INSTRUCTOR'S SUPPLEMENT PREVIEW BOOK

to accompany

FUNDAMENTALS OF PHYSICS

SIXTH EDITION

DAVID HALLIDAY
University of Pittsburgh

ROBERT RESNICK
Rensselaer Polytechnic Institute

JEARL WALKER
Cleveland State University

Includes Chapters 1-12 of the

- INSTRUCTOR'S MANUAL
  J. Richard Christman
  U.S. Coast Guard Academy, New London, CT
- TEST BANK
  J. Richard Christman
- INSTRUCTOR'S SOLUTIONS MANUAL
  James B. Whitenton
  Southern Polytechnic State University

JOHN WILEY & SONS, INC.
New York • Chichester • Weinheim • Brisbane • Singapore • Toronto

COVER PHOTO © Tsuyoshi Nishiinoue/Orion Press

ISBN 0-471-38396-1

Printed in the United States of America

10 9 8 7 6 5 4 3 2 1

Printed and bound by Bradford & Bigelow, Inc.

CONTENTS

PART I

Instructor's Manual

Chapters 1-12

PREFACE

This manual contains material we hope will be useful in the design of an introductory physics course based on the text *FUNDAMENTALS OF PHYSICS*, sixth edition, by David Halliday, Robert Resnick, and Jearl Walker. It may be used with either the extended or regular versions of the text. In Section One, we have included material to help instructors choose topics and design courses. In Section Two, we have also provided lecture notes, outlining the important topics of each chapter, and have suggested demonstration and laboratory experiments, computer software, video cassettes, and video disks.

Because some instructors avoid assigning problems that are discussed in *A Student's Companion* and in the *Student Solution Manual*, while others desire to include a few of these in many assignments, Section Three of the manual contains a list of problems in the two student supplements. Sections Four, Five, and Six contain answers to checkpoints, answers to end-of-chapter questions, and answers to end-of-chapter problems.

To help ease the transition from the fifth to the sixth edition of the text, Section Seven of the manual cross references end-of-chapter problems between the two editions.

Some instructors include in their courses computational exercises to be carried out by students on computers, using either spreadsheet software or their own programs. To aid these instructors, the final section of the manual contains a selection of computer projects.

The principal author is grateful to Stanley Williams, who co-authored the first edition of the instructor manual for *Fundamentals of Physics*. Much of his material has been retained in this manual. He is also grateful to Walter Eppenstein, who helped with suggestions for demonstration and laboratory experiments. Jearl Walker helped significantly by supplying answers to checkpoint questions, end-of-chapter questions, and new and revised end-of-chapter problems.

The author is indebted to Joan Kalkut who provided much good advice and managed many aspects of the production side of this project. Many good people at Wiley helped in a host of ways. Special thanks go to Stuart Johnson, Cliff Mills, Catherine Donovan, and Tom Hempstead. Karen Christman carefully read earlier editions of the manuscript and made many useful suggestions. Her fine work is gratefully noted. The unfailing support of Mary Ellen Christman is joyfully acknowledged.

J. Richard Christman
U.S. Coast Guard Academy
New London, Connecticut 06320

TABLE OF CONTENTS

SECTION ONE
PATHWAYS

Fundamentals of Physics, sixth edition, follows the sequence of topics found in most introductory courses. In fact, earlier editions of this text were instrumental in establishing the sequence. It is, however, extremely flexible in regard to both the range of topics and the depth of coverage. As a result, it can be used for a two, three, or four term course along traditional lines. It can also be used with many of the innovative courses that are presently being designed and taught. In many instances sections that discuss fundamental principles and give applications are followed by other sections that go deeper into the physics. Some instructors prefer to cover fewer topics than others but treat the topics they do cover in great depth. Others prefer to cover more topics with less depth. Courses of both types can easily be accommodated by selecting appropriate sections of the text.

By carefully choosing sections of the text to be included, your course might be a two-term, in-depth study of the fundamentals of classical mechanics and electromagnetism. With the addition of another term you might include more applications and the thermodynamics and optics chapters. In a three-term course, you might also forgo thermodynamics and optics but include Chapter 38 (Relativity) and some of the quantum mechanics chapters added in the extended version.

When designing the course, some care must be taken in the selection of topics because many discussions in later chapters presume coverage of prior material. Here are some comments you might find useful in designing your course. Also refer to the *Lecture Notes* section of this manual.

Mechanics. The central concepts of classical mechanics are covered in Chapters 1 through 12. Some minor changes that are possible, chiefly in the nature of postponements, are mentioned in the Lecture Notes. For example, the scalar product can be postponed until the discussion of work in Chapter 7 and the vector product can be postponed until the discussion of torque in Chapter 12.

Coverage of Chapter 5 can be shortened to two lectures or elongated to over four, depending on the time spent on applications. Other sections in the first twelve chapters that can be used to adjust the length of the course are 2–7, 3–6, 4–8, 4–9, 6–3, 9–7, 9–8, 10–5, and 11–3. However, you should consider covering section 9–8 if you include the thermodynamics chapters in your course. Coverage of Chapter 10 (Collisions) can be shortened significantly with safety. Section 11–7, which deals with the calculation of the rotational inertias of extended bodies, can be covered in detail or can be shortened by simply stating results once the definition as a sum over particles has been discussed. The parallel axis theorem is needed to solve some end-of-chapter problems in this chapter and in Chapter 16 and it should be covered if those problems are assigned.

The order of the chapters should be retained. For example, difficulties arise if you precede dynamics with statics as is sometimes done in other texts. To do so, you would need to discuss torque, introduced in Chapter 11, and explain its relation to angular acceleration. This involves considerable effort and is of questionable value.

SUGGESTED COURSES

A bare bones two-semester course (about 90 meetings) can be constructed around Chapters 1 through 12, 16, 17, and 22 through 34, with the omission of sections 32–3 through 32–8. The course can be adjusted to the proper length by the inclusion or omission of supplementary material and optional topics. If four to eight additional meetings are available each term, Chapter 14 or 15 (or perhaps both) can be inserted after Chapter 12 and one or more of the optics chapters can be inserted after Chapter 34. As an alternative, you might consider including sections on dielectrics, magnetic properties, semiconductors, and superconductors to emphasize properties of materials.

A three-term course (about 135 meetings) can be constructed by adding the thermodynamics chapters (19 through 21) and some or all of the modern physics chapters (38 through 45) to those mentioned above. If the needs of the class dictate a section on alternating current, some modern physics material can be replaced by Chapter 33.

ESTIMATES OF TIME

The following chart gives estimates of the time required to cover all of each chapter, in units of 50 minute periods. The second and fifth columns of the chart contain estimates of the number of lecture periods needed and includes the time needed to perform demonstrations and discuss the main points of the chapter. The third and sixth columns contain estimates of the number of recitation periods required and includes the time needed to go over problem solutions, answers to end-of-chapter questions, and points raised by students. If your course is organized differently, you may wish to add the two numbers to obtain the total estimated time for each chapter.

Use the chart as a rough guide when planning the syllabus for a semester, quarter, or year course. If you omit parts of chapters, reduce the estimated time accordingly.

Text Chapter	Number of Lectures	Number of Recitations	Text Chapter	Number of Lectures	Number of Recitations
1	0.3	0.2	24	1.8	1.8
2	2.0	2.0	25	1.8	1.8
3	1.0	1.0	26	1.5	2.0
4	2.0	2.5	27	1.0	1.0
5	2.0	2.0	28	2.0	2.3
6	2.0	2.0	29	2.0	1.8
7	1.8	1.5	30	2.0	1.2
8	2.0	2.0	31	2.5	2.5
9	2.0	1.5	32	1.5	1.8
10	2.0	1.6	33	2.5	2.5
11	2.0	1.5	34	2.9	2.7
12	2.0	2.0	35	2.5	2.5
13	1.0	2.0	36	2.0	2.0
14	2.3	2.0	37	2.0	2.0
15	2.0	2.0	38	2.5	2.0
16	2.5	1.8	39	2.0	2.0
17	2.5	2.0	40	2.0	2.0
18	2.5	2.0	41	2.2	2.0
19	2.5	2.5	42	2.0	2.0
20	2.0	1.4	43	1.8	2.0
21	1.5	1.6	44	2.0	2.0
22	1.0	1.0	45	2.0	2.0
23	1.6	1.3			

SECTION TWO
LECTURE NOTES

Lecture notes for each chapter of the text are grouped under the headings BASIC TOPICS and SUGGESTIONS.

BASIC TOPICS contains the main points of the chapter in outline form. In addition, one or two demonstrations are recommended to show the main theme of the chapter. You may wish to pattern your lectures after the notes, suitably modified, or simply use them as a check on the completeness of your own notes.

The SUGGESTIONS sections recommend end-of-chapter questions and problems, video cassettes, video disks, computer software, computer projects, alternate demonstrations, and other material that might be useful for the course. Many of the questions concentrate on points that seem to give students trouble, and it is worthwhile dealing with some of them before students tackle a problem assignment. Some questions and problems might be incorporated into the lectures while some might be assigned and used to generate discussion by students in small recitation sections. Answers to the questions appear in Section Five of this manual and answers to the problems appear in Section Six.

Each chapter contains several semi-quantitative questions, called checkpoints. Encourage students to use them to check their understanding of the concepts and relationships discussed in the chapter. Go over some or all of them in recitation classes or lectures. Answers to the checkpoint questions are given in Section Four of this manual.

If funds are available, consider setting up an interactive class room or lecture hall in which students can be polled remotely. The checkpoints and end-of-chapter questions are excellent for this purpose. A system called Classtalk, which utilizes student input via individual computers, is available from Better Education, Inc., 4824 George Washington Memorial Highway, Yorktown, Va 23692.

General Two excellent books that deal with teaching the introductory calculus-based course are

Teaching Introductory Physics; Arnold B. Arons; John Wiley (1997).

Teaching Introductory Physics (A Sourcebook); Clifford E. Swartz and Thomas Miner; Springer-Verlag (1998).

Both of these provide well thought-out explanations of some of the concepts that perplex students and give help with teaching those concepts. They are also excellent sources of demonstration and laboratory experiments that illuminate the important ideas of the introductory physics course.

Over the past ten years or so the field of physics education research has grown tremendously. Many research projects focus on the troubles students have in learning physics and analyze proposed remedies. Lillian McDermot and Edward Redish have compiled an extensive resource letter that lists books and journal articles in the field. It appeared in the September 1999 issue of the American Journal of Physics and is highly recommended as a source of material for improvement of the course.

Video. All of the video cassette and video disk items listed in the SUGGESTIONS sections are short, well done, and highly pertinent to the chapter. It is not possible to review all available material and there are undoubtedly many other fine video cassettes and disks that are not listed. Video might be incorporated into the lectures, shown during laboratory periods, or set up in a special room for more informal viewing.

An excellent set of video cassettes and disks, *THE MECHANICAL UNIVERSE*, can be obtained from The Annenberg CPB Collection, PO Box 2345, South Burlington, VT 05407–2345. The set consists of 52 half-hour segments dealing with nearly all the important concepts of introductory physics. Historical information and animated graphics are used to present the concepts in an imaginative and engaging fashion. Some physics departments run appropriate segments throughout the course in special viewing rooms. Accompanying textbooks, teacher manuals, and study guides are also available.

Many time-tested films originally from Encyclopaedia Britannica, PSSC, Project Physics, and elsewhere have been transferred to video disk by the AAPT Instructional Media Center and are available under the title *Physics: Cinema Classics* from Ztek Co., PO Box 11768, Lexington, KY 40577–1768. The films cover a host of topics in mechanics, thermodynamics, electricity and magnetism, optics, and modern physics. Other short films that have been transferred to video are the AAPT Collections 1 and 2 and the Miller Collection. These and many other video tapes and disks are referenced in the SUGGESTIONS section of the Lecture Notes.

Computer Software. Computers have made significant contributions to the teaching of physics. They are widely used in lectures to provide animated illustrations, with parameters under the control of the lecturer; they also provide tutorials and drills that students can work through on their own. Specialized programs are listed in appropriate SUGGESTION sections of the Lecture Notes in this manual. In addition, several available software packages cover large portions of an introductory course. Four of them are:

Core Concepts in Physics; CD-ROM; Macintosh, Windows; Saunders College Publishing, The Public Ledger Building, Suite 1250, 150 South Independence Mall West, Philadelphia, PA 19106–3412. A great many animations and live videos, laboratory demonstrations, and graphics. Most are interactive. Many step-by-step solutions are given to example problems. Covers a large portion of the introductory calculus-based course.

Interactive Journey Through Physics; CD-Rom; Macintosh, Windows; Cindy Schwartz and Bob Beicher; Prentice-Hall, 240 Frisch Ct., Paramus, NJ 07652-5240. A large number of animations and simulations, many with audio descriptions. Self-check quizzes are associated with the simulations. Over fifty narrated video segments. Covers most of the topics of the introductory course.

Multimedia Enhanced Physics Instruction; CD ROM; Macintosh, Windows; Maha Ashour-Abdalla; McGraw-Hill, P.O. Box 182604, Columbus, OH 43272–303143004. Extensive illustrated concepts sections, which give the important ideas of the introductory course, along with applications and examples, often by way of video. Numerous problems with interactive help. Each concept module also has a self-check quiz.

Interactive Physics; Macintosh, Windows; Knowledge Revolution, 15 Bush Pl., San Francisco, CA 94103). Animations and graphs for a wide variety of mechanical phenomena. The user can set up "experiments" with massive objects, strings, springs, dampers, and constant forces. Parameters can easily be changed. For Macintosh and Windows computers. Reviewed in The Physics Teacher, September 1991.

You might consider setting aside a room or portion of a lab, equip it with several computers, and make tutorial, drill, and simulation programs available to students. If you have sufficient hardware (and software), you might base some assignments on computer materials.

Computers and top-of-the line graphing calculators might also be used by students to perform calculations. Properly selected computer projects can add greatly to the students' understanding of physics. Projects involving the investigation of some physical system of interest might be assigned to individuals or might be carried out by a laboratory class. Some computer projects are outlined in Section Eight of this manual and are referenced in the Lecture Notes of this manual. A large number of suitable problems and projects can also be found in the book *Introduction to Computational Physics* by Marvin L. De Jong (Addison-Wesley, 1991).

Commercial spreadsheet programs can facilitate problem solving. PSI-Plot (Windows; Poly Software International, P.O. Box 1457, Sandy, UT 84091) and $f(g)$ Scholar (Macintosh, Windows; Future Graph, Inc., 75 James Way, Southampton, PA 18966) are high-end spreadsheet programs that incorporate many science and engineering problem-solving and graphing capabilities. Commercial problem-solving programs such as *MathCAD* (Windows; MathSoft, Inc., 101 Main Street, Cambridge, MA 02142-1521), *DERIVE* (Windows; Soft Warehouse; 3660 Waialae Avenue, Suite 304, Honolulu, HI 96815), *MAPLE* (Macintosh, Windows; Waterloo Maple, 57 Erb Street W., Waterloo, Ontario, Canada N2L 6C2), *Pro Solve* (Macintosh, Windows; Problem Solving Concepts, Inc., 1980 E. 116th Street, Ste 220, Carmel, IN 46032), and *Mathematica* (Macintosh, Windows; Wolfram Research, Inc., 100 Trade Center Drive, Champaign, IL 61820–7237) can easily be used by students to solve problems and graph results. All these programs allow students to set up a problem generically, then view solutions for various values of input parameters. For example, the range or maximum height of a projectile can be found as a function of initial speed or firing angle, even if air resistance is taken into account.

A number of computer programs allow you to view digitized video on a computer monitor and mark the position of an object in each frame. The coordinates of the object can be listed and plotted. They can then be used to find the velocity and acceleration of the object, either within the program itself or by exporting the data to a spreadsheet. Three of these are: *Videopoint* (Windows, CD-ROM; Pasco Scientific, 10101 Foothills Blvd., Roseville, CA 95678), *VideoGraph* (Macintosh; Physics Academic Software, Box 8202, North Carolina State University, Raleigh, NC 27695–8202), and *World in Motion* (Windows; Physics Curriculum and Instruction, 22585 Woodhill Drive, Lakeville, MN 55044). All of these come with an assortment of video clips. Home-made videos can also be used. The capabilities of the programs are different. Check carefully before purchasing.

Demonstrations. Notes for most of the chapters are developed around demonstration experiments. Generally speaking, these use relatively inexpensive, readily available equipment, yet clearly demonstrate the main ideas of the chapter. The choice of demonstrations, however, is highly personal and you may wish to substitute others for those suggested here or you may wish to present the same ideas using chalkboard diagrams. Several excellent books give many other examples of demonstration experiments. The following are available from the American Association of Physics Teachers, One Physics Ellipse, College Park MD 20740–3845:

A Demonstration Handbook for Physics, G.D. Freier and F.J. Anderson, 320 pages (1981). Contains over 800 demonstrations, including many that use everyday materials and that can be constructed with minimal expense. Line drawings are used to illustrate the demonstrations.

String and Sticky Tape Experiments, Ronald Edge, 448 pages (1987). Contains a large number of illuminating experiments that can be constructed from inexpensive, readily available materials.

Apparatus for Teaching Physics, edited by Karl C. Mamola. A collection of articles from The Physics Teacher that describe laboratory and demonstration apparatus.

How Things Work, H. Richard Crane, 114 pages, 1992. A collection of 20 articles from The Physics Teacher.

The following is currently out of print but is available in many college libraries and physics departments:

Physics Demonstration Experiments, H.F. Meiners, ed. An excellent source of ideas, information, and construction details on a large number of experiments, with over 2000 line drawings and photographs. It also contains some excellent articles on the philosophical aspects of lecture demonstrations, the use of shadow projectors, TV, films, overhead projectors, and stroboscopes.

Appropriate demonstrations described in Freier and Anderson are listed in the SUGGESTIONS sections of the notes. This book does not give any construction details, but more information about most demonstrations can be obtained from the book edited by Meiners.

The *Physics InfoMall* CD-ROM (The Learning Team, 84 Business Park Drive, Armonk, NY 10504), a searchable database of over 1000 demonstrations, is another excellent source. There are both Windows and Macintosh versions. The CD also contains articles and abstracts, problems with solutions, whole reference books, and a physics calendar. The following three books, all available from the AAPT, are also sources of ideas for demonstrations and examples:

Physics of Sports; edited by C. Frohlich. Contains reprints and a resource letter.

Amusement Park Physics; edited by Carole Escobar. In workbook form. The activities described are perhaps more appropriate for a high school class but some can be used in college level lectures as examples.

Potpourri of Physics Teaching Ideas; edited by Donna Berry; reprints of articles on apparatus from The Physics Teacher.

A computer can also be used for data acquisition during demonstrations. Photogate timers, temperature probes, strain gauges, voltage probes, and other devices can be input directly into the computer and results can be displayed as tables or graphs. The screen can be shown to a large class by using a large monitor, a TV projection system, or an overhead projector adapter. Inexpensive software and hardware can be purchased from Vernier Software (8565 SW Beaverton-Hillsdale Hwy., Portland, OR 97225-2429). PASCO Scientific (PO Box 619011, 10101 Foothills Boulevard, Roseville, CA 95661–9011) has data acquisition software and an extensive variety of probes for both Macintosh and Windows computers. If more sophisticated software is desired, consider the commercial package Labview (National Instruments Corporation, 11500 N. Mopac Expwy., Austin, TX 78759–3504). The monograph *Photodetectors* by Jon W. McWane, J. Edward Neighbor, and Robert F. Tinker (available from the AAPT) is a good source of technical information about photodetectors.

Laboratories. Hands-on experience with actual equipment is an extremely important element of an introductory physics course. There are many different views as to the objectives of the physics laboratory and the final decision on the types of experiments to be used has to be made by the individual instructor or department. This decision is usually based on financial and personnel considerations as well as on the pedagogical objectives of the laboratory.

Existing laboratories vary widely. Some use strictly cookbook type experiments while others allow the students to experiment freely, with practically no instructions. The equipment ranges from very simple apparatus to rather complex and sophisticated equipment. Physical phenomena may be observed directly or simulated on a computer. Data may be taken by the students or fed into a computer.

The equipment described above can be used for data acquisition in a student lab. Even if data acquisition software is not used, consider having students use computers and spreadsheet programs to analyze and graph data.

Many physics departments have written their own notes or laboratory manuals and relatively few physics laboratory texts are on the market. Two such books, both available from John Wiley & Sons, are

Laboratory Physics, second edition, H.F. Meiners, W. Eppenstein, R.A. Oliva, and T. Shannon. (1987).

Laboratory Experiments in College Physics, seventh edition, C.H. Bernard and C.D. Epp. (1994).

Experiments from these books are listed in the SUGGESTIONS section of the Lecture Notes. Meiners is used to designate the Meiners, Eppenstein, Oliva, and Shannon book while Bernard is used to designate the Bernard and Epp book. Both books contain excellent introductory sections explaining laboratory procedures to students. Meiners also contains a large amount of material on the use of microprocessors in the lab.

Student supplements. Several supplements, all available from Wiley, might be recommended to the students:

A Student's Companion to Fundamentals of Physics. A study guide. The basic concepts of each chapter are reviewed in a workbook format that helps students focus their attention on the important ideas and their relationships to each other. Each copy contains a password for an internet site that provides the students with hints for about 650 end-of-chapter exercises and problems. The list of exercises and problems on the site are given in Section Three of this manual.

Student Solution Manual. Contains fully worked solutions to about 880 end-of-chapter problems and exercises.

CD Physics. A CD ROM version of the text and supplements for Windows and Macintosh. This contains the complete text, the *Student Solution Manual*, the interactive tutorials, the interactive simulations, and a glossary. It is extensively hyperlinked.

Instructor aids. In addition to this *Instructor Manual* Wiley provides several other aids for instructors:

Instructor's Solution Manual Contains fully worked solutions to all the end-of-chapter problems.

Test Bank. Contains over 2800 multiple choice questions (with answers) for use on exams and quizzes. Both quantitative and qualitative questions are included. In each chapter, some of the questions are modeled after the checkpoints and end-of-chapter questions, as well as after the end-of-chapter problems and exercises. There is also a computerized version of the *Test bank* on a CD ROM, which also contains a program for constructing exams (along with answer keys).

Instructor's Resource CD. A CD ROM for Windows. Contains the *Instructor's Solution Manual* (as both LaTEX source files and in PDF form), reproductions of illustrations from the text (in JPEG form), and the *Test Bank*. There is a computer program that allows instructors to generate exams from the test bank questions.

WebCT. A collections instructor resources that can be downloaded from the Wiley server. It includes the set of text illustrations, the interactive simulations, and test questions.

Chapter 1 MEASUREMENT

BASIC TOPICS

I. Base and derived units.
   A. Explain that standards are associated with base units and that measurement of a physical quantity takes place by means of comparison with a standard. Discuss qualitatively the SI standards for time, length, and mass. Show a 1 kg mass and a meter stick. Show the simple well-known procedure for measuring length with a meter stick.
   B. Explain that derived units are combinations of base units. Emphasize that the speed of light is now a defined unit and the meter is a derived unit. Discuss an experiment in which the time taken for light to travel a certain distance is measured. Example: the reflection of a light signal from the Moon. Use a clock and a meter stick to find your walking speed in m/s.
   C. This is a good place to review area, volume, and mass density. Use simple geometric figures (circle, rectangle, triangle, cube, sphere, cylinder, etc.) as examples.

II. Systems of units.
   A. Explain what a system of units is. Give the 1971 SI base units (Table 1–1). Stress that they will be used extensively.
   B. Point out the SI prefixes (Table 1–2). The important ones for this course are mega, kilo, centi, milli, micro, nano, and pico. Discuss powers of ten arithmetic and stress the simplicity of the notation. This might be a good place to say something about significant digits.
   C. Discuss unit arithmetic and unit conversion.
   D. Most of the students' experience is with the British system. Relate the inch to the centimeter and the slug to the kilogram. Discuss unit conversion. Use speed as an example: convert 50 mph and 3 mph to km/h and m/s. Point out the conversion tables in Appendix D.

III. Properties of standards.
   A. Discuss accessibility and invariability as desirable properties of standards.
   B. Discuss secondary standards such as the meter stick used earlier.

IV. Measurements.
   A. Stress the wide range of magnitudes measured. See Tables 1–3, 1–4, and 1–5. Explain the atomic mass unit. One atom of ^{12}C has a mass of exactly 12 u. 1 u is approximately 1.661×10^{-27} kg.
   B. Discuss indirect measurements.

SUGGESTIONS

1. Assignments
   a. To emphasize SI prefixes assign problems 1 and 5.
   b. Unit conversion is covered in many problems. Choose some, such as 2, 3, 4, that deal with unfamiliar units.
   c. According to the needs of the class, assign one or more problems that deal with area and volume calculations, such as 5, 7, and 8.
2. Demonstrations
   Examples of "standards" and measuring instruments: Freier and Anderson Ma1 — 3.
3. Books and Monographs
   a. *Frequency and Time Measurements*, edited by Christine Hackman and Donald B. Sullivan; available from AAPT, One Physics Ellipse, College Park MD 20740–3845.

b. *SI: The International System of Units*; edited by Robert A. Nelson; available from AAPT (see above for address).

c. *Connecting Time and Space*; edited by Harry E. Bates; available from AAPT (see above for address). Reprints that discuss measurements of the speed of light and the redefinition of the meter. Students will not be able to understand much of this material at this stage of the course but it is nevertheless useful for background.

4. Audio/Visual

a. *Time and Place, Measuring Short Distances*; Side A: Mechanics (I) of Cinema Classics; video disk; available from Ztek Co., PO Box 11768, Lexington, KY 40577–1768.

b. *Powers of Ten* from the Films of Charles and Ray Eames; video disk; available from Ztek Co. (see above for address).

5. Laboratory

a. Meiners Experiment 7-1: *Measurement of Length, Area, and Volume.* Gives students experience using the vernier caliper, micrometer, and polar planimeter. Good introduction to the determination of error limits (random and least count) and calculation of errors in derived quantities (volume and area).

b. Bernard Experiments 1 and 2: *Determination of Length, Mass, and Density* and *Measurements, Measurement Errors, and Graphical Analysis.* Roughly the same as the Meiners experiment, but a laboratory balance is added to the group of instruments and the polar planimeter is not included. Graphs of mass versus radius and radius squared for a collection of disks made of the same material, with the same thickness, are used to establish the quadratic dependence of mass on radius.

c. Meiners Experiment 7-3: *The Simple Pendulum* and Bernard Experiment 3: *The Period of a Pendulum — An Application of the Experimental Method.* Students time simple pendulums of different lengths, then use the data and graphs (including a logarithmic plot) to determine the relationship between length and period. They calculate the acceleration due to gravity. This is an exercise in finding functional relationships and does not require knowledge of dynamics.

Chapter 2 MOTION ALONG A STRAIGHT LINE

BASIC TOPICS

I. Position and displacement.

A. Move a toy cart with constant velocity along a table top. Select an origin, place a meter stick and clock on the table, and demonstrate how x(t) is measured in principle. Emphasize that x is always measured from the origin; it is not the cart's displacement during any time interval.

B. Draw a graph of $x(t)$ and point out that it is a straight line. Show what the graph looks like if the cart is not moving. Point out that the line has a greater slope if the cart is going faster. Move the cart so its speed increases with time and show what the curve $x(t)$ looks like. Do with same for a cart that is slowing down.

C. Some students think of a coordinate as distance. Distinguish between these concepts. Point out that a coordinate defines a position on an axis and can be positive or negative. Demonstrate a negative velocity, both with the cart and on a graph. As another example, throw a ball into the air, pick a coordinate axis (positive in the upward direction, say), and point out when the velocity is positive and when it is negative. Draw the graph of the coordinate as a function of time. Repeat with the positive direction upward.

D. Define the displacement of an object during a time interval. Emphasize that only the initial and final coordinates enter and that an object may have many different motions between these while still having the same displacement. Point out that the displacement is zero if the initial and final coordinates are the same.

II. Velocity.

A. Define average velocity over an interval. Stress the meaning of the sign. Go over Sample Problem 2–1. Draw a graph of x versus t for an object that is accelerating. Pick an interval and draw the line between the end points on the graph. Observe that the average velocity in the interval is the slope of the line. Figs. 2–3 and 2–4 may also be used. Show how to calculate average velocity if the function $x(t)$ is given in algebraic form.

B. Define instantaneous velocity. Demonstrate the limiting process. Use a graph of x versus t for an accelerating cart to demonstrate that the line used to find the average velocity becomes tangent to the curve in the limit as Δt vanishes. Remark that the slope of the tangent line gives the instantaneous velocity. Show a plot of v versus t that corresponds to the x versus t graph used previously. Show how to calculate the instantaneous velocity if the function $x(t)$ is given in algebraic form. See Sample Problem 2–4. Stress that a value of the instantaneous velocity is associated with each instant of time. Some students think of velocity as being associated with a time interval rather than an instant of time.

C. Define instantaneous speed as the magnitude of the velocity. Compare to the average speed in an interval, which is the total path length divided by the time. Remark that the average speed is not necessarily the same as the magnitude of the average velocity.

D. Note that many calculus texts use a prime to denote a derivative. They also define the derivative of x with respect to time by the limit of $[x(t + \Delta t) - x(t)]/\Delta t$ rather than by the limit of $\Delta x/\Delta t$. Mention the different notations in class so students can relate their physics and calculus texts.

III. Acceleration.

A. Define average and instantaneous acceleration. Show the previous v versus t graph and point out the lines used to find the average acceleration in an interval and the instantaneous acceleration at a given time. Show how to calculate the average and instantaneous acceleration if either $x(t)$ or $v(t)$ is given in algebraic form.

B. Interpret the sign of the acceleration. Give examples of objects with acceleration in the same direction as the velocity (speeding up) and in the opposite direction (slowing down). Be sure to include both directions of velocity. Emphasize that a positive acceleration does not necessarily imply speeding up and a negative acceleration does not necessarily imply slowing down.

C. Use graphs of $x(t)$ and $v(t)$ to point out that an object may simultaneously have zero velocity and non-zero acceleration. Explain that if the direction of motion reverses the object must have zero velocity at some instant. Give the position as a function of time as $x(t) = At^2$, for example, and show that the velocity is 0 at $t = 0$ but the acceleration is not 0. Illustrate the function with a graph.

IV. Motion in one dimension with constant acceleration.

A. Derive the kinematic equations for $x(t)$ and $v(t)$. If students know about integration, use methods of the integral calculus (as in Section 2–7). In any event, show that $v(t)$ is the derivative of $x(t)$ and that a is the derivative of $v(t)$.

B. Discuss kinematics problems in terms of a set of simultaneous equations to be solved. Examples: use equations for $x(t)$ and $v(t)$ to algebraically eliminate the time and to algebraically eliminate the acceleration. The equations of constant acceleration motion are listed in Table 2–1. Some instructors teach students to use the table. Others ask students to always start with Eqs. 2–11 and 2–15, then use algebra to obtain the equations needed for a particular problem. See Sample Problem 2–5.

C. To help students see the influence of the initial conditions, sketch graphs of $v(t)$ and $x(t)$ for various initial conditions but the same acceleration. Include both positive and negative initial velocities. Draw a different set of graphs for positive and negative acceleration. Point out where the particle has zero velocity and when it returns to its initial position.

V. Free fall.

A. Give the value for g in SI units. Point out that the free-fall acceleration is due to gravity and that it is directed toward the center of Earth. Say that locally Earth's surface is essentially flat and the free-fall acceleration may be taken to be in the same direction at slightly different points. Explain that $a = +g$ if down is taken to be the positive direction and $a = -g$ if up is the positive direction. Do examples using both choices. Throw a ball into the air and emphasize that its acceleration is g throughout its motion, even at the top of its trajectory.

B. Drop a small ball through two photogates, one near the top to turn on a timer and one further down to turn it off. Repeat for various distances and plot the position of the ball as a function of time. Explain that the curve is parabolic and indicates a constant acceleration.

C. Explain that all objects at the same place have the same free-fall acceleration. In reality, different objects may have different accelerations because air influences their motions differently. This can be demonstrated by placing a coin and a wad of cotton in a glass cylinder about 1 m long. Turn the cylinder over and note that the coin reaches the bottom first. Now use a vacuum pump to partially evacuate the cylinder and repeat the experiment. Repeat again with as much air as possible pumped out.

D. Point out that free-fall problems are special cases of constant acceleration kinematics and the methods described earlier can be used. Work a few examples. For an object thrown into the air, calculate the time to reach the highest point, the height of the highest point, the time to return to the initial height, and its velocity when it returns, all in terms of the initial velocity.

SUGGESTIONS

1. Assignments

a. To help students obtain some qualitative understanding of velocity and acceleration, ask them to discuss questions 1, 2, 3, and 4. Some aspects of motion with constant acceleration are covered in questions 5 and 6. Free fall is covered in questions 8 and 9.

b. To make more use of the calculus assign some of problems 7, 11, 12, 15, 19, and 21. Problems 12 and 19 can also be used to discuss differences between average and instantaneous velocity.

c. To emphasize the interpretation of graphs assign a few of problems 6, 7, 10, 12, 13, 16, and 18. Some of these require students to draw graphs after performing calculations.

d. Ask students to solve a few problems dealing with motion with constant acceleration. Consider problems 25, 26, 34, and 35. For a little more challenge, consider 37.

e. Problems 42, 45, 46, 50, and 55 are good problems to test understanding of free-fall motion. Problem 62 is more challenging.

2. Demonstrations

Uniform velocity and acceleration, velocity as a limiting process: Freier and Anderson Mb10 — 13, 15, 18, 21, 22.

3. Audio/Visual
   a. *Acceleration due to Gravity*; from AAPT collection 1 of single-concept films; video tape; available from Ztek Co., PO Box 11768, Lexington, KY 40577–1768.
   b. *One Dimensional Motion*; *Distance, Time & Speed*; *One Dimensional Acceleration*; *Constant Velocity & Uniform Acceleration*; from AAPT collection 2 of single-concept films; video tape; available from Ztek Co. (see above for address).
   c. *Uniform Motion*, *Free Fall*; Side A: Mechanics (I) of Cinema Classics; video disk; available from Ztek Co. (see above for address).
   d. *Acceleration*; VHS video tape (8 min); Films for the Humanities & Sciences, PO Box 2053, Princeton, NJ 08543–2053.
   e. *Falling Motion*; VHS video tape (8 min); Films for the Humanities & Sciences (see above for address).
   f. *Uniform and Accelerated Motion*; *Gravitational Acceleration*; Physics Demonstrations in Mechanics, Part I; VHS video tape; ≈3 min each; Physics Curriculum & Instruction, 22585 Woodhill Drive, Lakeville, MN 55044.
   g. *Graphical Analysis of Motion*; Demonstrations in Mechanics, Part III; VHS video tape or video disk; ≈3 min; Physics Curriculum & Instruction (see above for address).
4. Computer Software
   a. *Graphs and Tracks*; David Trowbridge; Windows, Macintosh; available from Physics Academic Software, North Carolina State University, PO Box 8202, Raleigh, NC 27690–0739. A ball rolls on a series of connected inclines. In one part the student is given graphs of the position, velocity, and acceleration and is asked to adjust the tracks to produce the graphed motion. In a second part the student is shown the motion and asked to sketch the graphs. Complements lab experiments with a sonic ranger.
   b. *Newtonian Sandbox*; Judah Schwartz; Windows, Macintosh; available from Physics Academic Software (see above for address). Generates the motion of a point particle in one and two dimensions. Plots trajectories, coordinates, velocity components, radial and angular positions, and phase space trajectories.
   c. *Objects in Motion*; Peter Cramer; Windows, Macintosh; available from Physics Academic Software (see above for address). Simulates the motion of an object under various conditions and plots graphs of the position, velocity and accelerations. Situations considered are: uniform acceleration along a straight line, projectile motion, relative motion, circular motion, planetary motion, and elastic collisions.
   d. *Physics Demonstrations*; Julien C. Sprott; Windows; available from Physics Academic Software (see above for address). Ten simulations of motion and sound demonstrations. Includes "the monkey and the coconut", "ballistics cat", "flame pipe", "Doppler effect".
   e. *Motion*; Macintosh, Windows; Cross Educational Software, Inc., 508 E. Kentucky Avenue, PO Box 1536, Ruston, LA 71270. Contains sections on graphing motion and motion with constant acceleration.
   f. *Accelerated Motion*; Windows; Cross Educational Software, Inc. (see above for address).
   g. *Conceptual Kinematics*; Frank Griffin and Louis Turner; Macintosh, Windows; available from Physics Academic Software (see above for address). An interactive, animated tutorial, with quiz questions for self-testing.
   h. *Dynamic Analyzer*; Roger F. Sipson; available from Physics Academic Software (see above for address).
   i. *Motion in One Dimension*; Socrates Software, 6187 Rosecommon Drive, Norcross, GA 30092. Uses animation and graphical analysis to demonstrate the relationships between displacement, velocity, and acceleration.

5. Computer Projects
   a. Use a spreadsheet or your own computer program to demonstrate the limiting processes used to define velocity and acceleration. Given the functional form of $x(t)$, have the computer calculate and display the coordinate for some time t and a succession of later times, closer and closer to t. For each interval, have it calculate and display the average velocity. Be careful to refrain from displaying non-significant figures and be sure to stop the process before all significance is lost.
   b. Have students use the root finding capability of a commercial math program or their own computer programs to solve kinematic problems for which $x(t)$ and $v(t)$ are given functions. Nearly all of them can be set up as problems that involve finding the root of either the coordinate or velocity as a function of time, followed perhaps by substitution of the root into another kinematic equation. Problems need not be limited to those involving constant acceleration. Air resistance, for example, can be taken into account. The same program can be used to solve rotational kinematic problems in Chapter 11.
6. Laboratory
   a. Motion detectors. As a student moves toward and away from a sonar-like ranging device, his position, velocity, or acceleration is plotted on a computer monitor. A graph can be designed by the instructor and the student can be asked to duplicate it by moving in front of the ranger. Several sonic rangers are reviewed in The Physics Teacher of January 1988. An extremely popular model is available from Vernier Software, 8565 SW Beaverton-Hillsdale Hwy., Portland, OR 97225-2429.
   b. Meiners Experiment 7–5: *Analysis of Rectilinear Motion.* Students measure the position as a function of time for various objects rolling down an incline, then use the data to plot speeds and accelerations as functions of time. No knowledge of rotational motion is required. This experiment emphasizes the definitions of velocity and acceleration as differences over a time interval.
   c. Meiners Experiment 8–1: *Motion in One Dimension* (omit the part dealing with conservation of energy). Essentially the same experiment except pucks sliding on a nearly frictionless surface are used. This experiment may be done with dry ice pucks or on an air table or air track.
   d. Bernard Experiment 7: *Uniformly Accelerated Motion.* The same technique as the Meiners experiments but a variety of setups are described: the standard free fall apparatus, the free fall apparatus with an Atwood attachment, an inclined plane, an inclined air track, and a horizontal air track with a pulley attachment.

Chapter 3 VECTORS

BASIC TOPICS

I. Definition.
   A. Explain that vectors have magnitude and direction, and that they obey certain rules of addition.
   B. Example of a vector: displacement. Give the definition of displacement and point out that a displacement does not describe the path of the object. Give the definition and physical interpretation of the sum of two displacements. Demonstrate vector addition by walking along two sides of the room. Point out the two displacements and their sum. Note that the distance traveled is not the magnitude of the displacement. Go back to your original position and point out that the displacement is now zero.
   C. Compare vectors with scalars and present a list of each.

D. Go over vector notation and insist that students use it to identify vectors clearly. In this text a vector is indicated by placing an arrow over an algebraic symbol. The italic version of the symbol, without the arrow, indicates the magnitude of the vector. Point out that many other texts use boldface type to indicate vectors.

II. Vector addition and subtraction by the graphical method.

A. Draw two vectors tail to head, draw the resultant, and point out its direction. Explain how the magnitude of the resultant can be measured with a ruler and the orientation can be measured with a protractor. Explain how a scale is used to draw the original vectors and find the magnitude of the resultant.

B. Define the negative of a vector and define vector subtraction as $\vec{a}-\vec{b}=\vec{a}+(-\vec{b})$. Graphically show that if $\vec{a}+\vec{b}=\vec{c}$ then $\vec{a}=\vec{c}-\vec{b}$.

C. Show that vector addition is both commutative and associative.

III. Vector addition and subtraction by the analytic method.

A. Derive expressions for the components of a vector, given its magnitude and the angles it makes with the coordinate axes. In preparation for the analysis of forces, find the x component of a vector in the xy plane in terms of the angles it makes with the positive and negative x axis and also in terms of the angles it makes with the positive and negative y axis.

B. Point out that the components depend on the choice of coordinate system, and compare the behavior of vector components with the behavior of a scalar when the orientation of the coordinate system is changed. Find the components of a vector using two differently oriented coordinate systems. Point out that it is possible to orient the coordinate system so that only one component of a given vector is not zero. Remark that a pure translation of a vector (or coordinate system) does not change the components.

C. Define the unit vectors along the coordinate axes. Give the form used to write a vector in terms of its components and the unit vectors. Explain that unit vectors are unitless so they can be used to write any vector quantity.

D. Vector addition. Give the expressions for the components of the resultant in terms of the components of the addends. Demonstrate the equivalence of the graphical and analytic methods of finding a vector sum. See the diagram to the right.

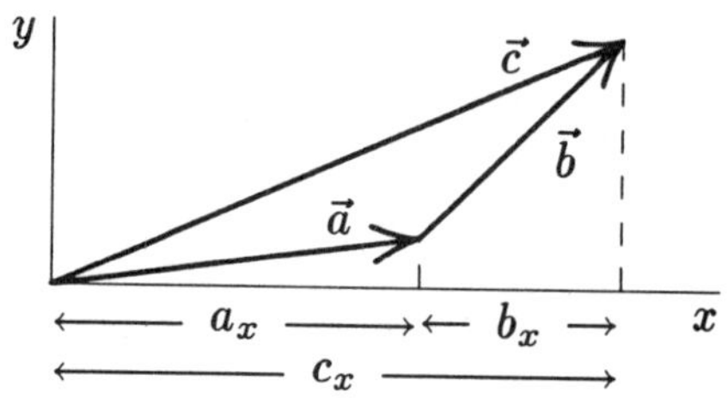

E. Give the expression for vector subtraction in terms of components. You may also wish to demonstrate the equivalence of the graphical and analytical methods of vector subtraction.

F. Show how to find the magnitude and angles with the coordinate axes, given the components. Explain that calculators give only one of the two possible values for the inverse tangent and show how to determine the correct angle for a given situation.

G. State that two vectors are equal only if their corresponding components are equal. State that many physical laws are written in terms of vectors and that many take the form of an equality between two vectors.

IV. Multiplication involving vectors.

A. Multiplication by a scalar. Give examples of both positive and negative scalars multiplying a vector. Give the components of the resulting vector as well as its magnitude and direction. Remark that division of a vector by a scalar is equivalent to multiplication by the reciprocal of the scalar.

B. (May be postponed until Chapter 7.) Scalar product of two vectors. Emphasize that the product is a scalar. Give the expression for the product in terms of the magnitudes of the vectors and the angle between them. To determine the angle, the vectors must be drawn with their tails at the same point. Point out that $\vec{a} \cdot \vec{b}$ is the magnitude of $\vec{a}$ multiplied by the component of $\vec{b}$ along an axis in the direction of $\vec{a}$. Explain that $\vec{a} \cdot \vec{b} = 0$ if $\vec{a}$ is perpendicular to $\vec{b}$.

C. Either derive the expression for a scalar product in terms of cartesian components or else assign Problem 30. See the discussion leading to Eq. 3–23. Specialize the expression to show that $\vec{a} \cdot \vec{a} = a^2$. Show how to use the scalar product to calculate the angle between two vectors if their components are known. Consider problem 31.

D. Vector product of two vectors (may be postponed until Chapter 12). Emphasize that the product is a vector. Give the expression for the magnitude of the product and the right hand rule for determining the direction. Explain that $\vec{a} \times \vec{b} = 0$ if $\vec{a}$ and $\vec{b}$ are parallel. Point out that $|\vec{a} \times \vec{b}|$ is the magnitude of $\vec{a}$ multiplied by the component of $\vec{b}$ along an axis perpendicular to $\vec{a}$ and in the plane of $\vec{a}$ and $\vec{b}$. Show that $\vec{b} \times \vec{a} = -\vec{a} \times \vec{b}$.

E. Either derive the expression for a vector product in terms of cartesian components or else assign Problem 32. See the discussion leading to Eq. 3–30. Give students the useful mnemonic for the vector products of the unit vectors $\hat{\imath}$, $\hat{\jmath}$, and $\hat{k}$, written in that order clockwise around a circle. One starts with the first named vector in the vector product and goes around the circle toward the second named vector. If the direction of travel is clockwise the result, is the third vector. If it is counterclockwise, the result is the negative of the third vector.

SUGGESTIONS

1. Assignments
   a. Use questions 1, 2, and 3 to discuss properties of vectors. Questions 4 and 5 deal with vector addition and subtraction. Question 6 deals with the signs of components.
   b. Ask students to use graphical representations of vectors to think about problems such as 1 and 2.
   c. Problems 3, 5, and 6 cover the fundamentals of vector components. Problems 7 and 8 stress the physical meaning of vector components. Some good problems to test understanding of analytic vector addition and subtraction are 12, 13, 15, 18, and 19.

2. Demonstrations
   Vector addition: Freier and Anderson Mb2, 3.

3. Audio/Visual
   a. *Vector Addition — Velocity of a Boat*; from AAPT collection 1 of single-concept films; video tape; available from Ztek Co., PO Box 11768, Lexington, KY 40577–1768.
   b. *Vectors*; Side A: Mechanics (I) of Cinema Classics; video disk; available from Ztek Co. (see above for address).
   c. *Vector Addition*; Physics Demonstrations in Mechanics, Part III; VHS video tape or video disk; ≈3 min; Physics Curriculum & Instruction, 22585 Woodhill Drive, Lakeville, MN 55044.

4. Computer Software
   a. *Math Methods in Physics*; Macintosh, Windows; Cross Educational Software, Inc., 508 E. Kentucky Avenue, PO Box 1536, Ruston, LA 71270. Contains a section on vector components, vector addition, vector and scalar products. Also contains sections on numbers and units, algebra, geometry, trigonometry, and calculus.
   b. *Vectors*; Richard R. Silbar; available from Physics Academic Software, Box 8202, North Carolina State University, Raleigh, NC 27695–8202.

5. Computer Project
   Have students use a commercial math program or write their own computer programs to carry out conversions between polar and cartesian forms of vectors, vector addition, scalar and vector products.
6. Laboratory
   Bernard Experiment 4: *Composition and Resolution of Coplanar Concurrent Forces.* Students mathematically determine a force that balances 2 or 3 given forces, then check the calculation using a commercial force table. They need not know the definition of a force, only that the forces in the experiment are vectors along the strings used, with magnitudes proportional to the weights hung on the strings. The focus is on resolving vectors into components and finding the magnitude and direction of a vector, given its components.

Chapter 4 MOTION IN TWO AND THREE DIMENSIONS

BASIC TOPICS

I. Definitions.

   A. Draw a curved particle path. Show the position vector for several times and the displacement vector for several intervals. Define average velocity over an interval. Write the definition in both vector and component form.

   B. Define velocity as $d\vec{r}/dt$. Write the definition in both vector and component form. Point out that the velocity vector is tangent to the path. Define speed of the magnitude of the velocity.

   C. Define acceleration as $d\vec{v}/dt$. Write the definition in both vector and component form. Point out that $\vec{a}$ is not zero if either the magnitude or direction of $\vec{v}$ changes with time.

   D. Show that the particle is speeding up only if $\vec{a}\cdot\vec{v}$ is positive. If $\vec{a}\cdot\vec{v}$ is negative, the particle is slowing down, and if $\vec{a}\cdot\vec{v} = 0$, its speed is not changing.

   E. Remark that sometimes the magnitude and direction of the acceleration are given, rather than its components. Remind students how to find the components if such is the case.

   F. Go over Sample Problem 4–4 or a similar problem of your own devising. It shows how to find and use the components of the acceleration.

II. Projectile motion.

   A. Demonstrate projectile motion by using a spring gun to fire a ball onto a surface at the firing height. Use various firing angles, including 45°, and point out that the maximum range occurs for a firing angle of 45°. Remark on the symmetry of the range as a function of firing angle. Mention that the maximum range occurs for a different angle when the ball is fired onto a surface at a different height.

   B. Draw the trajectory of a projectile, show the direction of the initial velocity, and derive its components in terms of the initial speed and firing angle.

   C. Write down the kinematic equations for $x(t)$, $y(t)$, $v_x(t)$, and $v_y(t)$. At first, include both a_x and a_y but then specialize to $a_x = 0$ and $a_y = -g$ for positive y up. Stress that these form two sets of one dimensional equations, linked by the common variable t and are to be solved simultaneously. Note that a_x affects only v_x, not v_y or v_z. Make similar statements about the other components. Throw a ball vertically, then catch it. Repeat while walking with constant velocity across the room. Ask students to observe the motion of the ball relative to the chalkboard and to describe its motion relative to your hand.

   D. Point out that the acceleration is the same at all points of the trajectory, even the highest point. Also point out that the horizontal component of the velocity is constant.

E. Work examples. Use punted footballs, hit baseballs, or thrown basketballs according to season.
   1. Find the time for the projectile to reach its highest point, then find the coordinates of the highest point.
   2. Find the time for the projectile to hit the ground, at the same level as the firing point. Then, find the horizontal range and the velocity components just before landing.
   3. Show that maximum range over level ground is achieved when the firing angle is 45°.
   4. Show how to work problems for which the landing point is not at the same level as the firing point.

F. Point out that all projectiles follow some piece of the full parabolic trajectory. For example, A to D could be the trajectory of a ball thrown at an upward angle from a roof to the street; B to D could be the trajectory of a ball thrown horizontally; C to D could be the trajectory of a ball thrown downward.

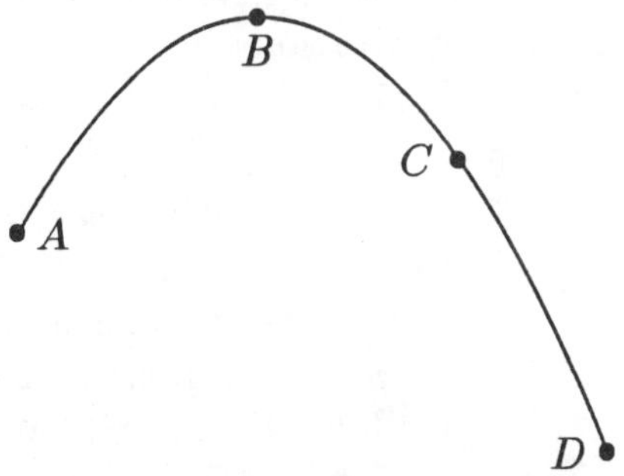

G. Explain how to find the speed and direction of travel for any time. Specialize to the time of impact on level ground and show that the speed is the same as the firing speed but that the vertical component of the velocity has changed sign. Remark that this result is true only because air resistance has been neglected.

H. Work some sample problems. Consider Sample Problems 4–7 and 4–8 or others of your own devising.

III. Circular motion.

A. Draw the path and describe uniform circular motion, emphasizing that the speed remains constant. Remind students that the acceleration must be perpendicular to the velocity. By drawing the velocity vector at two times, argue that it must be directed inward. On the diagram show the velocity and acceleration vectors for several positions of the particle.

B. Derive $a = v^2/r$. As an alternative to the derivation given in the text, write the equations for the particle coordinates as functions of time, then differentiate twice.

C. Example: calculate the speed of an Earth satellite, given the orbit radius and the acceleration to due to gravity at the orbit. Emphasize that the acceleration is toward Earth.

IV. Relative motion.

A. Material in this section is used in Chapter 5 to discuss inertial frames and in Chapter 12 to discuss rolling without slipping. It is also useful as a prelude to relativity.

B. Relate the position of a particle as given in coordinate system A to the position as given in coordinate system B by $\vec{r}_{\rm PA} = \vec{r}_{\rm PB} + \vec{r}_{\rm BA}$, where $\vec{r}_{\rm BA}$ is the position of the origin of B relative to the origin of A. Differentiate to show that $\vec{v}_{\rm PA} = \vec{v}_{\rm PB} + \vec{v}_{\rm BA}$ and $\vec{a}_{\rm PA} = \vec{a}_{\rm PB} + \vec{a}_{\rm BA}$, where $\vec{v}_{\rm BA}$ and $\vec{a}_{\rm BA}$ are the velocity and acceleration, respectively, of B relative to A.

C. Discuss examples of a ball thrown or rolled in accelerating and non-accelerating trains. The discussion may be carried out for either one- or two-dimensional motion.

D. Remark that $\vec{a}_{\rm PB} = \vec{a}_{\rm PA}$ if the two coordinate systems are not accelerating with respect to each other. This is an important point for the discussion of inertial reference frames in Chapter 5.

E. Work several problems dealing with airplanes flying in the wind and boats sailing in moving water. Emphasize that relative motion problems are chiefly exercises in vector addition. To help students understand some of the problems explain that a boat's "heading" is its direction of motion in a frame attached to the water, while its direction of travel is its direction of motion in a frame attached to the ground.

SUGGESTIONS

1. Assignments
   a. Assign problems 1, 10, and 15 to have students think about the analysis of motion in two dimensions.
   b. Use questions 3 through 10 to generate discussions of ideal projectile motion.
   c. Ask questions 12 and 13 in connection with centripetal acceleration.
   d. Have students work several of problems 17 through 26 and 28 through 41. Some of these deal with sports. See, for example, problems 39, 40, and 41.
   e. Assign problems 46 and 48 in connection with uniform circular motion.
   f. Assign one or two problems that deal with relative motion. Good examples are 61 and 64.
2. Demonstrations
   Projectile motion: Freier and Anderson Mb14, 16, 17, 19, 20, 23, 24, 28.
3. Audio/Visual
   a. *A Matter of Relative Motion*, *Galilean Relativity — Ball Dropped from Mast of Ship*; *Object Dropped from Aircraft*, *Projectile Fired Vertically*; from AAPT collection 1 of single-concept films; video tape; available from Ztek Co., PO Box 11768, Lexington, KY 40577–1768.
   b. *Projectile Motion*, *Circular Motion*; Side B: Mechanics (II) of Cinema Classics; video disk; available from Ztek Co. (see above for address).
   c. *Projectile Motion*; VHS video tape (15 min); Films for the Humanities & Sciences, PO Box 2053, Princeton, NJ 08543–2053.
   d. *Circular Motion*; VHS video tape (10 min); Films for the Humanities & Sciences (see above for address).
   e. *Introduction to Relative Motion*; VHS video tape (8 min); Films for the Humanities & Sciences (see above for address).
   f. *The Ape in the Tree*; Demonstrations of Physics: Motion; VHS video tape (3:13); Media Design Associates, Inc., Box 3189, Boulder, CO 80307–3190.
   g. *Reference Frames* from Skylab Physics Videodisc; video disk; available from Ztek Co. (see above for address).
   h. *Projectile Motion*; Physics Demonstrations in Mechanics, Part I; VHS video tape; ≈3 min; Physics Curriculum & Instruction, 22585 Woodhill Drive, Lakeville, MN 55044.
   i. *Circular Motion*; Physics Demonstrations in Mechanics, Part I; VHS video tape; ≈3 min; Physics Curriculum & Instruction (see above for address).
   j. *Velocity and Acceleration Vectors*; *Frame of Reference*; Physics Demonstrations in Mechanics, Part III; VHS video tape or video disk; ≈3 min each; Physics Curriculum & Instruction (see above for address).
   k. *Projectile Motion*; *Circular Motion*; Physics Demonstrations in Mechanics, Part V; VHS video tape or video disk; ≈3 min; Physics Curriculum & Instruction (see above for address).
4. Computer Software
   a. *Adding Velocities*; Windows; Cross Educational Software, Inc., 508 E. Kentucky Avenue, PO Box 1536, Ruston, LA 71270. Relative velocity is considered in both one and two dimensions.
   b. *Projectiles*; Vernier Software, 8565 S.W. Beaverton-Hillside Hwy., Portland, OR 97225–2429. A simulation program that allows students to experiment with projectile motion.
   c. *Mechanics in Motion*; Stephen Saxon; available from Physics Academic Software, Box 8202, North Carolina State University, Raleigh, NC 27695–8202. Contains projectile, pendulum, and collision simulators. Can also be used to demonstrate conservation of energy and rotational motion.
   d. *Newtonian Sandbox*. See Chapter 2 SUGGESTIONS.

e. *Objects in Motion*. See Chapter 2 SUGGESTIONS.
f. *Physics Demonstrations*. See Chapter 2 SUGGESTIONS.
g. *Dynamic Analyzer*. See Chapter 2 SUGGESTIONS.

5. Computer Projects
   a. Have students use a commercial math program or their own root finding programs to solve projectile motion problems.
   b. Have students use a spreadsheet or write a computer program to tabulate the coordinates and velocity components of a projectile as functions of time. Have them change the initial velocity and observe changes in the coordinates of the highest point and in the range. Ask them to find the firing angle for the greatest horizontal coordinate when the landing point is above or below the firing point.
6. Laboratory
   a. Meiners Experiment 7–9: *Ballistic Pendulum — Projectile Motion* (use only the first method in connection with this chapter). Students find the initial velocity of a ball shot from a spring gun by measuring its range. Emphasizes the use of kinematic equations.
   b. *Inelastic Impact and the Velocity of a Projectile* (use only Procedure B with this chapter). In addition to using range data to find the initial velocity, students plot the range as a function of firing angle.

Chapter 5 FORCE AND MOTION — I

BASIC TOPICS

I. Overview
   A. Explain that objects may interact with each other and their velocities change as a result. State that the strength of an interaction depends on properties of the objects and their relative positions. Gravitational mass is responsible for gravitational interactions, electric charge is responsible for electric and magnetic interactions.
   B. Explain that we split the problem into two parts and say that each body exerts a force on the other and that the net force on a body changes its velocity. Remark that an equation that gives the force in terms of the properties of the objects and their positions is called a force law. Force laws are discussed throughout the course. The dominant theme of this chapter, however, is the relationship between the net force and the acceleration it produces.

II. Newton's first law.
   A. State the law: if an object does not interact with any other objects, its acceleration is zero.
   B. Point out that the acceleration depends on the reference frame used to measure it and that the first law can be true for only a select set of frames. Cover the essential parts of the relative motion section of Chapter 4, if they were not covered earlier. Define an inertial frame. Tell students that an inertial frame can be constructed, in principle, by finding an object that is not interacting with other objects and then attaching a reference frame to it. Any frame that moves with constant velocity relative to an inertial frame is also an inertial frame, but one that is accelerating relative to an inertial frame is not.
   C. Explain that we may take a reference frame attached to Earth as an inertial frame for the description of most laboratory phenomena but we cannot for the description of ocean and wind currents, space probes, and astronomical phenomena.

III. Newton's second law.
   A. Explain that the environment influences the motion of an object and that force measures the extent of the interaction. The result of the interaction is an *acceleration*. Place a cart at rest on the air track. Push it to start it moving and note that it continues at constant

velocity. After it is moving, push it to increase its speed, then push it to decrease its speed. In each case note the direction of the force and the direction of the acceleration. Also give an eraser a shove across a table and note that it stops. Point out that the table top exerts a force of friction while the eraser is moving. Push the eraser at constant velocity and explain that the force of your hand and the force of friction sum to zero.

B. Define force in terms of the acceleration imparted to the standard 1 kg mass. Explain how this definition can be used to calibrate a spring, for example. Point out that force is a vector, in the same direction as the acceleration. If two or more forces act on the standard mass, its acceleration is the same as when a force equal to the resultant acts. Unit: newton. Explain that 1 N is 1 kg·m/s^2.

C. Have three students pull on a rope, knotted together as shown. Ask one to increase his or her pull and ask the others to report what they had to do to remain stationary.

D. Define mass in terms of the ratio of the acceleration imparted to the standard mass and to the unknown mass, with the same force acting. Attach identical springs to two identical carts, one empty and the other containing a lead brick. Pull with the same force (same elongation of the springs) and observe the difference in acceleration. Unit: kilogram.

E. State the second law. Stress that the force that appears is the net or resultant force. Explain that the law holds in inertial frames. Point out that this is an experimentally established law and does not follow as an identity from the definitions of force and mass. Emphasize that $m\vec{a}$ is not a force.

F. Discuss examples: calculate the constant force required to stop an object in a given time, given the mass and initial velocity; calculate the force required to keep an object in uniform circular motion, given its speed and the radius of its orbit. Calculate the acceleration of an object being pushed by two forces in opposite directions and note that the acceleration vanishes if the forces have equal magnitudes. Emphasize that the forces continue to act but their sum vanishes. Some students believe that the forces literally cancel each other and no longer act.

IV. Newton's third law.

A. State the law. Stress that the two forces in question act on different bodies and each helps to determine the acceleration of the body on which it acts. Explain that the third law describes a characteristic of force laws. State that the two forces in an action-reaction pair are of the same type: gravitational, for example.

B. Discuss examples. Hold a book stationary in your hand, identify action-reaction pairs (hand-book, book-Earth). Now allow your hand and the book to accelerate downward with an acceleration less than g and again identify action-reaction pairs. Note that you can control the acceleration of the book by means of the force you exert but once you exert a given force you cannot control the force that the book exerts on you.

C. Attach a force probe to each of two air-track carts. Use a computer to plot the force that each exerts on the other as the carts collide. Point out that at each instant the forces have the same magnitude and are in opposite directions.

V. Applications of Newton's laws involving a single object.

A. Go over the steps used to solve a one-body problem: identify the body and all forces acting on it, draw a free-body diagram, choose a coordinate system, write the second law in component form, and finally solve for the unknown.

B. Some special forces should be explained. They are important for many of the problems but are rarely mentioned explicitly. Warn students they must take these forces into account if they act.

1. Point out that the magnitude of the gravitational force is mg, where g is the local acceleration due to gravity and m is the mass of the object. It is directed toward the center of Earth. Explain that the magnitude of this force is the weight of the object. Explain that weight varies with altitude and slightly from place to place on the surface of Earth, but mass does not vary. Emphasize that the appearance of g in the formula for the gravitational force does not imply that the acceleration of the body is g.
2. Point out that a massless rope transmits force unaltered in magnitude and that the magnitude of the force it exerts on objects at each end is called the tension force. If a person pulls an object by exerting a force on a string attached to the object, the motion is as if the person pulled directly on the object. The string serves to define the direction of the force. A frictionless, massless pulley serves to change the direction but not the magnitude of the tension force of the rope passing over it.
3. Explain that the normal force of a surface on an object originates in elastic and ultimately electric forces. It prevents the object from moving through the surface. State that it is perpendicular to the surface. If the surface is at rest, the normal force adjusts so the acceleration component perpendicular to the surface vanishes. More generally, the object and the surface have the same perpendicular acceleration component. Place a book on the table and press on it. State that the normal force is greater than when you were not pressing. Hold the book against the wall by pressing on it and mention that the normal force is horizontal.

C. Set up the situation described in Sample Problem 5–7 using an inclined air track but attach a calibrated spring scale to the support at the top of the incline and tie the other end of the scale to the block. Calculate the tension force of the string and compare the result to the reading on the scale. Cut the string, then calculate the acceleration.

D. Consider a person standing on a scale in an elevator. State that the scale measures the normal force and calculate its value for an elevator at rest, one accelerating upward, one accelerating downward with $a < g$, and one in free fall. See Sample Problem 5–8.

VI. Applications of Newton's laws involving more than one object.

A. Explain that when two or more objects are involved, a free-body diagram must be drawn for each. A Newton's second law equation, in component form, is also written for each object. Point out that differently oriented coordinate systems may be used for different bodies. Show how to invoke the third law when necessary. Explain that the same symbol should be used for the magnitude of the two forces of an action-reaction pair and that their opposing directions are taken into account when drawing the free-body diagram and in writing the second law equations.

B. Explain that in some cases both objects can be considered as a single object. Say that the objects must have the same acceleration and that the forces they exert on each other must not be requested. The mass of the single object is then the sum of the masses of the constituent objects and internal forces are not included in the analysis.

C. Use examples to show how rods, strings, and pulleys relate the motions of bodies in various cases. Explain that, in addition to the second law equations, there will be equations relating the accelerations of the objects. Show that these equations depend on the choice of coordinate systems.

D. Consider several examples, carefully explaining each step. If you have not developed an application of your own, work Sample Problem 5–9 in the text. If possible, give a demonstration.

SUGGESTIONS

1. Assignments
   a. Use questions 1 through 6 to help students think about the influence of forces on bodies. Assign one or two of problems 1, 2, and 3.
   b. Use question 8 to help students think about normal forces.
   c. Use questions 7 and 9 and problem 9 to help students with tensions in ropes.
   d. Discuss question 11 in connection with Newton's third law of motion.
   e. Discuss question 12 in connection with inclined plane problems.
   f. Assign problem 1 to emphasize the definition of force and problem 6 or 7 to demonstrate Newton's second law.
   g. Use problems 11 and 12 to discuss mass and weight.
   h. Use problems 27 and 31 to discuss Newton's third law.
   i. Assign a few applications problems from the group 13 through 56, according to the needs and interests of the class.
   j. As a prelude to Chapter 9 (where the center of mass and conservation of momentum are discussed) assign problem 25.
2. Demonstrations
   a. Inertia: Freier and Anderson Mc1 — 5, Me1.
   b. $\vec{F} = m\vec{a}$: Freier and Anderson Md2, Ml1.
   c. Third-law pairs: Freier and Anderson Md1, 3, 4.
   d. Mass and weight: Freier and Anderson Mf1, 2.
   e. Tension in a string: Freier and Anderson Ml1.
3. Books and Monographs

   Resource Letters, Book Four; American Association of Physics Teachers, One Physics Ellipse, College Park, MD 20740–3845. Contains a resource letter on mechanics.
4. Audio/Visual
   a. *Frames of Reference Videodisc*; video disk; available from Ztek Co., PO Box 11768, Lexington, KY 40577–1768.
   b. *Human Mass Measurement* from Skylab Physics Videodisc; video disk; available from Ztek Co. (see above for address).
   c. *Newton's First and Second Laws*; *Newton's Third Law*; *Inertial Forces*; *Translational Acceleration*; from AAPT collection 2 of single-concept films; video tape; available from Ztek Co., PO Box 11768, Lexington, KY 40577–1768.
   d. *Inertial Forces – Translational Acceleration*; from AAPT Miller collection of single-concept films; video tape; available from Ztek Co. (see above for address).
   e. *Forces, Newton's Laws*; Side A: Mechanics (I) of Cinema Classics; video disk; available from Ztek Co. (see above for address).
   f. *Newton's First Law*; Demonstrations of Physics: Motion; VHS video tape (5:40); Media Design Associates, Inc., Box 3189, Boulder, CO 80307–3190.
   g. *Newton's Second Law*; Demonstrations of Physics: Motion; VHS video tape (6:31); Media Design Associates, Inc. (see above for address).
   h. *Inertia*; VHS video tape (10 min); Films for the Humanities & Sciences, PO Box 2053, Princeton, NJ 08543–2053.
   i. *Mechanics in Action*; VHS video tape (25 min); Films for the Humanities & Sciences (see above for address).
   j. *Energy and Force: Part 1*; VHS video tape (24 min); Films for the Humanities & Sciences (see above for address).
   k. *The Laws of Motion*; VHS video tape (21 min); Films for the Humanities & Sciences (see above for address).

l. *The Laws of Motion Applied*; VHS video tape (21 min); Films for the Humanities & Sciences (see above for address).

m. *Newton's 1st Law*; *Newton's 2nd Law*; *Newton's 3rd Law*; Physics Demonstrations in Mechanics, Part II; VHS video tape; ≈3 min each; Physics Curriculum & Instruction, 22585 Woodhill Drive, Lakeville, MN 55044.

n. *Newton's 1st Law*; Physics Demonstrations in Mechanics, Part III; VHS video tape or video disk; ≈3 min; Physics Curriculum & Instruction (see above for address).

o. *Fundamental Forces*; *Mass and Weight*; *Newton's 2nd Law*; *Newton's 2nd and 3rd Laws*; *Force Components*; Physics Demonstrations in Mechanics, Part IV; VHS video tape or video disk; ≈3 min each; Physics Curriculum & Instruction (see above for address).

5. Computer Software

   a. *Freebody*; Graham Oberum; available from Physics Academic Software, Box 8202, North Carolina State University, Raleigh, NC 27695–8202. Students draw force vectors and can change the length and orientation of the vectors in response to questions. The screen gives the components.

   b. *Force and Motion Microworld*; Ping-Kee L. Tao and Ming-Wai Tse; available from Physics Academic Software (see above for address). Uses velocity graphs to display the effects of force on the motion of an object. Includes drag forces.

   c. *Dynamic Analyzer*; Roger Sipson; Windows; available from Physics Academic Software (see above for address). The program allows students to explore one- and two-dimensional motion, harmonic motion, and motion in electric and magnetic fields. The motions are simulated on the screen and various graphs are drawn side-by-side with the simulations. The user can change the parameters of the motion.

   d. *Newtonian Sandbox*. See Chapter 2 SUGGESTIONS.

   e. *Motion*; Macintosh, Windows; Cross Educational Software, Inc., 508 E. Kentucky Avenue, PO Box 1536, Ruston, LA 71270. Contains sections on second law problems, including projectiles, friction, and uniform circular motion.

6. Laboratory

   Meiners Experiment 8–2: *Concept of Mass: Newton's Second Law of Motion.* Students measure the accelerations of two pucks that interact via a spring on a nearly frictionless surface and compare the ratio to the ratio of their masses. This experiment may be done with dry ice pucks or on an air table or air track.

Chapter 6 FORCE AND MOTION — II

BASIC TOPICS

I. Frictional forces.

   A. Place a large massive wooden block on the lecture table. Attach a spring scale, large enough to be read easily. If necessary, tape sandpaper to the table under the block. Pull weakly on the scale and note that the reading is not zero although the block does not move. Pull slightly harder and note that the reading increases but the block still does not move. Remark that there must be a force of friction opposing the pull and that the force of friction increases as the pull increases. Now increase your pull until the block moves and note the reading just before it starts to move. Pull the block at constant speed and note the reading. Have the students repeat the experiment in a qualitative manner, using books resting on their chair arms. To show that the phenomenon depends on the nature of the surface, the demonstration can be repeated after waxing the wooden block and table top.

B. Give a brief qualitative discussion about the source of frictional forces. Stress that the force of static friction has whatever magnitude and direction are required to hold the two bodies in contact at rest relative to each other, up to a certain limit in magnitude. Define the coefficient of static friction and explain the use of $f_s < \mu_s N$. In particular, explain that if the surface is stationary the force of static friction is determined by the condition that the object on it has zero acceleration. To test if an object remains at rest, the frictional force required to produce zero relative acceleration is calculated and compared with $\mu_s N$.

C. Define the coefficient of kinetic friction and explain that $f_k = \mu_k N$ gives the frictional force as long as the object is sliding on the surface. Also explain that if the surface is stationary the force of kinetic friction is directed opposite to the velocity of the object sliding on it.

D. Work some examples:

1. Find the angle of an inclined plane for which sliding starts; find the angle at which the body slides at constant speed. These examples can be analyzed in association with a demonstration and the students can use the data to find the coefficients of friction.
2. Analyze an object resting on the floor, with a person applying a force that is directed at some angle above the horizontal. Find the minimum applied horizontal force that will start the object moving and point out that it is a function of the angle between the applied force and the horizontal.
3. Consider the same situation but with the object moving. Find its acceleration. This and the previous example demonstrate the dependence of the normal force and the force of friction on the externally applied force.
4. To give an illuminating variant, consider a book being held against the wall by a horizontal force. Calculate the minimum applied force that will keep the book from falling.

II. Drag forces and terminal speed.

A. Make or buy a small toy parachute. Drop two weights side by side and note they reach the floor at the same time. Attach the parachute to one and repeat. Explain that the force of the air reduces the acceleration.

B. State that for turbulent flow of air around an object the magnitude of the drag force is given by $D = \frac{1}{2}C\rho Av^2$, where A is the effective cross-sectional area, ρ is the density of air, and v is the speed of the object relative to the air. C is a drag coefficient, usually determined by experiment. Remark that a parachute increases the cross-sectional area. State that the drag force is directed opposite to the velocity.

C. Explain that as an object falls its speed approaches terminal speed as a limit. Write down Newton's second law for a falling object and point out that the drag and gravitational forces are in opposite directions. Suppose the object is dropped from rest and point out that the acceleration is g at first but as the object gains speed its acceleration decreases in magnitude. At terminal speed the acceleration is zero and remains zero, so the velocity no longer changes. Show that zero acceleration leads to $v_t = \sqrt{2mg/C\rho A}$. Point out Table 6–1, which gives some terminal speeds.

D. Remark that if an object is thrown downward with a speed that is greater than terminal speed it slows down until terminal speed is reached.

E. Qualitatively discuss projectile motion with drag. The horizontal component of the velocity tends to zero while the vertical component tends to the terminal speed. Contrast the trajectory with one in the absence of air resistance.

III. Uniform circular motion.

A. Point out that for uniform circular motion to occur there must be a radially inward force of constant magnitude and that something in the environment of the body supplies the force. Whirl a mass tied to a string around your head and explain that the string supplies the force. Set up a loop-the-loop with a ball or toy cart on a track and explain that the combination of the normal force of the track and the force of gravity supplies the centripetal force. Have students identify the source of the force in examples and problems as they are discussed.

B. Point out that $F = mv^2/r$ is just $F = ma$ with the expression for centripetal acceleration substituted for a.

C. Discuss problem solving strategy. After identifying the forces, find the radial component of the resultant and equate it to mv^2/r.

D. Examples:
   1. Find the speed and period of a conical pendulum.
   2. Find the speed with which a car can round an unbanked curve, given the coefficient of static friction.
   3. Find the angle of banking required to hold a car on a curve without aid of friction.
   4. Analyze the loop-the-loop and point out that the ball leaves the track when the normal force vanishes. Show that the critical speed at the top is given by $v^2/r = g$.

SUGGESTIONS

1. Assignments
   a. Discuss some or all of questions 1, 3, 4, 5, and 6 in connection with the force of static friction and the onset of sliding. Kinetic friction is the subject of questions 2, 7, and 8. Consider asking question 8 after assigning problem 25.
   b. Ways in which the coefficient of static friction is used are emphasized in problems 1, 2, 8, 9, and 11. Problem 25 is more challenging. To help students understand the role played by the normal force in the onset of sliding, assign problem 12. Problem 16 deals with the role of the normal force in kinetic friction.
   c. Problems 17, 26, and 27 provide some interesting applications of the laws of friction.
   d. Use questions 9 and 10 in your discussion of centripetal acceleration and force. Problems 37 and 39 cover the role of friction in rounding a level curve. Also consider problems 44 and 45.
2. Demonstrations
   a. Friction: Freier and Anderson Mk.
   b. Inclined plane: Freier and Anderson Mj2.
   c. Centripetal acceleration: Freier and Anderson Mb29, 31, Mm1, 2, 4 — 8, Ms5.
3. Audio/Visual
   a. *Trajectories*; from AAPT collection 2 of single-concept films; video tape; available from Ztek Co., PO Box 11768, Lexington, KY 40577–1768.
   b. *Inertial Forces — Centripetal Acceleration*; from AAPT Miller collection of single-concept films; video tape; available from Ztek Co. (see above for address).
   c. *Uniform Circular Motion*; Demonstrations of Physics: Motion; VHS video tape (5:57); Media Design Associates, Inc., Box 3189, Boulder, CO 80307–3190.
   d. *Circular Motion*; VHS video tape (10 min); Films for the Humanities & Sciences, PO Box 2053, Princeton, NJ 08543–2053.
   e. *Terminal Velocity*; Physics Demonstrations in Mechanics, Part II; VHS video tape; ≈3 min; Physics Curriculum & Instruction, 22585 Woodhill Drive, Lakeville, MN 55044.
4. Computer Software
   a. *Dynamic Analyzer*. See Chapter 5 SUGGESTIONS.
   b. *Motion*; Macintosh, Windows; Cross Educational Software, Inc., 508 E. Kentucky Avenue, PO Box 1536, Ruston, LA 71270. Contains sections on second law problems, including projectiles, friction, and uniform circular motion.

5. Computer Projects
   a. Have students use a computer program to investigate objects that are subjected to time dependent forces. To check the program first have them consider a constant force and compare machine generated functions with the known kinematic equations.
   b. Have students modify the program to integrate Newton's second law for velocity dependent forces, then have them investigate the motion of an object subjected to a force that is proportional to v or v^2. It is instructive to have them plot the velocity components as functions of time for a projectile fired straight up or down, subject to air resistance. Consider initial velocities that are both greater and less than the terminal velocity. Also have them study the maximum height and range of projectiles with various coefficients of air resistance. Some projects are given in the Computer Projects section of this manual.
6. Laboratory
   a. Meiners Experiment 7–6: *Coefficient of Friction — The Inclined Plane.* Students determine the coefficients of static and sliding friction for three blocks on an inclined plane. They devise their own experimental procedures.
   b. Meiners Experiment 7–7: *Radial Acceleration* (Problem I only). The centripetal force and the speed of a ball on a string, executing uniform circular motion, are measured for various orbit radii. Essentially a verification of $F = mv^2/r$.
   c. Meiners Experiment 7–8: *Investigation of Uniform Circular Motion*, or Bernard Experiment 13: *Centripetal Force.* Students measure the force acting on a body undergoing uniform circular motion, with the centripetal force provided by a spring.
   d. Meiners Experiment 8–3: *Centripetal Force.* Students measure the speed of a puck undergoing uniform circular motion on a nearly frictionless surface. The data is used to calculate the centripetal force.

Chapter 7 KINETIC ENERGY AND WORK

BASIC TOPICS

I. Kinetic energy and the work-kinetic energy theorem.
   A. Define kinetic energy for a particle. Remind students that kinetic energy is a scalar and depends on the speed but not on the direction of the velocity. Point out that $v^2 = v_x^2 + v_y^2$ for two-dimensional motion and remark that the appearance of velocity components in the expression does *not* mean K has components.
   B. Consider a ball thrown into the air. Neglect air resistance and point out that during the upward part of the motion the force of gravity slows the ball and the kinetic energy decreases. As the ball falls, the force of gravity speeds the ball and the kinetic energy increases. Remind students that for a constant force (and acceleration) $v^2 = v_0^2 + 2a\,\Delta x$ (which was derived in the study of kinematics). Multiply by $m/2$ to obtain $K = K_0 + F\,\Delta x$. Say that for a constant force acting on a particle that moves in one dimension $W = F\,\Delta x$ is the work done by the force F as the particle travels through the displacement Δx. State that $K = K_0 + W$ is an example of the work-kinetic energy theorem: the change in the kinetic energy of a particle during a given interval equals the work done on the particle by the total force during the interval.
   C. Point out that only the component of a force parallel or antiparallel to the velocity changes the speed. Other components change the direction of motion. Positive total work results in an increase in kinetic energy and speed, negative total work results in a decrease. Remind students of previous examples in which the object moved with constant speed (including uniform circular motion). The total work was zero and the kinetic energy did not change. Avoid quantitative calculations involving frictional forces.

D. Explain that the work-kinetic energy theorem can be applied only to particles and objects that can be treated as particles. To give an example in which it cannot be applied directly, consider a car crashing into a rigid barrier: the barrier does no work but the kinetic energy of the car decreases.

E. Explain that observers in different inertial frames will measure different values of the net work done and for the change in the kinetic energy but both will find $W_{\text{net}} = \Delta K$.

F. Use Newton's second law to prove the theorem for motion in one dimension. If the students are mathematically sophisticated, extend the theorem to the general case. Stress that it is the total work (done by the resultant force) that enters the theorem.

II. Work done by a constant force.

A. Write down $W = \vec{F} \cdot \vec{d} = Fd\cos\phi$ and point out ϕ on a diagram. Explain that this is the work done *on* a particle *by* the constant force $\vec{F}$ as the particle undergoes a displacement $\vec{d}$. Explain that work can be calculated for each individual force and that the total work done on the particle is the work done by the resultant force. Point out that work is a scalar quantity. Also point out that work is zero for a force that is perpendicular to the displacement and that, in general, only the component of $\vec{F}$ tangent to the path contributes to the work. The force does no work if the displacement is zero. Emphasize that work can be positive or negative, depending on the relative orientation of $\vec{F}$ and $\vec{d}$. For a constant force, the work depends only on the displacement, not on details of the path. Unit: joule.

B. Calculate the work done by the force of gravity as a mass falls a distance h and as it rises a distance h. Emphasize the sign. Calculate the work done by a non-horizontal force used to pull a box across a horizontal floor. Point out that the work done by the normal force and the work done by the force of gravity are zero. Consider both an accelerating box and one moving with constant velocity. Repeat the calculation for a crate being pulled up an incline by a force applied parallel to the incline. Show the work done by gravity is $-mgh$, where h is the change in the height of the crate.

III. Work done by a variable force.

A. For motion in one dimension, discuss the integral form for work as the limit of a sum over infinitesimal path segments. Explain that the sum can be carried out by a computer even if the integral cannot be evaluated analytically.

B. Examples: derive expressions for the work done by an ideal spring and a force of the form k/x^2. If you have not yet discussed the force of an ideal spring, do so now as a preface to the calculation of work. Explain how the spring constant can be found by hanging a mass from the spring and measuring the extension. Demonstrate changes in the spring length during which the spring does positive work and during which the spring does negative work.

C. As an example consider a stone dropped onto a vertical spring and calculate the maximum compression of the spring, given the mass of the stone, the height from which it is dropped, and the spring constant of the spring.

D. For motion in more than one dimension, write down the expression for the work in the form $\int_i^f \vec{F} \cdot d\vec{r}$ and explain its interpretation as the limit of a sum over infinitesimal path segments. Explain that this is the general definition of work. Calculate the work done by the applied force, the force of gravity, and the tension in the string as a simple pendulum is pulled along its arc until it is displaced vertically through a height h by a horizontal applied force $\vec{F}$.

IV. Power.

A. Define power as $P = dW/dt$. Unit: watt.

B. Show that $P = \vec{F} \cdot \vec{v}$. Explain that the work done over a time interval is given by $\int P\,dt$.

SUGGESTIONS

1. Assignments
   a. The idea of kinetic energy is covered in question 1. Assign one or more of problems 1, 2, 3, and 5 to exercise the concept.
   b. The idea of work is covered in questions 2, 3, and 6. Problems 6, 7, and 10 are good examples of the quantitative aspects. The work done by the net force, as opposed to individual forces, is emphasized in question 3 and problems 11 and 13.
   c. To discuss the work done by gravity, ask questions 7 and 8, then assign problems 14, 16, and 17. To discuss the force exerted by an ideal spring and the work done by it, ask questions 10 and 11, then assign problem 21.
   d. Questions 4, 5, 9, 12, and 13 and problem 10 are a good introduction to the work-kinetic energy theorem. Also assign problem 22 or 23.
   e. Assign problem 24 or 29 in connection with the work done by a variable force.
   f. Assign one or more of problems 30, 32, 34, and 36 in connection with power.
2. Demonstrations
   Work: Freier and Anderson Mv1.
3. Audio/Visual
   a. *Work and Energy*; Side E: Conservation Laws of Cinema Classics; video disk; available from Ztek Co., PO Box 11768, Lexington, KY 40577–1768.
   b. *Energy and Force: Part 2*; VHS video tape (25 min); Films for the Humanities & Sciences, PO Box 2053, Princeton, NJ 08543–2053.
   c. *Potential Energy and Kinetic Energy*; VHS video tape (10 min); Films for the Humanities & Sciences (see above for address).
   d. *Work and Energy*; Physics Demonstrations in Mechanics, Part VI; VHS video tape or video disk; ≈3 min; Physics Curriculum & Instruction, 22585 Woodhill Drive, Lakeville, MN 55044.
4. Computer Project
   Have students use a commercial math program or write a program to numerically evaluate the integral for work, then use the program to calculate the work done by various forces, given as functions of position. Include a nonconservative force and use the program to show the work done on a round trip does not vanish. Some projects can be found in the Computer Projects section of this manual.
5. Laboratory
   a. Meiners Experiment 7–16: *Elongation of an Elastomer.* Students measure the elongation of an elastomer for a succession of applied forces and use a polar planimeter to calculate the work done by the force. The experiment may also be done in connection with Chapter 13.
   b. Bernard Experiment 10: *Mechanical Advantage and Efficiency of Simple Machines.* This experiment can be used to broaden the course to include these topics. A lever, an inclined plane, a pulley system, and a wheel and axle are studied. In each case the force output is measured for a given force input and the work input is compared to the work output.

Chapter 8 POTENTIAL ENERGY AND CONSERVATION OF ENERGY

BASIC TOPICS

I. Potential energy, conservative and nonconservative forces
   A. Explain that potential energy is an energy of configuration. The potential energy of a

system of objects depends on the relative positions of the objects. A system consisting of an object and Earth has a potential energy that depends on the separation of Earth and the object, for example.

B. State that a potential energy can be associated with a force only if that force is conservative and explain that a force is conservative if the work done by the force when the system starts and ends with the same configuration is zero, no matter what the configurations and no matter what motions occur between the beginning and end of the interval. Show that this implies that the work done by the force between any given starting and ending configurations is the same no matter what intervening configurations are assumed by the system.

C. Discuss the force of gravity and the force of an ideal spring as examples. For either or both of these, show that the work done depends only on the end points and not on the path between, then argue that the work vanishes for a round trip. Point out that on some parts of the path the force does positive work while on other parts it does negative work. Demonstrate that the work done by a spring is independent of the path by considering two different motions with the same end points. For the first motion, have the mass go directly from the initial point to the final point; for the second, have it first go away from the final point before going there.

D. Use a force of friction with constant magnitude as an example of a nonconservative force. Consider a block on a horizontal table top and argue that the work done by the force cannot vanish over a round trip since it is negative for each segment. Suppose the block moves around a circular path and friction is the only force that does work. Argue that the object returns to its initial position with less kinetic energy than it had when it started. State that a potential energy cannot be associated with a frictional force.

E. Use a cart on a linear air track to demonstrate these ideas. Couple each end of the cart, via a spring, to a support at the corresponding end of the air track. Give the cart an initial velocity and tell students to observe its speed each time it returns to its initial position. Point out that the kinetic energy returns to nearly the same value and that the springs do zero work during a round trip. Reduce or eliminate air flow to show the influence of a nonconservative force. If this is done rapidly and skillfully, you can cause the cart to stop far from the starting point.

II. Potential energy.

A. Give the definition of potential energy in terms of work for motion in one dimension. See Eq. 8–6 and emphasize that the change in potential energy is the negative of the work done *by* the force responsible for the potential energy.

B. Discuss the following properties:

1. The zero is arbitrary. Only potential energy differences have physical meaning.
2. The potential energy is a *scalar* function of position.
3. The force is given by $F = -\mathrm{d}U/\mathrm{d}x$ in one dimension.
4. Unit: joule.

C. Derive expressions for the potential energy functions associated with the force of gravity (uniform gravitational field) and the force of an ideal spring. Stress that the potential energy is a property of the object-Earth or spring-mass system and depends on the *configuration* of the system.

D. Use the work-kinetic energy theorem to show that $W = \Delta U$ is the work that must be applied by an external agent to increase the potential energy by ΔU if the kinetic energy does not change. Show that ΔU is recovered as kinetic energy when the external agent is removed. Example: raising an object in a gravitational field.

III. Conservation of energy.

A. Explain that if all the forces acting between the objects of a system are conservative and the net work done by external forces on objects of the system is zero then $K + U =$ constant. This follows from the work-kinetic energy theorem with the work of the conservative forces represented by the change in potential energy. The negative sign in Eq. 8–6 is essential to obtain this result. Emphasize that U is the sum of the individual potential energies if more than one conservative force acts. Define the total mechanical energy as $E_{\rm mec} = K + U$.

B. Discuss the conversion of kinetic to potential energy and vice versa. Drop a superball on a rigid table top and point out when the potential and kinetic energies are maximum and when they are minimum. The question of elasticity can be glossed over by saying that to a good approximation the ball rebounds with unchanged speed. Also discuss the energy in a spring-mass system. Return to the cart on the air track and discuss its motion in terms of $K + U =$ constant. To avoid later confusion in the students' minds, start the motion with neither K nor U equal to zero. Emphasize that the energy remains in the system but changes its form during the motion. The agent of the change is the work done by the forces of the springs.

C. Show how to calculate the total energy for a spring-mass system from the initial conditions. Write the conservation principle in the form $\frac{1}{2}mv^2 + U(x) = \frac{1}{2}mv_0^2 + U(x_0)$. Use conservation of energy to find expressions for the maximum speed, maximum extension, and maximum compression, given the total energy.

D. Use the example of a ball thrown upward to demonstrate that conservation of energy must be applied to a system rather than to a single particle. Remark that Earth does work on the ball, the ball does work on Earth, and the change in potential energy is the negative of the sum. Show it is mgh, where h is the change in their separation. Remark that both kinetic energies change and the total change is the negative of the change in the potential energy. Explain that because Earth is so massive the change in its kinetic energy is small and may be neglected.

E. Discuss potential energy curves. Use the curve for a spring-mass system, then a more general one, and show how to calculate the kinetic energy and speed from the coordinate and total energy. Point out the turning points on the curves and discuss their physical significance. Use $F = -{\rm d}U/{\rm d}x$ to argue that the particle turns around at a turning point. For an object on a frictionless roller coaster track, find the speed at various points and identify the turning points.

F. Define stable, unstable, and neutral equilibrium. Use a potential energy curve (a frictionless roller coaster, say) to illustrate. Emphasize that ${\rm d}U/{\rm d}x = 0$ at an equilibrium point.

IV. Potential energy in two and three dimensions.

A. Define potential energy as a line integral and explain that it is the limit of a sum over infinitesimal path segments. Remark that conservation of energy leads to $\frac{1}{2}mv^2 + U(x, y, z) =$ constant. Explain that $v^2 = v_x^2 + v_y^2 + v_z^2$ and that v^2 is a scalar.

B. Example: simple pendulum. Since the gravitational potential energy depends on height, in the absence of nonconservative forces the pendulum has the same swing on either side of the equilibrium point and always returns to the same turning points. Demonstrate with a pendulum hung near a blackboard and mark the end points of the swing on the board. For a more adventurous demonstration, suspend a bowling ball pendulum from the ceiling and release the ball from rest in contact with your nose. Stand very still while it completes its swing and returns to your nose.

V. External work and thermal energy.

A. Explain that when forces due to objects external to the system do work W on the system the energy equation becomes $\Delta K + \Delta U = W$ if the internal forces are conservative. K is the total kinetic energy of all objects in the system and U is the total potential energy of their interactions with each other.

B. Show that if the external force is conservative the system can be enlarged to include the (previously) external agent. Then, $W = 0$ and U must be augmented to include the new interactions. Give the example of a ball thrown upward in Earth's gravitational field.

C. Explain that if some or all of the internal forces are nonconservative then the total energy must include an thermal energy term to take account of energy that enters or leaves some or all of the objects and contributes to the energy (kinetic and potential) of the particles that make up the objects. Distinguish between thermal energy and the energy associated with the motion and interactions of the object as a whole. Write $\Delta K + \Delta U + \Delta E_{\rm th} = W$.

D. Refer back to the block sliding on the horizontal table top, discussed earlier. Explain that when the block stops all the original kinetic energy has been converted to thermal energy. As an example assume the table top exerts a constant frictional force of magnitude f. Explain that $K_i = fd$, where K_i is the initial kinetic energy and d is the distance the block slides before stopping. Say that if the system consists of only the block, then $-K_i + \Delta E_{\rm th} = W$, where W is the work done by friction on the block and $\Delta E_{\rm therma}$ is the change in the thermal energy of the block. Note that $-fd$ is not the work done by friction. Now take the system to consist of the both the block and table top. Then $-K_i + \Delta E_{\rm th} = 0$, where $\Delta E_{\rm th}$ is the total change in the thermal energies of the block and table top. Argue that $\Delta E_{\rm th} = fd$. Explain that the division of thermal energy between the block and the table top cannot be calculated without a detailed model of friction.

E. Explain that the quantity $\int \vec{f} \cdot d\vec{r}$ along the path of an object does NOT give the work done by friction but it does contribute to the change in kinetic energy of the object, along with similar contributions from other forces, if any. For a block sliding on a table top, the work done by friction is algebraically greater than the value of the integral and the difference is the increase in the thermal energies of the block and table top.

SUGGESTIONS

1. Assignments
   a. Use question 1 to discuss the idea of a conservative force.
   b. Problems 2, 3, 4, and 7 test basic understanding of gravitational potential energy. Use question 2 and problem 1 to test basic understanding of elastic potential energy.
   c. Test for understanding of the conservation of mechanical energy by asking questions 3, 4, and 5 and assigning some of problems 9, 10, 11, 12, 16, 17, 18, 19, 21, 27, and 34. Some of these are related to previous problems. Some combine gravitational and elastic potential energies.
   d. Draw several potential energy curves and have the class analyze the particle motion for various values of the total energy. This can provide particularly useful feedback as to how well the students have mastered the idea of energy conservation. Also ask question 6 and assign problems 36 and 38.
   e. Assign a few problems dealing with applied and dissipative forces. Consider 41, 42, 43, 45, 50, 51, 56, 61, and 65.

2. Demonstrations
   a. Conservation of energy: Freier and Anderson Mn1 — 3, 6.
   b. Nonconservative forces: Freier and Anderson Mw1.

3. Books and Monographs
   The Bicycle by Phillip DiLavore; available from AAPT, One Physics Ellipse, College Park MD 20740–3845. Deals chiefly with energy. Students do not need to know about rotational motion.

4. Audio/Visual
   a. *Gravitational Potential Energy*; *Conservation of Energy — Pole Vault and Aircraft Take-off*; from AAPT collection 1 of single-concept films; video tape; available from Ztek Co., PO Box 11768, Lexington, KY 40577–1768.
   b. *Energy Conservation*; Side E: Conservation Laws of Cinema Classics; video disk; available from Ztek Co. (see above for address).
   c. *Conservation of Energy*; *Work and Conservation of Energy*; Physics Demonstrations in Mechanics, Part I; VHS video tape; ≈3 min each; Physics Curriculum & Instruction, 22585 Woodhill Drive, Lakeville, MN 55044.
5. Computer Software
   Conservation Laws; Macintosh, Windows; Cross Educational Software, Inc., 508 E. Kentucky Avenue, PO Box 1536, Ruston, LA 71270. Includes a section on conservation of energy.
6. Laboratory
   Bernard Experiment 9: *Work, Energy, and Friction.* A string is attached to a car on an incline and passes over a pulley at the top of the incline. Weights on the free end of the string are adjusted so the car rolls down the incline at constant speed. The work done by gravity on the weights and on the car is calculated and used to find the change in mechanical energy due to friction. The coefficient of friction is computed. The experiment is repeated for the car rolling up the incline and for various angles of incline. It is also repeated with the car sliding on its top and the coefficients of static and kinetic friction are found.

Chapter 9 SYSTEMS OF PARTICLES

BASIC TOPICS

I. Center of mass.
   A. Spin a chalkboard eraser as you toss it. Point out that, if the influence of air can be neglected, one point (the center of mass) follows the parabolic trajectory of a projectile although the motions of other points are more complicated.
   B. Define the center of mass by giving its coordinates in terms of the coordinates of the individual particles in the system. As an example, consider a system consisting of three discrete particles and calculate the coordinates of the center of mass, given the masses and coordinates of the particles. Point out that no particle need be at the center of mass.
   C. Extend the definition to include a continuous mass distribution. Note that if the object has a point, line, or plane of symmetry, the center of mass must be at that point, on that line, or in that plane. Examples: a uniform sphere or spherical shell, a uniform cylinder, a uniform square, a rectangular plate, or a triangular plate. Show how to compute the coordinates of the center of mass of a complex object comprised of a several simple parts, a table for example. Each part is replaced by a particle with mass equal to the mass of the part, positioned at the center of mass of the part. The center of mass of the particles is then found. Explain how to find the center of mass of a simple shape, such as a rectangular or circular plate, with a hole cut in it.
   D. Explain that the general motion of a rigid body may be described by giving the motion of the center of mass and the motion of the object around the center of mass.
   E. Derive expressions for the velocity and acceleration of the center of mass in terms of the velocities and accelerations of the particles in the system.

F. Derive $\vec{F}_{net} = M\vec{a}_{com}$ and emphasize that $\vec{F}_{net}$ is the net *external* force on all objects of the system. As an example, consider a two-particle system with external forces acting on both particles and each particle interacting with the other. Invoke Newton's third law to show that the internal forces cancel when all forces are summed.

G. State that if $\vec{F}_{net} = 0$ and the center of mass is initially at rest, then it remains at the same point no matter how individual parts of the system move. Refer to the two carts of C above.Work a sample problem. For example, consider a person running from one end to the other of a slab that is free to slide on a horizontal frictionless surface or two skaters who pull themselves toward each other with a rope or pole. Ask how far the cart and running person or the two skaters each move.

II. Momentum.

A. Define momentum for a single particle. If you are including modern physics topics in the course, also give the relativistic definition of momentum.

B. Show that Newton's second law can be written $\vec{F}_{net} = d\vec{p}/dt$ for a particle. Emphasize that the mass of the particle is constant and that this form of the law does not imply that a new term $\vec{v}\, dm/dt$ has been added to $\vec{F}_{net} = m\vec{a}$.

C. State that the total momentum of a system of particles is the *vector* sum of the individual momenta and show that $\vec{P} = M\vec{v}_{com}$.

D. Show that Newton's second law for the center of mass can be written $\vec{F}_{net} = d\vec{P}/dt$, where $\vec{P}$ is the total momentum of the system. Stress that $\vec{F}_{net}$ is the net external force and that this equation is valid only if the mass of the system is constant.

III. Conservation of linear momentum.

A. Point out that $\vec{P}$ = constant if $\vec{F}_{net} = 0$. Stress that one examines the *external* forces to see if momentum is conserved in any particular situation. Point out that one component of $\vec{P}$ may be conserved when others are not.

B. Put two carts, connected by a spring, on a horizontal air track and set them in oscillation by pulling them apart and releasing them from rest. Explain that the center of mass does not accelerate and the total momentum of the system is constant. Use the conservation of momentum principle to derive an expression for the velocity of one cart in terms of the velocity of the other. Push one cart and explain that the center of mass is now accelerating and the total momentum is changing.

C. Consider a projectile that splits in two and find the velocity of one part, given the velocity of the other. Point out that mechanical energy is conserved for the cart-spring system but is not for the fragmenting projectile. The exploding projectile idea can be demonstrated with an air track and two carts, one more massive than the other. Attach a brass tube to one cart and a tapered rubber stopper to the other. Arrange so that the tube is horizontal and the stopper fits in its end. The tube has a small hole in its side, through which a firecracker fuse fits. Start the carts at rest and light a firecracker in the tube. The carts rapidly separate, strike the ends of the track, come back together again, and stop. Arrange the initial placement so the carts strike the ends of the track simultaneously. Explain that $\vec{P} = 0$ throughout the motion. For a less dramatic demonstration, tie two carts together with a compressed spring between them, then cut the string.

D. Explain that observers in two different inertial frames will measure different values of the momentum for a system but they will agree on the conservation of momentum. That is, if the net external force is found to vanish in one inertial frame, it vanishes in all inertial frames.

E. Illustrate the use of conservation of momentum to solve problems by considering the firing of a cannon initially resting on a frictionless surface. Assume the barrel is horizontal and calculate the recoil velocity of the cannon. Explain that muzzle velocity is measured relative to the cannon and that we must use the velocity of the cannonball relative to Earth.

IV. Variable mass systems.

A. Derive the rocket equations, Eqs. 9–42 and 9–43. Emphasize that we must consider the rocket and fuel to be a single constant mass system. Derive expressions for the momentum before and after a small amount of fuel is expelled, in terms of the mass of the rocket and fuel together, the mass of the fuel expelled, the initial velocity of the rocket, the change in its velocity, and the relative velocity of the expelled fuel. Assume no external forces act on the rocket-fuel system and equate the two expressions for the total momentum.

B. To demonstrate, screw several hook eyes into a toy CO_2 propelled rocket, run a line through the eyes, and string the line across the lecture hall. Start the rocket from rest and have the students observe its acceleration as it crosses the hall.

C. As a second example, consider the loading of sand on a conveyor belt and calculate the force required to keep the belt moving at constant velocity.

V. External forces and internal energy.

A. Show that if the system is replaced by a single particle with mass equal to the total mass of the system, located at the system center of mass, and acted on by a force equal to the total external force acting on the system, then the work done by the force equals the change in the kinetic energy associated with the motion of the center of mass (defined by $\frac{1}{2}mv_{\text{com}}^2$).

B. Remark that the integral $\int \vec{F}_{\text{net}} \cdot d\vec{s}$ over the path of the center of mass is NOT necessarily the work done by the net external force. In particular, it is not when the motions of the center of mass and the point of application of the force are different. In an extreme case the point of application of the external force does not move so the force does zero work, but the center of mass accelerates and the kinetic energy changes. Remind students of the energy equation $\Delta K + \Delta U + \Delta E_{\text{int}} = W$ and stress that W is the true work done by the external force. State that internal energy may be converted to center of mass kinetic energy or vice versa by the action of an external force.

C. Discuss the example of an accelerating car. Point out that the force of the road on the tires accelerates the car but does no work if the tires do not skid. Internal energy is converted to center of mass external energy. Discuss the example of a skater on ice pushing herself away from a wall and observe that the force of the wall on her accounts for the acceleration of her center of mass but this force does not work.

D. State that for a closed system $\Delta E_{\text{int}} + \Delta E_{\text{mec}} = 0$ and apply this equation to the skater pushing from a wall and to the accelerating automobile.

SUGGESTIONS

1. Assignments

a. Use question 1, along with problems 3, 4, and 6 to generate discussion about the position of the center of mass. To present a challenge, assign problem 9.

b. Questions 3, 4, 5, and 6 are good tests of understanding of the motion of the center of mass. Discuss them as an introduction to the problems. Assign some of problems 11, 14, 15, and 16.

c. Assign some problems in which the center of mass does not move: 10, 12, 18, and 19, for example.

d. To emphasize the vector nature of momentum, assign problems 23 and 24.

e. Use questions 7, 8, and 9 to discuss conservation of momentum, then assign problems such as 27, 29, 32, and 34. To emphasize the difference between conservation of energy and conservation of momentum assign problem 36.

f. Assign a problem or two such as 41, 44, 46, and 47, which are concerned with variable mass systems.

g. To generate a discussion of the role played by internal forces assign problems 52, 55, and 58.

2. Demonstrations

   a. Center of mass, center of gravity: Freier and Anderson Mp7, 12, 13.

   b. Motion of center of mass: Freier and Anderson Mp1, 2, 16 — 19.

   c. Conservation of momentum: Freier and Anderson Mg4, 5, Mi2.

   d. Rockets: Freier and Anderson Mh.

3. Books and Monographs

   Rockets by David Keeports; available from AAPT, One Physics Ellipse, College Park MD 20740–3845.

4. Audio/Visual

   a. *Finding the Speed of a Rifle Bullet*; from AAPT collection 1 of single-concept films; video tape; available from Ztek Co., PO Box 11768, Lexington, KY 40577–1768.

   b. *Center of Mass*, *Conservation of Linear/Angular Momentum*, *Conservation of Momentum*; from AAPT collection 2 of single-concept films; video tape; available from Ztek Co. (see above for address).

   c. *Linear Momentum*; Side E: Conservation Laws of Cinema Classics; video disk; available from Ztek Co. (see above for address).

   d. *Momentum*; VHS video tape (9 min); Films for the Humanities & Sciences, PO Box 2053, Princeton, NJ 08543–2053.

   e. *Human Momenta, No Initial Motion*, *Human Momenta, Initial Translation* from Skylab Physics Videodisc; video disk; available from Ztek Co. (see above for address).

   f. *Center of Mass Motion*; Physics Experiments; VHS video tape (15 min); Films for the Humanities & Sciences (see above for address).

   g. *Human Momenta*; from AAPT Skylab videotape; American Association of Physics Teachers, One Physics Ellipse, College Park, MD 20740–3845.

   h. *Motion of Center of Mass*; *Conservation of Momentum*; Physics Demonstrations in Mechanics, Part II; VHS video tape; ≈3 min; Physics Curriculum & Instruction, 22585 Woodhill Drive, Lakeville, MN 55044.

   i. *Impulse and Momentum*;*Conservation of Momentum*; Physics Demonstrations in Mechanics, Part V; VHS video tape or video disk; ≈3 min; Physics Curriculum & Instruction (see above for address).

   j. *Motion of Center of Mass*; Physics Demonstrations in Mechanics, Part V; VHS video tape or video disk; ≈3 min; Physics Curriculum & Instruction (see above for address).

5. Computer Software

   Conservation Laws; Macintosh, Windows; Cross Educational Software, Inc., 508 E. Kentucky Avenue, PO Box 1536, Ruston, LA 71270. Includes a section on conservation of momentum.

6. Computer Project

   Have students use a spreadsheet or write a computer program to follow individual particles in a two or three particle system, given the force law for the forces they exert on each other. The program should integrate Newton's second law for each particle. Have the students use their data to verify the conservation of momentum. See the Computer Projects section of this manual.

7. Laboratory
   a. Meiners Experiment 7–9: *Ballistic Pendulum — Projectile Motion.* A ball is shot into a trapping mechanism at the end of a pendulum. The initial speed of the ball is found by applying conservation of momentum to the collision and conservation of energy to the subsequent swing of the pendulum. Also see Bernard Experiment 11: *Inelastic Impact and the Velocity of a Projectile.*
   b. Meiners Experiment 8–6: *Center of Mass Motion.* Two pucks are connected by a rubber band or spring and move toward each other on a nearly frictionless surface. A spark timer is used to record their positions as functions of time. Students calculate and study the position of the center of mass as a function of time. They also find the center of mass velocity. Can be performed with dry ice pucks or on an air table or air track.
   c. Meiners Experiment 8–7: *Linear Momentum.* Essentially the same as 8–6 but data is analyzed to give the individual momenta and total momentum as functions of time. Kinetic energy is also analyzed.

Chapter 10 COLLISIONS

BASIC TOPICS

I. Properties of collisions
   A. Set up a collision between two carts on an air track. Point out the interaction interval and the intervals before and after the interaction.
   B. Explain that for the collisions considered two bodies interact with each other over a short period of time and that the times before and after the collision are well defined. The force of interaction is great enough that external forces can be ignored during the interaction time.

II. Impulse.
   A. Define the impulse of a force as the integral over time of the force. Note that it is a vector. Clearly distinguish between impulse (integral over time) and work (integral over path). Draw a force versus time graph for the force of one body on the other during a one-dimensional collision and point out the impulse is the area under the curve.
   B. Define the time averaged force and show that the impulse is the product $\vec{F}_{\text{avg}}\Delta t$, where Δt is the time of interaction. Remark that we can use the average force to estimate the strength of the interaction during the collision.
   C. Use Newton's second law to show that the impulse on a body equals the change in its momentum. Remark that the change in momentum depends not only on the force but also on the duration of the interaction. Refer to the air track collision and point out that it is the impulse of one body on the other that changes the momentum of the second body. Repeat the air track collision. Measure the velocity of one cart before and after the collision and calculate the change in its momentum. Equate this to the impulse the other cart exerts. Estimate the collision time and calculate the average force exerted on the cart.
   D. Use the third law to show that two bodies in a collision exert equal and opposite impulses on each other and show that, if external impulses can be ignored, then total momentum is conserved. Refer to the air track collision. Again stress that external forces are neglected during the collisions considered here.
   E. Force probes, with input to a computer, can be used to plot the forces acting during a collision as functions of time. The curves can be integrated to find the impulse.

III. Two-body collisions in one dimension.
   A. Define the terms "elastic", "inelastic", and "completely inelastic". Distinguish between the

transfer of kinetic energy from one colliding object to the another and the loss of kinetic energy to internal energy.

B. Two-body completely inelastic collisions.

1. Derive an expression for the velocity of the bodies after the collision in terms of their masses and initial velocities.
2. Demonstrate the collision on an air track, using carts with velcro bumpers. Point out that the kinetic energy of the bodies is not conserved and calculate the energy loss. Remark that the energy is dissipated by the mechanism that binds the objects to each other. Some goes to internal energy, some to deformation energy. Note that $\frac{1}{2}(m_1 + m_2)v_{com}^2$ is retained. If we use a reference frame attached to the center of mass to describe the collision, we would find the combined bodies at rest after the collision and all kinetic energy lost.

C. Two-body elastic collisions.

1. Derive expressions for the final velocities in terms of the masses and initial velocities.
2. Specialize the general result to the case of equal masses and one body initially at rest. Demonstrate this collision on the air track using carts with spring bumpers. Point out that the carts exchange velocities.
3. Specialize the general result to the case of a light body, initially at rest, struck by a heavy body. Demonstrate this collision on the air track. Point out that the velocity of the heavy body is reduced only slightly and that the light body shoots off at high speed. Relate to a bowling ball hitting a pin.
4. Specialize the general result to the case of a heavy body, initially at rest, struck by a light body. Demonstrate this collision on the air track. Point out the low speed acquired by the heavy body and the rebound of the light body. Relate to a ball rebounding from a wall. A nearly elastic collision can be obtained with a superball.
5. Point out that, although the total kinetic energy does not change, kinetic energy is usually transferred from one body to the other. Consider a collision in which one body is initially at rest and calculate the fraction that is transferred. Show that the fraction is small if either mass is much greater than the other and that the greatest fraction is transferred if the two masses are the same. This is important, for example, in deciding what moderator to use to thermalize neutrons from a fission reactor.

D. Point out that while the greatest energy loss occurs when the interaction is completely inelastic, there are many other inelastic collisions in which less than the maximum energy loss occurs. Note that it is possible to have a collision in which kinetic energy increases (an explosive impact, for example).

IV. Two-body collisions in two dimensions.

A. Write down the equations for the conservation of momentum, in component form, for a collision with one body initially at rest. Mention that these can be solved for two unknowns.

B. Consider an elastic collision for which one body is at rest initially and the initial velocity of the second is given. Write the conservation of kinetic energy and conservation of momentum equations. Point out that the outcome is not determined by the initial velocities but that the impulse of one body on the other must be known to determine the velocities of the two bodies after the collision. State that the impulse is usually not known and in practice physicists observe one of the outgoing particles to determine its direction of motion. Carry out the calculation: assume the final direction of motion of one body is known and calculate the final direction of motion of the other and both final speeds.

C. State, or perhaps prove, that if the particles have the same mass then their directions of motion after the collision are perpendicular to each other.

D. Consider a completely inelastic collision for which the two bodies do not move along the same line initially. Mention that the outcome of this type collision *is* determined by the initial velocities. Calculate the final velocity. Calculate the fraction of energy dissipated.

SUGGESTIONS

1. Assignments
   a. To help students with the concept of impulse and the impulse-momentum theorem ask question 1. Also consider problems 3, 5, 9, 14, and 16.
   b. Use questions 3, 4, and 5 to discuss the motion of the center of mass during a collision. Use questions 2 and 6 to discuss conservation of momentum.
   c. Assign some problems dealing with inelastic collisions, such as 23 and 28. Assign and discuss some problems for which the mechanism of kinetic energy loss is given explicitly. See problems 31 and 33.
   d. Assign a problem, such as 35 or 52, that ask students to determine if a collision is elastic.
   e. Ask question 7 in support of the discussion of elastic collisions. Assign problems 37, 39, and 44. For some fun, carry out the demonstration described in question 8 and assign problem 45.
   f. Demonstrate the ballistic pendulum and show how it can be used to measure the speed of a bullet. Assign problems 22 and 24.
   g. If you intend to cover the thermodynamics chapters, assign and discuss problem 13. Also consider problem 12. They will prove helpful for the discussion of the microscopic basis of pressure.
   h. Assign some problems that deal with two-dimensional collisions: the group 46 through 55. Include both elastic and inelastic collisions.
2. Demonstrations
   Freier and Anderson Mg1 — 3, Mi1, 3, 4, Mw3, 4.
3. Audio/Visual
   a. *One-Dimensional Collisions*, *Two-Dimensional Collisions*, *Scattering of a Cluster of Objects*, *Dynamics of a Billiard Ball*, *Inelastic One-Dimensional Collisions*, *Inelastic Two-Dimensional Collisions*, and *Colliding Freight Cars*; from AAPT collection 1 of single-concept films; video tape; available from Ztek Co., PO Box 11768, Lexington, KY 40577–1768.
   b. *Drops and Splashes*, *Collisions in Two Dimensions*, *Inelastic Collisions*; from AAPT collection 2 of single-concept films; video tape; available from Ztek Co. (see above for address).
   c. *Elastic Collisions*, *Inelastic Collisions*, *Collisions*; Side E: Conservation Laws of Cinema Classics; video disk; available from Ztek Co. (see above for address).
   d. *Air Track Collisions*; Demonstrations of Physics: Energy and Momentum; VHS video tape (7:24); Media Design Associates, Inc., Box 3189, Boulder, CO 80307–3190.
   e. *The Dynamics of a Karate Punch*; Demonstrations of Physics: Energy and Momentum; VHS video tape (4:55); Media Design Associates, Inc. (see above for address).
   f. *Collisions* from Skylab Physics Videodisc; video disk; available from Ztek Co. (see above for address).
   g. *Physics and Automobile Collisions*; interactive video disk by Dean Zollman (John Wiley, 1984). The disk shows collisions of cars with fixed barriers and two car collisions (head-on, at 90°, and at 60°). One sequence shows the influence of bumper design, others show the influence of air bags and shoulder straps on manikins. All are slow motion movies of manufacturers' tests and many show grids and clocks. Students can stop the action to take measurements, then make calculations of momentum and energy transfers. For most exercises, a standard disk player is satisfactory; for a few, a computer-controlled player is required.

h. *Characteristics of Collisions*; *Elastic Collision*; Physics Demonstrations in Mechanics, Part V; VHS video tape or video disk; ≈3 min; Physics Curriculum & Instruction, 22585 Woodhill Drive, Lakeville, MN 55044.

4. Computer Software
   Objects in Motion. See Chapter 2 SUGGESTIONS.

5. Computer Project
   Have students use a commercial math program or write a program to graph the total final kinetic energy as a function of the final velocity of one object in a two-body, one-dimensional collision, given the initial velocities and masses of the two objects. Ask them to run the program for specific initial conditions and identify elastic, inelastic, completely inelastic, and explosive collisions on their graphs. Details of this and other projects are given in the Computer Projects section of this manual.

6. Laboratory
   a. Meiners Experiment 7–10: *Impulse and Momentum.* Students use a microprocessor to measure the force as a function of time as a toy truck hits a force transducer. They numerically integrate the force to find the impulse, then compare the result with the change in momentum, found by measuring the velocity before and after the collision.
   b. Bernard Experiment 8: *Impulse, Momentum, and Energy.* Part A deal with a mass that is hung on a string passing over a pulley and attached to an air track glider. The glider accelerates from rest for a known time and a spark timer is used to find its velocity at the end of the time. The impulse is calculated and compared with the momentum. In part B a glider is launched by a stretched rubber band and a spark record of its position as a function of time is made while it is in contact with the rubber band. A static technique is used to measure the force of the rubber band for each of the recorded glider positions and the impulse is approximated. The result is again compared with the final momentum of the glider.
   c. Meiners Experiment 7–11: *Scattering* (for advanced groups). The deflection of pellets from a stationary disk is used to investigate the scattering angle as a function of impact parameter and to find the radius of the disk.
   d. Meiners Experiment 8–5: *One-Dimensional Collisions.* A puck moving on a nearly frictionless surface collides with a stationary puck. A spark timer is used to record the positions of the pucks as functions of time. Students calculate the velocities, momenta, and energies before and after the collision. May be performed with dry ice pucks or on an air table or track.
   e. Meiners Experiment 8–8: *Two-Dimensional Collisions.* Same as Meiners 8–5 but the pucks are allowed to scatter out of the original line of motion. Students must measure angles and calculate components of the momenta. The experiment may be performed with dry ice pucks or on an air table.
   f. Bernard Experiment 12: *Elastic Collision — Momentum and Energy Relations in Two Dimensions.* A ball rolls down an incline on a table top and strikes a target ball initially at rest at the edge of the table. The landing points of the balls on the floor are used to find their velocities just after the collision. The experiment is run without a target ball to find the velocity of the incident ball just before the collision. Data is used to check for conservation of momentum and energy. Both head-on and grazing collisions are investigated. A second experiment, similar to Meiners 8–8, is also described.

Chapter 11 ROTATION

BASIC TOPICS

I. Rotation about a fixed axis.

A. Spin an irregular object on a fixed axis. A bicycle wheel or spinning platform with the object attached can be used. Draw a rough diagram, looking along the rotation axis. Explain that each point in the body has a circular orbit and that, for any selected point, the radius of the orbit is the perpendicular distance from the point to the rotation axis. Contrast to a body that is simultaneously rotating and translating. See Figs. 11–2 and 11–3.

B. Define angular position θ (in radians and revolutions), angular displacement $\Delta\theta$, angular velocity ω (in rad/s, deg/s, and rev/s), and angular acceleration α (in rad/s^2, deg/s^2, and rev/s^2). Treat both average and instantaneous quantities but emphasize that the instantaneous quantities are most important for us. Remind students of radian measure.

C. Use Fig. 11–4 to show how an angular displacement is measured. Note that θ is positive for an angular position in one direction from the fixed axis and negative for an angular position in the other. By convention in this text position angles are positive in the counterclockwise direction. Remark that as the body rotates θ continues increase beyond 2π rad.

D. Interpret the signs of ω and α. Give examples of spinning objects for which ω and α have the same sign and for which they have opposite signs.

E. Point out the analogy to one-dimensional linear motion. θ corresponds to x, ω to v, and α to a.

F. Point out that ω and α can be thought of as the components of vectors $\vec{\omega}$ and $\vec{\alpha}$, respectively. For fixed axis rotation, the vectors lie along the rotation axis, with the direction of $\vec{\omega}$ determined by a right hand rule: if the fingers curl in the direction of rotation, then the thumb points in the direction of $\vec{\omega}$. If $d\omega/dt > 0$, then $\vec{\alpha}$ is in the same direction; if $d\omega/dt < 0$, then it is in the opposite direction. Use Fig. 11–7 to explain that a vector cannot be associated with a finite angular displacement because displacements do not add as vectors.

II. Rotation with constant angular acceleration.

A. Emphasize that the discussion here is restricted to rotation about a fixed axis but that the same equations can be used when the rotation axis is in linear translation. This type motion will be discussed in the next chapter.

B. Write down the kinematic equations for $\theta(t)$ and $\omega(t)$. Make a comparison with the analogous equations for linear motion (see Table 11–1).

C. Point out that the problems of rotational kinematics are similar to those for one-dimensional linear kinematics and that the same strategies are used for their solution.

D. Go over examples. Calculate the time and number of revolutions for an object to go from some initial angular velocity to some final angular velocity, given the angular acceleration. If time permits, consider both a body that is speeding up and one that is slowing down. For the latter, calculate the time to stop and the number of revolutions made while stopping. Calculate the time to rotate a given number of revolutions and the final angular velocity, again given the angular acceleration.

III. Linear speed and acceleration of a point rotating about a fixed axis.

A. Write down $s = \theta r$ for the arc length. Explain that it is a rearrangement of the defining equation for the radian and that θ must be in radians for it to be valid.

B. Wrap a string on a large spool that is free to rotate about a fixed axis. Mark the spool so the angle of rotation can be measured. Slowly pull out the string and explain that the length of string pulled out is equal to the arc length through which a point on the rim moves. Compare the string length to θr for $\theta = \pi/2$, π, $3\pi/2$, and 2π rad. Show that $s = \theta r$ reduces to the familiar result for $\theta = 2\pi$ rad.

C. Differentiate $s = \theta r$ to obtain $v = r\omega$ and $a_t = \alpha r$. Emphasize that radian measure *must* be used. Point out that v gives the speed and a_t gives the acceleration of the string as it is pulled provided it does not slip on the spool. Point out that all points in a rotating rigid body have the same value of ω and the same value of α but points that are different distances from the rotation axis have different values of v and different values of a_t.

D. Point out that the velocity is tangent to the circular orbit but that the total acceleration is not. a_t gives the tangential component while $a_r = v^2/r = \omega^2$ gives the radial component. The tangential component is not zero only when the point on the rim speeds up or slows down in its rotational motion while the radial component is not zero as long as the object is turning. For students who have forgotten, reference the derivation of $a_r = v^2/r$, given in Chapter 4.

E. Explain how to find the magnitude and direction of the total acceleration in terms of ω, α, and r.

IV. Kinetic energy of rotation and rotational inertia.

A. By substituting $v = r\omega$ into $K = \frac{1}{2}mv^2$, show that $K = \frac{1}{2}mr^2\omega^2$ for a particle moving around a circle with angular velocity ω and, by summing over all particles in a rigid body, show that $K = \frac{1}{2}I\omega^2$, where $I = \sum m_i r_i^2$ if the body is rotating about a fixed axis. Explain that I is called the rotational inertia of the body. Mention that many texts call it the moment of inertia.

B. Point out that rotational inertia depends on the distribution of mass and on the position and orientation of the rotation axis. Explain that two bodies may have the same mass but quite different rotational inertias. State that Table 11–2 gives the rotational inertia for various objects and axes. Particularly point out the rotational inertia of a hoop rotating about the axis through its center and perpendicular to its plane. Note that all its mass is the same distance from the rotation axis. Also point out the rotational inertias of a cylinder rotating about its axis and a sphere rotating about a diameter. Note that the mass is now distributed through a range of distances from the rotation axis and the rotational inertia is less than that for a hoop with the same mass and radius.

C. Optional: show how to convert the sum for I to an integral. Use the integral to find the rotational inertia for a thin rod rotating about an axis through its center and perpendicular to its length. If the students have experience with volume integrals using spherical coordinates, derive the expression for the rotational inertia of a sphere.

D. Prove the parallel axis theorem. The proof can be carried out using a sum for I rather than an integral. Explain its usefulness for finding the rotational inertia when the rotation axis is not through the center of mass. Emphasize that the actual axis and the axis through the center of mass must be parallel for the theorem to be valid. Use the parallel axis theorem to obtain the rotational inertia for the rotation of a thin rod about one end from the rotational inertia for rotation about the center, given in Table 11–2.

V. Torque.

A. Define torque for a force acting on a single particle. Consider forces that lie in planes perpendicular to the axis of rotation and take $\tau = rF\sin\phi$, where $\vec{r}$ is a vector that is perpendicular to the rotation axis and points from the axis to the point of application of the force. ϕ is the angle between $\vec{r}$ and $\vec{F}$ when they are drawn with their tails at the same point. The definition will be generalized in the next chapter. Explain that $\tau = rF_t = r_\perp F$, where F_t is the tangential component of $\vec{F}$ and $r_\perp$ is the moment arm.

B. Explain that the torque vanishes if $\vec{F}$ is along the same line as $\vec{r}$ and that only the component of $\vec{F}$ that is perpendicular to $\vec{r}$ produces a torque. This is a mechanism for picking out the part of the force that produces angular acceleration, as opposed to the part that produces centripetal acceleration. Also explain that the same force can produce a larger torque if it is applied at a point farther from the rotation axis.

C. Use a wrench tightening a bolt as an example. The force is applied perpendicular to the wrench arm and long moment arms are used to obtain large torques.

D. Explain the sign convention for torques applied to a body rotating about a fixed axis. For example, torques tending to give the body a counterclockwise (positive) angular acceleration are positive while those tending to give the body a clockwise angular acceleration are negative. Remark that the convention is arbitrary and the opposite convention may be convenient for some problems.

VI. Newton's second law for rotation.

A. Use a single particle on a circular orbit to introduce the topic. Start with $F_t = ma_t$ and show that $\tau = I\alpha$, where $I = mr^2$. Explain that this equation also holds for extended bodies, although I is then the sum given above.

B. Remark that problems are solved similarly to linear second law problems. Tell students to identify torques, draw a force diagram, choose the direction of positive rotation, and substitute the total torque into $\tau_{\text{net}} = I\alpha$. Remark that the point of application of a force is important for rotation, so the object cannot be represented by a dot on a force diagram. Tell students to sketch the object and place the tails of force vectors at the application points.

C. Wrap a string around a cylinder, free to rotate on a fixed horizontal axis. Attach the free end of the string to a mass and allow the mass to fall from rest. Note that its acceleration is less than g, perhaps by dropping a free mass beside it. See Sample Problem 11–7.

VII. Work-kinetic energy theorem for rotation.

A. Use $dW = \vec{F} \cdot d\vec{s}$ to show that the work done by a torque is given by $W = \int \tau \, d\theta$ and that the power delivered is given by $P = \tau\omega$.

B. Use $\tau \, d\theta = I\alpha \, d\theta = \frac{1}{2} I \, d(\omega^2)$ to show that $W = \frac{1}{2} I(\omega_f^2 - \omega_i^2)$.

C. For the situation of Sample Problem 11–7 use conservation of energy to find the angular velocity of the cylinder after the mass has fallen a distance h. Use rotational kinematics and the value for the angular acceleration found in the text to check the answer.

SUGGESTIONS

1. Assignments

a. Use questions 1 and 2 to discuss graphical interpretations of angular position and velocity.

b. Use techniques of the calculus to derive the kinematic equations for constant angular acceleration. That is, integrate $\alpha =$ constant twice with respect to time. Assign one or two problems from the group 1, 4, and 5.

c. Assign some problems that deal with rotation with constant angular acceleration: 12, 15, 17, and 18, for example. Also consider question 4.

d. To discuss the relationship between angular and linear variables, assign some of problems 22, 23, 24, and 29. Questions 5 and 6 might help with the ideas of the radial and tangential components of the linear acceleration.

e. Use question 7 to guide students through a qualitative discussion of rotational inertia. Assign problems 37, 39, and 41. Use problem 43 to discuss the radius of gyration, if desired.

f. Use problem 47 or 48 to discuss the calculation of torque.

g. To help students think about torque and $\tau_{\text{net}} = I\alpha$, discuss questions 8, 9, 10, and 11. Assign some of problems 52, 56, and 57. To deal with a situation in which the dynamics of more than one object is important, demonstrate the Atwood machine and discuss problem 55.

h. Use question 12 to discuss the work done by a torque and changes in rotational kinetic energy. Discuss conservation of mechanical energy and assign problems 65 and 66.

2. Demonstrations
   a. Rotational dynamics: Freier and Anderson Ms7, Mt 5, 6, Mo5.
   b. Rotational work and energy: Freier and Anderson Mv2, Mr5, Ms2.
3. Audio/Visual
   a. *Newton's Second Law for Rotational Motion*; Demonstrations of Physics: Motion; VHS video tape (8:35); Media Design Associates, Inc., Box 3189, Boulder, CO 80307–3190.
4. Computer Project
   Ask students to use a commercial math program or their own root finding programs to solve rotational kinematic problems.
5. Laboratory
   Meiners Experiment 7–14: *Rotational Inertia.* The rotational inertia of a disk is measured dynamically by applying a torque (a falling mass on a string wrapped around a flange on the disk). A microprocessor is used to measure the angular acceleration. Small masses are attached to the disk and their influence on the rotational inertia is studied. The acceleration of the mass can also be found by timing its fall through a measured distance. Then, $a_t = \alpha r$ is used to find the angular acceleration of the disk. Also see Bernard Experiment 14: *Moment of Inertia.*

Chapter 12 ROLLING, TORQUE, AND ANGULAR MOMENTUM

BASIC TOPICS

I. Rolling.

A. Remark that a rolling object can be considered to be rotating about an axis through the center of mass while the center of mass moves. The text considers the special case for which the axis of rotation does not change direction. Point out that the rotational motion obeys $\tau_{\text{net}} = I_{\text{com}}\alpha$ and the translational motion of the center of mass obeys $\vec{F}_{\text{net}} = m\vec{a}_{\text{com}}$, where τ_{net} is the sum of external torques and $\vec{F}_{\text{net}}$ is the sum of external forces. Emphasize that one of the forces acting may be the force of friction produced by the surface on which the object rolls.

B. Explain that the speed of a point at the top of a rolling object is $v_{\text{com}} + \omega R$ and the speed of a point at the bottom is $v_{\text{com}} - \omega R$. Specialize to the case of rolling without slipping. Point out that the point in contact with the ground has zero velocity, so $v_{\text{com}} = \omega R$. Use Fig. 12–4 as evidence. Also point out that tire tracks in the snow are clean (not smudged) if the tires do not slip.

C. Explain that a wheel rolling without slipping can be viewed as rotating about an axis through the point of contact with the ground. Use this and the parallel axis theorem to show that the kinetic energy is $\frac{1}{2}Mv_{\text{com}}^2 + \frac{1}{2}I_{\text{com}}\omega^2$.

D. Consider objects rolling down an inclined plane and show how to calculate the speed at the bottom using energy considerations. If time permits, carry out an analysis using the equations of motion and show how to find the frictional force that prevents slipping.

E. Roll a sphere, a hoop, and a cylinder, all with the same radius and mass, down an incline. Start the objects simultaneously at the same height and ask students to pick the winner.

Point out that the speed at the bottom is determined by the dimensionless parameter $\beta = I/MR^2$ and not by I, M, and R alone. All uniform cylinders started from rest reach the bottom in the same time and have the same speed when they get there.

F. Consider a ball striking a bat. Show how to find the point at which the ball should hit so the instantaneous center of rotation is at the place where the bat is held. The striking point is called the center of percussion. When the ball hits there the batter feels no sting.

II. Torque and angular momentum.

A. Define torque as $\vec{\tau} = \vec{r} \times \vec{F}$ and explain that this is the general definition. Review the vector product, give the expression for the magnitude ($\tau = rF\sin\phi$), and give the right hand rule for finding the direction. Explain that $\vec{\tau} = 0$ if $\vec{r} = 0$, $\vec{F} = 0$, or $\vec{r}$ is parallel (or antiparallel) to $\vec{F}$.

B. Consider an object going around a circle and suppose a force is applied tangentially. Take the origin to be on the rotation axis but not at the circle center and show the general definition reduces to the expression used in Chapter 11 for the component of the torque along the rotation axis: $\tau = F_t R$, where R is the radius of the circle (not the distance from the origin to the point of application).

C. Use vector notation to define angular momentum for a single particle ($\vec{\ell} = m\vec{r} \times \vec{v}$). Give the expression for the magnitude and the right-hand rule for the direction.

D. Derive the relationship $\ell = mr^2\omega$ between the magnitude of the angular momentum and the angular velocity for a particle moving on a circle centered at the origin. Also find the angular momentum if the origin is on a line through the circle center, perpendicular to the circle, but not at the center. Explain that the component along the rotation axis is $mr^2\omega$ and is independent of the position of the origin along the line and that the component perpendicular to the axis rotates with the particle.

E. To show that a particle may have angular momentum even if it is not moving in a circle, calculate the angular momentum of a particle moving with constant velocity along a line not through the origin. Point out that the angular momentum depends on the choice of origin. In preparation for G below you might want to find the time rate of change of $\vec{\ell}$.

F. Use Newton's second law to derive $\vec{\tau} = d\vec{\ell}/dt$ for a particle. Consider a particle moving in a circle, subjected to both centripetal and tangential forces. Take the origin to be at the center of the circle and show that $\vec{\tau} = d\vec{\ell}/dt$ reduces to $F_t = ma_t$, as expected. Take the origin to be on the line through the center, perpendicular to the circle, but not at the center. Show that the torque associated with the centripetal force produces the change in $\vec{\ell}$ expected from the discussion of D.

G. Show that the magnitude of the torque about the origin exerted by gravity on a falling mass is mgd, where d is the perpendicular distance from the line of fall to the origin. Write down the velocity as a function of time and show that the angular momentum is $mgtd$. Remark that $\tau_{net} = d\ell/dt$ by inspection. See Sample Problem 12–5.

III. Systems of particles.

A. Explain that the total angular momentum for a system of particle is the vector sum of the individual momenta.

B. Show that $\vec{\tau}_{net} = d\vec{L}/dt$ for a system of particles for which internal torques cancel. Emphasize that $\vec{\tau}_{net}$ is the result of summing all torques on all particles in the system and that $\vec{L}$ is the sum of all individual angular momenta. Demonstrate in detail the cancellation of internal torques for two particles that interact via central forces. Point out that this equation is the starting point for investigations of the rotational motion of bodies.

C. Show that the component along the rotation axis of the total angular momentum of a rigid body rotating about a fixed axis is $I\omega$. Use the example of a single particle to point out that the angular momentum vector is along the rotation axis if the body is symmetric about the axis but that otherwise it is not. Emphasize that for fixed axis rotation we are chiefly interested in the components of angular momentum and torque along the rotation axis.

D. Make a connection to material of the last chapter by showing that $L = I\omega$ and $\tau_{\text{net}} = \mathrm{d}L/\mathrm{d}t$ lead to $\tau_{\text{net}} = I\alpha$ for a rigid body rotating about a fixed axis. Here τ_{net} is the component of the total external torque along the rotation axis.

IV. Conservation of angular momentum.

A. Point out that $\vec{L}$ =constant if $\vec{\tau}_{\text{net}} = 0$. State that different objects in a system may change each other's angular momentum but the changes sum vectorially to zero. Also explain that the rotational inertia of an object may change while it is spinning. Then, $I_i\omega_i = I_f\omega_f$ if the net external torque vanishes.

B. As examples consider a mass dropped on the rim of a freely spinning platform, a person running tangent to the rim of a merry-go-round and jumping on, and a spinning skater whose rotational inertia is changed by dropping her arms.

C. The third example can be demonstrated easily if you have a rotating platform that can hold a person. Have a student hold weights in each hand to increase the rotational inertia. Start him spinning with arms extended, then have him bring his arms in toward the sides of his body. See Fig. 12–16. Also carry out the spinning bicycle wheel demonstration described in the text. See Fig. 12–19.

SUGGESTIONS

1. Assignments

   a. In connection with rolling without sliding ask questions 1 through 3 and assign problems 6, 7, and 11. To emphasize the condition for no slipping assign problem 8 or 14.

   b. Assign problems 20 and 26 to stress the importance of the origin in calculations of torque and angular momentum. Problem 25 asks students to calculate the angular momentum if the cartesian components of the position and momentum vectors are given. Problem 24 deals with both angular momentum and torque. Be sure students can calculate the angular momentum of an object moving along a straight line and the angular momentum of a rigid body rotating about a fixed axis. See problems 23, 27, and 35. Also consider discussing the angular momentum of a projectile.

   c. Use question 9 to start the discussion of conservation of angular momentum. Problems 43, 46, and 48 include motion along a straight line. Problems 39 and 44 deal with changes in rotational inertia. Problems 41 and 42 deal with inelastic rotational collisions. Assign one or more from each of these groups. To include a situation for which angular momentum is not conserved, assign problem 37.

2. Demonstrations

   a. Rolling: Freier and Anderson Mb4, 7, 30, Mo3, Mp3, Mr1, 4, Ms1, 3, 4, 6.

   b. Conservation of angular momentum: Freier and Anderson Mt1 — 4, 7, 8, Mu1.

   c. Gyroscopes: Freier and Anderson Mu2 — 18.

3. Audio/Visual

   a. *Human Momenta, Initial Translation and Rotation* from Skylab Physics Videodisc; video disk; available from Ztek Co., PO Box 11768, Lexington, KY 40577–1768.

   b. *Conservation of Linear/Angular Momentum*; from AAPT collection 2 of single-concept films; video tape; available from Ztek Co. (see above for address).

   c. *Angular Momentum*; Side F: Angular Momentum and Modern Physics of Cinema Classics; video disk; available from Ztek Co. (see above for address).

   d. *Conservation of Angular Momentum*; Demonstrations of Physics: Energy and Momentum; VHS video tape (5:52); Media Design Associates, Inc., Box 3189, Boulder, CO 80307–3190.

e. *Why a Spinning Top Doesn't Fall*; Demonstrations of Physics: Energy and Momentum; VHS video tape (4:24); Media Design Associates, Inc. (see above for address).
f. *Conservation Laws in Zero-G*; Skylab Demonstrations; VHS video tape (18 min); Media Design Associates, Inc. (see above for address).
g. *Rotational Dynamics*; Physics Demonstrations in Mechanics, Part VI; VHS video tape or video disk; ≈3 min; Physics Curriculum & Instruction, 22585 Woodhill Drive, Lakeville, MN 55044.
h. *Conservation of Angular Momentum*; *Center of Percussion*; Physics Demonstrations in Mechanics, Part II; VHS video tape; ≈3 min; Physics Curriculum & Instruction (see above for address).

4. Computer Software
   a. *Conservation Laws*; Macintosh, Windows; Cross Educational Software, Inc., 508 E. Kentucky Avenue, PO Box 1536, Ruston, LA 71270. Includes a section on conservation of angular momentum.
   b. *Circular Motion*; Macintosh, Windows; Cross Educational Software, Inc. (see above for address).
5. Computer Project
   Given the law for the torque between two rotating rigid bodies a computer program or spreadsheet can be used to integrate Newton's second law for rotation and tabulate the angular positions and angular velocities as functions of time. The data can be used to verify the conservation of angular momentum. See the Computer Projects section of this manual.
6. Laboratory
   a. Meiners Experiment 7–12: *Rotational and Translational Motion.* Students measure the center of mass acceleration of various bodies rolling down an incline and calculate the center of mass velocities at the bottom. Results are compared to measured velocities. It is also instructive to use energy methods to find the final speeds.
   b. Meiners Experiment 7–13: *Rotational Kinematics and Dynamics.* Students find the velocity and acceleration of a ball rolling around a loop-the-loop and analyze the forces acting on it.
   c. Meiners Experiment 8–9: *Conservation of Angular Momentum.* Uses the Pasco rotational dynamics apparatus. A ball rolls down a ramp and becomes coupled to the rim of a disk that is free to rotate on a vertical axis. Students measure the velocity of the ball before impact and the angular velocity of the disk-ball system after impact, then check for conservation of angular momentum.

SECTION THREE
PROBLEMS IN THE STUDENT SOLUTION MANUAL AND ON THE STUDENT'S COMPANION WEBSITE

The *Student Solution Manual* contains fully worked out solutions to many end-of-chapter problems and the *Student's Companion Website* contains guidelines and strategies for working many others. The *Student Solution Manual* is available to students as a print supplement. A password to the *Student's Companion Website* is given to students who purchase the study guide (*A Student's Companion*). The problems included in the *Student Solution Manual* and on the Student's Companion Website are listed here. A few complete solutions for each chapter are included in a free student website, with the address given in the text. They are underlined here.

As part of each assignment you may wish to have students study a few of the solutions in the *Student Solution Manual* before attempting their own solutions to other problems. You may also wish to include in the assignment several of the problems discussed on the *Student's Companion Website*. These will help students develop problem-solving strategies and learn effective problem-solving techniques without actually solving problems for them.

Chapter 1

Solution Manual: 1E, 3E, 5E, 7P, 9P, 11E, 13P, 15P, 17P, 19E, 21P, 23P
Website: 2E, 6E, 8P, 10E, 14P, 20P, 24P

Chapter 2

Solution Manual: 1E, 3E, 7P, 17E, 19P, 21P, 23E, 25E, 27E, 29E, 31E, 33P, 35P, 37P, 41E, 43E, 45E, 47P, 49P, 51P, 53P, 57P, 59P, 61P
Website: 5P, 8P, 11E, 12P, 18P, 20P, 26E, 28E, 32E, 36P, 38P, 44E, 52P

Chapter 3

Solution Manual: 3E, 5E, 7P, 9P, 11E, 13E, 15E, 17E, 21P, 23P, 25P, 27P, 29E, 31P, 33P, 35P
Website: 1E, 8P, 12E, 18P, 22P, 28E, 34P, 38P

Chapter 4

Solution Manual: 5E, 9E, 13P, 17E, 21E, 23E, 25P, 27P, 31P, 35P, 37P, 39P, 41P, 43E, 45E, 47P, 49P, 51P, 55P, 57E, 59P, 61P
Website: 4P, 8P, 11E, 14P, 16P, 20E, 28E, 30P, 36P, 42E, 44E, 48P, 52P 56E

Chapter 5

Solution Manual: 7P, 9E, 11E, 13E, 15E, 19E, 21E, 25P, 29P, 31P, 33P, 35P, 37P, 41P, 43P, 45P, 47P, 49P, 51P, 53P
Website: 2E, 6P, 12E, 16E, 22P, 26P, 34P, 36P, 40P, 44P, 48P, 50P, 56P

Chapter 6

Solution Manual: 1E, 3E, 5E, 7E, 9P, 11P, 13P, 15P, 19P, 21P, 23P, 27P, 29P, 31P, 33E, 37E, 41E, 43P, 45P, 47P

Website: 8E, 14P, 18P, 22P, 25P, 28P, 30P, 34E, 40P, 42P, 44P, 46P

Chapter 7

Solution Manual: 1E, 5P, 7E, 11P, 13P, 17P, 19P, 21E, 23P, 25E, 27P, 29P, 31E, 33P, 35P, 37P

Website: 6E, 8E, 10P, 15E, 16E, 18P, 22P, 24E, 26P, 28P, 32E, 34P, 36P, 40P

Chapter 8

Solution Manual: 1E, 3E, 5E, 7P, 9E, 11E, 13E, 15P, 17P, 21P, 25P, 27P, 29P, 31P, 33P*, 35P*, 37P, 41P, 43E, 51P, 53P, 55P, 61P

Website: 4E, 6P, 8P, 14P, 19P, 26P, 30P, 32P, 36P, 40E, 42P, 46E, 50P, 58P

Chapter 9

Solution Manual: 1E, 3E, 9P*, 11E, 15P, 17P, 19P, 27E, 29E, 33P, 37P, 39P, 41E, 43E, 47P, 55E

Website: 6P, 7P, 8P, 10E, 13E, 18P, 24P, 25P, 28E, 32E, 36P, 40P, 42E, 46E, 54P

Chapter 10

Solution Manual: 1E, 5E, 9P, 11P, 13P, 15P, 19P, 21E, 23E, 27P, 29P, 31P, 35E, 37E, 41P, 43P, 45P, 49E, 51P, 53P, 55P

Website: 2E, 8P, 14P, 16P, 20E, 26P, 33P, 34P, 40P, 48E, 50P, 52P

Chapter 11

Solution Manual: 3E, 5E, 9E, 13P, 19E, 21E, 25P, 27P, 29P, 31P, 33E, 35E, 39E, 41P, 43P, 45E, 47P, 49E, 51E, 55P, 63P, 65P, 67P

Website: 4E, 6P, 8E, 15P, 16P, 23E, 26P, 34P, 38E, 42P, 44P, 46E, 48P, 52E, 58E

Chapter 12

Solution Manual: 3E, 5E, 9P, 15E, 21E, 25E, 27P, 29E, 33E, 35E, 37P*, 39E, 41E, 43E, 45P, 51P*

Website: 2P, 7E, 8P, 12P, 14P, 20P, 24E, 26P, 33P, 36P, 38P, 42E, 46P, 48P, 52P, 58P

SECTION FOUR
ANSWERS TO CHECKPOINTS

The following are the answers to the Checkpoints that appear throughout the text.

Chapter 2

1. b and c
2. zero (zero displacement for the entire trip)
3. (check the derivative dx/dt) (a) 1 and 4; (b) 2 and 3
4. (see Tactic 5) (a) plus; (b) minus; (c) minus; (d) plus
5. 1 and 4 ($a = d^2x/dt^2$ must be a constant)
6. (a) plus (upward displacement on y axis); (b) minus (downward displacement on y axis); (c) $a = -g = -9.8\,\text{m/s}^2$

Chapter 3

1. (a) 7 m ($\vec{a}$ and $\vec{b}$ are in the same direction); (b) 1 m ($\vec{a}$ and $\vec{b}$ are in opposite directions)
2. c, d, f (components must be head-to-tail; $\vec{a}$ must extend from the tail of one component to the head of the other)
3. (a) +, +; (b) +, −; (c) +, + (draw vector from the tail of $\vec{d}_1$ to the head of $\vec{d}_2$)
4. (a) 90°; (b) 0 (vectors are parallel, in the same direction); (c) 180° (vectors are antiparallel, in opposite directions)
5. (a) 0° or 180°; (b) 90°

Chapter 4

1. (a) $(8\,\text{m})\hat{\imath} - (6\,\text{m})\hat{\jmath}$; (b) yes, the xy plane (no z component)
2. Draw $\vec{v}$ tangent to the path with its tail on the path. (a) first; (b) third
3. Take the second derivative with respect to time. (1) and (3): a_x and a_y are both constant and thus $\vec{a}$ is constant; (2) and (4): a_y is constant but a_x is not, thus $\vec{a}$ is not
4. $4\,\text{m/s}^3$, $-2\,\text{m/s}$, 3 m
5. (a) v_x constant; (b) v_y initially positive, decreases to zero, and then becomes progressively more negative; (c) $a_x = 0$ throughout; (d) $a_y = -g$ throughout
6. (a) $-(4\,\text{m/s})\hat{\imath}$; (b) $-(8\,\text{m/s}^2)\hat{\jmath}$
7. (a) 0, distance not changing; (b) +70 km/h, distance increasing; (c) +80 km/h, distance decreasing
8. (a) — (c) increase

Chapter 5

1. c, d, and e ($\vec{F}_1$ and $\vec{F}_2$ must be head-to-tail, $\vec{F}_{\text{net}}$ must be from the tail of one of them to the head of the other)
2. (a) and (b) 2 N, leftward (acceleration is zero in each situation)
3. (a) and (b) 1, 4, 3, 2
4. (a) equal; (b) greater (acceleration is upward, thus net force on body must be upward)
5. (a) equal; (b) greater; (c) less
6. (a) increase; (b) yes; (c) same; (d) yes
7. (a) $F\sin\theta$; (b) increase
8. 0 (because now $a = -g$)

Chapter 6

1. (a) zero (because there is no attempt at sliding); (b) 5 N; (c) no; (d) yes; (e) 8 N
2. (a) same (10 N); (b) decreases; (c) decreases (because N decreases)
3. greater (from Sample Problem 6–5, v_t depends on $\sqrt{R}$)
4. ($\vec{a}$ is directed toward the center of the circular path) (a) $\vec{a}$ downward, $\vec{N}$ upward; (b) $\vec{a}$ and $\vec{N}$ upward
5. (a) same (must still match the gravitational force on the rider); (b) increases ($N = mv^2/R$); (c) increases ($f_{s\,\max} = \mu_s N$)
6. (a) $4R_1$; (b) $4R_1$

Chapter 7

1. (a) decrease; (b) same; (c) negative, zero
2. d, c, b, a
3. (a) same; (b) smaller
4. (a) positive; (b) negative; (c) zero
5. zero

Chapter 8

1. no
2. 3, 1, 2
3. (a) all tie; (b) all tie
4. (a) CD, AB, BC (zero); (b) positive direction of x
5. all tie

Chapter 9

1. (a) origin; (b) fourth quadrant; (c) on y axis below origin; (d) origin; (e) third quadrant; (f) origin
2. (a) to (c) at the center of mass, still at the origin (their forces are internal to the system and cannot move the center of mass)
3. (a) 1, 3, then 2 and 4 tie (zero force); (b) 3
4. (a) 0; (b) no; (c) negative
5. (a) 500 km/h; (b) 2600 km/h; (c) 1600 km/h
6. (a) yes; (b) no

Chapter 10

1. (a) unchanged; (b) unchanged; (c) decreased
2. (a) zero; (b) positive; (c) positive direction of y
3. (a) $10\,\text{kg}\cdot\text{m/s}$; (b) $14\,\text{kg}\cdot\text{m/s}$; (c) $6\,\text{kg}\cdot\text{m/s}$
4. (a) $4\,\text{kg}\cdot\text{m/s}$; (b) $8\,\text{kg}\cdot\text{m/s}$; (c) 3 J
5. (a) $2\,\text{kg}\cdot\text{m/s}$; (b) $3\,\text{kg}\cdot\text{m/s}$

Chapter 11

1. b and c
2. a and d
3. (a) yes; (b) no; (c) yes; (d) yes
4. all tie
5. 1, 2, 4, 3
6. 1 and 3 tie, 4, then 2 and 5 tie (zero)
7. (a) downward in the figure; (b) less

Chapter 12

1. (a) same; (b) less
2. less
3. (a) $\pm z$; (b) $+y$; (c) $-x$
4. (a) 1 and 3 tie, then 2 and 4 tie, then 5 (zero); (b) 2 and 3
5. (a) 3, 1, and then 2 and 4 tie (zero); (b) 3
6. (a) all tie (same r, same t, thus same ΔL); (b) sphere, disk, hoop (reverse order of I)
7. (a) decreases; (b) same; (c) increases

SECTION FIVE
ANSWERS TO QUESTIONS

The following are the answers to the end-of-chapter questions.

Chapter 2

1. (a) all tie; (b) 4, tie of 1 and 2, then 3
2. (a) negative direction of x; (b) positive direction of x; (c) yes (when graphed line crosses the t axis); (d) positive; (e) constant
3. E
4. (a) 2, 3; (b) 1, 3; (c) 4
5. a and c
6. 60 km/h, not 0
7. $x = t^2$ and $x = 8(t-2) + (1.5)(t-2)^2$
8. (a) 3, 2, 1; (b) 1, 2, 3; (c) all tie; (d) 1, 2, 3
9. same

Chapter 3

1. $\vec{A}$ and $\vec{B}$
2. (a) — (c) yes (example: $5\hat{\imath}$ and $-2\hat{\jmath}$)
3. no, but $\vec{a}$ and $-\vec{b}$ are commutative: $\vec{a} + (-\vec{b}) = (-\vec{b}) + \vec{a}$
4. (a) yes; (b) yes; (c) no
5. (a) $\vec{a}$ and $\vec{b}$ are parallel; (b) $\vec{b} = 0$; (c) $\vec{a}$ and $\vec{b}$ are perpendicular
6. (a) — (d) negative
7. all but (e)
8. no (their orientations can differ)
9. (a) 0 (vectors are parallel); (b) 0 (vectors are antiparallel)
10. (a) $\vec{B}$ and $\vec{C}$, $\vec{D}$ and $\vec{E}$; (b) $\vec{D}$ and $\vec{E}$

Chapter 4

1. (a) $(7\,\text{m})\hat{\imath} + (1\,\text{m})\hat{\jmath} + (-2\,\text{m})\hat{k}$; (b) $(5\,\text{m})\hat{\imath} + (-3\,\text{m})\hat{\jmath} + (1\,\text{m})\hat{k}$; (c) $(-2\,\text{m})\hat{\imath}$
2. (a) 1 and 3: a_y is constant but a_x is not and thus $\vec{a}$ is not; 2: a_x is constant but a_y is not and thus $\vec{a}$ is not; 4: a_x and a_y are both constant and thus $\vec{a}$ is constant; (b) $-2\,\text{m/s}^2$, $-3\,\text{m/s}$
3. a, b, c
4. yes (the vertical component of $\vec{v}$ is downward)
5. (a) all tie; (b) 1 and 2 tie (the rocket is shot upward), then 3 and 4 tie (it is shot into the ground!)
6. $(2\,\text{m/s})\hat{\imath} - (4\,\text{m/s})\hat{\jmath}$
7. (a) 3, 2, 1; (b) 1, 2, 3; (c) all tie; (d) 6, 5, 4
8. (a) 0; (b) 350 km/h; (c) 350 km/h; (d) same nothing changed about the vertical motion
9. (a) less; (b) unanswerable; (c) equal; (d) unanswerable
10. (a) all tie; (b) all tie; (c) 3, 2, 1; (d) 3, 2, 1
11. (a) 2; (b) 3; (c) 1; (d) 2; (e) 3; (f) 1
12. 2, then 1 and 4 tie, then 3
13. (a) yes; (b) no; (c) yes

Chapter 5

1. (a) 5; (b) 7; (c) $(2\,\text{N})\hat{\imath}$; (d) $-(6\,\text{N})\hat{\jmath}$; (e) fourth; (f) fourth
2. (a) 2 and 3; (b) 2
3. (a) 2 and 4; (b) 2 and 4
4. increase
5. (a) 2, 3, 4; (b) 1, 3, 4; (c) 1, $+y$; 2, $+x$; 3, fourth quadrant; 4, third quadrant
6. 1, graphs a and e; 2, graphs b and d; 3, graphs b and f; 4, graphs c and f
7. (a) less; (b) greater
8. (a) increases from an initial value of mg; (b) decreases from mg to zero (after which the block moves up away from the floor)
9. (a) 20 kg; (b) 18 kg; (c) 10 kg; (d) all tie; (e) 3, 2, 1
10. (a) 17 kg; (b) 12 kg; (c) 10 kg; (d) all tie; (e) $\vec{F}$, $\vec{F}_{21}$, $\vec{F}_{32}$
11. (a) 4 or 5, choose 4; (b) 2; (c) 1; (d) 4 or 5, choose 5; (e) 3; (f) 6; (g) 3 and 6; 1, 2, and 5; (h) 3 and 6; (i) 1, 2, and 5
12. (a) increases; (b) increases; (c) decreases; (d) decreases; (e) a and b, 3; c and d, 2

Chapter 6

1. (a) F_1, F_2, F_3; (b) all tie

2. (a) rightward; (b) leftward; (c) decrease; (d) leftward; (e) rightward; (f) increase; (g) no
3. (a) same; (b) increases; (c) increases; (d) no
4. (a) upward; (b) horizontal, toward you; (c) no change; (d) increases; (e) increases
5. (a) decrease; (b) decrease; (c) increase; (d) increase; (e) increase
6. (a) decrease; (b) decrease; (c) decrease; (d) decrease; (e) decrease
7. (a) the block's mass m; (b) equal (they are a third-law pair); (c) that on the slab is in the direction of the applied force; that on the block is in the opposite direction; (d) the slab's mass M
8. (a) same; (b) less
9. 4, 3, then 1, 2, and 5 tie
10. (a) all tie; (b) all tie; (c) 2, 3, 1

Chapter 7

1. all tie
2. (a) positive; (b) negative; (c) negative
3. (a) positive; (b) zero; (c) negative; (d) negative; (e) zero; (f) positive
4. (a) 3 m; (b) 3 m; (c) 0 and 6 m; (d) negative direction of x
5. (a) A, B, C; (b) C, B, A; (c) C, B, A; (d) A, 2; B, 3; C, 1
6. b (positive work), a (zero work), c (negative work), d (more negative work)
7. all tie
8. (a) to (c) yes; (d) and (e) no
9. c, d, then a and b tie, then f, e
10. (a) A; (b) B
11. (a) $2F_1$; (b) $2W_1$
12. (a) 3 and 6; (b) 1, 4, then 2 and 5 tie (zero), then 6, then 3; (c) 1 and 4 tie, then 2 and 5 tie (zero), then 6, then 3
13. B, C, A

Chapter 8

1. (a) 12 J; (b) −2 J
2. c and d tie, then a and b tie
3. (a) all tie; (b) all tie
4. (a) and (b) all tie (same change in elevation)
5. (a) 4; (b) returns to its starting point and repeats the trip; (c) 1; (d) 1
6. (a) AB, CD, and then BC and DE tie (zero force); (b) 5 J; (c) 5 J; (d) 6 J; (e) FG; (f) DE
7. (a) fL; (b) 0.50; (c) 1.25; (d) 2.25; (e) b, center; c, right; d, left
8. (a) less; (b) equal
9. (a) increasing; (b) decreasing; (c) decreasing; (d) constant in AB and BC, decreasing in CD

Chapter 9

1. (a) to (d) at the origin
2. same (hang is a visual illusion, due primarily to skilled, rapid passing of the ball from hand to hand during the jump; so much occurs that the jump seems to last longer than normal; the illusion may also be enhanced if arms or legs are raised to flatten the body's path, as in Fig. 9–6.)
3. (a) at the center of the sled; (b) $L/4$, to the right; (c) not at all (no net external force); (d) $L/4$, to the left; (e) L; (f) $L/2$; (g) $L/2$
4. (a) less; (b) greater
5. (a) ac, cd, and bc; (b) bc; (c) bd and ad
6. (a) 2 N, rightward; (b) 2 N, rightward; (c) greater than 2 N, rightward
7. c, d, then a and b tie
8. (a) a, d, and f; (b) 2; (c) d, f, a
9. b, c, a

Chapter 10

1. all tie
2. 1B, $p_x = -2\,\text{kg}\cdot\text{m/s}$, $p_y = 4\,\text{kg}\cdot\text{m/s}$; 2C, $p_y = 1\,\text{kg}\cdot\text{m/s}$; 2D, $p_x = 11\,\text{kg}\cdot\text{m/s}$; 3E, $p_y = -2\,\text{kg}\cdot\text{m/s}$, $p_x = 4\,\text{kg}\cdot\text{m/s}$
3. b and c
4. (a) positive; (b) positive; (c) 2 and 3
5. (a) rightward; (b) rightward; (c) smaller
6. a, projectile momentum cannot increase, target cannot move in the negative x direction; b, projectile momentum cannot increase, total momentum is not conserved; e, target cannot move in the negative x direction, total momentum is not conserved; h, total momentum is not conserved; i, target cannot move in the negative x direction, total momentum is not conserved

7. (a) one stationary; (b) 2; (c) 5; (d) equal (pool player's result)
8. (a) lower (almost zero); (b) greater (baseball may smash into the ceiling)
9. (a) 2; (b) 1; (c) 3; (d) yes; (e) no
10. c

Chapter 11

1. (a) positive; (b) zero; (c) negative; (d) negative
2. (a) clockwise; (b) counterclockwise; (c) yes; (d) positive; (e) constant
3. finite angular displacements are not commutative
4. a and c
5. (a) c, a, then b and d tie; (b) b, then a and c tie, then d
6. (a) $+x$; (b) $+y$
7. 3, 1, 2
8. larger
9. 90°, then 70° and 110° tie
10. $\vec{F}_5$, $\vec{F}_4$, $\vec{F}_2$, $\vec{F}_1$, $\vec{F}_3$ (zero)
11. (a) decrease; (b) clockwise; (c) counterclockwise
12. (a) 3 and 6; (b) 1, 4, then 2 and 5 tie (zero), then 6, then 3; (c) 1 and 4 tie, then 2 and 5 tie (zero), then 6, then 3

Chapter 12

1. (a) same; (b) block; (c) block
2. (a) greater; (b) same
3. (a) L; (b) $1.5L$
4. (a) 0 or 180°; (b) 90°
5. b, then c and d tie, then a and e tie (zero)
6. (a) 0 ($\vec{r}$ and $\vec{F}$ are both radial); (b) same
7. a, then b and c tie, then e, d (zero)
8. (a) same; (b) greater (smaller rotational inertia)
9. (a) same; (b) increase; (c) decrease; (d) same, decrease, increase
10. (a) 4, 6, 7, 1, and then 2, 3, and 5 tie (zero); (b) 1, 4, and 7
11. (a) 1, 2, 3 (zero); (b) 1 and 2 tie, then 3; (c) 1 and 3 tie, then 2
12. (a) 5 and 6; (b) 1 and 4 tie, then the rest tie
13. (a) 3, 1, 2; (b) 3, 1, 2

SECTION SIX
ANSWERS TO PROBLEMS

The following are the answers to the end-of-chapter exercises and problems.

Chapter 1

1. (a) 10^9; (b) 10^{-4}; (c) 9.1×10^5
2. 403 L
3. (a) 160 rods; (b) 40 chains
4. (a) 23 points; (b) 1.9 picas
5. (a) 4.00×10^4 km; (b) $5.10 \times 10^8\ \text{km}^2$; (c) $1.08 \times 10^{12}\ \text{km}^3$
6. (a) 14.5 roods; (b) $1.47 \times 10^4\ \text{m}^2$
7. $1.9 \times 10^{22}\ \text{cm}^3$
8. (a) $1.0\ \text{m}^3$; (b) $6.0 \times 10^{-4}\ \text{m}^3$
9. 1.1×10^3 acre · feet
10. (a) 52.6 min; (b) 5.2%
11. (a) 0.98 ft/ns; (b) 0.30 mm/ps
12. (a) yes; (b) 8.6 s
13. C, D, A, B, E; the important criterion is the consistency of the daily variation, not its magnitude
14. (a) 495 s; (b) 141 s; (c) 198 s; (d) −245 s
15. 0.12 AU/min
16. 15°
17. 2.1 h
18. (a) 3.88×10^8; (b) 1557.806 448 872 75 s; (c) $\pm 3 \times 10^{-11}$ s
19. 9.0×10^{49}
20. (a) $1.430\ \text{m}^2$; (b) 72.84 km
21. (a) 10^3 kg; (b) 158 kg/s
22. 1.2×10^9 kg
23. (a) $1.18 \times 10^{-29}\ \text{m}^3$: (b) 0.282 nm
24. 0.260 kg
25. (a) 60.8 W; (b) 43.3 Z
26. (a) 22 pecks; (b) 5.5 bushels; (c) 200 L
27. 89 km
28. 1.2 m
29. $\approx 1 \times 10^{36}$

Chapter 2

1. 414 ms
2. 6.150 s
3. (a) +40 km/h; (b) 40 km/h
4. 1.29 s
5. (a) 73 km/h; (b) 68 km/h; (c) 70 km/h; (d) 0
6. (a) 1.74 m/s; (b) 2.14 m/s
7. (a) 0, −2, 0, 12 m/s; (b) +12 m; (c) +7 m/s
8. (a) 60 km
9. 1.4 m
10. (a) $2\ \text{s} < t < 4\ \text{s}$; (b) $0 < t < 3\ \text{s}$; (c) $3\ \text{s} < t < 7\ \text{s}$; (d) $t = 3\ \text{s}$
11. (a) −6 m/s; (b) negative x direction; (c) 6 m/s; (d) first smaller, then zero, and then larger; (e) yes ($t = 2$ s); (f) no
12. (a) 28.5 cm/s; (b) 18.0 cm/s; (c) 40.5, cm/s; (d) 28.1 cm/s; (e) 30.3 cm/s
13. 100 m
14. (e) situations (a), (b), and (d)
15. (a) velocity squared; (b) acceleration; (c) m^2/s^2, m/s^2
17. 20 m/s^2, in the direction opposite to its initial velocity
18. (a) 1.10 m/s; (b) 6.11 mm/s^2; (c) 1.47 m/s; (d) 6.11 mm/s^2
19. (a) 80 m/s; (b) 110 m/s; (c) 20 m/s^2
20. 5.9 m
21. (a) m/s^2, m/s^3; (b) 1.0 s; (c) 82 m; (d) −80 m; (e) 0, −12, −36, −72 m/s; (f) −6, −18, −30, −42 m/s^2
22. each, 0.28 m/s^2
23. 0.10 m
24. 0.556 s
25. (a) 1.6 m/s; (b) 18 m/s
26. 3.75 ms
27. (a) 3.1×10^6 s (1.2 months); (b) 4.6×10^{13} m
28. 2.8 m/s^2
29. $1.62 \times 10^{15}\ \text{m/s}^2$
30. 21g
31. 2.65 s
32. 4 m/s^2, positive x direction
33. (a) 3.56 m/s^2; (b) 8.43 m/s
34. yes; red train, 0; green train, 10 m/s
35. (a) 5.00 m/s; (b) 1.67 m/s^2; (c) 7.50 m
36. (a) 82 m; (b) 19 m/s
37. (a) 0.74 s; (b) −6.2 m/s^2

38. (a) $0.994\,m/s^2$
39. (a) 10.6 m; (b) 41.5 s
40. 183 m/s; no
41. (a) 29.4 m; (b) 2.45 s
42. (a) 1.54 s; (b) 27.1 m/s
43. (a) 31 m/s; (b) 6.4 s
44. (a) 5.44 s; (b) 53.3 m/s; (c) 5.80 m
45. (a) 3.2 s; (b) 1.3 s
46. (a) $v = \sqrt{v_0^2 + 2gh}$; (b) $t = [\sqrt{v_0^2 + 2gh} - v_0]/g$; (c) same as (a); (d) $t = [\sqrt{v_0^2 + 2gh} + v_0]/g$, greater
47. (a) 3.70 m/s; (b) 1.74 m/s; (c) 0.154 m
48. 34 m
49. 4.0m/s
50. 22.0 m/s
51. $857\,m/s^2$, upward
52. (a) 101 m; (b) 13.0 s
53. $+1.26 \times 10^3\,m/s^2$, upward
54. (a) 350 ms; (b) 82 ms (each includes both ascent and descent through the 15 cm)
55. 22 cm and 89 cm below the nozzle
56. (a) $8.0\,m/s^2$; (b) 20 m/s
57. 1.5 s
58. $4H$
59. (a) 5.4 s; (b) 41 m/s
60. (a) 12.3 m/s
61. (a) 76 m; (b) 4.2 s
62. (a) 8.85 m/s; (b) 1.00 m
63. (a) 1.23 cm; (b) 4 times, 9 times, 16 times, 25 times
64. (a) 17 s; (b) 290 m
65. 2.34 m

Chapter 3

1. The displacements should be (a) parallel, (b) antiparallel, (c) perpendicular
2. Walpole (where the state prison is located)
3. (a) −2.5 m, −6.9 m
4. (a) 0.349 rad; (b) 0.873 rad; (c) 1.75 rad; (d) 18.9°; (e) 120°; (f) 441°
5. (a) 47.2 m; (b) 122°
6. (a) 13 m; (b) 7.5 m
7. (a) 168 cm; (b) 32.5° above the floor
8. (a) 27.8 m; (b) 13.4 m
9. (a) 6.42 m; (b) no; (c) yes; (d) yes; (e) a possible answer: $(4.30\,m)\hat{\imath} + (3.70\,m)\hat{\jmath} + (3.00\,m)\hat{k}$; (f) 7.96 m
10. (a) 81 km; (b) 40° north of east
11. (a) 370 m; (b) 36° north of east; (c) 425 m; (d) the distance
12. (b) 3.2 km; (c) 41° south of west
13. (a) $(-9\,m)\hat{\imath} + (10\,m)\hat{\jmath}$; (b) 13 m; (c) +132°
14. (a) 12 m; (b) −5.8 m; (c) −2.8 m
15. (a) 4.2 m; (b) 40° east of north; (c) 8.0 m; (d) 24° north of west
16. (a) $(8.0\,m)\hat{\imath} + (2.0\,m\,\hat{\jmath}$; (b) 8.2 m; (c) 14°; (d) $(2.0\,m)\hat{\imath} - (6.0\,m)\hat{\jmath}$; (e) 6.3 m; (f) −72°
17. (a) $(3.0\,m)\hat{\imath} - (2.0\,m)\hat{\jmath} + (5.0\,m)\hat{k}$; (b) $(5.0\,m)\hat{\imath} - (4.0\,m)\hat{\jmath} - (3.0\,m)\hat{k}$; (c) $(-5.0\,m)\hat{\imath} + (4.0\,m)\hat{\jmath} + (3.0\,m)\hat{k}$
18. (a) 5.0 m; (b) −37°; (c) 10 m; (d) 53°; (e) 11 m; (f) 27°; (g) 11 m; (h) 80°; (i) 11 m; (j) 260°; (k) 180°
19. (a) 38 m; (b) 320°; (c) 130 m; (d) 1.2°; (e) 62 m; (f) 130°
20. (a) $(1.28\,m)\hat{\imath} + (6.60\,m)\hat{\jmath}$; (b) 6.72 m, at 79.1° and 1.38 rad
21. (a) 1.59 m; (b) 12.1 m; (c) 12.2 m; (d) 82.5°
22. (a) 26.6 m; (b) 209° or −151°
24. (a) $(10.0\,m)\hat{\imath} + (1.63\,m)\hat{\jmath}$; (b) 10.2 m; (c) 9.24° counterclockwise
26. (a) $(-3.18\,m)\hat{\imath} + (4.72\,m)\hat{\jmath}$; (b) 5.69 m; (c) +124°
27. (a) Put axes along cube edges with origin at one corner. Diagonals are $a\,\hat{\imath} + a\,\hat{\jmath} + a\,\hat{k}$, $a\,\hat{\imath} + a\,\hat{\jmath} - a\,\hat{k}$, $a\,\hat{\imath} - a\,\hat{\jmath} - a\,\hat{k}$, $a\,\hat{\imath} - a\,\hat{\jmath} + a\,\hat{k}$; (b) 54.7°; (c) $\sqrt{3}a$
28. (a) $(9.19\,m)\hat{\imath} + (7.71\,m)\hat{\jmath}$; (b) $(14.0\,m)\hat{\imath} + (3.41\,m)\hat{\jmath}$
29. (a) 30; (b) 52
31. 22°
34. $\vec{B} = -3.00\,\hat{\imath} - 3.00\,\hat{\jmath} - 4.00\,\hat{k}$
35. (b) $a^2 b \sin\phi$
36. 540
37. (a) 3.0 m; (b) 0; (c) 3.46 m; (d) 2.00 m; (e) −5.00 m; (f) 8.66 m; (g) −6.67; (h) 4.33
38. (a) 57°; (b) 2.2 m; (c) −4.5 m; (d) −2.2 m; (e) 4.5 m

Chapter 4

1. (a) $(-5.0\,m)\hat{\imath}+(8.0\,m)\hat{\jmath}$; (b) 9.4 m; (c) 122°; (e) $(8\,m)\hat{\imath} - (8\,m)\hat{\jmath}$; (f) 11 m; (g) −45°
2. (a) 6.2 m
3. (a) $(-7.0\,m)\hat{\imath} + (12\,m)\hat{\jmath}$; (b) xy plane
4. 1030 m, horizontally to the west

5. 7.59 km/h, 22.5° east of north
6. $(-0.70\,\text{m/s})\hat{\imath} + (1.4\,\text{m/s})\hat{\jmath} - (0.40\,\text{m/s})\hat{k}$
7. (a) $(3.00\,\text{m/s})\hat{\imath} - (8.00\,\text{m/s}^2)t\,\hat{\jmath}$; (b) $(3.00\,\text{m/s})\hat{\imath} - (16.0\,\text{m/s})\hat{\jmath}$; (c) 16.3 m/s; (d) −79.4°
8. (a) 63.1 km, 18.3° south of west; (b) 0.70 km/h, 18.3° south of east; (c) 1.56 km/h; (d) 1.20 km/h, 33.4° north of east
9. (a) $(8\,\text{m/s}^2)t\,\hat{\jmath} + (1\,\text{m/s})\hat{k}$; (b) $(8\,\text{m/s}^2)\hat{\jmath}$
10. (a) $(-1.5\,\text{m/s}^2)\hat{\imath} + (0.50\,\text{m/s}^2)\hat{k}$; (b) $1.6\,\text{m/s}^2$, in the xz plane, 162° from $+x$
11. (a) $(6.00\,\text{m})\hat{\imath} - (106\,\text{m})\hat{\jmath}$; (b) $(19.0\,\text{m/s})\hat{\imath} - (224\,\text{m/s})\hat{\jmath}$; (c) $(24.0\,\text{m/s}^2)\hat{\imath} - (336\,\text{m/s}^2)\hat{\jmath}$; (d) −85.2° to $+x$
12. $(-2.10\,\text{m/s}^2)\hat{\imath} + (2.81\,\text{m/s}^2)\hat{\jmath}$
13. (a) $(-1.5\,\text{m/s})\hat{\jmath}$; (b) $(4.5\,\text{m})\hat{\imath} - (2.25\,\text{m})\hat{\jmath}$
14. (a) $(-18\,\text{m/s}^2)\hat{\imath}$; (b) 0.75 s; (c) never; (d) 2.2 s
15. (a) 45 m; (b) 22 m/s
16. 60.0°
17. (a) 62 ms; (b) 480 m/s
18. (a) 0.495 s; (b) 3.07 m/s
19. (a) 0.205 s; (b) 0.205 s; (c) 20.5 cm; (d) 61.5 cm
20. (a) 18 cm; (b) 1.9 m
21. (a) 2.00 ns; (b) 2.00 ns; (c) 1.0×10^7 m/s; (d) 2.00×10^6 m/s
22. 25.9 cm
23. (a) 16.9 m; (b) 8.21 m; (c) 27.6 m; (d) 7.26 m; (e) 40.2 m; (f) 0
24. (a) 95 m; (b) 31 m
25. 4.8 cm
26. (a) 38 ft/s; (b) 32 ft/s; (c) 9.3 ft
28. (a) 12.0 m; (b) 19.2 m/s; (c) 4.80 m/s; (d) no
29. (a) 11 m; (b) 23 m; (c) 17 m/s; (d) 63° below horizontal
30. (a) 20 m/s; (b) 36 m/s; (c) 74 m
31. (a) 24 m/s; (b) 65° above the horizontal
32. (a) 1.60 m; (b) 6.86 m; (c) 2.86 m
33. (a) 10 s; (b) 897 m
34. 78.5°
35. the third
36. 5.8 m/s
37. (a) 202 m/s; (b) 806 m; (c) 161 m/s; (d) −171 m/s
38. (a) 20.3 m/s; (b) 21.7 m/s
39. (a) yes; (b) 2.56 m
40. (a) yes; (b) 20 cm; (c) no; (d) 0.86 cm
41. between the angles 31° and 63° above the horizontal
42. $4.0\,\text{m/s}^2$
43. (a) 7.49 km/s; (b) $8.00\,\text{m/s}^2$
44. (a) 0.94 m; (b) 19 m/s; (c) $2400\,\text{m/s}^2$; (d) 0.05 s
45. (a) 19 m/s; (b) 35 rev/min; (c) 1.7 s
46. (a) 7.32 m west of the center; (b) 7.32 m north of the center
47. (a) $0.034\,\text{m/s}^2$; (b) 84 min
48. (a) 7.3 km; (b) 80 km/h
49. (a) 12 s; (b) $4.1\,\text{m/s}^2$, down; (c) $4.1\,\text{m/s}^2$, up
50. (a) 1.3×10^5 m/s; (b) $7.9 \times 10^5\,\text{m/s}^2$; (c) increase
51. $160\,\text{m/s}^2$
52. (a) 4.2 m, 45°; (b) 5.5 m, 68°; (c) 6.0 m, 90°; (d) 4.2 m, 135°; (e) 0.85 m/s, 135°; (f) 0.94 m/s, 90°; (g) 0.94 m/s, 180°; (h) $0.30\,\text{m/s}^2$, 180°; (i) $0.30\,\text{m/s}^2$, 270°
53. (a) $13\,\text{m/s}^2$, eastward; (b) $13\,\text{m/s}^2$, eastward
54. (a) 5 km/h, upstream; (b) 1 km/h, downstream
55. 36 s, no
56. 130°
57. 60°
58. (a) $(80\,\text{km/h})\hat{\imath} - (60\,\text{km//h})\hat{\jmath}$; (b) $\vec{v}$ happens to be along the line of sight; (c) answers do not change
59. 32 m/s
60. (a) $(-32\,\text{km/h})\hat{\imath} - (46\,\text{km/h})\hat{\jmath}$; (b) $[(2.5\,\text{km}) - (32\,\text{km/h})t]\hat{\imath} + [(4.0\,\text{km}) - (46\,\text{km/h})t]\hat{\jmath}$; (c) 0.084 h; (d) ≈ 200 m
61. (a) 38 knots, 1.5° east of north; (b) 4.2 h; (c) 1.5° west of south
62. 93° from the car's direction of motion
63. (a) 37° west of north; (b) 62.6 s
64. (a) A: 10.1 km, 0.556 km; B: 12.1 km, 1.51 km; C: 14.3 km, 2.68 km; D: 16.4 km, 3.99 km; E: 18.5 km, 5.53 km; (b) the rocks form a curtain that curves upward and away from you

Chapter 5

1. (a) $F_x = 1.88\,\text{N}$; (b) $F_y = 0.684\,\text{N}$; (c) $(1.88\,\text{N})\hat{\imath} + (0.684\,\text{N})\hat{\jmath}$
2. (a) 0; (b) $(4\,\text{m/s}^2),\hat{\jmath}$; (c) $(3\,\text{m/s}^2)\hat{\imath}$
3. $2.9\,\text{m/s}^2$
4. $(-2\,\text{N})\hat{\imath} + (6\,\text{N})\hat{\jmath}$
5. $(3\,\text{N})\hat{\imath} - (11\,\text{N})\hat{\jmath} + (4\,\text{N})\hat{k}$
6. (a) $(0.86\,\text{m/s}^2)\hat{\imath} - (0.16\,\text{m/s}^2)\hat{\jmath}$; (b) $0.88\,\text{m/s}^2$; (c) $-11°$ from $+x$
7. (a) $(-32\,\text{N})\hat{\imath} - (21\,\text{N})\hat{\jmath}$; (b) 38 N; (c) 213° from $+x$
8. (a) $\vec{F}_2$ and $\vec{F}_3$ are in the $-x$ direction, $\vec{a} = 0$; (b) $\vec{F}_2$ and $\vec{F}_3$ are in the $-x$ direction, $\vec{a}$ is on the x axis, $a = 0.83\,\text{m/s}^2$; (c) $\vec{F}_2$ and $\vec{F}_3$ are 34° above and below the $-x$ direction; $\vec{a} = 0$
9. (a) 108 N; (b) 108 N; (c) 108 N
10. 2.0 N, down
11. (a) 11 N; (b) 2.2 kg; (c) 0; (d) 2.2 kg
12. (a) 740 N; (b) 290 N; (c) 0; (d) 75 kg at each location
13. 16 N
14. (a) 328 N, down; (b) 0; (c) 0; (d) 284 N, up
15. (a) 42 N; (b) 72 N; (c) $4.9\,\text{m/s}^2$
16. (a) 590 N, up; (b) 340 N, up; (c) 590 N, down
17. (a) $0.02\,\text{m/s}^2$; (b) 8×10^4 km; (c) 2×10^3 m/s
18. 310 N
19. 1.2×10^5 N
20. (a) 5500 N; (b) 2.7 s; (c) 4.0; (d) 2.0
21. 1.5 mm
22. 6.8×10^3 N
23. (a) $(285\,\text{N})\hat{\imath} + (705\,\text{N})\hat{\jmath}$; (b) $(285\,\text{N})\hat{\imath} - (115\,\text{N})\hat{\jmath}$; (c) 307 N; (d) $-22°$ from $+x$; (e) $3.67\,\text{m/s}^2$; (f) $-22°$ from $+x$
24. (a) 68 N; (b) 73 N
25. (a) $0.62\,\text{m/s}^2$; (b) $0.13\,\text{m/s}^2$; (c) 2.6 m
26. (a) $\cos\theta$; (b) $\sqrt{\cos\theta}$
27. (a) 494 N, up; (b) 494 N, down
28. (a) $(5\,\text{m/s})\hat{\imath} + (4.3\,\text{m/s})\hat{\jmath}$; (b) $(15\,\text{m})\hat{\imath} + (6.4\,\text{m})\hat{\jmath}$
29. (a) 2.2×10^{-3} N; (b) 3.7×10^{-3} N
30. (a) 68 N, up the slope; (b) 28 N, up the slope; (c) 12 N, down the slope
31. (a) 1.1 N
32. (a) 4.6×10^3 N; (b) 5.8×10^3 N
33. 1.8×10^4 N
34. 23 kg
35. (a) 620 N; (b) 580 N
36. (a) $0.970\,\text{m/s}^2$; (b) 11.6 N; (c) 34.9 N
37. (a) 3260 N; (b) 2.7×10^3 kg; (c) $1.2\,\text{m/s}^2$
38. (a) $0.74\,\text{m/s}^2$; (b) $7.3\,\text{m/s}^2$
39. (a) 180 N; (b) 640 N
40. 5.1 m/s
41. (a) 1.23 N; (b) 2.46 N; (c) 3.69 N; (d) 4.92 N; (e) 6.15 N; (f) 0.25 N
42. 839 kN
43. (a) $0.735\,\text{m/s}^2$; (b) downward; (c) 20.8 N
44. (a) 3.1 N; (b) 14.7 N
45. (a) 1.18 m; (b) 0.674 s; (c) 3.50 m/s
46. (a) $120\,\text{m/s}^2$; (b) $12g$; (c) 1.4×10^8 N; (d) 4.2 y
47. (a) $4.9\,\text{m/s}^2$; (b) $2.0\,\text{m/s}^2$; (c) upward; (d) 120 N
48. (a) 6800 N; (b) 21° to the barge's line of motion
49. (a) $2.18\,\text{m/s}^2$; (b) 116 N; (c) $21.0\,\text{m/s}^2$
50. (a) 466 N; (b) 527 N; (c) 931 N; (d) 1050 N; (e) 931 N; (f) 1050 N; (g) 1860 N; (h) 2110 N
51. (b) $F/(m+M)$; (c) $MF/(m+M)$; (d) $F(m+2M)/2(m+M)$
52. (a) 566 N; (b) 1130 N
53. $2Ma/(a+g)$
54. 18 kN
55. (a) 31.3 kN; (b) 24.3 kN
56. (a) 7.3 kg; (b) 89 N

Chapter 6

1. (a) 200 N; (b) 120 N
2. 2°
3. 0.61
4. 160 N
5. (a) 190 N; (b) $0.56\,\text{m/s}^2$
6. 20°
7. (a) 0.13 N; (b) 0.12
8. (b) 240 N; (c) 0.60
9. (a) no; (b) $(-12\,\text{N})\hat{\imath} + (5\,\text{N})\hat{\jmath}$
10. (a) 6.0 N, to the left; (b) 3.6 N, to the left; (c) 3.1 N, to the left
12. (a) 110 N; (b) 130 N; (c) no; (d) 46 N; (e) 17 N
13. (a) 300 N; (b) $1.3\,\text{m/s}^2$
14. 0.53
15. (a) 66 N; (b) $2.3\,\text{m/s}^2$
16. (a) 11 N; (b) $0.14\,\text{m/s}^2$, rightward
17. (b) 3.0×10^7 N

18. (a) 8.6 N; (b) 46 N; (c) 39 N
19. 100 N
20. (a) 17 N, up the incline; (b) 20 N, up the incline; (c) 15 N, up the incline
21. (a) 0; (b) 3.9 m/s^2 down the incline; (c) 1.0 m/s^2 down the incline
22. 3.3 kg
23. (a) 3.5 m/s^2; (b) 0.21 N; (c) blocks move independently
24. 8.5 N
25. 490 N
26. (a) 1.05 N, in tension; (b) 3.62 m/s^2 down the plane; (c) answers are the same except that the rod is under compression
27. (a) 6.1 m/s^2, leftward; (b) 0.98 m/s^2, leftward
28. (a) 3.0×10^5 N; (b) 1.2°
29. $g(\sin\theta - \sqrt{2}\mu_k \cos\theta)$
30. (a) 19°; (b) 3300 N
31. 9.9 s
32. (a) 320 km/h; (b) 650 km/h, no
33. 6200 N
34. 3.75
35. 2.3
36. 9.7g
37. about 48 km/h
38. (a) 3.7 kN, up; (b) 1.3 kN, down
39. 21 m
40. (a) 3.7 kN, up; (b) 2.3 kN, down
41. (a) $\sqrt{Mgr/m}$
42. (a) 275 N; (b) 877 N
43. (a) light; (b) 778 N; (c) 223 N
44. 12°
45. 2.2 km
46. (a) 210 N; (b) 44 m/s
47. (b) 8.74 N; (c) 37.9 N, radially inward; (d) 6.45 m/s
48. (a) 70 km/h; (b) 140 km/h; (c) yes

Chapter 7

1. 1.2×10^6 m/s
2. (a) 5×10^{14} J; (b) 0.1 megaton TNT; (c) 8 bombs
3. (a) 3610 J; (b) 1900 J; (c) 1.1×10^{10} J
4. (a) 2.4 m/s; (b) 4.8 m/s
5. (a) 2.9×10^7 m/s; (b) 2.1×10^{-13} J
6. 5000 J
7. (a) 590 J; (b) 0; (c) 0; (d) 590 J
8. 0.32 J
9. (a) 170 N; (b) 340 m; (c) -5.8×10^4 J; (d) 340 N, (e) 170 m; (f) -5.8×10^4 J
10. 530 J
11. (a) 1.50 J; (b) increases
12. 20 J
13. 15.3 J
14. (a) 36 kJ; (b) 200 J
15. (a) 98 N; (b) 4.0 cm; (c) 3.9 J; (d) −3.9 J
16. (a) 270 N; (b) −400 J; (c) 400 J; (d) 0; (e) 0
17. (a) 1.2×10^4 J; (b) -1.1×10^4 J; (c) 1100 J; (d) 5.4 m/s
18. (a) 8840 J; (b) 7840 J; (c) 6840 J
19. (a) $-3Mgd/4$; (b) Mgd; (c) $Mgd/4$; (d) $\sqrt{gd/2}$
20. 1250 J
21. (a) −0.043 J; (b) −0.13 J
22. (a) 0.29 J; (b) −1.8 J; (c) 3.5 m/s; (d) 23 cm
23. (a) 6.6 m/s; (b) 4.7 m
24. 25 J
25. 800 J
26. (a) 12 J; (b) 4.0 m; (c) 18 J
27. 0, by both methods
28. (a) 2.3 J; (b) 2.6 J
29. −6 J
30. 270 kW
31. 490 W
32. (a) 28 W; (b) (6 m/s) $\hat{j}$
33. (a) 0.83 J; (b) 2.5 J; (c) 4.2 J; (d) 5.0 W
34. (a) 900 J; (b) 113 W; (c) 225 W
35. 740 W
36. (a) 0; (b) −350 W
37. 68 kW
38. (a) 1.7 W; (b) 0; (c) −1.7 W
39. (a) 1.8×10^5 ft · lb; (b) 0.55 hp
40. (a) 100 J; (b) 67 W; (c) 33 W

Chapter 8

1. 89 N/cm
2. (a) 167 J; (b) −167 J; (c) 196 J; (d) 29 J (e) 167 J; (f) −167 J; (g) 296 J; (h) 129 J
3. (a) 4.31 mJ; (b) −4.31 mJ; (c) 4.31 mJ; (d) −4.31 mJ; (e) all increase
4. (a) 0; (b) $mgh/2$; (c) mgh; (d) $mgh/2$; (e) mgh; (f) increase
5. (a) mgL; (b) $-mgL$; (c) 0; (d) $-mgL$; (e) mgL; (f) 0; (g) same
6. (a) $4mgR$; (b) $3mgR$; (c) $5mgR$; (d) mgR; (e) $2mgR$; (f) all the same

7. (a) 184 J; (b) −184 J; (c) −184 J
8. (a) $mgL(1 - \cos\theta)$; (b) $-mgL(1 - \cos\theta)$; (c) $mgL(1 - \cos\theta)$; (d) all increase
9. (a) 2.08 m/s; (b) 2.08 m/s; (c) increase
10. (a) 12.9 m/s; (b) 12.9 m/s; (c) increase
11. (a) $\sqrt{2gL}$; (b) $2\sqrt{gL}$; (c) $\sqrt{2gL}$; (d) all the same
12. (a) v_0; (b) $(v_0^2 + gh)^{1/2}$; (c) $(v_0^2 + 2gh)^{1/2}$; (d) $v_0^2/2g + h$; (e) all the same
13. (a) 260 m; (b) same; (c) decrease
14. (a) 2.29 m/s; (b) same
15. (a) 21.0 m/s; (b) 21.0 m/s; (c) 21 m/s
16. (a) 784 N/m; (b) 62.7 J; (c) 62.7 J; (d) 80.0 cm
17. (a) 0.98 J; (b) −0.98 J; (c) 3.1 N/cm;
18. (a) $\sqrt{v_0^2 + 2gL(1 - \cos\theta_0)}$; (b) $\sqrt{2gL\cos\theta_0}$; (c) $\sqrt{gL(3 + 2\cos\theta_0)}$; (d) both decrease
19. (a) 39.2 J; (b) 39.2 J; (c) 4.00 m
20. (a) $8mg$ leftward; (b) mg downward; (c) $2.5R$
21. (a) 35 cm (b) 1.7 m/s
22. −320 J
23. (a) 4.8 m/s; (b) 2.4 m/s
24. (a) 4.4 m; (b) same
25. 10 cm
26. (a) no; (b) 930 N
27. 1.25 cm
28. (a) 7.2 J; (b) −7.2 J; (c) 86 cm; (d) 26 cm
30. (a) 1.4 m/s; (b) 1.9 m/s; (c) 28°
31. (a) $2\sqrt{gL}$: (b) $5mg$; (c) 71°
32. (a) 2.40 m/s; (b) 4.19 m/s
33. $mgL/32$
34. (a) $U_g = 0$, $U_e = 12.5$ J, $K = 0$; (b) $U_g = 2.0$ J, $U_e = 8.0$ J, $K = 2.5$ J; (c) $U_g = 4.0$ J, $U_e = 4.5$ J, $K = 4.0$ J; (d) $U_g = 6.0$ J, $U_e = 2.0$ J, $K = 4.5$ J; (e) $U_g = 8.0$ J, $U_e = 0.50$ J, $K = 4.0$ J; (f) $U_g = 10$ J, $U_e = 0$; $K = 2.5$ J; (g) $U_g = 12$ J, $U_e = 0$; $K = 0.50$ J; (h) 31.3 cm
36. (a) 4.8 N, positive x direction; (b) between $x \approx 1.5$ m and $x \approx 13.5$ m; (c) 3.5 m/s
37. (a) $1.12(A/B)^{1/6}$; (b) repulsive; (c) attractive
38. (a) −3.73 J; (c) 1.29 m; (d) 9.12 m; (e) 2.16 J; (f) 4.0 m; (g) $F(x) = (4 - x)e^{-x/4}$ N; (h) 4.0 m
39. (a) 5.6 J; (b) 3.5 J
40. 25 J
41. (a) 30.1 J; (b) 30.1 J; (c) 0.22;
42. (a) 560 J; (b) 560 J
43. (a) −2900 J; (b) 390 J; (c) 210 N
44. (a) −3800 J; (b) 31 kN
45. 11 kJ
46. 0.53 J
47. 20 ft · lb
48. (a) 2700 MJ; (b) 2700 MW; (c) 240 M$
49. (a) 1.5 MJ; (b) 0.51 MJ; (c) 1.0 MJ; (d) 63 m/s
50. 0.15
51. (a) 67 J; (b) 67 J; (c) 46 cm
52. 75 J
53. (a) 31.0 J; (b) 5.35 m/s; (c) conservative
54. 4.3 m
55. (a) 44 m/s; (b) 0.036
56. (a) 150 J; (b) 5.5 m/s
57. (a) −0.90 J; (b) 0.46 J; (c) 1.0 m/s
58. (a) 5.5 m/s; (b) 5.4 m; (c) same
59. 1.2 m
60. (a) 13 cm.; (b) 2.7 m; (c) both increase
62. (a) 10 m; (b) 50 N; (c) 4.1 m; (d) 120 N
63. in the center of the flat part
64. (a) 7.4 m/s; (b) 90 cm; (c) 2.8 m; (d) 15 m
65. (a) 216 J; (b) 1180 N; (c) 432 J; (d) motor also supplies thermal energy to crate and belt
66. (a) 3.0 mm; (b) 1.1 J; (d) yes; (e) 37 J; (f) no
67. (b) $\rho(L - x)/2$; (c) $v = v_0[2(\rho L + m_f)/(\rho L + 2m_f - \rho x)]^{1/2}$; (e) 35 m/s

---

Chapter 9

1. (a) 4600 km; (b) $0.73R_e$
2. 6.46×10^{-11} m, along the line joining the atoms
3. (a) 1.1 m; (b); 1.3 m; (c) shifts toward topmost particle
4. $0.2L$ down from the heavy rod, along symmetry axis
5. −0.25 m; (b) 0
6. in the iron, at midheight and midwidth, 2.7 cm from midlength
7. 6.8×10^{-12} m toward the nitrogen atom, along axis of symmetry
8. 20 cm; (b) 20 cm; (c) 16 cm

9. (a) $H/2$; (b) $H/2$; (c) descends to lowest point and then ascends to $H/2$; (d) $\frac{HM}{m}\left(\sqrt{1+\frac{m}{M}}-1\right)$
10. 6.2 m
11. 72 km/h
12. (a) down, $mv/(m+M)$; (b) balloon again stationary
13. (a) 28 cm; (b) 2.3 m/s
14. (a) 22 m; (b) 9.3 m/s
15. 53 m
16. $(-4.0\text{ m})\hat{\imath} + (4.0\text{ m})\hat{\jmath}$
17. (a) half way between the containers; (b) 26 mm toward the heavier container; (c) down; (d) -1.6×10^{-2} m/s^2
18. 58 kg
19. 4.2 m
20. (a) 52.0 km/h; (b) 28.8 km/h
21. 24 km/h
22. 4.9 kg · m/s
23. (a) 7.5×10^4 J; (b) 3.8×10^4 kg·m/s; (c) 38° south of east
24. (a) 30°; (b) $(-0.572\text{ kg}\cdot\text{m/s})\hat{\jmath}$
25. (a) $(-4.0 \times 10^4\text{ kg}\cdot\text{m/s})\hat{\imath}$; (b) west; (c) 0
26. (a) 5.0 kg · m/s; (b) 10 kg · m/s
27. 3.0 mm/s, away from the stone
28. 0.57 m/s, toward center of mass
29. increases by 4.4 m/s
30. $-1.4\text{ m/s})\hat{\imath}$
31. 4400 km/h
32. $wv_{\text{rel}}/(W+w)$
33. (a) 7290 m/s; (b) 8200 m/s; (c) 1.271×10^{10} J; (d) 1.275×10^{10} J
34. 3.5 m/s
35. (a) 1.4×10^{-22} kg · m/s; (b) 150°; (c) 120°; (d) 1.6×10^{-19} J
36. (a) 20 J; (b) 40 J
37. (a) 1014 m/s, 9.48° clockwise from the positive x direction; (b) 3.23 MJ
38. $mv^2/6$
39. 14 m/s, 135° from the other pieces
40. one chunk stops, the other moves ahead with a speed of 4.0 m/s
41. 108 m/s m/s
42. (a) 8.0×10^4 N; (b) 27 kg/s
43. (a) 1.57×10^6 N; (b) 1.35×10^5 kg; (c) 2.08 km/s
44. (a) 2.72; (b) 7.39
45. 2.2×10^{-3}
46. 28.8 N
47. (a) 46 N; (b) none
48. (a) 50 kg/s; (b) 160 kg/s
49. (a) 0.2 to 0.3 MJ; (b) same amount
50. (a) 7.8 MJ; (b) 6.2
51. (a) 8.8 m/s; (b) 2600 J; (c) 1.6 kW
52. 5.5×10^6 N
53. 24 W
54. 100 m
55. (a) 860 N; (b) 2.4 m/s
56. (a) 3.0×10^5 J; (b) 10 kW; (c) 20 kW
57. (a) 2.1×10^6 kg; (b) $\sqrt{100+1.5t}$ m/s; (c) $(1.5 \times 10^6)/\sqrt{100+1.5t}$ N; (d) 6.7 km
58. 69 hp
59. 1.5 cm downward (the bubbles rise but the layers descend)

Chapter 10

1. 2.5 m/s
2. 6.2×10^4 N
3. 3000 N
4. (a) 25 to 30 kN; (b) 1000 to 1200 kg · m/s
5. 67 m/s, in opposite direction
6. (a) 1.1 m; (b) 4.8×10^3 kg · m/s
7. (a) 42 N · s; (b) 2100 N
8. 5 N
9. (a) $(7.4 \times 10^3\text{ N}\cdot\text{s})\hat{\imath} - (7.4 \times 10^3\text{ N}\cdot\text{s})\hat{\jmath}$; (b) $(-7.4 \times 10^3\text{ N}\cdot\text{s})\hat{\imath}$; (c) 2.3×10^3 N; (d) 2.1×10^4 N; (e) −45°
10. 2.9×10^3 N
11. 10 m/s
12. (a) 0.48 g; (b) 7200 N
13. (a) 1.0 kg · m/s; (b) 250 J; (c) 10 N; (d) 1700 N (e) answer for (c) includes time between pellet collisions
14. 990 N
15. 41.7 cm/s
16. (a) 3.7 m/s; (b) 1.3 N · s; (c) 180 N
17. (a) 1.8 N·s, upward in the figure; (b) 180 N, downward in the figure
18. (a) 1.95×10^5 kg · m/s for each direction of thrust; (b) backward: −50.9 MJ, forward: +66.1 MJ, sideways: +7.61 MJ
19. (a) 9.0 kg · m/s; (b) 3000 N; (c) 4500 N; (d) 20 m/s
20. (a) 1.81 m/s; (b) 5.0 m/s
21. 3.0 m/s
22. 310 m/s

23. $\approx 2\,\text{mm/y}$
24. (a) 2.7 m/s; (b) 1400 m/s
25. (a) 4.6 m/s; (b) 3.9 m/s; (c) 7.5 m/s
26. (a) 721 m/s; (b) 937 m/s
27. (a) $mR(\sqrt{2gh} + gt)$; (b) 5.06 kg
28. (a) +2.0 m/s; (b) −1.3 J; (c) +40 J; (d) got energy from some source, such as a small explosion
29. 1.18×10^4 kg
30. 7.3 cm
31. (a) $mv_i/(m + M)$; (b) $M/(m + M)$
32. 120 J
33. 25 cm
34. 0.33 m
35. (a) 1.9 m/s, to the right; (b) yes; (c) no, total kinetic energy would have increased
36. 0.22%
37. (a) 99 g; (b) 1.9 m/s; (c) 0.93 m/s
38. 38 km/s
39. 7.8%
40. (a) 2.47 m/s; (b) 1.23 m/s
41. (a) 1.2 kg; (b) 2.5 m/s
42. (a) 0.60 cm; (b) 4.9 cm; (c) 9.0 cm; (d) 0
43. (a) 100 g; (b) 1.0 m/s
44. $m_1/3$
45. (a) 1/3; (b) $4h$
46. (a) $(10\,\text{m/s})\,\hat{\imath} + (15\,\text{m/s})\,\hat{\jmath}$; (b) 500 J lost
47. (a) 4.15×10^5 m/s; (b) 4.84×10^5 m/s
48. (a) 250 m/s; (b) 433 m/s
49. (a) 41°; (b) 4.76 m/s; (c) no
50. (a) 117° from the final direction of B
51. 120°
52. (a) 1.9 m/s, 30° to initial direction, on opposite side from first ball; (b) no
53. (a) 6.9 m/s, 30° to positive x direction; (b) 6.9 m/s, −30° to positive x direction; (c) 2.0 m/s, in negative x direction
54. 103°
56. (a) 4.50×10^{-3} kg · m/s; (b) 0.529 kg · m/s; (c) stroke
57. (a) $5mg$; (b) $7mg$; (c) 5 m

Chapter 11

1. (a) $a + 3bt^2 - 4ct^3$; (b) $6bt - 12ct^2$
2. (a) 0.105 rad/s; (b) 1.75×10^{-3} rad/s; (c) 1.45×10^{-4} rad/s
3. (a) 5.5×10^{15} s; (b) 26
4. (a) 4.0 rad/s; (b) 28 rad/s; (c) 12 rad/s^2; (d) 6.0 rad/s^2; (e) 18 rad/s^2
5. (a) 2 rad; (b) 0; (c) 130 rad/s; (d) 32 rad/s^2; (e) no
6. (a) 4.0 m/s: (b) no
7. 11 rad/s
8. (a) 9000 rev/min^2; (b) 420 rev
9. (a) −67 rev/min^2; (b) 8.3 rev
10. (a) 30 s; (b) 1800 rad
11. 200 rev/min
12. (a) 2.0 rad/s^2; (b) 5.0 rad/s; (c) 10 rad/s; (d) 75 rad
13. 8.0 s
14. (a) 13.5 s; (b) 27.0 rad/s
15. (a) 44 rad; (b) 5.5 s, 32 s; (c) −2.1 s, 40 s
16. (a) 1.0 rev/s^2; (b) 4.8 s; (c) 9.6 s; (d) 48 rev
17. (a) 340 s; (b) -4.5×10^{-3} rad/s^2; (c) 98 s
18. (a) 12 rad/s; (b) 3.0 s
19. 1.8 m/s^2, toward the center
20. (a) 3.5 rad/s; (b) 52 cm/s; (c) 26 cm/s
21. 0.13 rad/s
22. (a) 20.9 rad/s; (b) 12.5 m/s; (c) 800 rev/min^2; (d) 600 rev
23. (a) 3.0 rad/s; (b) 30 m/s; (c) 6.0 m/s^2; (d) 90 m/s^2
24. (a) 2.50×10^{-3} rad/s; (b) 20.2 m/s^2; (c) 0
25. (a) 3.8×10^3 rad/s; (b) 190 m/s
26. (a) −1.1 rev/min^2; (b) 9900 rev; (c) −0.99 mm/s^2; (d) 31 m/s^2
27. (a) 7.3×10^{-5} rad/s; (b) 350 m/s; (c) 7.3×10^{-5} rad/s; (d) 460 m/s
28. (a) 40.2 cm/s^2; (b) 2.36×10^3 m/s^2; (c) 83.2 m
29. 16 s
30. (a) 6.4 cm/s^2; (b) 2.6 cm/s^2
31. (a) -2.3×10^{-9} rad/s^2; (b) 2600 y; (c) 24 ms
32. (a) 73 cm/s^2; (b) 0.075; (c) 0.11
33. 12.3 kg · m^2
34. 6.75×10^{12} rad/s
35. (a) 1100 J; (b) 9700 J
36. (a) 221 kg · m^2; (b) 1.10×10^4 J
37. (a) $5md^2 + \frac{8}{3}Md^2$; (b) $(\frac{5}{2}m + \frac{4}{3}M)d^2\omega^2$
38. (a) 6490 kg · m^2; (b) 4.36 MJ
39. 0.097 kg · m^2
40. (a) 2.0 kg · m^2; (b) 6.0 kg · m^2; (c) 2.0 kg · m^2
41. $\frac{1}{3}M(a^2 + b^2)$
42. (a) 1300 g · cm^2; (b) 550 g · cm^2; (c) 1850 g · cm^2; (d) $A + B$
44. (a) 49 MJ; (b) 100 min

45. $4.6\,N \cdot m$
46. (a) $8.4\,N \cdot m$; (b) $17\,N \cdot m$; (c) 0
47. (a) $r_1 F_1 \sin\theta_1 - r_2 F_2 \sin\theta_2$; (b) $-3.8\,N \cdot m$
48. $12\,N \cdot m$
49. (a) $28.2\,rad/s^2$; (b) $338\,N \cdot m$
50. $1.28\,kg \cdot m^2$
51. (a) $155\,kg \cdot m^2$; (b) 64.4 kg
52. $9.7\,rad/s^2$, counterclockwise
53. 130 N
54. $4.6\,rad/s^2$
55. (a) $6.00\,cm/s^2$; (b) 4.87 N; (c) 4.54 N; (d) $1.20\,rad/s^2$; (e) $0.0138\,kg \cdot m^2$
56. (a) $420\,rad/s^2$; (b) 500 rad/s
57. (a) $1.73\,m/s^2$; (b) $6.92\,m/s^2$
58. (a) 1.4 m/s; (b) 1.4 m/s
59. $396\,N \cdot m$
60. (a) 19.8 kJ; (b) 1.32 kW
61. (a) $mL^2\omega^2/6$; (b) $L^2\omega^2/6g$
62. (a) $8.2 \times 10^{28}\,N \cdot m$; (b) $2.6 \times 10^{29}\,J$; (c) $3.0 \times 10^{21}\,kW$
63. 5.42 m/s
64. (a) $0.15\,kg \cdot m^2$; (b) 11 rad/s
65. $\sqrt{9g/4L}$
66. $\sqrt{\dfrac{2gh}{1 + (2M/3m) + (I/mr^2)}}$
67. (a) $[(3g/H)(1 - \cos\theta)]^1/2$; (b) $3g(1 - \cos\theta)$; (c) $(3/2)g\sin\theta$; (d) 41.8°
68. (a) 5.1 h; (b) 8.1 h
69. 17
70. (a) 0.32 rad/s; (b) 100 km/h

Chapter 12

1. (a) 59.3 rad/s; (b) $9.31\,rad/s^2$; (c) 70.7 m
2. (a) 0; (b) 0; (c) 22 m/s, positive x direction; (d) $1500\,m/s^2$; (e) 22 m/s, negative x direction; (f) $1500\,m/s^2$; (g) 22 m/s, in $+x$ direction; (h) 0; (i) 44 m/s; (j) $1500\,m/s^2$, in $+x$ direction; (k) 0; (l) $1500\,m/s^2$
3. −3.15 J
4. 1.00
5. 1/50
6. (a) $\frac{1}{2}mR^2$; (b) a solid circular cylinder
7. (a) 8.0°; (b) more
8. (a) 4.0 N, to the left; (b) $0.60\,kg \cdot m^2$
9. 4.8 m
10. (a) $mg(R - r)$; (b) 2/7; (c) $(17/7)mg$
11. (a) 63 rad/s; (b) 4.0 m
12. (a) $2.7R$; (b) $(50/7)mg$, to the left
13. (a) 8.0 J; (b) 3.0 m/s; (c) 6.9 J; (d) 1.8 m/s
14. (a) -0.11ω; (b) $-2.1\,m/s^2$; (c) $-47\,rad/s^2$; (d) 1.2 s; (e) 8.6 m; (f) 6.1 m/s
15. (a) $13\,cm/s^2$; (b) 4.4 s; (c) 55 cm/s; (d) $1.8 \times 10^{-2}\,J$; (e) 1.4 J; (f) 27 rev/s
16. (a) 0.89 s; (b) 9.4 J; (c) 1.4 m/s; (d) 0.12 J; (e) 440 rad/s; (f) 9.2 J
18. (a) $24\,N \cdot m$, in positive y direction; (b) $24\,N \cdot m$, in negative y direction; (c) $12\,N \cdot m$, in positive y direction; (d) $12\,N \cdot m$, in negative y direction
19. (a) $10\,N \cdot m$, parallel to yz plane, at 53° to $+y$; (b) $22\,N \cdot m$, $-x$
20. (a) $(-1.5\,N{\cdot}m)\hat{\imath} - (4.0\,N{\cdot}m)\hat{\jmath} - (1.0\,N{\cdot}m)\hat{k}$; (b) $(-1.5\,N{\cdot}m)\hat{\imath} - (4.0\,N{\cdot}m)\hat{\jmath} - (1.0\,N{\cdot}m)\hat{k}$
21. (a) $(50\,N \cdot m)\,\hat{k}$; (b) 90°
22. (a) $(6.0\,N{\cdot}m)\,\hat{\imath} - (3.0\,N{\cdot}m)\,\hat{\jmath} - (6.0\,N{\cdot}m)\,\hat{k})$; (b) $(26\,N \cdot m)\,\hat{\imath} + (3.0\,N \cdot m)\,\hat{\jmath} - (18\,N \cdot m)\,\hat{k}$; (c) $(32\,N \cdot m)\,\hat{\imath} - (24\,N \cdot m)\,\hat{k})$; (d) 0
23. $9.8\,kg \cdot m^2/s$
24. (a) $12\,kg \cdot m^2/s$, out of page; (b) $3.0\,N \cdot m$, out of page
25. (a) 0; (b) $(8.0\,N \cdot m)\,\hat{\imath} + (8.0\,N \cdot m)\,\hat{k}$
26. (a) $(600\,kg \cdot m^2/s)\hat{k}$; (b) $(720\,kg \cdot m^2/s)\,\hat{k}$
27. (a) mvd; (b) no; (c) 0, yes
28. (a) $(-32\,kg \cdot m^2/s)\,\hat{k}$; (b) $(-32\,kg \cdot m^2/s)\,\hat{k}$; (c) $(12\,N \cdot m)\,\hat{k}$; (d) 0
29. (a) $(-170\,kg \cdot m^2/s)\,\hat{k}$; (b) $(+56\,N \cdot m)\,\hat{k}$; (c) $(+56\,kg \cdot m^2/s)\,\hat{k}$
30. $4.5\,N \cdot m$, parallel to xy plane at −63° from $+x$
31. (a) 0; (b) $(8t)\,N \cdot m$, in $-z$ direction; (c) $(2/\sqrt{t})\,N \cdot m$, in $-z$ direction; (d) $(8/t^3)\,N \cdot m$, in $+z$ direction
32. (a) $(-24t^2\,kg \cdot m^2/s)\,\hat{k}$; (b) $(-48t\,N \cdot m)\,\hat{k}$; (c) $(+12t^2\,kg \cdot m^2/s)\,\hat{k}$, $(+24t\,N \cdot m)\,\hat{k}$
33. (a) $-1.47\,N \cdot m$; (b) 20.4 rad; (c) −29.9 J; (d) 19.9 W
34. (a) $0.53\,kg \cdot m^2/s$; (b) 4200 rev/min
35. (a) $14md^2$; (b) $4md^2\omega$; (c) $14md^2\omega$
37. $\omega_0 R_1 R_2 I_1/(I_1 R_2^2 + I_2 R_1^2)$
38. (a) $1.6\,kg \cdot m^2$; (b) $4.0\,kg \cdot m^2/s$
39. (a) 3.6 rev/s; (b) 3.0; (c) work done by man in moving bricks inward
40. 500 rev
41. (a) 267 rev/min; (b) 2/3

42. (a) 750rev/min; (b) 450 rev/min, in the original direction of the second disk
43. (a) $149 kg \cdot m^2$; (b) $158 kg \cdot m^2/s$; (c) 0.746rad/s
44. 3
45. $\frac{m}{M+m}\left(\frac{v}{R}\right)$
46. (a) they revolve in a circle of radius 1.5 m at an angular speed of 0.93 rad/s; (b) 98 J; (c) 8.4 rad/s; (d) 880 J; (e) energy is transferred from internal energy of the skaters to kinetic energy
47. (a) $(mRv - I\omega_0)/(I + mR^2)$; (b) no, energy transferred to internal energy of cockroach
48. (a) $mvR/(I + MR^2)$; (b) $mvR^2/(I + MR^2)$
49. 3.4 rad/s
50. 1300 m/s
51. (a) 0.148 rad/s; (b) 0.0123; (c) 181°
52. (a) $+0.14\omega_0$; (b) 1.14; (c) energy is transferred from internal energy of the cockroach to kinetic energy
53. the day would be longer by about 0.8 s
54. 2.6 rad/s
55. (a) 18 rad/s; (b) 0.92
56. 0.070 rad/s
57. (a) $0.24 kg \cdot m^2$; (b) 1800 m/s
58. 1.5 rad/s
59. $\theta = \cos^{-1}\left[1 - \frac{6m^2h}{d(2m+M)(3m+M)}\right]$

SECTION SEVEN
COMPARISON OF PROBLEMS WITH THE FIFTH EDITION

In the table below, the left column (in bold type) of each group gives the numbers of problems and exercises in the current (sixth) edition of *Fundamentals of Physics* and for each entry the right column gives the number of the same problem or exercise in the fifth edition. Any problem that did not appear in that edition is labeled "new". Significant changes in a problem are noted here. If units have been switched from British to SI units, the problem is labeled "SI". Some of the exercises and problems in the sixth edition are from the supplemental problem book published to accompany the fifth edition.

Chapter 1

1E	1–5E
2E	new
3E	1–3E
4E	new
5E	1–6E
6E	new
7P	1–12P
8P	new
9P	1–14P
10E	1–20E
11E	1–19E
12E	1–24E
13P	1–30P
14P	new
15P	1–26P
16P	1–27P
17P	1–28P
18P	new
19E	1–34E
20P	new
21P	1–37E
22P	1–35P
23P	1–39P
24P	1–38P
25	new
26	new
27	1–40
28	new
29	new

Chapter 2

1E	2–4E	
2E	new	
3E	2–8E	
4P	2–14P	
5P	2–11P	SI
6P	2–6P	SI
7P	2–12P	
8P	2–15P	
9P	1–29P	SI
10E	2–17E	
11E	2–16E	
12P	2–13P	
13P	2–19P	
14E	2–25E	
15E	2–26E	
16E	2–21E	
17E	2–20E	
18P	2–29P	
19P	2–27E	
20P	2–103	
21P	2–32P	
22E	2–35E	
23E	2–38E	
24E	2–33E	
25E	2–34E	
26E	2–102	
27E	2–36E	
28E	2–37E	
29E	2–39E	
30E	2–41E	
31E	2–44P	
32E	2–98	
33P	2–48P	
34P	2–58P	
35P	2–49P	
36P	2–56P	
37P	2–51P	
38P	2–57E	SI
39P	2–55P	
40E	2–62E	
41E	2–60E	
42E	2–64E	
43E	2–61E	
44E	2–65E	
45E	2–67E	
46P	2–74P	
47P	2–68P	
48P	2–111	

49P	2–69P
50P	2–97
51P	2–73P
52P	2–70P
53P	2–78P
54P	2–71P
55P	2–79P
56P	2–113
57P	2–85P
58P	2–76P
59P	2–87P
60P	2–83P
61P	2–88P
62P	2–77P
63P	2–75P
64P	2–84P
65P	2–90P

Chapter 3

1E	3–1E
2P	3–6P
3E	3–9E
4E	3–8E
5E	3–10E
6P	3–11E
7P	3–17P
8P	3–16P
9P	3–18P
10E	3–4E
11E	3–2E
12E	3–3E
13E	3–20E
14E	3–19E
15E	3–5P
16E	3–21E
17E	3–22E
18P	3–23E
19P	3–7P
20P	3–64
21P	3–27P
22P	3–66
23P	3–32P
24P	3–65
25P	3–33P
26P	3–63
27P	3–34P
28E	3–67
29E	3–42E
30E	3–46P
31P	3–48P
32P	3–49P
33P	3–55P
34P	3–61
35P	3–56P
36P	3–68
37P	3–58P
38P	3–51P

Chapter 4

1E	4–1E	
2E	4–2E	
3E	4–3E	
4P	3–29P	
5E	4–6E	
6E	4–8E	
7P	4–9E	
8P	4–104	
9E	4–11E	
10E	4–10E	
11E	4–12E	
12E	4–13E	
13P	4–15P	
14P	4–16P	
15P	4–14P	
16P	4–17P	
17E	4–19E	
18E	4–22E	
19E	4–24E	
20E	4–18E	
21E	4–21E	
22E	4–34P	
23E	4–26E	
24P	4–113	
25P	4–35P	
26P	4–109	
27P	4–36P	
28P	4–28E	
29P	4–37P	
30P	4–101	
31P	4–44P	
32P	4–108	
33P	4–48P	SI
34P	4–32P	
35P	4–50P	
36P	4–49P	
37P	4–51P	
38P	4–111	
39P	4–55P	SI
40P	4–53P	
41P	4–56P	
42E	4–58E	
43E	4–61E	
44E	4–63E	
45E	4–66E	
46P	4–114	
47P	4–67P	
48P	4–64E	
49P	4–68P	
50P	4–65E	
51P	4–71P	
52P	4–70P	
53E	4–75E	
54E	4–72E	
55P	4–73E	
56E	4–77E	
57E	4–79E	
58E	4–78E	
59P	4–82P	
60P	4–118	
61P	4–84P	
62P	4–86P	
63P	4–117	
64	new	

Chapter 5

1E	5–1E
2E	5–4E
3E	5–82
4E	5–5E
5E	5–6E
6P	5–10P
7P	5–8E
8P	5–11P
9E	5–18E
10E	5–78
11E	5–15E
12E	5–14E
13E	5–23E
14E	5–88
15E	5–27E
16E	5–89
17E	5–29E
18E	5–30E
19E	5–21E

20E	5–33E
21E	5–32E
22E	5–22E
23E	5–93
24P	5–94
25P	5–36P
26P	5–41P
27P	5–37P
28P	5–42P
29P	5–38P
30P	5–91
31P	5–40P
32P	5–45P
33P	5–43P
34P	5–49P
35P	5–51P
36P	5–48P
37P	5–54P
38P	5–39P
39P	5–77
40P	5–52P
41P	5–56P
42P	5–53P
43P	5–58P
44P	5–96
45P	5–60P
46P	5–66P
47P	5–64P
48P	5–69P
49P	5–68P
50P	5–74P
51P	5–73P
52P	5–63P
53P	5–70P
54P	5–65P
55P	5–50P
56P	5–62P

Chapter 6

1E	6–1E	
2E	6–3E	
3E	6–2E	
4E	6–5E	
5E	6–8E	
6E	6–13E	
7E	6–15E	
8E	6–12E	
9P	6–11E	
10P	6–76	
11P	6–18P	
12P	6–7E	
13P	6–23P	
14P	6–24P	
15P	6–25P	
16P	6–26P	
17P	6–29P	
18P	6–30P	
19P	6–31P	
20P	6–73	
21P	6–32P	
22P	6–33P	
23P	6–35P	
24P	6–75	
25P	6–38P	
26P	6–36P	
27P	6–39P	
28P	6–41P	
29P	6-40P	
30P	6–42P	
31P	6–43P	
32E	new	
33E	6–44E	
34P	6–45P	
35P	6–46P	
36E	6–50E	SI
37E	6–48E	
38E	6–91	
39E	6–49E	
40P	6–74	
41P	6–57P	
42P	6–61P	
43P	6–62P	
44P	6–69P	
45P	6–66P	
46P	6–92	
47P	6–70P	
48	new	

Chapter 7

1E	7–2E
2E	7–4E
3E	7–3E
4P	7–7P
5P	7–6P
6E	7–11E
7E	7–10E
8E	7–60
9E	7–9E
10P	7–17P
11P	7–16P
12P	7–58
13P	7–18P
14E	7–20E
15E	7–23E
16E	6–22E
17P	7–24P
18P	6–26P
19P	7–25P
20E	7–36E
21E	7–35E
22P	7–40P
23P	7–37P
24E	7–27E
25E	7–28E
26P	7–33P
27P	7–30P
28P	7–62
29P	7–31P
30E	7–41E
31E	7–45P
32E	7–44E
33P	7–48P
34P	7–65
35P	7–49P
36P	7–51P
37P	7–50P
38P	7–64
39P	7–46P
40P	7–47P

Chapter 8

1E	8–2E
2E	8–3E
3E	8–4E
4E	8–5E
5E	8–6E
6P	8–10P
7P	8–8P
8P	8–9P
9E	8–13E
10E	8–12E
11E	8–15E
12E	8–14E
13E	8–17E

14P	8–20P
15P	8–21P
16P	8–22P
17P	8–23P
18P	8–35P
19P	8–25P
20P	8–39P
21P	8–26P
22P	8–150
23P	8–33P
24P	8–110
25P	8–34P
26P	8–40P
27P	8–36P
28P	8–99
29P	8–41P
30P	8–100
31P	8–42P
32P	8–112
33P	8–43P
34P	8–113
35P	8–45P
36E	8–48E
37P	8–47P
38P	8–115
39E	8–50E
40E	8–51E
41P	8–54P
42P	8–52P
43E	8–66E
44E	8–68E
45E	8–102
46E	8–57E
47E	8–58E
48E	8–62E
49E	8–67E
50P	8–72P
51P	8–69P
52P	8–70P
53P	8–79P
54P	8–77P
55P	8–74P
56P	8–80P
57P	8–73P
58P	8–75P
59P	8–81P
60P	8–78P
61P	8–83P
62P	8–84P
63P	8–85P
64P	8–87P
65P	8–89P
66	8–68
67	8–136

Chapter 9

1E	9–1E
2E	9–2E
3E	9–3E
4E	9–4E
5E	9–5E
6P	9–7P
7P	9–8P
8P	9–9P
9P	9–11P
10E	9–12E
11E	9–13E
12E	9–14E
13P	9–18P
14P	9–19P
15P	9–21P
16P	9–82
17P	9–22P
18P	9–20P
19P	9–23P
20E	9–25E
21E	9–24E
22E	9–28E
23P	9–30P
24P	9–29P
25P	9–31P
26P	9–84
27E	9–34E
28E	9–35E
29E	9–37E
30P	9–78
31E	9–36E
32E	9–38E
33P	9–39P
34P	9–42P
35P	9–40P
36P	10–49P
37P	10–66P
38P	10–51P
39P	9–43P
40P	9–46P
41E	9–51E
42E	9–50E
43E	9–49E
44E	9–52E
45E	9–53E
46E	9–54E
47P	9–57P
48P	9–56P
49E	new
50E	9–59E
51E	9–62E
52E	9–63E
53E	9–65E
54E	9–67E
55E	9–69E
56P	9–71P
57P	9–73P
58P	9–74P
59	new

Chapter 10

1E	10–2E
2E	10–3E
3E	10–5E
4E	10–6E
5E	10–8E
6E	10–7E
7E	10–9E
8P	10–15P
9P	10–11P
10P	10–82
11P	10–12P
12P	10–16P
13P	10–13P
14P	10–19P
15P	10–24P
16P	10–20P
17P	10–21P
18P	10–22P
19P	10–26P
20E	10–42E
21E	10–41E
22E	10–44E
23E	10–40E
24E	10–45E
25P	10–47P
26P	10–50P
27P	10–52P
28P	10–46E

29P	10–53P
30P	10–89
31P	10–54P
32P	10–88
33P	10–55P
34P	10–56P
35E	10–28E
36E	10–30E
37E	10–31E
38E	10–27P
39E	10–32E
40P	10–34P
41P	10–33E
42P	new
43P	10–36P
44P	10–38P
45P	10–37P
46E	10–43E
47E	10–60P
48E	10–61E
49E	10–63E
50P	10–67P
51P	10–69P
52P	10–71P
53P	10–72P
54P	10–92
55P	10–68P
56	new
57	new

Chapter 11

1E	11–2E
2E	11–3E
3E	11–4E
4E	11–5E
5E	11–6E
6P	11–10P
7P	11–9P
8E	11–11E
9E	11–12E
10E	11–13E
11E	11–15E
12E	11–17E
13P	11–20P
14P	11–19P
15P	11–22P
16P	11–23P
17P	11–21P
18P	11–99
19E	11–25E
20E	11–26E
21E	11–27E
22E	11–28E
23E	11–32E
24E	11–33E
25P	11–38P
26P	11–35P
27P	11–40P
28P	11–36P
29P	11–41P
30P	11–96
31P	11–44P
32P	11–43P
33E	11–45E
34P	11–46P
35E	11–47E
36E	11–49E
37E	11–50E
38E	11–51E
39E	11–53E
40P	11–95
41P	11–56P
42P	11–57P
43P	11–58P
44P	11–59P
45E	11–61E
46E	11–60E
47P	11–63P
48P	11–64P
49E	11–66E
50E	11–65E
51E	11–69E
52E	11–67E
53P	11–70P
54P	11–105
55P	11–74P
56P	11–71P
57P	11–75P
58E	11–77E
59E	11–78E
60E	11–79E
61E	11–80E
62P	11–81P
63P	11–82P
64P	11–84P
65P	11–83P
66P	11–86P
67P	11–87P
68	11–31E
69	new
70	new

Chapter 12

1E	12–3E
2P	12–10P
3E	12–2E
4E	12–1E
5E	12–4E
6E	12–11E
7E	12–7E
8P	12–9P
9P	12–12P
10P	12–13P
11P	12–14P
12P	12–15P
13P	12–80
14P	12–16P
15E	12–17E
16P	12–18P
17E	12–20E
18E	12–21E
19E	12–22E
20P	12–23P
21P	12–26P
22P	12–24P
23E	12–27E
24E	12–29E
25E	12–31E
26P	12–34P
27P	12–33P
28P	12–83
29E	12–36E
30E	12–37E
31E	12–38E
32P	12–40P
33E	12–43E
34E	12–42E
35E	12–44E
36P	12–48P
37P	12–49P
38P	12–46P
39E	12–51E
40E	12–50E
41E	12–53E
42E	12–52E

43E	12–56E
44E	12–54E
45P	12–61P
46P	12–59P
47P	12–62P
48P	12–63P
49P	12–64P
50P	12–65P
51P	12–70P
52P	12–66P
53P	12–68P
54P	12–57E
55P	12–67P
56P	12–86
57P	12–87
58P	12–88
59P	12–69P

SECTION EIGHT
COMPUTER PROJECTS

Computational physics has become a part of many introductory courses. Its inclusion serves two purposes: students use their programming skills and they learn to apply the fundamental principles they are studying to situations that are more complicated than those presented in their physics text. The act of writing and debugging a computer program often helps students understand more precisely the equations they are working with. Perhaps more importantly from a pedagogical viewpoint, a computer can be programmed to solve the same problem over and over with different input data. With proper instruction, a student can use this technique to develop insight into the physics being studied.

To help in these endeavors some computer projects are suggested on the following pages. In addition, outlines of programs are supplied to help with the work on the projects. The projects and programs are keyed to the chapter of *Fundamentals of Physics* in which the physical principles used are discussed. You or the students must flesh out the programs to make them compatible with the programming language or spreadsheet that is used and to make the input and output routines work on the specific computer that is used. You may want to revise some projects to bring them more in line with your goals for the course.

Many of the projects require graphs to be drawn. Beginners may want to use a spreadsheet with graphing capability or else a simple BASIC or Pascal program to generate the required data, then draw the graphs manually. More advanced programmers will want to include programming steps to draw the plots on a monitor screen. Either technique is satisfactory.

Projects may be assigned as homework or carried out in groups during the laboratory portion of the course. The latter is especially beneficial since the projects can then be used to generate class discussions of the physics.

Chapter 2

A computer can be used to calculate the position and velocity of an object when its acceleration is a known function of time. Divide the time axis into a large number of small intervals, each of duration Δt. If Δt is sufficiently small the acceleration can be approximated by a constant in each interval, perhaps with a different value in different intervals. If v_b is the velocity at the beginning of an interval then $v_e = v_b + a\Delta t$ can be used to calculate the velocity at the end of the interval. If a is the average acceleration for the interval then this expression is exact. In most cases, however, the average acceleration is not known and so must be approximated. For the projects in this section it is approximated by $[a(t_b) + a(t_e)]/2$, where $a(t_b)$ is the acceleration at the beginning of the interval and $a(t_e)$ is the acceleration at the end. For each of the projects below the function $a(t)$ must be supplied.

If x_b is the coordinate of the object at the beginning of the interval, then the coordinate at the end of the interval is given by $x_e = x_b + v\Delta t$, where v is the average velocity in the interval. Approximate v by $(v_b + v_e)/2$. Smaller values for Δt make the approximations for v_e and x_e better. Δt, however, cannot be taken to be so small that significance is lost when the computer sums the terms in these equations. Note that the coordinate and velocity at the end of any interval are the coordinate and velocity at the beginning of the next interval. A skeleton program might look like this:

```
input initial values: t_0, x_0, v_0
input final time and interval width: t_f, Δt
set t_b = t_0, x_b = x_0, and v_b = v_0
calculate acceleration at beginning of first interval: a_b = a(t_b)
begin loop over intervals
    calculate time at end of interval: t_e = t_b + Δt
    calculate acceleration at end of interval: a_e = a(t_e)
    calculate "average" acceleration in interval: a = (a_b + a_e)/2
    calculate velocity at end of interval: v_e = v_b + aΔt
    calculate "average" velocity in interval: v = (v_b + v_e)/2
    calculate coordinate at end of interval: x_e = x_b + vΔt
  * if t_e ≥ t_f then
        print or display t_b, x_b, v_b
        print or display t_e, x_e, v_e
        exit loop
    end of if statement
    set t_b = t_e, x_b = x_e, v_b = v_e, a_b = a_e in preparation for next interval
end loop over intervals
stop
```

The line marked with an asterisk will be different for different applications. You will change this line, for example, if you want to display results and exit the loop when the velocity or coordinate have certain values.

Because Δt is arbitrary, the end of the last interval may not correspond exactly to t_f. The first line of output corresponds to a time just before t_f and the second line corresponds to a time just after. You can force $t = t_f$ at the end of the last interval by asking the computer to recalculate Δt near the beginning of the program. First use $(t_f - t_0)/\Delta t$ to estimate the number of intervals, then round the result to the nearest integer N. Finally take $\Delta t = (t_f - t_0)/N$.

Most programming languages allow you to write a separate section of the program to define the function $a(t)$. Then, the lines implementing $a_b = a(t_b)$ and $a_e = a(t_e)$ simply refer to the function definition. This is more efficient than writing the instructions for $a(t)$ twice, once for $t = t_b$ and once for $t = t_e$.

PROJECT 1. Test the program on a problem you can analyze analytically. Suppose the acceleration in m/s^2 is given by $12t$, for t in seconds. Take the initial coordinate (at $t = 0$) to be 5.0 m and the initial velocity to be -120 m/s. Find the coordinate and velocity at $t = 4.5$ s.

Start with $\Delta t = 0.5$ s and carry out the computation. Repeat several times, each time halving Δt, until you get the same answers to three significant figures on successive trials. Work the problem analytically to obtain an exact solution and carefully compare the answers with those obtained by the program. Check your program code if they differ.

PROJECT 2. Now use the program to determine when and where the particle is instantaneously stopped. You want the program to stop when the velocity changes sign in some interval. Use $v_b v_e \le 0$ as the condition for displaying results and exiting the loop. The correct answer is between the two results displayed. If they are the same to within the number of significant figures you desire you are finished. If they are not, run the program again with a smaller value of Δt. Try to obtain three significant figure accuracy.

Now find when the particle is at $x = -100$ m and its velocity when it is there. The condition for displaying results and exiting the loop is now $(x_b + 100)(x_e + 100) \le 0$. Obtain three significant figure accuracy.

PROJECT 3. Try a more complicated example. Suppose the particle starts at $x = 0$ with a velocity of $-120\,\mathrm{m/s}$ and has an acceleration that is given in $\mathrm{m/s^2}$ by $a(t) = 30e^{-t/8}$, where t is the time in seconds. Find the time at which it is instantaneously at rest and its position at that time. [ans: 5.55 s; −295 m]

You can use the program to generate a list of values ready to be plotted by hand or to plot values on the monitor screen. Suppose the time interval desired between displayed points is Δt_d. Input the value of Δt_d just after the values for t_f and Δt are read, then set $t_d = t_0 + \Delta t_d$. Replace the last statements of the program, from the if statement, with

if $t_e \geq t_d$ then
  print, display, or plot x_e or v_e
  increment t_d by Δt_d
end of if statement
set $t_b = t_e$, $x_b = x_e$, $v_b = v_e$, $a_b = a_e$ in preparation for next interval
if $t_e \geq t_f$ then exit loop
end loop over intervals
stop

Because the computer may produce values of t_e that are in error in the last place carried, the set of points you get may be somewhat different from what you anticipate. If this is intolerable, round the values of t_e so the position of the last non-zero digit is a few less than the number of figures carried by the machine.

PROJECT 4. A particle starts from rest at the origin and has an acceleration that is given in $\mathrm{m/s^2}$ by $a(t) = 30te^{-t}$, where t is the time in seconds. Draw graphs of its coordinate and velocity as functions of time from $t = 0$ to $t = 5\,\mathrm{s}$. Plot points every 0.5 s. You should obtain two significant figure accuracy with $N = 100$.

The acceleration is quite small near $t = 0$ and increases as t increases. This is the effect of the factor t. It reaches a maximum at $t = 1\,\mathrm{s}$, then decreases. This is the effect of the exponential factor. Your graph of $v(t)$ should have its greatest slope in the vicinity of $t = 1\,\mathrm{s}$, then approach a line with a slope of zero. The velocity has become nearly constant. What do you think the limiting value of the velocity is?

The graph of $x(t)$ should have a slope of zero at $t = 0$ (the initial velocity is zero) but it soon curves upward and eventually approaches a straight line, the slope of which is the limiting value of the velocity.

Chapter 4

The program outlined for Chapter 2 projects can be revised to investigate two-dimensional motion. You must now supply two components of the acceleration as functions of t and have the computer carry out calculations for each of the two components of the velocity and position vectors. The outline of a sample program might be:

input initial values: t_0, x_0, y_0, v_{0x}, v_{0y}
input final time and interval width: t_f, Δt
set $t_b = t_0$, $x_b = x_0$, $y_b = y_0$, $v_{xb} = v_{0x}$, $v_{yb} = v_{0y}$
calculate acceleration at beginning of first interval: $a_{xb} = a_x(t_b)$, $a_{yb} = a_y(t_b)$
begin loop over intervals
  calculate time at end of interval: $t_e = t_b + \Delta t$
  calculate acceleration at end of interval: $a_{xe} = a_x(t_e)$, $a_{ye} = a_y(t_e)$

calculate "average" acceleration: $a_x = (a_{xb} + a_{xe})/2$, $a_y = (a_{yb} + a_{ye})/2$
calculate velocity at end of interval: $v_{xe} = v_{xb} + a_x\Delta t$, $v_{ye} = v_{yb} + a_y\Delta t$
calculate "average" velocity : $v_x = (v_{xb} + v_{xe})/2$, $v_y = (v_{yb} + v_{ye})/2$
calculate coordinates at end of interval: $x_e = x_b + v_x\Delta t$, $y_e = y_b + v_y\Delta t$
* if $t_e \geq t_f$ then
    print or display t_b, x_b, y_b, v_{xb}, v_{yb}
    print or display t_e, x_e, y_e, v_{xe}, v_{ye}
    exit loop
end of if statement
set $t_b = t_e$, $x_b = x_e$, $y_b = y_e$, $v_{xb} = v_{xe}$, $v_{yb} = v_{ye}$, $a_{xb} = a_{xe}$, $a_{yb} = a_{ye}$
end loop over intervals
stop

As before, the line with the asterisk may be changed for other applications.

Instead of v_{0x} and v_{0y} you may wish to input the initial speed v_0 and angle ϕ_0 between the velocity and the x axis, then have the computer calculate v_{0x} and v_{0y} using $v_{0x} = v_0 \cos\phi_0$ and $v_{0y} = v_0 \sin\phi_0$. You may also wish to define $a_x(t)$ and $a_y(t)$ in a separate section of the program.

PROJECT 1. Start with a projectile motion problem that can be solved analytically. Suppose a projectile is fired over level ground at 350 m/s, at an angle of 25° above the horizontal. When does it reach the highest point? How high and how far down range is the highest point? When does it hit the ground and what is the range? Compare your answers with the analytic solutions.

If the x axis is horizontal and the y axis is positive in the upward direction then the components of the acceleration are given by $a_x = 0$ and $a_y = -9.8\,\mathrm{m/s^2}$. To find the highest point exit the loop when $v_{ye} \leq 0$. To find the range, exit the loop when $y_e \leq 0$. You must experiment a little to find an appropriate value for Δt. Start with 0.1 s and reduce it by a factor of 5 in successive calculations. Stop when you get the same results to three significant figures.

PROJECT 2. Now suppose the acceleration of the projectile has a horizontal component that varies with time, as a rocket might have. Take $a_x = 3.0t\,\mathrm{m/s^2}$ and $a_y = -9.8\,\mathrm{m/s^2}$, where t is in seconds. Use the same initial conditions (an initial velocity of 350 m/s, 25° above the horizontal) and find the time the projectile reaches its highest point and the coordinates of the highest point, assuming it is fired from the origin. Find the range over level ground and the velocity of the object just before it lands. [ans: highest point: $t = 15.1\,\mathrm{s}$, $x = 6.51 \times 10^3\,\mathrm{m}$, $y = 1.12 \times 10^3\,\mathrm{m}$; range: $t = 30.2\,\mathrm{s}$, $x = 2.33 \times 10^4\,\mathrm{m}$, $v_x = 1.68 \times 10^3\,\mathrm{m/s}$, $v_y = -148\,\mathrm{m/s}$]

PROJECT 3. Now suppose the horizontal component of the acceleration is given in $\mathrm{m/s^2}$ by $a_x(t) = 30te^{-t}$ while the y component is still $a_y = -9.8\,\mathrm{m/s^2}$. The x axis is horizontal and the y axis is positive in the upward direction. The initial velocity is still 350 m/s, 25° above the horizontal. At what time does the object reach the highest point on its trajectory and what are the coordinates of that point? At what time does it return to the level of the firing point and what is the range? What are the components of its velocity just before landing? [ans: highest point: $t = 15.1\,\mathrm{s}$, $x = 5.18 \times 10^3\,\mathrm{m}$, $y = 1.12 \times 10^3\,\mathrm{m}$; range: $t = 30.2\,\mathrm{s}$, $x = 1.04 \times 10^4\,\mathrm{m}$; velocity: $v_x = 347\,\mathrm{m/s}$, $v_y = -148\,\mathrm{m/s}$]

Chapter 6

The computer program outlined for the projects of Chapter 2 must be revised if the acceleration is a function of position or velocity. The average acceleration in an interval cannot be approximated by $a = (a_b + a_e)/2$ because a_e cannot be computed until v_e or x_e are known. You can, however, use the acceleration at the beginning of the interval and write $v_e = v_b + a_b\Delta t$. This is usually a poor approximation to the average acceleration, so compared to the program of Chapter 2, much smaller intervals must be used to obtain the same accuracy.

Errors arise when a large number of intervals are used because the computer normally carries only a small number of significant figures (eight or ten) and the last is often in error. These so-called truncation errors accumulate. You can decrease the effect significantly by carrying out the calculation in double precision.

The outline of a possible program for one-dimensional motion is:

input initial values: t_0, x_0, v_0
input final time and interval width: t_f, Δt
set $t_b = t_0$, $x_b = x_0$, $v_b = v_0$
begin loop over intervals
  calculate acceleration at beginning of interval: $a_b = a(t_b)$
  calculate velocity at end of interval: $v_e = v_b + a_b\Delta t$
  calculate "average" velocity: $v = (v_b + v_e)/2$
  calculate coordinate at end of interval: $x_e = x_b + v\Delta t$
  calculate time at end of interval: $t_e = t_b + \Delta t$
 * if $t_e \geq t_f$ then
   print or display t_b, x_b, v_b
   print or display t_e, x_e, v_e
   exit loop
  end of if statement
  set $t_b = t_e$, $x_b = x_e$, $v_b = v_e$
end loop over intervals
stop

You may want to revise the program so it displays values for a sequence of times. See the Computer Projects section for Chapter 2.

Sometimes air resistance must be taken into account when an object moves in the air. Consider an object that is moving vertically and take its acceleration to be given by $a = -g - (b/m)v$, where the positive direction is upward, m is the mass of the object, and b is a constant that depends on the interaction of the object with the air.

PROJECT 1. Suppose an object with $(b/m) = 0.100\,\text{s}^{-1}$ is fired upward from ground level with an initial speed of 100 m/s. On separate graphs plot the coordinate and velocity every 0.5 s from $t = 0$ to $t = 17$ s. First test the program to see what interval width it needs to obtain two significant figure accuracy over that range of time.

How much time does the projectile take to get to the highest point on its trajectory and how high is that point? How long does it take to get back to the ground and what is its velocity just before it reaches the ground? You might use your graph to obtain approximate answers, then use the program to refine them. [ans: highest point: 7.03 s, 311 m; ground: 16.2 s, 58.9 m/s]

Notice that the projectile takes longer to fall from the highest point than it does to reach that point and that it returns to ground level with a speed that is less than the firing speed. Compare your answers with those you would obtain if the object were in free fall. The time to reach the highest point is ______ and the highest point is ______ with air resistance than without. The total flight time is ______ and the speed on impact is ______ with air resistance than without.

PROJECT 2. Here's a problem for which the acceleration depends on position. Starting from rest at $x = 0$, a 3.5 kg box is dragged along the ground in the positive x direction by a constant force of 12 N acting horizontally. The ground is rougher toward larger x and the coefficient of kinetic friction increases according to $\mu_k = 0.070\sqrt{x}$. Use the program to find its position and velocity every second from $t = 0$ to the time it comes to rest again. At what time does the box come to rest? How far has it been dragged? [ans: 11.4 s; 56.2 m]

What is the maximum speed of the box? When does it have this speed? Where is it when it has this speed? [ans: 7.56 m/s; 4.84 s; 25.0 m from the starting point]

For the special case of an acceleration that is linear in the velocity, $a = (a_b + a_e)/2$ *can* be used to calculate the velocity at the end of the interval. Here's how. The acceleration at the beginning of the interval is given by $a_b = -g - (b/m)v_b$ and the acceleration at the end of the interval is given by $a_e = -g - (b/m)v_e$, so $a = (a_b + a_e)/2 = -g - (b/2m)v_b - (b/2m)v_e$. The velocity at the end of the interval is given by $v_e = v_b + a\Delta t = v_b - g\Delta t - (b\Delta t/2m)v_b - (b\Delta t/2m)v_e$. Solve this expression for v_e. The result is $v_e = [v_b(1 - b\Delta t/2m) - g\Delta t]/(1 + b\Delta t/2m)$. If $h = b\Delta t/2m$ this becomes $v_e = [v_b(1 - h) - g\Delta t]/(1 + h)$. An outline of a program is:

input initial values: t_0, x_0, v_0
input final time and interval width: t_f, Δt
calculate parameter: $h = b\Delta t/2m$
set $t_b = t_0$, $x_b = x_0$, $v_b = v_0$
begin loop over intervals
  calculate velocity at end of interval: $v_e = [v_b(1 - h) - g\Delta t]/(1 + h)$
  calculate "average" velocity: $v = (v_b + v_e)/2$
  calculate coordinate at end of interval: $x_e = x_b + v\Delta t$
  calculate time at end of interval: $t_e = t_b + \Delta t$
  * if $t_e \geq t_f$ then
    print or display t_b, x_b, v_b
    print or display t_e, x_e, v_e
    exit loop
  end of if statement
  set $t_b = t_e$, $x_b = x_e$, $v_b = v_e$
end loop over intervals
stop

PROJECT 3. Use this program to find the position and velocity of the projectile of the first project at the end of 17 s. Notice that far fewer intervals are required to obtain two significant figure accuracy.

Now use the program to investigate the influence of the air on a projectile fired straight upward. Make a table of the time to reach the highest point, the coordinate of the highest point, the time to reach the ground, and the velocity just before reaching ground, all as functions of the resistance parameter b/m. Try $b/m = 1.00\,\text{s}^{-1}$, $0.100\,\text{s}^{-1}$, and $0.0100\,\text{s}^{-1}$.

This exercise is designed to give you some idea of the effect of air resistance. A javelin has a small value of b/m, a basketball has a larger value, and a ping-pong ball has a still larger value. The larger b/m the greater the effects.

You might try to find a value for the resistance coefficient b/m of a ping-pong ball or other object. Shoot it into the air with a spring gun, measure the initial speed with a photogate timer, and time its return to the firing level. Then try various values of b in your computer program until you find the value that reproduces the experimentally determined time of flight.

PROJECT 4. You can use the program to investigate the approach to terminal velocity. Suppose an object with $b/m = 0.100\,\mathrm{s}^{-1}$ is dropped from a high cliff. Use the program to plot its velocity every 2 s over the first minute of its fall. The speed should approach the value $mg/b = 98\,\mathrm{m/s}$. Repeat for objects with $b/m = 0.0500\,\mathrm{s}^{-1}$ and $0.500\,\mathrm{s}^{-1}$.

The program can be modified to deal with a projectile moving in two dimensions, and subjected to a drag force that is proportional to its velocity. The acceleration components are now given by $a_x = -(b/m)v_x$ and $a_y = -g - (b/m)v_y$. The outline of a program is:

input initial values: t_0, x_0, y_0, v_{0x}, v_{0y}
input final time and interval width: t_f, Δt
calculate parameter: $h = b\Delta t/2m$
set $t_b = t_0$, $x_b = x_0$, $y_b = y_0$, $v_{xb} = v_{0x}$, $v_{yb} = v_{0y}$
begin loop over intervals
    calculate velocity at end of interval:
        $v_{xe} = v_{xb}(1 - h)/(1 + h)$
        $v_{ye} = [v_{yb}(1 - h) - g\Delta t]/(1 + h)$
    calculate "average" velocity: $v_x = (v_{xb} + v_{xe})/2$, $v_y = (v_{yb} + v_{ye})/2$
    calculate coordinates at end of interval: $x_e = x_b + v_x\Delta t$, $y_e = y_b + v_y\Delta t$
    calculate time at end of interval: $t_e = t_b + \Delta t$
  * if $t_e \geq t_f$ then
        print or display t_b, x_b, y_b, v_{xb}, v_{yb}
        print or display t_e, x_e, y_e, v_{xe}, v_{ye}
        exit loop
    end of if statement
    set $t_b = t_e$, $x_b = x_e$, $y_b = y_e$, $v_{xb} = v_{xe}$, $v_{yb} = v_{ye}$
end loop over intervals
stop

PROJECT 5. A projectile with $b/m = 0.100\,\mathrm{s}^{-1}$ is fired over level ground with an initial velocity of 100 m/s, 45° above the horizontal. Find its coordinates every 0.5 s from $t = 0$ (the time of firing) to $t = 12.5\,\mathrm{s}$. Plot its trajectory. First find the value of Δt required to obtain three significant figure accuracy over the entire time interval.

Find the time the projectile reaches the highest point and the coordinates of the highest point. Take the x axis to be horizontal and the y axis to be positive in the upward direction. [ans: $t = 5.43\,\mathrm{s}$, $x = 296\,\mathrm{m}$, $y = 175\,\mathrm{m}$]

Find the time of flight and range. What are its velocity components just before it lands? [ans: $t = 12.1\,\mathrm{s}$, $x = 495\,\mathrm{m}$, $v_x = 21.2\,\mathrm{m/s}$, $v_y = -47.5\,\mathrm{m/s}$]

PROJECT 6. If air resistance is significant, maximum range is obtained for a firing angle that is different from 45°. Consider a projectile with $b/m = 0.100\,\mathrm{s}^{-1}$, fired with an initial speed of 100 m/s over level ground. Find the range to three significant figures for firing angles of 30°, 35°, and 40°. You already know the range for 45°. The firing angle for maximum range is between ______ and ______. You may wish to refine the interval to obtain the firing angle to two significant figures. [ans: 510 m; 519 m; 514 m]

PROJECT 7. Suppose the projectile of the last project ($b/m = 0.100$) is fired from the edge of a high cliff instead of over level ground. Assume the same initial conditions (fired from the origin with a speed of 100 m/s, 45° above the horizontal) and plot its trajectory for the first minute of its flight.

Notice that the trajectory is not symmetric about the highest point but is blunted in the forward direction. Near the end of the time interval the projectile is falling nearly straight down. Both the horizontal and vertical components of the velocity approach limiting values. The limiting value of the horizontal component is ______ and the limiting value of the vertical component is ______.

Chapter 7

A force with x component $F(x)$ does work given by $W = \int_{x_i}^{x_f} F(x)\,dx$ as the object on which it acts moves from x_i to x_f along the x axis. A computer can be used to evaluate integrals of this form. One of the simplest techniques to use is Simpson's rule. An interval of the x axis, from x_b to x_e, is divided into two equal parts of width Δx. Let F_b be the value of the force at the beginning of the interval ($x = x_b$), F_m be the value at the middle ($x = x_b + \Delta x$), and F_e be the value of the end ($x = x_e$). Then according to Simpson's rule the integral over the interval can be approximated by

$$\int_{x_b}^{x_e} F(x)\,dx = \frac{\Delta x}{3}(F_b + 4F_m + F_e)\,.$$

This expression can be derived easily by fitting the function $F(x)$ to a quadratic of the form $F(x) = A_0 + A_1(x - x_b) + A_2(x - x_b)^2$. The coefficients are chosen so the quadratic yields the correct values of the function at $x = x_b$, $x = x_m$, and $x = x_e$. They are $A_0 = F_b$, $A_1 = -(F_e - 4F_m + 3F_b)/2\Delta x$, and $A_2 = (F_e - 2F_m + F_b)/2(\Delta x)^2$. The quadratic can be integrated easily. The approximation becomes better as the interval becomes narrower.

To evaluate an integral for an extended portion of the x axis divide the region from x_i to x_f into N intervals, where N is an even number. The interval width is given by $\Delta x = (x_f - x_i)/N$. Label the points x_0 ($= x_i$), x_1, x_2, ..., x_N ($= x_f$) and the corresponding values of the force F_0, F_1, F_2, ..., F_N. Apply the formula given above to each pair of intervals; for example, x_0 is the first point, x_1 is the second point, and x_2 is the third point in the first application; x_2 is the first point, x_3 is the second point, and x_4 is the third point in the next application. Except for x_0 and x_N, each point with an even label enters the final formula twice: once as the first point in an integration and once as a third point. Each point with an odd label enters as a midpoint of an integration. Thus

$$\int_{x_i}^{x_f} F(x)\,dx = \frac{\Delta x}{3}\Big[(F_N - F_0) + 2(F_0 + F_2 + F_4 + \ldots + F_{N-2}) + 4(F_1 + F_3 + \ldots + F_{N-1})\Big]$$

F_0 is subtracted at the beginning of the equation because it is also included in the sum over values with even labels. There it is multiplied by 2 but it should be included only once. F_N is not included in any of the sums but it must be included once in the final equation, so it is added at the beginning.

Write a computer program to evaluate a work integral for one-dimensional motion. Input the lower and upper limits of the integral and the number of intervals. You can force N to be an even integer by dividing the value read by 2, rounding the result to the nearest integer, then multiplying by 2. Calculate Δx. Now write a loop to sum all the force values with even labels and sum all the force values with odd labels. Instructions in the loop will be executed $N/2$ times. An outline might be:

input limits of integral: x_i, x_f
input number of intervals: N
replace N with nearest even integer
calculate interval width: $\Delta x = (x_f - x_i)/N$
initialize quantity to hold sum of values with even labels: $S_e = 0$
initialize quantity to hold sum of values with odd labels: $S_o = 0$
set $x = x_i$
begin loop over intervals: counter runs from 1 to $N/2$
    calculate $F(x)$ and add it to the sum of values with even labels:
      replace S_e with $S_e + F(x)$
    increment x by Δx
    calculate $F(x)$ and add it to the sum of values with odd labels:
      replace S_o with $S_o + F(x)$
    increment x by Δx
end loop
calculate force for upper and lower limits: $F_0 = F(x_i)$, $F_N = F(x_f)$
evaluate integral: $W = (\Delta x/3)(F_N - F_0 + 2S_e + 4S_o)$
display result
stop

You may want to define the force function $F(x)$ in a separate section of the program.

PROJECT 1. This problem can be solved analytically. Use it to check your program. A 2.00 kg block moves along the x axis, subjected to the force $\mathbf{F} = (6 - 4x)\,\mathbf{i}$, where $\mathbf{F}$ is in newtons and x is in meters. What work is done by this force as the block moves from $x = 0$ to $x = 1.00$ m? from $x = 0$ to $x = 5.00$ m? Try $N = 2$ and double its value on successive runs until you get the same first three significant figures. Check the answers by working the problem analytically. [ans: 4.00 J; −20.0 J]

If $\mathbf{F}$ is the total force acting on the block, then the work that it does is equal to the change in the kinetic energy of the block. Suppose the block has a speed of 6.00 m/s when it is at $x = 0$. What is its speed when it is at $x = 1.00$ m? at 5.00 m? [ans: 6.32 m/s; 4.00 m/s]

PROJECT 2. A certain non-ideal spring exerts a force that is given by $F(x) = -250x - 125xe^{-x}$, where x is its extension (if positive) or its compression (if negative). A 2.3 kg block is attached to one end and the other end is held fixed. The block is pulled out so the spring is extended by 25 cm, then it is released from rest. How much work does the spring do on the block as the block moves from its initial position to the position for which the spring is neither extended or compressed? If the force of the spring is the only force acting on the block, what is its speed as it passes this point? [ans: 11.1 J; 3.11 m/s]

The block continues to move, compressing the spring. By how much is it compressed when the block comes to rest instantaneously? Since the block is at rest both initially and finally the total work done by the spring over the entire trip is 0. Use trial and error to find the final value of x. Alternatively you might imbed the program in a loop in which x_f is incremented each time around and search for the values of x_f that straddle $W = 0$. [ans: −0.237 m]

If the force is given as a function of time and you are asked for the work it does over a given time interval, you must change your strategy somewhat. If the force is the total force, you can use one of the programs of Chapters 2 or 4 to find the velocity. The total work done, of course, is just the change in kinetic energy. If the force is not the total force, you can use the power equation $dW/dt = Fv$ to write $W = \int Fv\,dt$. The program is the same as before except the integrand is now Fv. You must know the velocity as a function of time to use this method.

PROJECT 3. The position as a function of time for a 25-kg crate sliding across the floor (along the x axis) is given by $x(t) = 3te^{-0.60t}$, where x is in meters and t is in seconds. Numerically evaluate the integral for the work done by the total force during the first 2.0 s. You must find $v(t)$ and $F(t)$ by differentiating $x(t)$ and using $F = ma$. Select an integration interval for three significant figure accuracy. [ans: −112 J]

Use the expression you found for $v(t)$ to calculate the change in the kinetic energy during the first 2.0 s. Your result should agree with the result of the numerical integration.

PROJECT 4. A 2.0-kg block starts from rest at the origin and moves along the x axis. The total force acting on it is given by $F(t) = (8.0 + 12t^2)$, where F is in newtons and t is in seconds. Use analytical means to show that $v(t) = 4.0t + 2.0t^3$ and $x(t) = 2.0t^2 + 0.50t^4$.

One of the forces acting on the block is given by $F_1(t) = 3.0t^2$. Find the work done by F_1 during the first 3.0 s of the motion. Do this by analytical evaluation of $\int Fv\,dt$ and by numerical integration. [ans: 972 J]

PROJECT 5. The velocity of a body with mass m dropped from rest near the surface of the earth is given by $v(t) = (mg/b)(1 - e^{-bt/m})$, where b is the drag coefficient and down is taken to be the positive direction. Suppose an object with mass $m = 5.0$ kg and drag coefficient $b = 75$ kg/s is dropped from the edge of a high cliff. Use numerical integration to find the work done by gravity during the first 5.0 s. [ans: 5280 J]

The work done by gravity, of course, is given by mgs, where s is the distance fallen. The expression for $v(t)$ can be integrated with respect to time, with the result $s(t) = (mg/b)[t - (m/b) + (m/b)e^{-bt/m}]$. Use this expression to calculate the distance fallen in 5.0 s, then use $W = mgs$ to calculate the work done by gravity. You should obtain agreement with your numerical calculation.

If the object moves in two dimensions, the integral for the work becomes $W = \int \mathbf{F} \cdot d\mathbf{s} = \int (F_x\,dx + F_y\,dy)$, where $d\mathbf{s} = \mathbf{i}\,dx + \mathbf{j}\,dy$ is an infinitesimal segment of the path. It is a vector tangent to the path. Because the object is on a specified path in the xy plane, the infinitesimals dx and dy are related to each other. You must know the path to determine the relationship.

There are several ways of specifying a path. One way is to give the functional relationship between the coordinates of points on the path: $y = g(x)$, where $g(x)$ is a specified function. Then, $dy = (dg/dx)\,dx$ and the work integral becomes $W = \int [F_x + (dg/dx)F_y]\,dx$. For example, a straight line with slope S and intercept A is given by $y = A + Sx$ and the work is given by $W = \int (F_x + SF_y)\,dx$. The limits of integration are the x coordinates of the points at the beginning and end of the path.

You must use the functional relationship between x and y in another way. The components of the force will be given as functions of both x and y but there can be only a single variable of integration. You must substitute the function $g(x)$ wherever y occurs in the expressions for the force components. The following is an example.

PROJECT 6. A particle moves along a straight line from $x = 1.5$ m, $y = 2.0$ m to $x = 3.0$ m, $y = 5.0$ m. One of the forces acting on it is given by $\mathbf{F} = 3x^2y^2\,\mathbf{i} + 2x^3y\,\mathbf{j}$. What work does this force do?

The path is given by $y = -1.0 + 2.0x$, so $dy = 2.0\,dx$. The force components are given by $F_x = 3x^2y^2 = 3x^2(-1 + 2x)^2$ and $F_y = 2x^3y = 2x^3(-1 + 2x)$. The work integral is $\int_{1.5}^{3.0}[3x^2(-1+2x)^2+4x^3(-1+2x)]\,dx$. The integration program can now be used to calculate the work done. [ans: 662 J]

The method does not work if the path is parallel to the y axis and it requires a large number of intervals if the path is nearly parallel to the y axis. Then, you should use y as the variable of integration. The expression for work becomes $\int[(dx/dy)F_x+F_y]\,dy$.

A path can also be specified by giving both x and y as functions of some parameter. The parameter need not have any physical significance. For the straight line path of the previous project, you might write $x=1.5+1.5s$ and $y=2.0+3.0s$. When the parameter $s=0$, then $x=1.5$ and $y=2.0$. When $s=1$, then $x=3.0$ and $y=5.0$. Now, $dx=1.5\,ds$ and $dy=3.0\,ds$. The integral for the work becomes $W=\int_0^1[(dx/ds)F_x+(dy/ds)F_y]\,ds$. You must now substitute for both x and y in terms of s in the expressions for F_x and F_y.

If the path is an arc of a circle, you might use the angle between the x axis and the radial line to the point as a parameter. Thus $x=R\cos\theta$ and $y=R\sin\theta$, where R is radius of the path. The integral for the work becomes $W=\int[-F_x\sin\theta+F_y\cos\theta]R\,d\theta$. Notice that $-F_x\sin\theta+F_y\cos\theta$ is the component of the force along a line that is tangent to the path and $R\,d\theta$ is the length of a path segment that subtends the angle $d\theta$ at the center of the circle.

PROJECT 7. A particle travels counterclockwise around a circular path with a radius of 2.0 m, centered at the origin. What work is done by the force $\mathbf{F}=3x^2y^2\,\mathbf{i}+2x^3y\,\mathbf{j}$ as the particle goes from $x=2.0$ m, $y=0$ to $x=-\sqrt{2.0}$ m, $y=\sqrt{2.0}$ m? This is an arc of $3\pi/4$ radians or 135°. [ans: −5.66 J]

Chapter 8

The force $\mathbf{F}=3x^2y^2\,\mathbf{i}+2x^3y\,\mathbf{j}$ is a conservative force. You can see what this means by computing the work it does as the particle on which it acts goes via different paths from the same initial point to the same final point.

PROJECT 1. Use the program of the previous chapter to calculate the work done by the force $\mathbf{F}=3x^2y^2\,\mathbf{i}+2x^3y\,\mathbf{j}$ as the particle goes from $x=2.0$ m, $y=0$ to $x=4.0$ m, $y=2.0$ m along each of the following paths.

a. Along the x axis from $x=2.0$ m to $x=4.0$ m, then along the line $x=4.0$ m from $y=0$ to $y=2.0$ m. Carry out the integration in two parts, one for each segment.
b. Along the line $x=2.0$ m from $y=0$ to $y=2.0$ m, then along the line $y=2.0$ m from $x=2.0$ m to $x=4.0$ m.
c. Along the straight line that joins the to points. You might take $x=2.0+2.0s$ and $y=2.0s$, where s is a parameter that varies between 0 and 1.
d. Along the perimeter of a 2.0-m radius circle centered at $x=4.0$ m, $y=0$. For this path $x=4.0-2.0\cos\theta$ and $y=2.0\sin\theta$, where θ varies from 0 to $\pi/2$ radians (90°).

Notice that in every case the work done is 256 J.

PROJECT 2. Since the work done by the force of the previous project is independent of the path, a potential energy function is associated with it. The value of this function at any point is the negative of work done by the force as the particle moves from some reference point to the point in question. The table below is actually a grid of points in the xy plane. The x coordinate is given at the top and the y coordinate is given at the left side. Take the origin to be the reference point ($U=0$ there) and calculate the potential energy associated with each of the other points. You may want to automate the process by reading in the coordinates x_f and y_f of the point. Use the path given by $x = x_f s$ and $y = y_f s$, where s is a parameter than varies from 0 to 1. Since $dx/ds = x_f$ and $dy/ds = y_f$ the potential energy is given by $U(x_f, y_f) = -\int_0^1 (F_x x_f + F_y y_f)\,ds$. You must write the force components in terms of s. Fill in the table below with values of U.

$y(m)$ \ $x(m)$	0	1	2	3	4
4					
3					
2					
1					
0	0				

The analytic form for the potential energy function is $U(x,y) = -x^3y^2$. You can easily check this since the derivative of U with respect to x must give the negative of the x component of the force and the derivative of U with respect to y must give the negative of the y component. Use the exact analytic expression to check the results of your numerical calculations. You might construct a program consisting of two loops, one over values of the x coordinate and the other over values of the y coordinate, to compute and display values of U.

Use the table to compute the following quantities:

a. The change in the potential energy when the particle moves from $x = 2\,\text{m}$, $y = 1\,\text{m}$ to $x = 3\,\text{m}$, $y = 3\,\text{m}$. [ans: $-235\,\text{J}$]

b. The work done by the force when the particle moves from $x = 3\,\text{m}$, $y = 4\,\text{m}$ to $x = 1\,\text{m}$, $y = 1\,\text{m}$. [ans: $-431\,\text{J}$]

c. The change in the kinetic energy of the particle when it moves from $x = 3\,\text{m}$, $y = 1\,\text{m}$ to $x = 2\,\text{m}$, $y = 4\,\text{m}$. Assume only one force acts on it. [ans: $+101\,\text{J}$]

PROJECT 3. By way of contrast, the force $\mathbf{F} = (3x^2 - 6)y^2\,\mathbf{i} + 2x^3y\,\mathbf{j}$ is not conservative. Use the program to calculate the work done by the force as the particle goes from $x = 2.0\,\text{m}$, $y = 0$ to $x = 4.0\,\text{m}$, $y = 2.0\,\text{m}$ along each of the following paths.

a. Along the x axis from $x = 2.0\,\text{m}$ to $x = 4.0\,\text{m}$, then along the line $x = 4.0\,\text{m}$ from $y = 0$ to $y = 2.0\,\text{m}$.

b. Along the line $x = 2.0\,\text{m}$ from $y = 0$ to $y = 2.0\,\text{m}$, then along the line $y = 2.0\,\text{m}$ from $x = 2.0\,\text{m}$ to $x = 4.0\,\text{m}$.

c. Along the straight line that joins the to points.

d. Along the perimeter of a 2.0-m radius circle centered at $x = 4.0\,\text{m}$, $y = 0$.

Notice that the work done is different for different paths. You can easily see that it is impossible to assign a potential energy to each point in space so that the work done by the force equals the difference in the values assigned to the end points of the path. [ans: 256 J; 208 J; 240 J; 224 J]

Chapter 9

A computer can be used to follow individual objects in a system as they interact with each other and to follow the center of mass of the system. As an example, consider two carts connected by an ideal spring on a horizontal frictionless air track. If the spring has a natural length ℓ_0 (when it is neither compressed or extended), then when its length is ℓ it exerts a force of magnitude $k|\ell - \ell_0|$ on each of the carts. Here k is the force constant for the spring. If the spring is extended ($\ell > \ell_0$), then both forces are toward the center of the spring. If the spring is compressed ($\ell < \ell_0$), then both forces are away from the center of the spring.

Let x_1 be the coordinate of one cart and x_2 be the coordinate of the other and take $x_2 > x_1$. Then, $\ell = x_2 - x_1$. The force on cart 1 is given by $F_1 = k(x_2 - x_1 - \ell_0)$ and the force on cart 2 is given by $F_2 = -k(x_2 - x_1 - \ell_0)$.

A program that will calculate the coordinate and velocity of each cart as functions of time can be modeled after the first program of Chapter 6. In addition, the program calculates the coordinate and velocity of the center of mass. Here's an outline.

input initial values: t_0, x_{10}, v_{10}, x_{20}, v_{20}
input final time and interval width: t_f, Δt
input display interval: Δt_d
set $t_b = t_0$, $t_d = t_0 + \Delta t_d$, $x_{1b} = x_{10}$, $v_{1b} = v_{10}$, $x_{2b} = x_{20}$, $v_{2b} = v_{20}$
calculate coordinate and velocity of center of mass:
    $x_{cm} = (m_1 x_{1b} + m_2 x_{2b})/(m_1 + m_2)$
    $v_{cm} = (m_1 v_{1b} + m_2 v_{2b})/(m_1 + m_2)$
display x_{1b}, x_{2b}, x_{cm}
begin loop over intervals
    calculate force on 1 at beginning of interval: $F_1 = k(x_{2b} - x_{1b} - \ell_0)$
    calculate accelerations at beginning of interval: $a_1 = F_1/m_1$, $a_2 = -F_1/m_2$
    calculate velocities at end of interval: $v_{1e} = v_{1b} + a_1\Delta t$, $v_{2e} = v_{2b} + a_2\Delta t$
    calculate "average" velocities: $v_1 = (v_{1b} + v_{1e})/2$, $v_2 = (v_{2b} + v_{2e})/2$
    calculate coordinates at end of interval: $x_{1e} = x_{1b} + v_1\Delta t$, $x_{2e} = x_{2b} + v_2\Delta t$
    calculate time at end of interval: $t_e = t_b + \Delta t$
    if $t_e \geq t_d$ then
        calculate coordinate and velocity of center of mass:
            $x_{cm} = (m_1 x_{1e} + m_2 x_{2e})/(m_1 + m_2)$
            $v_{cm} = (m_1 v_{1e} + m_2 v_{2e})/(m_1 + m_2)$
      * display x_{1e}, x_{2e}, x_{cm}
        increment t_d by Δt_d
    end of if statement
    if $t_e \geq t_f$ then exit loop
    set $t_b = t_f$, $x_{1b} = x_{1e}$, $v_{1b} = v_{1e}$, $x_{2b} = x_{2e}$, $v_{2b} = v_{2e}$
end loop over intervals
stop

For some applications the line marked with an asterisk is replaced by: display v_{1e}, v_{2e}, v_{cm}. You will need to use a large number of extremely small intervals, so truncation errors may be significant. If possible, use double precision variables.

PROJECT 1. Take $m_1 = 250\,\text{g}$, $m_2 = 600\,\text{g}$, $\ell_0 = 0.50\,\text{m}$, and $k = 5.00\,\text{N/m}$. Initially ($t = 0$), cart 1 is at rest at the origin and cart 2 is at rest at $x_2 = 0.70\,\text{m}$. Plot the coordinates of the carts and the coordinate of the center of mass every $0.10\,\text{s}$ from $t = 0$ to $t = 2.0\,\text{s}$. You will need to take $\Delta t = 0.02\,\text{s}$ to obtain two significant figure accuracy.

Notice that the center of mass is initially at rest and remains at rest even though the carts oscillate back and forth. Now repeat the calculation with $v_{20} = 0.30\,\mathrm{m/s}$. The carts again oscillate but as they do the center of mass moves with constant velocity in the positive x direction. Check that the calculation gives the same result as $v_{\mathrm{cm}} = (m_1 v_{10} + m_2 v_{20})/(m_1 + m_2)$ for the velocity of the center of mass.

For both of the situations considered, the total momentum of the system, given by $P = (m_1 + m_2)v_{\mathrm{cm}}$, is conserved. Now, suppose an external force of 0.50 N is applied to the first cart in the positive x direction. The acceleration of that cart is computed using $a_1 = (F_1 + 0.50)/m_1$, where F_1 is still the force of the spring on cart 1. The acceleration of cart 2 is $a_2 = -F_1/m_2$, as before. Plot the coordinates of the carts and the coordinate of the center of mass every 0.10 s from $t = 0$ to $t = 2.0$ s. The curve representing the coordinate of the center of mass should be parabolic. In fact, $x_{\mathrm{cm}} = \frac{1}{2}a_{\mathrm{cm}}t^2$, where $a_{\mathrm{cm}} = F_e/(m_1 + m_2)$. Here F_e is the external force, 0.50 N. Plot v_{cm} as a function of time and verify that it is a straight line with the correct slope.

A computer program can also be used to investigate the motion of a rocket. If the rocket and fuel have mass $M(t)$ and the fuel is ejected with velocity $\mathbf{u}$, measured relative to the rocket, then the acceleration of the rocket is given by

$$M\frac{d\mathbf{v}}{dt} = \mathbf{F} + \frac{dM}{dt}\mathbf{u}\,,$$

where $\mathbf{F}$ is the external force acting on the rocket (the force of gravity, for example).

First, consider a rocket that is fired straight upward and assume the acceleration due to gravity is uniform. If the positive y axis is upward, then $\mathbf{u} = -u\,\mathbf{j}$, $\mathbf{v} = v\,\mathbf{j}$, and $\mathbf{F} = -Mg\,\mathbf{j}$. If the fuel is expended uniformly, you may write $M(t) = M_0 - Kt$, where K is a constant. The differential equation becomes

$$(M_0 - Kt)\frac{dv}{dt} = uK - (M_0 - Kt)g\,.$$

This expression gives the acceleration as a function of time and can be used in conjunction with the computer program of Chapter 2 to calculate the position and velocity of the rocket at any time until the rocket runs out of fuel.

The differential equation above can also be solved analytically to obtain

$$v(t) = v_0 - gt + u\ln\frac{M_0}{M_0 - Kt}\,,$$

where v_0 is the velocity at $t = 0$. You will use this expression in the next computer project to check your program.

PROJECT 2. Consider a rocket that carries 80% of its original mass as fuel and ejects it uniformly for 5.0 s, at which time the fuel is gone and the engine shuts off. The rocket starts from rest and the speed of the fuel relative to the rocket is $u = 5000\,\mathrm{m/s}$. Use the program of Chapter 2 to find the speed and altitude of the rocket at burnout. First, you need to find a value for K. At the end of 5.0 s, M is the mass of the rocket alone and this is $0.20M_0$. Thus, $M_0 - 5.0K = 0.20M_0$, so $K = 0.16M_0$. While fuel is being expended, the acceleration of the rocket is given by $a(t) = -g + 0.16u/(1 - 0.16t)$. Use this expression in the program. Check your answer by using the analytic expression above to calculate the speed. [ans: 8.00×10^3 m/s; 1.48×10^4 m]

The next project deals with a rocket that moves in two dimensions. If the acceleration due to gravity is uniform, then

$$\frac{d\mathbf{v}}{dt} = -\frac{K\mathbf{u}}{M_0 - Kt} - g\mathbf{j},$$

where the y axis is positive in the upward direction. Since $\mathbf{u}$ is always opposite to $\mathbf{v}$, we may write $\mathbf{u} = -u\mathbf{v}/v$, where the unit vector $\mathbf{v}/v$ is in the direction of $\mathbf{v}$. Thus,

$$\frac{d\mathbf{v}}{dt} = \left[\frac{Ku}{M_0 - Kt}\right]\frac{\mathbf{v}}{v} - g\mathbf{j}.$$

This equation holds while fuel is being ejected. After burnout, the rocket becomes an ordinary projectile and $d\mathbf{v}/dt = -g\mathbf{j}$.

Write a program to investigate the motion of a rocket fired from rest over level ground. Since each component of the acceleration depends on both components of the velocity, you cannot use $\mathbf{s} = (\mathbf{a}_b + \mathbf{a}_e)/2$ to approximate the average acceleration in an interval. Use instead the acceleration at the beginning of the interval. For the first interval the program will have trouble with the unit vector $\mathbf{v}/v$ since it involves division by 0. For that interval replace v_x/v by $\cos\phi_0$ and v_y/v by $\sin\phi_0$, where ϕ_0 is the firing angle. At the end of the interval, after the velocity components have been computed, calculate $v = \sqrt{v_{xe}^2 + v_{ye}^2}$ and the ratios v_{xe}/v and v_{ye}/v, in preparation for the next interval. This can be handled using the unit vector components α_x and α_y as you will see in the outline below.

The program should provide for the possibility of different interval widths before and after burnout. Before burnout the acceleration may be great and both the speed and direction of travel may change rapidly. You will want to use a small interval. After burnout you will want to use a larger interval to save time. Here's an outline of a program.

input burnout time: t_{bo}
input firing angle: ϕ_0
input interval widths: $(\Delta t)_b$ for $t < t_{bo}$, $(\Delta t)_a$ for $t > t_{bo}$
set interval width: $\Delta t = (\Delta t)_b$
calculate unit vector components: $\alpha_x = \cos\phi_0$, $\alpha_y = \sin\phi_0$
set initial values: $t_b = 0$, $x_b = 0$, $v_{xb} = 0$, $y_b = 0$, $v_{yb} = 0$
begin loop over intervals
    calculate x component of acceleration at beginning of interval:
        if $t_b < t_{bo}$ then $a_x = \alpha_x Ku/(M_0 - Kt_b)$
        if $t_b \geq t_{bo}$ then $a_x = 0$
    calculate y component of acceleration at beginning of interval:
        if $t_b < t_{bo}$ then $a_y = -g + \alpha_y Ku/(M_0 - Kt_b)$
        if $t_b \geq t_{bo}$ then $a_y = -9.8$
    calculate velocity at end of interval: $v_{xe} = v_{xb} + a_x\Delta t$, $v_{ye} = v_{yb} + a_y\Delta t$
    calculate "average" velocity: $v_x = (v_{xb} + v_{xe})/2$, $v_y = (v_{yb} + v_{ye})/2$
    calculate x coordinates at end of interval: $x_e = x_b + v_x\Delta t$, $y_e = y_b + v_y\Delta t$
    calculate time at end of interval: $t_e = t_b + \Delta t$
  * if $t_e \geq t_f$ then
            display or print t_b, x_b, y_b, v_{xb}, v_{yb}
            display or print t_e, x_e, y_e, v_{xe}, v_{ye}
            exit loop
    end of if statement
    set $t_b = t_e$, $x_b = x_e$, $v_{xb} = v_{xe}$, $y_b = y_e$, $v_{yb} = v_{ye}$
    if $t_b < t_{bo}$ set $\Delta t = (\Delta t)_b$, if $t_b \geq t_{bo}$ set $\Delta t = (\Delta t)_a$

calculate unit vector components:

$v = \sqrt{v_{xe}^2 + v_{ye}^2}$, $\alpha_x = v_{xe}/v$, $\alpha_y = v_{ye}/v$

end loop over intervals when $y_e < 0$

stop

The line marked with an asterisk will be different for different applications. To calculate the coordinates and velocity for the highest point on the trajectory, end the loop when $v_{ye} \le 0$. To calculate the range over level ground, end the loop when $y_e \le 0$.

PROJECT 3. A toy rocket has a mass (without fuel) of 250 g. It carries 10 g of fuel and is fired from rest at 45° above the horizontal. The fuel leaves the rocket at 4000 m/s and burns out after 0.200 s of flight. Use the program to find the coordinates and velocity components at burnout ($t = 0.20$ s). For two significant figure accuracy use $\Delta t = 0.001$ s before burnout and $\Delta t = 0.01$ s after burnout. [ans: $x = 12.0$ m, $y = 10.7$ m, $v_x = 121$ m/s, $v_y = 108$ m/s]

Use the program to find the time the rocket reaches the highest point on its trajectory. Find its coordinates and speed when it is there. [ans: $t = 11.2$ s, $x = 1.35 \times 10^3$ m, $y = 605$ m, $v = 121$ m/s]

Use the program to find the time the rocket lands if it is fired over level ground. Find its coordinates and speed just before it lands. [ans: $t = 22.3$ s, $x = 2.70 \times 10^3$ m, $v_x = 121$ m/s, $v_y = -109$ m/s]

Chapter 10

Many details of a two-body collision are controlled by the impulse exerted by each object on the other. Consider the situation in which an object of mass m_1, moving with velocity v_{1b}, impinges on an object of mass m_2, initially at rest. Suppose that after the collision both objects move along the line of incidence. Let J be the impulse object 1 exerts on object 2. Then, the velocity of object 2 after the collision is given by $v_{2a} = J/m_2$. Object 2 exerts an impulse $-J$ on object 1 and its velocity after the collision is given by $v_{1a} = v_{1b} - J/m_1$. Both final velocities can be computed if the initial velocity of object 1 and the impulse J are known (along with the masses, of course).

Once the velocities are known, the change in the kinetic energy of the system can be found. Before the collision the kinetic energy is $K_b = \frac{1}{2}m_1v_{1b}^2$ and after the collision it is $K_a = \frac{1}{2}m_1v_{1a}^2 + \frac{1}{2}m_2v_{2a}^2$. The kinetic energy lost during the collision is given by $Q = K_b - K_a$. A computer can be used to investigate the outcome of a one-dimensional collision as a function of the impulse.

Write a program that calculates the final velocities and the change in kinetic energy for a given range of impulses. Input the masses and the initial velocity of object 1. Also input the limits of the range of impulses to be considered. The outline of a program is:

input masses, initial velocity: m_1, m_2, v_{1b}
calculate initial kinetic energy: $K_b = \frac{1}{2}m_1v_{1b}^2$
input limits of range of impulse: J_i, J_f
number of values of J: $N = 21$
calculate impulse increment: $\Delta J = (J_f - J_i)/(N - 1)$
set $J = J_i$
begin loop over impulse values; counter runs from 1 to N
  calculate final velocities: $v_{1a} = v_{1b} - J/m_1$, $v_{2a} = J/m_2$
  calculate final kinetic energy: $K_a = \frac{1}{2}m_1v_{1a}^2 + \frac{1}{2}m_2v_{2a}^2$
  calculate kinetic energy loss: $Q = K_b - K_a$
  print or display J, v_{1a}, v_{2a}, Q

increment J: replace J with $J + \Delta J$

end loop

stop

The number N was chosen to be 21 because 21 lines nicely fill the usual monitor screen. You may want to use a different value. Both J_i and J_f are included in the list of values considered.

PROJECT 1. An air-track cart of mass $m_1 = 250\,\text{g}$ impinges at $3.0\,\text{m/s}$ on a cart of mass $m_2 = 600\,\text{g}$, initially at rest. Use the computer program to list the final velocities and kinetic energy loss for values of the impulse J from 0 to $2\,\text{kg·m/s}$.

From the table you generated, pick out the two values of the impulse that straddle the condition $v_{1a} = v_{2a}$. This is the completely inelastic collision. Note that Q is a maximum in this region of the table. Rerun the program with a narrower range of impulse values. Repeat if necessary and find J, v_{1a}, and Q for a completely inelastic collision to three significant figure accuracy. [ans: $0.529\,\text{kg·m/s}$; $0.882\,\text{m/s}$; $0.794\,\text{J}$]

From the original table ($J = 0$ to $2\,\text{kg·m/s}$) pick out the two values that straddle the condition $v_{1a} = 0$. Find to three significant figures the impulse, the final velocity of cart 2, and the kinetic energy loss. [ans: $0.750\,\text{kg·m/s}$; $1.250\,\text{m/s}$; $0.656\,\text{J}$]

From the original table pick out the two values that straddle the condition $Q = 0$. This is the elastic collision. Find to three significant figures the impulse and final velocities. [ans: $1.06\,\text{kg·m/s}$; $-1.24\,\text{m/s}$; $1.77\,\text{m/s}$]

PROJECT 2. Now repeat the calculations for two carts with the same mass. Take $m_1 = m_2 = 250\,\text{g}$ and $v_{1b} = 3.0\,\text{m/s}$. Notice that for this special case the impulse that stops the incident cart is the same as the impulse that produces an elastic collision. [ans: completely inelastic: $J = 0.375\,\text{kg·m/s}$, $v_{1a} = v_{2a} = 1.50\,\text{m/s}$, $Q = 0.563\,\text{J}$; elastic: $J = 0.750\,\text{kg·m/s}$, $v_{1a} = 0$, $v_{2a} = 3.00\,\text{m/s}$, $Q = 0$]

Two-dimensional collisions are somewhat more difficult to analyze. Consider a collision between two pucks on an air table. One puck, with mass m_1 and speed v_{1b}, is incident along the x axis on a puck with mass m_2, initially at rest. Suppose the impulse exerted by the moving puck on the puck at rest has magnitude J and makes the angle α with the x axis. Puck 2 leaves the collision along the line of the impulse; its velocity makes the angle $\theta_2 = \alpha$ with the x axis. Its speed after the collision is given by $v_{2a} = J/m_2$. The impulse-momentum equations for the incident puck are $-J\cos\alpha = m_1 v_{1a}\cos\theta_1 - m_1 v_{1b}$ and $-J\sin\alpha = m_1 v_{1a}\sin\theta_1$. These can be solved for v_{1a} and θ_1, with the result $v_{1a} = \sqrt{(J^2 + m_1^2 v_{1b}^2 - 2m_1 v_{1b} J\cos\alpha)}\big/ m_1$ and $\theta_1 = -\arctan[J\sin\alpha/(m_1 v_{1b} - J\cos\alpha)]$. To conserve momentum the sign of θ_1 must be opposite that of θ_2; that is, the two pucks must leave the collision on opposite sides of the line of incidence. If the sign of θ_1 produced by the computer is not correct, add 180° to it or subtract 180° from it.

Write a program that accepts values for the masses, initial velocity of the incident puck, and angle of the impulse with the x axis, then calculates the magnitudes and directions of the velocities after the collision and the loss in kinetic energy, all as functions of the magnitude of the impulse. An outline might be:

input masses, initial velocity: m_1, m_2, v_{1b}
calculate initial kinetic energy: $K_a = \frac{1}{2}m_1 v_{1a}^2$
input angle of impulse: α
input limits of range of impulse: J_i, J_f
number of values of J: $N = 21$
calculate impulse increment: $\Delta J = (J_f - J_i)/(N-1)$
set $J = J_i$

begin loop over impulse values; counter runs from 1 to $N+1$

calculate final speeds:

$$v_{1a} = \sqrt{(J^2 + m_1^2 v_{1b}^2 - 2m_1 v_{1b} J\cos\alpha)} \Big/ m_1$$

$$v_{2a} = J/m_2$$

calculate angles:

$$\theta_1 = -\arctan[J\sin\alpha/(m_1 v_{1b} - J\cos\alpha)]$$

$$\theta_2 = \alpha$$

check sign of θ_1 and correct if necessary

calculate final kinetic energy: $K_a = \frac{1}{2}m_1 v_{1a}^2 + \frac{1}{2}m_2 v_{2a}^2$

calculate kinetic energy loss: $Q = K_b - K_a$

print or display J, v_{1a}, v_{2a}, θ_1, θ_2, Q

increment J: replace J with $J + \Delta J$

end loop

stop

PROJECT 3. Puck 1 has a mass of 250 g and is incident at 3.0 m/s on puck 2, which has a mass of 600 g and is initially at rest. The impulse of puck 1 on puck 2 makes an angle of 20° with the axis of incidence (the x axis). Use the program to construct a table of the final speeds, angles of motion, and kinetic energy losses for impulses with magnitudes in the range from 0 to 2 kg·m/s.

Use the table to find two values of the impulse magnitude that straddle the value for an elastic collision. Rerun the program to find the impulse for an elastic collision to three significant figures. What are the final speeds and angle of motion then? [ans: $J = 0.995$ kg·m/s; $\mathbf{v}_{1a} = 1.55$ m/s at $-119°$; $\mathbf{v}_{2a} = 1.66$ m/s at 20°]

Now find the magnitude of the impulse for which puck 2 is scattered through 90°. What are the velocities of the pucks after the collision and what kinetic energy is lost? [ans: $J = 0.798$ kg·m/s; $\mathbf{v}_{1a} = 1.09$ m/s at $-90°$; $\mathbf{v}_{2a} = 1.33$ m/s at 20°; $Q = 1.33$ J]

Now suppose the impulse makes an angle of 60° with the line of incidence. What magnitude of impulse produces an elastic collision? What are the final velocities? [ans: $J = 0.529$ kg·m/s; $\mathbf{v}_{1a} = 2.67$ m/s at $-43.4°$; $\mathbf{v}_{2a} = 0.882$ m/s at 60°]

For what magnitude of impulse is puck 1 scattered through 90°? What are the final velocities and the kinetic energy loss? [ans: $J = 1.50$ kg·m/s; $\mathbf{v}_{1a} = 5.20$ m/s at $-90°$; $\mathbf{v}_{2a} = 2.50$ m/s at 60°; $Q = -4.13$ J]

The negative value of Q in the last case means kinetic energy must be *added* during the collision (in an explosion, for example).

PROJECT 4. What happens if the incident puck has greater mass than the target puck? Take $m_1 = 600$ g, $v_{1b} = 3.0$ m/s, and $m_2 = 250$ g. For $\alpha = 20°$ and 60° calculate the magnitude of the impulse that will produce an elastic collision and find the velocities of the pucks after such a collision. [ans: 20°: $J = 0.995$ kg·m/s, $\mathbf{v}_{1a} = 1.55$ m/s at $-21.5°$, $\mathbf{v}_{2a} = 3.78$ m/s at 20°; 60°: $J = 0.529$ kg·m/s, $\mathbf{v}_{1a} = 2.67$ m/s at $-16.6°$, $\mathbf{v}_{2a} = 2.12$ m/s at 60°]

PROJECT 5. If the masses are equal, the pucks always leave an elastic collision with their lines of motion making an angle of 90° with each other. Use the program to verify this for $m_1 = m_2 = 250$ g, $v_{1b} = 3.0$ m/s, and $\alpha = 20°$. Also verify it for $\alpha = 60°$. [ans: 20°: $J = 0.705$ kg·m/s, $\mathbf{v}_{1a} = 1.03$ m/s at $-70.0°$, $\mathbf{v}_{2a} = 2.82$ m/s at 20°; 60°: $J = 0.375$ kg·m/s, $\mathbf{v}_{1a} = 2.60$ m/s at $-30.0°$, $\mathbf{v}_{2a} = 1.50$ m/s at 60°]

Chapter 12

Conservation of angular momentum problems usually involve a system of two or more parts that exert torques on each other. The net external torque, however, vanishes. The angular momentum of each part changes, but the sum of the changes vanishes. You can use a computer to examine details of the motion.

Consider two wheels that are free to rotate on the same axle and that exert torques on each other. If the torque on wheel 1 is τ, then the angular velocity ω_1 of that wheel obeys $\tau = I_1\, d\omega_1/dt$, where I_1 is the rotational inertia of the wheel. The torque on wheel 2 is $-\tau$ and the angular velocity of that wheel obeys $-\tau = I_2\, d\omega_2/dt$. If τ is given as a function of time, a computer can be used to solve for θ_1, ω_1, θ_2, and ω_2, all as functions of time. Here's an outline, modeled after the program given in Chapter 2. The program also contains instructions to compute the angular momenta of the wheels and the total angular momentum of the system so you can check on angular momentum conservation.

input initial angular velocities: ω_{10}, ω_{20}
input final time, interval width: t_f, Δt
input display interval: Δt_d
set $t_b = 0$, $t_d = \Delta t_d$, $\theta_{1b} = 0$, $\omega_{1b} = \omega_{10}$, $\omega_{2b} = \omega_{20}$
calculate torque on 1: $\tau_b = \tau(t_b)$
begin loop over intervals
  calculate time at end of interval: $t_e = t_b + \Delta t$
  calculate torque on 1 at end of interval: $\tau_e = \tau(t_e)$
  calculate "average" torque: $\tau = (\tau_b + \tau_e)/2$
  calculate angular velocities at end of interval:
    $\omega_{1e} = \omega_{1b} + \tau\Delta t/I_1$
    $\omega_{2e} = \omega_{2b} - \tau\Delta t/I_2$
  calculate "average" angular velocities:
    $\omega_1 = (\omega_{1b} + \omega_{1e})/2$
    $\omega_2 = (\omega_{2b} + \omega_{2e})/2$
  calculate angular positions at end of interval:
    $\theta_{1e} = \theta_{1b} + \omega_1\Delta t$
    $\theta_{2e} = \theta_{2b} + \omega_2\Delta t$
  if $t_e \geq t_d$ then
    calculate angular momenta: $L_1 = I_1\omega_{1e}$, $L_2 = I_2\omega_{2e}$, $L = L_1 + L_2$
    print or display t_e, θ_{1e}, ω_{1e}, θ_{2e}, ω_{2e}, L_1, L_2, L
    increment t_d by Δt_d
  end of if statement
  if $t_e \geq t_f$ then exit loop
  set $t_b = t_e$, $\theta_{1b} = \theta_{1e}$, $\omega_{1b} = \omega_{1e}$, $\theta_{2b} = \theta_{2e}$, $\omega_{2b} = \omega_{2e}$, $\tau_b = \tau_e$
end loop over intervals
stop

PROJECT 1. Suppose wheel 1 has a rotational inertia of $2.5\,\mathrm{kg\cdot m^2}$ and is initially rotating at 100 rad/s. Wheel 2 has a rotational inertia of $1.5\,\mathrm{kg\cdot m^2}$ and is initially at rest. The wheels come in contact and exert torques on each other, the torque of wheel 2 on wheel 1 being given by $\tau(t) = -20te^{-t/2}$, in N·m for t in seconds. Use the program to plot the angular velocities of the wheels from $t = 0$ to $t = 15$ s. Use an integration interval of 0.005 s and a display interval of 0.5 s.

What are the initial values of the angular momenta of the wheels and the total angular momentum of the system? What are the values after 15 s? [ans: $L_1 = 250\,\text{kg}\cdot\text{m}^2/\text{s}$, $L_2 = 0$, $L = 250\,\text{kg}\cdot\text{m}^2/\text{s}$; $L_1 = 170\,\text{kg}\cdot\text{m}^2/\text{s}$, $L_2 = 80\,\text{kg}\cdot\text{m}^2/\text{s}$, $L = 250\,\text{kg}\cdot\text{m}^2/\text{s}$]

Suppose that in addition to the torque that wheel 2 exerts on wheel 1 an external agent exerts a torque of 8.0 N·m on wheel 1, in the direction of its angular velocity. What now are the angular momenta of the wheels and the total angular momentum of the system after 15 s? The torque on wheel 1 is $8.0 - 20te^{-t/2}$ and the torque on wheel 2 is $20te^{-t/2}$. [ans: $L_1 = 290\,\text{kg}\cdot\text{m}^2/\text{s}$, $L_2 = 80\,\text{kg}\cdot\text{m}^2/\text{s}$, $L = 370\,\text{kg}\cdot\text{m}^2/\text{s}$]

Notice that the angular momenta of the individual wheels change with time in both cases. The total angular momentum, however, is conserved in the first situation but not in the second. Since $dL/dt = \tau_{\text{ext}}$ and in the second situation $\tau_{\text{ext}} = 8.0\,\text{N}\cdot\text{m}$ the total angular momentum at the end of 15 s should be $L_f = 8.0 \times 15 = 120\,\text{kg}\cdot\text{m}^2/\text{s}$. Is this result produced by the program?

PROJECT 2. Conservation of angular momentum is sometimes demonstrated by considering the inelastic "collision" between two wheels on the same axle. Suppose wheel 1 of the previous project starts with an angular velocity of 100 rad/s and exerts a constant torque of 8.0 N·m on wheel 2, which starts from rest. Wheel 2, of course, exerts a torque of −8.0 N·m on wheel 1. Use the program to find the time for which $\omega_1 = \omega_2$ and the value of the angular velocity then. Through what angle has each wheel rotated by the time they reach the same angular velocity? Compare the value obtained for the final angular velocity with that predicted by $I_1\omega_{10} = I_1\omega_1 + I_2\omega_2$. [ans: 11.7 s, 62.5 rad/s, 952 rad, 366 rad]

The time for the wheels to reach the same angular velocity and the angle through which they turn depend on the torque they exert on each other but the final value of the angular velocity does not. To verify this statement repeat the calculation for the same initial conditions but for a torque on wheel 1 that is given by $\tau = -20te^{-t/3}$, in N·m for t in seconds. [ans: 5.24 s, 62.5 rad/s, 438 rad, 144 rad]

PROJECT 3. The kinetic energy of the wheels is not conserved during the inelastic "collision" of the previous project. You can see this by calculating the initial and final kinetic energies, but you can also use a computer to calculate the work done by each torque. If a wheel turns through the small angle $\Delta\theta$ while the torque τ acts on it, the work done by the torque is given by $\Delta W = \tau\Delta\theta$. Modify the program so it computes the work done on each wheel. Just before the loop over intervals set $W_1 = 0$ and $W_2 = 0$. These variables will be used to sum the work done in the intervals. Just before displaying results have the computer increment W_1 by $\tau(\theta_{1e} - \theta_{1b})$ and W_2 by $-\tau(\theta_{2e} - \theta_{2b})$. Here τ is the average torque exerted by wheel 2 on wheel 1. Also have the computer calculate the total work $W = W_1 + W_2$ and display the result.

Take the initial conditions to be as before ($\omega_{10} = 100\,\text{rad/s}$, $\omega_{20} = 0$). Take the torque of wheel 1 on wheel 2 to be a constant 8.0 N·m. Find the work done on each wheel and the total work done from $t = 0$ to the time when the wheels have the same angular velocity. Compare the total work done to the change in the total kinetic energy: $\Delta K = \frac{1}{2}(I_1 + I_2)\omega_f^2 - \frac{1}{2}I_1\omega_{10}^2$, where ω_f is the final angular velocity. [ans: -7.62×10^3 J, 2.93×10^3 J, -4.69×10^3 J]

Now repeat the calculation for a torque on wheel 1 that is given by $\tau = -20te^{-t/3}$, in N·m for t in seconds. You should get the same answers.

Can the "collision" be elastic? Take the torque on wheel 1 to be −8.0 N·m and calculate the total work done as a function of time. Search for a time such that the total work is zero. You might use the condition $W \geq 0$ to exit the loop over intervals. If the wheels no longer interact after this time, the collision is elastic. What are the final angular velocities of the wheels? [ans: 23.4 s, 25.0 rad/s, 125 rad/s]

The text showed you how to analyze the change in angular velocity of a spinning ice skater as she brings her arms toward the center of her torso from an outstretched position. Before she brings them in, her rotational inertia is I_i and her angular velocity is ω_i. After she brings them in, her rotational inertia is I_f and her angular velocity is ω_f. Conservation of angular momentum leads to $I_i\omega_i = I_f\omega_f$, so $\omega_f = (I_i/I_f)\omega_i$.

Here you will use a computer program to investigate some of the details. In particular, you will be able to follow her angular velocity as she brings her arms in. Let I_T be the rotational inertia of her torso alone and let I_A be the rotational inertia of her arms, including any weights she might hold to heighten the effect. I_A is a function of time, determined by the rate with which she moves her arms. Angular momentum is conserved at each stage of the process so

$$\omega(t) = \frac{I_{Ai} + I_T}{I_A(t) + I_T}\omega_i\,,$$

where I_{Ai} is the initial value of the rotational inertia of her arms.

If $\omega(t)$ is known, a computer program can be used to calculate the angle $\theta(t)$ through which the skater has rotated since $t = 0$. The angle θ_e at the end of any interval of width Δt is related to the angle θ_b at the beginning by $\theta_e = \theta_b + \omega\Delta t$, where ω is the average angular velocity in the interval. As usual, we approximate the average angular velocity by $(\omega_b + \omega_e)/2$. Here is an outline of a program.

> input final time and interval width: t_f, Δt
> input display interval: Δt_d
> set $t_b = 0$, $t_d = \Delta t_d$, $\theta_b = 0$
> calculate angular velocity at beginning of first interval: $\omega_b = \omega(t_b)$
> **begin loop** over intervals
>   calculate time at end of interval: $t_e = t_b + \Delta t$
>   calculate angular velocity at end of interval: $\omega_e = \omega(t_e)$
>   calculate "average" angular velocity: $\omega = (\omega_b + \omega_e)/2$
>   calculate angle at end of interval: $\theta_e = \theta_b + \omega\Delta t$
>   if $t_e \geq t_d$ then
>   calculate torque at end of interval: τ_e
>   print or display t_e, θ_e, ω_e
>   increment t_d by Δt_d
>   end of if statement
>   if $t_e \geq t_f$ then exit loop
>   set $t_b = t_e$, $\theta_b = \theta_e$, $\omega_b = \omega_e$ in preparation for next interval
> **end loop** over intervals
> stop

Both the angular momentum of the skater's torso and the angular momentum of her arms change. Her torso exerts an torque on her arms and her arms exert a torque of equal magnitude but opposite direction on her torso. If τ represents the torque on her torso, then $\tau = I_T\,\mathrm{d}\omega/\mathrm{d}t$. Differentiation of the expression above for $\omega(t)$ produces

$$\tau = -\frac{I_T\omega_i(I_{Ai} + I_T)}{(I_A + I_T)^2}\frac{\mathrm{d}I_A}{\mathrm{d}t}$$

Use this expression in the program to calculate the torque just before t_e, θ_e, and ω_e are displayed, then display it along with the other quantities. A line for the instruction has already been indicated in the outline above.

PROJECT 4. A skater is initially turning at 12 rad/s. Her torso has a rotational inertia of 0.65 kg·m^2. She carries weights in her hands so that as she pulls her arms in their rotational inertia changes from 0.50 kg·m^2 to 0.04 kg·m^2. Suppose she moves her arms in such a way that their rotational inertia changes uniformly over 2.0 s. That is, $I_A(t) = I_{Ai} - Kt$, where $I_{Ai} = 0.50\,\text{kg·m}^2$ and K is a constant chosen so $I_A = 0.04\,\text{kg·m2}$ when $t = 2.0\,\text{s}$.

Use the program to plot her angular velocity and the torque acting on her torso as functions of time from $t = 0$ to $t = 2.0\,\text{s}$. What is her final angular velocity? Through what angle did she rotate during the process? At what time did she exert the maximum torque on her arms and what was the magnitude of that torque? [ans: 20.0 rad/s; 30.7 rad (4.88 rev); 1.73 N·m at 2.0 s]

Suppose now that all of the rotational inertia of her arms is actually due to the weights she is carrying and that she pulls them in at a constant rate. If m is the mass of one weight and r is it distance from the axis of rotation, then the rotational inertia of both weights together is $I_A = 2mr^2$. Since the weights are brought in at a constant rate, we may write $r = r_i + Kt$, where r_i is the initial distance and K is a constant. Take $m = 1.0\,\text{kg}$ and pick values of r_i and K so $I_{Ai} = 0.50\,\text{kg·m}^2$ and $I_A = 0.05\,\text{kg·m}^2$ at the end of 2.0 s. Then, use the program to plot her angular velocity and the torque acting on her torso as functions of time from $t = 0$ to $t = 2.0\,\text{s}$. What is her final angular velocity? Through what angle did she rotate during the process? At what time did she exert the maximum torque on her arms and what was the magnitude of that torque? [ans: 20.0 rad/s; 32.2 rad (5.12 rev); 2.82 N·m at 0.95 s]

PART II

Test Bank

Chapters 1-12

TO THE INSTRUCTOR

Over 2500 multiple choice problems have been culled from the many thousands accumulated over the years and are presented in this collection. They come from numerous sources, chiefly our own examinations and files and examinations sent to us by others. As far as we are aware, all are meant to be freely used. Most have been altered somewhat, either to meet the circumstances of our own courses or to improve them over repeated uses. Several hundred have been created expressly for this test bank to achieve uniformity of coverage. Users will, no doubt, be led to invent their own problems as natural follow-ups or as variants on those in this set.

Problems are organized according to the chapters of *Fundamentals of Physics*, fifth edition, and numbering is restarted with each chapter. Each problem is self-contained, with an appropriate figure if needed. Instructors can retype any desired problem or can photocopy selected pages, then cut and paste an examination. Answers to all problems are included.

The problems are also available from Wiley on computer disks for both Windows and Macintosh computers. The set of disks contains a program which will allow you to select and preview problems on a monitor screen, then print an exam and answer key. Tests can also be stored on disk for retrieval later and may be edited using a word processor.

Acknowledgements. Great thanks goes to Joan Kalkut who managed much of this and previous editions. She is a joy to work with. The contributions of the following people at Wiley is also greatly appreciated: Cliff Mills, Stuart Johnson, Catherine Donovan, and Thomas Hempstead. Their help speeded the project and smoothed the publication process. Karen Christman skillfully copy-edited the draft copy of a previous edition.

J. Richard Christman
Kenneth R. Brownstein

TABLE OF CONTENTS

Chapter 1: MEASUREMENT

1. The standard of time is based on:

   A. the daily rotation of the earth
   B. the frequency of light emitted by Kr^{86}
   C. the yearly revolution of the earth about the sun
   D. a precision pendulum clock
   E. none of these

   ans: E

2. A nanosecond is:

   A. 10^9 s
   B. 10^{-9} s
   C. 10^{-10} s
   D. 10^{-10} s
   E. 10^{-12} s

   ans: B

3. The standard of length is based on:

   A. the distance from the north pole to the equator along a meridian passing through Paris
   B. wavelength of light emitted by Hg^{198}
   C. wavelength of light emitted by Kr^{86}
   D. a precision meter stick in Paris
   E. the speed of light

   ans: E

4. In 1866, the U. S. Congress defined the U. S. yard as exactly 3600/3937 international meter. This was done primarily because:

   A. length can be measured more accurately in meters than in yards
   B. the meter is more stable than the yard
   C. this definition relates the common U. S. length units to a more widely used system
   D. there are more wavelengths in a yard than in a meter
   E. the members of this Congress were exceptionally intelligent

   ans: C

5. Which of the following is closest to your height?

   A. 0.02 m
   B. 0.2 m
   C. 2 m
   D. 200 m
   E. 2000 m

   ans: C

6. There is no SI base unit for area because:

A. an area has no thickness; hence no physical standard can be built
B. we live in a three (not a two) dimensional world
C. it is impossible to express ft^2 in terms of m
D. area can be expressed in terms of m^2
E. area is not an important physical quantity

ans: D

7. The SI base unit for mass is:

A. gram
B. pound
C. kilogram
D. ounce
E. kilopound

ans: C

8. A gram is:

A. 10^{-6} kg
B. 10^{-3} kg
C. 1 kg
D. 10^{3} kg
E. 10^{6} kg

ans: B

9. Which of the following is closest to your mass?

A. 0.06 kg
B. 0.6 kg
C. 6 kg
D. 60 kg
E. 600 kg

ans: D

10. $5.0 \times 10^4 \times 3.0 \times 10^6 =$

A. 1.5×10^{9}
B. 1.5×10^{10}
C. 1.5×10^{11}
D. 1.5×10^{12}
E. 1.5×10^{13}

ans: C

11. $5.0 \times 10^4 \times 3.0 \times 10^{-6} =$

A. 1.5×10^{-3}
B. 1.5×10^{-1}
C. 1.5×10^{1}
D. 1.5×10^{3}
E. 1.5×10^{5}

ans: B

12. $5.0 \times 10^5 + 3.0 \times 10^6 =$

A. 8.0×10^5
B. 8.0×10^6
C. 5.3×10^5
D. 3.5×10^5
E. 3.5×10^6

ans: E

13. $7.0 \times 10^6 / 2.0 \times 10^{-6} =$

A. 3.5×10^{-12}
B. 3.5×10^{-6}
C. 3.5
D. 3.5×10^6
E. 3.5×10^{12}

ans: E

14. The number of significant figures in 0.00150 is:

A. 2
B. 3
C. 4
D. 5
E. 6

ans: B

15. The number of significant figures in 15.0 is:

A. 1
B. 2
C. 3
D. 4
E. 5

ans: C

16. $3.2 \times 2.7 =$

A. 9
B. 8
C. 8.6
D. 8.64
E. 8.640

ans: C

17. 1 mi is equivalent to 1609 m so 55 mph is:

A. 15 m/s
B. 25 m/s
C. 66 m/s
D. 88 m/s
E. 1500 m/s

ans: B

18. A sphere with a radius of 1.7 cm has a volume of:

A. 2.1×10^{-5} m^3
B. 9.1×10^{-4} m^3
C. 3.6×10^{-3} m^3
D. 0.11 m^3
E. 21 m^3

ans: A

19. A sphere with a radius of 1.7 cm has a surface area of:

A. 2.1×10^{-5} m^2
B. 9.1×10^{-4} m^2
C. 3.6×10^{-3} m^2
D. 0.11 m^2
E. 36 m^2

ans: C

20. A right circular cylinder with a radius of 2.3 cm and a height of 1.4 m has a volume of:

A. 0.20 m^3
B. 0.14 m^3
C. 9.3×10^{-3} m^3
D. 2.3×10^{-3} m^3
E. 7.4×10^{-4} m^3

ans: D

21. A right circular cylinder with a radius of 2.3 cm and a height of 1.4 cm has a total surface area of:

A. 1.7×10^{-3} m^2
B. 3.2×10^{-3} m^2
C. 2.0×10^{-3} m^3
D. 5.3×10^{-3} m^2
E. 7.4×10^{-3} m^2

ans: D

22. A cubic box with an edge of exactly 1 cm has a volume of:

A. 10^{-9} m^3
B. 10^{-6} m^3
C. 10^{-3} m^3
D. 10^{3} m^3
E. 10^{6} m^3

ans: B

23. A square with an edge of exactly 1 cm has an area of:

A. 10^{-6} m^2
B. 10^{-4} m^2
C. 10^2 m^2
D. 10^4 m^2
E. 10^6 m^2

ans: B

24. 1 m is equivalent to 3.281 ft. A cube with an edge of 1.5 ft has a volume of:

A. 1.2×10^2 m^3
B. 9.6×10^{-2} m^3
C. 10.5 m^3
D. 9.5×10^{-2} m^3
E. 0.21 m^3

ans: B

25. During a short interval of time the velocity v in m/s of an automobile is given by $v = at^2 + bt^3$, where the time t is in seconds. The units of a and b are respectively:

A. $\mathrm{m \cdot s^2}$; $\mathrm{m \cdot s^4}$
B. $\mathrm{s^3/m}$; $\mathrm{s^4/m}$
C. $\mathrm{m/s^2}$; $\mathrm{m/s^3}$
D. $\mathrm{m/s^3}$; $\mathrm{m/s^4}$
E. $\mathrm{m/s^4}$; $\mathrm{m/s^5}$

ans: D

26. Suppose $A = BC$, where A has the dimensions L/M and C has the dimensions L/T. Then B has dimension:

A. T/M
B. $\mathrm{L^2/TM}$
C. $\mathrm{TM/L^2}$
D. $\mathrm{L^2T/M}$
E. $\mathrm{M/L^2T}$

ans: A

27. Suppose $A = B^nC^m$, where A has dimensions LT, B has dimensions $\mathrm{L^2T^{-1}}$, and C has dimensions $\mathrm{LT^2}$. Then the exponents n and m have the values:

A. 2/3; 1/3
B. 2; 3
C. 4/5; −1/5
D. 1/5; 3/5
E. 1/2; 1/2

ans: D

Chapter 2: MOTION ALONG A STRAIGHT LINE

1. A particle moves along the x axis from x_i to x_f. Of the following values of the initial and final coordinates, which results in the displacement with the largest magnitude?
   A. $x_i = 4\,\text{m}$, $x_f = 6\,\text{m}$
   B. $x_i = -4\,\text{m}$, $x_f = -8\,\text{m}$
   C. $x_i = -4\,\text{m}$, $x_f = 2\,\text{m}$
   D. $x_i = 4\,\text{m}$, $x_f = -2\,\text{m}$
   E. $x_i = -4\,\text{m}$, $x_f = 4\,\text{m}$
   ans: E

2. A particle moves along the x axis from x_i to x_f. Of the following values of the initial and final coordinates, which results in a negative displacement?
   A. $x_i = 4\,\text{m}$, $x_f = 6\,\text{m}$
   B. $x_i = -4\,\text{m}$, $x_f = -8\,\text{m}$
   C. $x_i = -4\,\text{m}$, $x_f = 2\,\text{m}$
   D. $x_i = -4\,\text{m}$, $x_f = -2\,\text{m}$
   E. $x_i = -4\,\text{m}$, $x_f = 4\,\text{m}$
   ans: B

3. The average speed of a moving object during a given interval of time is always:
   A. the magnitude of its average velocity over the interval
   B. the distance covered during the time interval divided by the time interval
   C. one-half its speed at the end of the interval
   D. its acceleration multiplied by the time interval
   E. one-half its acceleration multiplied by the time interval.
   ans: B

4. Two automobiles are 150 kilometers apart and traveling toward each other. One automobile is moving at 60 km/h and the other is moving at 40 km/h mph. In how many hours will they meet?
   A. 2.5
   B. 2.0
   C. 1.75
   D. 1.5
   E. 1.25
   ans: D

5. A car travels 40 kilometers at an average speed of 80 km/h and then travels 40 kilometers at an average speed of 40 km/h. The average speed of the car for this 80-km trip is:
   A. 40 km/h
   B. 45 km/h
   C. 48 km/h
   D. 53 km/h
   E. 80 km/h
   ans: D

6. A car starts from Hither, goes 50 km in a straight line to Yon, immediately turns around, and returns to Hither. The time for this round trip is 2 hours. The magnitude of the average velocity of the car for this round trip is:

   A. 0
   B. 50 km/hr
   C. 100 km/hr
   D. 200 km/hr
   E. cannot be calculated without knowing the acceleration

   ans: A

7. A car starts from Hither, goes 50 km in a straight line to Yon, immediately turns around, and returns to Hither. The time for this round trip is 2 hours. The average speed of the car for this round trip is:

   A. 0 km/h
   B. 50 km/h
   C. 100 km/h
   D. 200 km/h
   E. cannot be calculated without knowing the acceleration

   ans: B

8. The coordinate of a particle in meters is given by $x(t) = 12t - 3.0t^2$, where the time t is in seconds. The particle is momentarily at rest at $t =$

   A. 2.0 s
   B. 3.0 s
   C. 4.0 s
   D. 5.0 s
   E. 6.0 s

   ans: A

9. A drag racing car starts from rest at $t = 0$ and moves along a straight line with velocity given by $v = bt^2$, where b is a constant. The expression for the distance traveled by this car from its position at $t = 0$ is:

   A. bt^3
   B. $bt^3/3$
   C. $4bt^2$
   D. $3bt^2$
   E. $bt^{3/2}$

   ans: B

10. A ball rolls up a slope. At the end of three seconds its velocity is 20 cm/s; at the end of eight seconds its velocity is 0 cm/s. What is the average acceleration (in cm/s^2) from the third to the eighth second?

   A. 2.5
   B. 4.0
   C. 5.0
   D. 6.0
   E. 6.67

   ans: B

11. The coordinate of an object is given as a function of time by $x = 7t - 3t^2$, where x is in meters and t is in seconds. Its average velocity over the interval from $t = 0$ to $t = 4\,\text{s}$ is:

A. 5 m/s
B. −5 m/s
C. 11 m/s
D. −11 m/s
E. −14.5 m/s

ans: B

12. The velocity of an object is given as a function of time by $v = 4t - 3t^2$, where v is in m/s and t is in seconds. Its average velocity over the interval from $t = 0$ to $t = 2\,\text{s}$ is:

A. 0
B. −2 m/s
C. 2 m/s
D. −4 m/s
E. can't tell unless the initial position is given

ans: A

13. The coordinate of an object is given as a function of time by $x = 4t^2 - 3t^3$, where x is in meters and t is in seconds. Its average acceleration over the interval from $t = 0$ to $t = 2\,\text{s}$ is:

A. −4 m/s
B. 4 m/s
C. −10 m/s
D. 10 m/s
E. −13 m/s

ans: C

14. Each of four particles move along an x axis. Their coordinates (in meters) as functions of time (in seconds) are given by

particle 1: $x(t) = 3.5 - 2.7t^3$
particle 2: $x(t) = 3.5 + 2.7t^3$
particle 3: $x(t) = 3.5 + 2.7t^2$
particle 4: $x(t) = 3.5 - 3.4t - 2.7t^2$

Which of these particles have constant acceleration?

A. All four
B. Only 1 and 2
C. Only 2 and 3
D. Only 3 and 4
E. None of them

ans: D

15. Each of four particles move along an x axis. Their coordinates (in meters) as functions of time (in seconds) are given by

particle 1: $x(t) = 3.5 - 2.7t^3$
particle 2: $x(t) = 3.5 + 2.7t^3$
particle 3: $x(t) = 3.5 + 2.7t^2$
particle 4: $x(t) = 3.5 - 3.4t - 2.7t^2$

Which of these particles is speeding up for $t > 0$?

A. All four
B. Only 1
C. Only 2 and 3
D. Only 2, 3, and 4
E. None of them

ans: A

16. An object starts from rest at the origin and moves along the x axis with a constant acceleration of $4\,\text{m/s}^2$. Its average velocity as it goes from $x = 2\,\text{m}$ to $x = 8\,\text{m}$ is:

A. $1\,\text{m/s}$
B. $2\,\text{m/s}$
C. $3\,\text{m/s}$
D. $5\,\text{m/s}$
E. $6\,\text{m/s}$

ans: E

17. Of the following situations, which one is impossible?

A. A body having velocity east and acceleration east
B. A body having velocity east and acceleration west
C. A body having zero velocity and non-zero acceleration
D. A body having constant acceleration and variable velocity
E. A body having constant velocity and variable acceleration

ans: E

18. Throughout a time interval, while the speed of a particle increases as it moves along the x axis, its velocity and acceleration might be:

A. positive and negative, respectively
B. negative and positive, respectively
C. negative and negative, respectively
D. negative and zero, respectively
E. positive and zero, respectively

ans: C

19. A particle moves on the x axis. When its acceleration is positive and increasing:

A. its velocity must be positive
B. its velocity must be negative
C. it must be slowing down
D. it must be speeding up
E. none of the above must be true

ans: E

20. The position y of a particle moving along the y axis depends on the time t according to the equation $y = at - bt^2$. The dimensions of the quantities a and b are respectively:

A. L^2/T, L^3/T^2
B. L/T^2, L^2/T
C. L/T, L/T^2
D. L^3/T, T^2/L
E. none of these

ans: C

21. A particle moves along the x axis according to the equation $x = 6t^2$, where x is in meters and t is in seconds. Therefore:

A. the acceleration of the particle is 6 m/s^2
B. t cannot be negative
C. the particle follows a parabolic path
D. each second the velocity of the particle changes by 9.8 m/s
E. none of the above

ans: E

22. Over a short interval near time $t = 0$ the coordinate of an automobile in meters is given by $x(t) = 27t - 4.0t^3$, where t is in seconds. At the end of 1.0 s the acceleration of the auto is:

A. 27 m/s^2
B. 4.0 m/s^2
C. -4.0 m/s^2
D. -12 m/s^2
E. -24 m/s^2

ans: E

23. Over a short interval, starting at time $t = 0$, the coordinate of an automobile in meters is given by $x(t) = 27t - 4.0t^3$, where t is in seconds. The magnitudes of the initial (at $t = 0$) velocity and acceleration of the auto respectively are:

A. 0; 12 m/s^2
B. 0; 24 m/s^2
C. 27 m/s; 0
D. 27 m/s; 12 m/s^2
E. 27 m/s; 24 m/s^2

ans: C

24. At time $t = 0$ a car has a velocity of 16 m/s. It slows down with an acceleration given by $-0.50t$, in m/s^2 for t in seconds. It stops at $t =$

A. 64 s
B. 32 s
C. 16 s
D. 8.0 s
E. 4.0 s

ans: D

25. At time $t = 0$ a car has a velocity of 16 m/s. It slows down with an acceleration given by $-0.50t$, in m/s^2 for t in seconds. At the end of 4.0 s it has traveled:

A. 0
B. 12 m
C. 14 m
D. 25 m
E. 59 m

ans: E

26. At time $t = 0$ a car has a velocity of 16 m/s. It slows down with an acceleration given by $-0.50t$, in m/s^2 for t in seconds. By the time it stops it has traveled:

A. 15 m
B. 31 m
C. 62 m
D. 85 m
E. 100 m

ans: D

27. Starting at time $t = 0$, an object moves along a straight line with velocity in m/s given by $v(t) = 98 - 2t^2$, where t is in seconds. When it momentarily stops its acceleration is:

A. 0
B. $-4.0\ m/s^2$
C. $-9.8\ m/s^2$
D. $-28\ m/s^2$
E. $49\ m/s^2$

ans: D

28. Starting at time $t = 0$, an object moves along a straight line. Its coordinate in meters is given by $x(t) = 75t - 1.0t^3$, where t is in seconds. When it momentarily stops its acceleration is:

A. 0
B. $-73\ m/s^2$
C. $-30\ m/s^2$
D. $-9.8\ m/s^2$
E. $9.2 \times 10^3\ m/s^2$

ans: C

29. A car, initially at rest, travels 20 m in 4 s along a straight line with constant acceleration. The acceleration of the car (in m/s^2) is:

A. 0.4
B. 1.3
C. 2.5
D. 4.9
E. 9.8

ans: C

30. A racing car traveling with constant acceleration increases its speed from 10 m/s to 30 m/s over a distance of 80 m. How long does this take?

A. 2.0 s
B. 4.0 s
C. 5.0 s
D. 8.0 s
E. The time cannot be calculated since the speed is not constant

ans: B

31. A car starts from rest and goes down a slope with a constant acceleration of 5 m/s^2. After 5 s the car reaches the bottom of the hill. Its speed at the bottom of the hill, in meters per second, is:

A. 1
B. 12.5
C. 25
D. 50
E. 160

ans: C

32. A car moving with an initial velocity of 25 m/s north has a constant acceleration of 3 m/s^2 south. After 6 seconds its velocity will be:

A. 7 m/s north
B. 7 m/s south
C. 43 m/s north
D. 20 m/s north
E. 20 m/s south

ans: A

33. An object with an initial velocity of 12 m/s west experiences a constant acceleration of 4 m/s^2 west for 3 seconds. During this time the object travels a distance of:

A. 12 m
B. 24 m
C. 36 m
D. 54 m
E. 144 m

ans: D

34. How far does a car travel in 6 s if its initial velocity is 2 m/s and its acceleration is 2 m/s^2 in the forward direction?

A. 12 m
B. 14 m
C. 24 m
D. 36 m
E. 48 m

ans: E

35. At a stop light, a truck traveling at 15 m/s passes a car as it starts from rest. The truck travels at constant velocity and the car accelerates at 3 m/s^2. How many seconds will it take for the car to catch up to the truck?

A. 5
B. 10
C. 15
D. 20
E. 25

ans: B

36. A ball is in free fall. Its acceleration is:

A. downward during both ascent and descent
B. downward during ascent and upward during descent
C. upward during ascent and downward during descent
D. upward during both ascent and descent
E. downward at all times except at the very top, when it is zero

ans: A

37. A ball is in free fall. Upward is taken to be the positive direction. The displacement of the ball during a short time interval is:

A. positive during both ascent and descent
B. negative during both ascent and descent
C. negative during ascent and positive during descent
D. positive during ascent and negative during descent
E. none of the above

ans: D

38. A baseball is thrown vertically into the air. The acceleration of the ball at its highest point is:

A. 9.8 m/s^2 down
B. 9.8 m/s^2 up
C. changing suddenly from 9.8 m/s^2 up to 9.8 m/s^2 down
D. zero
E. cannot be calculated without knowing the initial velocity

ans: A

39. Which one of the following statements is correct for an object released from rest?

A. The average velocity during the first second of time is 4.9 m/s
B. During each second the object falls 9.8 m
C. The acceleration changes by 9.8 m/s^2 every second
D. The object falls 9.8 m during the first second of time
E. The acceleration of the object is proportional to its weight

ans: A

40. A freely falling body has a constant acceleration of 9.8 m/s^2. This means that:
  A. the body falls 9.8 m during each second
  B. the body falls 9.8 m during the first second only
  C. the speed of the body increases by 9.8 m/s during each second
  D. the acceleration of the body increases by 9.8 m/s^2 during each second
  E. the acceleration of the body decreases by 9.8 m/s^2 during each second

  ans: C

41. An object is shot vertically upward. While it is rising:
  A. its velocity and acceleration are both upward
  B. its velocity is upward and its acceleration is downward
  C. its velocity and acceleration are both downward
  D. its velocity is downward and its acceleration is upward
  E. its velocity and acceleration are both decreasing

  ans: B

42. An object is thrown straight up from ground level with a speed of 50 m/s. If $g = 10$ m/s^2 its distance above ground level 1.0 s later is:
  A. 40 m
  B. 45 m
  C. 50 m
  D. 55 m
  E. 60 m

  ans: B

43. An object is thrown straight up from ground level with a speed of 50 m/s. If $g = 10$ m/s^2 its distance above ground level 6.0 s later is:
  A. 0.00 m
  B. 270 m
  C. 330 m
  D. 480 m
  E. none of these

  ans: E

44. At a location where $g = 9.80$ m/s^2, an object is thrown vertically down with an initial speed of 1.00 m/s. After 5.00 s the object will have traveled:
  A. 125 m
  B. 127.5 m
  C. 245 m
  D. 250 m
  E. 255 m

  ans: B

45. An object is thrown vertically upward at 35 m/s. Taking $g = 10$ m/s^2, the velocity of the object 5 s later is:

A. 7.0 m/s up
B. 15 m/s down
C. 15 m/s up
D. 85 m/s down
E. 85 m/s up

ans: B

46. A feather, initially at rest, is released in a vacuum 12 m above the surface of the earth. Which of the following statements is correct?

A. The maximum velocity of the feather is 9.8 m/s
B. The acceleration of the feather decreases until terminal velocity is reached
C. The acceleration of the feather remains constant during the fall
D. The acceleration of the feather increases during the fall
E. The acceleration of the feather is zero

ans: C

47. An object is released from rest and falls a distance H during the first second of time. How far will it fall during the next second of time?

A. H
B. $2H$
C. $3H$
D. $4H$
E. H^2

ans: C

48. A heavy ball falls freely, starting from rest. Between the third and fourth second of time it travels a distance of:

A. 4.9 m
B. 9.8 m
C. 29.4 m
D. 34.3 m
E. 39.8 m

ans: D

49. As a rocket is accelerating vertically upward at 9.8 m/s^2 near Earth's surface, it releases a projectile. Immediately after release the acceleration (in m/s^2) of the projectile is:

A. 9.8 down
B. 0
C. 9.8 up
D. 19.6 up
E. none of the above

ans: A

50. A stone is released from a balloon that is descending at a constant speed of 10 m/s. Neglecting air resistance, after 20 s the speed of the stone is:

A. 2160 m/s
B. 1760 m/s
C. 206 m/s
D. 196 m/s
E. 186 m/s

ans: C

51. An object dropped from a stationary balloon hits the ground in 12.0 s. If its acceleration is 9.80 m/s^2, the height of the balloon is:

A. 29.4 m
B. 58.8 m
C. 118 m
D. 353 m
E. 706 m

ans: E

52. Neglecting the effect of air resistance a stone dropped off a 175-m high building lands on the ground in:

A. 3 s
B. 4 s
C. 6 s
D. 18 s
E. 36 s

ans: C

53. A stone is thrown vertically upward with an initial speed of 19.5 m/s. It will rise to a maximum height of:

A. 4.9 m
B. 9.8 m
C. 19.4 m
D. 38.8 m
E. none of these

ans: C

54. A baseball is hit straight up and is caught by the catcher 2.0 s later. The maximum height of the ball during this interval is:

A. 4.9 m
B. 7.4 m
C. 9.8 m
D. 12.6 m
E. 19.6 m

ans: A

55. An object is thrown straight down with an initial speed of 4 m/s from a window which is 8 m above the ground. The time it takes the object to reach the ground is:

A. 0.80 s
B. 0.93 s
C. 1.3 s
D. 1.7 s
E. 2.0 s

ans: B

56. A stone is released from rest from the edge of a building 190 m above the ground. Neglecting air resistance, the speed of the stone, just before striking the ground, is:

A. 43 m/s
B. 61 m/s
C. 120 m/s
D. 190 m/s
E. 1400 m/s

ans: B

57. An object is thrown vertically upward with a certain initial velocity in a world where the acceleration due to gravity is 19.6 m/s^2. The height to which it rises is ___ that to which the object would rise if thrown upward with the same initial velocity on the Earth. Neglect friction.

A. half
B. $\sqrt{2}$ times
C. twice
D. four times
E. cannot be calculated from the given data

ans: A

58. A projectile is shot vertically upward with a given initial velocity. It reaches a maximum height of 100 m. If, on a second shot, the initial velocity is doubled then the projectile will reach a maximum height of:

A. 70.7 m
B. 141.4 m
C. 200 m
D. 241 m
E. 400 m

ans: E

59. One object is thrown vertically upward with an initial velocity of 100 m/s and another object with an initial velocity of 10 m/s. The maximum height reached by the first object will be ___ that of the other.

A. 10 times
B. 100 times
C. 1000 times
D. 10,000 times
E. none of these

ans: B

60. The area under a velocity-time graph represents:
  A. acceleration
  B. change in acceleration
  C. speed
  D. change in velocity
  E. displacement

  ans: E

61. Displacement can be obtained from:
  A. the slope of an acceleration-time graph
  B. the slope of a velocity-time graph
  C. the area under an acceleration-time graph
  D. the area under a velocity-time graph
  E. the slope of an acceleration-time graph

  ans: D

62. An object has a constant acceleration of 3 m/s^2. The coordinate versus time graph for this object has a slope:
  A. that increases with time
  B. that is constant
  C. that decreases with time
  D. of 3 m/s
  E. of 3 m/s^2

  ans: A

63. The coordinate-time graph of an object is a straight line with a positive slope. The object has:
  A. constant displacement
  B. steadily increasing acceleration
  C. steadily decreasing acceleration
  D. constant velocity
  E. steadily increasing velocity

  ans: D

64. Which of the following five coordinate versus time graphs represents the motion of an object moving with a constant nonzero speed?

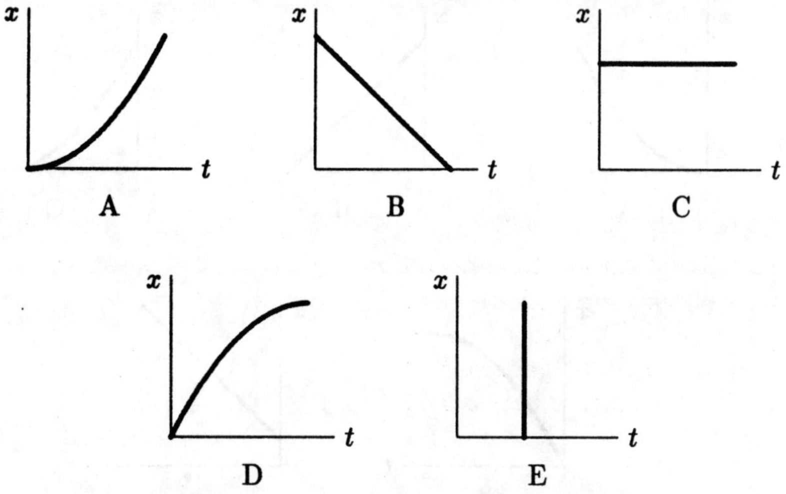

ans: B

65. Which of the following five acceleration versus time graphs is correct for an object moving in a straight line at a constant velocity of 20 m/s?

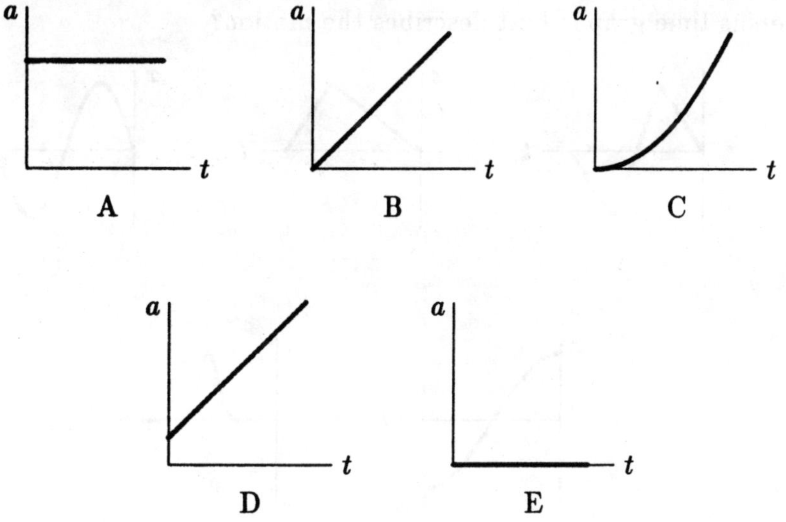

ans: E

66. Which of the following five coordinate versus time graphs represents the motion of an object whose speed is increasing?

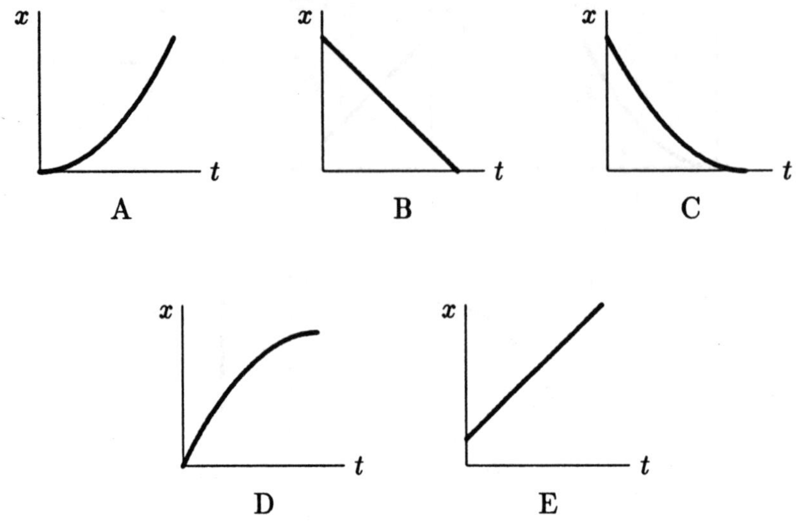

ans: A

67. A car accelerates from rest on a straight road. A short time later, the car decelerates to a stop and then returns to its original position in a similar manner. Which of the following five coordinate versus time graphs best describes the motion?

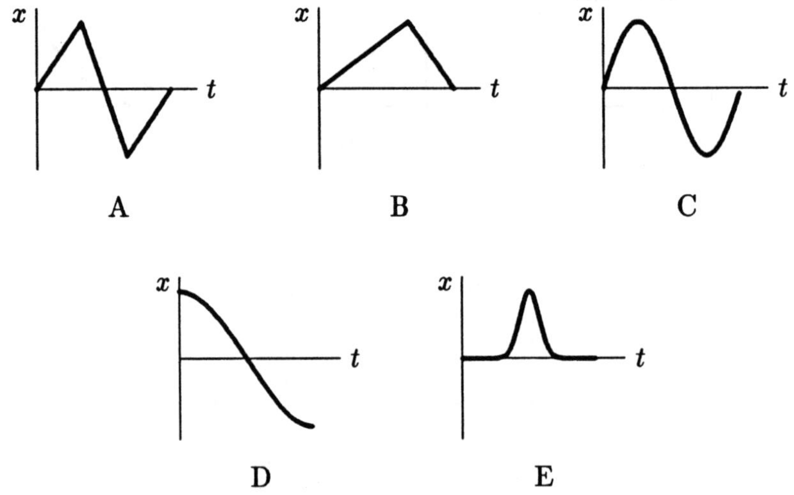

ans: E

68. The acceleration of an object, starting from rest, is shown in the graph below. Other than at $t = 0$, when is the velocity of the object equal to zero?

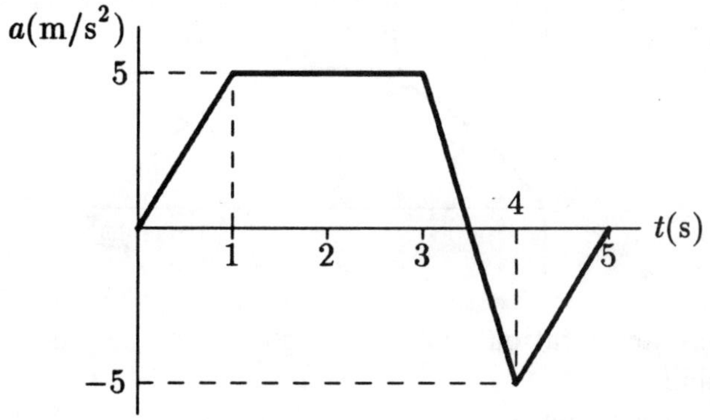

A. During the interval from 1.0 s to 3.0 s
B. At $t = 3.5\,\text{s}$
C. At $t = 4.0\,\text{s}$
D. At $t = 5.0\,\text{s}$
E. At no other time on this graph

ans: E

69. An elevator is moving upward with constant acceleration. The dashed curve shows the position y of the ceiling of the elevator as a function of the time t. At the instant indicated by the dot, a bolt breaks loose and drops from the ceiling. Which curve best represents the position of the bolt as a function of time?

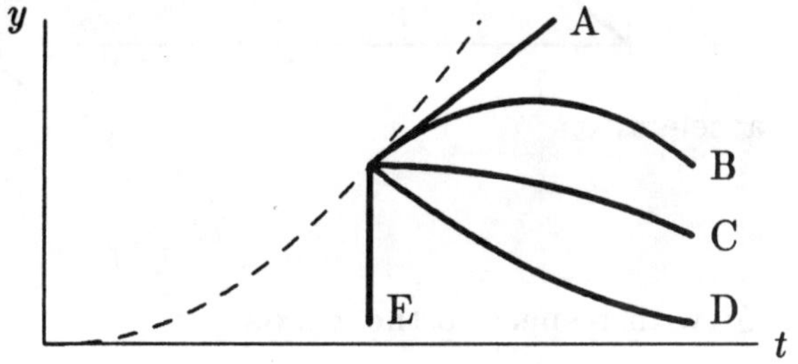

ans: B

70. The diagram shows a velocity-time graph for a car moving in a straight line. At point Q the car must be:

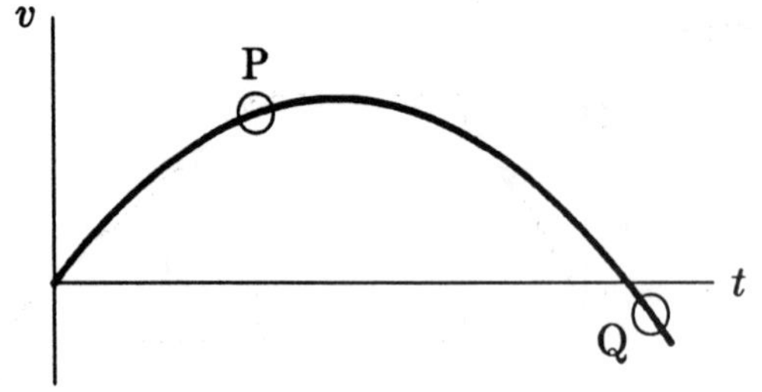

A. moving with zero acceleration
B. traveling downhill
C. traveling below ground-level
D. reducing speed
E. traveling in the reverse direction to that at point P

ans: E

71. The diagram shows a velocity-time graph for a car moving in a straight line. At point P the car must be:

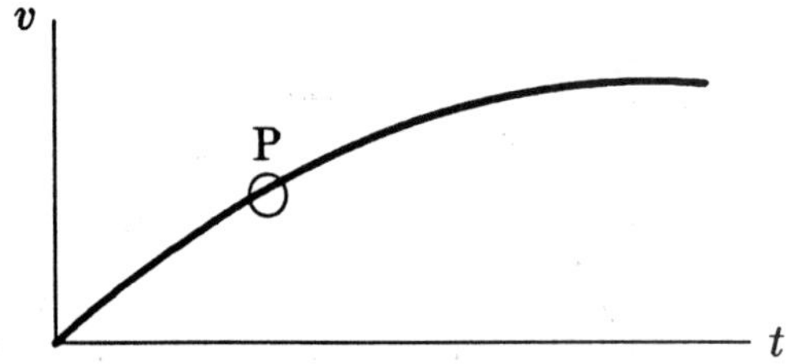

A. moving with zero acceleration
B. climbing the hill
C. accelerating
D. stationary
E. moving at about 45° with respect to the x axis

ans: C

72. The graph represents the straight line motion of a car. How far does the car travel between $t = 2$ s and $t = 5$ s?

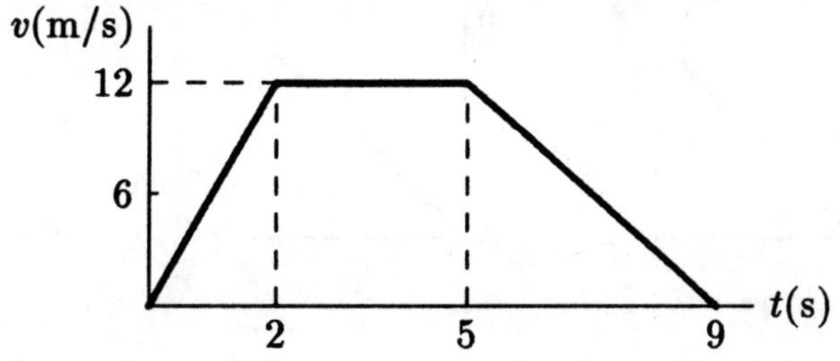

A. 4 m
B. 12 m
C. 24 m
D. 36 m
E. 60 m

ans: D

73. The diagram represents the straight line motion of a car. Which of the following statements is true?

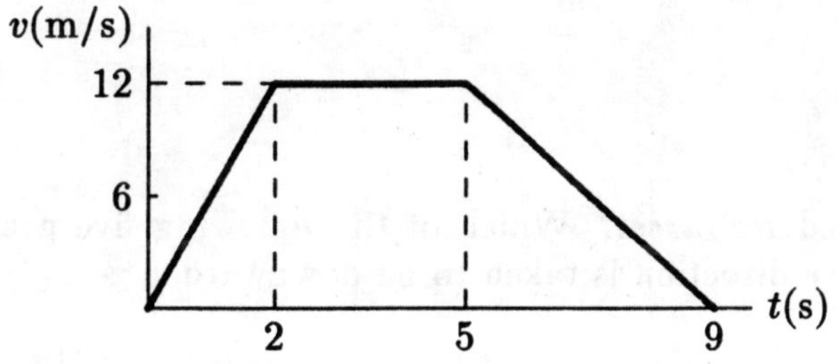

A. The car accelerates, stops, and reverses
B. The car accelerates at 6 m/s^2 for the first 2 s
C. The car is moving for a total time of 12 s
D. The car decelerates at 12 m/s^2 for the last 4 s
E. The car returns to its starting point when $t = 9$ s

ans: B

74. Consider the following five graphs (note the axes carefully). Which of these represents motion at constant speed?

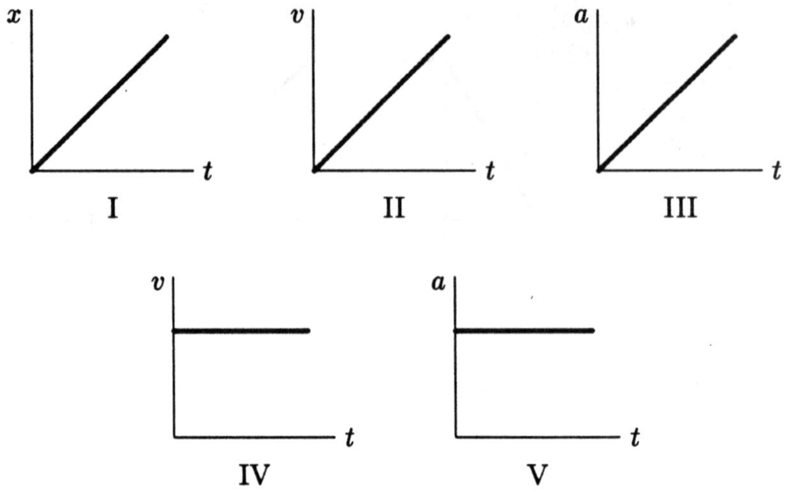

A. IV only
B. IV and V only
C. I; II, and III only
D. I and II only
E. I and IV only

ans: E

75. An object is dropped from rest. Which of the following five graphs correctly represents its motion? The positive direction is taken to be downward.

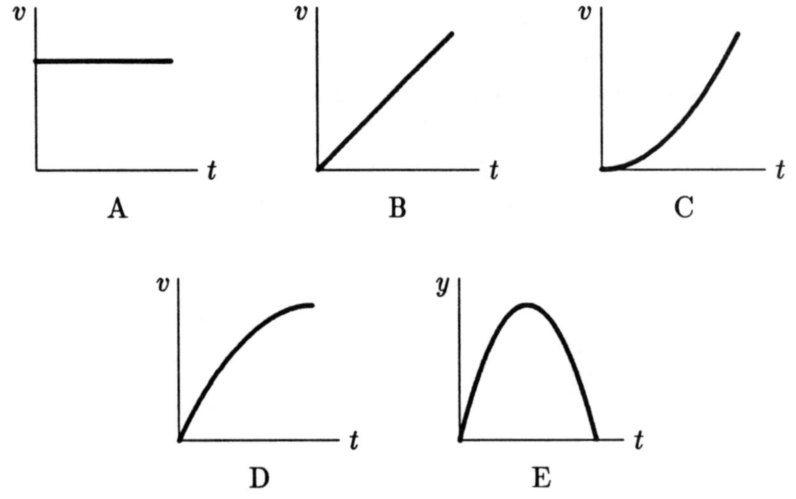

ans: B

76. **A stone is dropped from a cliff. The graph (carefully note the axes) which best represents its motion while it falls is:**

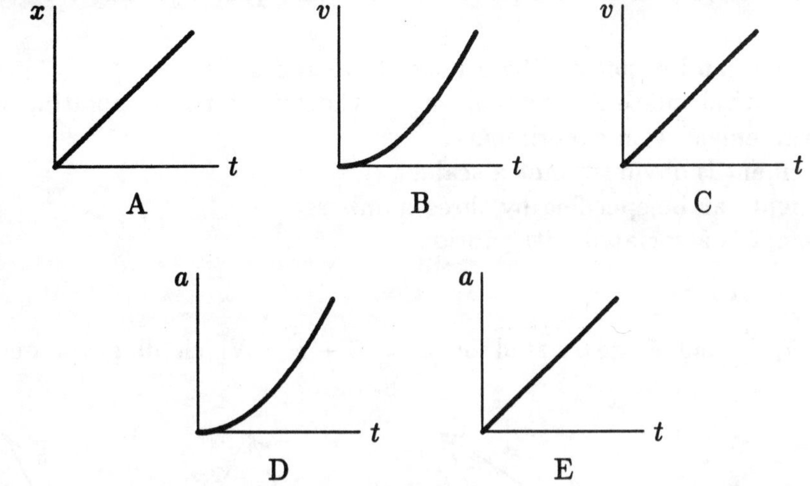

ans: C

77. **An object is thrown vertically into the air. Which of the following five graphs represents the velocity (v) of the object as a function of the time (t)? The positive direction is taken to be upward.**

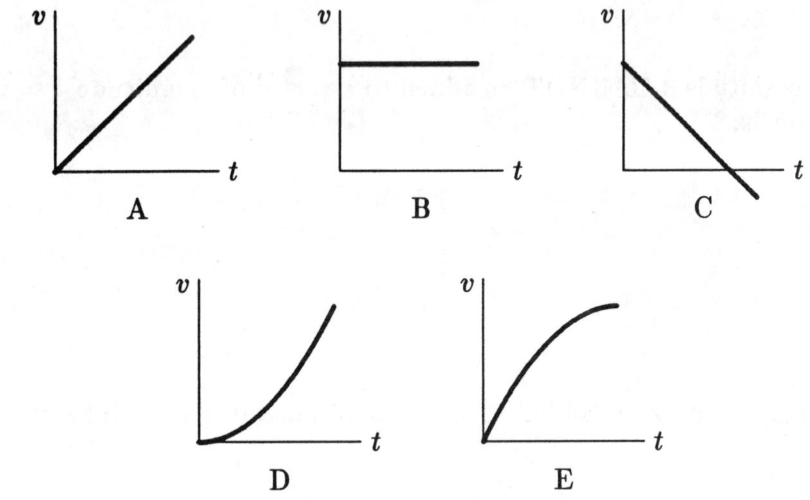

ans: C

Chapter 3: VECTORS

1. We say that the displacement of a particle is a vector quantity. Our best justification for this assertion is:

   A. displacement can be specified by a magnitude and a direction
   B. operating with displacements according to the rules for manipulating vectors leads to results in agreement with experiments
   C. a displacement is obviously not a scalar
   D. displacement can be specified by three numbers
   E. displacement is associated with motion

   ans: B

2. The vectors $\vec{a}$, $\vec{b}$, and $\vec{c}$ are related by $\vec{c} = \vec{a} + \vec{b}$. Which diagram below illustrates this relationship?

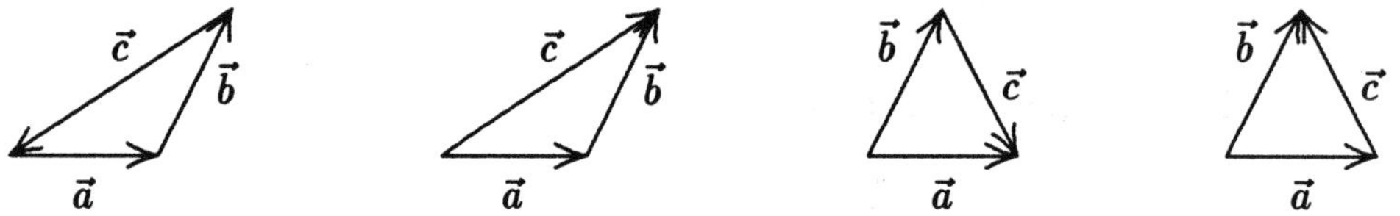

A B C D

   E. None of these

   ans: B

3. A vector of magnitude 3 CANNOT be added to a vector of magnitude 4 so that the magnitude of the resultant is:

   A. zero
   B. 1
   C. 3
   D. 5
   E. 7

   ans: A

4. A vector of magnitude 20 is added to a vector of magnitude 25. The magnitude of this sum might be:

   A. zero
   B. 3
   C. 12
   D. 47
   E. 50

   ans: C

5. A vector $\vec{S}$ of magnitude 6 and another vector $\vec{T}$ have a sum of magnitude 12. The vector $\vec{T}$:

A. must have a magnitude of at least 6 but no more than 18
B. may have a magnitude of 20
C. cannot have a magnitude greater than 12
D. must be perpendicular to $\vec{S}$
E. must be perpendicular to the vector sum

ans: A

6. The vector $-\vec{A}$ is:

A. longer than $\vec{A}$
B. shorter than $\vec{A}$
C. in the same direction as $\vec{A}$
D. in the direction opposite to $\vec{A}$
E. perpendicular to $\vec{A}$

ans: D

7. The vector $\vec{V}_3$ in the diagram is equal to:

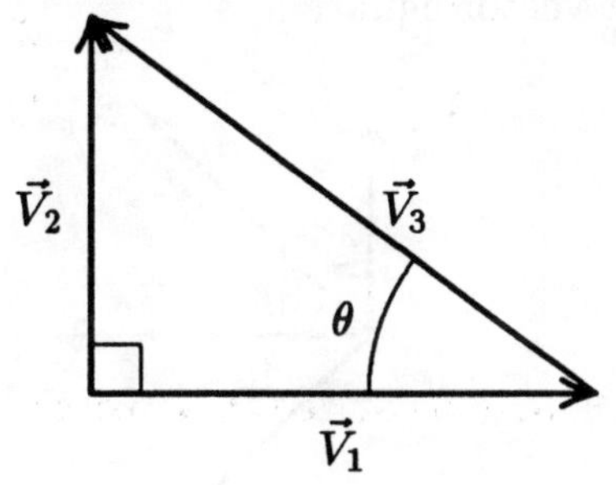

A. $\vec{V}_1 - \vec{V}_2$
B. $\vec{V}_1 + \vec{V}_2$
C. $\vec{V}_2 - \vec{V}_1$
D. $\vec{V}_1 \cos\theta$
E. $\vec{V}_1/(\cos\theta)$

ans: C

8. If $|\vec{A} + \vec{B}|^2 = A^2 + B^2$, then:

A. $\vec{A}$ and $\vec{B}$ must be parallel and in the same direction
B. $\vec{A}$ and $\vec{B}$ must be parallel and in opposite directions
C. either $\vec{A}$ or $\vec{B}$ must be zero
D. the angle between $\vec{A}$ and $\vec{B}$ must be 60°
E. none of the above is true

ans: E

9. If $|\vec{A} + \vec{B}| = A + B$ and neither $\vec{A}$ nor $\vec{B}$ vanish, then:

A. $\vec{A}$ and $\vec{B}$ are parallel and in the same direction
B. $\vec{A}$ and $\vec{B}$ are parallel and in opposite directions
C. the angle between $\vec{A}$ and $\vec{B}$ is 45°
D. the angle between $\vec{A}$ and $\vec{B}$ is 60°
E. $\vec{A}$ is perpendicular to $\vec{B}$

ans: A

10. If $|\vec{A} - \vec{B}| = A + B$ and neither $\vec{A}$ nor $\vec{B}$ vanish, then:

A. $\vec{A}$ and $\vec{B}$ are parallel and in the same direction
B. $\vec{A}$ and $\vec{B}$ are parallel and in opposite directions
C. the angle between $\vec{A}$ and $\vec{B}$ is 45°
D. the angle between $\vec{A}$ and $\vec{B}$ is 60°
E. $\vec{A}$ is perpendicular to $\vec{B}$

ans: B

11. Four vectors ($\vec{A}$, $\vec{B}$, $\vec{C}$, $\vec{D}$) all have the same magnitude. The angle θ between adjacent vectors is 45° as shown. The correct vector equation is:

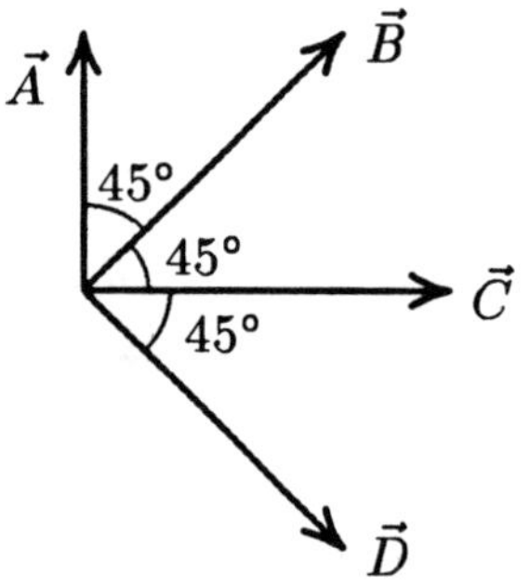

A. $\vec{A} - \vec{B} - \vec{C} + \vec{D} = 0$
B. $\vec{B} + \vec{D} - \sqrt{2}\vec{C} = 0$
C. $\vec{A} + \vec{B} = \vec{B} + \vec{D}$
D. $\vec{A} + \vec{B} + \vec{C} + \vec{D} = 0$
E. $(\vec{A} + \vec{C})/\sqrt{2} = -\vec{B}$

ans: B

12. Vectors $\vec{A}$ and $\vec{B}$ lie in the xy plane. We can deduce that $\vec{A} = \vec{B}$ if:

A. $A_x^2 + A_y^2 = B_x^2 + B_y^2$
B. $A_x + A_y = B_x + B_y$
C. $A_x = B_x$ and $A_y = B_y$
D. $A_y/A_x = B_y/B_x$
E. $A_x = A_y$ and $B_x = B_y$

ans: C

13. A vector has a magnitude of 12. When its tail is at the origin it lies between the positive x axis and the negative y axis and makes an angle of 30° with the x axis. Its y component is:

A. $6/\sqrt{3}$
B. $-6\sqrt{3}$
C. 6
D. -6
E. 12

ans: D

14. If the x component of a vector $\vec{A}$, in the xy plane, is half as large as the magnitude of the vector, its y component is:

A. $A/2$
B. $2A$
C. $3A/4$
D. $\sqrt{3}A/2$
E. $\sqrt{5}A/2$

ans: D

15. If $\vec{A} = (6\,\mathrm{m})\,\hat{\imath} - (8\,\mathrm{m})\,\hat{\jmath}$ then $4\vec{A}$ has magnitude:

A. 10 m
B. 20 m
C. 30 m
D. 40 m
E. 50 m

ans: D

16. A vector has a component of 10 m in the $+x$ direction, a component of 10 m in the $+y$ direction, and a component of 5 m in the $-z$ direction. The magnitude of this vector is:

A. zero
B. 15 m
C. 20 m
D. 25 m
E. 225 m

ans: B

17. Let $\vec{V} = (2.00\,\mathrm{m})\,\hat{\imath} + (6.00\,\mathrm{m})\,\hat{\jmath} - (3.00\,\mathrm{m})\,\hat{k}$. The magnitude of $\vec{V}$ is:

A. 5.00 m
B. 5.57 m
C. 7.00 m
D. 7.42 m
E. 8.54 m

ans: C

18. A vector in the xy plane has a magnitude of 25 m and an x component of 12 m. The angle it makes with the positive x axis is:

A. 26°
B. 29°
C. 61°
D. 64°
E. 241°

ans: C

19. The angle between $\vec{A} = (25\,\text{m})\,\hat{\imath} + (45\,\text{m})\,\hat{\jmath}$ and the positive x axis is:

A. 29°
B. 61°
C. 151°
D. 209°
E. 241°

ans: B

20. The angle between $\vec{A} = (-25\,\text{m})\,\hat{\imath} + (45\,\text{m})\,\hat{\jmath}$ and the positive x axis is:

A. 29°
B. 61°
C. 119°
D. 151°
E. 209°

ans: C

21. Let $\vec{A} = (2\,\text{m})\,\hat{\imath}+(6\,\text{m})\,\hat{\jmath}-(3\,\text{m})\,\hat{k}$ and $\vec{B} = (4\,\text{m})\,\hat{\imath}+(2\,\text{m})\,\hat{\jmath}+(1\,\text{m})\,\hat{k}$. The vector sum $\vec{S} = \vec{A}+\vec{B}$ equals:

A. $(6\,\text{m})\,\hat{\imath} + (8\,\text{m})\,\hat{\jmath} - (2\,\text{m})\,\hat{k}$
B. $(-2\,\text{m})\,\hat{\imath} + (4\,\text{m})\,\hat{\jmath} - (4\,\text{m})\,\hat{k}$
C. $(2\,\text{m})\,\hat{\imath} - (4\,\text{m})\,\hat{\jmath} + (4\,\text{m})\,\hat{k}$
D. $(8\,\text{m})\,\hat{\imath} + (12\,\text{m})\,\hat{\jmath} - (3\,\text{m})\,\hat{k}$
E. none of these

ans: A

22. Let $\vec{A} = (2\,\text{m})\,\hat{\imath} + (6\,\text{m})\,\hat{\jmath} - (3\,\text{m})\,\hat{k}$ and $\vec{B} = (4\,\text{m})\,\hat{\imath} + (2\,\text{m}\,\hat{\jmath} + (1\,\text{m})\,\hat{k}$. The vector difference $\vec{D} = \vec{A} - \vec{B}$ is:

A. $(6\,\text{m})\,\hat{\imath} + (8\,\text{m})\,\hat{\jmath} - (2\,\text{m})\,\hat{k}$
B. $(-2\,\text{m})\,\hat{\imath} + (4\,\text{m})\,\hat{\jmath} - (4\,\text{m})\,\hat{k}$
C. $(2\,\text{m})\,\hat{\imath} - (4\,\text{m})\,\hat{\jmath} + (4\,\text{m})\,\hat{k}$
D. $(8\,\text{m})\,\hat{\imath} + (12\,\text{m})\,\hat{\jmath} - (3\,\text{m})\,\hat{k}$
E. none of these

ans: B

23. If $\vec{A} = (2\,\mathrm{m})\,\hat{\imath} - (3\,\mathrm{m})\,\hat{\jmath}$ and $\vec{B} = (1\,\mathrm{m})\,\hat{\imath} - (2\,\mathrm{m})\,\hat{\jmath}$, then $\vec{A} - 2\vec{B} =$

A. $(1\,\mathrm{m})\,\hat{\jmath}$
B. $(-1\,\mathrm{m})\,\hat{\jmath}$
C. $(4\,\mathrm{m})\,\hat{\imath} - (7\,\mathrm{m})\,\hat{\jmath}$
D. $(4\,\mathrm{m})\,\hat{\imath} + (1\,\mathrm{m})\,\hat{\jmath}$
E. $(-4\,\mathrm{m})\,\hat{\imath} + (7\,\mathrm{m})\,\hat{\jmath}$

ans: A

24. In the diagram, $\vec{A}$ has magnitude 12 m and $\vec{B}$ has magnitude 8 m. The x component of $\vec{A} + \vec{B}$ is about:

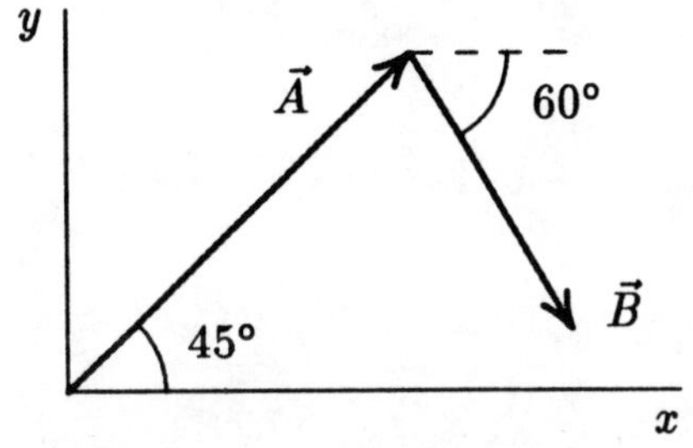

A. 5.5 m
B. 7.6 m
C. 12 m
D. 14 m
E. 15 m

ans: C

25. A certain vector in the xy plane has an x component of 4 m and a y component of 10 m. It is then rotated in the xy plane so its x component is doubled. Its new y component is about:

A. 20 m
B. 7.2 m
C. 5.0 m
D. 4.5 m
E. 2.2 m

ans: B

26. Vectors $\vec{A}$ and $\vec{B}$ each have magnitude L. When drawn with their tails at the same point, the angle between them is 30°. The value of $\vec{A} \cdot \vec{B}$ is:

A. zero
B. L^2
C. $\sqrt{3}L^2/2$
D. $2L^2$
E. none of these

ans: C

27. Let $\vec{A} = (2\,\text{m})\,\hat{\imath} + (6\,\text{m})\,\hat{\jmath} - (3\,\text{m})\,\hat{k}$ and $\vec{B} = (4\,\text{m})\,\hat{\imath} + (2\,\text{m})\,\hat{\jmath} + (1\,\text{m})\,\hat{k}$. Then $\vec{A} \cdot \vec{B} =$

A. $(8\,\text{m})\,\hat{\imath} + (12\,\text{m})\,\hat{\jmath} - (3\,\text{m})\,\hat{k}$
B. $(12\,\text{m})\,\hat{\imath} - (14\,\text{m})\,\hat{\jmath} - (20\,\text{m})\,\hat{k}$
C. $23\,\text{m}^2$
D. $17\,\text{m}^2$
E. none of these

ans: D

28. Two vectors have magnitudes of 10 m and 15 m. The angle between them when they are drawn with their tails at the same point is 65°. The component of the longer vector along the line of the shorter is:

A. 0
B. 4.2 m
C. 6.3 m
D. 9.1 m
E. 14 m

ans: C

29. Let $\vec{S} = (1\,\text{m})\,\hat{\imath} + (2\,\text{m})\,\hat{\jmath} + (2\,\text{m})\,\hat{k}$ and $\vec{T} = (3\,\text{m})\,\hat{\imath} + (4\,\text{m})\,\hat{k}$. The angle between these two vectors is given by:

A. $\cos^{-1}(14/15)$
B. $\cos^{-1}(11/225)$
C. $\cos^{-1}(104/225)$
D. $\cos^{-1}(11/15)$
E. cannot be found since $\vec{S}$ and $\vec{T}$ do not lie in the same plane

ans: D

30. Two vectors lie with their tails at the same point. When the angle between them is increased by 20° their scalar product has the same magnitude but changes from positive to negative. The original angle between them was:

A. 0
B. 60°
C. 70°
D. 80°
E. 90°

ans: D

31. If the magnitude of the sum of two vectors is less than the magnitude of either vector, then:

A. the scalar product of the vectors must be negative
B. the scalar product of the vectors must be positive
C. the vectors must be parallel and in opposite directions
D. the vectors must be parallel and in the same direction
E. none of the above

ans: A

32. If the magnitude of the sum of two vectors is greater than the magnitude of either vector, then:
    A. the scalar product of the vectors must be negative
    B. the scalar product of the vectors must be positive
    C. the vectors must be parallel and in opposite directions
    D. the vectors must be parallel and in the same direction
    E. none of the above

    ans: E

33. Vectors $\vec{A}$ and $\vec{B}$ each have magnitude L. When drawn with their tails at the same point, the angle between them is 30°. The magnitude of $\vec{A} \times \vec{B}$ is:
    A. $L^2/2$
    B. L^2
    C. $\sqrt{3}L^2/2$
    D. $2L^2$
    E. none of these

    ans: A

34. Two vectors lie with their tails at the same point. When the angle between them is increased by 20° the magnitude of their vector product doubles. The original angle between them was about:
    A. 0
    B. 18°
    C. 25°
    D. 45°
    E. 90°

    ans: B

35. Two vectors have magnitudes of 10 m and 15 m. The angle between them when they are drawn with their tails at the same point is 65°. The component of the longer vector along the line perpendicular to the shorter vector, in the plane of the vectors, is:
    A. 0
    B. 4.2 m
    C. 6.3 m
    D. 9.1 m
    E. 14 m

    ans: E

36. The two vectors $(3\,\text{m})\hat{\imath} - (2\,\text{m})\hat{\jmath}$ and $(2\,\text{m})\hat{\imath} + (3\,\text{m})\hat{\jmath} - (2\,\text{m})\hat{k}$ define a plane. It is the plane of the triangle with both tails at one vertex and each head at one of the other vertices. Which of the following vectors is perpendicular to the plane?
    A. $(4\,\text{m})\hat{\imath} + (6\,\text{m})\hat{\jmath} + (13\,\text{m})\hat{k}$
    B. $(-4\,\text{m})\hat{\imath} + (6\,\text{m})\hat{\jmath} + (13\,\text{m})\hat{k}$
    C. $(4\,\text{m})\hat{\imath} - (6\,\text{m})\hat{\jmath} + (13\,\text{m})\hat{k}$
    D. $(4\,\text{m})\hat{\imath} + (6\,\text{m}\hat{\jmath} - (13\,\text{m})\hat{k}$
    E. $(4\,\text{m})\hat{\imath} + (6\,\text{m})\hat{\jmath}$

    ans: A

37. Let $\vec{R} = \vec{S} \times \vec{T}$ and $\theta \neq 90°$, where θ is the angle between $\vec{S}$ and $\vec{T}$ when they are drawn with their tails at the same point. Which of the following is NOT true?

A. $|\vec{R}| = |\vec{S}||\vec{T}| \sin \theta$
B. $-\vec{R} = \vec{T} \times \vec{S}$
C. $\vec{R} \cdot \vec{S} = 0$
D. $\vec{R} \cdot \vec{T} = 0$
E. $\vec{S} \cdot \vec{T} = 0$

ans: E

38. The value of $\hat{\imath} \cdot (\hat{\jmath} \times \hat{k})$ is:

A. zero
B. +1
C. −1
D. 3
E. $\sqrt{3}$

ans: B

39. The value of $\hat{k} \cdot (\hat{k} \times \hat{\imath})$ is:

A. zero
B. +1
C. −1
D. 3
E. $\sqrt{3}$

ans: A

Chapter 4: MOTION IN TWO AND THREE DIMENSIONS

1. Velocity is defined as:
   - A. rate of change of position with time
   - B. position divided by time
   - C. rate of change of acceleration with time
   - D. a speeding up or slowing down
   - E. change of position

   ans: A

2. Acceleration is defined as:
   - A. rate of change of position with time
   - B. speed divided by time
   - C. rate of change of velocity with time
   - D. a speeding up or slowing down
   - E. change of velocity

   ans: C

3. Which of the following is a scalar quantity?
   - A. Speed
   - B. Velocity
   - C. Displacement
   - D. Acceleration
   - E. None of these

   ans: A

4. Which of the following is a vector quantity?
   - A. Mass
   - B. Density
   - C. Speed
   - D. Temperature
   - E. None of these

   ans: E

5. Which of the following is NOT an example of accelerated motion?
   - A. Vertical component of projectile motion
   - B. Circular motion at constant speed
   - C. A swinging pendulum
   - D. Earth's motion about sun
   - E. Horizontal component of projectile motion

   ans: E

6. A particle goes from $x = -2\,\text{m}$, $y = 3\,\text{m}$, $z = 1\,\text{m}$ to $x = 3\,\text{m}$, $y = -1\,\text{m}$, $z = 4\,\text{m}$. Its displacement is:
   A. $(1\,\text{m})\,\hat{\imath} + (2\,\text{m})\,\hat{\jmath} + (5\,\text{m})\,\hat{k}$
   B. $(5\,\text{m})\,\hat{\imath} - (4\,\text{m})\,\hat{\jmath} + (3\,\text{m})\,\hat{k}$
   C. $-(5\,\text{m})\,\hat{\imath} + (4\,\text{m})\,\hat{\jmath} - (3\,\text{m})\,\hat{k}$
   D. $-(1\,\text{m})\,\hat{\imath} - (2\,\text{m})\,\hat{\jmath} - (5\,\text{m})\,\hat{k}$
   E. $-(5\,\text{m})\,\hat{\imath} - (2\,\text{m})\,\hat{\jmath} + (3\,\text{m})\,\hat{k}$

   ans: B

7. A jet plane in straight horizontal flight passes over your head. When it is directly above you, the sound seems to come from a point behind the plane in a direction 30° from the vertical. The speed of the plane is:
   A. the same as the speed of sound
   B. half the speed of sound
   C. three-fifths the speed of sound
   D. 0.866 times the speed of sound
   E. twice the speed of sound

   ans: B

8. A plane traveling north at 200 m/s turns and then travels south at 200 m/s. The change in its velocity is:
   A. zero
   B. 200 m/s north
   C. 200 m/s south
   D. 400 m/s north
   E. 400 m/s south

   ans: E

9. Two bodies are falling with negligible air resistance, side by side, above a horizontal plane. If one of the bodies is given an additional horizontal acceleration during its descent, it:
   A. strikes the plane at the same time as the other body
   B. has the vertical component of its velocity altered
   C. has the vertical component of its acceleration altered
   D. follows a hyperbolic path
   E. follows a straight line path along the resultant acceleration vector

   ans: A

10. The velocity of a projectile equals its initial velocity added to:
   A. a constant horizontal velocity
   B. a constant vertical velocity
   C. a constantly increasing horizontal velocity
   D. a constantly increasing downward velocity
   E. a constant velocity directed at the target

   ans: D

11. A stone thrown from the top of a tall building follows a path that is:

A. circular
B. made of two straight line segments
C. hyperbolic
D. parabolic
E. a straight line

ans: D

12. Identical guns fire identical bullets horizontally at the same speed from the same height above level planes, one on the Earth and one on the Moon. Which of the following three statements is/are true?

I. The horizontal distance traveled by the bullet is greater for the Moon.
II. The flight time is less for the bullet on the Earth.
III. The velocity of the bullets at impact are the same.

A. III only
B. I and II only
C. I and III only
D. II and III only
E. I, II, III

ans: B

13. A stone is thrown horizontally and follows the path XYZ shown. The direction of the acceleration of the stone at point Y is:

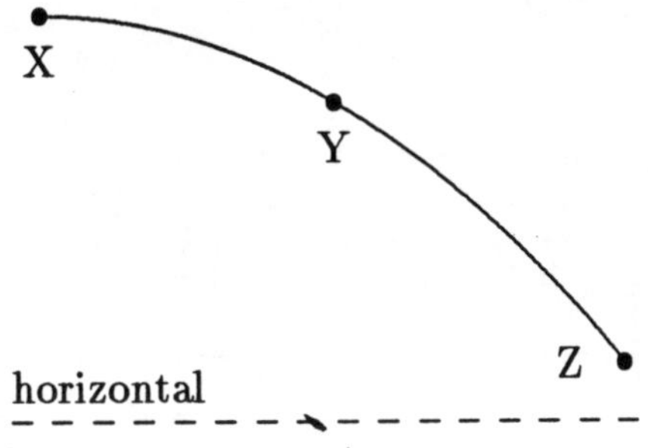

A. ↓
B. →
C. ↘
D. ↙
E. ↗

ans: A

14. A bullet shot horizontally from a gun:
   A. strikes the ground much later than one dropped vertically from the same point at the same instant
   B. never strikes the ground
   C. strikes the ground at approximately the same time as one dropped vertically from the same point at the same instant
   D. travels in a straight line
   E. strikes the ground much sooner than one dropped from the same point at the same instant

   ans: C

15. A bomber flying in level flight must release its bomb before it is over the target. Neglecting air resistance, which one of the following is NOT true?
   A. The bomber will be over the target when the bomb strikes
   B. The acceleration of the bomb is constant
   C. The horizontal velocity of the plane equals the vertical velocity of the bomb when it hits the target
   D. The bomb travels in a curved path
   E. The time of flight of the bomb is independent of the horizontal speed of the plane

   ans: C

16. The airplane shown is in level flight at an altitude of 0.50 km and a speed of 150 km/h. At what distance d should it release a heavy bomb to hit the target X? Take $g = 10\,\mathrm{m/s^2}$.

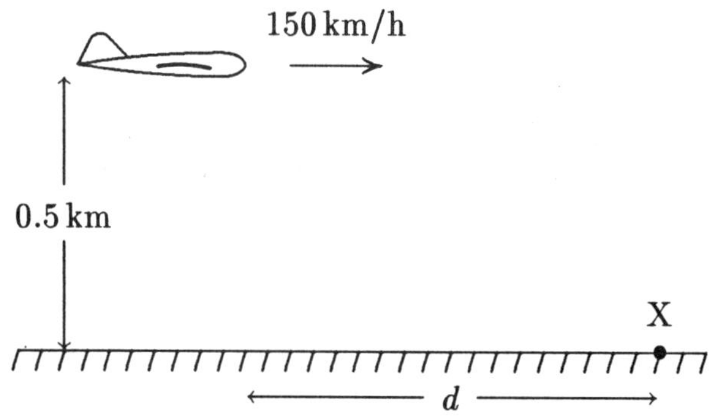

   A. 150 m
   B. 295 m
   C. 420 m
   D. 2550 m
   E. 15,000 m

   ans: C

17. An object is shot from the back of a truck moving at 40 km/h on a straight horizontal road. The launcher is aimed upward, perpendicular to the bed of the truck. The object falls:

A. in front of the truck
B. behind the truck
C. on the truck
D. depends on the initial speed of the object
E. depends on the value of g

ans: C

18. A stone is thrown horizontally from the top of a 20-m high hill. It strikes the ground at an angle of 45°. With what speed was it thrown?

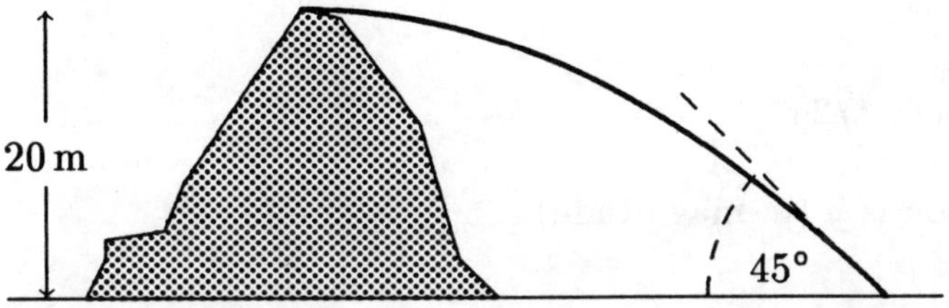

A. 14 m/s
B. 20 m/s
C. 28 m/s
D. 32 m/s
E. 40 m/s

ans: B

19. A stone is thrown outward from the top of a 59.4-m high cliff with an upward velocity component of 19.5 m/s. How many seconds will the stone be in the air?

A. 4
B. 5
C. 6
D. 7
E. 8

ans: C

20. A large cannon is fired from ground level over level ground at an angle of 30° above the horizontal. The muzzle speed is 980 m/s. Neglecting air resistance, the projectile will travel what horizontal distance before striking the ground?

A. 4300 m
B. 8500 m
C. 43,000 m
D. 85,000 m
E. 170,000 m

ans: D

21. A projectile has an initial velocity $\vec{v}_0$ at an angle θ_0 above the horizontal. It reaches the highest point P of its trajectory in a time T after launch. The highest point is a vertical distance h_{max} and a horizontal distance d from the firing point and the speed of the projectile at the highest point is v. All of the following statements are true except:

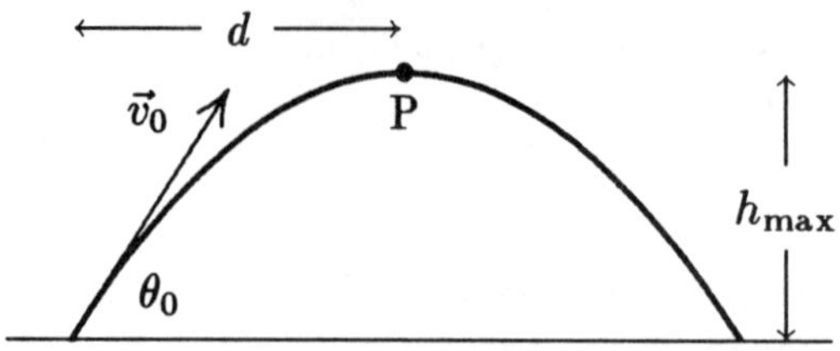

A. $T = (v_0 \cos\theta_0)/g$
B. $d = v_0 T \cos\theta_0$
C. $h_{max} = (v_0 \sin\theta_0)^2/2g$
D. $v = v_0 \cos\theta_0$
E. the acceleration is g (in magnitude)

ans: A

22. A boy on the edge of a vertical cliff 20 m high throws a stone horizontally outwards with a speed of 20 m/s. It strikes the ground at what horizontal distance from the foot of the cliff? Use $g = 10\,\mathrm{m/s^2}$.

A. 10 m
B. 40 m
C. 50 m
D. $50\sqrt{5}$ m
E. none of these

ans: B

23. Which of the curves on the graph below best represents the vertical component v_y of the velocity versus the time t for a projectile fired at an angle of 45° above the horizontal?

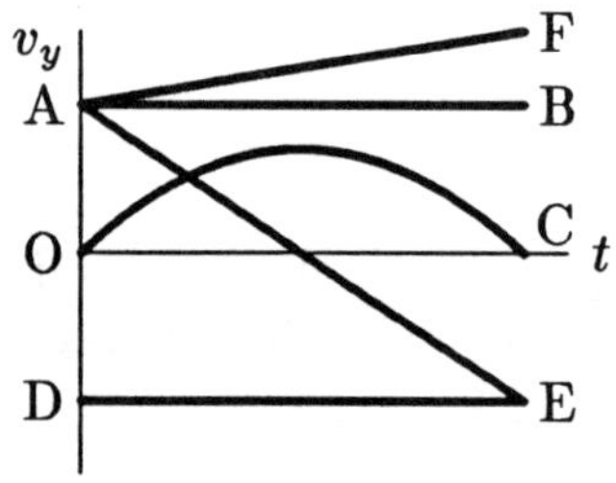

A. OC
B. DE
C. AB
D. AE
E. AF

ans: D

24. A cannon fires a projectile as shown. The dashed line shows the trajectory in the absence of gravity; points MNOP correspond to one second intervals. Using $g = 10\,\mathrm{m/s^2}$, the lengths X,Y,Z are:

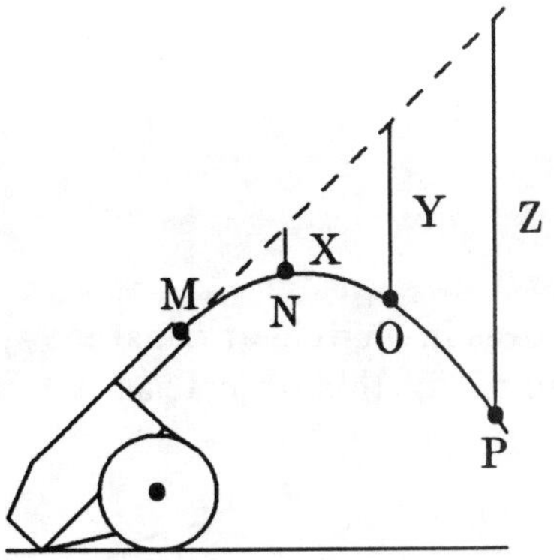

A. 5 m, 10 m, 15 m
B. 5 m, 20 m, 45 m
C. 10 m, 40 m, 90 m
D. 10 m, 20 m, 30 m
E. 0.2 m, 0.8 m, 1.8 m

ans: B

25. A dart is thrown horizontally toward X at 20 m/s as shown. It hits Y 0.1 s later. The distance XY is:

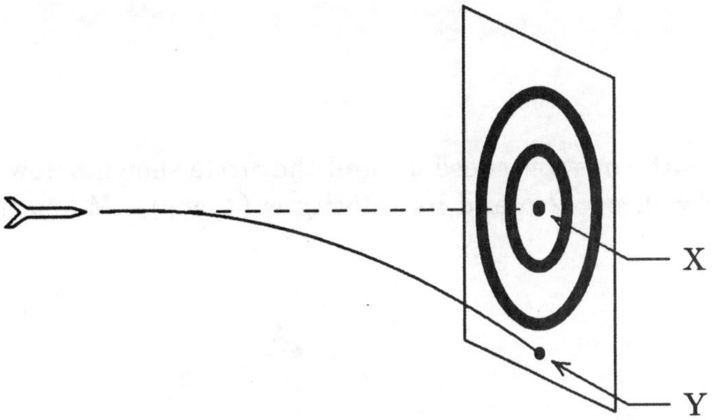

A. 2 m
B. 1 m
C. 0.5 m
D. 0.1 m
E. 0.05 m

ans: E

26. A projectile is fired from ground level over level ground with an initial velocity that has a vertical component of 20 m/s and a horizontal component of 30 m/s. Using $g = 10\,\mathrm{m/s^2}$, the distance from launching to landing points is:

A. 40 m
B. 60 m
C. 80 m
D. 120 m
E. 180 m

ans: D

27. An object, tied to a string, moves in a circle at constant speed on a horizontal surface as shown. The direction of the displacement of this object, as it travels from W to X is:

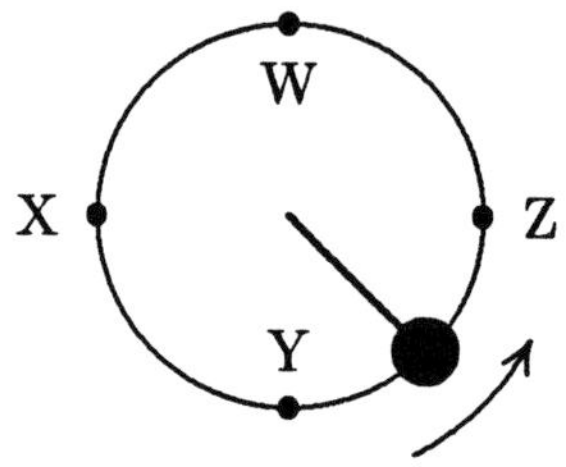

A. ←
B. ↓
C. ↑
D. ↗
E. ↙

ans: E

28. A particle moves with constant speed around the circle shown below. When it is at point A its coordinates are $x = 0$, $y = 2\,\mathrm{m}$ and its velocity is $(4\,\mathrm{m/s})\,\hat{\imath}$. When it is at point B its velocity and acceleration are:

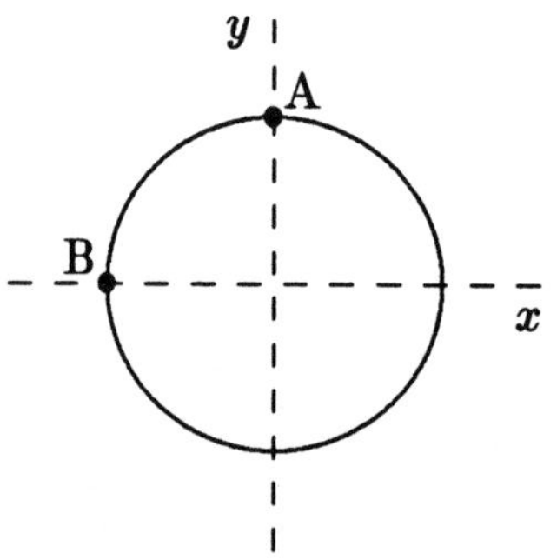

A. $-(4\,\mathrm{m/s})\,\hat{\jmath}$ and $(8\,\mathrm{m/s^2})\,\hat{\imath}$, respectively
B. $(4\,\mathrm{m/s})\,\hat{\imath}$ and $-(8\,\mathrm{m/s^2})\,\hat{\imath}$, respectively
C. $(4\,\mathrm{m/s})\,\hat{\jmath}$ and $(8\,\mathrm{m/s^2})\,\hat{\imath}$, respectively
D. $(4\,\mathrm{m/s})\,\hat{\imath}$ and $(2\,\mathrm{m/s^2})\,\hat{\jmath}$, respectively
E. $(4\,\mathrm{m/s})\,\hat{\jmath}$ and 0, respectively

ans: C

29. An airplane makes a gradual 90° turn while flying at a constant speed of 200 m/s. The process takes 20.0 seconds to complete. For this turn the magnitude of the average acceleration of the plane is:

A. zero
B. $40\,m/s^2$
C. $20\,m/s^2$
D. $14\,m/s^2$
E. $10\,m/s^2$

ans: D

30. An airplane is flying north at 500 km/h. It makes a gradual 180° turn at constant speed, changing its direction of travel from north through east to south. The process takes 40 s. The average acceleration of the plane for this turn (in km/h·s) is:

A. 12.5 (north)
B. 12.5 (east)
C. 12.5 (south)
D. 25 (north)
E. 25 (south)

ans: E

31. An object is moving on a circular path of radius π meters at a constant speed of 4.0 m/s. The time required for one revolution is:

A. $2/\pi^2$ s
B. $\pi^2/2$ s
C. $\pi/2$ s
D. $\pi^2/4$
E. $2/\pi$ s

ans: B

32. A particle moves at constant speed in a circular path. The instantaneous velocity and instantaneous acceleration vectors are:

A. both tangent to the circular path
B. both perpendicular to the circular path
C. perpendicular to each other
D. opposite to each other
E. none of the above

ans: C

33. A stone is tied to a string and whirled at constant speed in a horizontal circle. The speed is then doubled without changing the length of the string. Afterward, the magnitude of the acceleration of the stone is:

A. the same
B. twice as great
C. four times as great
D. half as great
E. one-fourth as great

ans: C

34. Two objects are traveling around different circular orbits with constant speed. They both have the same acceleration but object A is traveling twice as fast as object B. The orbit radius for object A is ______ the orbit radius for object B.

A. one-fourth
B. one-half
C. the same as
D. twice
E. four times

ans: E

35. A stone is tied to a 0.50-m string and whirled at a constant speed of 4.0 m/s in a vertical circle. Its acceleration in m/s^2 at the top of the circle is:

A. 9.8, up
B. 9.8, down
C. 8.0, down
D. 32, up
E. 32, down

ans: E

36. A stone is tied to a 0.50-m string and whirled at a constant speed of 4.0 m/s in a vertical circle. The magnitude of its acceleration in m/s^2 at the bottom of the circle is:

A. 9.8, up
B. 9.8, down
C. 8.0, up
D. 32, up
E. 32, down

ans: D

37. A car rounds a 20-m radius curve at 10 m/s. The magnitude of its acceleration in m/s^2 is:

A. 0
B. 0.20
C. 5.0
D. 40
E. 400

ans: C

38. For a biological sample in a 1.0-m radius centrifuge to have a centripetal acceleration of $25g$ its speed in m/s must be:

A. 11
B. 16
C. 50
D. 122
E. 245

ans: B

39. A girl on a merry-go-round moves horizontally in a circle at constant speed. She travels one fourth of a revolution, a distance of 25 m along the circumference of the circle, in 5.0 s. The magnitude of her acceleration is:

A. $0.31\,\text{m/s}^2$
B. $1.3\,\text{m/s}^2$
C. $1.6\,\text{m/s}^2$
D. $3.9\,\text{m/s}^2$
E. $6.3\,\text{m/s}^2$

ans: C

40. A stone is tied to the end of a string and is swung with constant speed around a horizontal circle with a radius of 1.5 m. If it makes two complete revolutions each second, its acceleration is:

A. $0.24\,\text{m/s}^2$
B. $2.4\,\text{m/s}^2$
C. $24\,\text{m/s}^2$
D. $240\,\text{m/s}^2$
E. $2400\,\text{m/s}^2$

ans: D

41. A Ferris wheel with a radius of 8.0 m makes 1 revolution every 10 s. When a passenger is at the top, essentially a diameter above the ground, he releases a ball. How far from the point on the ground directly under the release point does the ball land?

A. 0
B. 1.0 m
C. 8.0 m
D. 9.1 m
E. 16 m

ans: D

42. A boat is able to move through still water at 20 m/s. It makes a round trip to a town 3.0 km downstream. If the river flows at 5 m/s, the time required for this round trip is:

A. 120 s
B. 150 s
C. 200 s
D. 300 s
E. 320 s

ans: E

43. A boat is traveling upstream at 14 km/h with respect to a river that is flowing at 6 km/h (with respect to the ground). A man runs directly across the boat, from one side to the other, at 6 km/h (with respect to the boat). The speed of the man with respect to the ground is:

A. 10 km/h
B. 14 km/h
C. 18.5 km/h
D. 21 km/h
E. 26 km/h

ans: A

44. A ferry boat is sailing at 12 km/h 30° W of N with respect to a river that is flowing at 6.0 km/h E. As observed from the shore, the ferry boat is sailing:

A. 30° E of N
B. due N
C. 30° W of N
D. 45° E of N
E. none of these

ans: B

45. A boy wishes to row across a river in the shortest possible time. He can row at 2 m/s in still water and the river is flowing at 1 m/s. At what angle θ should he point the bow (front) of his boat?

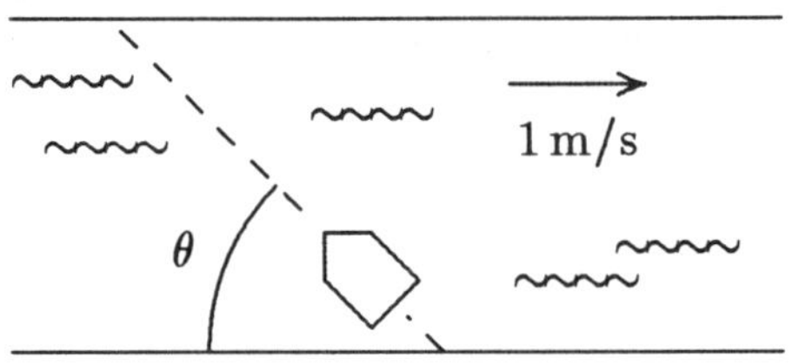

A. 30°
B. 45°
C. 60°
D. 63°
E. 90°

ans: E

46. A girl wishes to row across a river to a point directly opposite as shown. She can row at 2 m/s in still water and the river is flowing at 1 m/s. At what angle with respect to the line joining the starting and finishing points should she point the bow (front) of her boat?

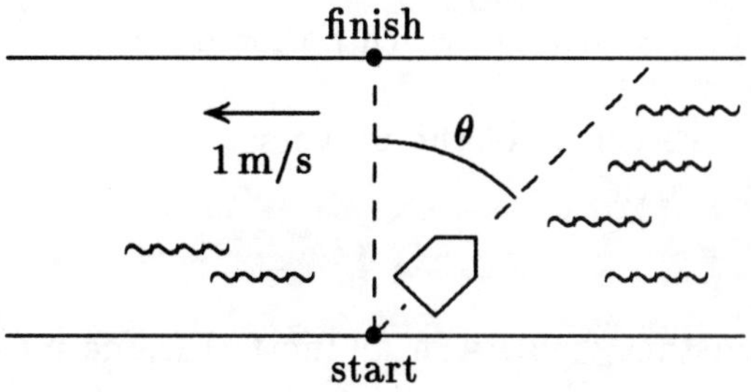

A. 30°
B. 45°
C. 60°
D. 63°
E. 90°

ans: A

47. A motor boat can travel at 10 km/h in still water. A river flows at 5 km/h west. A boater wishes to cross from the south bank to a point directly opposite on the north bank. At what angle must the boat be headed?

A. 27° E of N
B. 30° E of N
C. 45° E of N
D. 60° E of N
E. depends on the width of the river

ans: B

48. Two projectiles are in flight at the same time. The acceleration of one relative to the other:

A. is always $9.8\,\mathrm{m/s^2}$
B. can be as large as $19.8\,\mathrm{m/s^2}$
C. can be horizontal
D. is zero
E. none of these

ans: D

Chapter 5: FORCE AND MOTION — I

1. An example of an inertial reference frame is:
   A. any frame that is not accelerating
   B. a frame attached to a particle on which no forces act
   C. any frame that is at rest
   D. a frame attached to the center of the universe
   E. a frame attached to Earth

   ans: B

2. An object moving at constant velocity in an inertial frame must:
   A. have a net force acting on it
   B. eventually stop due to gravity
   C. not have any force of gravity acting on it
   D. have zero net force acting on it
   E. have no frictional force acting on it

   ans: D

3. In principle, a force is measured by measuring the ______, when the force is applied to it.
   A. velocity of the standard kilogram
   B. speed of the standard kilogram
   C. velocity of any object
   D. acceleration of the standard kilogram
   E. acceleration of any object

   ans: D

4. Which of the following quantities is NOT a vector?
   A. Mass
   B. Displacement
   C. Weight
   D. Acceleration
   E. Force

   ans: A

5. A newton is the force:
   A. of gravity on a 1 kg body
   B. of gravity on a 1 g body
   C. that gives a 1 g body an acceleration of $1\,\mathrm{cm/s^2}$
   D. that gives a 1 kg body an acceleration of $1\,\mathrm{m/s^2}$
   E. that gives a 1 kg body an acceleration of $9.8\,\mathrm{m/s^2}$

   ans: D

6. The unit of force called the newton is:
   A. $9.8\,\mathrm{kg \cdot m/s^2}$
   B. $1\,\mathrm{kg \cdot m/s^2}$
   C. defined by means of Newton's third law
   D. 1 kg of mass
   E. 1 kg of force

   ans: B

7. A force of 1 N is:
   A. $1\,\mathrm{kg/s}$
   B. $1\,\mathrm{kg \cdot m/s}$
   C. $1\,\mathrm{kg \cdot m/s^2}$
   D. $1\,\mathrm{kg \cdot m^2/s}$
   E. $1\,\mathrm{kg \cdot m^2/s^2}$

   ans: C

8. The standard 1-kg mass is attached to a compressed spring and the spring is released. If the mass initially has an acceleration of $5.6\,\mathrm{m/s^2}$, the force of the spring has a magnitude of:
   A. 2.8 N
   B. 5.6 N
   C. 11.2 N
   D. 0
   E. an undetermined amount

   ans: B

9. Acceleration is always in the direction:
   A. of the displacement
   B. of the initial velocity
   C. of the final velocity
   D. of the net force
   E. opposite to the frictional force

   ans: D

10. The term "mass" refers to the same physical concept as:
    A. weight
    B. inertia
    C. force
    D. acceleration
    C. volume

    ans: B

11. The inertia of a body tends to cause the body to:
   A. speed up
   B. slow down
   C. resist any change in its motion
   D. fall toward Earth
   E. decelerate due to friction

   ans: C

12. A heavy ball is suspended as shown. A quick jerk on the lower string will break that string but a slow pull on the lower string will break the upper string. The first result occurs because:

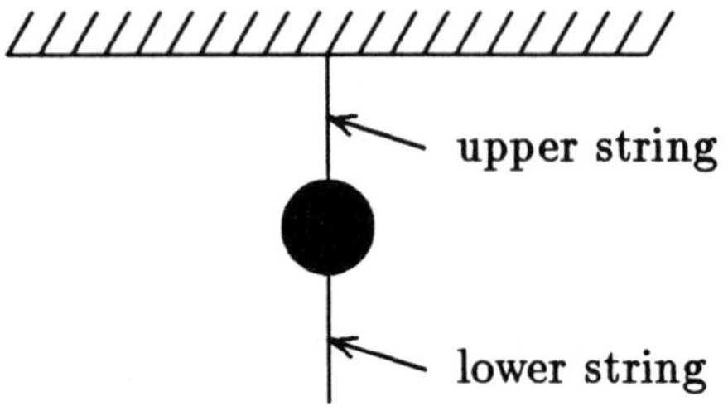

   A. the force is too small to move the ball
   B. action and reaction
   C. the ball has inertia
   D. air friction holds the ball back
   E. the ball has too much energy

   ans: C

13. When a certain force is applied to the 1-kg standard mass its acceleration is $5.0\,\mathrm{m/s^2}$. When the same force is applied to another object its acceleration is one-fifth as much. The mass of the object is:
   A. 0.2 kg
   B. 0.5 kg
   C. 1.0 kg
   D. 5.0 kg
   E. 10 kg

   ans: D

14. Mass differs from weight in that:
   A. all objects have weight but some lack mass
   B. weight is a force and mass is not
   C. the mass of an object is always more than its weight
   D. mass can be expressed only in the metric system
   E. there is no difference

   ans: B

15. The mass of a body:
   A. is slightly different at different places on Earth
   B. is a vector
   C. is independent of the free-fall acceleration
   D. is the same for all bodies of the same volume
   E. can be measured most accurately on a spring scale

   ans: C

16. The mass and weight of a body:
   A. differ by a factor of 9.8
   B. are identical
   C. are the same physical quantities expressed in different units
   D. are both a direct measure of the inertia of the body
   E. have the same ratio as that of any other body placed at that location

   ans: E

17. An object placed on an equal-arm balance requires 12 kg to balance it. When placed on a spring scale, the scale reads 12 kg. Everything (balance, scale, set of masses and object) is now transported to the moon where the free-fall acceleration is one-sixth that on Earth. The new readings of the balance and spring scale (respectively) are:
   A. 12 kg, 12 kg
   B. 2 kg, 2 kg
   C. 12 kg, 2 kg
   D. 2 kg, 12 kg
   E. 12 kg, 72 kg

   ans: C

18. Two objects, one having three times the mass of the other, are dropped from the same height in a vacuum. At the end of their fall, their velocities are equal because:
   A. anything falling in vacuum has constant velocity
   B. all objects reach the same terminal velocity
   C. the acceleration of the larger object is three times greater than that of the smaller object
   D. the force of gravity is the same for both objects
   E. none of the above

   ans: E

19. A feather and a lead ball are dropped from rest in vacuum on the moon. The acceleration of the feather is:
   A. more than that of the lead ball
   B. the same as that of the lead ball
   C. less than that of the lead ball
   D. $9.8\,\mathrm{m/s^2}$
   E. zero since it floats in a vacuum

   ans: B

20. The block shown moves with constant velocity on a horizontal surface. Two of the forces acting on the block are shown. A frictional force exerted by the surface is the only other horizontal force acting on the block. The frictional force is:

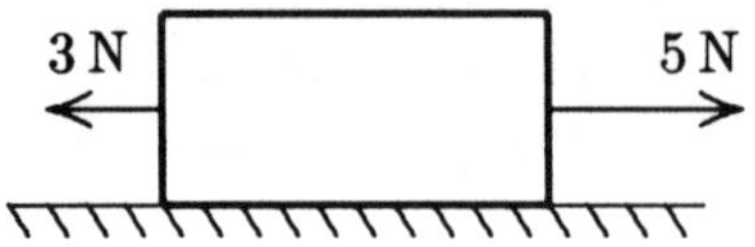

A. 0
B. 2 N, leftward
C. 2 N, rightward
D. slightly more than 2 N, leftward
E. slightly less than 2 N, leftward

ans: B

21. Two forces, one with a magnitude of 3 N and the other with a magnitude of 5 N, are applied to an object. For which orientations of the forces shown in the diagrams is the magnitude of the acceleration of the object the least?

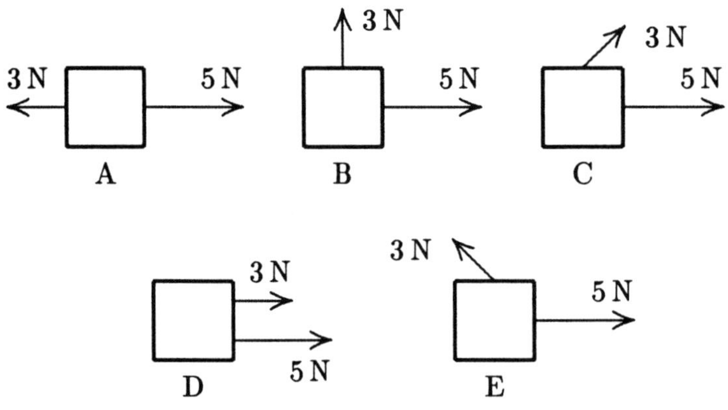

ans: A

22. A crate rests on a horizontal surface and a woman pulls on it with a 10-N force. Rank the situations shown below according to the magnitude of the normal force exerted by the surface on the crate, least to greatest.

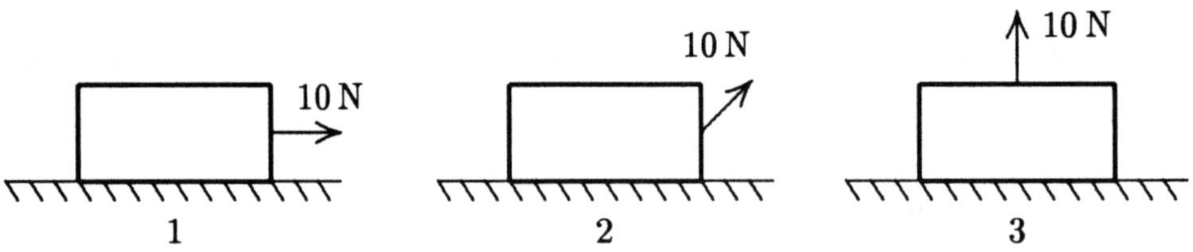

A. 1, 2, 3
B. 2, 1, 3
C. 2, 3, 1
D. 1, 3, 2
E. 3, 2, 1

ans: E

23. A heavy wooden block is dragged by a force $\vec{F}$ along a rough steel plate, as shown in the diagrams for two cases. The magnitude of the applied force $\vec{F}$ is the same for both cases. The normal force in (ii), as compared with the normal force in (i) is:

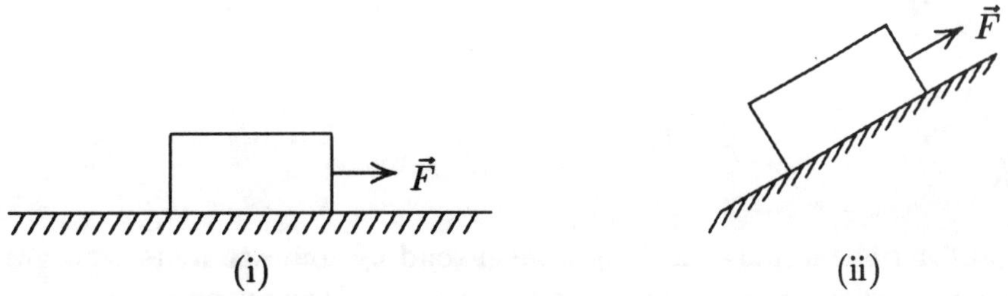

A. the same
B. greater
C. less
D. less for some angles of the incline and greater for others
E. less or greater, depending on the magnitude of the applied force $\vec{F}$.

ans: C

24. Equal forces $\vec{F}$ act on isolated bodies A and B. The mass of B is three times that of A. The magnitude of the acceleration of A is:

A. three times that of B
B. 1/3 that of B
C. the same as B
D. nine times that of B
E. 1/9 that of B

ans: A

25. A car travels east at constant velocity. The net force on the car is:

A. east
B. west
C. up
D. down
E. zero

ans: E

26. A constant force of 8.0 N is exerted for 4.0 s on a 16-kg object initially at rest. The change in speed of this object will be:

A. 0.5 m/s
B. 2 m/s
C. 4 m/s
D. 8 m/s
E. 32 m/s

ans: B

27. A 6-kg object is moving south. A net force of 12 N north acting on it will result in the object having an acceleration of:
   A. $2\,m/s^2$, north
   B. $2\,m/s^2$, south
   C. $6\,m/s^2$, north
   D. $18\,m/s^2$, north
   E. $18\,m/s^2$, south

   ans: A

28. A 9000-N automobile is pushed along a level road by four students who apply a total forward force of 500 N. Neglecting friction, the acceleration of the automobile is:
   A. $0.055\,m/s^2$
   B. $0.54\,m/s^2$
   C. $1.8\,m/s^2$
   D. $9.8\,m/s^2$
   E. $18\,m/s^2$

   ans: B

29. An object rests on a horizontal frictionless surface. A horizontal force of magnitude F is applied. This force produces an acceleration:
   A. only if F is larger than the weight of the object
   B. only while the mass suddenly changes from rest to motion
   C. always
   D. only if the inertia of the object decreases
   E. only if F is increasing

   ans: C

30. A 25-kg chair is pushed across a frictionless horizontal floor with a force of 20 N, directed 20° below the horizontal. The acceleration of the chair is:
   A. $0.27\,m/s^2$
   B. $0.75\,m/s^2$
   C. $0.80\,m/s^2$
   D. $170\,m/s^2$
   E. $470\,m/s^2$

   ans: B

31. A 1.5-N ball is thrown at an angle of 30° above the horizontal with an initial speed of 12 m/s. At its highest point, the net force on the ball is:
   A. 9.8 N, 30° below horizontal
   B. zero
   C. 9.8 N, up
   D. 9.8 N, down
   E. 1.5 N, down

   ans: E

32. Two forces are applied to a 5.0-kg object; one is 6.0 N to the north and the other is 8.0 N to the west. The magnitude of the acceleration of the object is:

A. $0.50\,\mathrm{m/s^2}$
B. $2.0\,\mathrm{m/s^2}$
C. $2.8\,\mathrm{m/s^2}$
D. $10\,\mathrm{m/s^2}$
E. $50\,\mathrm{m/s^2}$

ans: B

33. In a tug-of-war, two men each pull on the rope with 400 N forces, in opposite directions. The tension in the rope is:

A. 400 N
B. 800 N
C. zero
D. 200 N
E. 560 N

ans: A

34. A heavy steel ball B is suspended by a cord from a block of wood W. The entire system is dropped through the air. Neglecting air resistance, the tension in the cord is:

A. zero
B. the difference in the masses of B and W
C. the difference in the weights of B and W
D. the weight of B
E. none of these

ans: A

35. A circus performer of weight W is walking along a "high wire" as shown. The tension in the wire:

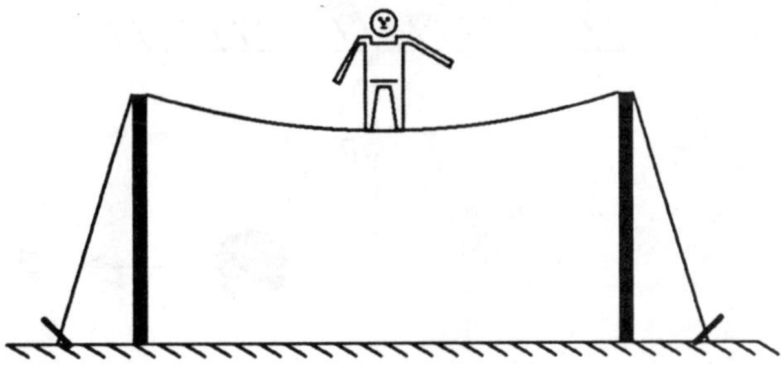

A. is approximately W
B. is approximately $W/2$
C. is much less than W
D. is much more than W
E. depends on whether he stands on one foot or two feet

ans: D

36. A 1000-kg elevator is rising and its speed is increasing at $3\,\text{m/s}^2$. The tension force of the cable on the elevator is:

A. 6800 N
B. 1000 N
C. 3000 N
D. 9800 N
E. 12800 N

ans: E

37. A 700-kg elevator accelerates downward at $3.0\,\text{m/s}^2$. The tension force of the cable on the elevator is:

A. 2.1 kN, up
B. 2.1 kN, down
C. 4.8 kN, up
D. 4.8 kN, down
E. 9.0 kN, up

ans: C

38. A crane operator lowers a 16,000-N steel ball with a downward acceleration of $3\,\text{m/s}^2$. The tension force of the cable is:

A. 4900 N
B. 11,000 N
C. 16,000 N
D. 21,000 N
E. 48,000 N

ans: B

39. A 1-N pendulum bob is held at an angle θ from the vertical by a 2-N horizontal force F as shown. The tension in the string supporting the pendulum bob (in newtons) is:

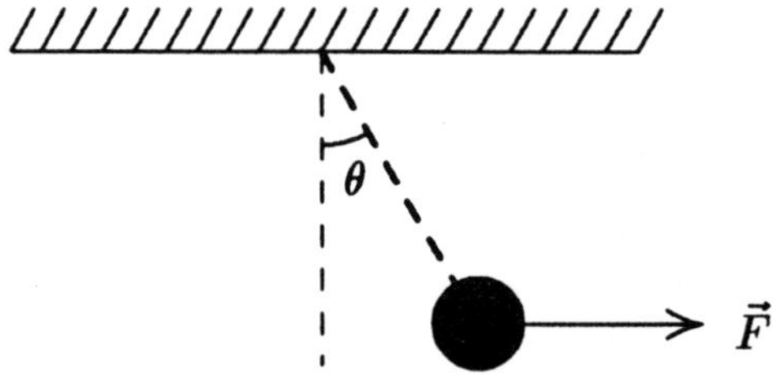

A. $\cos\theta$
B. $2/\cos\theta$
C. $\sqrt{5}$
D. 1
E. none of these

ans: C

40. A car moves horizontally with a constant acceleration of $3\,m/s^2$. A ball is suspended by a string from the ceiling of the car. The ball does not swing, being at rest with respect to the car. What angle does the string make with the vertical?

A. 17°
B. 35°
C. 52°
D. 73°
E. Cannot be found without knowing string length

ans: A

41. A man weighing 700 Nb is in an elevator that is accelerating upward at $4\,m/s^2$. The force exerted on him by the elevator floor is:

A. 71 N
B. 290 N
C. 410 N
D. 700 N
E. 990 N

ans: E

42. A physics textbook is suspended on a spring scale in an elevator. Of the following, the scale shows the highest reading when the elevator:

A. moves upward with increasing speed
B. moves upward with decreasing speed
C. remains stationary
D. moves downward with increasing speed
E. moves downward at constant speed

ans: A

43. A physics textbook is suspended on a spring scale in an elevator. Of the following, the scale shows the highest reading when the elevator:

A. moves downward with increasing speed
B. moves downward with decreasing speed
C. remains stationary
D. moves upward with decreasing speed
E. moves upward at constant speed

ans: B

44. A 25-kg chair is pushed across a frictionless horizontal floor with a force of 200 N, directed 20° below the horizontal. The magnitude of the normal force of the floor on the chair is:

A. 25 N
B. 68 N
C. 180 N
D. 250 N
E. 310 N

ans: E

45. A block slides down a frictionless plane that makes an angle of 30° with the horizontal. The acceleration of the block (in cm/s^2) is:

A. 980
B. 566
C. 849
D. zero
E. 490

ans: E

46. A 25-N crate slides down a frictionless incline that is 25° above the horizontal. The magnitude of the normal force of the incline on the crate is:

A. 11 N
B. 23 N
C. 25 N
D. 100 N
E. 220 N

ans: B

47. A 25-N crate is held at rest on a frictionless incline by a force that is parallel to the incline. If the incline is 25° above the horizontal the magnitude of the applied force is:

A. 4.1 N
B. 4.6 N
C. 8.9 N
D. 11 N
E. 23 N

ans: D

48. A 25-N crate is held at rest on a frictionless incline by a force that is parallel to the incline. If the incline is 25° above the horizontal the magnitude of the normal force of the incline on the crate is:

A. 4.1 N
B. 4.6 N
C. 8.9 N
D. 11 N
E. 23 N

ans: E

49. A 32-N force, parallel to the incline, is required to push a certain crate at constant velocity up a frictionless incline that is 30° above the horizontal. The mass of the crate is:

A. 3.3 kg
B. 3.8 kg
C. 5.7 kg
D. 6.5 kg
E. 160 kg

ans: D

50. When a 40-N force, parallel to the incline and directed up the incline, is applied to a crate on a frictionless incline that is 30° above the horizontal, the acceleration of the crate is $2.0\,\mathrm{m/s^2}$, down the incline. The mass of the crate is:

A. 3.8 kg
B. 4.1 kg
C. 5.8 kg
D. 6.2 kg
E. 10 kg

ans: E

51. When a 40-N force, parallel to the incline and directed up the incline, is applied to a crate on a frictionless incline that is 30° above the horizontal, the acceleration of the crate is $2.0\,\mathrm{m/s^2}$, up the incline. The mass of the crate is:

A. 3.8 kg
B. 4.1 kg
C. 5.8 kg
D. 6.2 kg
E. 10 kg

ans: C

52. The "reaction" force does not cancel the "action" force because:

A. the action force is greater than the reaction force
B. they act on different bodies
C. they are in the same direction
D. the reaction force exists only after the action force is removed
E. the reaction force is greater than the action force

ans: B

53. A book rests on a table, exerting a downward force on the table. The reaction to this force is:

A. the force of Earth on the book
B. the force of the table on the book
C. the force of Earth on the table
D. the force of the book on Earth
E. the inertia of the book

ans: B

54. A lead block is suspended from your hand by a string. The reaction to the force of gravity on the block is the force exerted by:

A. the string on the block
B. the block on the string
C. the string on the hand
D. the hand on the string
E. the block on Earth

ans: E

55. A 5-kg concrete block is lowered with a downward acceleration of $2.8\,\mathrm{m/s^2}$ by means of a rope. The force of the block on the rope is:

A. 14 N, up
B. 14 N, down
C. 35 N, up
D. 35 N, down
E. 49 N, up

ans: D

56. A 90-kg man stands in an elevator that is moving up at a constant speed of $5.0\,\mathrm{m/s}$. The force exerted by him on the floor is about:

A. zero
B. 90 N
C. 880 N
D. 450 N
E. 49 N

ans: C

57. A 90-kg man stands in an elevator that has a downward acceleration of $1.4\,\mathrm{m/s^2}$. The force exerted by him on the floor is about:

A. zero
B. 90 N
C. 760 N
D. 880 N
E. 1010 N

ans: C

58. A 5-kg concrete block is lowered with a downward acceleration of $2.8\,\mathrm{m/s^2}$ by means of a rope. The force of the block on Earth is:

A. 14 N, up
B. 14 N, down
C. 35 N, up
D. 35 N, down
E. 49 N, up

ans: E

59. Two blocks are connected by a string and pulley as shown. Assuming that the string and pulley are massless, the magnitude of the acceleration of each block is:

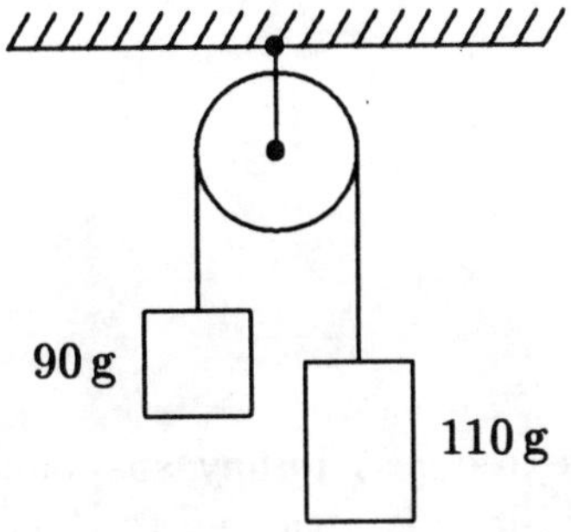

A. $0.049\,m/s^2$
B. $0.020\,m/s^2$
C. $0.0098\,m/s^2$
D. $0.54\,m/s^2$
E. $0.98\,m/s^2$
ans: E

60. A 70-N block and an 35-N block are connected by a string as shown. If the pulley is massless and the surface is frictionless, the magnitude of the acceleration of the 70-N block is:

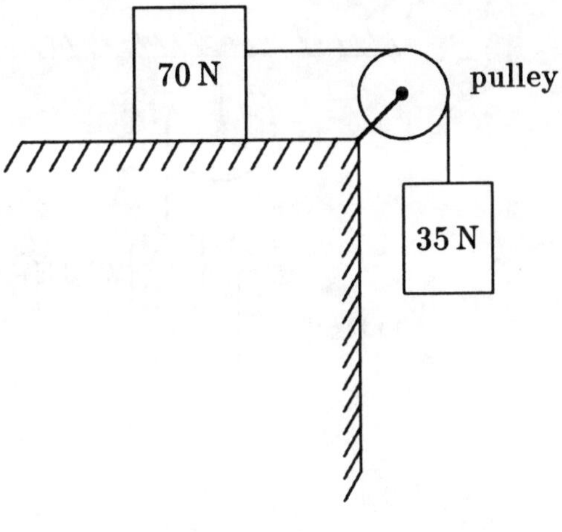

A. $1.6\,m/s^2$
B. $3.3\,m/s^2$
C. $4.9\,m/s^2$
D. $6.7\,m/s^2$
E. $9.8\,m/s^2$
ans: B

61. A 13-N weight and a 12-N weight are connected by a massless string over a massless, frictionless pulley. The 13-N weight has a downward acceleration equal to that of a freely falling body times:

A. 1
B. 1/12
C. 1/13
D. 1/25
E. 13/25

ans: D

62. A massless rope passes over a massless pulley suspended from the ceiling. A 4-kg block is attached to one end and a 5-kg block is attached to the other end. The acceleration of the 5-kg block is:

A. $g/4$
B. $5g/9$
C. $4g/9$
D. $g/5$
E. $g/9$

ans: E

63. Two blocks, weighing 250 N and 350 N, respectively, are connected by a string that passes over a massless pulley as shown. The tension in the string is:

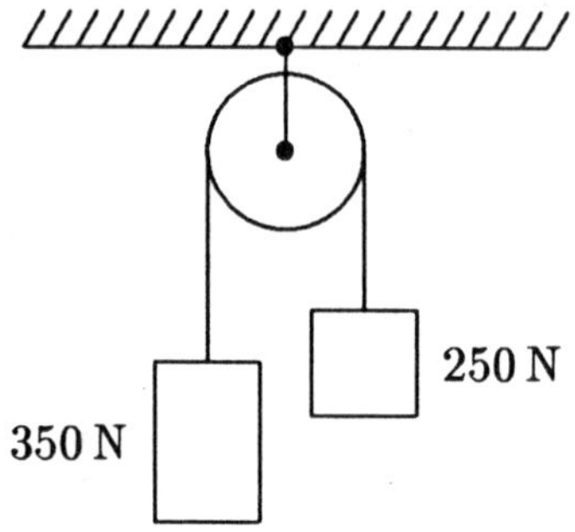

A. 210 N
B. 290 N
C. 410 N
D. 500 N
E. 4900 N

ans: B

64. Three books (X, Y, and Z) rest on a table. The weight of each book is indicated. The net force acting on book Y is:

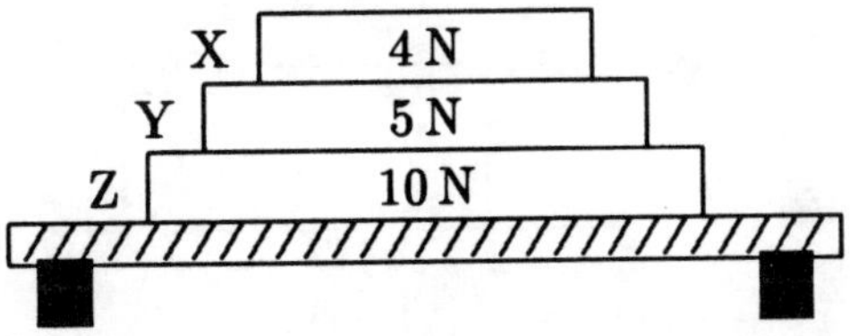

A. 4 N down
B. 5 N up
C. 9 N down
D. zero
E. none of these

ans: D

65. Three books (X, Y, and Z) rest on a table. The weight of each book is indicated. The force of book Z on book Y is:

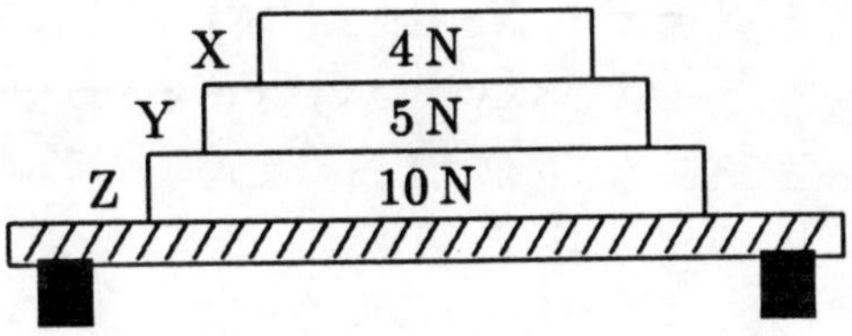

A. 0
B. 5 N
C. 9 N
D. 14 N
E. 19 N

ans: C

66. Three blocks (A,B,C), each having mass M, are connected by strings as shown. Block C is pulled to the right by a force $\vec{F}$ that causes the entire system to accelerate. Neglecting friction, the net force acting on block B is:

A. zero
B. $\vec{F}/3$
C. $\vec{F}/2$
D. $2\vec{F}/3$
E. $\vec{F}$

ans: B

67. Two blocks with masses m and M are pushed along a horizontal frictionless surface by a horizontal applied force $\vec{F}$ as shown. The magnitude of the force that either of these blocks exerts on the other is:

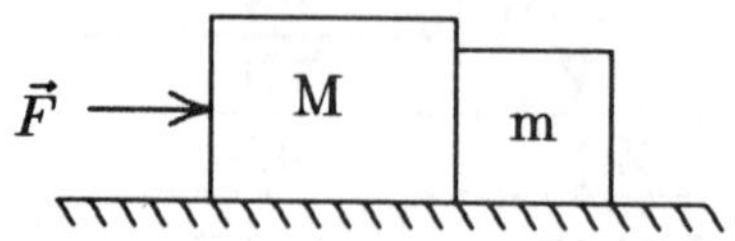

A. $mF/(m+M)$
B. mF/M
C. $mF/(M-m)$
D. $MF/(M+m)$
E. MF/m

ans: A

68. Two blocks (X and Y) are in contact on a horizontal frictionless surface. A 36-N constant force is applied to X as shown. The force exerted by X on Y is:

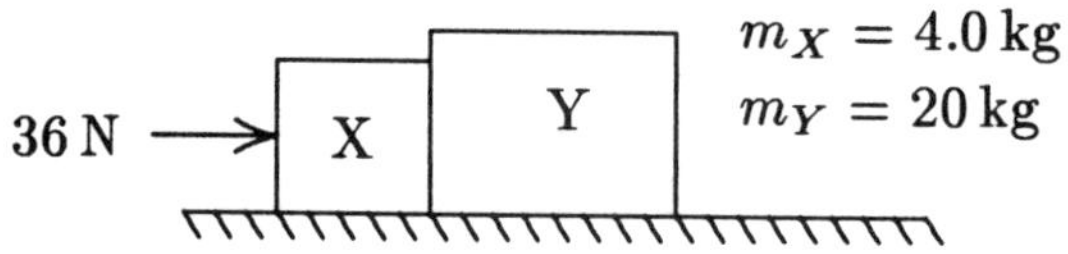

A. 1.5 N
B. 6.0 N
C. 29 N
D. 30 N
E. 36 N

ans: D

69. Two blocks (X and Y) are in contact on a horizontal frictionless surface. A 36-N constant force is applied to X as shown. The force exerted by Y on X is:

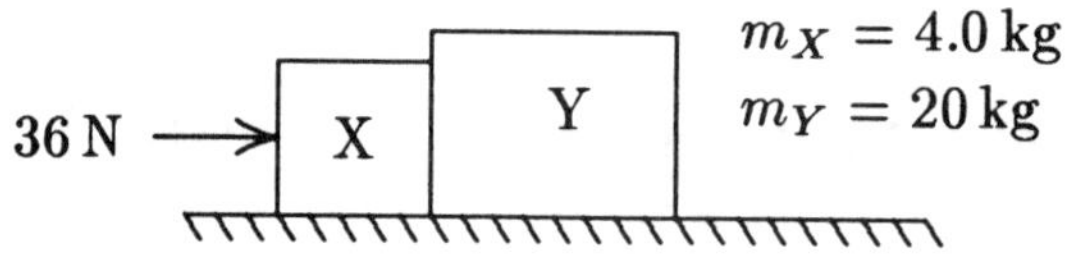

A. 1.5 N
B. 6.0 N
C. 29 N
D. 30 N
E. 36 N

ans: D

70. A short 10-g string is used to pull a 50-g toy across a frictionless horizontal surface. If a 3.0×10^{-2}-N force is applied horizontally to the free end, the force of the string on the toy, at the other end, is:

A. 0.15 N
B. 6.0×10^{-3} N
C. 2.5×10^{-2} N
D. 3.0×10^{-2} N
E. 3.5×10^{-2} N

ans: C

Chapter 6: FORCE AND MOTION — II

1. A brick slides on a horizontal surface. Which of the following will increase the frictional force on it?

   A. Putting a second brick on top
   B. Decreasing the surface area of contact
   C. Increasing the surface area of contact
   D. Decreasing the mass of the brick
   E. None of the above

   ans: A

2. The coefficient of kinetic friction:

   A. is in the direction of the frictional force
   B. is in the direction of the normal force
   C. is the ratio of force to area
   D. can have units of newtons
   E. is none of the above

   ans: E

3. When the brakes of an automobile are applied, the road exerts the greatest retarding force:

   A. while the wheels are sliding
   B. just before the wheels start to slide
   C. when the automobile is going fastest
   D. when the acceleration is least
   E. at the instant when the speed begins to change

   ans: B

4. A forward horizontal force of 12 N is used to pull a 240-N sled at constant velocity on a frozen pond. The coefficient of friction is:

   A. 0.5
   B. 0.05
   C. 2
   D. 0.2
   E. 20

   ans: B

5. The velocity of a 4.0-N hockey puck, sliding across a level ice surface, decreases at the rate of $0.61\,\mathrm{m/s^2}$. The coefficient of kinetic friction between the puck and ice is:

   A. 0.062
   B. 0.41
   C. 0.62
   D. 1.2
   E. 9.8

   ans: A

6. A crate rests on a horizontal surface and a woman pulls on it with a 10-N force. No matter what the orientation of the force, the crate does not move. Rank the situations shown below according to the magnitude of the frictional force exerted by the surface on the crate, least to greatest.

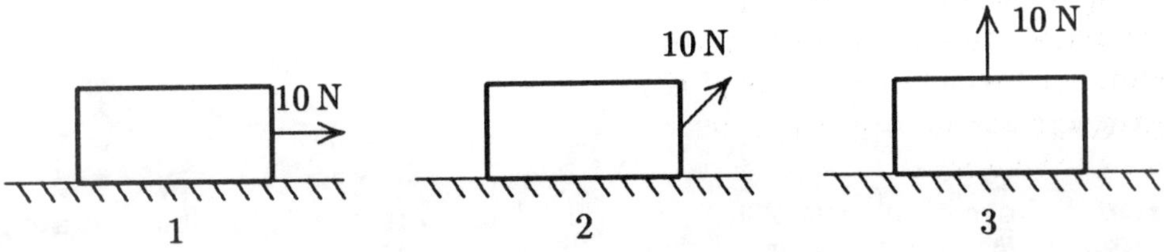

A. 1, 2, 3
B. 2, 1, 3
C. 2, 3, 1
D. 1, 3, 2
E. 3, 2, 1

ans: E

7. A box with a weight of 50 N rests on a horizontal surface. A person pulls horizontally on it with a force of 10 N and it does not move. To start it moving, a second person pulls vertically upward on the box. If the coefficient of static friction is 0.4, what is the smallest vertical force for which the box moves?

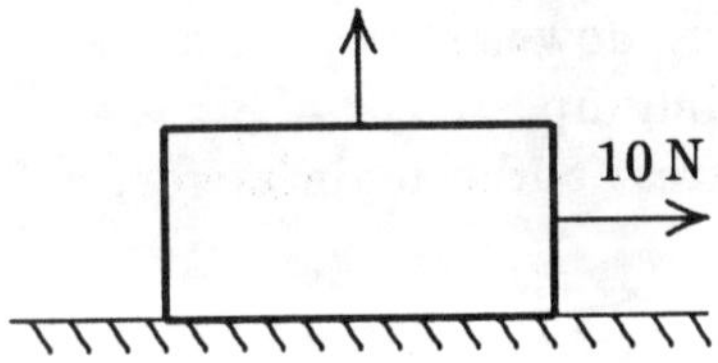

A. 4 N
B. 10 N
C. 14 N
D. 25 N
E. 35 N

ans: D

8. A 12-N horizontal force is applied to a 40-N block on a rough horizontal surface. The block is initially at rest. If $\mu_s = 0.5$ and $\mu_k = 0.4$, the frictional force on the block is:

A. 8 N
B. 12 N
C. 16 N
D. 20 N
E. 40 Nb

ans: B

9. A 24-N horizontal force is applied to a 40-N block initially at rest on a rough horizontal surface. If $\mu_s = 0.5$ and $\mu_k = 0.4$, the frictional force on the block is:
   A. 8 N
   B. 12 N
   C. 16 N
   D. 20 N
   E. 400 N

   ans: C

10. A horizontal shove of at least 200 N is required to start moving a 800-N crate initially at rest on a horizontal floor. The coefficient of static friction is:
   A. 0.25
   B. 0.125
   C. 0.50
   D. 4.00
   E. none of these

   ans: A

11. A force $\vec{F}$ (larger than the largest possible force of static friction) is applied to the left to an object moving to the right on a horizontal surface. Then:
   A. the object must be moving at constant speed
   B. $\vec{F}$ and the friction force act in opposite directions
   C. the object must be slowing down
   D. the object must be speeding up
   E. the object must come to rest and remain at rest

   ans: C

12. A block rests on a rough horizontal surface ($\mu_s = 0.50$, $\mu_k = 0.40$). A constant horizontal force, just sufficient to start the block in motion, is applied. The acceleration of the block, in m/s^2, is:
   A. 0
   B. 0.98
   C. 3.3
   D. 4.5
   E. 8.9

   ans: B

13. A car is traveling at 15 m/s on a horizontal road. The brakes are applied and the car skids to a stop in 4.0 s. The coefficient of kinetic friction between the tires and road is:
   A. 0.38
   B. 0.69
   C. 0.76
   D. 0.92
   E. 1.11

   ans: A

14. A boy pulls a wooden box along a rough horizontal floor at constant speed by means of a force $\vec{P}$ as shown. In the diagram f is the magnitude of the force of friction, N is the magnitude of the normal force, and F_g is the magnitude of the force of gravity. Which of the following must be true?

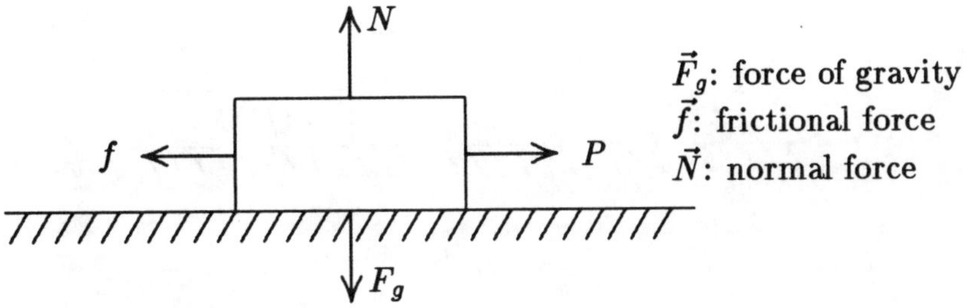

A. $P = f$ and $N = F_g$
B. $P = f$ and $N > F_g$
C. $P > f$ and $N < F_g$
D. $P > f$ and $N = F_g$
E. none of these

ans: A

15. A boy pulls a wooden box along a rough horizontal floor at constant speed by means of a force $\vec{P}$ as shown. In the diagram f is the magnitude of the force of friction, N is the magnitude of the normal force, and F_g is the magnitude of the force of gravity. Which of the following must be true?

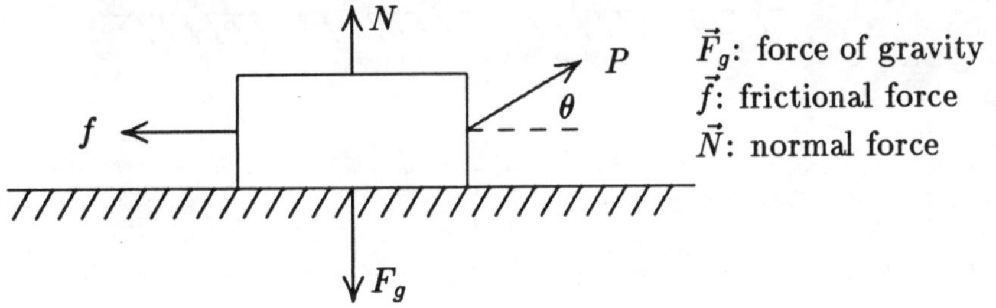

A. $P = f$ and $N = F_g$
B. $P = f$ and $N > F_g$
C. $P > f$ and $N < F_g$
D. $P > f$ and $N = F_g$
E. none of these

ans: C

16. A 400-N block is dragged along a rough ($\mu_k = 0.4$) horizontal surface by an applied force $\vec{F}$ as shown. The block moves at constant velocity. The magnitude of $\vec{F}$ is:

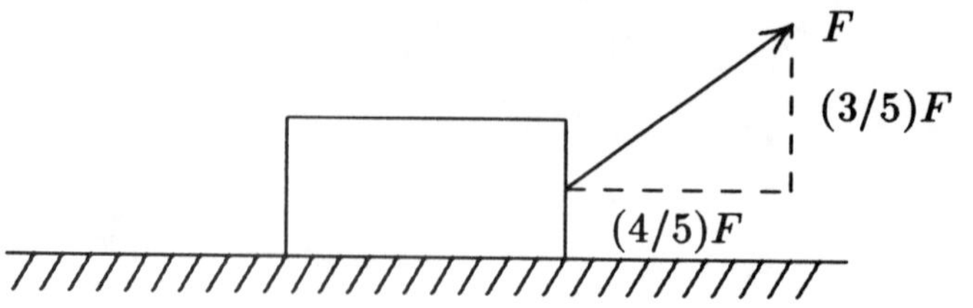

A. 100 N
B. 150 N
C. 200 N
D. 290 N
E. 400 Nb

ans: B

17. A block of mass m is pulled at constant velocity along a rough horizontal floor by an applied force $\vec{T}$ as shown. The frictional force is:

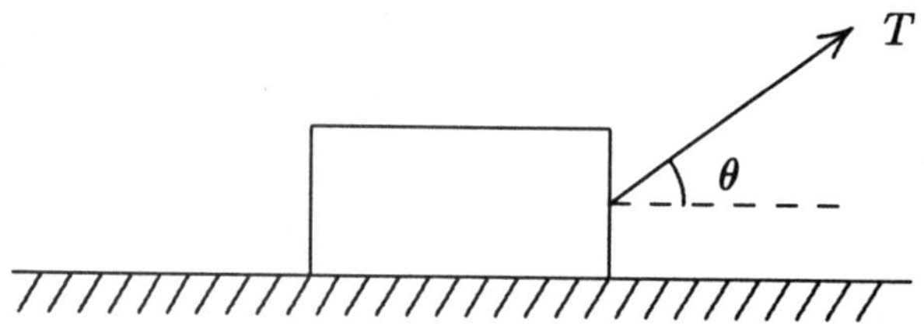

A. $T\cos\theta$
B. $T\sin\theta$
C. zero
D. mg
E. $mg\cos\theta$

ans: A

18. A block of mass m is pulled along a rough horizontal floor by an applied force $\vec{T}$ as shown. The vertical component of the force exerted on the block by the floor is:

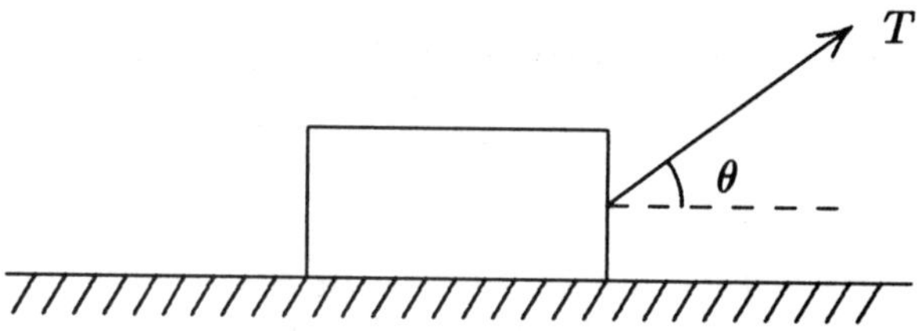

A. mg
B. $mg - T\cos\theta$
C. $mg + T\cos\theta$
D. $mg - T\sin\theta$
E. $mg + T\sin\theta$

ans: D

19. A 12-kg crate rests on a horizontal surface and a boy pulls on it with a force that is 30° above the horizontal. If the coefficient of static friction is 0.40, the minimum magnitude force he needs to start the crate moving is:

A. 44 N
B. 47 N
C. 54 N
D. 56 N
E. 71 N

ans: A

20. A crate resting on a rough horizontal floor is to be moved horizontally. The coefficient of static friction is 0.40. To start the crate moving with the weakest possible applied force, in what direction should the force be applied?

A. Horizontal
B. 24° below the horizontal
C. 22° above the horizontal
D. 24° above the horizontal
E. 66° below the horizontal

ans: C

21. A 50-N force is applied to a crate on a horizontal rough floor, causing it to move horizontally. If the coefficient of kinetic friction is 0.50, in what direction should the force be applied to obtain the greatest acceleration?

A. Horizontal
B. 60° above the horizontal
C. 30° above the horizontal
D. 27° above the horizontal
E. 30° below the horizontal

ans: D

22. A professor holds an eraser against a vertical chalkboard by pushing horizontally on it. He pushes with a force that is much greater than is required to hold the eraser. The force of friction exerted by the board on the eraser increases if he:

A. pushes with slightly greater force
B. pushes with slightly less force
C. stops pushing
D. raises his elbow so the force he exerts is slightly downward but has the same magnitude
E. lowers his elbow so the force he exerts is slightly upward but has the same magnitude

ans: D

23. A horizontal force of 12 N pushes a 0.5-kg block against a vertical wall. The block is initially at rest. If $\mu_s = 0.6$ and $\mu_k = 0.8$ which of the following is true?

A. The frictional force is 4.9 N
B. The frictional force is 7.2 N
C. The normal force is 4.9 N
D. The block will start moving and accelerate
E. If started moving downward, the block will decelerate

ans: A

24. A horizontal force of 5.0 N pushes a 0.50-kg block against a vertical wall. The block is initially at rest. If $\mu_s = 0.6$ and $\mu_k = 0.80$, the frictional force is:

A. 0
B. 4.9 N
C. 3.0 N
D. 5.0 N
E. 4.0 N

ans: E

25. A horizontal force of 12 N pushes a 0.50-kg block against a vertical wall. The block is initially at rest. If $\mu_s = 0.6$ and $\mu_k = 0.80$, the acceleration of the block in m/s^2 is:

A. 0
B. 9.4
C. 9.8
D. 14.4
E. 19.2

ans: A

26. A horizontal force of 5.0 N pushes a 0.50-kg block against a vertical wall. The block is initially at rest. If $\mu_s = 0.60$ and $\mu_k = 0.80$, the acceleration of the block in m/s^2 is:

A. 0
B. 1.8
C. 6.0
D. 8.0
E. 9.8

ans: B

27. A heavy wooden block is dragged by a force $\vec{F}$ along a rough steel plate, as shown below for two possible situations. The magnitude of $\vec{F}$ is the same for the two situations. The frictional force in (ii), as compared with that in (i) is:

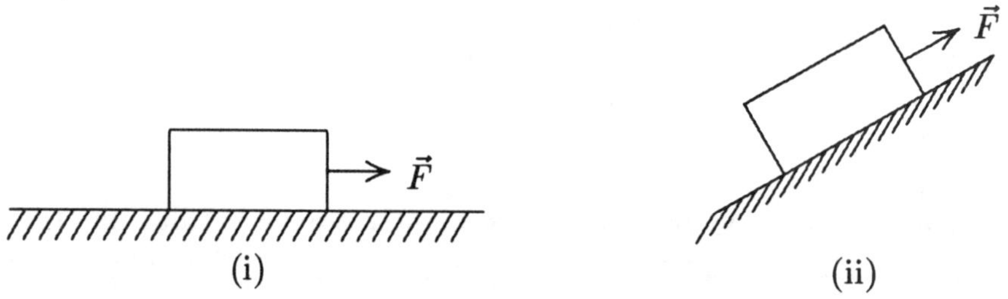

A. the same
B. greater
C. less
D. less for some angles and greater for others
E. can be less or greater, depending on the magnitude of the applied force.

ans: C

28. A block is first placed on its long side and then on its short side on the same inclined plane, as shown. The block slides down the plane on its short side but remains at rest on its long side. A possible explanation is:

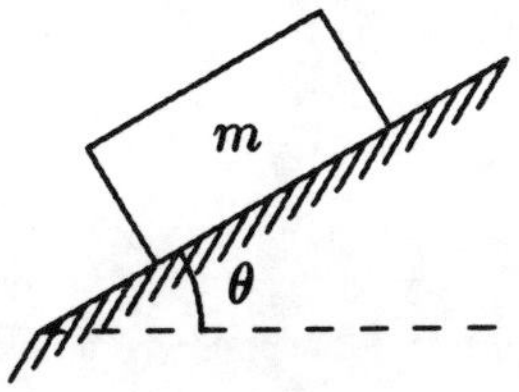

(ii)

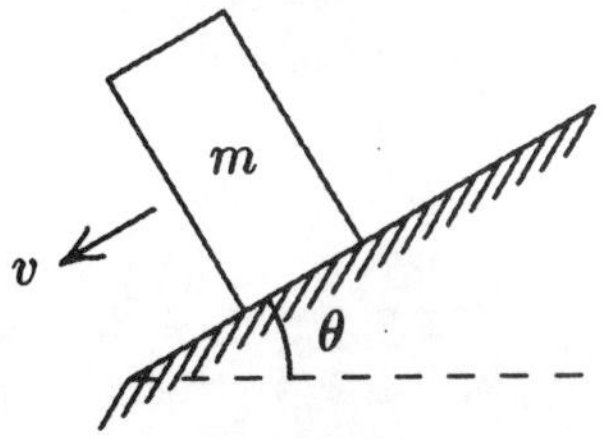

(ii)

A. the short side is smoother
B. the frictional force is less because the contact area is less
C. the center of gravity is higher in the second case
D. the normal force is less in the second case
E. the force of gravity is more nearly down the plane in the second case

ans: A

29. A box rests on a rough board 10 meters long. When one end of the board is slowly raised to a height of 6 meters above the other end, the box begins to slide. The coefficient of static friction is:

A. 0.8
B. 0.25
C. 0.4
D. 0.6
E. 0.75

ans: E

30. A block is placed on a rough wooden plane. It is found that when the plane is tilted 30° to the horizontal, the block will slide down at constant speed. The coefficient of kinetic friction of the block with the plane is:

A. 0.500
B. 0.577
C. 1.73
D. 0.866
E. 4.90

ans: B

31. A crate is sliding down an incline that is 35° above the horizontal. If the coefficient of kinetic friction is 0.40, the acceleration of the crate is:

A. 0
B. $2.4\,m/s^2$
C. $5.8\,m/s^2$
D. $8.8\,m/s^2$
E. $10.3\,m/s^2$

ans: B

32. A 5.0-kg crate is resting on a horizontal plank. The coefficient of static friction is 0.50 and the coefficient of kinetic friction is 0.40. After one end of the plank is raised so the plank makes an angle of 25° with the horizontal, the force of friction is:

A. 0
B. 18 N
C. 21 N
D. 22 N
E. 44 N

ans: C

33. A 5.0-kg crate is resting on a horizontal plank. The coefficient of static friction is 0.50 and the coefficient of kinetic friction is 0.40. After one end of the plank is raised so the plank makes an angle of 30° with the horizontal, the force of friction is:

A. 0
B. 18 N
C. 21 N
D. 22 N
E. 44 N

ans: B

34. A 5.0-kg crate is on an incline that makes an angle of 30° with the horizontal. If the coefficient of static friction is 0.50, the minimum force that can be applied parallel to the plane to hold the crate at rest is:

A. 0
B. 3.3 N
C. 30 N
D. 46 N
E. 55 N

ans: B

35. A 5.0-kg crate is on an incline that makes an angle of 30° with the horizontal. If the coefficient of static friction is 0.5, the maximum force that can be applied parallel to the plane without moving the crate is:

A. 0
B. 3.3 N
C. 30 N
D. 46 N
E. 55 N

ans: D

36. Block A, with mass m_A, is initially at rest on a horizontal floor. Block B, with mass m_B, is initially at rest on the horizontal top surface of A. The coefficient of static friction between the two blocks is μ_s. Block A is pulled with an increasing force. It begins to slide out from under B when its acceleration reaches:

A. g
B. $\mu_s g$
C. $m_B \mu_s g$
D. $(m_A/m_B)\mu_s g$
E. $(m_B/m_A)\mu_s g$

ans: B

37. The system shown remains at rest. Each block weighs 20 N. The force of friction on the upper block is:

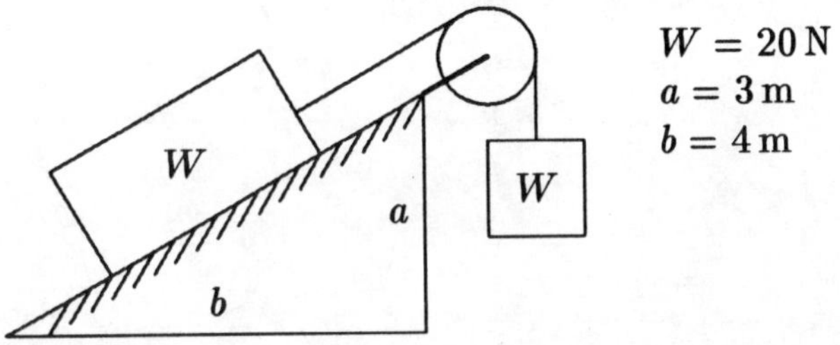

A. 4 N
B. 8 N
C. 12 N
D. 16 N
E. 20 N

ans: B

38. Block A, with a mass of 50 kg, rests on a horizontal table top. The coefficient of static friction is 0.40. A horizontal string is attached to A and passes over a massless, frictionless pulley as shown. The smallest mass m_B that will start A moving when it is attached to the other end of the string is:

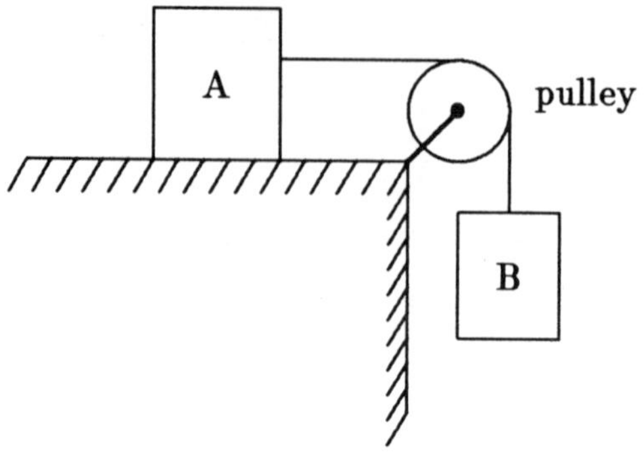

A. 20 kg
B. 30 kg
C. 40 kg
D. 50 kg
E. 70 kg

ans: A

39. Block A, with a mass of 10 kg, rests on a 35° incline. The coefficient of static friction is 0.40. An attached string is parallel to the incline and passes over a massless, frictionless pulley at the top. The smallest mass m_B, attached to the dangling end, for which A remains at rest is:

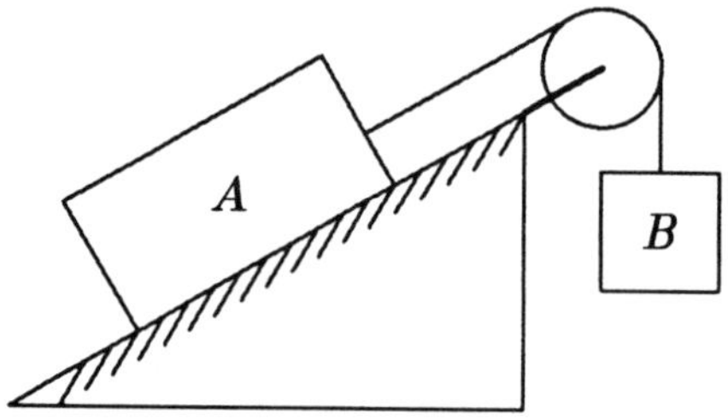

A. 2.5 kg
B. 3.5 kg
C. 5.9 kg
D. 9.0 kg
E. 10.5 kg

ans: A

40. Block A, with a mass of 10 kg, rests on a 35° incline. The coefficient of static friction is 0.40. An attached string is parallel to the incline and passes over a massless, frictionless pulley at the top. The largest mass m_B, attached to the dangling end, for which A remains at rest is:

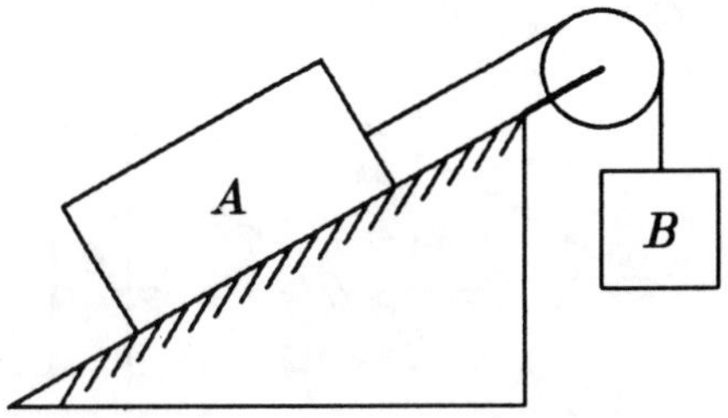

A. 2.5 kg
B. 3.5 kg
C. 5.9 kg
D. 9.0 kg
E. 10.5 kg

ans: D

41. Block A, with a mass of 10 kg, rests on a 30° incline. The coefficient of kinetic friction is 0.20. The attached string is parallel to the incline and passes over a massless, frictionless pulley at the top. Block B, with a mass of 8.0 kg, is attached to the dangling end of the string. The acceleration of B is:

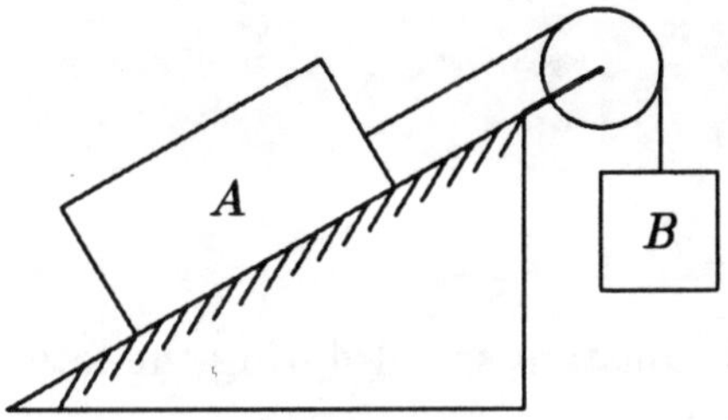

A. $0.69\,\mathrm{m/s^2}$, up
B. $0.69\,\mathrm{m/s^2}$, down
C. $2.6\,\mathrm{m/s^2}$, up
D. $2.6\,\mathrm{m/s^2}$, down
E. 0

ans: B

42. Block A, with a mass of 10 kg, rests on a 30° incline. The coefficient of kinetic friction is 0.20. The attached string is parallel to the incline and passes over a massless, frictionless pulley at the top. Block B, with a mass of 3.0 kg, is attached to the dangling end of the string. The acceleration of B is:

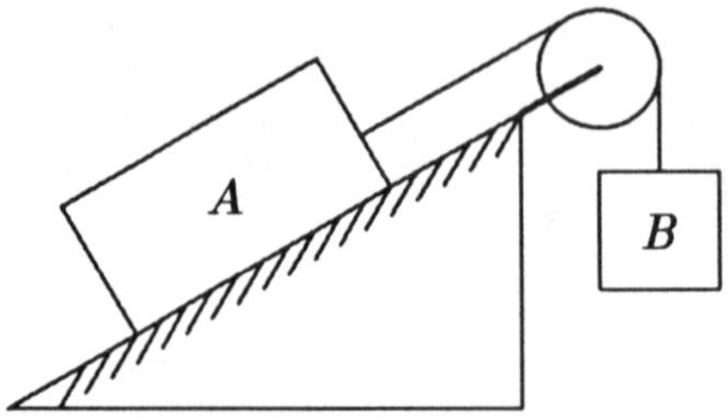

A. $0.20\,\mathrm{m/s^2}$, up
B. $0.20\,\mathrm{m/s^2}$, down
C. $2.8\,\mathrm{m/s^2}$, up
D. $2.8\,\mathrm{m/s^2}$, down
E. 0

ans: A

43. A 1000-kg airplane moves in straight flight at constant speed. The force of air friction is 1800 N. The net force on the plane is:

A. zero
B. 11800 N
C. 1800 N
D. 9800 N
E. none of these

ans: A

44. Why do raindrops fall with constant speed during the later stages of their descent?

A. The gravitational force is the same for all drops
B. Air resistance just balances the force of gravity
C. The drops all fall from the same height
D. The force of gravity is negligible for objects as small as raindrops
E. Gravity cannot increase the speed of a falling object to more than 9.8 m/s

ans: B

45. A ball is thrown downward from the edge of a cliff with an initial speed that is greater than the terminal speed. Initially its acceleration is

A. 0
B. upward
C. downward and greater than g
D. downward and less than g
E. downward and equal to g

ans: B

46. A ball is thrown upward into the air with a speed that is greater than terminal speed. On the way up it slows down and, after its speed equals the terminal speed but before it gets to the top of its trajectory:

A. its speed is constant
B. it continues to slow down
C. it speeds up
D. its motion becomes jerky
E. none of the above

ans: B

47. A ball is thrown upward into the air with a speed that is greater than terminal speed. It lands at the place where it was thrown. During its flight the force of air resistance is the greatest:

A. just after it is thrown
B. halfway up
C. at the top of its trajectory
D. halfway down
E. just before it lands.

ans: A

48. Uniform circular motion is the direct consequence of:

A. Newton's third law
B. a force that is always tangent to the path
C. an acceleration tangent to the path
D. conservation of energy and momentum
E. an acceleration always directed toward the same point

ans: E

49. An object moving in a circle at constant speed:

A. must have only one force acting on it
B. is not accelerating
C. is held to its path by centrifugal force
D. has an acceleration of constant magnitude
E. has an acceleration that is tangent to the circle

ans: D

50. An object of mass m and another object of mass $2m$ are each forced to move along a circle of radius 1.0 m at a constant speed of 1.0 m/s. The magnitudes of their accelerations are:

A. equal
B. in the ratio of $\sqrt{2} : 1$
C. in the ratio of $2 : 1$
D. in the ratio of $4 : 1$
E. zero

ans: A

51. The magnitude of the force (in newtons) required to cause a 0.04-kg object to move at 0.6 m/s in a circle of radius 1.0 m is:

A. 2.4×10^{-2}
B. 1.4×10^{-2}
C. $1.4\pi \times 10^{-2}$
D. $2.4\pi^2 \times 10^{-2}$
E. 3.13

ans: B

52. A 0.2-kg stone is attached to a string and swung in a circle of radius 0.6 m on a horizontal and frictionless surface. If the stone makes 150 revolutions per minute, the tension force of the string is:

A. 0.03 N
B. 0.2 N
C. 0.9 N
D. 1.96 N
E. 30 N

ans: E

53. Which of the following five graphs is correct for a particle moving in a circle of radius r at a constant speed of 10 m/s?

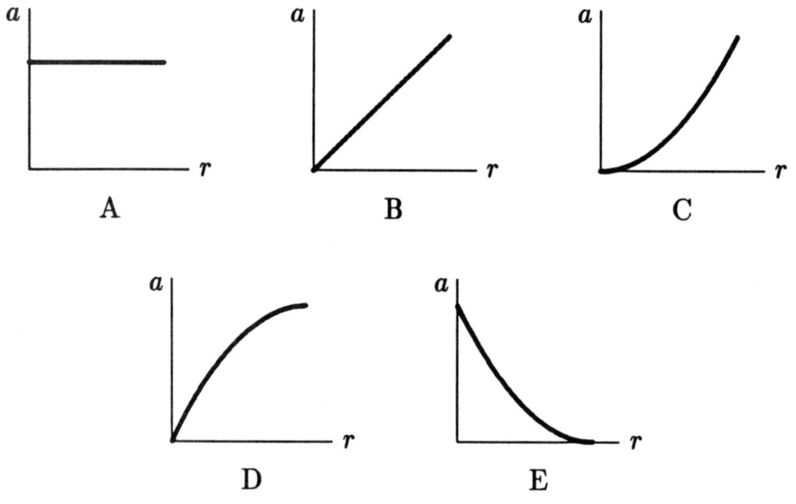

ans: E

54. An object moves in a circle. If the radius is doubled keeping the speed the same then the centripetal force must be:

A. twice as great
B. half as great
C. four times as great
D. one-fourth as great
E. the same

ans: B

55. An object moves in a circle. If the mass is tripled, the speed halved, and the radius unchanged, then the centripetal force must change by a factor of:

A. 3/2
B. 3/4
C. 9/4
D. 6
E. 12

ans: B

56. If a satellite moves above Earth's atmosphere in a circular orbit with constant speed, then:

A. its acceleration and velocity are always in the same direction
B. the net force on it is zero
C. its velocity is constant
D. it will fall back to Earth when its fuel is used up
E. its acceleration is toward the Earth

ans: E

57. A 800-N passenger in a car presses against the car door with a 200 N force when the car makes a left turn at 13 m/s. The (faulty) door will pop open under a force of 800 N. Of the following, the least speed for which the passenger is thrown out of the car is:

A. 14 m/s
B. 19 m/s
C. 20 m/s
D. 26 m/s
E. 54 m/s

ans: D

58. If a certain car, going with speed v_1, rounds a level curve with a radius R_1, it is just on the verge of skidding. If its speed is now doubled, the radius of the tightest curve on the same road that it can round without skidding is:

A. $2R_1$
B. $4R_1$
C. $R_1/2$
D. $R_1/4$
E. R_1

ans: B

59. An automobile moves on a level horizontal road in a circle of radius 30 m. The coefficient of friction between tires and road is 0.50. The maximum speed with which this car can round this curve is:

A. 3.0 m/s
B. 4.9 m/s
C. 9.8 m/s
D. 12 m/s
E. 13 m/s

ans: D

60. The driver of a 1000-kg car tries to turn through a circle of radius 100 m on an unbanked curve at a speed of 10 m/s. The actual frictional force between the tires and slippery road is 900 N. The car will:

A. slide into the inside of the curve
B. make the turn
C. slow down due to the frictional force
D. make the turn only if it went faster
E. slide off to the outside of the curve

ans: E

61. A car rounds a 75-m radius curve at a constant speed of 18 m/s. A ball is suspended by a string from the ceiling the car and moves with the car. The angle between the string and the vertical is:

A. 0
B. 1.4°
C. 24°
D. 90°
E. cannot be found without knowing the mass of the ball

ans: C

62. A giant wheel, having a diameter of 40 m, is fitted with a cage and platform on which a man of mass m stands. The wheel is rotated in a vertical plane at such a speed that the force exerted by the man on the platform is equal to his weight when the cage is at X, as shown. The net force on the man at point X is:

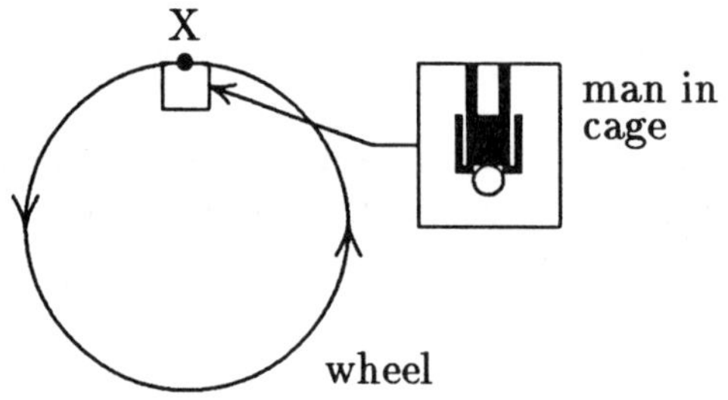

A. zero
B. mg, down
C. mg, up
D. $2mg$, down
E. $2mg$, up

ans: D

63. A giant wheel, 40 m in diameter, is fitted with a cage and platform on which a man can stand. The wheel rotates at such a speed that when the cage is at X (as shown) the force exerted by the man on the platform is equal to his weight. The speed of the man (in m/s) is:

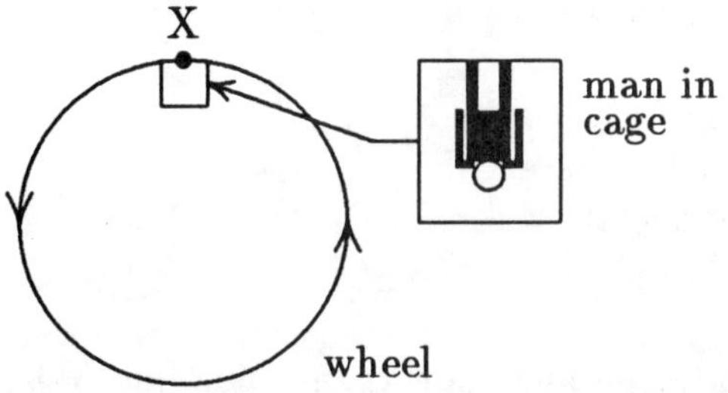

A. 14
B. 20
C. 28
D. 80
E. 120

ans: B

64. A person riding a Ferris wheel is strapped into her seat by a seat belt. The wheel is spun so that the centripetal acceleration is g. Select the correct combination of forces that act on her when she is at the top. In the table F_g = force of gravity, down; F_b = seat belt force, down; and F_s = seat force, up.

	F_g	F_b	F_s
A.	0	mg	0
B.	mg	0	0
C.	0	0	mg
D.	mg	mg	0
E.	mg	0	mg

ans: B

65. One end of a 1.0-m long string is fixed, the other end is attached to a 2.0-kg stone. The stone swings in a vertical circle, passing the bottom point at 4.0 m/s. The tension force of the string (in newtons) at this point is about:

A. 0
B. 12
C. 20
D. 32
E. 52

ans: E

66. One end of a 1.0-m string is fixed, the other end is attached to a 2.0-kg stone. The stone swings in a vertical circle, passing the top point at 4.0 m/s. The tension force of the string (in newtons) at this point is about:

A. 0
B. 12
C. 20
D. 32
E. 52

ans: B

67. A coin is placed on a horizontal phonograph turntable. Let N be the normal force exerted by the turntable on the coin, f be the frictional force exerted by the turntable on the coin, and $f_{s,\ max}$ be the maximum possible force of static friction. The speed of the turntable is increased in small steps. If the coin does not slide, then

A. N increases, f increases, and $f_{s,\ max}$ stays the same
B. N increases, f increases, and $f_{s,\ max}$ increases
C. f increases and both N and $f_{s,\ max}$ stay the same
D. N, f, and $f_{s,\ max}$ all stay the same
E. N, f, and $f_{s,\ max}$ all increase

ans: C

68. The iron ball shown is being swung in a vertical circle at the end of a 0.7-m long string. How slowly can the ball go through its top position without having the string go slack?

A. 1.3 m/s
B. 2.6 m/s
C. 3.9 m/s
D. 6.9 m/s
E. 9.8 m/s

ans: B

69. On some roller coasters each car is suspended below the track by means of a rod that is free to swivel in any direction. When such a car rounds a 45-m radius horizontal curve at 22 m/s (about 50 mph), what angle does the rod make with the vertical?

A. 0
B. 25°
C. 48°
D. 65°
E. 90°

ans: C

70. Circular freeway entrance and exit ramps are commonly banked to handle a car moving at 13 m/s. To design a similar ramp for 26 m/s one should:

A. increase radius by factor of 2
B. decrease radius by factor of 2
C. increase radius by factor of 4
D. decrease radius by factor of 4
E. increase radius by factor of $\sqrt{2}$

ans: C

71. At what angle should the roadway on a curve with a 50 m radius be banked to allow cars to negotiate the curve at 12 m/s even if the roadway is icy (and the frictional force is zero)?

A. 0
B. 16°
C. 18°
D. 35°
E. 73°

ans: B

Chapter 7: KINETIC ENERGY AND WORK

1. Which of the following is NOT a correct unit for work?

   A. erg
   B. ft·lb
   C. watt
   D. newton·meter
   E. joule

   ans: C

2. Which of the following groups does NOT contain a scalar quantity?

   A. velocity, force, power
   B. displacement, acceleration, force
   C. acceleration, speed, work
   D. energy, work, distance
   E. pressure, weight, time

   ans: B

3. A boy holds a 40-N weight at arm's length for 10 s. His arm is 1.5 m above the ground. The work done by the force of the boy on the weight while he is holding it is:

   A. 0
   B. 6.1 J
   C. 40 J
   D. 60 J
   E. 90 J

   ans: A

4. A crate moves to the right on a horizontal surface as a woman pulls on it with a 10-N force. Rank the situations shown below according to the work done by the 10-N force, least to greatest. The displacement is the same for all cases.

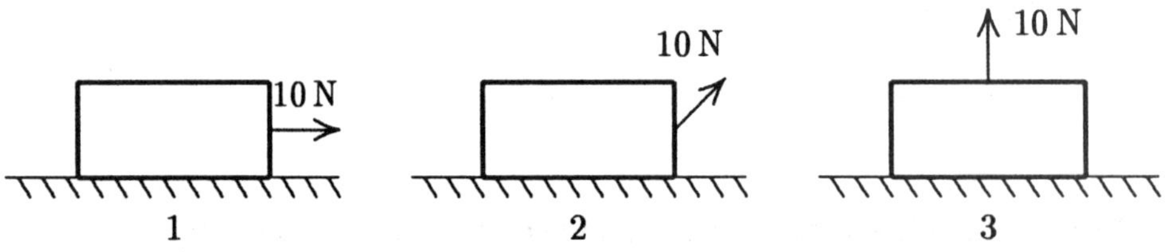

   A. 1, 2, 3
   B. 2, 1, 3
   C. 2, 3, 1
   D. 1, 3, 2
   E. 3, 2, 1

   ans: E

5. An object moves in a circle at constant speed. The work done by the centripetal force is zero because:
   A. the displacement for each revolution is zero
   B. the average force for each revolution is zero
   C. there is no friction
   D. the magnitude of the acceleration is zero
   E. the centripetal force is perpendicular to the velocity

   ans: E

6. An object of mass 1 g is whirled in a horizontal circle of radius 0.5 m at a constant speed of 2 m/s. The work done on the object during one revolution is:
   A. 0
   B. 1 J
   C. 2 J
   D. 4 J
   E. 16 J

   ans: A

7. The work done by gravity during the descent of a projectile:
   A. is positive
   B. is negative
   C. is zero
   D. depends for its sign on the direction of the y axis
   E. depends for its sign on the direction of both the x and y axes

   ans: A

8. A baseball is hit high into the upper bleachers of left field. Over its entire flight the work done by gravity and the work done by air resistance, respectively, are:
   A. positive; positive
   B. positive; negative
   C. negative; positive
   D. negative; negative
   E. unknown since vital information is lacking

   ans: D

9. A fly ball to left field is caught at the same height as it was originally hit. Over its entire flight the work done by gravity and the work done by air resistance, respectively, are:
   A. zero; positive
   B. zero; negative
   C. positive; negative
   D. negative; positive
   E. negative; negative

   ans: B

10. A 2-kg object is moving at 3 m/s. A 4-N force is applied in the direction of motion and then removed after the object has traveled an additional 5 m. The work done by this force is:
    A. 12 J
    B. 15 J
    C. 18 J
    D. 20 J
    E. 38 J

    ans: D

11. A sledge (including load) weighs 5000 N. It is pulled on level snow by a dog team exerting a horizontal force on it. The coefficient of kinetic friction between sledge and snow is 0.05. How much work is done by the dog team pulling the sledge 1000 m at constant speed?
    A. 2.5×10^4 J
    B. 2.5×10^5 J
    C. 5.0×10^5 J
    D. 2.5×10^6 J
    E. 5.0×10^6 J

    ans: B

12. Camping equipment weighing 6000 N is pulled across a frozen lake by means of a horizontal rope. The coefficient of kinetic friction is 0.05. The work done by the campers in pulling the equipment 1000 m at constant velocity is:
    A. 3.1×10^4 J
    B. 1.5×10^5 J
    C. 3.0×10^5 J
    D. 2.9×10^6 J
    E. 6.0×10^6 J

    ans: C

13. Camping equipment weighing 6000 N is pulled across a frozen lake by means of a horizontal rope. The coefficient of kinetic friction is 0.05. How much work is done by the campers in pulling the equipment 1000 m if its speed is decreasing at the constant rate of $0.20\,\text{m/s}^2$?
    A. -1.2×10^6 J
    B. 1.8×10^5 J
    C. 3.0×10^5 J
    D. 4.2×10^5 J
    E. 1.2×10^6 J

    ans: B

14. A 1-kg block is lifted vertically 1 m by a boy. The work done by the boy is about:
    A. 1 ft · lb
    B. 1 J
    C. 10 J
    D. 0.1 J
    E. zero

    ans: C

15. A 0.50-kg object moves in a horizontal circular track with a radius of 2.5 m. An external force of 3.0 N, always tangent to the track, causes the object to speed up as it goes around. The work done by the external force as the mass makes one revolution is:

A. 24 J
B. 47 J
C. 59 J
D. 94 J
E. 120 J

ans: B

16. A man pulls a 100-N crate up a frictionless 30° slope 5 m high, as shown. Assuming that the crate moves at constant speed, the work done by the man is:

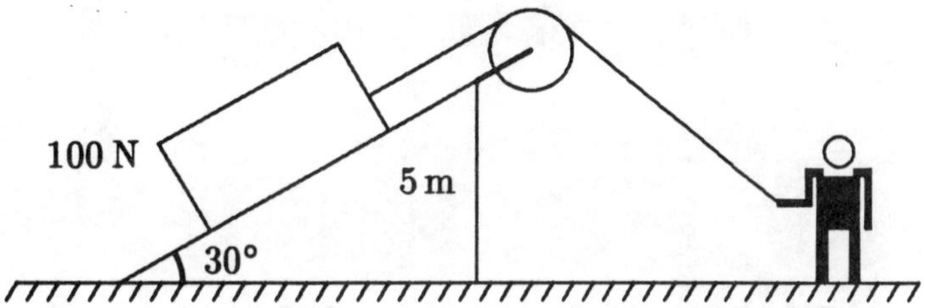

A. −500 J
B. −250 J
C. 0
D. 250 J
E. 500 J

ans: E

17. A man pushes an 80-N crate a distance of 5.0 m upward along a frictionless slope that makes an angle of 30° with the horizontal. The force he exerts is parallel to the slope. If the speed of the crate increases at a rate of $1.5\,\mathrm{m/s^2}$, then the work done by the man is:

A. −200 J
B. 61 J
C. 140 J
D. 200 J
E. 260 J

ans: E

18. A man pushes an 80-N crate a distance of 5.0 m upward along a frictionless slope that makes an angle of 30° with the horizontal. The force he exerts is parallel to the slope. If the speed of the crate is constant, then the work done by the man is:

A. −200 J
B. 61 J
C. 140 J
D. 200 J
E. 260 J

ans: D

19. A man pushes an 80-N crate at constant speed a distance of 5.0 m upward along a rough slope that makes an angle of 30° with the horizontal. The work done by the force of gravity is:

A. −400 J
B. −200 J
C. −69 J
D. 200 J
E. 400 J

ans: B

20. A man pulls a sled along a rough horizontal surface by applying a constant force $\vec{F}$ at an angle θ above the horizontal. In pulling the sled a horizontal distance d, the work done by the man is:

A. Fd
B. $Fd\cos\theta$
C. $Fd\sin\theta$
D. $Fd/\cos\theta$
E. $Fd/\sin\theta$

ans: B

21. A man wishes to pull a crate 15 m across a rough floor by exerting a force of 100 N. The coefficient of kinetic friction is 0.25. For the man to do the least work, the angle between the force and the horizontal should be:

A. 0
B. 14°
C. 43°
D. 66°
E. 76°

ans: A

22. A particle moves 5 m in the $+x$ direction while being acted upon by a constant force $\vec{F} = (4\,\mathrm{N})\hat{\imath} + (2\,\mathrm{N})\hat{\jmath} - (4\,\mathrm{N})\hat{k}$. The work done on the particle by this force is:

A. 20 J
B. 10 J
C. −20 J
D. 30 J
E. impossible to calculate without knowing other forces

ans: A

23. A block is attached to the end of an ideal spring and moved from coordinate x_i to coordinate x_f. The relaxed position is at $x = 0$. The work done by spring is positive if:

A. $x_i = 2\,\mathrm{cm}$ and $x_f = 4\,\mathrm{cm}$
B. $x_i = -2\,\mathrm{cm}$ and $x_f = 4\,\mathrm{cm}$
C. $x_i = -2\,\mathrm{cm}$ and $x_f = -4\,\mathrm{cm}$
D. $x_i = 2\,\mathrm{cm}$ and $x_f = -4\,\mathrm{cm}$
E. $x_i = -4\,\mathrm{cm}$ and $x_f = -2\,\mathrm{cm}$

ans: E

24. An ideal spring, with a pointer attached to its end, hangs next to a scale. With a 100-N weight attached, the pointer indicates "40" on the scale as shown. Using a 200-N weight instead results in "60" on the scale. Using an unknown weight X instead results in "30" on the scale. The weight of X is:

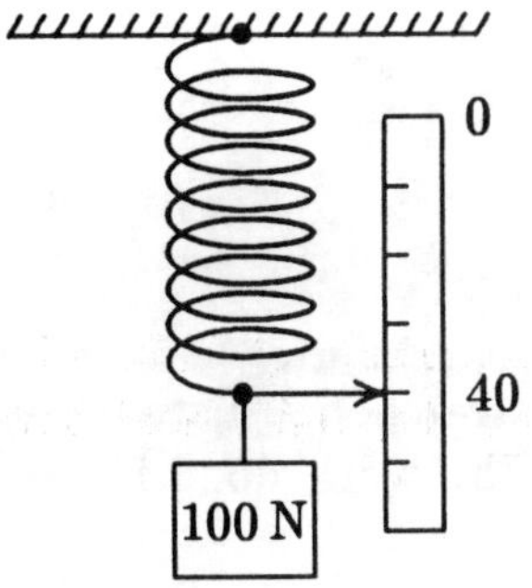

A. 10 N
B. 20 N
C. 30 N
D. 40 N
E. 50 N

ans: E

25. Three identical ideal springs (X,Y,Z) are arranged as shown. When a 4.0-kg mass is hung on X, the mass descends 3.0 cm. When a 6.0-kg mass is hung on Y, the mass descends:

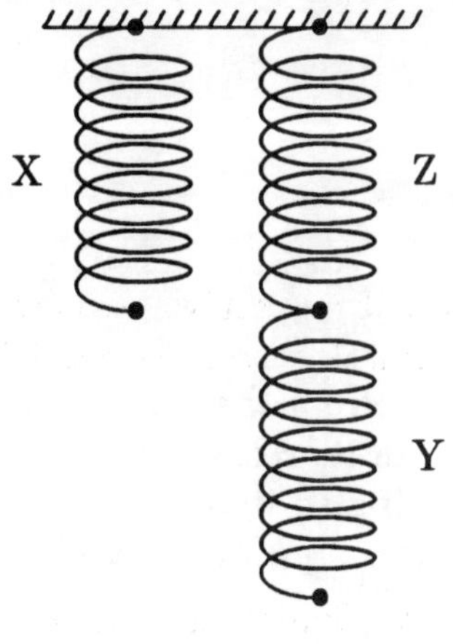

A. 2.0 cm
B. 4.0 cm
C. 4.5 cm
D. 6.0 cm
E. 9.0 cm

ans: E

26. When a certain rubber band is stretched a distance x, it exerts a restoring force of magnitude $F = Ax$, where A is a constant. The work done by a person in stretching this rubber band from $x = 0$ to $x = L$ is:

A. AL^2
B. $A + 2L$
C. $A + 2L^2$
D. A/L
E. $AL^2/2$

ans: E

27. When a certain rubber band is stretched a distance x, it exerts a restoring force $F = ax + bx^2$, where a and b are constants. The work done in stretching this rubber band from $x = 0$ to $x = L$ is:

A. $aL^2 + bLx^3$
B. $aL + 2bL^2$
C. $a + 2bL$
D. bL
E. $aL^2/2 + bL^3/3$

ans: E

28. An ideal spring is hung vertically from the ceiling. When a 2.0-kg mass hangs at rest from it the spring is extended 6.0 cm from its relaxed length. A downward external force is now applied to the mass to extend the spring an additional 10 cm. While the spring is being extended by the force, the work done by the spring is:

A. −3.6 J
B. −3.3 J
C. -3.4×10^{-5} J
D. 3.3 J
E. 3.6 J

ans: A

29. An ideal spring is hung vertically from the ceiling. When a 2.0-kg mass hangs at rest from it the spring is extended 6.0 cm from its relaxed length. A upward external force is then applied to the mass to compress the spring so it is 10 cm shorter than its equilibrium length. While the spring is being compressed the work done by the spring is:

A. −1.0 J
B. −0.52 J
C. −0.26 J
D. 0.52 J
E. 1.0 J

ans: A

30. Which of the following bodies has the largest kinetic energy?

A. Mass $3M$ and speed V
B. Mass $3M$ and speed $2V$
C. Mass $2M$ and speed $3V$
D. Mass M and speed $4V$
E. All four of the above have the same kinetic energy

ans: C

31. Two trailers, X with mass 500 kg and Y with mass 2000 kg, are being pulled at the same speed. The ratio of the kinetic energy of Y to that of X is:

A. 1:1
B. 2:1
C. 4:1
D. 9:1
E. 1500:1

ans: C

32. A 8000-N car is traveling at 12 m/s along a horizontal road when the brakes are applied. The car skids to a stop in 4.0 s. How much kinetic energy does the car lose in this time?

A. 4.8×10^4 J
B. 5.9×10^4 J
C. 1.2×10^5 J
D. 5.8×10^5 J
E. 4.8×10^6 J

ans: B

33. The velocity of a particle moving along the x axis changes from v_i to v_f. For which values of v_i and v_f is the total work done on the particle negative?

A. $v_i = 2\,\text{m/s}$, $v_f = 5\,\text{m/s}$
B. $v_i = -2\,\text{m/s}$, $v_f = 5\,\text{m/s}$
C. $v_i = -2\,\text{m/s}$, $v_f = -5\,\text{m/s}$
D. $v_i = 2\,\text{m/s}$, $v_f = -5\,\text{m/s}$
E. $v_i = -5\,\text{m/s}$, $v_f = 2\,\text{m/s}$

ans: E

34. An object is constrained by a cord to move in a circular path of radius 0.5 m on a horizontal frictionless surface. The cord will break if its tension exceeds 16 N. The maximum kinetic energy the object can have is:

A. 4 J
B. 8 J
C. 16 J
D. 32 J
E. 64 J

ans: A

35. The weight of an object on the moon is one-sixth of its weight on Earth. The ratio of the kinetic energy of a body on Earth moving with speed V to that of the same body moving with speed V on the moon is:

A. 6:1
B. 36:1
C. 1:1
D. 1:6
E. 1:36

ans: C

36. Which of the following is the correct combination of dimensions for the quantity "energy"?

A. MLT
B. LT^2/m
C. ML^2/T^2
D. M^2L^3T
E. ML/T^2

ans: C

37. The amount of work required to stop a moving object is equal to:

A. the velocity of the object
B. the kinetic energy of the object
C. the mass of the object times its acceleration
D. the mass of the object times its velocity
E. the square of the velocity of the object

ans: B

38. A 5.0-kg cart is moving horizontally at 6.0 m/s. In order to change its speed to 10.0 m/s, the net work done on the cart must be:

A. 40 J
B. 90 J
C. 160 J
D. 400 J
E. 550 J

ans: C

39. A particle is initially at rest on a horizontal frictionless table. It is acted upon by a constant horizontal force F. Which of the following five graphs is a correct plot of work W as a function of particle speed v?

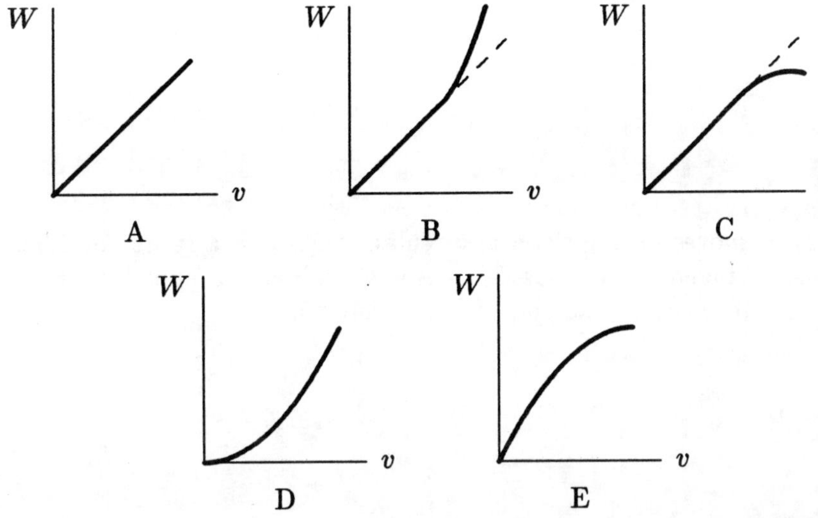

ans: D

40. An 8-N block slides down an incline. It has an initial speed of 7 m/s. The work done by the resultant force on this block is:
   A. 3 J
   B. 6 J
   C. 56 J
   D. impossible to calculate without more information
   E. none of these

   ans: D

41. A 4-kg cart starts up an incline with a speed of 3 m/s and comes to rest 2 m up the incline. The total work done on the car is:
   A. 6 J
   B. 8 J
   C. 12 J
   D. 18 J
   E. impossble to calculate without more information

   ans: D

42. Two objects with masses of m_1 and m_2 have the same kinetic energy and are both moving to the right. The same constant force $\vec{F}$ is applied to the left to both masses. If $m_1 = 4m_2$, the ratio of the stopping distance of m_1 to that of m_2 is:
   A. 1:4
   B. 4:1
   C. 1:2
   D. 2:1
   E. 1:1

   ans: E

43. A Boston Red Sox baseball player catches a ball of mass m that is moving toward him with speed v. While bringing the ball to rest, his hand moves back a distance d. Assuming constant deceleration, the horizontal force exerted on the ball by his hand is:

A. mv/d
B. mvd
C. mv^2/d
D. $2mv/d$
E. $mv^2/(2d)$

ans: E

44. A 0.50-kg object moves in a horizontal circular track with a radius of 2.5 m. An external force of 3.0 N, always tangent to the track, causes the object to speed up as it goes around. If it starts from rest its speed at the end of one revolution is:

A. 9.8 m/s
B. 14 m/s
C. 15 m/s
D. 19 m/s
E. 21 m/s

ans: B

45. A 0.50-kg object moves on a horizontal frictionless circular track with a radius of 2.5 m. An external force of 3.0 N, always tangent to the track, causes the object to speed up as it goes around. If it starts from rest, then at the end of one revolution the radial component of the force of the track on it is:

A. 19 N
B. 38 N
C. 47 N
D. 75 N
E. 96 N

ans: B

46. A 2-kg particle on the end of an ideal spring is pulled out 0.5 m and released from rest. The spring constant is 200 N/m. When the particle passes the point where the spring force is zero its speed is:

A. 0
B. 0.05 m/s
C. 5 m/s
D. 10 m/s
E. 100 m/s

ans: C

47. At time $t = 0$ a particle starts moving along the x axis. If its kinetic energy increases uniformly with t the net force acting on it must be:

A. constant
B. proportional to t
C. inversely proportional to t
D. proportional to $\sqrt{t}$
E. proportional to $1/\sqrt{t}$

ans: E

48. At time $t = 0$ a 2-kg particle has a velocity of $(4\,\text{m/s})\,\hat{\imath} - (3\,\text{m/s})\,\hat{\jmath}$. At $t = 3\,\text{s}$ its velocity is $(2\,\text{m/s})\,\hat{\imath} + (3\,\text{m/s})\,\hat{\jmath}$. During this time the work done on it was:

A. $4\,\text{J}$
B. $-4\,\text{J}$
C. $-12\,\text{J}$
D. $-40\,\text{J}$
E. $(4\,\text{J})\hat{\imath} + (36\,\text{J})\hat{\jmath}$

ans: C

49. A particle starts from rest at time $t = 0$ and moves along the x axis. If the net force on it is proportional to t, its kinetic energy is proportional to:

A. t
B. t^2
C. t^4
D. $1/t^2$
E. none of the above

ans: C

50. A 1.5-kg crate falls from a height of 2.0 m onto an industrial spring scale with a spring constant of $1.5 \times 10^5\,\text{N/m}$. At its greatest compression the reading on the scale is:

A. 15 N
B. 30 N
C. $1.5 \times 10^3\,\text{N}$
D. $2.1 \times 10^3\,\text{N}$
E. $3.0 \times 10^3\,\text{N}$

ans: E

51. A particle moving along the x axis is acted upon by a single force $F = F_0 e^{-kx}$, where F_0 and k are constants. The particle is released from rest at $x = 0$. It will attain a maximum kinetic energy of:

A. F_0/k
B. F_0/e^k
C. kF_0
D. $1/2(kF_0)^2$
E. ke^kF_0

ans: A

52. The mechanical advantage of any machine is:

A. the efficiency of the machine
B. the work done by the machine
C. the ratio of the work done by the machine to the work expended on it
D. the ratio of the force exerted by the machine to the force applied to it
E. the ratio of the force applied to the machine to the force exerted by it

ans: D

53. In raising an object to a given height by means of an inclined plane, as compared with raising the object vertically, there is a reduction in:

A. work required
B. distance pushed
C. friction
D. force required
E. value of the acceleration due to gravity

ans: D

54. A watt is:

A. $kg \cdot m/s^3$
B. $kg \cdot m^2/s$
C. $kg \cdot m^2/s^3$
D. $kg \cdot m/s$
E. $kg \cdot m^2/s^2$

ans: C

55. Power has the dimensions of:

A. ML^2/T^2
B. MT/L^2
C. ML/T^2
D. ML^2/T^3
E. none of these

ans: D

56. Which of the following five units represents a quantity that is NOT the same as the other four?

A. joule
B. erg
C. watt
D. foot·pound
E. newton·meter

ans: C

57. Which of the following five quantities is NOT an expression for energy? Here m is a mass, g is the acceleration due to gravity, h and d are distances, F is a force, v is a speed, a is an acceleration, P is power, and t is time.

A. mgh
B. Fd
C. $1/2mv^2$
D. ma
E. Pt

ans: D

58. A watt·second is a unit of:

A. force
B. power
C. displacement
D. speed
E. energy

ans: E

59. A watt per hour is a unit of:

A. energy
B. power
C. force
D. acceleration
E. none of these

ans: E

60. A kilowatt·hour is a unit of:

A. power
B. energy/time
C. work
D. power/time
E. force/distance

ans: C

61. A man moves the 10-g object shown in a vertical plane from position X to position Y along a circular track of radius 20 m. The process takes 0.75 min. The work done by the man is about:

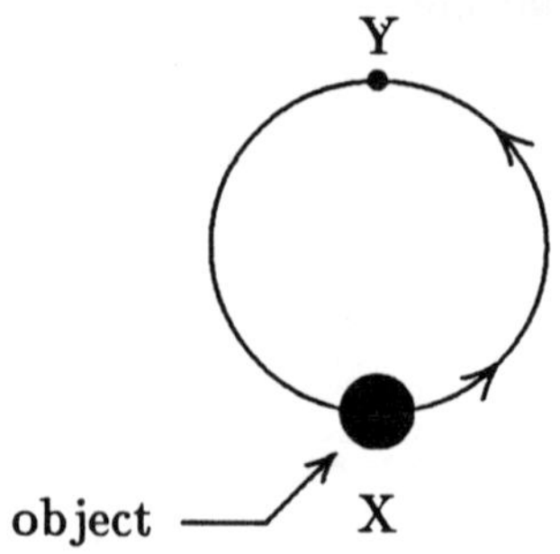

A. 1 J
B. 2 J
C. 4 J
D. 6 J
E. 12 J

ans: C

62. A woman lifts a barbell 2.0 m in 5.0 s. If she lifts it the same distance in 10 s, the work done by her is:

A. four times as great
B. two times as great
C. the same
D. half as great
E. one-fourth as great

ans: C

63. An escalator is used to move 20 people (60 kg each) per minute from the first floor of a department store to the second floor, 5 m above. Neglecting friction, the power required is approximately:

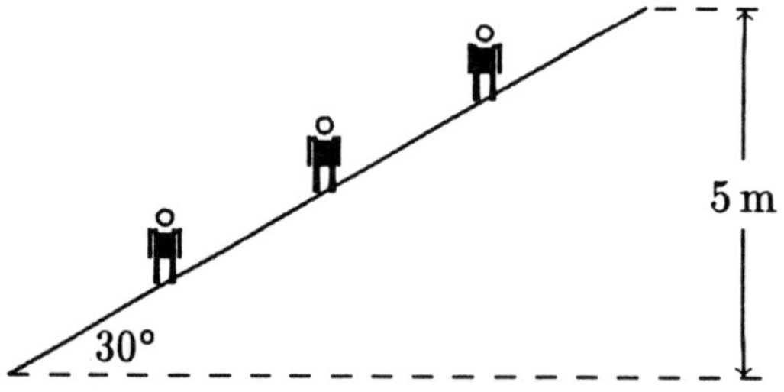

A. 100 W
B. 200 W
C. 1000 W
D. 2000 W
E. 60,000 W

ans: C

64. A person holds an 80-N weight 2 m above the floor for 30 seconds. The power required to do this is:

A. 80 W
B. 40 W
C. 20 W
D. 10 W
E. none of these

ans: E

65. A 50-N force acts on a 2-kg crate that starts from rest. When the force has been acting for 2 s the rate at which it is doing work is:

A. 75 W
B. 100 W
C. 1000 W
D. 2500 W
E. 5000 W

ans: D

66. A 50-N force acts on a 2-kg crate that starts from rest. At the instant the particle has gone 2 m the rate at which the force is doing work is:

A. 2.5 W
B. 25 W
C. 75 W
D. 100 W
E. 500 W

ans: E

67. A particle starts from rest and is acted on by a net force that does work at a rate that is proportional to the time t. The speed of the particle is proportional to:

A. $\sqrt{t}$
B. t
C. t^2
D. $1/\sqrt{t}$
E. $1/t$

ans: A

Chapter 8: POTENTIAL ENERGY AND CONSERVATION OF ENERGY

1. If a force acting is conservative then:
   A. its work is zero when the particle moves exactly once around any closed path
   B. its work equals the change in the kinetic energy of the particle
   C. it obeys Newton's second law
   D. it obeys Newton's third law
   E. it is not a frictional force

   ans: A

2. A nonconservative force:
   A. violates Newton's second law
   B. violates Newton's third law
   C. cannot do any work
   D. must be perpendicular to the velocity of the particle on which it acts
   E. none of the above

   ans: E

3. The sum of the kinetic and potential energies of a system of objects is conserved:
   A. only when no external force acts on the objects
   B. only when the objects move along closed paths
   C. only when the work done by the resultant external force is zero
   D. always
   E. none of the above

   ans: E

4. A force acting on a particle is conservative if:
   A. its work equals the change in the kinetic energy of the particle
   B. it obeys Newton's second law
   C. it obeys Newton's third law
   D. its work depends on the end points of the motion, not the path between
   E. it is not a frictional force

   ans: D

5. Two particles interact by conservative forces and no other forces act. They complete round trips, ending at the points where they started. Over this trip:
   A. the total kinetic energy might have a different value at the beginning and end
   B. the potential energy might have a different value at the beginning and end
   C. the total mechanical energy might have a different value at the beginning and end
   D. the total linear momentum might have different values at the beginning and end
   E. none of the above

   ans: E

6. Two objects interact with each other and with no other objects. Initially object A has a speed of 5 m/s and object B has a speed of 10 m/s. In the course of their motion they return to their initial positions. Then A has a speed of 4 m/s and B has a speed of 7 m/s. We can conclude:
   A. the potential energy changed from the beginning to the end of the trip
   B. mechanical energy was increased by nonconservative forces
   C. mechanical energy was decreased by nonconservative forces
   D. mechanical energy was increased by conservative forces
   E. mechanical energy was decreased by conservative forces

   ans: C

7. A good example of kinetic energy is provided by:
   A. a woundup clock spring
   B. the raised weights of a grandfather's clock
   C. a tornado
   D. a gallon of gasoline
   E. an automobile storage battery

   ans: C

8. No kinetic energy is possessed by:
   A. a shooting star
   B. a rotating propeller on a moving airplane
   C. a pendulum at the bottom of its swing
   D. an elevator standing at the fifth floor
   E. a cyclone

   ans: D

9. The woundup spring of a clock possesses:
   A. kinetic but no potential energy
   B. potential but no kinetic energy
   C. both potential and kinetic energy
   D. neither potential nor kinetic energy
   E. depends on the system of units

   ans: B

10. A body at rest is capable of doing work if:
   A. the potential energy is positive
   B. the potential energy is negative
   C. it is free to move in such a way as to decrease its kinetic energy
   D. it is free to move in such a way as to decrease the potential energy
   E. it is free to move in such a way as to increase the potential energy

   ans: D

11. Which one of the following five quantities CANNOT be used as a unit of potential energy?

A. watt·second
B. gram·cm/s^2
C. joule
D. kg·m^2/s^2
E. ft·lb

ans: B

12. Suppose that the fundamental dimensions are taken to be: force (F), velocity (V) and time (T). The dimensions of potential energy are then:

A. F/T
B. FVT
C. FV/T
D. F/T^2
E. FV^2/T^2

ans: B

13. The graphs below show the force acting on a particle as the particle moves along the positive x axis from the origin to $x = x_1$. The force is parallel to the x axis and is conservative. The maximum magnitude F_1 has the same value for all graphs. Rank the situations according to the change in the potential energy associated with the force, least (or most negative) to greatest (or most positive).

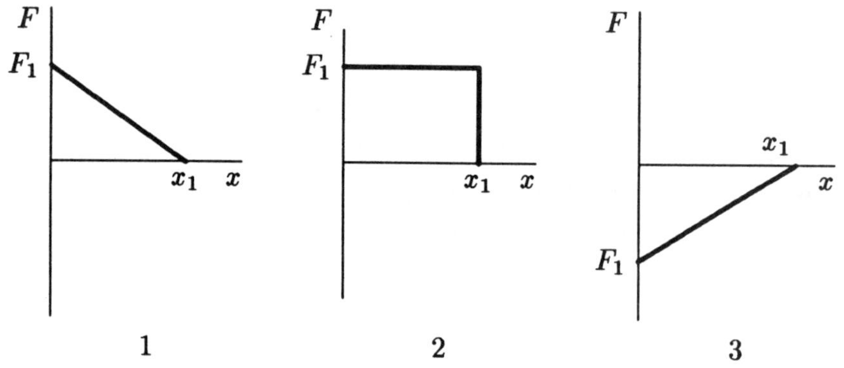

A. 1, 2, 3
B. 1, 3, 2
C. 2, 3, 1
D. 3, 2, 1
E. 2, 1, 3

ans: E

14. A golf ball is struck by a golf club and falls on a green three meters above the tee. The potential energy of the Earth-ball system is greatest:

A. just before the ball is struck
B. just after the ball is struck
C. just after the ball lands on the green
D. when the ball comes to rest on the green
E. when the ball reaches the highest point in its flight

ans: E

15. A ball is held at a height H above a floor. It is then released and falls to the floor. If air resistance can be ignored, which of the five graphs below correctly gives the mechanical energy , E of the Earth-ball system as a function of the altitude y of the ball?

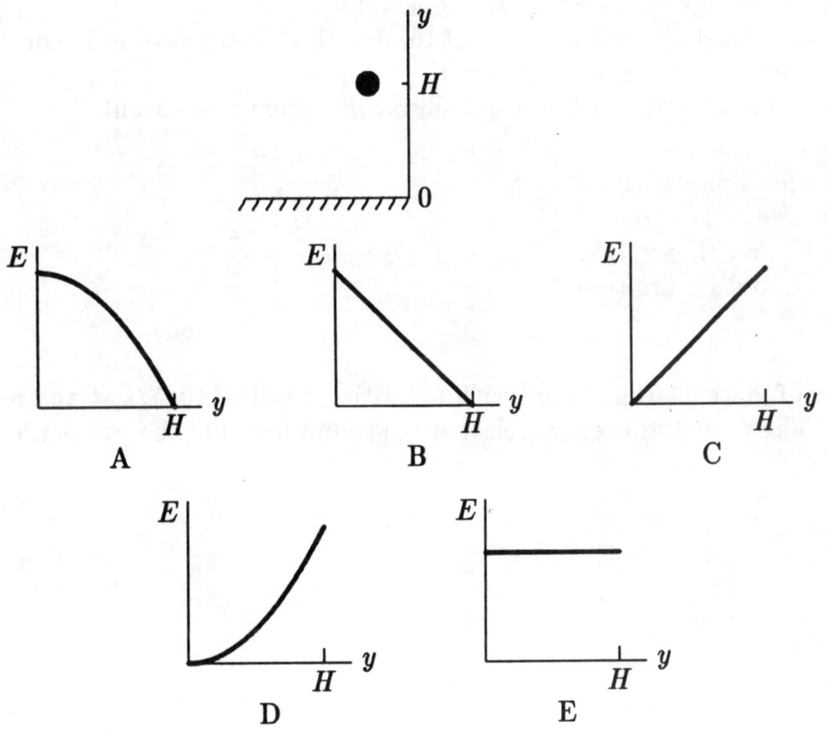

ans: E

16. A 6.0-kg block is released from rest 80 m above the ground. When it has fallen 60 m its kinetic energy is approximately:

A. 4800 J
B. 3500 J
C. 1200 J
D. 120 J
E. 60 J

ans: B

17. A 2-kg block is thrown upward from a point 20 m above Earth's surface. At what height above Earth's surface will the gravitational potential energy of the Earth-block system have increased by 500 J?

A. 5 m
B. 25 m
C. 46 m
D. 70 m
E. 270 m

ans: C

18. An elevator is rising at constant speed. Consider the following statements:

I. the upward cable force is constant
II. the kinetic energy of the elevator is constant
III. the gravitational potential energy of the Earth-elevator system is constant
IV. the acceleration of the elevator is zero
V. the mechanical energy of the Earth-elevator system is constant

A. all five are true
B. only II and V are true
C. only IV and V are true
D. only I, II, and III are true
E. only I, II, and IV are true

ans: E

19. A projectile of mass 0.50 kg is fired with an initial speed of 10 m/s at an angle of 60° above the horizontal. The potential energy (relative to ground level) of the projectile at its highest point is:

A. 25 J
B. 18.75 J
C. 12.5 J
D. 6.25 J
E. none of these

ans: B

20. For a block of mass m to slide without friction up the rise of height h shown, it must have a minimum initial speed of:

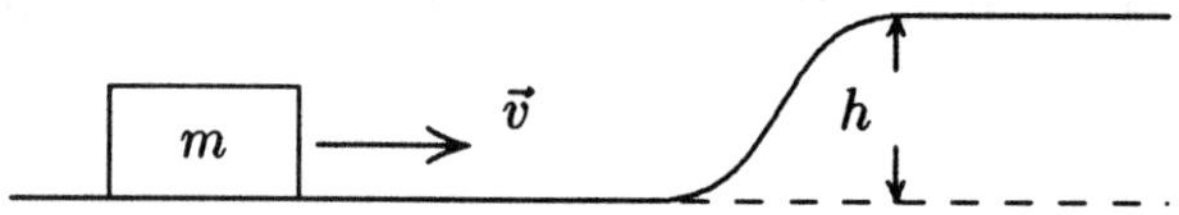

A. $1/2\sqrt{gh}$
B. $\sqrt{gh/2}$
C. $\sqrt{2gh}$
D. $2\sqrt{2gh}$
E. $2\sqrt{gh}$

ans: C

21. A 2.2-kg block starts from rest on a rough inclined plane that makes an angle of 25° with the horizontal. The coefficient of kinetic friction is 0.25. As the block goes 2.0 m down the plane, the mechanical energy of the Earth-block system changes by:

A. 0
B. −9.8 J
C. 9.8 J
D. −18 J
E. 18 J

ans: B

22. A simple pendulum consists of a 2.0-kg mass attached to a string. It is released from rest at X as shown. Its speed at the lowest point Y is about:

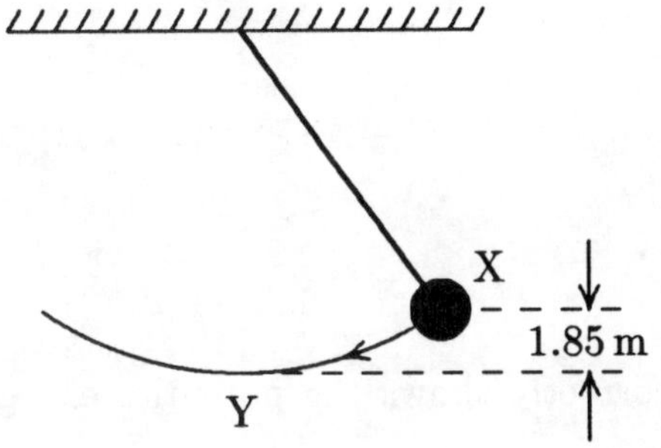

A. 0.90 m/s
B. $\sqrt{3.6}$ m/s
C. 3.6 m/s
D. 6.0 m/s
E. 36 m/s

ans: D

23. The long pendulum shown is drawn aside until the ball has risen 0.5 m. It is then released from rest. The speed of the ball at its lowest position:

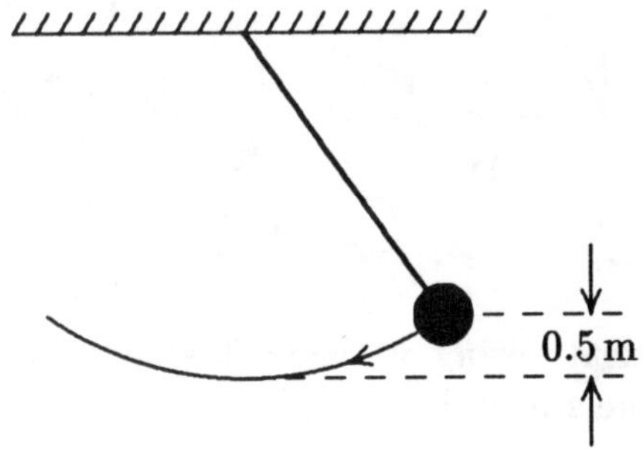

A. 3.1 m/s
B. 4.2 m/s
C. 5.8 m/s
D. 20 m/s
E. cannot be determined unless the mass is known

ans: A

24. A particle moves along the x axis under the influence of a stationary object. The force exerted on the particle is given by $F = 8x^3$, where F is in newtons and x is in meters. If $x = 0$ is taken to be the zero of potential energy, then the potential energy in joules is given by:

A. $2x^4$
B. $-2x^4$
C. $24x^2$
D. $-24x^2$
E. $5 - 2x^4$

ans: B

25. A 0.20-kg particle moves along the x axis under the influence of a stationary object. The potential energy is given by $U(x) = 8x^2 + 2x^4$, where U is in joules and x is in meters. If the particle has a speed of 5.0 m/s when it is at $x = 1.0$ m, its speed when it is at the origin is:

A. 0
B. 2.5 m/s
C. 5.7 m/s
D. 7.9 m/s
E. 11 m/s

ans: E

26. Which of the five graphs correctly shows the potential energy of a spring as a function of its elongation x?

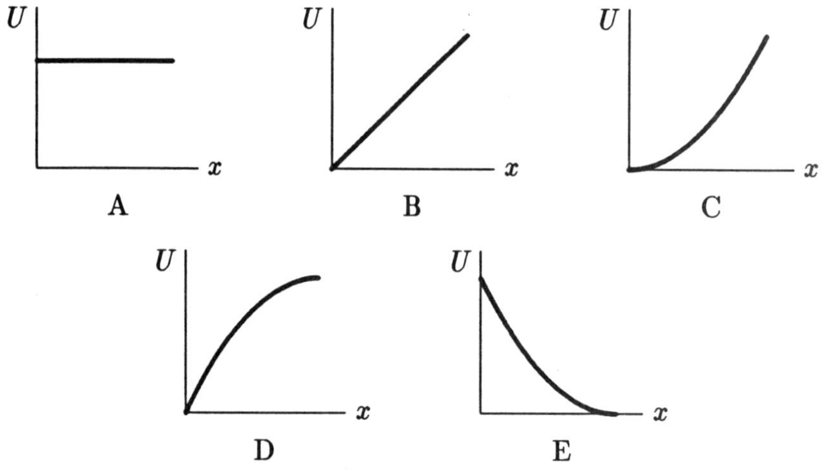

ans: C

27. A force of 10 N holds an ideal spring with a 20-N/m spring constant in compression. The potential energy stored in the spring is:

A. 0.5 J
B. 2.5 J
C. 5 J
D. 10 J
E. 200 J

ans: B

28. An ideal spring is used to fire a 15.0-g block horizontally across a frictionless table top. The spring has a spring constant of 20 N/m and is initially compressed by 7.0 cm. The speed of the block as it leaves the spring is:

A. 0
B. 1.9×10^{-3} m/s
C. 2.6×10^{-2} m/s
D. 0.39 m/s
E. 2.6 m/s

ans: E

29. A 0.50-kg block attached to an ideal spring with a spring constant of 80 N/m oscillates on a horizontal frictionless surface. The total mechanical energy is 0.12 J. The greatest extension of the spring from its equilibrium length is:

A. 1.5×10^{-3} m
B. 3.0×10^{-3} m
C. 0.039 m
D. 0.054 m
E. 18 m

ans: D

30. A 0.50-kg block attached to an ideal spring with a spring constant of 80 N/m oscillates on a horizontal frictionless surface. The total mechanical energy is 0.12 J. The greatest speed of the block is:

A. 0.15 m/s
B. 0.24 m/s
C. 0.49 m/s
D. 0.69 m/s
E. 1.46 m/s

ans: D

31. A 0.50-kg block attached to an ideal spring with a spring constant of 80 N/m oscillates on a horizontal frictionless surface. When the spring is 4.0 cm shorter than its equilibrium length, the speed of the block is 0.50 m/s. The greatest speed of the block is:

A. 0.23 m/s
B. 0.32 m/s
C. 0.55 m/s
D. 0.71 m/s
E. 0.93 m/s

ans: D

32. A 0.5-kg block slides along a horizontal frictionless surface at 2 m/s. It is brought to rest by compressing a very long spring of spring constant 800 N/m. The maximum spring compression is:

A. .1 cm
B. 3 cm
C. 5 cm
D. 80 cm
E. 80 m

ans: C

33. A block of mass m is initially moving to the right on a horizontal frictionless surface at a speed v. It then compresses a spring of spring constant k. At the instant when the kinetic energy of the block is equal to the potential energy of the spring, the spring is compressed a distance of:

A. $v\sqrt{m/2k}$
B. $(1/2)mv^2$
C. $(1/4)mv^2$
D. $mv^2/4k$
E. $(1/4)\sqrt{mv/k}$

ans: A

34. A 700-N man jumps out of a window into a fire net 10 m below. The net stretches 2 m before bringing the man to rest and tossing him back into the air. The maximum potential energy of the net is:

A. 300 J
B. 710 J
C. 850 J
D. 7000 J
E. 8400 J

ans: E

35. A toy cork gun contains a spring whose spring constant is 10.0 N/m. The spring is compressed 5 cm and then used to propel a 6.00-g cork. The cork, however, sticks to the spring for 1.00 cm beyond its unstretched length before separation occurs. The muzzle velocity of this cork is:

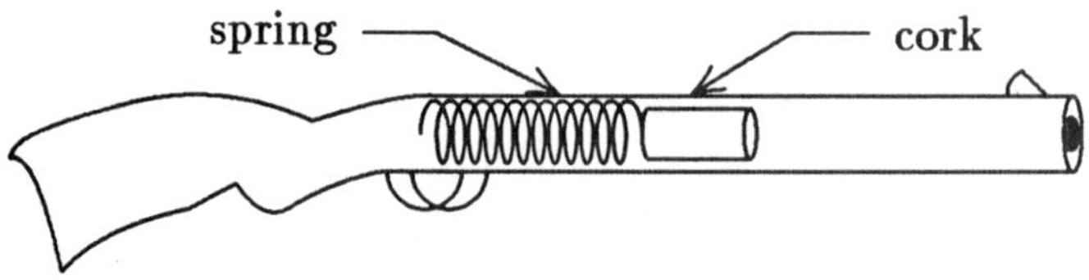

A. 1.02 m/s
B. 1.41 m/s
C. 2.00 m/s
D. 2.04 m/s
E. 4.00 m/s

ans: C

36. A small object of mass m, on the end of a light cord, is held horizontally at a distance r from a fixed support as shown. The object is then released. What is the tension force of the cord when the object is at the lowest point of its swing?

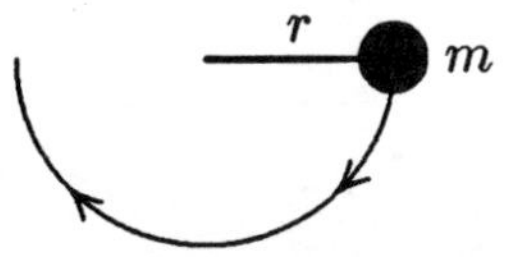

A. $mg/2$
B. mg
C. $2mg$
D. $3mg$
E. mgr

ans: D

37. The string in the figure is 50 cm long. When the ball is released from rest, it will swing along the dotted arc. How fast, in m/s, will it be going at the lowest point in its swing?

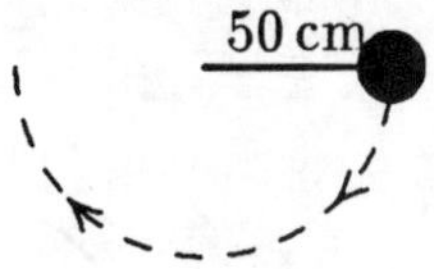

A. 2.0
B. 2.2
C. 3.1
D. 4.4
E. 6.0

ans: C

38. A block is released from rest at point P and slides along the frictionless track shown. At point Q, its speed is:

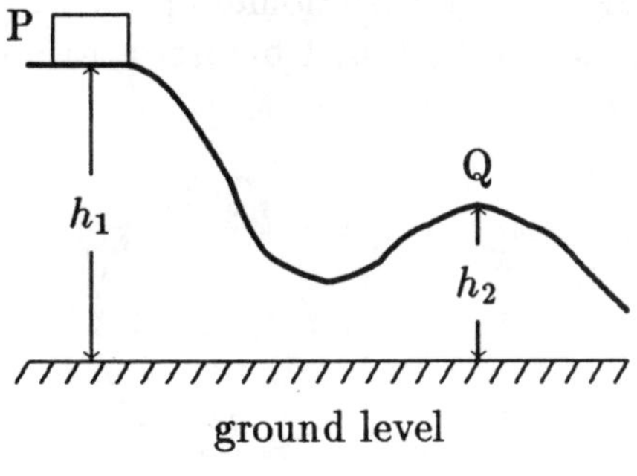

A. $2g\sqrt{h_1 - h_2}$
B. $2g(h_1 - h_2)$
C. $(h_1 - h_2)/2g$
D. $\sqrt{2g(h_1 - h_2)}$
E. $(h_1 - h_2)^2/2g$

ans: D

39. A small object of mass m starts from rest at the position shown and slides along the frictionless loop-the-loop track of radius R. What is the smallest value of y such that the object will slide without losing contact with the track?

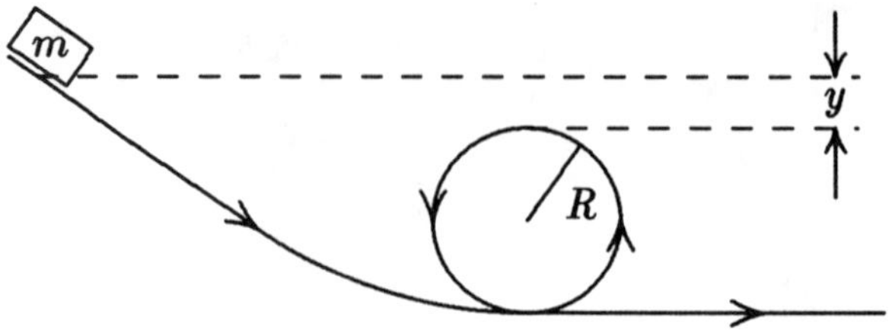

A. $R/4$
B. $R/2$
C. R
D. $2R$
E. zero

ans: B

40. A small object slides along the frictionless loop-the-loop with a diameter of 3 m. What minimum speed must it have at the top of the loop?

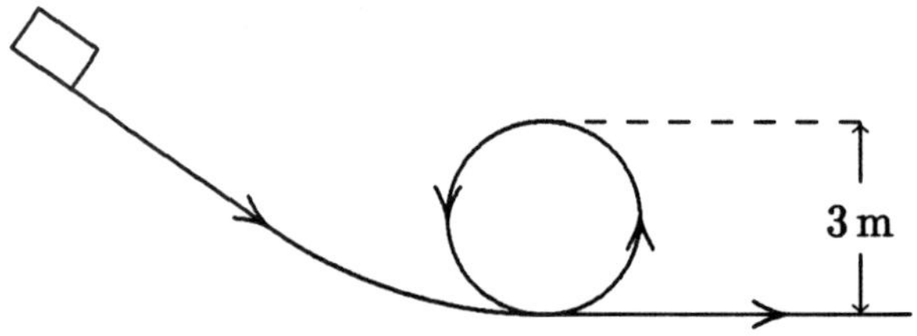

A. 1.9 m/s
B. 3.8 m/s
C. 5.4 m/s
D. 15 m/s
E. 29 m/s

ans: B

41. A rectangular block is moving along a frictionless path when it encounters the circular loop as shown. The block passes points 1, 2, 3, 4, 1 before returning to the horizontal track. At point 3:

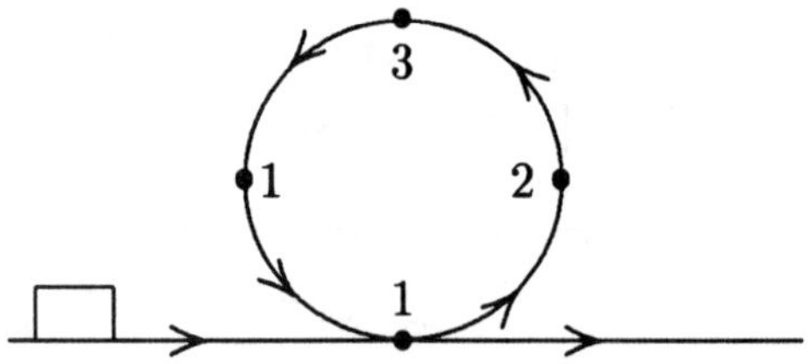

A. its mechanical energy is a minimum
B. the forces on it are balanced
C. it is not accelerating
D. its speed is a minimum
E. it experiences a net upward force

ans: D

42. A ball of mass m, at one end of a string of length L, rotates in a vertical circle just fast enough to prevent the string from going slack at the top of the circle. The speed of the ball at the bottom of the circle is:

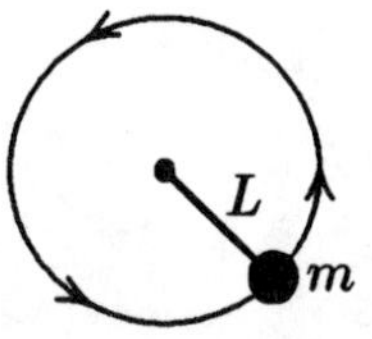

A. $\sqrt{2gL}$
B. $\sqrt{3gL}$
C. $\sqrt{4gL}$
D. $\sqrt{5gL}$
E. $\sqrt{7gL}$

ans: D

43. A particle is released from rest at the point $x = a$ and moves along the x axis subject to the potential energy function $U(x)$ shown. The particle:

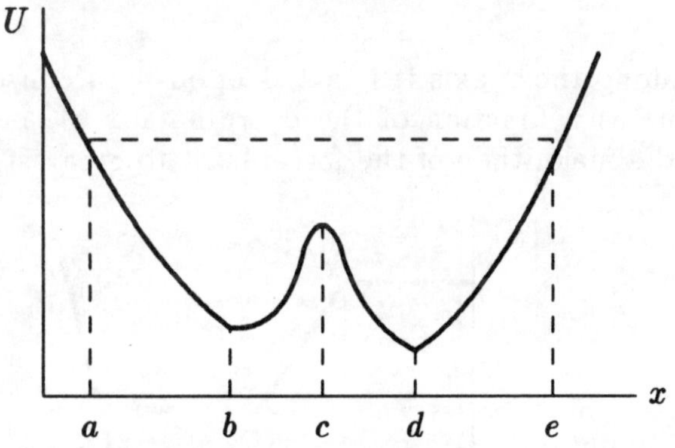

A. oscillates back and forth between $x = a$ and $x = e$
B. oscillates back and forth between $x = a$ and $x = c$
C. moves to infinity at varying speed
D. moves to $x = e$, where it remains at rest
E. moves to $x = b$, where it remains at rest

ans: A

44. The potential energy of a particle moving along the x axis is given by $U(x) = 8x^2 + 2x^4$, where U is in joules and x is in meters. If the total mechanical energy is 9.0 J, the limits of motion are:

A. -0.96 m; $+0.96$ m
B. -2.2 m; $+2.2$ m
C. -1.6 m; $+1.6$ m
D. -0.96 m; $+2.2$ m
E. -0.96 m; $+1.6$ m

ans: A

45. The potential energy of a 0.20-kg particle moving along the x axis is given by $U(x) = 8x^2 + 2x^4$, where U is in joules and x is in meters. When the particle is at $x = 1.0\,\text{m}$ it is traveling in the positive x direction with a speed of $5.0\,\text{m/s}$. It next stops momentarily to turn around at $x =$

A. 0
B. $-1.1\,\text{m}$
C. $1.1\,\text{m}$
D. $-2.3\,\text{m}$
E. $2.3\,\text{m}$

ans: C

46. Given a potential energy function $U(x)$, the corresponding force $\vec{F}$ is in the positive x direction if:

A. U is positive
B. U is negative
C. U is an increasing function of x
D. U is a decreasing function of x
E. it is impossible to obtain the direction of $\vec{F}$ from U

ans: D

47. As a particle moves along the x axis it is acted upon by a conservative force. The potential energy is shown below as a function of the coordinate x of the particle. Rank the labeled regions according to the magnitude of the force, least to greatest.

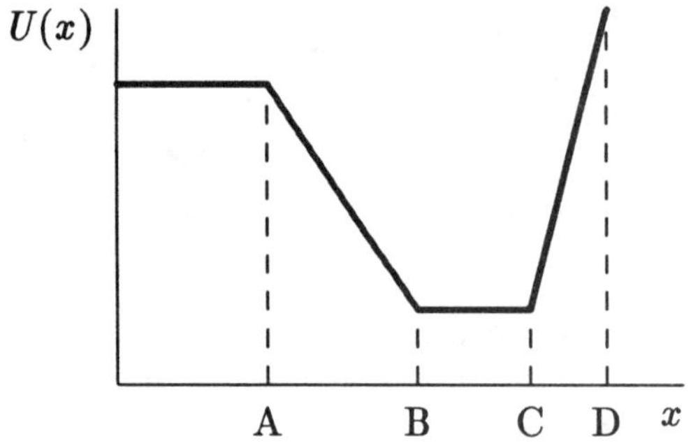

A. AB, BC, CD
B. AB, CD, BC
C. BC, CD, AB
D. BC, AB, CD
E. CD, BC, AB

ans: D

48. **The first graph shows the potential energy $U(x)$ for a particle moving on the x axis. Which of the other five graphs correctly gives the force F exerted on the particle?**

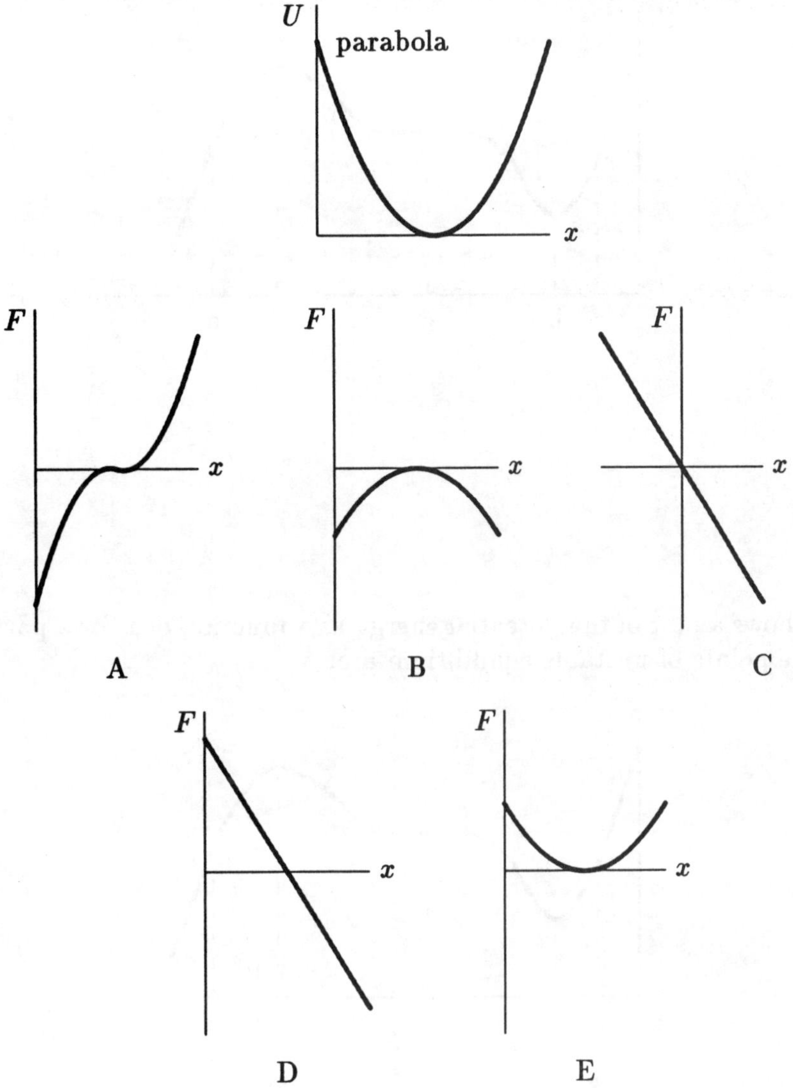

ans: D

49. The diagram shows a plot of the potential energy as a function of x for a particle moving along the x axis. The points of stable equilibrium are:

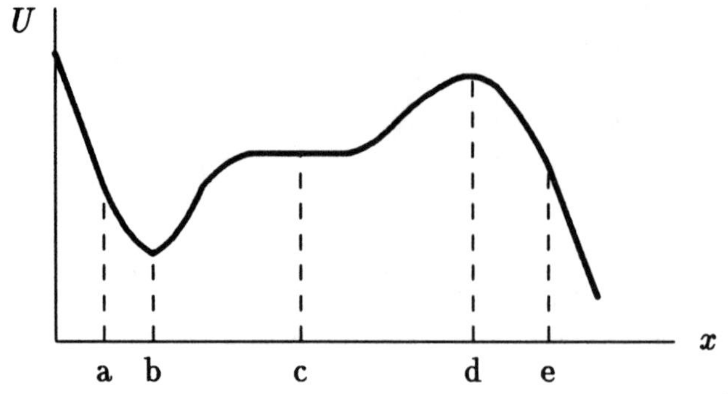

A. only a
B. only b
C. only c
D. only d
E. b and d

ans: B

50. The diagram shows a plot of the potential energy as a function of x for a particle moving along the x axis. The points of unstable equilibrium are:

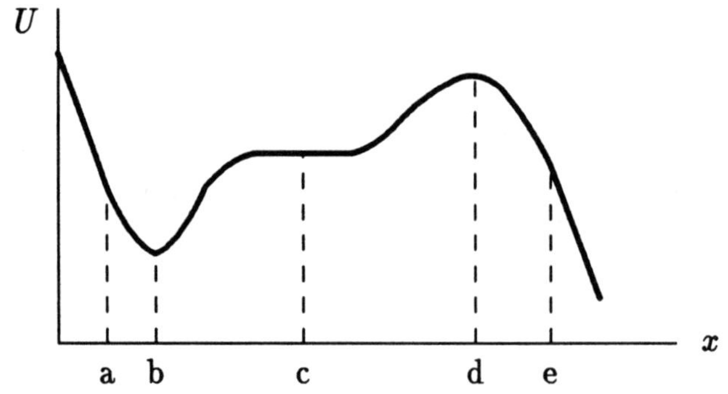

A. only a
B. only b
C. only c
D. only d
E. b and d

ans: D

51. The diagram shows a plot of the potential energy as a function of x for a particle moving along the x axis. Of the labeled points, the points of neutral equilibrium are:

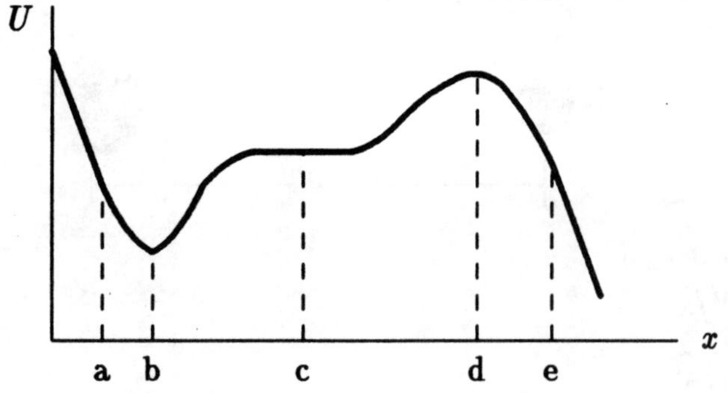

A. only a
B. only b
C. only c
D. only d
E. b and d

ans: C

52. The potential energy of a body of mass m is given by $U = -mgx + \frac{1}{2}kx^2$. The corresponding force is:

A. $-mgx^2/2 + kx^3/6$
B. $mgx^2/2 - kx^3/6$
C. $-mg + kx/2$
D. $-mg + kx$
E. $mg - kx$

ans: E

53. The potential energy of a 0.20-kg particle moving along the x axis is given by $U(x) = 8x^2 - 2x^4$, where U is in joules and x is in meters. When the particle is at $x = 1.0\,\mathrm{m}$ its acceleration is:

A. 0
B. $-8\,\mathrm{m/s^2}$
C. $8\,\mathrm{m/s^2}$
D. $-40\,\mathrm{m/s^2}$
E. $40\,\mathrm{m/s^2}$

ans: D

54. The potential energy for the interaction between the two atoms in a diatomic molecule is $U = A/x^{12} - B/x^6$, where A and B are constants and x is the interatomic distance. The force that one atom exerts on the other is:

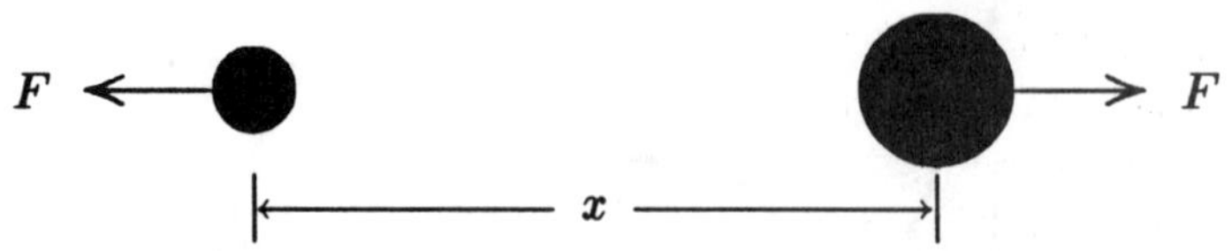

A. $12A/x^{13} - 6B/x^7$
B. $-13A/x^{13} + 7B/x^7$
C. $-11A/x^{11} + 5B/x^5$
D. $72A/x^{12} - 72B/x^6$
E. $A/x^{13} - B/x^7$

ans: A

55. The internal energy of a system consisting of a thrown ball, Earth, and the air is most closely associated with:

A. the gravitational interaction of Earth and the ball
B. the kinetic energy of the ball as a whole
C. motions of the individual particles within the ball
D. motions of individual particles within the ball and the air
E. the kinetic energy of Earth as a whole

ans: D

56. Three identical blocks move either on a horizontal surface, up a plane, or down a plane, as shown below. They all start with the same speed and continue to move until brought to rest by friction. Rank the three situations according to the mechanical energy dissipated by friction, least to greatest.

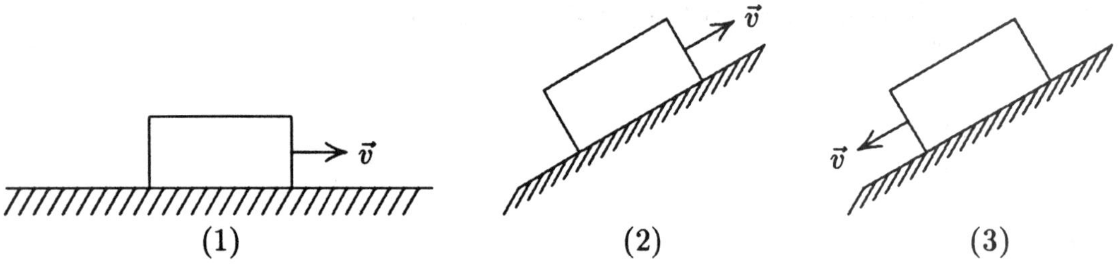

A. The same for all cases
B. 1, 2, 3
C. 1, then 2 and 3 tie
D. 3, 2, 1
E. 2, 1, 3

ans: E

57. Objects A and B interact with each other via both conservative and nonconservative forces. Let K_A and K_B be the kinetic energies, U be the potential energy, and E_{int} be the internal energy. If no external agent does work on the objects then:

A. $K_A + U$ is conserved
B. $K_A + U + E_{\text{int}}$ is conserved
C. $K_A + K_B + E_{\text{int}}$ is conserved
D. $K_A + K_B + U$ is conserved
E. $K_A + K_B + U + E_{\text{int}}$ is conserved

ans: E

58. A block slides across a rough horizontal table top. The work done by friction changes:

A. only the kinetic energy
B. only the potential energy
C. only the internal energy
D. only the kinetic and potential energies
E. only the kinetic and internal energies

ans: E

59. A 25-g ball is released from rest 80 m above the surface of Earth. Just before it hits the surface its speed is 20 m/s. During the fall the internal energy of the ball and air changed by:

A. +15 J
B. −15 J
C. +5 J
D. −5 J
E. 0

ans: A

60. A 5-kg projectile is fired over level ground with a velocity of 200 m/s at an angle of 25° above the horizontal. Just before it hits the ground its speed is 150 m/s. Over the entire trip the change in the internal energy of the projectile and air is:

A. +19,000 J
B. −19,000 J
C. +44,000 J
D. −44,000 J
E. 0

ans: C

61. A 0.75-kg block slides on a rough horizontal table top. Just before it hits a horizontal ideal spring its speed is 3.5 m/s. It compresses the spring 5.7 cm before coming to rest. If the spring constant is 1200 N/m, the internal energy of the block and the table top must have increased by:

A. 0
B. 1.9 J
C. 2.6 J
D. 4.6 J
E. 6.5 J

ans: C

Chapter 9: SYSTEMS OF PARTICLES

1. Which one of the following statements is true?
   A. the center of mass of an object must lie within the object
   B. all the mass of an object is actually concentrated at its center of mass
   C. the center of mass of an object cannot move if there is zero net force on the object
   D. the center of mass of a cylinder must lie on its axis
   E. none of the above

   ans: E

2. The x, y coordinates in meters of the center of mass of the three-particle system shown below are:

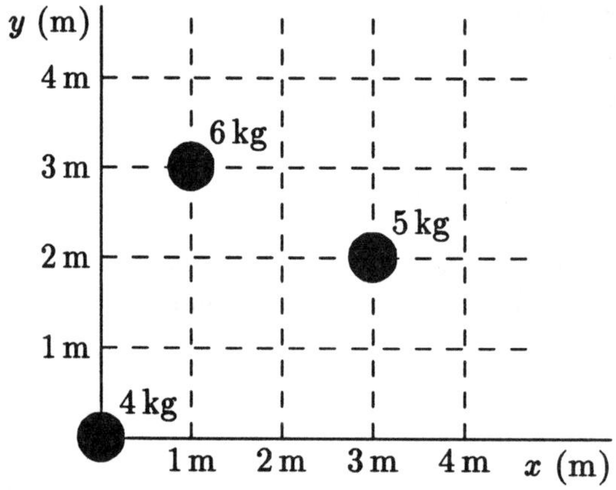

   A. 0,0
   B. 1.3 m, 1.7 m
   C. 1.4 m, 1.9 m
   D. 1.9 m, 2.5 m
   E. 1.4 m, 2.5 m

   ans: C

3. The center of mass of a uniform disk of radius R is located:
   A. on the rim
   B. a distance $R/2$ from the center
   C. a distance $R/3$ from the center
   D. a distance $2R/3$ from the center
   E. at the center

   ans: E

4. The center of mass of the system consisting of Earth, the Sun, and the planet Mars is:
   A. closer to Earth than to either of the other bodies
   B. closer to the Sun than to either of the other bodies
   C. closer to Mars than to either of the other bodies
   D. at the geometric center of the triangle formed by the three bodies
   E. at the center of the line joining Earth and Mars

   ans: B

5. The center of mass of Earth's atmosphere is:
   A. a little less than halfway between Earth's surface and the outer boundary of the atmosphere
   B. near the surface of Earth
   C. near the outer boundary of the atmosphere
   D. near the center of Earth
   E. none of the above

   ans: D

6. A thick uniform wire is bent into the shape of the letter "U" as shown. Which point indicates the location of the center of mass of this wire?

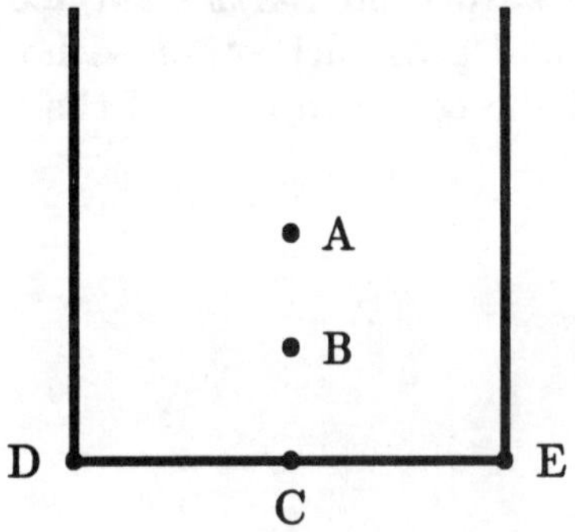

   ans: B

7. A machinist starts with three identical square plates but cuts one corner from one of them, two corners from the second, and three corners from the third. Rank the three plates according to the x coordinate of their centers of mass, from smallest to largest.

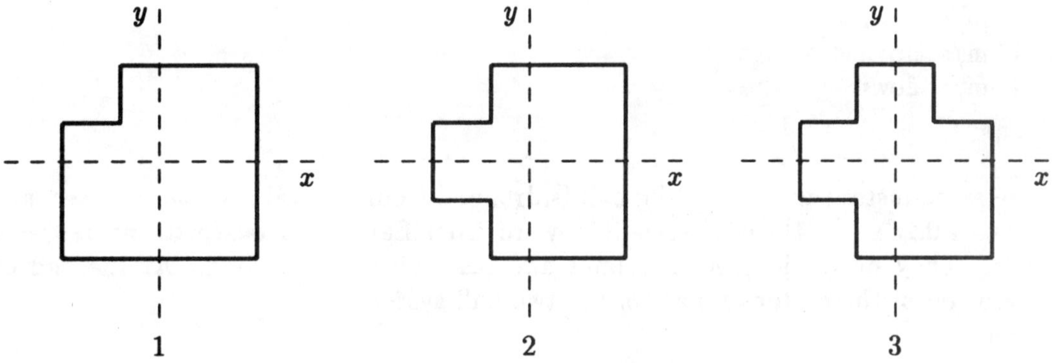

   A. 1, 2, 3
   B. 1 and 2 tie, then 3
   C. 1, then 2 and 3 tie
   D. 3, 2, 1
   E. 1 and 3 tie, then 2

   ans: E

8. Block A, with a mass of 4 kg, is stationary while block B, with a mass of 8 kg, is moving at 3 m/s. The center of mass of the two block system has a speed of:
   A. 0
   B. 1.5 m
   C. 2 m
   D. 3 m
   E. 12 m

   ans: C

9. At the same instant that a 0.50-kg ball is dropped from 25 m above Earth, a second ball, with a mass of 0.25 kg, is thrown straight upward from Earth's surface with an initial velocity of 15 m/s. They move along nearby lines and pass each other without colliding. At the end of 2.0 s the height above Earth's surface of the center of mass of the two-ball system is:
   A. 2.9 m
   B. 4.0 m
   C. 5.0 m
   D. 7.1 m
   E. 10.4 m

   ans: D

10. At the same instant that a 0.50-kg ball is dropped from 25 m above Earth, a second ball, with a mass of 0.25 kg, is thrown straight upward from Earth's surface with an initial velocity of 15 m/s. They move along nearby lines and pass without colliding. At the end of 2.0 s the velocity of the center of mass of the two-ball system is:
    A. 11 m/s, down
    B. 11 m/s, up
    C. 15 m/s, down
    D. 15 m/s, up
    E. 20 m/s, down

    ans: C

11. At the same instant that a 0.50-kg ball is dropped from 25 m above Earth, a second ball, with a mass of 0.25 kg, is thrown straight upward from Earth's surface with an initial velocity of 15 m/s. They move along nearby lines and pass without colliding. At the end of 2.0 s the acceleration of the center of mass of the two-ball system is:
    A. $0.25g$
    B. $0.50g$
    C. $0.75g$
    D. g
    E. $g/0.75$

    ans: D

12. At the same instant that a 0.50-kg ball is dropped from a high building, a second ball, with a mass of 0.25 kg, is thrown straight upward from Earth's surface with an initial velocity of 19.6 m/s. They move along nearby lines and pass without colliding. When the second ball is at its highest point the velocity of the center of mass of the two-ball system is:

A. 0
B. 13 m/s, down
C. 20 m/s, down
D. 27 m/s, down
E. 40 m/s, down

ans: B

13. A light rope passes over a light frictionless pulley attached to the ceiling. A large mass is tied to one end and a smaller mass is tied to the other. Starting from rest the larger mass moves downward and the smaller moves upward with the same magnitude acceleration. Which of the following statements is true for the system consisting of the two masses?

A. The center of mass remains at rest.
B. The net external force is zero.
C. The velocity of the center of mass is a constant.
D. The acceleration of the center of mass is g, downward.
E. None of the above statements are true.

ans: E

14. Two 4.0-kg blocks are tied together with a compressed spring between them. They are thrown from the ground with an initial velocity of 35 m/s, 45° above the horizontal. At the highest point of the trajectory they become untied and spring apart. About how far below the highest point is the center of mass of the two-block system 2.0 s later, before either fragment has hit the ground?

A. 12 m
B. 20 m
C. 31 m
D. Can't tell because the velocities of the fragments are not given.
E. Can't tell because the coordinates of the highest point are not given.

ans: B

15. The center of mass of a system of particles has a constant velocity if:

A. the forces exerted by the particles on each other sum to zero
B. the external forces acting on particles of the system sum to zero
C. the velocity of the center of mass is initially zero
D. the particles are distributed symmetrically around the center of mass
E. the center of mass is at the geometric center of the system

ans: B

16. The center of mass of a system of particles remains at the same place if:

A. it is initially at rest and the external forces sum to zero
B. it is initially at rest and the internal forces sum to zero
C. the sum of the external forces is less than the maximum force of static friction
D. no friction acts internally
E. none of the above

ans: A

17. A man sits in the back of a canoe in still water. He then moves to the front of the canoe and sits there. Afterwards the canoe:

A. is forward of its original position and moving forward
B. is forward of its original position and moving backward
C. is rearward of its original position and moving forward
D. is rearward of its original position and moving backward
E. is rearward of its original position and not moving

ans: E

18. A 640-N hunter gets a rope around a 3200-N polar bear. They are stationary, 20 m apart, on frictionless level ice. When the hunter pulls the polar bear to him, the polar bear will move:

A. 1.0 m
B. 3.3 m
C. 10 m
D. 12 m
E. 17 m

ans: B

19. Two boys, with masses of 40 kg and 60 kg, respectively, stand on a horizontal frictionless surface holding the ends of a light 10-m long rod. The boys pull themselves together along the rod. When they meet the 40-kg boy will have moved what distance?

A. 4 m
B. 5 m
C. 6 m
D. 10 m
E. need to know the forces they exert

ans: C

20. The center of mass of a system of particles obeys an equation similar to Newton's second law $\vec{F} = m\vec{a}_{\text{com}}$, where:

A. $\vec{F}$ is the net internal force and m is the total mass of the system
B. $\vec{F}$ is the net internal force and m is the mass acting on the system
C. $\vec{F}$ is the net external force and m is the total mass of the system
D. $\vec{F}$ is the force of gravity and m is the mass of Earth
E. $\vec{F}$ is the force of gravity and m is the total mass of the system

ans: C

21. A large wedge rests on a horizontal frictionless surface, as shown. A block starts from rest and slides down the inclined surface of the wedge, which is rough. During the motion of the block the center of mass of the block and wedge:

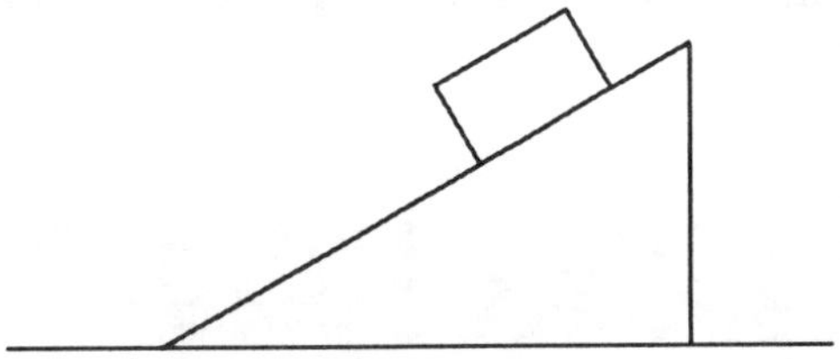

A. does not move
B. moves horizontally with constant speed
C. moves horizontally with increasing speed
D. moves vertically with increasing speed
E. moves both horizontally and vertically

ans: D

22. A large wedge with a mass of 10 kg rests on a horizontal frictionless surface, as shown. A block with a mass of 5.0 kg starts from rest and slides down the inclined surface of the wedge, which is rough. At one instant the wedge is moving to the left with a speed of 3.0 m/s. At that instant the horizontal component of the velocity of the block is:

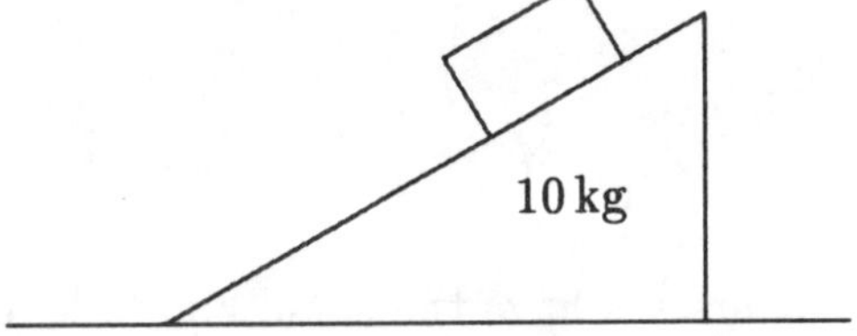

A. 3.0 m/s left
B. 3.0 m/s, right
C. 6.0 m/s, right
D. 6.0 m/s, left
E. 17 m/s, right

ans: C

23. A 2.0-kg block is attached to one end of a spring with a spring constant of 100 N/m and a 4.0-kg block is attached to the other end. The blocks are placed on a horizontal frictionless surface and set into motion. At one instant the 2.0-kg block is observed to be traveling to the right with a speed of 0.50 m/s and the 4.0-kg block is observed to be traveling to the left with a speed of 0.30 m/s. Since the only forces acting are the force of gravity, the normal force of the surface, and the force of the spring, we conclude that:

A. the spring is compressed at the time of the observation
B. the spring is not compressed at the time of observation
C. the motion was started with the masses at rest
D. the motion was started with at least one of masses moving
E. the motion was started by compressing the spring

ans: D

24. A 2.0-kg mass is attached to one end of a spring with a spring constant of 100 N/m and a 4.0-kg mass is attached to the other end. The masses are placed on a horizontal frictionless surface and the spring is compressed 10 cm. The spring is then released with the masses at rest and the masses oscillate. When the spring has its equilibrium length the speeds of the masses are:

A. 0 and 0.29 m/s
B. 0 and 0.58 m/s
C. 0.29 and 0.58 m/s
D. 0.14 and 0.29 m/s
E. 0.58 and 1.2 m/s

ans: C

25. Momentum may be expressed in:

A. kg/m
B. gram·s
C. N·s
D. kg/(m·s)
E. N/s

ans: C

26. The momentum of an object at a given instant is independent of its:

A. inertia
B. mass
C. speed
D. velocity
E. acceleration

ans: E

27. If two objects with different masses have the same kinetic energy, their momenta are:

A. proportional to their masses
B. proportional to the squares of their masses
C. proportional to the square roots of their masses
D. inversely proportional to their masses
E. inversely proportional to the square roots of their masses

ans: C

28. Two bodies, A and B, have equal kinetic energies. The mass of A is nine times that of B. The ratio of the momentum of A to that of B is:

A. 1:9
B. 1:3
C. 1:1
D. 3:1
E. 9:1

ans: D

29. Two objects, P and Q, have the same momentum. Q has more kinetic energy than P if it:

A. weighs more than P
B. is moving faster than P
C. weighs the same as P
D. is moving slower than P
E. is moving at the same speed as P

ans: B

30. Which of the following five objects requires the greatest change in momentum to stop moving?

	object	mass (kg)	speed (m/s)
A.	electron	10^{-30}	10^{7}
B.	oil tanker	10^{8}	10^{-1}
C.	rain drop	10^{-4}	10
D.	snail	10^{-2}	10^{-4}
E.	satellite	10	10^{4}

ans: B

31. Which of the following quantities has a numerical value that is independent of the frame of reference in which it is evaluated?

A. Velocity
B. Momentum
C. Kinetic energy
D. Displacement
E. None of these

ans: E

32. A particle moves along the x axis. Its momentum is graphed below as a function of time. Rank the numbered regions according to the magnitude of the force acting on the particle, least to greatest.

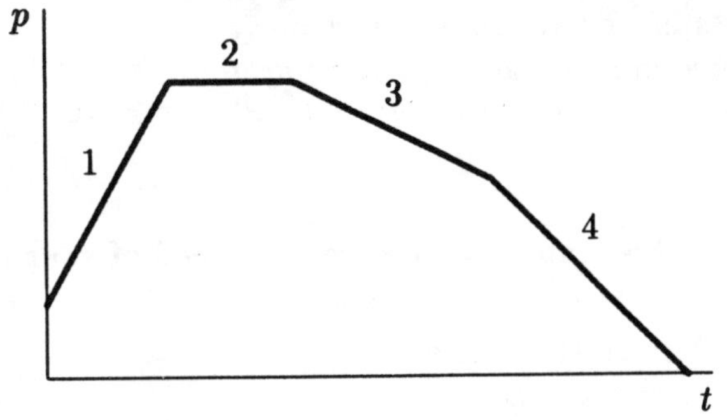

A. 1, 2, 3, 4
B. 2, 3, 4, 1
C. 1, 4, 3, 2
D. 1, 3, 4, 2
E. 2, 4, 3, 1

ans: B

33. A 0.20-kg rubber ball is dropped from the window of a building. It strikes the sidewalk below at 30 m/s and rebounds at 20 m/s. The change in momentum of the ball as a result of the collision with the sidewalk is (in kg·m/s):

A. 10
B. 6.0
C. 4.0
D. 2.0
E. 1.0

ans: A

34. A 1.0-kg ball moving at 2.0 m/s perpendicular to a wall rebounds from the wall at 1.5 m/s. The change in the momentum of the ball is:

A. zero
B. 0.5 N · s away from wall
C. 0.5 N · s toward wall
D. 3.5 N · s away from wall
E. 3.5 N · s toward wall

ans: D

35. A ball is thrown into the air. As it rises, there is an increase in its:

A. acceleration
B. velocity
C. kinetic energy
D. potential energy
E. momentum

ans: D

36. If the total momentum of a system is changing:

A. particles of the system must be exerting forces on each other
B. the system must be under the influence of gravity
C. the center of mass must have constant velocity
D. a net external force must be acting on the system
E. none of the above

ans: D

37. When you step on the accelerator to increase the speed of your car, the force that accelerates the car is:

A. the force of your foot on the accelerator
B. the force of friction of the road on the tires
C. the force of the engine on the drive shaft
D. the normal force of the road on the tires
E. none of the above

ans: B

38. When you step on the brake pedal to stop your car, the force that decelerates the car is:
    A. the force of your foot on the brake pedal
    B. the force of the brake shoes on the brake disk
    C. the force of friction of the road on the tires
    D. the normal force of the road on the tires
    E. none of the above

    ans: C

39. A 2.5-kg stone is released from rest and falls toward Earth. After 4.0 s, the magnitude of its momentum (in kg·m/s) is:
    A. 98
    B. 78
    C. 39
    D. 24
    E. zero

    ans: A

40. In a tug-of-war game the winning team stands its ground. It pulls the rope through its hands without moving along the ground. The losing team does not let the rope slip but drags its feet along the ground as it moves with constant velocity. For the system composed of the two teams and the rope, which of the following statements is true while the losing team is moving?
    A. External forces are doing positive work.
    B. External forces are doing zero work.
    C. Internal forces are doing positive work
    D. Internal forces are doing zero work
    E. None of the above statements are true.

    ans: C

41. A 1000-kg automobile starts from rest and accelerates to 20 m/s in a distance of 500 m, over a horizontal road. If the tires do not skid on the road which of the following statements is NOT true?
    A. The work done by the road on the car is at least 2.0×10^5 J.
    B. The work done by the engine is at least 2.0×10^5 J.
    C. The work done by the normal force of the road on the tires is zero.
    D. The total frictional force of the road on the tires is at least 400 N.
    E. The total frictional force of the tires on the road is at least 400 N.

    ans: A

42. A man, standing at rest on a horizontal frictionless floor, might get himself moving by:
    A. walking
    B. rolling
    C. exhaling vertically
    D. crawling slowly
    E. throwing a shoe horizontally

    ans: E

43. A 64-kg woman stands on frictionless level ice. She kicks a 0.10-kg stone backwards with her foot so that the stone acquires a velocity of 1.1 m/s. The velocity (in m/s) acquired by the woman is:

A. 1.1 forward
B. 0.0017 backward
C. 0.0017 forward
D. 1.1 backward
E. none of these

ans: C

44. A man is marooned at rest on level frictionless ice. In desperation, he hurls his shoe to the right at 15 m/s. If the man weighs 720 N and the shoe weighs 4.0 N, the man moves to the left at approximately:

A. 0
B. 2.1×10^{-2} m/s
C. 8.3×10^{-2} m/s
D. 15 m/s
E. 2.7×10^{3} m/s

ans: C

45. A rifle of mass M is initially at rest but free to recoil. It fires a bullet of mass m and velocity v (relative to the ground). After firing, the velocity of the rifle (relative to the ground) is:

A. $-mv$
B. $-Mv/m$
C. $-mv/M$
D. $-v$
E. mv/M

ans: C

46. Bullets from two revolvers are fired with the same velocity. The bullet from gun #1 is twice as heavy as the bullet from gun #2. Gun #1 weighs three times as much as gun #2. The ratio of the momentum imparted to gun #1 to that imparted to gun #2 is:

A. 2:3
B. 3:2
C. 2:1
D. 3:1
E. 6:1

ans: C

47. A 5-kg object can move along the x axis. It is subjected to a force $\vec{F}$ in the positive x direction; a graph of F as a function of time t is shown below. Over the time the force is applied the change in the velocity of the object is:

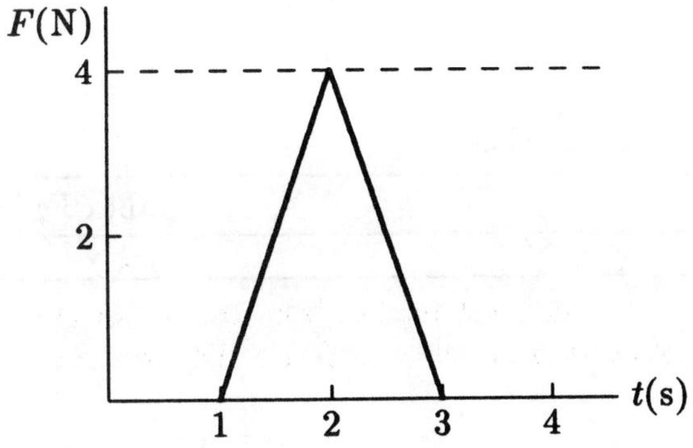

A. 0.8 m/s
B. 1.1 m/s
C. 1.6 m/s
D. 2.3 m/s
E. 4.0 m/s

ans: A

48. Force:

A. equals the negative integral (with respect to distance) of the potential energy function
B. is the ability to do work
C. is the rate of change of doing work
D. equals the time rate of change of momentum
E. has dimensions of momentum multiplied by time

ans: D

49. Cart A, with a mass of 0.2 kg, travels on a horizontal air track at 3 m/s and hits cart B, which has a mass of 0.4 kg and is initially at rest. After the collision the center of mass of the two cart system has a speed in m/s of:

A. 0
B. 1
C. 2
D. 3
E. can't tell without knowing the energy

ans: B

50. A 500-kg sack of coal is dropped on a 2000-kg railroad flatcar which was initially moving at 3 m/s as shown. After the sack rests on the flatcar, the speed of the flatcar is:

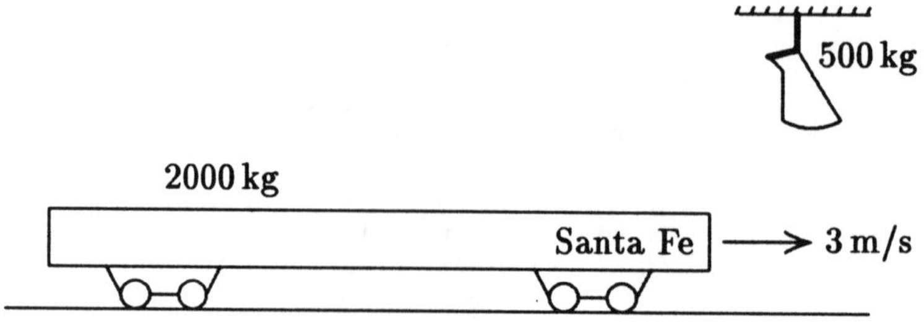

A. 0.6 m/s
B. 1.2 m/s
C. 1.8 m/s
D. 2.4 m/s
E. 3.6 m/s

ans: D

51. Sand is dropped straight down on a moving conveyor belt at the rate of 3.0 kg/s. If friction in the bearings can be ignored the power that must be expended to keep the belt moving at 2.0 m/s is:

A. 0
B. 3.0 W
C. 6.0 W
D. 9.0 W
E. 12 W

ans: E

52. A cart loaded with sand slides along a horizontal frictionless track. As the cart moves, sand trickles out at a constant rate through a hole in the back of the cart. The speed of the cart will:

A. decrease at a constant rate
B. increase at a constant rate
C. remain the same
D. decrease at a variable rate
E. increase at a variable rate

ans: C

53. The thrust of a rocket is:

A. a gravitational force acting on the rocket
B. the force of the exiting fuel gases on the rocket
C. any force that is external to the rocket-fuel system
D. a force that arises from the reduction in mass of the rocket-fuel system
E. none of the above

ans: B:

54. At one instant of time a rocket is traveling in outer space at 2500 m/s and is exhausting fuel at a rate of 100 kg/s. If the speed of the fuel as it leaves the rocket is 1500 m/s, relative to the rocket, the thrust is:

A. 0
B. 1.0×10^5 N
C. 1.5×10^5 N
D. 2.9×10^5 N
E. 2.5×10^5 N

ans: C

55. A rocket exhausts fuel with a velocity of 1500 m/s, relative to the rocket. It starts from rest in outer space with fuel comprising 80 per cent of the total mass. When all the fuel has been exhausted its speed is:

A. 3600 m/s
B. 2400 m/s
C. 1200 m/s
D. 880 m/s
E. 400 m/s

ans: B

56. A 1000-kg space probe is motionless in space. To start moving, its main engine is fired for 5 s during which time it ejects exhaust gases at 5000 m/s. At the end of this process it is moving at 20 m/s. The approximate mass of the ejected gas is:

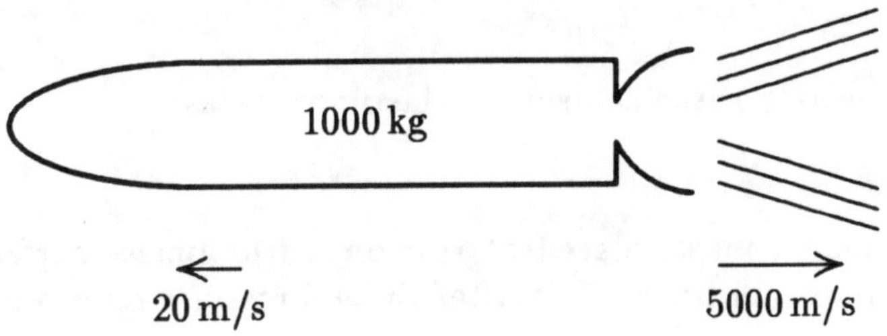

A. 0.8 kg
B. 4 kg
C. 5 kg
D. 20 kg
E. 25 kg

ans: B

57. Why does a rocket accelerate when it is fired horizontally?

A. Exhaust gases push against outside air
B. Exhaust gases, which the rocket pushes backwards, push the rocket forward
C. Outside air exerts more force on the back of the rocket than on the front
D. Exhaust speed is greater than the rocket speed
E. Atmospheric pressure at the back of the rocket is reduced

ans: B

Chapter 10: COLLISIONS

1. The physical quantity "impulse" has the same dimensions as that of:
   A. force
   B. power
   C. energy
   D. momentum
   E. work

   ans: D

2. The law of conservation of momentum applies to a system of colliding objects only if:
   A. there is no change in kinetic energy of the system
   B. the coefficient of restitution is one
   C. the coefficient of restitution is zero
   D. the net external impulse is zero
   E. the collisions are all elastic

   ans: D

3. Sphere X, of mass 2 kg, is moving to the right at 10 m/s. Sphere Y, of mass 4 kg, is moving to the left at 10 m/s. The two spheres collide head-on. The ratio of the magnitude of the impulse exerted by X on Y to that exerted by Y on X is:
   A. $1/4$
   B. $1/2$
   C. $1/\sqrt{2}$
   D. 1
   E. need to know whether the collision is elastic or inelastic

   ans: D

4. Two bodies of unequal mass, placed at rest on a frictionless surface, are acted on by equal horizontal forces for equal times. Just after these forces are removed, the body of greater mass will have:
   A. the greater speed
   B. the greater acceleration
   C. the smaller momentum
   D. the greater momentum
   E. the same momentum as the other body

   ans: E

5. A 0.2-kg rubber ball is dropped from the window of a building. It strikes the sidewalk below at 30 m/s and rebounds up at 20 m/s. The magnitude of the impulse due to the collision with the sidewalk is:
   A. $10\,\mathrm{N \cdot s}$
   B. $6.0\,\mathrm{N \cdot s}$
   C. $2.0\,\mathrm{N \cdot s}$
   D. $19.6\,\mathrm{N \cdot s}$
   E. $9.8\,\mathrm{N \cdot s}$

   ans: A

6. A 10-kg block of ice is at rest on a frictionless horizontal surface. A 1.0-N force is applied in an easterly direction for 1.0 s. During this time interval, the block:

A. acquires a speed of 1 m/s
B. moves 10 cm
C. acquires a momentum of 1.0 kg · m/s
D. acquires a kinetic energy of 0.1 J
E. none of the above

ans: C

7. A uniform narrow bar, resting on ice, is given a transverse horizontal impulse $\vec{J}$ at one end as shown. The center of mass of the bar COM will then:

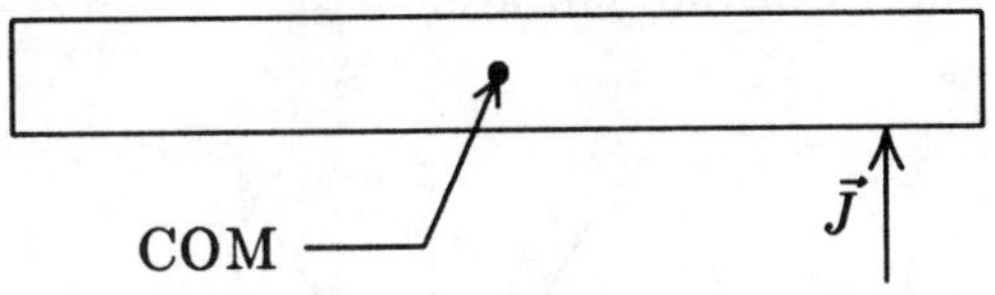

A. remain at rest
B. move in a circle
C. move in a straight line
D. move in a parabola
E. move along some other curve

ans: C

8. What impulse will give a 2.0-kg object a momentum change of + 50 kg · m/s?

A. +25 N · s
B. −25 N · s
C. +50 N · s
D. −50 N · s
E. +100 N · s

ans: C

9. A 2-kg object is acted upon by a single force in the x direction in a manner described by the graph shown. The momentum acquired by the object is:

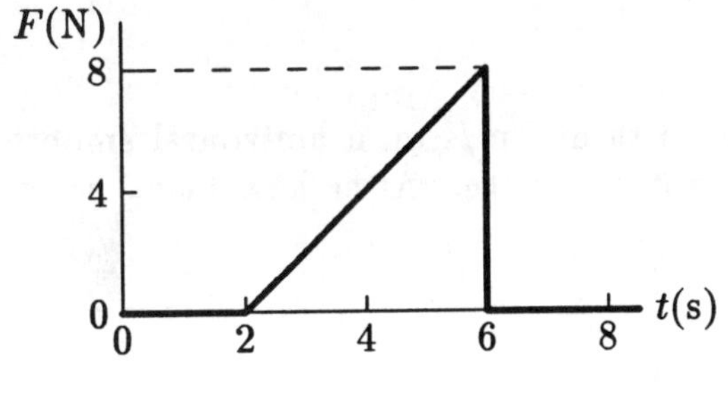

A. 16 N · s
B. 30 N · s
C. 32 N · s
D. 40 N · s
E. 48 N · s

ans: A

10. A student's life was saved in an automobile accident because an airbag expanded in front of his head. If the car had not been equipped with an airbag, the windshield would have stopped the motion of his head in a much shorter time. Compared to the windshield, the airbag:

A. causes a much smaller change in momentum
B. exerts a much smaller impulse
C. causes a much smaller change in kinetic energy
D. exerts a much smaller force
E. does much more work

ans: D

11. A ball hits a wall and rebounds with the same speed, as diagramed below. The changes in the components of the momentum of the ball are:

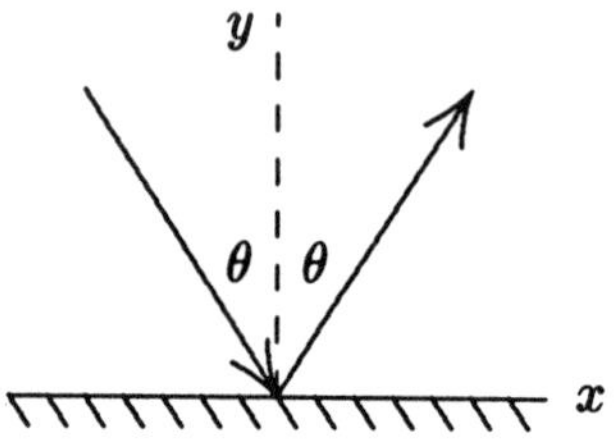

A. $\Delta p_x > 0$, $\Delta p_y > 0$
B. $\Delta p_x < 0$, $\Delta p_y > 0$
C. $\Delta p_x = 0$, $\Delta p_y > 0$
D. $\Delta p_x = 0$, $\Delta p_y < 0$
E. $\Delta p_x > 0$, $\Delta p_y < 0$

ans: C

12. A golf ball of mass m is hit by a golf club so that the ball leaves the tee with speed v. The club is in contact with the ball for time T. The average force on the club on the ball during the time T is:

A. mvT
B. mv/T
C. $(1/2)mv^2T$
D. $mv^2/(2T)$
E. $mT^2/(2v)$

ans: B

13. A 2-kg puck is moving north at 3 m/s on a horizontal smooth ice surface. The puck is then pushed east by a constant 4-N force. After how many seconds is the puck moving exactly northeast?

A. $1.5/\sqrt{2}$
B. $1.5\sqrt{2}$
C. 1.5
D. 2.75
E. 4.25

ans: C

14. A 640-N acrobat falls 5.0 m from rest into a net. The net tosses him back up with the same speed he had just before he hit the net. The average upward force exerted on him by the net during this collision is:

A. 32 N
B. 64 N
C. 320 N
D. 640 N
E. impossible to determine from given data

ans: E

15. Whenever an object strikes a stationary object of equal mass:

A. the two objects cannot stick together
B. the collision must be elastic
C. the first object must stop
D. momentum is not necessarily conserved
E. none of the above

ans: E

16. In which case is the momentum of the CAPITALIZED SYSTEM conserved?

A. A BALL falling freely in a vacuum
B. A RUBBER BALL as it bounces from the floor
C. A CAR making a turn at constant speed
D. TWO BALLS colliding at right angles
E. A BULLET AND THE GUN from which it was fired when the gun is firmly held

ans: D

17. An event in which a particle at rest decays into two different particles, only one of which remains at rest, would be extremely surprising because:

A. energy cannot be conserved in such a decay
B. mass cannot be conserved in such a decay
C. angular momentum cannot be conserved in such a decay
D. momentum cannot be conserved in such a decay
E. electric charge cannot be conserved in such a decay

ans: D

18. For a two-body collision, a frame of reference for which the total momentum is zero is one whose velocity, from the point of view of the laboratory, has the same value as:

A. the velocity of the center of mass
B. the velocity of the incident object
C. the velocity of the target object
D. the average velocity of the two bodies, before the collision
E. the average velocity of the two bodies, after the collision

ans: A

19. An inelastic collision is one in which:

A. momentum is not conserved but kinetic energy is conserved
B. total mass is not conserved but momentum is conserved
C. neither kinetic energy nor momentum is conserved
D. momentum is conserved but kinetic energy is not conserved
E. the total impulse is equal to the change in kinetic energy

ans: D

20. A 4000-kg freight car, moving at 4.0 m/s, collides and couples with an 6000-kg freight car, which was initially at rest. The common final speed of these two cars is:

A. 1.0 m/s
B. 1.3 m/s
C. 1.6 m/s
D. 2.5 m/s
E. 4.0 m/s

ans: C

21. A 4.0-N puck is traveling at 3.0 m/s. It strikes a 8.0-N puck, which is stationary. The two pucks stick together. Their common final speed is:

A. 1.0 m/s
B. 1.5 m/s
C. 2.0 m/s
D. 2.3 m/s
E. 3.0 m/s

ans: A

22. A 3.00-g bullet traveling horizontally at 400 m/s hits a 3.00-kg wooden block, which is initially at rest on a smooth horizontal table. The bullet buries itself in the block without passing through. The speed of the block after the collision is:

A. 1.33 m/s
B. 0.40 m/s
C. 12.0 m/s
D. 40.0 m/s
E. 160 m/s

ans: B

23. A cart of mass m, traveling on a horizontal air track with speed v, collides with a stationary cart of mass $2m$. The carts stick together. The impulse exerted by one cart on the other has magnitude:

A. 0
B. $2mv/3$
C. mv
D. $3mv/2$
E. $2mv$

ans: B

24. A 3-g bullet is fired horizontally into a 10-kg block of wood suspended by a rope from the ceiling. The block swings in an arc, rising 3 mm above its lowest position. The kinetic energy of the block at the bottom of its swing, just after the collision, is approximately:

A. 0.3 J
B. 0.0003 J
C. unknown since the heat generated in the collision was not given
D. zero
E. 30 J

ans: A

25. A 3-g bullet is fired horizontally into a 10-kg block of wood suspended by a rope from the ceiling. The block swings in an arc, rising 3 mm above its lowest position. The velocity of the bullet was:

A. unknown since the heat generated in the collision was not given
B. 8.0×10^2 m/s
C. 24.0 m/s
D. 8.00 m/s
E. 2.4×10^4 m/s

ans: B

26. A 3-kg and a 2-kg cart approach each other on a horizontal air track in such a way that their center of mass has a speed of 4 m/s. They collide and stick together. After the collision their total kinetic energy in joules is:

A. 16
B. 20
C. 24
D. 40
E. can't tell from the given data

ans: D

27. Blocks A and B are moving toward each other. A has a mass of 2.0 kg and a velocity of 50 m/s, while B has a mass of 4.0 kg and a velocity of −25 m/s. They suffer a completely inelastic collision. The kinetic energy dissipated during the collision is:

A. 0
B. 1250 J
C. 3750 J
D. 5000 J
E. 5600 J

ans: C

28. For a completely inelastic two-body collision the kinetic energy retained by the objects is the same as:

A. the total kinetic energy before the collision
B. the difference in the kinetic energies of the objects before the collision
C. $\frac{1}{2}Mv_{com}^2$, where M is the total mass and v_{com} is the velocity of the center of mass
D. the kinetic energy of the more massive body before the collision
E. the kinetic energy of the less massive body before the collision

ans: C

29. Two spacemen are floating together with zero speed in a gravity-free region of space. The mass of spaceman A is 120 kg and that of spaceman B is 90 kg. Spaceman A pushes B away from him with B attaining a final speed of 0.5 m/s. The final recoil speed of A is:

A. zero
B. 0.38 m/s
C. 0.5 m/s
D. 0.67 m/s
E. 1.0 m/s

ans: B

30. A 75-kg man is riding in a 30-kg cart at 2.0 m/s. He jumps off in such a way as to land on the ground with no horizontal velocity. The resulting change in speed of the cart is:

A. zero
B. 2.0 m/s
C. 3.0 m/s
D. 5.0 m/s
E. 7.0 m/s

ans: D

31. An unstable nucleus has mass M and is initially at rest. It ejects a particle of mass m with speed v. The recoil speed of the remaining nucleus is:

A. v
B. mv/M
C. $mv/(m+M)$
D. $(m+M)v/m$
E. $mv/(M-m)$

ans: E

32. A horizontal moving walkway is moving at 3 m/s. On the average, during each second, four stationary people step onto it and four people step off it. Assuming each person's mass is 60 kg, what average driving force must be exerted on the walkway to keep it moving?

A. 45 N
B. 80 N
C. 720 N
D. 1080 N
E. 2400 N

ans: C

33. An elastic collision is one in which:

A. momentum is not conserved but kinetic energy is conserved
B. total mass is not conserved but momentum is conserved
C. kinetic energy and momentum are both conserved
D. momentum is conserved but kinetic energy is not conserved
E. the total impulse is equal to the change in kinetic energy

ans: C

34. Object A strikes the stationary object B head-on in an elastic collision. The mass of A is fixed, you may choose the mass of B appropriately. Then:

A. for B to have the greatest recoil speed, choose $m_B = m_A$
B. for B to have the greatest recoil momentum, choose $m_B \ll m_A$
C. for B to have the greatest recoil kinetic energy, choose $m_B \ll m_A$
D. for B to have the least recoil speed, choose $m_B = m_A$
E. for B to have the greatest recoil kinetic energy, choose $m_B = m_A$

ans: E

35. Block A, with a mass of 2.0 kg, moves along the x axis with a velocity of 5.0 m/s in the positive x direction. It suffers an elastic collision with block B, initially at rest, and the blocks leave the collision along the x axis. If B is much more massive than A, the velocity of A after the collision is:

A. 0
B. +5.0 m/s
C. −5.0 m/s
D. +10 m/s
E. −10 m/s

ans: C

36. Block A, with a mass of 2.0 kg, moves along the x axis with a velocity of 5.0 m/s in the positive x direction. It suffers an elastic collision with block B, which initially has a velocity of −2.0 m/s (in the negative x direction). The blocks leave the collision along the x axis. If B is much more massive than A, the velocity of A after the collision is:

A. 0
B. −3.0 m/s
C. −5.0 m/s
D. −7.0 m/s
E. −9.0 m/s

ans: E

37. A very massive object traveling at 10 m/s strikes a very light object, initially at rest, and the light object moves off in the direction of travel of the heavy object. If the collision is elastic, the speed of the lighter object is:

A. 5.0 m/s
B. 10 m/s
C. 15 m/s
D. 20 m/s
E. Can't tell from the information given.

ans: D

38. A 5.0-g ball is moving downward at 100 cm/s. It undergoes an elastic collision with a heavy horizontal plate, which is fixed to Earth. The change in momentum of the ball due to this collision is:

A. 500 gm · cm/s, up
B. 1000 gm · cm/s, up
C. zero
D. 500 gm · cm/s, down
E. 1000 gm · cm/s, down

ans: B

39. Sphere A has mass m and is moving with velocity v. It makes a head-on elastic collision with a stationary sphere B of mass $2m$. After the collision their speeds (v_A and v_B) are:

A. 0, $v/2$
B. $-v/3$, $2v/3$
C. $-v$, v
D. $-2v/3$, $v/3$
E. none of these

ans: B

40. Blocks A and B are moving toward each other along the x axis. A has a mass of 2.0 kg and a velocity of 50 m/s (in the positive x direction), while B has a mass of 4.0 kg and a velocity of −25 m/s (in the negative x direction). They suffer an elastic collision and move off along the x axis. After the collision the velocities of A and B, respectively, are:

A. −50 and 25 m/s
B. 50 and −25 m/s
C. −25 and 50 m/s
D. 25 and −50 m/s
E. −25 and −50 m/s

ans: A

41. Blocks A and B move toward each other along the x axis. A has a mass of 2.0 kg and a velocity of 10 m/s (in the positive x direction); B has a mass of 3.0 kg and a velocity of −5.0 m/s (in the negative x direction). They suffer an elastic collision and go off along the x axis. After the collision the velocities of A and B, respectively, are:

A. −10 and +5.0 m/s
B. −8.0 and +7.0 m/s
C. −9.0 and +6.0 m/s
D. −5.0 and +10 m/s
E. +5.0 and −10 m/s

ans: B

42. Blocks A and B are moving toward each other along the x axis. A has a mass of 2.0 kg and a velocity of 50 m/s, while B has a mass of 4.0 kg and a velocity of -25 m/s. They suffer an elastic collision and move off along the x axis. The kinetic energy transferred from A to B during the collision is:

A. 0
B. 2500 J
C. 5000 J
D. 7500 J
E. 10000 J

ans: A

43. When a particle suffers a head-on elastic collision with another particle, initially at rest, the greatest fraction of kinetic energy is transferred if:

A. the incident particle is initially traveling very fast
B. the incident particle is traveling very slowly
C. the incident particle is much more massive than the target particle
D. the incident particle is much less massive than the target particle
E. the incident and target particle have the same mass

ans: E

44. Two objects, X and Y, are held at rest on a horizontal frictionless surface. A spring is compressed between X and Y. The mass of X is 2/5 times the mass of Y. When the objects are released, the ratio of the kinetic energy of X to that of Y is:

A. 2/5
B. 4/25
C. 25/4
D. 5/2
E. 1

ans: D

45. Two carts (A and B), having spring bumpers, collide as shown. Cart A has a mass of 2 kg and is initially moving to the right. Cart B has a mass of 3 kg and is initially stationary. When the separation between the carts is a minimum:

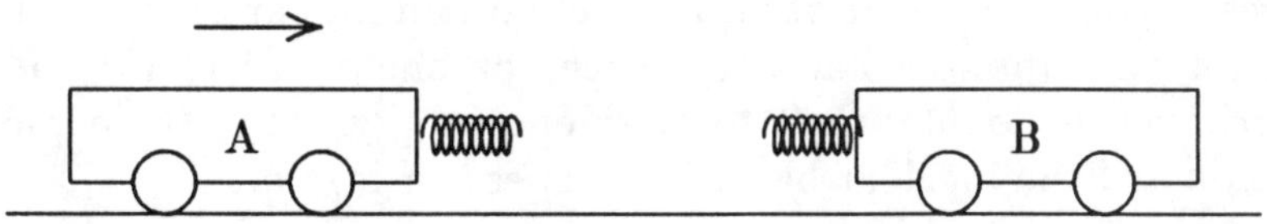

A. cart B is still at rest
B. cart A has come to rest
C. both carts have the same momentum
D. both carts have the same kinetic energy
E. the kinetic energy of the system is at a minimum

ans: E

46. Two blocks (1 kg and 3 kg) are joined by a thread and travel together over a horizontal frictionless surface at speed v as shown. The thread is then cut. Their final speeds (V_1 and V_3) are:

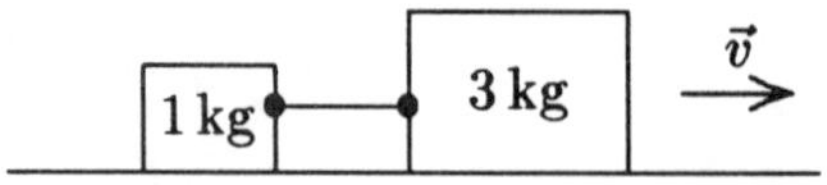

A. $V_1 = 2v/3$, $V_3 = v/3$
B. $V_1 = v/3$, $V_3 = 2v/3$
C. $V_1 = v/4$, $V_3 = 3v/4$
D. $V_1 = 3v/4$, $V_3 = v/4$
E. none of these

ans: E

47. Two identical carts travel at 1 m/s on a common surface. They collide head-on and are reported to rebound, each with a speed of 2 m/s. Then:

A. momentum was not conserved; therefore, the report must be false
B. if some other form of energy were changed to kinetic during the collision, the report could be true
C. if the collision were elastic, the report could be true
D. if the surface were inclined, the report could be true
E. if the duration of the collision were short enough, the report could be true

ans: B

48. A projectile in flight explodes into several fragments. The total momentum of the fragments immediately after this explosion:

A. is the same as the momentum of the projectile immediately before the explosion
B. has been changed into kinetic energy of the fragments
C. is less than the momentum of the projectile immediately before the explosion
D. is more than the momentum of the projectile immediately before the explosion
E. has been changed into radiant energy

ans: A

49. A block moves at 5 m/s in the positive x direction and hits an identical block, initially at rest. A small amount of gunpowder had been placed on one of the blocks. The explosion does not harm the blocks but it doubles their total kinetic energy. After the explosion the blocks move along the x axis and the incident block has a speed in m/s of:

A. 1.8
B. 5.0
C. 6.8
D. 7.1
E. 11.8

ans: A

50. A He^4 nucleus (atomic weight 4) moving with speed v breaks up into a neutron (atomic weight 1) and a He^3 nucleus. The neutron moves off at right angles to the original He^4 as shown. If the neutron speed is $3v$, the final speed of the He^3 nucleus is:

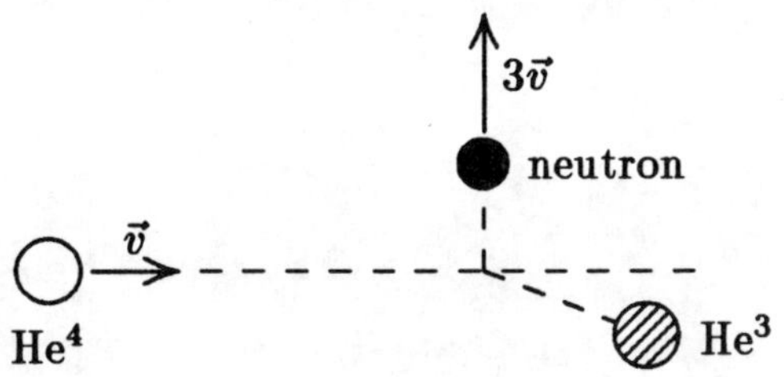

A. zero
B. v
C. $4v/3$
D. $5v/3$
E. none of these

ans: D

51. A stream of gas consists of n molecules. Each molecule has mass m and speed v. The stream is reflected elastically from a rigid surface as shown. The magnitude of the change in the total momentum of the stream is:

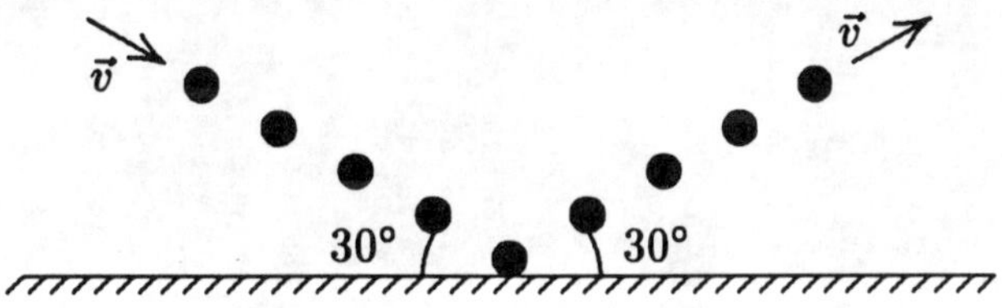

A. $2mnv$
B. $2mnv \sin 30°$
C. $mnv \sin 30°$
D. $mnv \cos 30°$
E. mnv

ans: B

Chapter 11: ROTATION

1. A radian is about:

   A. 25°
   B. 37°
   C. 45°
   D. 57°
   E. 90°

   ans: D

2. One revolution is about the same as:

   A. 1 rad
   B. 57 rad
   C. $\pi/2$ rad
   D. π rad
   E. 2π rad

   ans: E

3. One revolution per minute is about:

   A. 0.0524 rad/s
   B. 0.105 rad/s
   C. 0.95 rad/s
   D. 1.57 rad/s
   E. 6.28 rad/s

   ans: B

4. If a wheel turns with constant angular speed then:

   A. each point on its rim moves with constant velocity
   B. each point on its rim moves with constant acceleration
   C. the wheel turns through equal angles in equal times
   D. the angle through which the wheel turns in each second increases as time goes on
   E. the angle through which the wheel turns in each second decreases as time goes on

   ans: C

5. If a wheel is turning at 3.0 rad/s, the time it takes to complete one revolution is about:

   A. 0.33 s
   B. 0.67 s
   C. 1.0 s
   D. 1.3 s
   E. 2.1 s

   ans: E

6. If wheel turning at a constant rate completes 100 revolutions in 10 s its angular speed is:
   A. 0.31 rad/s
   B. 0.63 rad/s
   C. 10 rad/s
   D. 31 rad/s
   E. 63 rad/s

   ans: E

7. The angular speed in rad/s of the second hand of a watch is:
   A. $\pi/1800$
   B. $\pi/60$
   C. $\pi/30$
   D. 2π
   E. 60

   ans: C

8. The angular speed in rad/s of the minute hand of a watch is:
   A. $60/\pi$
   B. $1800/\pi$
   C. π
   D. $\pi/1800$
   E. $\pi/60$

   ans: D

9. A flywheel, initially at rest, has a constant angular acceleration. After 9 s the flywheel has rotated 450 rad. Its angular acceleration in rad/s^2 is:
   A. 100
   B. 1.77
   C. 50
   D. 11.1
   E. 15.9

   ans: D

10. Ten seconds after an electric fan is turned on, the fan rotates at 300 rev/min. Its average angular acceleration is:
   A. 3.14 rad/s^2
   B. 30 rad/s^2
   C. 30 rev/s^2
   D. 50 rev/min^2
   E. 1800 rev/s^2

   ans: A

11. A wheel rotates with a constant angular acceleration of $\pi\,\text{rad/s}^2$. During the time interval from t_1 to t_2 its angular displacement is π rad. At time t_2 its angular velocity is 2π rad/s. Its angular velocity in rad/s at time t_1 is:

A. zero
B. 1
C. π
D. $\pi\sqrt{2}$
E. 2π

ans: D

12. A flywheel rotating at 12 rev/s is brought to rest in 6 s. The magnitude of the average angular acceleration in rad/s^2 of the wheel during this process is:

A. $1/\pi$
B. 2
C. 4
D. 4π
E. 72

ans: D

13. A phonograph turntable, rotating at 0.75 rev/s, slows down and stops in 30 s. The magnitude of its average angular acceleration in rad/s^2 for this process is:

A. 1.5
B. 1.5π
C. $\pi/40$
D. $\pi/20$
E. 0.75

ans: D

14. The angular velocity of a rotating wheel increases by 2 rev/s every minute. The angular acceleration in rad/s^2 of this wheel is:

A. $4\pi^2$
B. 2π
C. 1/30
D. $\pi/15$
E. 4π

ans: D

15. A wheel initially has an angular velocity of 18 rad/s but it is slowing at a rate of $2.0\,\text{rad/s}^2$. The time it takes to stop is:

A. 3.0 s
B. 6.0 s
C. 9.0 s
D. 12 s
E. never stops

ans: C

16. A wheel initially has an angular velocity of 36 rad/s but after 6.0 s its angular velocity is 24 rad/s. If its angular acceleration is constant its value is:

A. $2.0\,\mathrm{rad/s^2}$
B. $-2.0\,\mathrm{rad/s^2}$
C. $3.0\,\mathrm{rad/s^2}$
D. $-3.0\,\mathrm{rad/s^2}$
E. $6.0\,\mathrm{rad/s^2}$

ans: B

17. A wheel initially has an angular velocity of −36 rad/s but after 6.0 s its angular velocity is −24 rad/s. If its angular acceleration is constant the value is:

A. $2.0\,\mathrm{rad/s^2}$
B. $-2.0\,\mathrm{rad/s^2}$
C. $3.0\,\mathrm{rad/s^2}$
D. $-3.0\,\mathrm{rad/s^2}$
E. $-6.0\,\mathrm{rad/s^2}$

ans: A

18. A wheel initially has an angular velocity of 18 rad/s but it is slowing at a rate of $2.0\,\mathrm{rad/s^2}$. By the time it stops it will have turned through:

A. 13 rev
B. 26 rev
C. 39 rev
D. 52 rev
E. 65 rev

ans: A

19. A wheel starts from rest and has an angular acceleration of $4.0\,\mathrm{rad/s^2}$. When it has made 10 rev its angular velocity is:

A. 16 rad/s
B. 22 rad/s
C. 32 rad/s
D. 250 rad/s
E. 500 rad/s

ans: B

20. A wheel starts from rest and has an angular acceleration of $4.0\,\mathrm{rad/s^2}$. The time it takes to make 10 rev is:

A. 0.50 s
B. 0.71 s
C. 2.2 s
D. 2.8 s
E. 5.6 s

ans: E

21. A wheel starts from rest and has an angular acceleration that is given by $\alpha(t) = 6t^2$, where t is in seconds and α is in radians per second-squared. The angle through which it turns in time t is given by:

A. $(1/8)t^4$
B. $(1/4)t^4$
C. $(1/2)t^4$
D. t^4
E. 12

ans: C

22. A wheel starts from rest and has an angular acceleration that is given by $\alpha(t) = 6t^2$, where t is in seconds and α is in radians per second-squared. The time it takes to make 10 rev is:

A. 2.8 s
B. 3.3 s
C. 4.0 s
D. 4.7 s
E. 5.3 s

ans: B

23. A wheel starts from rest and has an angular acceleration that is given by $\alpha(t) = 6t^2$, where t is in seconds and α is in radians per second-squared. After it has turned through 10 rev its angular velocity is:

A. 63 rad/s
B. 75 rad/s
C. 89 rad/s
D. 130 rad/s
E. 210 rad/s

ans: B

24. A wheel is spinning at 27 rad/s but is slowing with an angular acceleration that has a magnitude given by $3t^2$, in rad/s^2 for t in seconds. It stops in a time of:

A. 1.7 s
B. 2.6 s
C. 3.0 s
D. 4.4 s
E. 7.3 s

ans: C

25. If the angular velocity vector of a spinning body points out of the page then, when viewed from above the page, the body is spinning:

A. clockwise about an axis that is perpendicular to the page
B. counterclockwise about an axis that is perpendicular to the page
C. about an axis that is parallel to the page
D. about an axis that is changing orientation
E. about an axis that is getting longer

ans: B

26. The angular velocity vector of a spinning body points out of the page. If the angular acceleration vector points into the page then:

A. the body is slowing down
B. the body is speeding up
C. the body is starting to turn in the opposite direction
D. the axis of rotation is changing orientation
E. none of the above

ans: A

27. A child, riding on a large merry-go-round, travels a distance of 3000 m in a circle of diameter 40 m. The total angle in radians through which she revolves is:

A. 50
B. 75
C. 150
D. 314
E. none of these

ans: C

28. The figure shows a cylinder of radius 0.7 m rotating about its axis at 10 rad/s. The speed of the point P is:

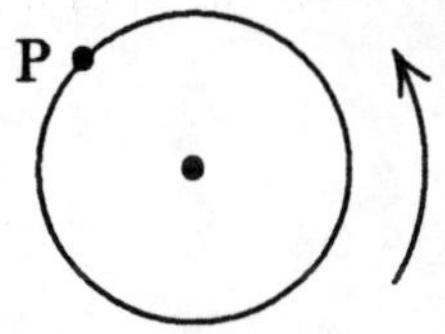

A. 7.0 m/s
B. 14π rad/s
C. 7.0π rad/s
D. 0.70 m/s
E. none of these

ans: A

29. The fan shown has been turned on and is speeding up as it rotates clockwise. The direction of the acceleration of the point X on the fan tip could be:

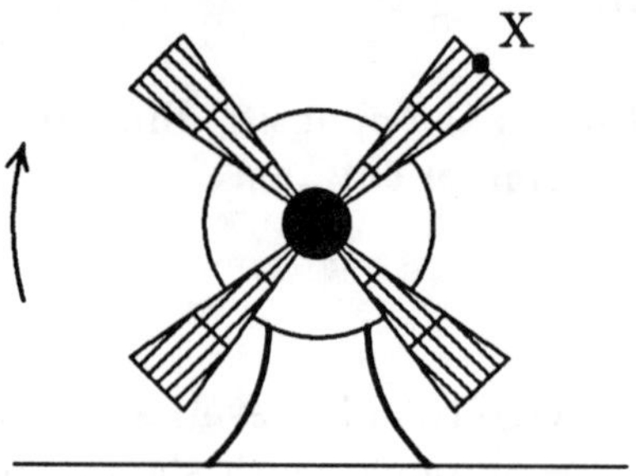

A. ↙
B. ↘
C. ↓
D. ←
E. →

ans: C

30. A wheel of diameter 3.0 cm has a 4.0-m cord wrapped around its periphery. Starting from rest, the wheel is given a constant angular acceleration of $2.0\,\mathrm{rad/s^2}$. The cord will unwind in:

A. 0.82 s
B. 2.0 s
C. 8.0 s
D. 16 s
E. 130 s

ans: D

31. A particle moves in a circular path of radius 0.10 m with a constant angular speed of 5 rev/s. The acceleration of the particle is:

A. $0.10\pi\,\mathrm{m/s^2}$
B. $0.50\,\mathrm{m/s^2}$
C. $500\pi\,\mathrm{m/s^2}$
D. $1000\pi^2\,\mathrm{m/s^2}$
E. $10\pi^2\,\mathrm{m/s^2}$

ans: E

32. A car travels north at constant velocity. It goes over a piece of mud, which sticks to the tire. The initial acceleration of the mud, as it leaves the ground, is:

A. vertically upward
B. horizontally to the north
C. horizontally to the south
D. zero
E. upward and forward at 45° to the horizontal

ans: A

33. String is wrapped around the periphery of a 5.0-cm radius cylinder, free to rotate on its axis. If the string is pulled out at a constant rate of 10 cm/s and does not slip on the cylinder, the angular velocity of the cylinder is:

A. 2.0 rad/s
B. 5.0 rad/s
C. 10 rad/s
D. 25 rad/s
E. 50 rad/s

ans: A

34. String is wrapped around the periphery of a 5.0-cm radius cylinder, free to rotate on its axis. The string is pulled straight out at a constant rate of 10 cm/s and does not slip on the cylinder. As each small segment of string leaves the cylinder, its acceleration changes by:

A. 0
B. 0.010 m/s^2
C. 0.020 m/s^2
D. 0.10 m/s^2
E. 0.20 m/s^2

ans: E

35. A flywheel of diameter 1.2 m has a constant angular acceleration of 5.0 rad/s^2. The tangential acceleration of a point on its rim is:

A. 5.0 rad/s^2
B. 3.0 m/s^2
C. 5.0 m/s^2
D. 6.0 m/s^2
E. 12 m/s^2

ans: B

36. For a wheel spinning with constant angular acceleration on an axis through its center, the ratio of the speed of a point on the rim to the speed of a point halfway between the center and the rim is:

A. 1
B. 2
C. 1/2
D. 4
E. 1/4

ans: B

37. For a wheel spinning on an axis through its center, the ratio of the tangential acceleration of a point on the rim to the tangential acceleration of a point halfway between the center and the rim is:

A. 1
B. 2
C. 1/2
D. 4
E. 1/4

ans: B

38. For a wheel spinning on an axis through its center, the ratio of the radial acceleration of a point on the rim to the radial acceleration of a point halfway between the center and the rim is:

A. 1
B. 2
C. 1/2
D. 4
E. 1/4

ans: B

39. Two wheels are identical but wheel B is spinning with twice the angular velocity of wheel A. The ratio of the radial acceleration of a point on the rim of B to the radial acceleration of a point on the rim of A is:

A. 1
B. 2
C. 1/2
D. 4
E. 1/4

ans: D

40. A wheel starts from rest and spins with a constant angular acceleration. As time goes on the acceleration vector for a point on the rim:

A. decreases in magnitude and becomes more nearly tangent to the rim
B. decreases in magnitude and becomes more early radial
C. increases in magnitude and becomes more nearly tangent to the rim
D. increases in magnitude and becomes more nearly radial
E. increases in magnitude but retains the same angle with the tangent to the rim

ans: D

41. The acceleration of a point on a spinning wheel is increased by a factor of 4 if:

A. the angular velocity and the angular acceleration are each increased by a factor of 4
B. the angular velocity is increased by a factor of 4 and the angular acceleration is not changed
C. the angular velocity and the angular acceleration are each increased by a factor of 2
D. the angular velocity is increased by a factor of 2 and the angular acceleration is not changed
E. the angular velocity is increased by a factor of 2 and the angular acceleration is increased by a factor of 4

ans: E

42. Three identical balls are tied by light strings to the same rod and rotate around it, as shown below. Rank the balls according to their rotational inertia, least to greatest.

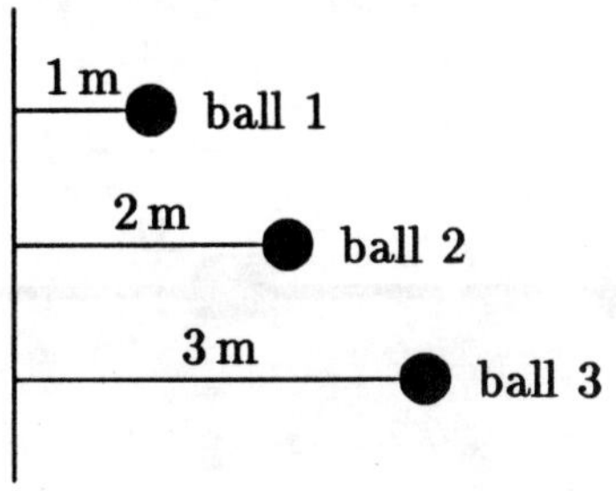

A. 1, 2, 3
B. 3, 2, 1
C. 3, then 1 and 2 tie
D. 1, 3, 2
E. All are the same

ans: A

43. Four identical particles, each with mass m, are arranged in the x, y plane as shown. They are connected by light sticks to form a rigid body. If $m = 2.0\,\text{kg}$ and $a = 1.0\,\text{m}$, the rotational inertia of this array about the y axis is:

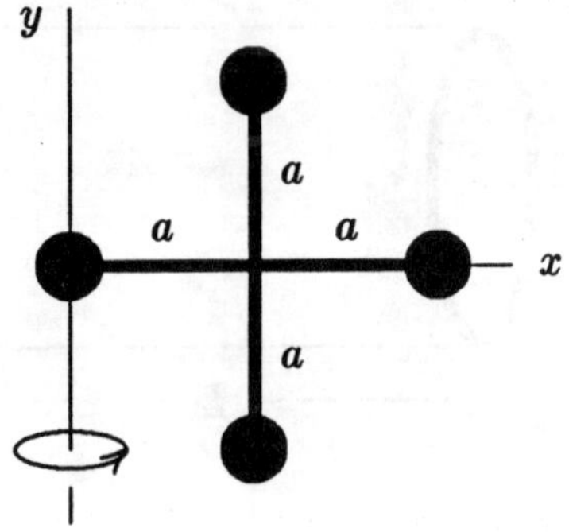

A. $4.0\,\text{kg}\cdot\text{m}^2$
B. $12\,\text{kg}\cdot\text{m}^2$
C. $9.6\,\text{kg}\cdot\text{m}^2$
D. $4.8\,\text{kg}\cdot\text{m}^2$
E. none of these

ans: B

44. Three identical objects, each of mass M, are fastened to a massless rod of length L as shown. The rotational inertia about one end of the rod of this array is:

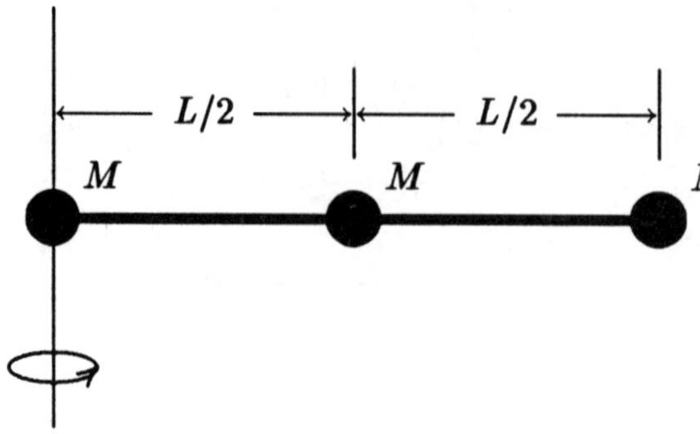

A. $ML^2/2$
B. ML^2
C. $3ML^2/2$
D. $5ML^2/4$
E. $3ML^2$

ans: D

45. The rotational inertia of a thin cylindrical shell of mass M, radius R, and length L about its central axis (X–X′) is:

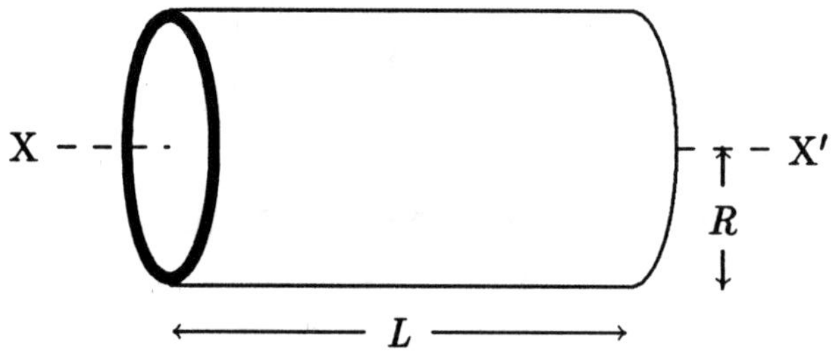

A. $MR^2/2$
B. $ML^2/2$
C. ML^2
D. MR^2
E. none of these

ans: D

46. The rotational inertia of a wheel about its axle does not depend upon its:

A. diameter
B. mass
C. distribution of mass
D. speed of rotation
E. material composition

ans: D

47. Consider four objects, each having the same mass and the same radius:

1. a solid sphere
2. a hollow sphere
3. a flat disk in the x, y plane
4. a hoop in the x, y plane

The order of increasing rotational inertia about an axis through the center of mass and parallel to the z axis is:

A. 1, 2, 3, 4
B. 4, 3, 2, 1
C. 1, 3, 2, 4
D. 4, 2, 3, 1
E. 3, 1, 2, 4

ans: C

48. A and B are two solid cylinders made of aluminum. Their dimensions are shown. The ratio of the rotational inertia of B to that of A about the common axis X–X′ is:

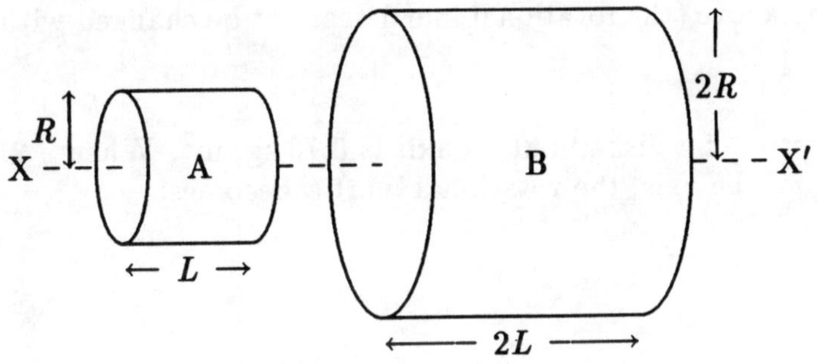

A. 2
B. 4
C. 8
D. 16
E. 32

ans: E

49. Two circular disks having the same mass and the same thickness are made from different materials. The disk with the smaller rotational inertia is:

A. the one made from the more dense material
B. the one made from the less dense material
C. neither — both rotational inertias are the same
D. the disk with the larger angular velocity
E. the disk with the larger torque

ans: A

50. A solid cylinder made of lead has the same mass and the same length as a solid cylinder made of wood. The rotational inertia of the lead cylinder compared to the wooden one is:

A. greater
B. less
C. same
D. unknown unless the radii are given
E. unknown unless both the masses and the radii are given

ans: B

51. To increase the rotational inertia of a solid disk about its axis without changing its mass:

A. drill holes near the rim and put the material near the axis
B. drill holes near the axis and put the material near the rim
C. drill holes at points on a circle near the rim and put the material at points between the holes
D. drill holes at points on a circle near the axis and put the material at points between the holes
E. do none of the above (the rotational inertia cannot be changed without changing the mass)

ans: B

52. The rotational inertia of a disk about its axis is $0.70\,\mathrm{kg}\cdot\mathrm{m}^2$. When a 2.0-kg weight is added to its rim, 0.40 m from the axis, the rotational inertia becomes:

A. $0.38\,\mathrm{kg}\cdot\mathrm{m}^2$
B. $0.54\,\mathrm{kg}\cdot\mathrm{m}^2$
C. $0.70\,\mathrm{kg}\cdot\mathrm{m}^2$
D. $0.86\,\mathrm{kg}\cdot\mathrm{m}^2$
E. $1.0\,\mathrm{kg}\cdot\mathrm{m}^2$

ans: E

53. When a thin uniform stick of mass M and length L is pivoted about its midpoint, its rotational inertia is $ML^2/12$. When pivoted about a parallel axis through one end, its rotational inertia is:

A. $ML^2/12$
B. $ML^2/6$
C. $ML^2/3$
D. $7ML^2/12$
E. $13ML^2/12$

ans: C

54. The rotational inertia of a solid uniform sphere about a diameter is $(5/2)MR^2$, where M is its mass and R is its radius. If the sphere is pivoted about an axis that is tangent to its surface, its rotational inertia is:

A. MR^2
B. $(1/2)MR^2$
C. $(3/2)MR^2$
D. $(5/2)MR^2$
E. $(7/2)MR^2$

ans: E

55. A solid uniform sphere of radius R and mass M has a rotational inertia about a diameter that is given by $(5/2)MR^2$. A light string of length $2R$ is attached to the surface and used to suspend the sphere from the ceiling. Its rotational inertia about the point of attachment at the ceiling is:

A. $(5/2)MR^2$
B. $4MR^2$
C. $(9/2)MR^2$
D. $(13/2)MR^2$
E. $(23/2)MR^2$

ans: E

56. A force with a given magnitude is to be applied to a wheel. The torque can be maximized by:

A. applying the force near the axle, radially outward from the axle
B. applying the force near the rim, radially outward from the axle
C. applying the force near the axle, parallel to a tangent to the wheel
D. applying the force at the rim, tangent to the rim
E. applying the force at the rim, at 45° to the tangent

ans: D

57. The meter stick shown below rotates about an axis through the point marked •, 20 cm from one end. Five forces act on the stick: one at each end, one at the pivot point, and two 40 cm from one end, as shown. The magnitudes of the forces are all the same. Rank the forces according to the magnitudes of the torques they produce about the pivot point, least to greatest.

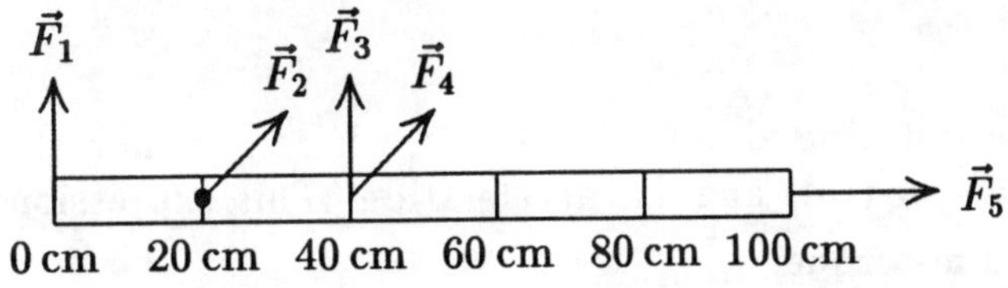

A. $\vec{F}_1, \vec{F}_2, \vec{F}_3, \vec{F}_4, \vec{F}_5$
B. $\vec{F}_1$ and $\vec{F}_2$ tie, then $\vec{F}_3, \vec{F}_4, \vec{F}_5$
C. $\vec{F}_2$ and $\vec{F}_5$ tie, then $\vec{F}_4, \vec{F}_1, \vec{F}_3$
D. $\vec{F}_2, \vec{F}_5, \vec{F}_1$ and $\vec{F}_3$ tie, then $\vec{F}_4$
E. $\vec{F}_2$ and $\vec{F}_5$ tie, then $\vec{F}_4$, then $\vec{F}_1$ and $\vec{F}_3$ tie

ans: E

58. A rod is pivoted about its center. A 5-N force is applied 4 m from the pivot and another 5-N force is applied 2 m from the pivot, as shown. The magnitude of the total torque about the pivot (in N·m) is:

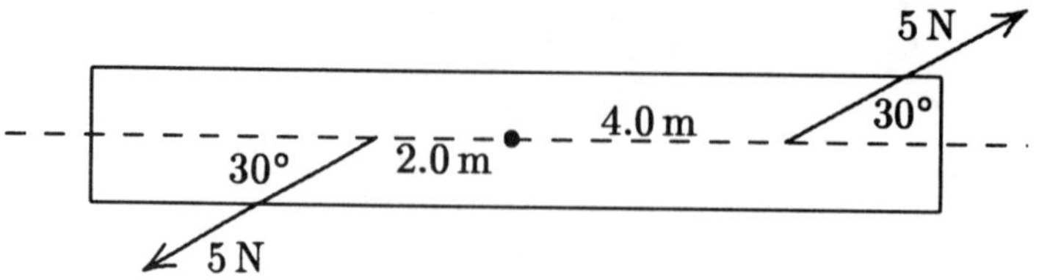

A. 0
B. 5
C. 8.7
D. 15
E. 26

ans: D

59. As a particle starts from rest and moves with constant angular acceleration around a circular orbit, the point of application of the force acting on it does not change. This must mean:

A. the torque acting on it is increasing in magnitude
B. the torque acting on it is decreasing in magnitude
C. the force acting on it is increasing in magnitude
D. the force acting on it is decreasing in magnitude
E. none of the above

ans: C

60. $\tau = I\alpha$ for an object rotating about a fixed axis, where τ is the net torque acting on it, I is its rotational inertia, and α is its angular acceleration. This expression:

A. is the definition of torque
B. is the definition of rotational inertia
C. is the definition of angular acceleration
D. follows directly from Newton's second law
E. depends on a principle of physics that is unrelated to Newton's second law

ans: D

61. A meter stick on a horizontal frictionless table top is pivoted at the 80-cm mark. It is initially at rest. A horizontal force $\vec{F_1}$ is applied perpendicularly to the end of the stick at 0 cm, as shown. A second horizontal force $\vec{F_2}$ (not shown) is applied at the 100-cm end of the stick. If the stick does not rotate:

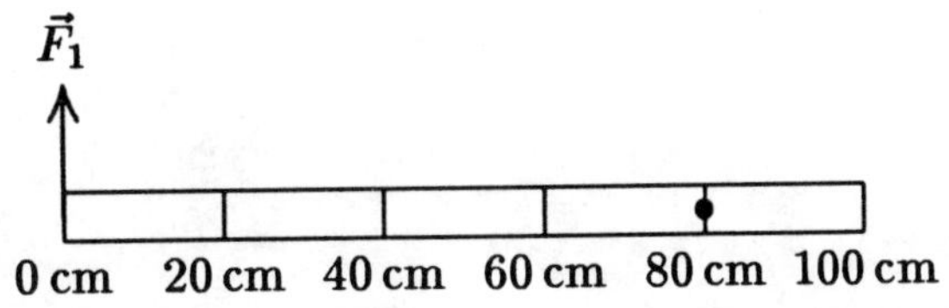

A. $|\vec{F_2}| > |\vec{F_1}|$ for all orientations of $\vec{F_2}$
B. $|\vec{F_2}| < |\vec{F_1}|$ for all orientations of $\vec{F_2}$
C. $|\vec{F_2}| = |\vec{F_1}|$ for all orientations of $\vec{F_2}$
D. $|\vec{F_2}| > |\vec{F_1}|$ for some orientations of $\vec{F_2}$ and $|\vec{F_2}| < |\vec{F_1}|$ for others
E. $|\vec{F_2}| > |\vec{F_1}|$ for some orientations of $\vec{F_2}$ and $|\vec{F_2}| = |\vec{F_1}|$ for others

ans: A

62. A uniform disk, a thin hoop, and a uniform sphere, all with the same mass and same outer radius, are each free to rotate about a fixed axis through its center. Assume the hoop is connected to the rotation axis by light spokes. With the objects starting from rest, identical forces are simultaneously applied to the rims, as shown. Rank the objects according to their angular velocities after a given time t, least to greatest.

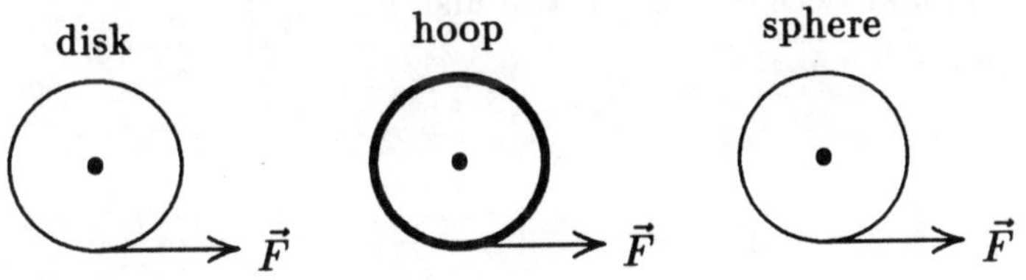

A. disk, hoop, sphere
B. hoop, disk, sphere
C. hoop, sphere, disk
D. hoop, disk, sphere
E. sphere, disk, hoop

ans: D

63. A disk is free to rotate on a fixed axis. A force of given magnitude F, in the plane of the disk, is to be applied. Of the following alternatives the greatest angular acceleration is obtained if the force is:

A. applied tangentially halfway between the axis and the rim
B. applied tangentially at the rim
C. applied radially halfway between the axis and the rim
D. applied radially at the rim
E. applied at the rim but neither radially nor tangentially

ans: B

64. A cylinder is 0.10 m in radius and 0.20 m in length. Its rotational inertia, about the cylinder axis on which it is mounted, is $0.020\,\mathrm{kg\cdot m^2}$. A string is wound around the cylinder and pulled with a force of 1.0 N. The angular acceleration of the cylinder is:

A. $2.5\,\mathrm{rad/s^2}$
B. $5.0\,\mathrm{rad/s^2}$
C. $10\,\mathrm{rad/s^2}$
D. $15\,\mathrm{rad/s^2}$
E. $20\,\mathrm{rad/s^2}$

ans: B

65. A disk with a rotational inertia of $5.0\,\mathrm{kg\cdot m^2}$ and a radius of 0.25 m rotates on a frictionless fixed axis perpendicular to the disk and through its center. A force of 8.0 N is applied tangentially to the rim. The angular acceleration of the disk is:

A. $0.40\,\mathrm{rad/s^2}$
B. $0.60\,\mathrm{rad/s^2}$
C. $1.0\,\mathrm{rad/s^2}$
D. $2.5\,\mathrm{rad/s^2}$
E. $10\,\mathrm{rad/s^2}$

ans: A

66. A disk with a rotational inertia of $5.0\,\mathrm{kg\cdot m^2}$ and a radius of 0.25 m rotates on a frictionless fixed axis perpendicular to the disk and through its center. A force of 8.0 N is applied parallel to the axis. The angular acceleration of the disk is:

A. 0
B. $0.40\,\mathrm{rad/s^2}$
C. $0.60\,\mathrm{rad/s^2}$
D. $1.0\,\mathrm{rad/s^2}$
E. $2.5\,\mathrm{rad/s^2}$

ans: A

67. A disk with a rotational inertia of $5.0\,\mathrm{kg\cdot m^2}$ and a radius of 0.25 m rotates on a frictionless fixed axis perpendicular to the disk and through its center. A force of 8.0 N is applied tangentially to the rim. If the disk starts at rest, then after it has turned through half a revolution its angular velocity is:

A. 0.57 rad/s
B. 0.64 rad/s
C. 0.80 rad/s
D. 1.6 rad/s
E. 3.2 rad/s

ans: D

68. A thin circular hoop of mass 1.0 kg and radius 2.0 m is rotating about an axis through its center and perpendicular to its plane. It is slowing down at the rate of $7.0\,\text{rad/s}^2$. The net torque acting on it is:

A. 7.0 N · m
B. 14.0 N · m
C. 28.0 N · m
D. 44.0 N · m
E. none of these

ans: C

69. A certain wheel has a rotational inertia of $12\,\text{kg}\cdot\text{m}^2$. As it turns through 5.0 rev its angular velocity increases from 5.0 rad/s to 6.0 rad/s. If the net torque is constant its value is:

A. 0.016 N · m
B. 0.18 N · m
C. 0.57 N · m
D. 2.1 N · m
E. 3.6 N · m

ans: D

70. A 16-kg block is attached to a cord that is wrapped around the rim of a flywheel of diameter 0.40 m and hangs vertically, as shown. The rotational inertia of the flywheel is $0.50\,\text{kg}\cdot\text{m}^2$. When the block is released and the cord unwinds, the acceleration of the block is:

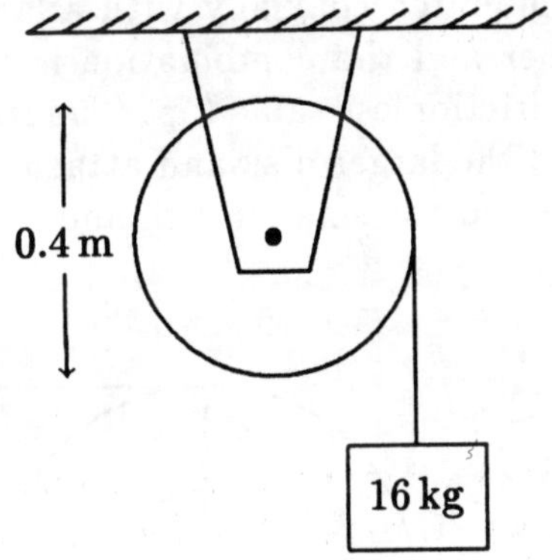

A. $0.15g$
B. $0.56g$
C. $0.84g$
D. g
E. $1.3g$

ans: B

71. A 8.0-cm radius disk with a rotational inertia of $0.12\,\text{kg}\cdot\text{m}^2$ is free to rotate on a horizontal axis. A string is fastened to the surface of the disk and a 10-kg mass hangs from the other end. The mass is raised by using a crank to apply a 9.0-N·m torque to the disk. The acceleration of the mass is:

A. $0.50\,\text{m/s}^2$
B. $1.7\,\text{m/s}^2$
C. $6.2\,\text{m/s}^2$
D. $12\,\text{m/s}^2$
E. $20\,\text{m/s}^2$

ans: A

72. A 0.70-kg disk with a rotational inertia given by $MR^2/2$ is free to rotate on a fixed horizontal axis suspended from the ceiling. A string is wrapped around the disk and a 2.0-kg mass hangs from the free end. If the string does not slip, then as the mass falls and the cylinder rotates, the suspension holding the cylinder pulls up on the cylinder with a force of:

A. 6.9 N
B. 9.8 N
C. 16 N
D. 26 N
E. 29 N

ans: B

73. A small disk of radius R_1 is mounted coaxially with a larger disk of radius R_2. The disks are securely fastened to each other and the combination is free to rotate on a fixed axle that is perpendicular to a horizontal frictionless table top. The rotational inertia of the combination is I. A string is wrapped around the larger disk and attached to a block of mass m, on the table. Another string is wrapped around the smaller disk and is pulled with a force $\vec{F}$ as shown. The acceleration of the block is:

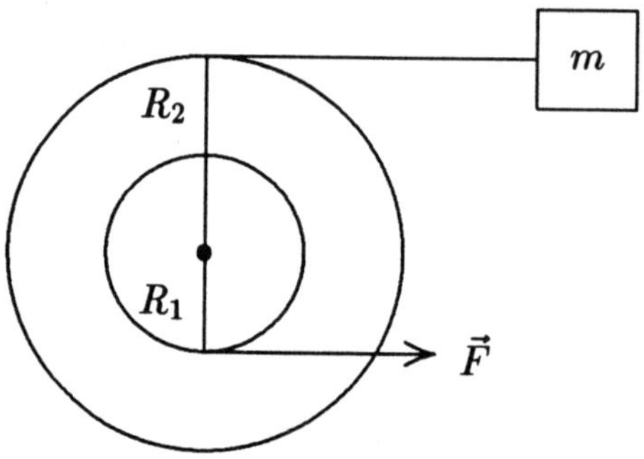

A. R_1F/mR_2
B. $R_1R_2F/(I - mR_2^2)$
C. $R_1R_2F/(I + mR_2^2)$
D. $R_1R_2F/(I - mR_1R_2)$
E. $R_1R_2F/(I + mR_1R_2)$

ans: C

74. A small disk of radius R_1 is fastened coaxially to a larger disk of radius R_2. The combination is free to rotate on a fixed axle, which is perpendicular to a horizontal frictionless table top. The rotational inertia of the combination is I. A string is wrapped around the larger disk and attached to a block of mass m, on the table. Another string is wrapped around the smaller disk and is pulled with a force $\vec{F}$ as shown. The tension in the string pulling the block is:

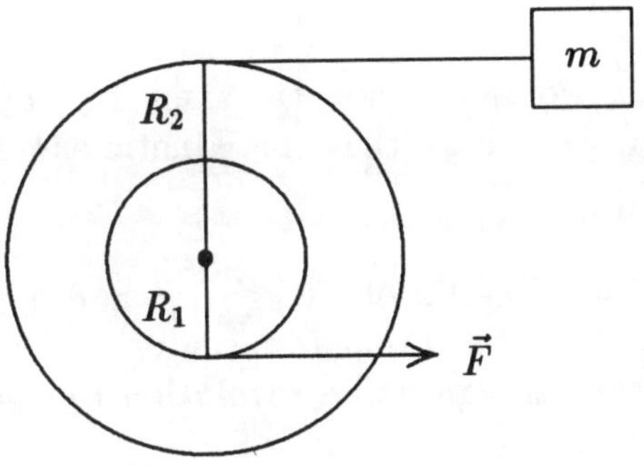

A. R_1F/R_2
B. $mR_1R_2F/(I - mR_2^2)$
C. $mR_1R_2F/(I + mR_2^2)$
D. $mR_1R_2F/(I - mR_1R_2)$
E. $mR_1R_2F/(I + mR_1R_2)$

ans: C

75. A block is attached to each end of a rope that passes over a pulley suspended from the ceiling. The blocks do not have the same mass. If the rope does not slip on the pulley, then at any instant after the blocks start moving, the rope:

A. pulls on both blocks, but exerts a greater force on the heavier block
B. pulls on both blocks, but exerts a greater force on the lighter block
C. pulls on both blocks and exerts the same magnitude force on both
D. does not pull on either block
E. pulls only on the lighter block

ans: A

76. A pulley with a radius of 3.0 cm and a rotational inertia of $4.5 \times 10^{-3}\,\mathrm{kg \cdot m^2}$ is suspended from the ceiling. A rope passes over it with a 2.0-kg block attached to one end and a 4.0-kg block attached to the other. The rope does not slip on the pulley. When the velocity of the heavier block is 2.0 m/s the total kinetic energy of the pulley and blocks is:

A. 2.0 J
B. 4.0 J
C. 14 J
D. 22 J
E. 28 J

ans: D

77. A pulley with a radius of 3.0 cm and a rotational inertia of $4.5 \times 10^{-3}\,\mathrm{kg \cdot m^2}$ is suspended from the ceiling. A rope passes over it with a 2.0-kg block attached to one end and a 4.0-kg block attached to the other. The rope does not slip on the pulley. At any instant after the blocks start moving, the object with the greatest kinetic energy is:

A. the heavier block
B. the lighter block
C. the pulley
D. either block (the two blocks have the same kinetic energy)
E. none (all three objects have the same kinetic energy)

ans: C

78. A disk with a rotational inertia of $5.0\,\mathrm{kg \cdot m^2}$ and a radius of 0.25 m rotates on a fixed axis perpendicular to the disk and through its center. A force of 2.0 N is applied tangentially to the rim. As the disk turns through half a revolution the work done by the force is:

A. 1.6 J
B. 2.5 J
C. 6.3 J
D. 10 J
E. 40 J

ans: A

79. A grinding wheel, used to sharpen tools, is powered by a motor. A knife held against the wheel exerts a torque of $0.80\,\mathrm{N \cdot m}$. If the wheel rotates with a constant angular velocity of 20 rad/s the work done on the wheel by the motor in 1.0 min is:

A. 0
B. 480 J
C. 960 J
D. 1400 J
E. 1800 J

ans: C

80. A disk has a rotational inertia of $6.0\,\mathrm{kg \cdot m^2}$ and a constant angular acceleration of $2.0\,\mathrm{rad/s^2}$. If it starts from rest the work done during the first 5.0 s by the net torque acting on it is:

A. 0
B. 30 J
C. 60 J
D. 300 J
E. 600 J

ans: D

81. A disk starts from rest and rotates around a fixed axis, subject to a constant net torque. The work done by the torque during the second 5 s is ______ as the work done during the first 5 s.

A. the same
B. twice as much
C. half as much
D. four times as much
E. one-fourth as much

ans: D

82. A disk starts from rest and rotates about a fixed axis, subject to a constant net torque. The work done by the torque during the second revolution is ____ as the work done during the first revolution.

A. the same
B. twice as much
C. half as much
D. four times as much
E. one-fourth as much

ans: A

Chapter 12: ROLLING, TORQUE, AND ANGULAR MOMENTUM

1. A wheel rolls without sliding along a horizontal road as shown. The velocity of the center of the wheel is represented by $\longrightarrow$. Point P is painted on the rim of the wheel. The instantaneous velocity of point P is:

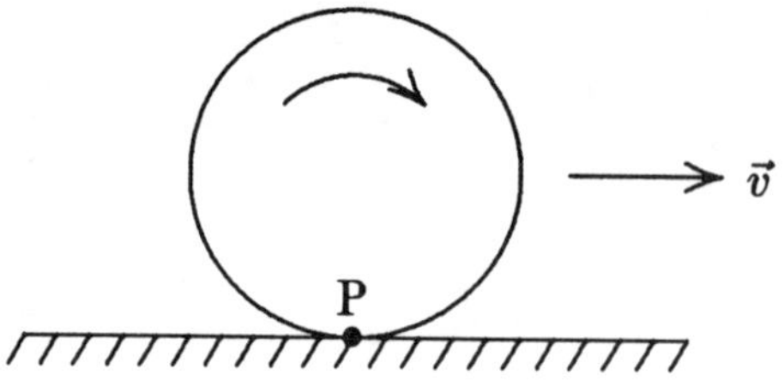

A. $\rightarrow$
B. $\leftarrow$
C. $\uparrow$
D. $\nearrow$
E. zero

ans: E

2. A wheel of radius 0.5 m rolls without sliding on a horizontal surface as shown. Starting from rest, the wheel moves with constant angular acceleration $6\,\text{rad/s}^2$. The distance in meters traveled by the center of the wheel from $t = 0$ to $t = 3\,\text{s}$ is:

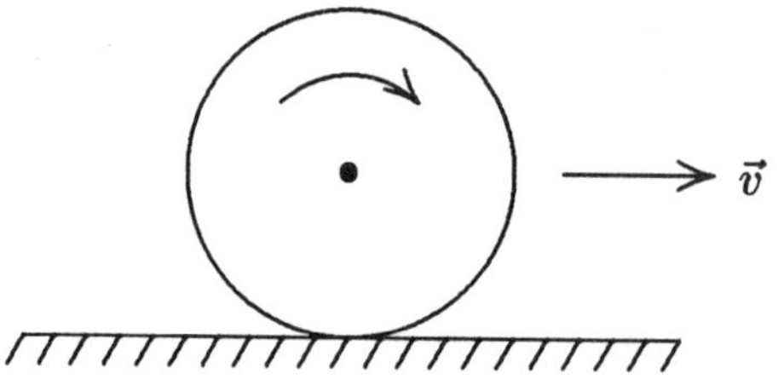

A. zero
B. 27
C. 13.5
D. 18
E. none of these

ans: C

3. Two wheels roll side-by-side without sliding, at the same speed. The radius of wheel 2 is twice the radius of wheel 1. The angular velocity of wheel 2 is:

A. twice the angular velocity of wheel 1
B. the same as the angular velocity of wheel 1
C. half the angular velocity of wheel 1
D. more than twice the angular velocity of wheel 1
E. less than half the angular velocity of wheel 1

ans: C

4. A forward force acting on the axle accelerates a rolling wheel on a horizontal surface. If the wheel does not slide the frictional force of the surface on the wheel is:

   A. zero
   B. in the forward direction and does zero work on the wheel
   C. in the forward direction and does positive work on the wheel
   D. in the backward direction and does zero work on the wheel
   E. in the backward direction and does positive work on the wheel

   ans: D

5. When the speed of a rear-drive car is increasing on a horizontal road the direction of the frictional force on the tires is:

   A. forward for all tires
   B. backward for all tires
   C. forward for the front tires and backward for the rear tires
   D. backward for the front tires and forward for the rear tires
   E. zero

   ans: D

6. A 3.0-kg solid wheel, rolling without sliding on a horizontal surface, has a rotational inertia about its axis given by $MR^2/2$, where M is its mass and R is its radius. A force is applied to the axle and the center of mass has an acceleration of $2.0\,\mathrm{m/s^2}$. The magnitudes of the applied force and the frictional force of the surface, respectively, are:

   A. 6.0 N, 0
   B. 0, 6.0 N
   C. 3.0 N, 9.0 N
   D. 9.0 N, 3.0 N
   E. 12 N, 3.0 N

   ans: D

7. The coefficient of static friction between a certain cylinder and a horizontal floor is 0.40. If the rotational inertia of the cylinder about its symmetry axis is given by $I = (1/2)MR^2$, then the maximum acceleration the cylinder can have without sliding is:

   A. $0.1g$
   B. $0.2g$
   C. $0.4g$
   D. $0.8g$
   E. g

   ans: D

8. A thin-walled hollow tube rolls without sliding along the floor. The ratio of its translational kinetic energy to its rotational kinetic energy (about an axis through its center of mass) is:

   A. 1
   B. 2
   C. 3
   D. 1/2
   E. 1/3

   ans: A

9. A sphere and a cylinder of equal mass and radius are simultaneously released from rest on the same inclined plane and roll without sliding down the incline. Then:

   A. the sphere reaches the bottom first because it has the greater inertia
   B. the cylinder reaches the bottom first because it picks up more rotational energy
   C. the sphere reaches the bottom first because it picks up more rotational energy
   D. they reach the bottom together
   E. none of the above are true

   ans: E

10. A hoop, a uniform disk, and a uniform sphere, all with the same mass and outer radius, start with the same speed and roll without sliding up identical inclines. Rank the objects according to how high they go, least to greatest.

   A. hoop, disk, sphere
   B. disk, hoop, sphere
   C. sphere, hoop, disk
   D. sphere, disk, hoop
   E. hoop, sphere, disk

   ans: A

11. A hoop has a mass of 200 g and a radius of 25 cm. It rolls without sliding along level ground at 5.0 m/s. Its total kinetic energy is:

   A. 2.5 J
   B. 5 J
   C. 10 J
   D. 250 J
   E. need to know the angular velocity

   ans: B

12. Two identical disks, with rotational inertia I $(= \frac{1}{2}MR^2)$, roll without sliding across a horizontal floor with the same speed and then up inclines. Disk A rolls up its incline without sliding. On the other hand, disk B rolls up a frictionless incline. Otherwise the inclines are identical. Disk A reaches a height 12 cm above the floor before rolling down again. Disk B reaches a height above the floor of:

   A. 24 cm
   B. 18 cm
   C. 12 cm
   D. 8 cm
   E. 6 cm

   ans: D

13. A yo-yo, arranged as shown, rests on a frictionless surface. When a force $\vec{F}$ is applied to the string as shown, the yo-yo:

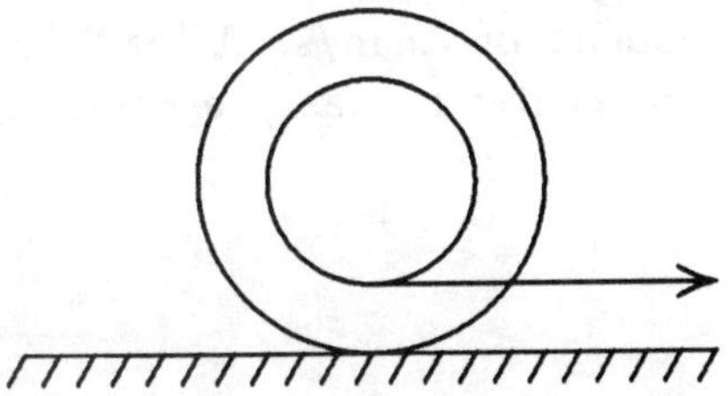

A. moves to the left and rotates counterclockwise
B. moves to the right and rotates counterclockwise
C. moves to the left and rotates clockwise
D. moves to the right and rotates clockwise
E. moves to the right and does not rotate

ans: B

14. We may apply conservation of energy to a cylinder rolling down an incline without sliding and exclude friction because:

A. there is no friction present
B. the angular velocity of the center of mass about the point of contact is zero
C. the coefficient of kinetic friction is zero
D. the linear velocity of the point of contact (relative to the inclined surface) is zero
E. the coefficient of static and kinetic friction are equal

ans: D

15. Two uniform cylinders have different masses and different rotational inertias. They simultaneously start from rest at the top of an inclined plane and roll without sliding down the plane. The cylinder that gets to the bottom first is:

A. the one with the larger mass
B. the one with the smaller mass
C. the one with the larger rotational inertia
D. the one with the smaller rotational inertia
E. neither (they arrive together)

ans: E

16. A 5.0-kg ball rolls without sliding from rest down an inclined plane. A 4.0-kg block, mounted on roller bearings totaling 100 g, rolls from rest down the same plane. At the bottom, the block has the:

A. greater speed
B. smaller speed
C. same speed as the ball
D. depends on the angle of inclination
E. depends on the radius of the ball

ans: A

17. A cylinder of radius $R = 6.0\,\text{cm}$ is resting on a rough horizontal surface. The coefficient of kinetic friction between the cylinder and the surface is 0.30 and the rotational inertia for rotation about the axis is given by $MR^2/2$, where M is its mass. Initially it is not rotating but its center of mass has a velocity of 7.0 m/s. After 2.0 s the velocity of its center of mass and its angular velocity about its center of mass, respectively, are:

A. 1.1 m/s, 0
B. 1.1 m/s, 19 rad/s
C. 1.1 m/s, 98 rad/s
D. 1.1 m/s, 200 rad/s
E. 5.9 m/s, 98 rad/s

ans: D

18. The fundamental dimensions of angular momentum are:

A. MLT^{-1}
B. $ML^{-2}T^{-2}$
C. ML^2T^{-1}
D. ML^2T^{-2}
E. none of these

ans: C

19. Possible units of angular momentum are:

A. kg·m/s
B. kg·m^2/s^2
C. kg·m/s^2
D. kg·m^2/s
E. none of these

ans: D

20. The unit kg·m^2/s can be used for:

A. angular momentum
B. rotational kinetic energy
C. rotational inertia
D. torque
E. power

ans: A

21. The newton·second is a unit of:

A. work
B. angular momentum
C. power
D. linear momentum
E. none of these

ans: D

22. A 2.0-kg block travels around a 0.50-m radius circle with an angular velocity of 12 rad/s. Its angular momentum about the center of the circle is:

A. $6.0\,\text{kg}\cdot\text{m}^2/\text{s}$
B. $12\,\text{kg}\cdot\text{m}^2/\text{s}$
C. $48\,\text{kg}/\text{m}^2\cdot\text{s}$
D. $72\,\text{kg}\cdot\text{m}^2/\text{s}^2$
E. $576\,\text{kg}/\text{m}^2\cdot\text{s}^2$

ans: A

23. The angular momentum vector of Earth about its rotation axis, due to its daily rotation, is directed:

A. tangent to the equator toward the east
B. tangent to the equator toward the west
C. north
D. south
E. toward the Sun

ans: C

24. A 6.0-kg particle moves to the right at 4.0 m/s as shown. Its angular momentum about the point O is:

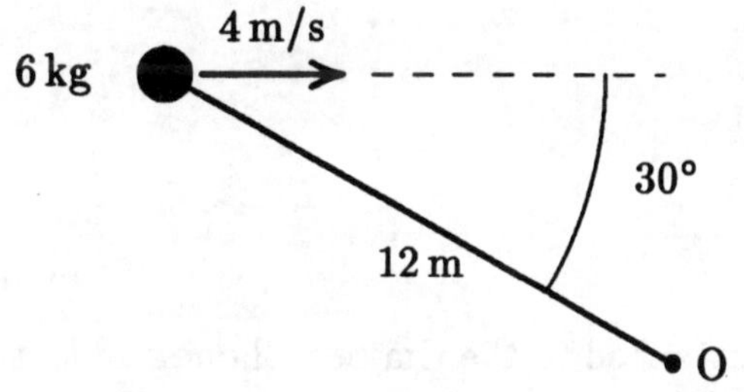

A. zero
B. $288\,\text{kg}\cdot\text{m}^2/\text{s}$
C. $144\,\text{kg}\cdot\text{m}^2/\text{s}$
D. $24\,\text{kg}\cdot\text{m}^2/\text{s}$
E. $249\,\text{kg}\cdot\text{m}^2/\text{s}$

ans: C

25. Two objects are moving in the x, y plane as shown. The magnitude of their total angular momentum (about the origin O) is:

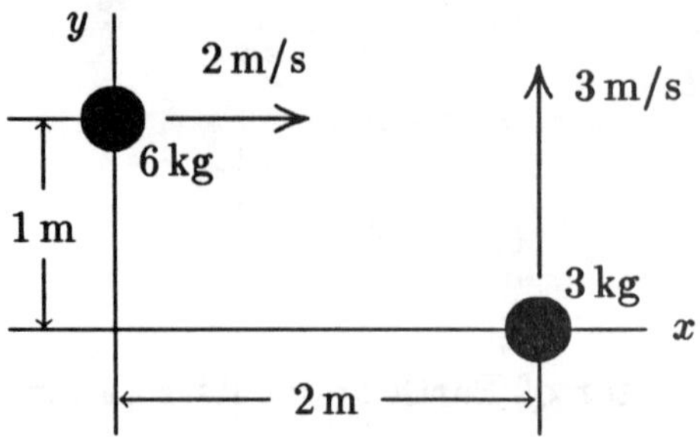

A. zero
B. $6\,\mathrm{kg\cdot m^2/s}$
C. $12\,\mathrm{kg\cdot m^2/s}$
D. $30\,\mathrm{kg\cdot m^2/s}$
E. $78\,\mathrm{kg\cdot m^2/s}$

ans: B

26. A 2.0-kg block starts from rest on the positive x axis 3.0 m from the origin and thereafter has an acceleration given by $\vec{a} = (4.0\,\mathrm{m/s^2})\,\hat{\imath} - (3.0\,\mathrm{m/s^2})\,\hat{\jmath}$. At the end of 2.0 s its angular momentum about the origin is:

A. 0
B. $(-36\,\mathrm{kg\cdot m^2/s})\,\hat{k}$
C. $(+48\,\mathrm{kg\cdot m^2/s})\,\hat{k}$
D. $(-96\,\mathrm{kg\cdot m^2/s})\,\hat{k}$
E. $(+96\,\mathrm{kg\cdot m^2/s})\,\hat{k}$

ans: B

27. A 15-g paper clip is attached to the rim of a phonograph record with a radius of 30 cm, spinning at 3.5 rad/s. Its angular momentum is:

A. $1.4 \times 10^{-3}\,\mathrm{kg\cdot m^2/s}$
B. $4.7 \times 10^{-3}\,\mathrm{kg\cdot m^2/s}$
C. $1.6 \times 10^{-2}\,\mathrm{kg\cdot m^2/s}$
D. $3.2 \times 10^{-1}\,\mathrm{kg\cdot m^2/s}$
E. $1.1\,\mathrm{kg\cdot m^2/s}$

ans: B

28. A 2.0-kg block travels around a 0.50-m radius circle with an angular speed of 12 rad/s. The circle is parallel to the xy plane and is centered on the z axis, 0.75 m from the origin. The magnitude of the angular momentum around the origin is:

A. $6.0\,\mathrm{kg\cdot m^2/s}$
B. $9.0\,\mathrm{kg\cdot m^2/s}$
C. $11\,\mathrm{kg\cdot m^2/s}$
D. $14\,\mathrm{kg\cdot m^2/s}$
E. $20\,\mathrm{kg\cdot m^2/s}$

ans: C

29. A 2.0-kg block travels around a 0.50-m radius circle with an angular speed of 12 rad/s. The circle is parallel to the xy plane and is centered on the z axis, a distance of 0.75 m from the origin. The z component of the angular momentum around the origin is:

A. $6.0\,\text{kg}\cdot\text{m}^2/\text{s}$
B. $9.0\,\text{kg}\cdot\text{m}^2/\text{s}$
C. $11\,\text{kg}\cdot\text{m}^2/\text{s}$
D. $14\,\text{kg}\cdot\text{m}^2/\text{s}$
E. $20\,\text{kg}\cdot\text{m}^2/\text{s}$

ans: A

30. A 2.0-kg block travels around a 0.50-m radius circle with an angular speed of 12 rad/s. The circle is parallel to the xy plane and is centered on the z axis, 0.75 m from the origin. The component in the xy plane of the angular momentum around the origin has magnitude:

A. 0
B. $6.0\,\text{kg}\cdot\text{m}^2/\text{s}$
C. $9.0\,\text{kg}\cdot\text{m}^2/\text{s}$
D. $11\,\text{kg}\cdot\text{m}^2/\text{s}$
E. $14\,\text{kg}\cdot\text{m}^2/\text{s}$

ans: C

31. A uniform disk has radius R and mass M. When it is spinning with angular velocity ω about an axis through its center and perpendicular to its face its angular momentum is $I\omega$. When it is spinning with the same angular velocity about a parallel axis a distance h away its angular momentum is:

A. $I\omega$
B. $(I + Mh^2)\omega$
C. $I - Mh^2)\omega$
D. $(I + MR^2)\omega$
E. $(I - MR^2)\omega$

ans: B

32. A pulley with radius R and rotational inertia I is free to rotate on a horizontal fixed axis through its center. A string passes over the pulley. Mass m_1 is attached to one end and mass m_2 is attached to the other. At one time m_1 is moving downward with speed v. If the string does not slip on the pulley, the magnitude of the total angular momentum, about the pulley center, of the masses and pulley, considered as a system, is given by:

A. $(m_1 - m_2)vR + Iv/R$
B. $(m_1 + m_2)vR + Iv/R$
C. $(m_1 - m_2)vR - Iv/R$
D. $(m_1 + m_2)vR - Iv/R$
E. none of the above

ans: B

33. A single force acts on a particle situated on the positive x axis. The torque about the origin is in the negative z direction. The force might be:

A. in the positive y direction
B. in the negative y direction
C. in the positive x direction
D. in the negative x direction
E. in the positive z direction

ans: B

34. A rod rests on frictionless ice. Forces that are equal in magnitude and opposite in direction are then simultaneously applied to its ends as shown. The quantity that vanishes is its:

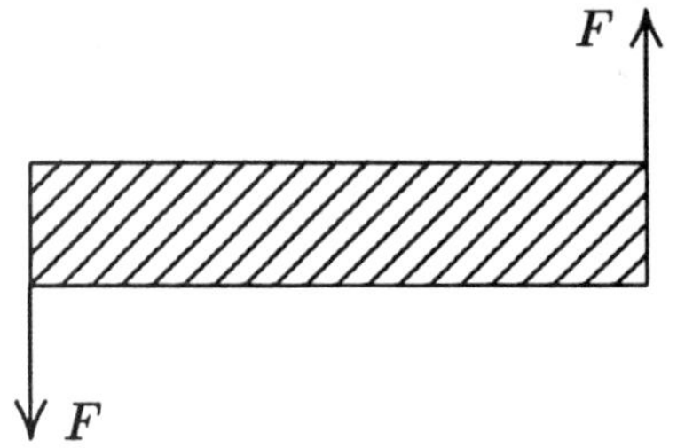

A. angular momentum
B. angular acceleration
C. total linear momentum
D. kinetic energy
E. rotational inertia

ans: C

35. A 2.0-kg stone is tied to a 0.50-m long string and swung around a circle at a constant angular velocity of 12 rad/s. The net torque on the stone about the center of the circle is:

A. 0
B. $6.0\,\mathrm{N \cdot m}$
C. $12\,\mathrm{N \cdot m}$
D. $72\,\mathrm{N \cdot m}$
E. $140\,\mathrm{N \cdot m}$

ans: A

36. A 2.0-kg stone is tied to a 0.50-m long string and swung around a circle at a constant angular velocity of 12 rad/s. The circle is parallel to the xy plane and is centered on the z axis, 0.75 m from the origin. The magnitude of the torque about the origin is:

A. 0
B. $6.0\,\mathrm{N \cdot m}$
C. $14\,\mathrm{N \cdot m}$
D. $72\,\mathrm{N \cdot m}$
E. $108\,\mathrm{N \cdot m}$

ans: E

37. A 2.0-kg block starts from rest on the positive x axis 3.0 m from the origin and thereafter has an acceleration given by $\vec{a} = (4.0\,\mathrm{m/s^2})\,\hat{\imath} - (3.0\,\mathrm{m/s^2})\,\hat{\jmath}$. The torque, relative to the origin, acting on it at the end of 2.0 s is:

A. 0
B. $(-18\,\mathrm{N\cdot m})\,\hat{k}$
C. $(+24\,\mathrm{N\cdot m})\,\hat{k}$
D. $(-144\,\mathrm{N\cdot m})\,\hat{k}$
E. $(+144\,\mathrm{N\cdot m})\,\hat{k}$

ans: B

38. A uniform disk, a thin hoop, and a uniform sphere, all with the same mass and same outer radius, are each free to rotate about a fixed axis through its center. Assume the hoop is connected to the rotation axis by light spokes. With the objects starting from rest, identical forces are simultaneously applied to the rims, as shown. Rank the objects according to their angular momenta after a given time t, least to greatest.

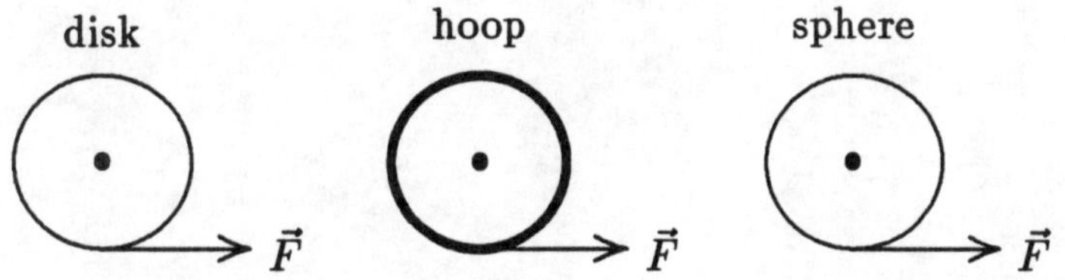

A. They are all the same
B. disk, hoop, sphere
C. hoop, disk, sphere
D. hoop, sphere, disk
E. hoop, disk, sphere

ans: A

39. A single force acts on a particle P. Rank each of the orientations of the force shown below according to the magnitude of the time rate of change of the particle's angular momentum about the point O, least to greatest.

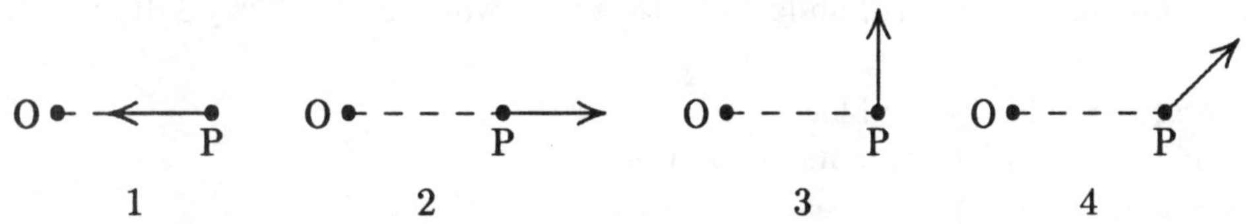

A. 1, 2, 3, 4
B. 1 and 2 tie, then 3, 4
C. 1 and 2 tie, then 4, 3
D. 1 and 2 tie, then 3 and 4 tie
E. All are the same

ans: C

40. A pulley with radius R is free to rotate on a horizontal fixed axis through its center. A string passes over the pulley. Mass m_1 is attached to one end and mass m_2 is attached to the other. The portion of the string attached to m_1 has tension T_1 and the portion attached to m_2 has tension T_2. The magnitude of the total external torque, about the pulley center, acting on the masses and pulley, considered as a system, is given by:

A. $|m_1 - m_2|gR$
B. $(m_1 + m_2)gR$
C. $|m_1 - m_2|gR + (T_1 + T_2)R$
D. $(m_1 + m_2)gR + (T_1 - T_2)R$
E. $|m_1 - m_2|gR + (T_2 - T_1)R$

ans: A

41. An ice skater with rotational inertia I_0 is spinning with angular speed ω_0. She pulls her arms in, decreasing her rotational inertia to $I_0/3$. Her angular speed becomes:

A. $\omega_0/3$
B. $\omega_0/\sqrt{3}$
C. ω_0
D. $\sqrt{3}\omega_0$
E. $3\omega_0$

ans: E

42. A man, with his arms at his sides, is spinning on a light frictionless turntable. When he extends his arms:

A. his angular velocity increases
B. his angular velocity remains the same
C. his rotational inertia decreases
D. his rotational kinetic energy increases
E. his angular momentum remains the same

ans: E

43. A man, holding a weight in each hand, stands at the center of a horizontal frictionless rotating turntable. The effect of the weights is to double the rotational inertia of the system. As he is rotating, the man opens his hands and drops the two weights. They fall outside the turntable. Then:

A. his angular velocity doubles
B. his angular velocity remains about the same
C. his angular velocity is halved
D. the direction of his angular momentum vector changes
E. his rotational kinetic energy increases

ans: B

44. A uniform sphere of radius R rotates about a diameter with angular momentum L. Under the action of internal forces the sphere collapses to a uniform sphere of radius $R/2$. Its new angular momentum is:

A. $L/4$
B. $L/2$
C. L
D. $2L$
E. $4L$

ans: C

45. When a man on a frictionless rotating stool extends his arms horizontally, his rotational kinetic energy:

A. must increase
B. must decrease
C. must remain the same
D. may increase or decrease depending on his initial angular velocity
E. may increase or decrease depending on his gravitational potential energy

ans: B

46. When a woman on a frictionless rotating turntable extends her arms out horizontally, her angular momentum:

A. must increase
B. must decrease
C. must remain the same
D. may increase or decrease depending on her initial angular velocity
E. tilts away from the vertical

ans: C

47. Two disks are mounted on low-friction bearings on a common shaft. The first disc has rotational inertia I and is spinning with angular velocity ω. The second disc has rotational inertia $2I$ and is spinning in the same direction as the first disc with angular velocity 2ω as shown. The two disks are slowly forced toward each other along the shaft until they couple and have a final common angular velocity of:

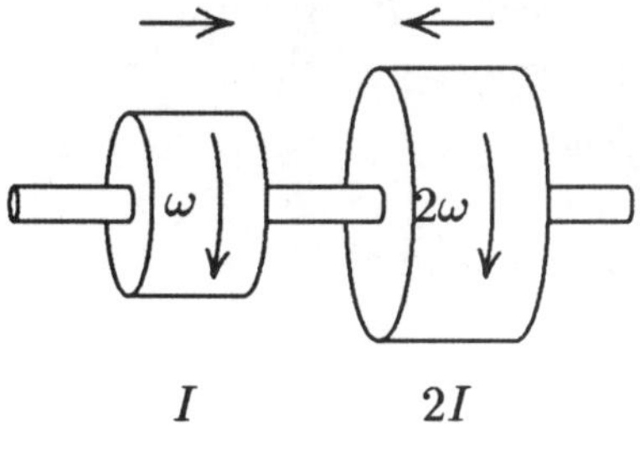

A. $5\omega/3$
B. $\omega\sqrt{3}$
C. $\omega\sqrt{7/3}$
D. ω
E. 3ω

ans: A

48. A wheel, mounted on a vertical shaft of negligible rotational inertia, is rotating at 500 rpm. Another identical (but not rotating) wheel is suddenly dropped onto the same shaft as shown. The resultant combination of the two wheels and shaft will rotate at:

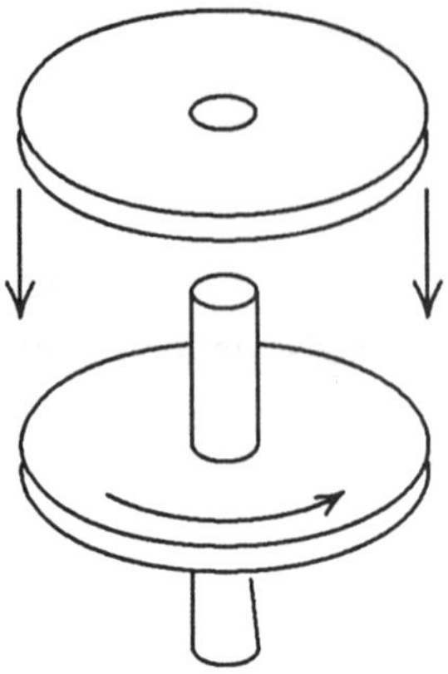

A. 250 rpm
B. 354 rpm
C. 500 rpm
D. 707 rpm
E. 1000 rpm

ans: A

49. A phonograph record is dropped onto a freely spinning turntable. Then:

A. neither angular momentum nor mechanical energy is conserved because of the frictional forces between record and turntable
B. the frictional force between record and turntable increases the total angular momentum
C. the frictional force between record and turntable decreases the total angular momentum
D. the total angular momentum remains constant
E. the sum of the angular momentum and rotational kinetic energy remains constant

ans: D

50. A playground merry-go-round has a radius of 3.0 m and a rotational inertia of $600\,\mathrm{kg \cdot m^2}$. When the merry-go-round is at rest, a 20-kg child runs at 5.0 m/s along a line tangent to the rim and jumps on. The angular velocity of the merry-go-round is then:

A. 0.38 rad/s
B. 0.45 rad/s
C. 0.71 rad/s
D. 0.56 rad/s
E. 1.2 rad/s

ans: A

51. A playground merry-go-round has a radius of 3.0 m and a rotational inertia of $600\,\mathrm{kg\cdot m^2}$. It is initially spinning at 0.80 rad/s when a 20-kg child crawls from the center to the rim. When the child reaches the rim the angular velocity of the merry-go-round is:

A. 0.62 rad/s
B. 0.73 rad/s
C. 0.80 rad/s
D. 0.89 rad/s
E. 1.1 rad/s

ans: A

52. Two pendulum bobs of unequal mass are suspended from the same fixed point by strings of equal length. The lighter bob is drawn aside and then released so that it collides with the other bob on reaching the vertical position. The collision is elastic. What quantities are conserved in the collision?

A. Both kinetic energy and angular momentum of the system
B. Only kinetic energy
C. Only angular momentum
D. Angular speed of lighter bob
E. None of the above

ans: A

53. A particle, held by a string whose other end is attached to a fixed point C, moves in a circle on a horizontal frictionless surface. If the string is cut, the angular momentum of the particle about the point C:

A. increases
B. decreases
C. does not change
D. changes direction but not magnitude
E. none of these

ans: C

54. A block with mass M, on the end of a string, moves in a circle on a horizontal frictionless table as shown. As the string is slowly pulled through a small hole in the table:

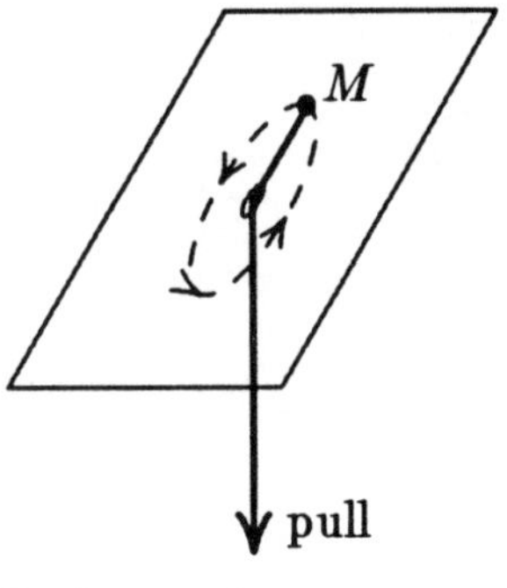

A. the angular momentum of the block remains constant
B. the angular momentum of the block decreases
C. the kinetic energy of the block remains constant
D. the kinetic energy of the block decreases
E. none of the above

ans: A

PART III

Instructor's Solutions Manual

Chapters 1-12

Preface

This booklet includes the solutions relevant to the EXERCISES & PROBLEMS sections of the 6th edition of **Fundamentals of Physics**, by Halliday, Resnick, and Walker. We also include solutions to problems in the Problems Supplement. We have not included solutions or discussions which pertain to the QUESTIONS sections.

I am very grateful for helpful input from J. Richard Christman, Meighan Dillon, Barbara Moore, and Jearl Walker regarding the development of this document.

TABLE OF CONTENTS

Chapter 1

1. The metric prefixes (micro, pico, nano, ...) are given for ready reference on the inside front cover of the textbook (see also Table 1-2).

   (a) Since $1\,\text{km} = 1 \times 10^3\,\text{m}$ and $1\,\text{m} = 1 \times 10^6\,\mu\text{m}$,
   $$1\,\text{km} = 10^3\,\text{m} = (10^3\,\text{m})(10^6\,\mu\text{m/m}) = 10^9\,\mu\text{m}\,.$$
   The given measurement is 1.0 km (two significant figures), which implies our result should be written as $1.0 \times 10^9\ \mu\text{m}$.

   (b) We calculate the number of microns in 1 centimeter. Since $1\,\text{cm} = 10^{-2}\,\text{m}$,
   $$1\,\text{cm} = 10^{-2}\,\text{m} = (10^{-2}\,\text{m})(10^6\,\mu\text{m/m}) = 10^4\,\mu\text{m}\,.$$
   We conclude that the fraction of one centimeter equal to $1.0\,\mu\text{m}$ is 1.0×10^{-4}.

   (c) Since $1\,\text{yd} = (3\,\text{ft})(0.3048\,\text{m/ft}) = 0.9144\,\text{m}$,
   $$1.0\,\text{yd} = (0.91\,\text{m})(10^6\,\mu\text{m/m}) = 9.1 \times 10^5\,\mu\text{m}\,.$$

2. The customer expects $20 \times 7056\ \text{in}^3$ and receives $20 \times 5826\ \text{in}^3$, the difference being 24600 cubic inches, or
   $$(24600\ \text{in}^3)\left(\frac{2.54\ \text{cm}}{1\ \text{inch}}\right)^3\left(\frac{1\ \text{L}}{1000\ \text{cm}^3}\right) = 403\ \text{L}$$
   where Appendix D has been used (see also Sample Problem 1-2).

3. Using the given conversion factors, we find

   (a) the distance d *in rods* to be
   $$d = 4.0\,\text{furlongs} = \frac{(4.0\,\text{furlongs})(201.168\,\text{m/furlong})}{5.0292\,\text{m/rod}} = 160\,\text{rods}\,,$$

   (b) and that distance *in chains* to be
   $$d = \frac{(4.0\,\text{furlongs})(201.168\,\text{m/furlong})}{20.117\,\text{m/chain}} = 40\,\text{chains}\,.$$

4. (a) Recalling that 2.54 cm equals 1 inch (exactly), we obtain
   $$(0.80\ \text{cm})\left(\frac{1\ \text{inch}}{2.54\ \text{cm}}\right)\left(\frac{6\ \text{picas}}{1\ \text{inch}}\right)\left(\frac{12\ \text{points}}{1\ \text{pica}}\right) \approx 23\ \text{points}\ ,$$

   (b) and
   $$(0.80\ \text{cm})\left(\frac{1\ \text{inch}}{2.54\ \text{cm}}\right)\left(\frac{6\ \text{picas}}{1\ \text{inch}}\right) \approx 1.9\ \text{picas}\ .$$

5. Various geometric formulas are given in Appendix E.

   (a) Substituting

$$R = (6.37 \times 10^6\,\text{m})\,(10^{-3}\,\text{km/m}) = 6.37 \times 10^3\,\text{km}$$

   into *circumference* $= 2\pi R$, we obtain 4.00×10^4 km.

   (b) The surface area of Earth is

$$4\pi R^2 = 4\pi\,(6.37 \times 10^3\,\text{km})^2 = 5.10 \times 10^8\ \text{km}^2\ .$$

   (c) The volume of Earth is

$$\frac{4\pi}{3}R^3 = \frac{4\pi}{3}\,(6.37 \times 10^3\,\text{km})^3 = 1.08 \times 10^{12}\ \text{km}^3\ .$$

6. (a) Using the fact that the area A of a rectangle is width×length, we find

$$\begin{aligned} A_{\text{total}} &= (3.00\ \text{acre}) + (25.0\,\text{perch})(4.00\,\text{perch}) \\ &= (3.00\,\text{acre})\left(\frac{(40\,\text{perch})(4\,\text{perch})}{1\ \text{acre}}\right) + 100\,\text{perch}^2 \\ &= 580\ \text{perch}^2\ . \end{aligned}$$

   We multiply this by the perch^2 → rood conversion factor (1 rood/40 perch^2) to obtain the answer: $A_{\text{total}} = 14.5$ roods.

   (b) We convert our intermediate result in part (a):

$$A_{\text{total}} = (580\ \text{perch}^2)\left(\frac{16.5\ \text{ft}}{1\ \text{perch}}\right)^2 = 1.58 \times 10^5\ \text{ft}^2\ .$$

   Now, we use the feet → meters conversion given in Appendix D to obtain

$$A_{\text{total}} = (1.58 \times 10^5\ \text{ft}^2)\left(\frac{1\ \text{m}}{3.281\ \text{ft}}\right)^2 = 1.47 \times 10^4\ \text{m}^2\ .$$

7. The volume of ice is given by the product of the semicircular surface area and the thickness. The semicircle area is $A = \pi r^2/2$, where r is the radius. Therefore, the volume is

$$V = \frac{\pi}{2}r^2\,z$$

   where z is the ice thickness. Since there are 10^3 m in 1 km and 10^2 cm in 1 m, we have

$$r = (2000\,\text{km})\left(\frac{10^3\ \text{m}}{1\,\text{km}}\right)\left(\frac{10^2\ \text{cm}}{1\,\text{m}}\right) = 2000 \times 10^5\ \text{cm}\ .$$

   In these units, the thickness becomes

$$z = (3000\,\text{m})\left(\frac{10^2\ \text{cm}}{1\,\text{m}}\right) = 3000 \times 10^2\ \text{cm}\ .$$

   Therefore,

$$V = \frac{\pi}{2}\,(2000 \times 10^5\ \text{cm})^2\,(3000 \times 10^2\ \text{cm}) = 1.9 \times 10^{22}\ \text{cm}^3\ .$$

8. The total volume V of the real house is that of a triangular prism (of height $h = 3.0$ m and base area $A = 20 \times 12 = 240\ \text{m}^2$) in addition to a rectangular box (height $h' = 6.0$ m and same base). Therefore,

$$V = \frac{1}{2}hA + h'A = \left(\frac{h}{2} + h'\right)A = 1800\ \text{m}^3\ .$$

(a) Each dimension is reduced by a factor of 1/12, and we find

$$V_{\text{doll}} = (1800 \text{ m}^3)\left(\frac{1}{12}\right)^3 \approx 1.0 \text{ m}^3 .$$

(b) In this case, each dimension (relative to the real house) is reduced by a factor of 1/144. Therefore,

$$V_{\text{miniature}} = (1800 \text{ m}^3)\left(\frac{1}{144}\right)^3 \approx 6.0 \times 10^{-4} \text{ m}^3 .$$

9. We use the conversion factors found in Appendix D.

$$1 \text{ acre} \cdot \text{ft} = (43,560 \text{ ft}^2) \cdot \text{ft} = 43,560 \text{ ft}^3 .$$

Since 2 in. = (1/6) ft, the volume of water that fell during the storm is

$$V = (26 \text{ km}^2)(1/6 \text{ ft}) = (26 \text{ km}^2)(3281 \text{ ft/km})^2(1/6 \text{ ft}) = 4.66 \times 10^7 \text{ ft}^3 .$$

Thus,

$$V = \frac{4.66 \times 10^7 \text{ ft}^3}{4.3560 \times 10^4 \text{ ft}^3/\text{acre} \cdot \text{ft}} = 1.1 \times 10^3 \text{ acre} \cdot \text{ft} .$$

10. The metric prefixes (micro (μ), pico, nano, ...) are given for ready reference on the inside front cover of the textbook (also, Table 1-2).

$$\begin{aligned} 1\,\mu\text{century} &= (10^{-6} \text{ century})\left(\frac{100 \text{ y}}{1 \text{ century}}\right)\left(\frac{365 \text{ day}}{1 \text{ y}}\right)\left(\frac{24 \text{ h}}{1 \text{ day}}\right)\left(\frac{60 \text{ min}}{1 \text{ h}}\right) \\ &= 52.6 \text{ min} . \end{aligned}$$

The percent difference is therefore

$$\frac{52.6 \text{ min} - 50 \text{ min}}{50 \text{ min}} = 5.2\% .$$

11. We use the conversion factors given in Appendix D and the definitions of the SI prefixes given in Table 1-2 (also listed on the inside front cover of the textbook). Here, "ns" represents the nanosecond unit, "ps" represents the picosecond unit, and so on.

(a) 1 m = 3.281 ft and 1 s = 10^9 ns. Thus,

$$3.0 \times 10^8 \text{ m/s} = \left(\frac{3.0 \times 10^8 \text{ m}}{\text{s}}\right)\left(\frac{3.281 \text{ ft}}{\text{m}}\right)\left(\frac{\text{s}}{10^9 \text{ ns}}\right) = 0.98 \text{ ft/ns} .$$

(b) Using 1 m = 10^3 mm and 1 s = 10^{12} ps, we find

$$\begin{aligned} 3.0 \times 10^8 \text{ m/s} &= \left(\frac{3.0 \times 10^8 \text{ m}}{\text{s}}\right)\left(\frac{10^3 \text{ mm}}{\text{m}}\right)\left(\frac{\text{s}}{10^{12} \text{ ps}}\right) \\ &= 0.30 \text{ mm/ps} . \end{aligned}$$

12. The number of seconds in a year is 3.156×10^7. This is listed in Appendix D and results from the product

$$(365.25 \text{ day/y})(24 \text{ h/day})(60 \text{ min/h})(60 \text{ s/min}) .$$

(a) The number of shakes in a second is 10^8; therefore, there are indeed more shakes per second than there are seconds per year.

(b) Denoting the age of the universe as 1 u-day (or 86400 u-sec), then the time during which humans have existed is given by

$$\frac{10^6}{10^{10}} = 10^{-4} \text{ u-day} ,$$

which we may also express as

$$(10^{-4} \text{ u-day}) \left(\frac{86400 \text{ u-sec}}{1 \text{ u-day}} \right) = 8.6 \text{ u-sec} .$$

13. None of the clocks advance by exactly 24 h in a 24-h period but this is not the most important criterion for judging their quality for measuring time intervals. What is important is that the clock advance by the same amount in each 24-h period. The clock reading can then easily be adjusted to give the correct interval. If the clock reading jumps around from one 24-h period to another, it cannot be corrected since it would impossible to tell what the correction should be. The following gives the corrections (in seconds) that must be applied to the reading on each clock for each 24-h period. The entries were determined by subtracting the clock reading at the end of the interval from the clock reading at the beginning.

CLOCK	Sun. -Mon.	Mon. -Tues.	Tues. -Wed.	Wed. -Thurs.	Thurs. -Fri.	Fri. -Sat
A	−16	−16	−15	−17	−15	−15
B	−3	+5	−10	+5	+6	−7
C	−58	−58	−58	−58	−58	−58
D	+67	+67	+67	+67	+67	+67
E	+70	+55	+2	+20	+10	+10

Clocks C and D are both good timekeepers in the sense that each is consistent in its daily drift (relative to WWF time); thus, C and D are easily made "perfect" with simple and predictable corrections. The correction for clock C is less than the correction for clock D, so we judge clock C to be the best and clock D to be the next best. The correction that must be applied to clock A is in the range from 15 s to 17s. For clock B it is the range from −5 s to +10 s, for clock E it is in the range from −70 s to −2 s. After C and D, A has the smallest range of correction, B has the next smallest range, and E has the greatest range. From best the worst, the ranking of the clocks is C, D, A, B, E.

14. The time on any of these clocks is a straight-line function of that on another, with slopes $\neq 1$ and y-intercepts $\neq 0$. From the data in the figure we deduce

$$\begin{aligned} t_C &= \frac{2}{7} t_B + \frac{594}{7} \\ t_B &= \frac{33}{40} t_A - \frac{662}{5} . \end{aligned}$$

These are used in obtaining the following results.

(a) We find

$$t'_B - t_B = \frac{33}{40} (t'_A - t_A) = 495 \text{ s}$$

when $t'_A - t_A = 600$ s.

(b) We obtain

$$t'_C - t_C = \frac{2}{7} (t'_B - t_B) = \frac{2}{7}(495) = 141 \text{ s} .$$

(c) Clock B reads $t_B = (33/40)(400) - (662/5) \approx 198$ s when clock A reads $t_A = 400$ s.

(d) From $t_C = 15 = (2/7)t_B + (594/7)$, we get $t_B \approx -245$ s.

15. We convert meters to astronomical units, and seconds to minutes, using

$$\begin{aligned} 1000\,\text{m} &= 1\,\text{km} \\ 1\,\text{AU} &= 1.50 \times 10^8\,\text{km} \\ 60\,\text{s} &= 1\,\text{min}\,. \end{aligned}$$

Thus, 3.0×10^8 m/s becomes

$$\left(\frac{3.0 \times 10^8\,\text{m}}{\text{s}}\right)\left(\frac{1\,\text{km}}{1000\,\text{m}}\right)\left(\frac{\text{AU}}{1.50 \times 10^8\,\text{km}}\right)\left(\frac{60\,\text{s}}{\text{min}}\right) = 0.12\,\text{AU/min}\,.$$

16. Since a change of longitude equal to 360° corresponds to a 24 hour change, then one expects to change longitude by $360°/24 = 15°$ before resetting one's watch by 1.0 h.

17. The last day of the 20 centuries is longer than the first day by

$$(20\,\text{century})(0.001\,\text{s/century}) = 0.02\,\text{s}\,.$$

The average day during the 20 centuries is $(0 + 0.02)/2 = 0.01$ s longer than the first day. Since the increase occurs uniformly, the cumulative effect T is

$$\begin{aligned} T &= (\text{average increase in length of a day})(\text{number of days}) \\ &= \left(\frac{0.01\,\text{s}}{\text{day}}\right)\left(\frac{365.25\,\text{day}}{\text{y}}\right)(2000\,\text{y}) \\ &= 7305\text{ s} \end{aligned}$$

or roughly two hours.

18. We denote the pulsar rotation rate f (for frequency).

$$f = \frac{1\text{ rotation}}{1.55780644887275 \times 10^{-3}\,\text{s}}$$

(a) Multiplying f by the time-interval $t = 7.00$ days (which is equivalent to 604800 s, if we ignore *significant figure* considerations for a moment), we obtain the number of rotations:

$$N = \left(\frac{1\,\text{rotation}}{1.55780644887275 \times 10^{-3}\,\text{s}}\right)(604800\,\text{s}) = 388238218.4$$

which should now be rounded to 3.88×10^8 rotations since the time-interval was specified in the problem to three significant figures.

(b) We note that the problem specifies the *exact* number of pulsar revolutions (one million). In this case, our unknown is t, and an equation similar to the one we set up in part (a) takes the form

$$\begin{aligned} N &= ft \\ 1 \times 10^6 &= \left(\frac{1\text{ rotation}}{1.55780644887275 \times 10^{-3}\,\text{s}}\right) t \end{aligned}$$

which yields the result $t = 1557.80644887275$ s (though students who do this calculation on their calculator might not obtain those last several digits).

(c) Careful reading of the problem shows that the time-uncertainty *per revolution* is $\pm 3 \times 10^{-17}$ s. We therefore expect that as a result of one million revolutions, the uncertainty should be $(\pm 3 \times 10^{-17})(1 \times 10^6) = \pm 3 \times 10^{-11}$ s.

19. If M_E is the mass of Earth, m is the average mass of an atom in Earth, and N is the number of atoms, then $M_E = Nm$ or $N = M_E/m$. We convert mass m to kilograms using Appendix D ($1\,\text{u} = 1.661 \times 10^{-27}$ kg). Thus,

$$N = \frac{M_E}{m} = \frac{5.98 \times 10^{24}\,\text{kg}}{(40\,\text{u})(1.661 \times 10^{-27}\,\text{kg/u})} = 9.0 \times 10^{49} \ .$$

20. To organize the calculation, we introduce the notion of density (which the students have probably seen in other courses):

$$\rho = \frac{m}{V} \ .$$

(a) We take the volume of the leaf to be its area A multiplied by its thickness z. With density $\rho = 19.32$ g/cm^3 and mass $m = 27.63$ g, the volume of the leaf is found to be

$$V = \frac{m}{\rho} = 1.430 \text{ cm}^3 \ .$$

We convert the volume to SI units:

$$(1.430 \text{ cm}^3)\left(\frac{1 \text{ m}}{100 \text{ cm}}\right)^3 = 1.430 \times 10^{-6} \text{ m}^3 \ .$$

And since $V = A\,z$ where $z = 1 \times 10^{-6}$ m (metric prefixes can be found in Table 1-2), we obtain

$$A = \frac{1.430 \times 10^{-6}\,\text{m}^3}{1 \times 10^{-6}\,\text{m}} = 1.430 \text{ m}^2 \ .$$

(b) The volume of a cylinder of length ℓ is $V = A\ell$ where the cross-section area is that of a circle: $A = \pi r^2$. Therefore, with $r = 2.500 \times 10^{-6}$ m and $V = 1.430 \times 10^{-6}$ m^3, we obtain

$$\ell = \frac{V}{\pi r^2} = 7.284 \times 10^4 \text{ m} \ .$$

21. We introduce the notion of density (which the students have probably seen in other courses):

$$\rho = \frac{m}{V}$$

and convert to SI units: $1\,\text{g} = 1 \times 10^{-3}$ kg.

(a) For volume conversion, we find $1\,\text{cm}^3 = (1 \times 10^{-2}\,\text{m})^3 = 1 \times 10^{-6}\,\text{m}^3$. Thus, the density in kg/m^3 is

$$1\,\text{g/cm}^3 = \left(\frac{1\,\text{g}}{\text{cm}^3}\right)\left(\frac{10^{-3}\,\text{kg}}{\text{g}}\right)\left(\frac{\text{cm}^3}{10^{-6}\,\text{m}^3}\right) = 1 \times 10^3 \text{ kg/m}^3 \ .$$

Thus, the mass of a cubic meter of water is 1000 kg.

(b) We divide the mass of the water by the time taken to drain it. The mass is found from $M = \rho V$ (the product of the volume of water and its density):

$$M = (5700\,\text{m}^3)(1 \times 10^3\,\text{kg/m}^3) = 5.70 \times 10^6 \text{ kg} \ .$$

The time is $t = (10\,\text{h})(3600\,\text{s/h}) = 3.6 \times 10^4$ s, so the *mass flow rate* R is

$$R = \frac{M}{t} = \frac{5.70 \times 10^6\,\text{kg}}{3.6 \times 10^4\,\text{s}} = 158 \text{ kg/s} \ .$$

22. The volume of the water that fell is

$$
\begin{aligned}
V &= (26\,\text{km}^2)(2.0\,\text{in.}) \\
&= (26\,\text{km}^2)\left(\frac{1000\,\text{m}}{1\,\text{km}}\right)^2 (2.0\,\text{in.})\left(\frac{0.0254\,\text{m}}{1\,\text{in.}}\right) \\
&= (26\times 10^6\,\text{m}^2)(0.0508\,\text{m}) \\
&= 1.3\times 10^6\ \text{m}^3\ .
\end{aligned}
$$

We write the mass-per-unit-volume (density) of the water as:

$$\rho = \frac{m}{V} = 1\times 10^3\ \text{kg/m}^3\ .$$

The mass of the water that fell is therefore given by $m = \rho V$:

$$
\begin{aligned}
m &= \left(1\times 10^3\,\text{kg/m}^3\right)(1.3\times 10^6\,\text{m}^3) \\
&= 1.3\times 10^9\ \text{kg}\ .
\end{aligned}
$$

23. We introduce the notion of density (which the students have probably seen in other courses):

$$\rho = \frac{m}{V}$$

and convert to SI units: $1000\,\text{g} = 1\,\text{kg}$, and $100\,\text{cm} = 1\ \text{m}$.

(a) The density ρ of a sample of iron is therefore

$$\rho = \left(7.87\,\text{g/cm}^3\right)\left(\frac{1\,\text{kg}}{1000\,\text{g}}\right)\left(\frac{100\,\text{cm}}{1\,\text{m}}\right)^3$$

which yields $\rho = 7870\ \text{kg/m}^3$. If we ignore the empty spaces between the close-packed spheres, then the density of an individual iron atom will be the same as the density of any iron sample. That is, if M is the mass and V is the volume of an atom, then

$$V = \frac{M}{\rho} = \frac{9.27\times 10^{-26}\,\text{kg}}{7.87\times 10^3\,\text{kg/m}^3} = 1.18\times 10^{-29}\ \text{m}^3\ .$$

(b) We set $V = 4\pi R^3/3$, where R is the radius of an atom (Appendix E contains several geometry formulas). Solving for R, we find

$$R = \left(\frac{3V}{4\pi}\right)^{1/3} = \left(\frac{3(1.18\times 10^{-29}\,\text{m}^3)}{4\pi}\right)^{1/3} = 1.41\times 10^{-10}\,\text{m}\,.$$

The center-to-center distance between atoms is twice the radius, or 2.82×10^{-10} m.

24. The metric prefixes (micro (μ), pico, nano, ...) are given for ready reference on the inside front cover of the textbook (see also Table 1-2). The surface area A of each grain of sand of radius $r = 50\ \mu\text{m} = 50\times 10^{-6}$ m is given by $A = 4\pi(50\times 10^{-6})^2 = 3.14\times 10^{-8}\ \text{m}^2$ (Appendix E contains a variety of geometry formulas). We introduce the notion of density (which the students have probably seen in other courses):

$$\rho = \frac{m}{V}$$

so that the mass can be found from $m = \rho V$, where $\rho = 2600\ \text{kg/m}^3$. Thus, using $V = 4\pi r^3/3$, the mass of each grain is

$$m = \left(\frac{4\pi\,(50\times 10^{-6}\,\text{m})^3}{3}\right)\left(2600\,\frac{\text{kg}}{\text{m}^3}\right) = 1.36\times 10^{-9}\ \text{kg}\ .$$

We observe that (because a cube has six equal faces) the indicated surface area is 6 m^2. The number of spheres (the grains of sand) N which have a total surface area of 6 m^2 is given by

$$N = \frac{6\,\mathrm{m}^2}{3.14 \times 10^{-8}\,\mathrm{m}^2} = 1.91 \times 10^8 \ .$$

Therefore, the total mass M is given by

$$M = Nm = \left(1.91 \times 10^8\right)\left(1.36 \times 10^{-9}\ \mathrm{kg}\right) = 0.260\,\mathrm{kg}\ .$$

25. From the Figure we see that, regarding differences in positions Δx, 212 S is equivalent to 258 W and 180 S is equivalent to 156 Z. Whether or not the origin of the Zelda path coincides with the origins of the other paths is immaterial to consideration of Δx.

(a)

$$\Delta x = (50.0\ \mathrm{S})\left(\frac{258\ \mathrm{W}}{212\ \mathrm{S}}\right) = 60.8\ \mathrm{W}$$

(b)

$$\Delta x = (50.0\ \mathrm{S})\left(\frac{156\ \mathrm{Z}}{180\ \mathrm{S}}\right) = 43.3\ \mathrm{Z}$$

26. The first two conversions are easy enough that a *formal* conversion is not especially called for, but in the interest of *practice makes perfect* we go ahead and proceed formally:

(a)

$$(11\ \mathrm{tuffet})\left(\frac{2\ \mathrm{peck}}{1\ \mathrm{tuffet}}\right) = 22\ \mathrm{peck}$$

(b)

$$(11\ \mathrm{tuffet})\left(\frac{0.50\ \mathrm{bushel}}{1\ \mathrm{tuffet}}\right) = 5.5\ \mathrm{bushel}$$

(c)

$$(5.5\ \mathrm{bushel})\left(\frac{36.3687\ \mathrm{L}}{1\ \mathrm{bushel}}\right) \approx 200\ \mathrm{L}$$

27. We make the assumption that the clouds are directly overhead, so that Figure 1-3 (and the calculations that accompany it) apply. Following the steps in Sample Problem 1-4, we have

$$\frac{\theta}{360^\circ} = \frac{t}{24\ \mathrm{h}}$$

which, for $t = 38$ min $= 38/60$ h yields $\theta = 9.5^\circ$. We obtain the altitude h from the relation

$$d^2 = r^2 \tan^2\theta = 2rh$$

which is discussed in that Sample Problem, where $r = 6.37 \times 10^6$ m is the radius of the earth. Therefore, $h = 8.9 \times 10^4$ m.

28. In the simplest approach, we set up a ratio for the total increase in *horizontal depth* x (where $\Delta x = 0.05$ m is the increase in horizontal depth per step)

$$x = N_{\text{steps}} \Delta x = \left(\frac{4.57}{0.19}\right)(0.05) = 1.2 \text{ m} .$$

However, we can approach this more carefully by noting that if there are $N = 4.57/.19 \approx 24$ rises then under normal circumstances we would expect $N - 1 = 23$ runs (horizontal pieces) in that staircase. This would yield $(23)(0.05) = 1.15$ m, which – to two significant figures – agrees with our first result.

29. Abbreviating wapentake as "wp" and assuming a hide to be 110 acres, we set up the ratio 25 wp/11 barn along with appropriate conversion factors:

$$\frac{(25\,\text{wp})\left(\frac{100\,\text{hide}}{1\,\text{wp}}\right)\left(\frac{110\,\text{acre}}{1\,\text{hide}}\right)\left(\frac{4047\,\text{m}^2}{1\,\text{acre}}\right)}{(11\,\text{barn})\left(\frac{1\times10^{-28}\,\text{m}^2}{1\,\text{barn}}\right)} \approx 1 \times 10^{36} .$$

30. It is straightforward to compute how many seconds in a year (about 3×10^7). Now, if we estimate roughly one breath per second (or every two seconds, or three seconds – it won't affect the result) then to within an order of magnitude, a person takes 10^7 breaths in a year.

31. A day is equivalent to 86400 seconds and a meter is equivalent to a million micrometers, so

$$\frac{(3.7\,\text{m})(10^6\,\mu\text{m/m})}{(14\,\text{day})(86400\,\text{s/day})} = 3.1\ \mu\text{m/s} .$$

32. The mass in kilograms is

$$(28.9\,\text{piculs})\left(\frac{100\,\text{gin}}{1\,\text{picul}}\right)\left(\frac{16\,\text{tahil}}{1\,\text{gin}}\right)\left(\frac{10\,\text{chee}}{1\,\text{tahil}}\right)\left(\frac{10\,\text{hoon}}{1\,\text{chee}}\right)\left(\frac{0.3779\,\text{g}}{1\,\text{hoon}}\right)$$

which yields 1.747×10^6 g or roughly 1750 kg.

33. (a) In atomic mass units, the mass of one molecule is $16 + 1 + 1 = 18$ u. Using Eq. 1-9, we find

$$(18\,\text{u})\left(\frac{1.6605402 \times 10^{-27}\,\text{kg}}{1\,\text{u}}\right) = 3.0 \times 10^{-26}\ \text{kg} .$$

(b) We divide the total mass by the mass of each molecule and obtain the (approximate) number of water molecules:

$$\frac{1.4 \times 10^{21}}{3.0 \times 10^{-26}} \approx 5 \times 10^{46} .$$

34. (a) We find the volume in cubic centimeters

$$(193\,\text{gal})\left(\frac{231\,\text{in}^3}{1\,\text{gal}}\right)\left(\frac{2.54\,\text{cm}}{1\,\text{in}}\right)^3 = 7.31 \times 10^5\ \text{cm}^3$$

and subtract this from 1×10^6 cm^3 to obtain 2.69×10^5 cm^3. The conversion gal $\rightarrow$ in^3 is given in Appendix D (immediately below the table of Volume conversions).

(b) The volume found in part (a) is converted (by dividing by $(100\,\text{cm/m})^3$) to 0.731 m^3, which corresponds to a mass of

$$(1000\,\text{kg/m}^3)\,(0.731\,\text{m}^2) = 731\ \text{kg}$$

using the density given in the problem statement. At a rate of 0.0018 kg/min, this can be filled in

$$\frac{731\,\text{kg}}{0.0018\,\text{kg/min}} = 4.06 \times 10^5\ \text{min}$$

which we convert to 0.77 y, by dividing by the number of minutes in a year (365 days)(24 h/day)(60 min/h).

35. (a) When θ is measured in radians, it is equal to the arclength divided by the radius. For very large radius circles and small values of θ, such as we deal with in this problem, the arcs may be approximated as straight lines – which for our purposes correspond to the diameters d and D of the Moon and Sun, respectively. Thus,

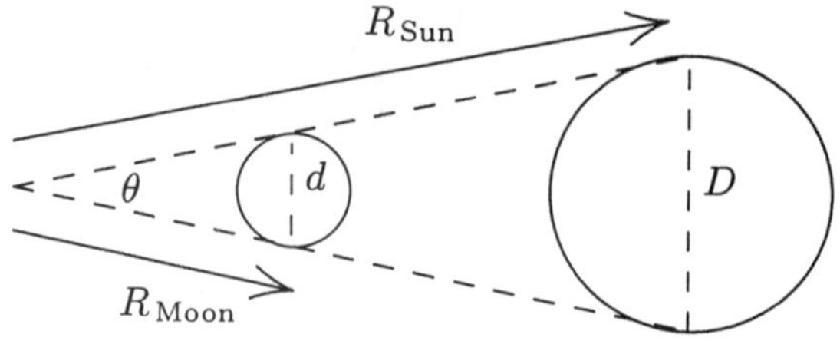

$$\theta = \frac{d}{R_{\text{Moon}}} = \frac{D}{R_{\text{Sun}}} \implies \frac{R_{\text{Sun}}}{R_{\text{Moon}}} = \frac{D}{d}$$

which yields $D/d = 400$.

(b) Various geometric formulas are given in Appendix E. Using r_s and r_m for the radius of the Sun and Moon, respectively (noting that their ratio is the same as D/d), then the Sun's volume divided by that of the Moon is

$$\frac{\frac{4}{3}\pi r_s^3}{\frac{4}{3}\pi r_m^3} = \left(\frac{r_s}{r_m}\right)^3 = 400^3 = 6.4 \times 10^7 \ .$$

(c) The angle should turn out to be roughly 0.009 rad (or about half a degree). Putting this into the equation above, we get

$$d = \theta R_{\text{Moon}} = (0.009)\left(3.8 \times 10^5\right) \approx 3.4 \times 10^3 \text{ km} \ .$$

36. (a) For the minimum (43 cm) case, 9 cubit converts as follows:

$$(9\,\text{cubit})\left(\frac{0.43\,\text{m}}{1\,\text{cubit}}\right) = 3.9 \text{ m} \ .$$

And for the maximum (43 cm) case we obtain

$$(9\,\text{cubit})\left(\frac{0.53\text{ m}}{1\,\text{cubit}}\right) = 4.8 \text{ m} \ .$$

(b) Similarly, with $0.43\,\text{m} \rightarrow 430\,\text{mm}$ and $0.53\,\text{m} \rightarrow 530\,\text{mm}$, we find 3.9×10^3 mm and 4.8×10^3 mm, respectively.

(c) We can convert length and diameter first and then compute the volume, or first compute the volume and then convert. We proceed using the latter approach (where d is diameter and ℓ is length).

$$\begin{aligned} V_{\text{cylinder, min}} &= \frac{\pi}{4}\,\ell\, d^2 = 28 \text{ cubit}^3 \\ &= \left(28\,\text{cubit}^3\right)\left(\frac{0.43\,\text{m}}{1\,\text{cubit}}\right)^3 \\ &= 2.2 \text{ m}^3 \ . \end{aligned}$$

Similarly, with 0.43 m replaced by 0.53 m, we obtain $V_{\text{cylinder, max}} = 4.2\,\text{m}^3$.

37. (a) Squaring the relation $1\,\text{ken} = 1.97\,\text{m}$, and setting up the ratio, we obtain

$$\frac{1\,\text{ken}^2}{1\,\text{m}^2} = \frac{1.97^2\,\text{m}^2}{1\,\text{m}^2} = 3.88 \ .$$

(b) Similarly, we find

$$\frac{1\,\text{ken}^3}{1\,\text{m}^3} = \frac{1.97^3\,\text{m}^3}{1\,\text{m}^3} = 7.65\ .$$

(c) The volume of a cylinder is the circular area of its base multiplied by its height. Thus,

$$\pi r^2 h = \pi(3.00)^2(5.50) = 155.5\ \text{ken}^3\ .$$

(d) If we multiply this by the result of part (b), we determine the volume in cubic meters: $(155.5)(7.65) = 1.19 \times 10^3\ \text{m}^3$.

38. Although we can look up the distance from Cleveland to Los Angeles, we can just as well (for an order of magnitude calculation) assume it's some relatively small fraction of the circumference of Earth – which suggests that (again, for an order of magnitude calculation) we can estimate the distance to be roughly r, where $r \approx 6 \times 10^6$ m is the radius of Earth. If we take each toilet paper sheet to be roughly 10 cm (0.1 m) then the number of sheets needed is roughly $6 \times 10^6/0.1 = 6 \times 10^7 \approx 10^8$.

39. Using the (exact) conversion 2.54 cm = 1 in. we find that 1 ft = (12)(2.54)/100 = 0.3048 m (which also can be found in Appendix D). The volume of a cord of wood is $8 \times 4 \times 4 = 128\ \text{ft}^3$, which we convert (multiplying by 0.3048^3) to $3.6\ \text{m}^3$. Therefore, one cubic meter of wood corresponds to $1/3.6 \approx 0.3$ cord.

40. (a) When θ is measured in radians, it is equal to the arclength s divided by the radius R. For a very large radius circle and small value of θ, such as we deal with in Fig. 1-9, the arc may be approximated as the straight line-segment of length 1 AU. First, we convert $\theta = 1$ arcsecond to radians:

$$(1\,\text{arcsecond})\left(\frac{1\,\text{arcminute}}{60\,\text{arcsecond}}\right)\left(\frac{1^\circ}{60\,\text{arcminute}}\right)\left(\frac{2\pi\,\text{radian}}{360^\circ}\right)$$

which yields $\theta = 4.85 \times 10^{-6}$ rad. Therefore, one parsec is

$$R_o = \frac{s}{\theta} = \frac{1\,\text{AU}}{4.85 \times 10^{-6}} = 2.06 \times 10^5\ \text{AU}\ .$$

Now we use this to convert $R = 1$ AU to parsecs:

$$R = (1\,\text{AU})\left(\frac{1\,\text{pc}}{2.06 \times 10^5\,\text{AU}}\right) = 4.9 \times 10^{-6}\ \text{pc}\ .$$

(b) Also, since it is straightforward to figure the number of seconds in a year (about 3.16×10^7 s), and (for constant speeds) distance = speed×time, we have

$$1\,\text{ly} = (186{,}000\,\text{mi/s})\,(3.16 \times 10^7\ \text{s})\ 5.9 \times 10^{12}\ \text{mi}$$

which we convert to AU by dividing by 92.6×10^6 (given in the problem statement), obtaining 6.3×10^4 AU. Inverting, the result is $1\,\text{AU} = 1/6.3 \times 10^4 = 1.6 \times 10^{-5}$ ly.

(c) As found in the previous part, $1\,\text{ly} = 5.9 \times 10^{12}$ mi.

(d) We now know what one parsec is in AU (denoted above as R_o), and we also know how many miles are in an AU. Thus, one parsec is equivalent to

$$(92.9 \times 10^6\,\text{mi/AU})\,(2.06\ \times 10^5\ \text{AU}) = 1.9 \times 10^{13}\ \text{mi}\ .$$

41. We reduce the stock amount to British teaspoons:

$$\begin{aligned} 1\ \text{breakfastcup} &= 2 \times 8 \times 2 \times 2 = 64\ \text{teaspoons} \\ 1\ \text{teacup} &= 8 \times 2 \times 2 = 32\ \text{teaspoons} \\ 6\ \text{tablespoons} &= 6 \times 2 \times 2 = 24\ \text{teaspoons} \\ 1\ \text{dessertspoon} &= 2\ \text{teaspoons} \end{aligned}$$

which totals to 122 teaspoons – which corresponds (since liquid measure is being used) to 122 U.S. teaspoons. Since each U.S cup is 48 teaspoons, then upon dividing $122/48 \approx 2.54$, we find this amount corresponds to two-and-a-half U.S. cups plus a remainder of precisely 2 teaspoons. For the nettle tops, one-half quart is still one-half quart. For the rice, one British tablespoon is 4 British teaspoons which (since dry-goods measure is being used) corresponds to 2 U.S. teaspoons. Finally, a British saltspoon is $\frac{1}{2}$ British teaspoon which corresponds (since dry-goods measure is again being used) to 1 U.S. teaspoon.

42. (a) *Mega*phone.
    (b) *micro*phone.
    (c) *deka*card ("deck of cards").
    (d) *Giga*low ("gigalo").
    (e) *tera*bull ("terrible").
    (f) *deci*mate.
    (g) *centi*pede.
    (h) *nano*nannette. ("No No Nanette").
    (i) *pico*boo ("peek-a-boo").
    (j) *atto*boy ("at-a-boy").
    (k) Two *hecto*withit ("to heck with it").
    (l) Two *kilo*mockingbird ("to kill a mockingbird").

43. The volume removed in one year is

$$V = \left(75 \times 10^4\,\text{m}^2\right)(26\,\text{m}) \approx 2 \times 10^7\ \text{m}^3$$

which we convert to cubic kilometers:

$$V = \left(2 \times 10^7\,\text{m}^3\right)\left(\frac{1\,\text{km}}{1000\,\text{m}}\right)^3 = 0.020\ \text{km}^3\ .$$

44. (a) Using Appendix D, we have $1\,\text{ft} = 0.3048\,\text{m}$, $1\,\text{gal} = 231\,\text{in.}^3$, and $1\,\text{in.}^3 = 1.639 \times 10^{-2}$ L. From the latter two items, we find that $1\,\text{gal} = 3.79$ L. Thus, the quantity $460\text{ft}^2/\text{gal}$ becomes

$$\left(\frac{460\,\text{ft}^2}{\text{gal}}\right)\left(\frac{1\,\text{m}}{3.28\,\text{ft}}\right)^2\left(\frac{1\,\text{gal}}{3.79\,\text{L}}\right) = 11.3\ \text{m}^2/\text{L}\ .$$

(b) Also, since 1 m^3 is equivalent to 1000 L, our result from part (a) becomes

$$\left(\frac{11.3\,\text{m}^2}{\text{L}}\right)\left(\frac{1000\,\text{L}}{1\,\text{m}^3}\right) = 1.13 \times 10^4\ \text{m}^{-1}\ .$$

(c) The inverse of the original quantity is $(460\,\text{ft}^2/\text{gal})^{-1} = 2.17 \times 10^{-3}\,\text{gal}/\text{ft}^2$, which is the volume of the paint (in gallons) needed to cover a square foot of area. From this, we could also figure the paint thickness (it turns out to be about a tenth of a millimeter, as one sees by taking the reciprocal of the answer in part (b)).

Chapter 2

1. Assuming the horizontal velocity of the ball is constant, the horizontal displacement is

$$\Delta x = v\Delta t$$

where Δx is the horizontal distance traveled, Δt is the time, and v is the (horizontal) velocity. Converting v to meters per second, we have $160\,\text{km/h} = 44.4\,\text{m/s}$. Thus

$$\Delta t = \frac{\Delta x}{v} = \frac{18.4\,\text{m}}{44.4\,\text{m/s}} = 0.414\,\text{s}\,.$$

The velocity-unit conversion implemented above can be figured "from basics" (1000 m = 1 km, 3600 s = 1 h) or found in Appendix D.

2. Converting to SI units, we use Eq. 2-3 with d for distance.

$$\begin{aligned} s_{\text{avg}} &= \frac{d}{t} \\ (110.6\ \text{km/h})\left(\frac{1000\,\text{m/km}}{3600\ \text{s/h}}\right) &= \frac{200.0\,\text{m}}{t} \end{aligned}$$

which yields $t = 6.510$ s. We converted the speed km/h $\rightarrow$ m/s by converting each unit (km $\rightarrow$ m, h $\rightarrow$ s) individually. But we mention that the "one-step" conversion can be found in Appendix D ($1\,\text{km/h} = 0.2778\,\text{m/s}$).

3. We use Eq. 2-2 and Eq. 2-3. During a time t_c when the velocity remains a positive constant, speed is equivalent to velocity, and distance is equivalent to displacement, with $\Delta x = v\,t_c$.

   (a) During the first part of the motion, the displacement is $\Delta x_1 = 40$ km and the time interval is

   $$t_1 = \frac{(40\,\text{km})}{(30\,\text{km/h})} = 1.33\,\text{h}\,.$$

   During the second part the displacement is $\Delta x_2 = 40\,\text{km}$ and the time interval is

   $$t_2 = \frac{(40\,\text{km})}{(60\,\text{km/h})} = 0.67\,\text{h}\,.$$

   Both displacements are in the same direction, so the total displacement is $\Delta x = \Delta x_1 + \Delta x_2 = 40\,\text{km} + 40\,\text{km} = 80\,\text{km}$. The total time for the trip is $t = t_1 + t_2 = 2.00\,\text{h}$. Consequently, the average velocity is

   $$v_{\text{avg}} = \frac{(80\,\text{km})}{(2.0\,\text{h})} = 40\,\text{km/h}\,.$$

   (b) In this example, the numerical result for the average speed is the same as the average velocity 40 km/h.

(c) In the interest of saving space, we briefly describe the graph (with kilometers and hours understood): two contiguous line segments, the first having a slope of 30 and connecting the origin to $(t_1, x_1) = (1.33, 40)$ and the second having a slope of 60 and connecting (t_1, x_1) to $(t, x) = (2.00, 80)$. The average velocity, from the graphical point of view, is the slope of a line drawn from the origin to (t, x).

4. If the plane (with velocity v) maintains its present course, and if the terrain continues its upward slope of 4.3°, then the plane will strike the ground after traveling

$$\Delta x = \frac{h}{\tan\theta} = \frac{35\,\text{m}}{\tan 4.3°} = 465.5 \text{ m} \approx 0.465 \text{ km} .$$

This corresponds to a time of flight found from Eq. 2-2 (with $v = v_{\text{avg}}$ since it is constant)

$$t = \frac{\Delta x}{v} = \frac{0.465\,\text{km}}{1300\,\text{km/h}} = 0.000358 \text{ h} \approx 1.3 \text{ s} .$$

This, then, estimates the time available to the pilot to make his correction.

5. (a) Denoting the travel time and distance from San Antonio to Houston as T and D, respectively, the average speed is

$$s_{\text{avg }1} = \frac{D}{T} = \frac{(55\,\text{km/h})\frac{T}{2} + (90\,\text{km/h})\frac{T}{2}}{T} = 72.5 \text{ km/h}$$

which should be rounded to 73 km/h.

(b) Using the fact that time = distance/speed while the speed is constant, we find

$$s_{\text{avg }2} = \frac{D}{T} = \frac{D}{\frac{D/2}{55\,\text{km/h}} + \frac{D/2}{90\,\text{km/h}}} = 68.3 \text{ km/h}$$

which should be rounded to 68 km/h.

(c) The total distance traveled ($2D$) must not be confused with the net displacement (zero). We obtain for the two-way trip

$$s_{\text{avg}} = \frac{2D}{\frac{D}{72.5\,\text{km/h}} + \frac{D}{68.3\,\text{km/h}}} = 70 \text{ km/h} .$$

(d) Since the net displacement vanishes, the average velocity for the trip in its entirety is zero.

(e) In asking for a *sketch*, the problem is allowing the student to arbitrarily set the distance D (the intent is *not* to make the student go to an Atlas to look it up); the student can just as easily arbitrarily set T instead of D, as will be clear in the following discussion. In the interest of saving space, we briefly describe the graph (with kilometers-per-hour understood for the slopes): two contiguous line segments, the first having a slope of 55 and connecting the origin to $(t_1, x_1) = (T/2, 55T/2)$ and the second having a slope of 90 and connecting (t_1, x_1) to (T, D) where $D = (55 + 90)T/2$. The average velocity, from the graphical point of view, is the slope of a line drawn from the origin to (T, D).

6. (a) Using the fact that time = distance/velocity while the velocity is constant, we find

$$v_{\text{avg}} = \frac{73.2\,\text{m} + 73.2\,\text{m}}{\frac{73.2\,\text{m}}{1.22\,\text{m/s}} + \frac{73.2\,\text{m}}{3.05\,\text{m}}} = 1.74\,\text{m/s} .$$

(b) Using the fact that distance $= vt$ while the velocity v is constant, we find

$$v_{\text{avg}} = \frac{(1.22\,\text{m/s})(60\,\text{s}) + (3.05\,\text{m/s})(60\,\text{s})}{120\,\text{s}} = 2.14\,\text{m/s} .$$

(c) The graphs are shown below (with meters and seconds understood). The first consists of two (solid) line segments, the first having a slope of 1.22 and the second having a slope of 3.05. The slope of the dashed line represents the average velocity (in both graphs). The second graph also consists of two (solid) line segments, having the same slopes as before – the main difference (compared to the first graph) being that the stage involving higher-speed motion lasts much longer.

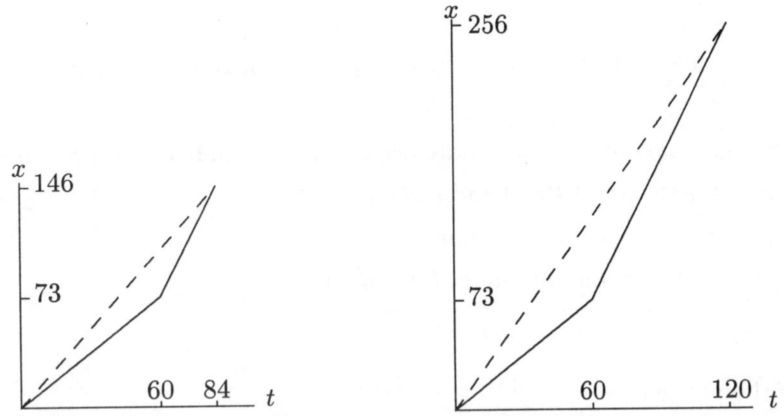

7. Using $x = 3t - 4t^2 + t^3$ with SI units understood is efficient (and is the approach we will use), but if we wished to make the units explicit we would write $x = (3\,\mathrm{m/s})t - (4\,\mathrm{m/s}^2)t^2 + (1\,\mathrm{m/s}^3)t^3$. We will quote our answers to one or two significant figures, and not try to follow the significant figure rules rigorously.

(a) Plugging in $t = 1$ s yields $x = 0$. With $t = 2$ s we get $x = -2$ m. Similarly, $t = 3$ s yields $x = 0$ and $t = 4$ s yields $x = 12$ m. For later reference, we also note that the position at $t = 0$ is $x = 0$.

(b) The position at $t = 0$ is subtracted from the position at $t = 4$ s to find the displacement $\Delta x = 12$ m.

(c) The position at $t = 2$ s is subtracted from the position at $t = 4$ s to give the displacement $\Delta x = 14$ m. Eq. 2-2, then, leads to

$$v_{\text{avg}} = \frac{\Delta x}{\Delta t} = \frac{14}{2} = 7 \text{ m/s} .$$

(d) The horizontal axis is $0 \le t \le 4$ with SI units understood.

Not shown is a straight line drawn from the point at $(t, x) = (2, -2)$ to the highest point shown (at $t = 4$ s) which would represent the answer for part (c).

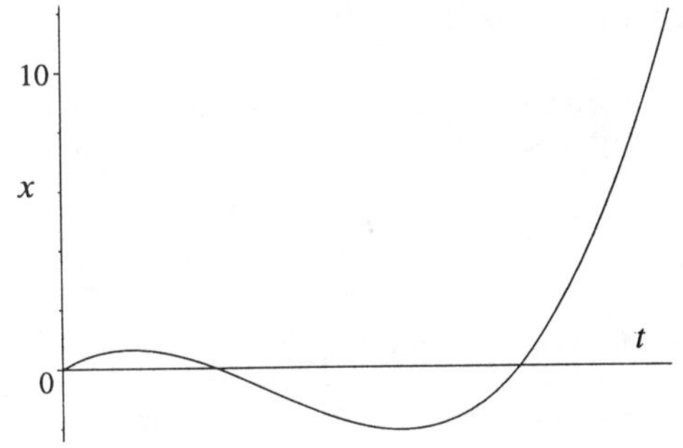

8. Recognizing that the gap between the trains is closing at a constant rate of 60 km/h, the total time which elapses before they crash is $t = (60\,\mathrm{km})/(60\,\mathrm{km/h}) = 1.0$ h. During this time, the bird travels a distance of $x = vt = (60\,\mathrm{km/h})(1.0\,\mathrm{h}) = 60$ km.

9. Converting to seconds, the running times are $t_1 = 147.95$ s and $t_2 = 148.15$ s, respectively. If the runners were equally fast, then

$$s_{\text{avg }1} = s_{\text{avg }2} \implies \frac{L_1}{t_1} = \frac{L_2}{t_2} .$$

From this we obtain

$$L_2 - L_1 = \left(\frac{148.15}{147.95} - 1\right) L_1 \approx 1.35 \text{ m}$$

where we set $L_1 \approx 1000$ m in the last step. Thus, if L_1 and L_2 are no different than about 1.35 m, then runner 1 is indeed faster than runner 2. However, if L_1 is shorter than L_2 than 1.4 m then runner 2 is actually the faster.

10. The velocity (both magnitude and sign) is determined by the slope of the x versus t curve, in accordance with Eq. 2-4.

    (a) The armadillo is to the left of the coordinate origin on the axis between $t = 2.0$ s and $t = 4.0$ s.

    (b) The velocity is negative between $t = 0$ and $t = 3.0$ s.

    (c) The velocity is positive between $t = 3.0$ s and $t = 7.0$ s.

    (d) The velocity is zero at the graph minimum (at $t = 3.0$ s).

11. We use Eq. 2-4.

    (a) The velocity of the particle is

    $$v = \frac{dx}{dt} = \frac{d}{dt}\left(4 - 12t + 3t^2\right) = -12 + 6t \ .$$

    Thus, at $t = 1$ s, the velocity is $v = (-12 + (6)(1)) = -6$ m/s.

    (b) Since $v < 0$, it is moving in the negative x direction at $t = 1$ s.

    (c) At $t = 1$ s, the *speed* is $|v| = 6$ m/s.

    (d) For $0 < t < 2$ s, $|v|$ decreases until it vanishes. For $2 < t < 3$ s, $|v|$ increases from zero to the value it had in part (c). Then, $|v|$ is larger than that value for $t > 3$ s.

    (e) Yes, since v smoothly changes from negative values (consider the $t = 1$ result) to positive (note that as $t \to +\infty$, we have $v \to +\infty$). One can check that $v = 0$ when $t = 2$ s.

    (f) No. In fact, from $v = -12 + 6t$, we know that $v > 0$ for $t > 2$ s.

12. We use Eq. 2-2 for average velocity and Eq. 2-4 for instantaneous velocity, and work with distances in centimeters and times in seconds.

    (a) We plug into the given equation for x for $t = 2.00$ s and $t = 3.00$ s and obtain $x_2 = 21.75$ cm and $x_3 = 50.25$ cm, respectively. The average velocity during the time interval $2.00 \leq t \leq 3.00$ s is

    $$v_{\text{avg}} = \frac{\Delta x}{\Delta t} = \frac{50.25 \text{ cm} - 21.75 \text{ cm}}{3.00 \text{ s} - 2.00 \text{ s}}$$

    which yields $v_{\text{avg}} = 28.5$ cm/s.

    (b) The instantaneous velocity is $v = \frac{dx}{dt} = 4.5t^2$, which yields $v = (4.5)(2.00)^2 = 18.0$ cm/s at time $t = 2.00$ s.

    (c) At $t = 3.00$ s, the instantaneous velocity is $v = (4.5)(3.00)^2 = 40.5$ cm/s.

    (d) At $t = 2.50$ s, the instantaneous velocity is $v = (4.5)(2.50)^2 = 28.1$ cm/s.

    (e) Let t_m stand for the moment when the particle is midway between x_2 and x_3 (that is, when the particle is at $x_m = (x_2 + x_3)/2 = 36$ cm). Therefore,

    $$x_m = 9.75 + 1.5t_m^3 \quad \Longrightarrow \quad t_m = 2.596$$

    in seconds. Thus, the instantaneous speed at this time is $v = 4.5(2.596)^2 = 30.3$ cm/s.

(f) The answer to part (a) is given by the slope of the straight line between $t = 2$ and $t = 3$ in this x-vs-t plot. The answers to parts (b), (c), (d) and (e) correspond to the slopes of tangent lines (not shown but easily imagined) to the curve at the appropriate points.

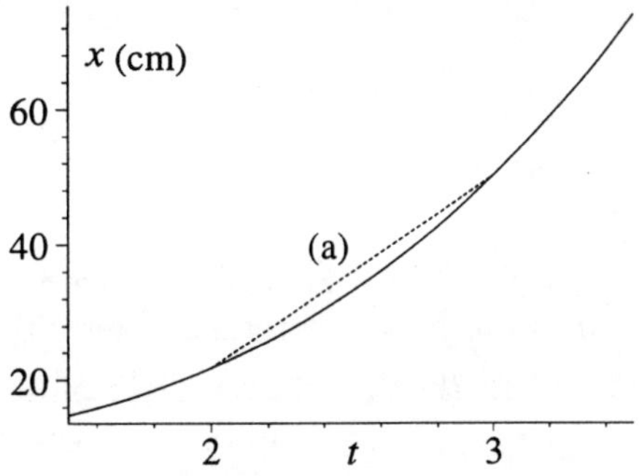

13. Since $v = \frac{dx}{dt}$ (Eq. 2-4), then $\Delta x = \int v\,dt$, which corresponds to the area under the v vs t graph. Dividing the total area A into rectangular (base×height) and triangular ($\frac{1}{2}$base×height) areas, we have

$$\begin{aligned} A &= A_{0<t<2} + A_{2<t<10} + A_{10<t<12} + A_{12<t<16} \\ &= \frac{1}{2}(2)(8) + (8)(8) + \left((2)(4) + \frac{1}{2}(2)(4)\right) + (4)(4) \end{aligned}$$

with SI units understood. In this way, we obtain $\Delta x = 100$ m.

14. From Eq. 2-4 and Eq. 2-9, we note that the sign of the velocity is the sign of the slope in an x-vs-t plot, and the sign of the acceleration corresponds to whether such a curve is concave up or concave down. In the interest of saving space, we indicate sample points for parts (a)-(d) in a single figure; this means that all points are not at $t = 1$ s (which we feel is an acceptable modification of the problem – since the datum $t = 1$ s is not used).

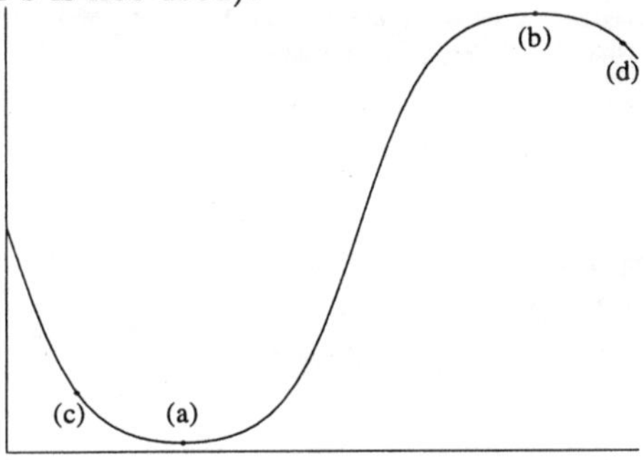

Any change from zero to non-zero values of $\vec{v}$ represents increasing $|\vec{v}|$ (speed). Also, $\vec{v} \parallel \vec{a}$ implies that the particle is going faster. Thus, points (a), (b) and (d) involve increasing speed.

15. We appeal to Eq. 2-4 and Eq. 2-9.

(a) This is v^2 – that is, the velocity squared.

(b) This is the acceleration a.

(c) The SI units for these quantities are $(\mathrm{m/s})^2 = \mathrm{m^2/s^2}$ and $\mathrm{m/s^2}$, respectively.

16. Eq. 2-9 indicates that acceleration is the slope of the v-vs-t graph.

Based on this, we show here a sketch of the acceleration (in $\mathrm{m/s^2}$) as a function of time. The values along the acceleration axis should not be taken too seriously.

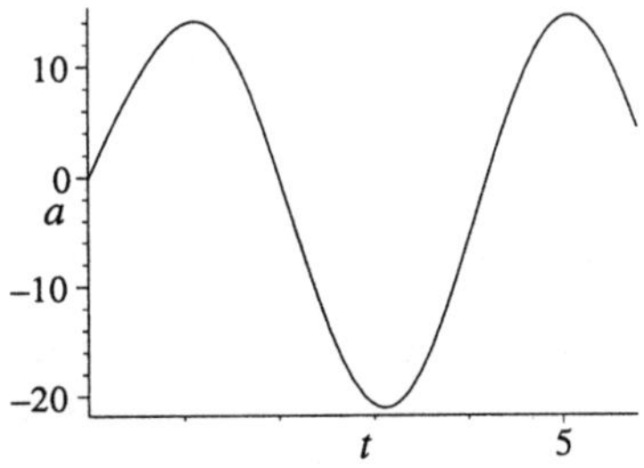

17. We represent its initial direction of motion as the $+x$ direction, so that $v_0 = +18$ m/s and $v = -30$ m/s (when $t = 2.4$ s). Using Eq. 2-7 (or Eq. 2-11, suitably interpreted) we find

$$a_{\text{avg}} = \frac{(-30) - (+18)}{2.4} = -20 \text{ m/s}^2$$

which indicates that the average acceleration has magnitude 20 m/s^2 and is in the opposite direction to the particle's initial velocity.

18. We use Eq. 2-2 (average velocity) and Eq. 2-7 (average acceleration). Regarding our coordinate choices, the initial position of the man is taken as the origin and his direction of motion during $5\,\text{min} \le t \le 10\,\text{min}$ is taken to be the positive x direction. We also use the fact that $\Delta x = v\Delta t'$ when the velocity is constant during a time interval $\Delta t'$.

(a) Here, the entire interval considered is $\Delta t = 8 - 2 = 6$ min which is equivalent to 360 s, whereas the sub-interval in which he is *moving* is only $\Delta t' = 8 - 5 = 3$ min $= 180$ s. His position at $t = 2$ min is $x = 0$ and his position at $t = 8$ min is $x = v\Delta t' = (2.2)(180) = 396$ m. Therefore,

$$v_{\text{avg}} = \frac{396\,\text{m} - 0}{360\,\text{s}} = 1.10 \text{ m/s} .$$

(b) The man is at rest at $t = 2$ min and has velocity $v = +2.2$ m/s at $t = 8$ min. Thus, keeping the answer to 3 significant figures,

$$a_{\text{avg}} = \frac{2.2\,\text{m/s} - 0}{360\,\text{s}} = 0.00611 \text{ m/s}^2 .$$

(c) Now, the entire interval considered is $\Delta t = 9 - 3 = 6$ min (360 s again), whereas the sub-interval in which he is moving is $\Delta t' = 9 - 5 = 4$ min $= 240$ s). His position at $t = 3$ min is $x = 0$ and his position at $t = 9$ min is $x = v\Delta t' = (2.2)(240) = 528$ m. Therefore,

$$v_{\text{avg}} = \frac{528\,\text{m} - 0}{360\,\text{s}} = 1.47 \text{ m/s} .$$

(d) The man is at rest at $t = 3$ min and has velocity $v = +2.2$ m/s at $t = 9$ min. Consequently, $a_{\text{avg}} = 2.2/360 = 0.00611$ m/s^2 just as in part (b).

(e) The horizontal line near the bottom of this x-vs-t graph represents the man standing at $x = 0$ for $0 \le t < 300$ s and the linearly rising line for $300 \le t \le 600$ s represents his constant-velocity motion. The dotted lines represent the answers to part (a) and (c) in the sense that their slopes yield those results.

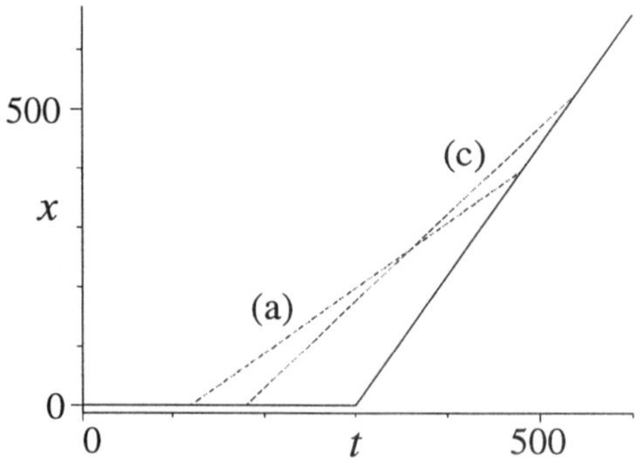

The graph of v-vs-t is not shown here, but would consist of two horizontal "steps" (one at $v = 0$ for $0 \le t < 300$ s and the next at $v = 2.2$ m/s for $300 \le t \le 600$ s). The indications of the average accelerations found in parts (b) and (d) would be dotted lines connected the "steps" at the appropriate t values (the slopes of the dotted lines representing the values of a_{avg}).

19. In this solution, we make use of the notation $x(t)$ for the value of x at a particular t. Thus, $x(t) = 50t + 10t^2$ with SI units (meters and seconds) understood.

   (a) The average velocity during the first 3 s is given by

$$v_{\text{avg}} = \frac{x(3) - x(0)}{\Delta t} = \frac{(50)(3) + (10)(3)^2 - 0}{3} = 80 \text{ m/s} .$$

   (b) The instantaneous velocity at time t is given by $v = dx/dt = 50 + 20t$, in SI units. At $t = 3.0$ s, $v = 50 + (20)(3.0) = 110$ m/s.

   (c) The instantaneous acceleration at time t is given by $a = dv/dt = 20 \text{ m/s}^2$. It is constant, so the acceleration at any time is $20 \, \text{m/s}^2$.

   (d) and (e) The graphs below show the coordinate x and velocity v as functions of time, with SI units understood. The dotted line marked (a) in the first graph runs from $t = 0$, $x = 0$ to $t = 3.0$ s, $x = 240$ m. Its slope is the average velocity during the first 3 s of motion. The dotted line marked (b) is tangent to the x curve at $t = 3.0$ s. Its slope is the instantaneous velocity at $t = 3.0$ s.

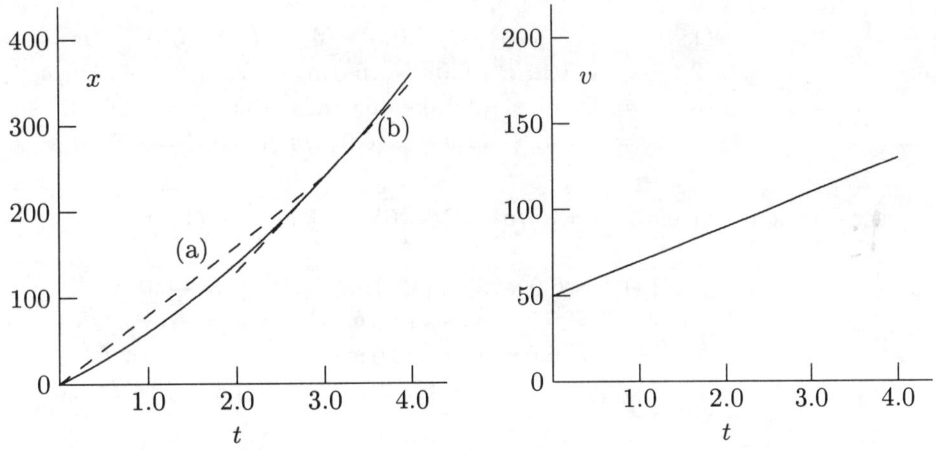

20. Using the general property $\frac{d}{dx} \exp(bx) = b \exp(bx)$, we write

$$v = \frac{dx}{dt} = \left(\frac{d\,(19t)}{dt}\right) \cdot e^{-t} + (19t) \cdot \left(\frac{d\,e^{-t}}{dt}\right) .$$

If a concern develops about the appearance of an argument of the exponential ($-t$) apparently having units, then an explicit factor of $1/T$ where $T = 1$ second can be inserted and carried through the computation (which does not change our answer). The result of this differentiation is

$$v = 16(1 - t)e^{-t}$$

with t and v in SI units (s and m/s, respectively). We see that this function is zero when $t = 1$ s. Now that we know *when* it stops, we find out *where* it stops by plugging our result $t = 1$ into the given function $x = 16te^{-t}$ with x in meters. Therefore, we find $x = 5.9$ m.

21. In this solution, we make use of the notation $x(t)$ for the value of x at a particular t. The notations $v(t)$ and $a(t)$ have similar meanings.

   (a) Since the unit of ct^2 is that of length, the unit of c must be that of length/time^2, or m/s^2 in the SI system. Since bt^3 has a unit of length, b must have a unit of length/time^3, or m/s^3.

(b) When the particle reaches its maximum (or its minimum) coordinate its velocity is zero. Since the velocity is given by $v = dx/dt = 2ct - 3bt^2$, $v = 0$ occurs for $t = 0$ and for

$$t = \frac{2c}{3b} = \frac{2(3.0\,\mathrm{m/s^2})}{3(2.0\,\mathrm{m/s^3})} = 1.0\,\mathrm{s}\,.$$

For $t = 0$, $x = x_0 = 0$ and for $t = 1.0\,\mathrm{s}, x = 1.0\,\mathrm{m} > x_0$. Since we seek the maximum, we reject the first root ($t = 0$) and accept the second ($t = 1\,\mathrm{s}$).

(c) In the first 4 s the particle moves from the origin to $x = 1.0\,\mathrm{m}$, turns around, and goes back to

$$x(4\,\mathrm{s}) = (3.0\,\mathrm{m/s^2})(4.0\,\mathrm{s})^2 - (2.0\,\mathrm{m/s^3})(4.0\,\mathrm{s})^3 = -80\,\mathrm{m}\,.$$

The total path length it travels is $1.0\,\mathrm{m} + 1.0\,\mathrm{m} + 80\,\mathrm{m} = 82\,\mathrm{m}$.

(d) Its displacement is given by $\Delta x = x_2 - x_1$, where $x_1 = 0$ and $x_2 = -80\,\mathrm{m}$. Thus, $\Delta x = -80\,\mathrm{m}$.

(e) The velocity is given by $v = 2ct - 3bt^2 = (6.0\,\mathrm{m/s^2})t - (6.0\,\mathrm{m/s^3})t^2$. Thus

$$\begin{aligned}
v(1\,\mathrm{s}) &= (6.0\,\mathrm{m/s^2})(1.0\,\mathrm{s}) - (6.0\,\mathrm{m/s^3})(1.0\,\mathrm{s})^2 = 0 \\
v(2\,\mathrm{s}) &= (6.0\,\mathrm{m/s^2})(2.0\,\mathrm{s}) - (6.0\,\mathrm{m/s^3})(2.0\,\mathrm{s})^2 = -12\,\mathrm{m/s} \\
v(3\,\mathrm{s}) &= (6.0\,\mathrm{m/s^2})(3.0\,\mathrm{s}) - (6.0\,\mathrm{m/s^3})(3.0\,\mathrm{s})^2 = -36.0\,\mathrm{m/s} \\
v(4\,\mathrm{s}) &= (6.0\,\mathrm{m/s^2})(4.0\,\mathrm{s}) - (6.0\,\mathrm{m/s^3})(4.0\,\mathrm{s})^2 = -72\,\mathrm{m/s}\,.
\end{aligned}$$

(f) The acceleration is given by $a = dv/dt = 2c - 6b = 6.0\,\mathrm{m/s^2} - (12.0\,\mathrm{m/s^3})t$. Thus

$$\begin{aligned}
a(1\,\mathrm{s}) &= 6.0\,\mathrm{m/s^2} - (12.0\,\mathrm{m/s^3})(1.0\,\mathrm{s}) = -6.0\,\mathrm{m/s^2} \\
a(2\,\mathrm{s}) &= 6.0\,\mathrm{m/s^2} - (12.0\,\mathrm{m/s^3})(2.0\,\mathrm{s}) = -18\,\mathrm{m/s^2} \\
a(3\,\mathrm{s}) &= 6.0\,\mathrm{m/s^2} - (12.0\,\mathrm{m/s^3})(3.0\,\mathrm{s}) = -30\,\mathrm{m/s^2} \\
a(4\,\mathrm{s}) &= 6.0\,\mathrm{m/s^2} - (12.0\,\mathrm{m/s^3})(4.0\,\mathrm{s}) = -42\,\mathrm{m/s^2}\,.
\end{aligned}$$

22. For the automobile $\Delta v = 55 - 25 = 30\,\mathrm{km/h}$, which we convert to SI units:

$$a = \frac{\Delta v}{\Delta t} = \frac{(30\,\mathrm{km/h})\left(\frac{1000\,\mathrm{m/km}}{3600\,\mathrm{s/h}}\right)}{(0.50\,\mathrm{min})(60\,\mathrm{s/min})} = 0.28\,\mathrm{m/s^2}\,.$$

The change of velocity for the bicycle, for the same time, is identical to that of the car, so its acceleration is also $0.28\,\mathrm{m/s^2}$.

23. The constant-acceleration condition permits the use of Table 2-1.

(a) Setting $v = 0$ and $x_0 = 0$ in $v^2 = v_0^2 + 2a(x - x_0)$, we find

$$x = -\frac{1}{2}\frac{v_0^2}{a} = -\frac{1}{2}\left(\frac{5.00 \times 10^6}{-1.25 \times 10^{14}}\right) = 0.100\ \mathrm{m}\,.$$

Since the muon is slowing, the initial velocity and the acceleration must have opposite signs.

(b) Below are the time-plots of the position x and velocity v of the muon from the moment it enters the field to the time it stops. The computation in part (a) made no reference to t, so that other equations from Table 2-1 (such as $v = v_0 + at$ and $x = v_0t + \frac{1}{2}at^2$) are used in making these plots.

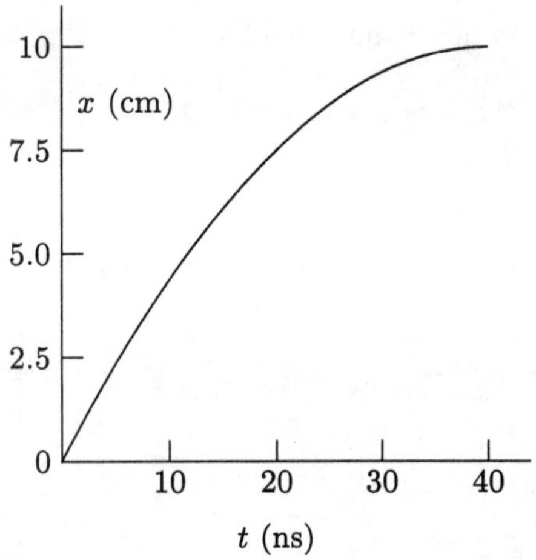

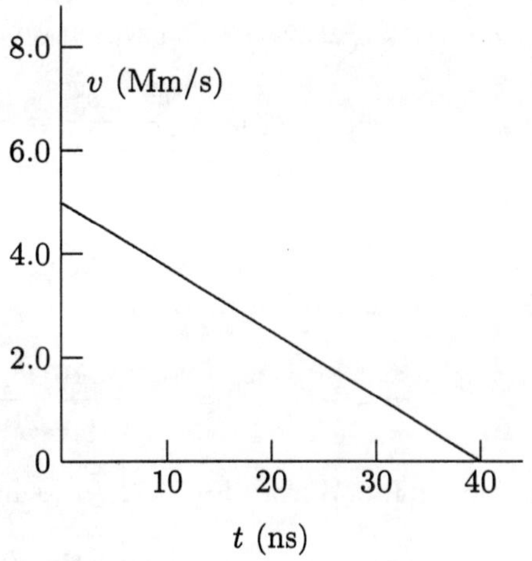

24. The time required is found from Eq. 2-11 (or, suitably interpreted, Eq. 2-7). First, we convert the velocity change to SI units:

$$\Delta v = (100\,\text{km/h})\left(\frac{1000\,\text{m/km}}{3600\,\text{s/h}}\right) = 27.8 \text{ m/s} .$$

Thus, $\Delta t = \Delta v/a = 27.8/50 = 0.556$ s.

25. We use $v = v_0 + at$, with $t = 0$ as the instant when the velocity equals +9.6 m/s.

(a) Since we wish to calculate the velocity for a time *before* $t = 0$, we set $t = -2.5$ s. Thus, Eq. 2-11 gives

$$v = (9.6\,\text{m/s}) + \left(3.2\,\text{m/s}^2\right)(-2.5\,\text{s}) = 1.6 \text{ m/s} .$$

(b) Now, $t = +2.5$ s and we find

$$v = (9.6\,\text{m/s}) + \left(3.2\,\text{m/s}^2\right)(2.5\,\text{s}) = 18 \text{ m/s} .$$

26. The bullet starts at rest ($v_0 = 0$) and after traveling the length of the barrel ($\Delta x = 1.2$ m) emerges with the given velocity ($v = 640$ m/s), where the direction of motion is the positive direction. Turning to the constant acceleration equations in Table 2-1, we use

$$\Delta x = \frac{1}{2}(v_0 + v)\,t \quad .$$

Thus, we find $t = 0.00375$ s (about 3.8 ms).

27. The constant acceleration stated in the problem permits the use of the equations in Table 2-1.

(a) We solve $v = v_0 + at$ for the time:

$$t = \frac{v - v_0}{a} = \frac{\frac{1}{10}\left(3.0 \times 10^8\,\text{m/s}\right)}{9.8\,\text{m/s}^2} = 3.1 \times 10^6 \text{ s}$$

which is equivalent to 1.2 months.

(b) We evaluate $x = x_0 + v_0 t + \frac{1}{2}at^2$, with $x_0 = 0$. The result is

$$x = \frac{1}{2}\left(9.8\,\text{m/s}^2\right)\left(3.1 \times 10^6\,\text{s}\right)^2 = 4.7 \times 10^{13}\,\text{m} .$$

28. From Table 2-1, $v^2 - v_0^2 = 2a\Delta x$ is used to solve for a. Its minimum value is

$$a_{\min} = \frac{v^2 - v_0^2}{2\Delta x_{\max}} = \frac{(360\,\text{km/h})^2}{2(1.80\,\text{km})} = 36000\ \text{km/h}^2$$

which converts to $2.78\ \text{m/s}^2$.

29. Assuming constant acceleration permits the use of the equations in Table 2-1. We solve $v^2 = v_0^2 + 2a(x - x_0)$ with $x_0 = 0$ and $x = 0.010$ m. Thus,

$$a = \frac{v^2 - v_0^2}{2x} = \frac{(5.7\times 10^5)^2 - (1.5\times 10^5)^2}{2(0.01)} = 1.62\times 10^{15}\ \text{m/s}^2\ .$$

30. The acceleration is found from Eq. 2-11 (or, suitably interpreted, Eq. 2-7).

$$a = \frac{\Delta v}{\Delta t} = \frac{(1020\,\text{km/h})\left(\frac{1000\,\text{m/km}}{3600\,\text{s/h}}\right)}{1.4\,\text{s}} = 202.4\ \text{m/s}^2\ .$$

In terms of the gravitational acceleration g, this is expressed as a multiple of $9.8\ \text{m/s}^2$ as follows:

$$a = \frac{202.4}{9.8}\,g = 21g\ .$$

31. We choose the positive direction to be that of the initial velocity of the car (implying that $a < 0$ since it is slowing down). We assume the acceleration is constant and use Table 2-1.

(a) Substituting $v_0 = 137\,\text{km/h} = 38.1\,\text{m/s}$, $v = 90\,\text{km/h} = 25\,\text{m/s}$, and $a = -5.2\,\text{m/s}^2$ into $v = v_0 + at$, we obtain

$$t = \frac{25\,\text{m/s} - 38\,\text{m/s}}{-5.2\,\text{m/s}^2} = 2.5\ \text{s}\ .$$

(b) We take the car to be at $x = 0$ when the brakes are applied (at time $t = 0$). Thus, the coordinate of the car as a function of time is given by

$$x = (38)t + \frac{1}{2}(-5.2)t^2$$

in SI units. This function is plotted from $t = 0$ to $t = 2.5$ s on the graph to the right. We have not shown the v-vs-t graph here; it is a descending straight line from v_0 to v.

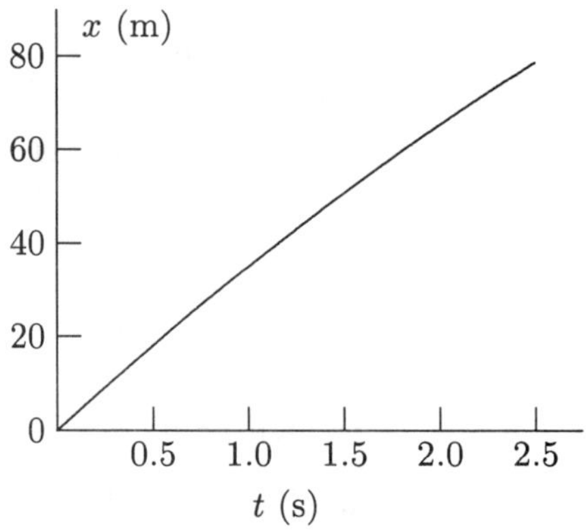

32. From the figure, we see that $x_0 = -2.0$ m. From Table 2-1, we can apply $x - x_0 = v_0 t + \frac{1}{2}at^2$ with $t = 1.0$ s, and then again with $t = 2.0$ s. This yields two equations for the two unknowns, v_0 and a. SI units are understood.

$$\begin{aligned} 0.0 - (-2.0) &= v_0\,(1.0) + \frac{1}{2}a(1.0)^2 \\ 6.0 - (-2.0) &= v_0\,(2.0) + \frac{1}{2}a(2.0)^2\ . \end{aligned}$$

Solving these simultaneous equations yields the results $v_0 = 0.0$ and $a = 4.0\ \text{m/s}^2$. The fact that the answer is positive tells us that the acceleration vector points in the $+x$ direction.

33. The problem statement (see part (a)) indicates that $a =$ constant, which allows us to use Table 2-1.

(a) We take $x_0 = 0$, and solve $x = v_0 t + \frac{1}{2}at^2$ (Eq. 2-15) for the acceleration: $a = 2(x - v_0 t)/t^2$. Substituting $x = 24.0\,\text{m}$, $v_0 = 56.0\,\text{km/h} = 15.55\,\text{m/s}$ and $t = 2.00\,\text{s}$, we find

$$a = \frac{2\,(24.0\,\text{m} - (15.55\,\text{m/s})(2.00\,\text{s}))}{(2.00\,\text{s})^2} = -3.56\,\text{m/s}^2 .$$

The negative sign indicates that the acceleration is opposite to the direction of motion of the car. The car is slowing down.

(b) We evaluate $v = v_0 + at$ as follows:

$$v = 15.55\,\text{m/s} - \left(3.56\,\text{m/s}^2\right)(2.00\,\text{s}) = 8.43\ \text{m/s}$$

which is equivalent to 30.3 km/h.

34. We take the moment of applying brakes to be $t = 0$. The deceleration is constant so that Table 2-1 can be used. Our primed variables (such as $v'_o = 72$ km/h $= 20$ m/s) refer to one train (moving in the $+x$ direction and located at the origin when $t = 0$) and unprimed variables refer to the other (moving in the $-x$ direction and located at $x_0 = +950$ m when $t = 0$). We note that the acceleration vector of the unprimed train points in the *positive* direction, even though the train is slowing down; its initial velocity is $v_o = -144$ km/h $= -40$ m/s. Since the primed train has the lower initial speed, it should stop sooner than the other train would (were it not for the collision). Using Eq 2-16, it should stop (meaning $v' = 0$) at

$$x' = \frac{(v')^2 - (v'_o)^2}{2a'} = \frac{0 - 20^2}{-2} = 200\ \text{m} .$$

The speed of the other train, when it reaches that location, is

$$v = \sqrt{v_o^2 + 2a\Delta x} = \sqrt{(-40)^2 + 2(1.0)(200 - 950)} = \sqrt{100} = 10\ \text{m/s}$$

using Eq 2-16 again. Specifically, its velocity at that moment would be -10 m/s since it is still traveling in the $-x$ direction when it crashes. If the computation of v had failed (meaning that a negative number would have been inside the square root) then we would have looked at the possibility that there was no collision and examined how far apart they finally were. A concern that can be brought up is whether the primed train collides before it comes to rest; this can be studied by computing the time it stops (Eq. 2-11 yields $t = 20$ s) and seeing where the unprimed train is at that moment (Eq. 2-18 yields $x = 350$ m, still a good distance away from contact).

35. The acceleration is constant and we may use the equations in Table 2-1.

(a) Taking the first point as coordinate origin and time to be zero when the car is there, we apply Eq. 2-17 (with SI units understood):

$$x = \frac{1}{2}(v + v_0)\,t = \frac{1}{2}(15 + v_0)\,(6) .$$

With $x = 60.0$ m (which takes the direction of motion as the $+x$ direction) we solve for the initial velocity: $v_0 = 5.00$ m/s.

(b) Substituting $v = 15$ m/s, $v_0 = 5$ m/s and $t = 6$ s into $a = (v - v_0)/t$ (Eq. 2-11), we find $a = 1.67$ m/s^2.

(c) Substituting $v = 0$ in $v^2 = v_0^2 + 2ax$ and solving for x, we obtain

$$x = -\frac{v_0^2}{2a} = -\frac{5^2}{2(1.67)} = -7.50\ \text{m} .$$

(d) The graphs require computing the time when $v = 0$, in which case, we use $v = v_0 + at' = 0$. Thus,

$$t' = \frac{-v_0}{a} = \frac{-5}{1.67} = -3.0 \text{ s}$$

indicates the moment the car was at rest. SI units are assumed.

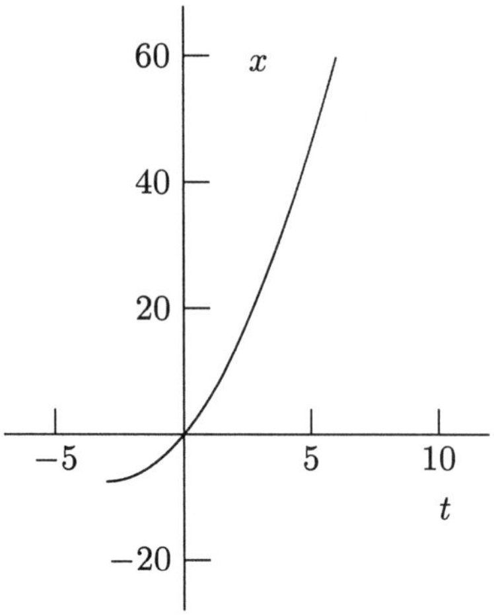

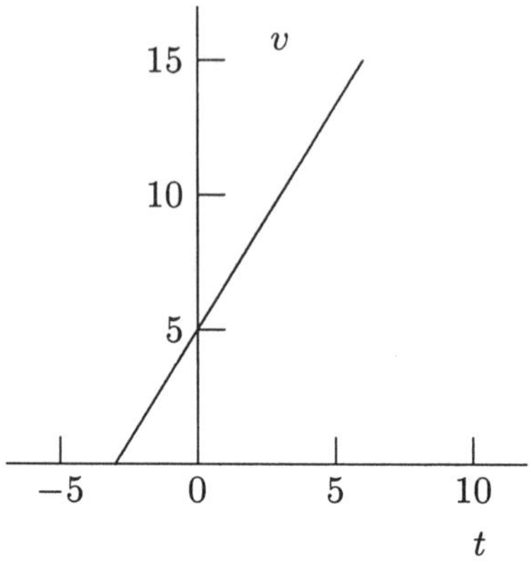

36. We denote the required time as t, assuming the light turns green when the clock reads zero. By this time, the distances traveled by the two vehicles must be the same.

(a) Denoting the acceleration of the automobile as a and the (constant) speed of the truck as v then

$$\Delta x = \left(\frac{1}{2}at^2\right)_{\text{car}} = (vt)_{\text{truck}}$$

which leads to

$$t = \frac{2v}{a} = \frac{2(9.5)}{2.2} = 8.6 \text{ s} .$$

Therefore,

$$\Delta x = vt = (9.5)(8.6) = 82 \text{ m} .$$

(b) The speed of the car at that moment is

$$v_{\text{car}} = at = (2.2)(8.6) = 19 \text{ m/s} .$$

37. We denote t_r as the reaction time and t_b as the braking time. The motion during t_r is of the constant-velocity (call it v_0) type. Then the position of the car is given by

$$x = v_0 t_r + v_0 t_b + \frac{1}{2}at_b^2$$

where v_0 is the initial velocity and a is the acceleration (which we expect to be negative-valued since we are taking the velocity in the positive direction and we know the car is decelerating). *After* the brakes are applied the velocity of the car is given by $v = v_0 + at_b$. Using this equation, with $v = 0$, we eliminate t_b from the first equation and obtain

$$x = v_0 t_r - \frac{v_0^2}{a} + \frac{1}{2}\frac{v_0^2}{a} = v_0 t_r - \frac{1}{2}\frac{v_0^2}{a} .$$

We write this equation for each of the initial velocities:

$$x_1 = v_{01} t_r - \frac{1}{2}\frac{v_{01}^2}{a}$$

and

$$x_2 = v_{02}t_r - \frac{1}{2}\frac{v_{02}^2}{a} \ .$$

Solving these equations simultaneously for t_r and a we get

$$t_r = \frac{v_{02}^2 x_1 - v_{01}^2 x_2}{v_{01}v_{02}(v_{02} - v_{01})}$$

and

$$a = -\frac{1}{2}\frac{v_{02}v_{01}^2 - v_{01}v_{02}^2}{v_{02}x_1 - v_{01}x_2} \ .$$

Substituting $x_1 = 56.7\,\text{m}$, $v_{01} = 80.5\,\text{km/h} = 22.4\,\text{m/s}$, $x_2 = 24.4$ m and $v_{02} = 48.3\,\text{km/h} = 13.4$ m/s, we find

$$t_r = \frac{13.4^2(56.7) - 22.4^2(24.4)}{(22.4)(13.4)(13.4 - 22.4)} = 0.74 \text{ s}$$

and

$$a = -\frac{1}{2}\frac{(13.4)22.4^2 - (22.4)13.4^2}{(13.4)(56.7) - (22.4)(24.4)} = -6.2 \text{ m/s}^2 \ .$$

The *magnitude* of the deceleration is therefore 6.2 m/s^2. Although rounded off values are displayed in the above substitutions, what we have input into our calculators are the "exact" values (such as $v_{02} = \frac{161}{12}$ m/s).

38. In this solution we elect to wait until the last step to convert to SI units. Constant acceleration is indicated, so use of Table 2-1 is permitted. We start with Eq. 2-17 and denote the train's initial velocity as v_t and the locomotive's velocity as v_ℓ (which is also the final velocity of the train, if the rear-end collision is barely avoided). We note that the distance Δx consists of the original gap between them D as well as the forward distance traveled during this time by the locomotive $v_\ell t$. Therefore,

$$\frac{v_t + v_\ell}{2} = \frac{\Delta x}{t} = \frac{D + v_\ell t}{t} = \frac{D}{t} + v_\ell \ .$$

We now use Eq. 2-11 to eliminate time from the equation. Thus,

$$\frac{v_t + v_\ell}{2} = \frac{D}{(v_\ell - v_t)/a} + v_\ell$$

leads to

$$a = \left(\frac{v_t + v_\ell}{2} - v_\ell\right)\left(\frac{v_\ell - v_t}{D}\right) = -\frac{1}{2D}(v_\ell - v_t)^2 \ .$$

Hence,

$$a = -\frac{1}{2(0.676\,\text{km})}\left(29\,\frac{\text{km}}{\text{h}} - 161\,\frac{\text{km}}{\text{h}}\right)^2 = -12888 \text{ km/h}^2$$

which we convert as follows:

$$a = \left(-12888\,\text{km/h}^2\right)\left(\frac{1000\,\text{m}}{1\text{ km}}\right)\left(\frac{1\text{ h}}{3600\,\text{s}}\right)^2 = -0.994 \text{ m/s}^2$$

so that its *magnitude* is 0.994 m/s^2. A graph is shown below for the case where a collision is just avoided (x along the vertical axis is in meters and t along the horizontal axis is in seconds). The top (straight) line shows the motion of the locomotive and the bottom curve shows the motion of the passenger train.

The other case (where the collision is not quite avoided) would be similar except that the slope of the bottom curve would be greater than that of the top line at the point where they meet.

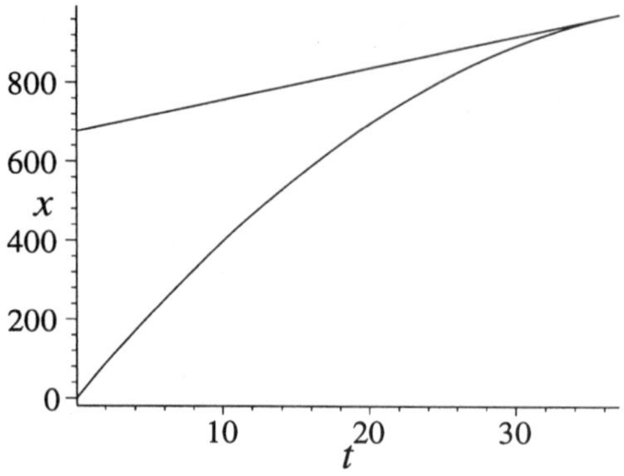

39. We assume the periods of acceleration (duration t_1) and deceleration (duration t_2) are periods of constant a so that Table 2-1 can be used. Taking the direction of motion to be $+x$ then $a_1 = +1.22$ m/s^2 and $a_2 = -1.22$ m/s^2. We use SI units so the velocity at $t = t_1$ is $v = 305/60 = 5.08$ m/s.

    (a) We denote Δx as the distance moved during t_1, and use Eq. 2-16:

    $$v^2 = v_0^2 + 2a_1\Delta x \implies \Delta x = \frac{5.08^2}{2(1.22)}$$

    which yields $\Delta x = 10.59 \approx 10.6$ m.

    (b) Using Eq. 2-11, we have

    $$t_1 = \frac{v - v_0}{a_1} = \frac{5.08}{1.22} = 4.17 \text{ s} .$$

    The deceleration time t_2 turns out to be the same so that $t_1 + t_2 = 8.33$ s. The distances traveled during t_1 and t_2 are the same so that they total to $2(10.59) = 21.18$ m. This implies that for a distance of $190 - 21.18 = 168.82$ m, the elevator is traveling at constant velocity. This time of constant velocity motion is

    $$t_3 = \frac{168.82\,\text{m}}{5.08\,\text{m/s}} = 33.21 \text{ s} .$$

    Therefore, the total time is $8.33 + 33.21 \approx 41.5$ s.

40. Neglect of air resistance justifies setting $a = -g = -9.8$ m/s^2 (where *down* is our $-y$ direction) for the duration of the fall. This is constant acceleration motion, and we may use Table 2-1 (with Δy replacing Δx).

    (a) Using Eq. 2-16 and taking the negative root (since the final velocity is downward), we have

    $$v = -\sqrt{v_0^2 - 2g\Delta y} = -\sqrt{0 - 2(9.8)(-1700)} = -183$$

    in SI units. Its magnitude is therefore 183 m/s.

    (b) No, but it is hard to make a convincing case without more analysis. We estimate the mass of a raindrop to be about a gram or less, so that its mass and speed (from part (a)) would be less than that of a typical bullet, which is good news. But the fact that one is dealing with *many* raindrops leads us to suspect that this scenario poses an unhealthy situation. If we factor in air resistance, the final speed is smaller, of course, and we return to the relatively healthy situation with which we are familiar.

41. We neglect air resistance, which justifies setting $a = -g = -9.8$ m/s^2 (taking *down* as the $-y$ direction) for the duration of the fall. This is constant acceleration motion, which justifies the use of Table 2-1 (with Δy replacing Δx).

(a) Starting the clock at the moment the wrench is dropped ($v_o = 0$), then $v^2 = v_o^2 - 2g\Delta y$ leads to

$$\Delta y = -\frac{(-24)^2}{2(9.8)} = -29.4 \text{ m}$$

so that it fell through a height of 29.4 m.

(b) Solving $v = v_0 - gt$ for time, we find:

$$t = \frac{v_0 - v}{g} = \frac{0 - (-24)}{9.8} = 2.45 \text{ s} .$$

(c) SI units are used in the graphs, and the initial position is taken as the coordinate origin. In the interest of saving space, we do not show the acceleration graph, which is a horizontal line at $-9.8\,\text{m/s}^2$.

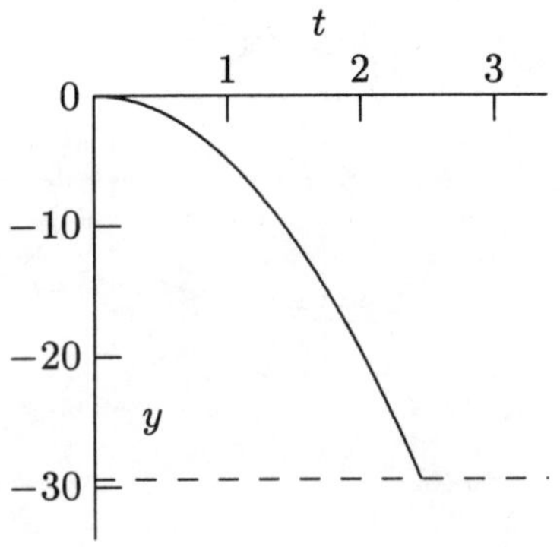

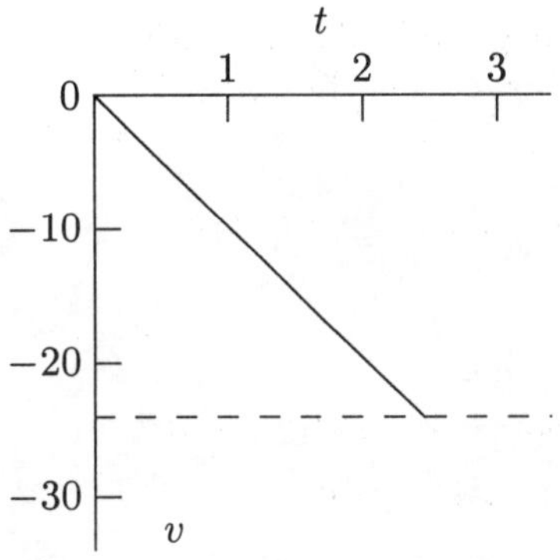

42. We neglect air resistance, which justifies setting $a = -g = -9.8 \text{ m/s}^2$ (taking *down* as the $-y$ direction) for the duration of the fall. This is constant acceleration motion, which justifies the use of Table 2-1 (with Δy replacing Δx).

(a) Noting that $\Delta y = y - y_0 = -30$ m, we apply Eq. 2-15 and the quadratic formula (Appendix E) to compute t:

$$\Delta y = v_0 t - \frac{1}{2}gt^2 \implies t = \frac{v_0 \pm \sqrt{v_0^2 - 2g\Delta y}}{g}$$

which (with $v_0 = -12$ m/s since it is downward) leads, upon choosing the positive root (so that $t > 0$), to the result:

$$t = \frac{-12 + \sqrt{(-12)^2 - 2(9.8)(-30)}}{9.8} = 1.54 \text{ s} .$$

(b) Enough information is now known that any of the equations in Table 2-1 can be used to obtain v; however, the one equation that does *not* use our result from part (a) is Eq. 2-16:

$$v = \sqrt{v_0^2 - 2g\Delta y} = 27.1 \text{ m/s}$$

where the positive root has been chosen in order to give *speed* (which is the magnitude of the velocity vector).

43. We neglect air resistance for the duration of the motion (between "launching" and "landing"), so $a = -g = -9.8 \text{ m/s}^2$ (we take downward to be the $-y$ direction). We use the equations in Table 2-1 (with Δy replacing Δx) because this is $a =$ constant motion.

(a) At the highest point the velocity of the ball vanishes. Taking $y_0 = 0$, we set $v = 0$ in $v^2 = v_0^2 - 2gy$, and solve for the initial velocity: $v_0 = \sqrt{2gy}$. Since $y = 50\,\text{m}$ we find $v_0 = 31$ m/s.

(b) It will be in the air from the time it leaves the ground until the time it returns to the ground ($y = 0$). Applying Eq. 2-15 to the entire motion (the rise and the fall, of total time $t > 0$) we have

$$y = v_0 t - \frac{1}{2} g t^2 \implies t = \frac{2v_0}{g}$$

which (using our result from part (a)) produces $t = 6.4$ s. It is possible to obtain this without using part (a)'s result; one can find the time just for the rise (from ground to highest point) from Eq. 2-16 and then double it.

(c) SI units are understood in the x and v graphs shown. In the interest of saving space, we do not show the graph of a, which is a horizontal line at $-9.8\,\mathrm{m/s^2}$.

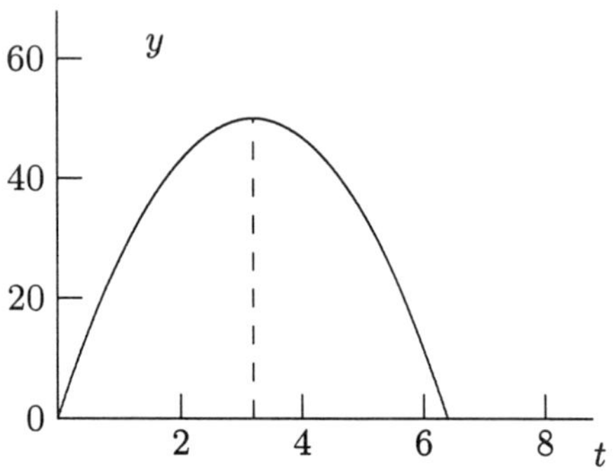

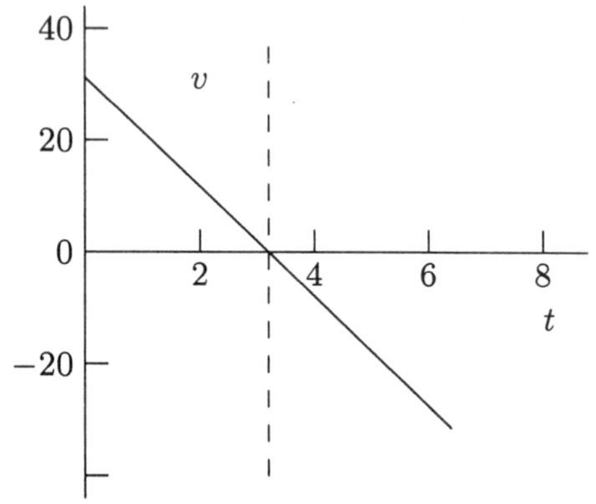

44. There is no air resistance, which makes it quite accurate to set $a = -g = -9.8$ m/s^2 (where downward is the $-y$ direction) for the duration of the fall. We are allowed to use Table 2-1 (with Δy replacing Δx) because this is constant acceleration motion; in fact, when the acceleration changes (during the process of catching the ball) we will again assume constant acceleration conditions; in this case, we have $a_2 = +25g = 245$ m/s^2.

(a) The time of fall is given by Eq. 2-15 with $v_0 = 0$ and $y = 0$. Thus,

$$t = \sqrt{\frac{2y_0}{g}} = \sqrt{\frac{2(145)}{9.8}} = 5.44 \text{ s} .$$

(b) The final velocity for its free-fall (which becomes the initial velocity during the catching process) is found from Eq. 2-16 (other equations can be used but they would use the result from part (a)).

$$v = -\sqrt{v_0^2 - 2g\,(y - y_0)} = -\sqrt{2gy_0} = -53.3 \text{ m/s}$$

where the negative root is chosen since this is a downward velocity.

(c) For the catching process, the answer to part (b) plays the role of an *initial* velocity ($v_0 = -53.3$ m/s) and the final velocity must become zero. Using Eq. 2-16, we find

$$\Delta y_2 = \frac{v^2 - v_0^2}{2a_2} = \frac{-(-53.3)^2}{2(245)} = -5.80 \text{ m}$$

where the negative value of Δy_2 signifies that the distance traveled while arresting its motion is downward.

45. Taking the $+y$ direction *downward* and $y_0 = 0$, we have $y = v_0 t + \frac{1}{2} g t^2$ which (with $v_0 = 0$) yields $t = \sqrt{2y/g}$.

(a) For this part of the motion, $y = 50$ m so that

$$t = \sqrt{\frac{2(50)}{9.8}} = 3.2 \text{ s} .$$

(b) For this next part of the motion, we note that the total displacement is $y = 100$ m. Therefore, the total time is

$$t = \sqrt{\frac{2(100)}{9.8}} = 4.5 \text{ s} .$$

The difference between this and the answer to part (a) is the time required to fall through that second 50 m distance: $4.5 - 3.2 = 1.3$ s.

46. We neglect air resistance, which justifies setting $a = -g = -9.8 \text{ m/s}^2$ (taking down as the $-y$ direction) for the duration of the motion. We are allowed to use Table 2-1 (with Δy replacing Δx) because this is constant acceleration motion. The ground level is taken to correspond to $y = 0$.

(a) With $y_0 = h$ and v_0 replaced with $-v_0$, Eq. 2-16 leads to

$$v = \sqrt{(-v_0)^2 - 2g(y - y_0)} = \sqrt{v_0^2 + 2gh} .$$

The positive root is taken because the problem asks for the speed (the *magnitude* of the velocity).

(b) We use the quadratic formula to solve Eq. 2-15 for t, with v_0 replaced with $-v_0$,

$$\Delta y = -v_0 t - \frac{1}{2}gt^2 \implies t = \frac{-v_0 + \sqrt{(-v_0)^2 - 2g\Delta y}}{g}$$

where the positive root is chosen to yield $t > 0$. With $y = 0$ and $y_0 = h$, this becomes

$$t = \frac{\sqrt{v_0^2 + 2gh} - v_0}{g} .$$

(c) If it were thrown upward with that speed from height h then (in the absence of air friction) it would return to height h with that same downward speed and would therefore yield the same final speed (before hitting the ground) as in part (a). An important perspective related to this is treated later in the book (in the context of energy conservation) .

(d) Having to travel up before it starts its descent certainly requires more time than in part (b). The calculation is quite similar, however, except for now having $+v_0$ in the equation where we had put in $-v_0$ in part (b). The details follow:

$$\Delta y = v_0 t - \frac{1}{2}gt^2 \implies t = \frac{v_0 + \sqrt{v_0^2 - 2g\Delta y}}{g}$$

with the positive root again chosen to yield $t > 0$. With $y = 0$ and $y_0 = h$, we obtain

$$t = \frac{\sqrt{v_0^2 + 2gh} + v_0}{g} .$$

47. We neglect air resistance, which justifies setting $a = -g = -9.8 \text{ m/s}^2$ (taking *down* as the $-y$ direction) for the duration of the motion. We are allowed to use Table 2-1 (with Δy replacing Δx) because this is constant acceleration motion. The ground level is taken to correspond to the origin of the y axis.

(a) Using $y = v_0 t - \frac{1}{2}gt^2$, with $y = 0.544$ m and $t = 0.200$ s, we find

$$v_0 = \frac{y + \frac{1}{2}gt^2}{t} = \frac{0.544 + \frac{1}{2}(9.8)(0.200)^2}{0.200} = 3.70 \text{ m/s} .$$

(b) The velocity at $y = 0.544$ m is

$$v = v_0 - gt = 3.70 - (9.8)(0.200) = 1.74 \text{ m/s} .$$

(c) Using $v^2 = v_0^2 - 2gy$ (with different values for y and v than before), we solve for the value of y corresponding to maximum height (where $v = 0$).

$$y = \frac{v_0^2}{2g} = \frac{3.7^2}{2(9.8)} = 0.698 \text{ m} .$$

Thus, the armadillo goes $0.698 - 0.544 = 0.154$ m higher.

48. We neglect air resistance, which justifies setting $a = -g = -9.8$ m/s^2 (taking *down* as the $-y$ direction) for the duration of the motion. We are allowed to use Table 2-1 (with Δy replacing Δx) because this is constant acceleration motion. The ground level is taken to correspond to the origin of the y axis. The total time of fall can be computed from Eq. 2-15 (using the quadratic formula).

$$\Delta y = v_0 t - \frac{1}{2}gt^2 \implies t = \frac{v_0 + \sqrt{v_0^2 - 2g\Delta y}}{g}$$

with the positive root chosen. With $y = 0$, $v_0 = 0$ and $y_0 = h = 60$ m, we obtain

$$t = \frac{\sqrt{2gh}}{g} = \sqrt{\frac{2h}{g}} = 3.5 \text{ s} .$$

Thus, "1.2 s earlier" means we are examining where the rock is at $t = 2.3$ s:

$$y - h = v_0(2.3) - \frac{1}{2}g(2.3)^2 \implies y = 34 \text{ m}$$

where we again use the fact that $h = 60$ m and $v_0 = 0$.

49. The speed of the boat is constant, given by $v_b = d/t$. Here, d is the distance of the boat from the bridge when the key is dropped (12 m) and t is the time the key takes in falling. To calculate t, we put the origin of the coordinate system at the point where the key is dropped and take the y axis to be positive in the *downward* direction. Taking the time to be zero at the instant the key is dropped, we compute the time t when $y = 45$ m. Since the initial velocity of the key is zero, the coordinate of the key is given by $y = \frac{1}{2}gt^2$. Thus

$$t = \sqrt{\frac{2y}{g}} = \sqrt{\frac{2(45\,\text{m})}{9.8\,\text{m/s}^2}} = 3.03 \text{ s} .$$

Therefore, the speed of the boat is

$$v_b = \frac{12\,\text{m}}{3.03\,\text{s}} = 4.0 \text{ m/s} .$$

50. With $+y$ upward, we have $y_0 = 36.6$ m and $y = 12.2$ m. Therefore, using Eq. 2-18 (the last equation in Table 2-1), we find

$$y - y_0 = vt + \frac{1}{2}gt^2 \implies v = -22 \text{ m/s}$$

at $t = 2.00$ s. The term *speed* refers to the magnitude of the velocity vector, so the answer is $|v| = 22.0$ m/s.

51. We first find the velocity of the ball just before it hits the ground. During contact with the ground its average acceleration is given by

$$a_{\text{avg}} = \frac{\Delta v}{\Delta t}$$

where Δv is the change in its velocity during contact with the ground and $\Delta t = 20.0 \times 10^{-3}$ s is the duration of contact. Now, to find the velocity just *before* contact, we put the origin at the point where the ball is dropped (and take $+y$ upward) and take $t = 0$ to be when it is dropped. The ball strikes the ground at $y = -15.0$ m. Its velocity there is found from Eq. 2-16: $v^2 = -2gy$. Therefore,

$$v = -\sqrt{-2gy} = -\sqrt{-2(9.8)(-15.0)} = -17.1 \text{ m/s}$$

where the negative sign is chosen since the ball is traveling downward at the moment of contact. Consequently, the average acceleration during contact with the ground is

$$a_{\text{avg}} = \frac{0 - (-17.1)}{20.0 \times 10^{-3}} = 857 \text{ m/s}^2 .$$

The fact that the result is positive indicates that this acceleration vector points upward. In a later chapter, this will be directly related to the magnitude and direction of the force exerted by the ground on the ball during the collision.

52. The y axis is arranged so that ground level is $y = 0$ and $+y$ is upward.

(a) At the point where its fuel gets exhausted, the rocket has reached a height of

$$y' = \frac{1}{2}at^2 = \frac{(4.00)(6.00)^2}{2} = 72.0 \text{ m} .$$

From Eq. 2-11, the speed of the rocket (which had started at rest) at this instant is

$$v' = at = (4.00)(6.00) = 24.0 \text{ m/s} .$$

The additional height Δy_1 the rocket can attain (beyond y') is given by Eq. 2-16 with vanishing final speed: $0 = v'^2 - 2g\Delta y_1$. This gives

$$\Delta y_1 = \frac{v'^2}{2g} = \frac{(24.0)^2}{2(9.8)} = 29.4 \text{ m} .$$

Recalling our value for y', the total height the rocket attains is seen to be $72.0 + 29.4 = 101$ m.

(b) The time of free-fall flight (from y' until it returns to $y = 0$) after the fuel gets exhausted is found from Eq. 2-15:

$$-y' = v't - \frac{1}{2}gt^2 \implies -72.0 = (24.0)t - \frac{9.80}{2}t^2 .$$

Solving for t (using the quadratic formula) we obtain $t = 7.00$ s. Recalling the upward acceleration time used in part (a), we see the total time of flight is $7.00 + 6.00 = 13.0$ s.

53. The average acceleration during contact with the floor is given by $a_{\text{avg}} = (v_2 - v_1)/\Delta t$, where v_1 is its velocity just before striking the floor, v_2 is its velocity just as it leaves the floor, and Δt is the duration of contact with the floor (12×10^{-3} s). Taking the y axis to be positively upward and placing the origin at the point where the ball is dropped, we first find the velocity just before striking the floor, using $v_1^2 = v_0^2 - 2gy$. With $v_0 = 0$ and $y = -4.00$ m, the result is

$$v_1 = -\sqrt{-2gy} = -\sqrt{-2(9.8)(-4.00)} = -8.85 \text{ m/s}$$

where the negative root is chosen because the ball is traveling downward. To find the velocity just after hitting the floor (as it ascends without air friction to a height of 2.00 m), we use $v^2 = v_2^2 - 2g(y - y_0)$ with $v = 0$, $y = -2.00$ m (it ends up two meters *below* its initial drop height), and $y_0 = -4.00$ m. Therefore,

$$v_2 = \sqrt{2g(y - y_0)} = \sqrt{2(9.8)(-2.00 + 4.00)} = 6.26 \text{ m/s} .$$

Consequently, the average acceleration is

$$a_{avg} = \frac{v_2 - v_1}{\Delta t} = \frac{6.26 + 8.85}{12.0 \times 10^{-3}} = 1.26 \times 10^3 \text{ m/s}^2 .$$

The positive nature of the result indicates that the acceleration vector points upward. In a later chapter, this will be directly related to the magnitude and direction of the force exerted by the ground on the ball during the collision.

54. The height reached by the player is $y = 0.76$ m (where we have taken the origin of the y axis at the floor and $+y$ to be upward).

   (a) The initial velocity v_0 of the player is

$$v_0 = \sqrt{2gy} = \sqrt{2(9.8)(0.76)} = 3.86 \text{ m/s} .$$

   This is a consequence of Eq. 2-16 where velocity v vanishes. As the player reaches $y_1 = 0.76 - 0.15 = 0.61$ m, his speed v_1 satisfies $v_0^2 - v_1^2 = 2gy_1$, which yields

$$v_1 = \sqrt{v_0^2 - 2gy_1} = \sqrt{(3.86)^2 - 2(9.80)(0.61)} = 1.71 \text{ m/s} .$$

   The time t_1 that the player spends *ascending* in the top $\Delta y_1 = 0.15$ m of the jump can now be found from Eq. 2-17:

$$\Delta y_1 = \frac{1}{2}(v_1 + v)\, t_1 \implies t_1 = \frac{2(0.15)}{1.71 + 0} = 0.175 \text{ s}$$

   which means that the total time spend in that top 15 cm (both ascending and descending) is $2(0.17) = 0.35 \text{ s} = 350$ ms.

   (b) The time t_2 when the player reaches a height of 0.15 m is found from Eq. 2-15:

$$0.15 = v_0 t_2 - \frac{1}{2} g t_2^2 = (3.86) t_2 - \frac{9.8}{2} t_2^2 ,$$

   which yields (using the quadratic formula, taking the smaller of the two positive roots) $t_2 = 0.041$ s $= 41$ ms, which implies that the total time spend in that bottom 15 cm (both ascending and descending) is $2(41) = 82$ ms.

55. We neglect air resistance, which justifies setting $a = -g = -9.8 \text{ m/s}^2$ (taking *down* as the $-y$ direction) for the duration of the motion. We are allowed to use Table 2-1 (with Δy replacing Δx) because this is constant acceleration motion. The ground level is taken to correspond to the origin of the y axis. The time drop 1 leaves the nozzle is taken as $t = 0$ and its time of landing on the floor t_1 can be computed from Eq. 2-15, with $v_0 = 0$ and $y_1 = -2.00$ m.

$$y_1 = -\frac{1}{2} g t_1^2 \implies t_1 = \sqrt{\frac{-2y}{g}} = \sqrt{\frac{-2(-2.00)}{9.8}} = 0.639 \text{ s} .$$

   At that moment,the fourth drop begins to fall, and from the regularity of the dripping we conclude that drop 2 leaves the nozzle at $t = 0.639/3 = 0.213$ s and drop 3 leaves the nozzle at $t = 2(0.213) = 0.426$ s. Therefore, the time in free fall (up to the moment drop 1 lands) for drop 2 is $t_2 = t_1 - 0.213 = 0.426$ s and the time in free fall (up to the moment drop 1 lands) for drop 3 is $t_3 = t_1 - 0.426 = 0.213$ s. Their positions at that moment are

$$\begin{aligned} y_2 &= -\frac{1}{2} g t_2^2 = -\frac{1}{2}(9.8)(0.426)^2 = -0.889 \text{ m} \\ y_3 &= -\frac{1}{2} g t_3^2 = -\frac{1}{2}(9.8)(0.213)^2 = -0.222 \text{ m} , \end{aligned}$$

   respectively. Thus, drop 2 is 89 cm below the nozzle and drop 3 is 22 cm below the nozzle when drop 1 strikes the floor.

56. The graph shows $y = 25$ m to be the highest point (where the speed momentarily vanishes). The neglect of "air friction" (or whatever passes for that on the distant planet) is certainly reasonable due to the symmetry of the graph.

(a) To find the acceleration due to gravity g_p on that planet, we use Eq. 2-15 (with $+y$ up)

$$y - y_0 = vt + \frac{1}{2}g_p t^2 \implies 25 - 0 = (0)(2.5) + \frac{1}{2}g_p(2.5)^2$$

so that $g_p = 8.0\ \mathrm{m/s^2}$.

(b) That same (max) point on the graph can be used to find the initial velocity.

$$y - y_0 = \frac{1}{2}(v_0 + v)\,t \implies 25 - 0 = \frac{1}{2}(v_0 + 0)\,(2.5)$$

Therefore, $v_0 = 20$ m/s.

57. Taking $+y$ to be upward and placing the origin at the point from which the objects are dropped, then the location of diamond 1 is given by $y_1 = -\frac{1}{2}gt^2$ and the location of diamond 2 is given by $y_2 = -\frac{1}{2}g(t-1)^2$. We are starting the clock when the first object is dropped. We want the time for which $y_2 - y_1 = 10$ m. Therefore,

$$-\frac{1}{2}g(t-1)^2 + \frac{1}{2}gt^2 = 10 \implies t = (10/g) + 0.5 = 1.5 \text{ s} .$$

58. We neglect air resistance, which justifies setting $a = -g = -9.8\ \mathrm{m/s^2}$ (taking *down* as the $-y$ direction) for the duration of the motion. We are allowed to use Table 2-1 (with Δy replacing Δx) because this is constant acceleration motion. When something is thrown straight up and is caught at the level it was thrown from (with a trajectory similar to that shown in Fig. 2-25), the time of flight t is half of its time of ascent t_a, which is given by Eq. 2-18 with $\Delta y = H$ and $v = 0$ (indicating the maximum point).

$$H = vt_a + \frac{1}{2}gt_a^2 \implies t_a = \sqrt{\frac{2H}{g}}$$

Writing these in terms of the total time in the air $t = 2t_a$ we have

$$H = \frac{1}{8}gt^2 \implies t = 2\sqrt{\frac{2H}{g}} .$$

We consider two throws, one to height H_1 for total time t_1 and another to height H_2 for total time t_2, and we set up a ratio:

$$\frac{H_2}{H_1} = \frac{\frac{1}{8}gt_2^2}{\frac{1}{8}gt_1^2} = \left(\frac{t_2}{t_1}\right)^2$$

from which we conclude that if $t_2 = 2t_1$ (as is required by the problem) then $H_2 = 2^2 H_1 = 4H_1$.

59. We neglect air resistance, which justifies setting $a = -g = -9.8\ \mathrm{m/s^2}$ (taking *down* as the $-y$ direction) for the duration of the motion. We are allowed to use Table 2-1 (with Δy replacing Δx) because this is constant acceleration motion. We placing the coordinate origin on the ground. We note that the initial velocity of the package is the same as the velocity of the balloon, $v_0 = +12$ m/s and that its initial coordinate is $y_0 = +80$ m.

(a) We solve $y = y_0 + v_0 t - \frac{1}{2}gt^2$ for time, with $y = 0$, using the quadratic formula (choosing the positive root to yield a positive value for t).

$$t = \frac{v_0 + \sqrt{v_0^2 + 2gy_0}}{g} = \frac{12 + \sqrt{12^2 + 2(9.8)(80)}}{9.8} = 5.4 \text{ s}$$

(b) If we wish to avoid using the result from part (a), we could use Eq. 2-16, but if that is not a concern, then a variety of formulas from Table 2-1 can be used. For instance, Eq. 2-11 leads to $v = v_0 - gt = 12 - (9.8)(5.4) = -41$ m/s. Its final *speed* is 41 m/s.

60. We neglect air resistance, which justifies setting $a = -g = -9.8$ m/s^2 (taking *down* as the $-y$ direction) for the duration of the motion. We are allowed to use Eq. 2-15 (with Δy replacing Δx) because this is constant acceleration motion. We use primed variables (except t) with the first stone, which has zero initial velocity, and unprimed variables with the second stone (with initial downward velocity $-v_0$, so that v_0 is being used for the initial *speed*). SI units are used throughout.

$$\begin{aligned}\Delta y' &= 0(t) - \frac{1}{2}gt^2 \\ \Delta y &= (-v_0)(t-1) - \frac{1}{2}g(t-1)^2\end{aligned}$$

Since the problem indicates $\Delta y' = \Delta y = -43.9$ m, we solve the first equation for t (finding $t = 2.99$ s) and use this result to solve the second equation for the initial speed of the second stone:

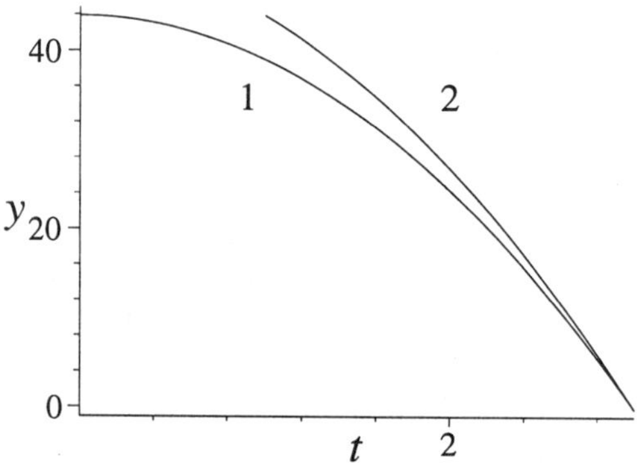

$$-43.9 = (-v_0)(1.99) - \frac{1}{2}(9.8)(1.99)^2$$

which leads to $v_0 = 12.3$ m/s.

61. We neglect air resistance, which justifies setting $a = -g = -9.8$ m/s^2 (taking *down* as the $-y$ direction) for the duration of the motion of the shot ball. We are allowed to use Table 2-1 (with Δy replacing Δx) because the ball has constant acceleration motion. We use primed variables (except t) with the constant-velocity elevator (so $v' = 20$ m/s), and unprimed variables with the ball (with initial velocity $v_0 = v' + 10 = 30$ m/s, relative to the ground). SI units are used throughout.

(a) Taking the time to be zero at the instant the ball is shot, we compute its maximum height y (relative to the ground) with $v^2 = v_0^2 - 2g(y - y_o)$, where the highest point is characterized by $v = 0$. Thus,

$$y = y_o + \frac{v_0^2}{2g} = 76 \text{ m}$$

where $y_o = y_o' + 2 = 30$ m (where $y_o' = 28$ m is given in the problem) and $v_0 = 30$ m/s relative to the ground as noted above.

(b) There are a variety of approaches to this question. One is to continue working in the frame of reference adopted in part (a) (which treats the ground as motionless and "fixes" the coordinate origin to it); in this case, one describes the elevator motion with $y' = y_o' + v't$ and the ball motion with Eq. 2-15, and solves them for the case where they reach the same point at the same time. Another is to work in the frame of reference of the elevator (the boy in the elevator might be oblivious to the fact the elevator is moving since it isn't accelerating), which is what we show here in detail:

$$\Delta y_e = v_{0e}t - \frac{1}{2}gt^2 \implies t = \frac{v_{0e} + \sqrt{v_{0e}^2 - 2g\Delta y_e}}{g}$$

where $v_{0e} = 20$ m/s is the initial velocity of the ball relative to the elevator and $\Delta y_e = -2.0$ m is the ball's displacement relative to the floor of the elevator. The positive root is chosen to yield a positive value for t ; the result is $t = 4.2$ s.

62. We neglect air resistance, which justifies setting $a = -g = -9.8\ \text{m/s}^2$ (taking *down* as the $-y$ direction) for the duration of the stone's motion. We are allowed to use Table 2-1 (with Δx replaced by y) because the ball has constant acceleration motion (and we choose $y_o = 0$).

(a) We apply Eq. 2-16 to both measurements, with SI units understood.

$$v_B^2 = v_0^2 - 2gy_B \quad \Longrightarrow \quad \left(\frac{1}{2}v\right)^2 + 2g(y_A + 3) = v_0^2$$
$$v_A^2 = v_0^2 - 2gy_A \quad \Longrightarrow \quad v^2 + 2gy_A = v_0^2$$

We equate the two expressions that each equal v_0^2 and obtain

$$\frac{1}{4}v^2 + 2gy_A + 2g(3) = v^2 + 2gy_A \quad \Longrightarrow \quad 2g(3) = \frac{3}{4}v^2$$

which yields $v = \sqrt{2g(4)} = 8.85$ m/s.

(b) An object moving upward at A with speed $v = 8.85$ m/s will reach a maximum height $y - y_A = v^2/2g = 4.00$ m above point A (this is again a consequence of Eq. 2-16, now with the "final" velocity set to zero to indicate the highest point). Thus, the top of its motion is 1.00 m above point B.

63. The object, once it is dropped ($v_0 = 0$) is in free-fall ($a = -g = -9.8\ \text{m/s}^2$ if we take *down* as the $-y$ direction), and we use Eq. 2-15 repeatedly.

(a) The (positive) distance D from the lower dot to the mark corresponding to a certain reaction time t is given by $\Delta y = -D = -\frac{1}{2}gt^2$, or $D = gt^2/2$. Thus for $t_1 = 50.0\,\text{ms}$

$$D_1 = \frac{(9.8\,\text{m/s}^2)(50.0\times 10^{-3}\,\text{s})^2}{2} = 0.0123\,\text{m} = 1.23\,\text{cm}\ .$$

(b) For $t_2 = 100\,\text{ms}$

$$D_2 = \frac{(9.8\,\text{m/s}^2)(100\times 10^{-3}\,\text{s})^2}{2} = 0.049\,\text{m} = 4D_1\ ;$$

for $t_3 = 150\,\text{ms}$

$$D_3 = \frac{(9.8\,\text{m/s}^2)(150\times 10^{-3}\,\text{s})^2}{2} = 0.11\,\text{m} = 9D_1\ ;$$

for $t_4 = 200\,\text{ms}$

$$D_4 = \frac{(9.8\,\text{m/s}^2)(200\times 10^{-3}\,\text{s})^2}{2} = 0.196\,\text{m} = 16D_1\ ;$$

and for $t_4 = 250\,\text{ms}$

$$D_5 = \frac{(9.8\,\text{m/s}^2)(250\times 10^{-3}\,\text{s})^2}{2} = 0.306\,\text{m} = 25D_1\ .$$

64. During free fall, we ignore the air resistance and set $a = -g = -9.8\ \text{m/s}^2$ where we are choosing *down* to be the $-y$ direction. The initial velocity is zero so that Eq. 2-15 becomes $\Delta y = -\frac{1}{2}gt^2$ where Δy represents the *negative* of the distance d she has fallen. Thus, we can write the equation as $d = \frac{1}{2}gt^2$ for simplicity.

(a) The time t_1 during which the parachutist is in free fall is (using Eq. 2-15) given by

$$d_1 = 50\,\text{m} = \frac{1}{2}gt_1^2 = \frac{1}{2}\left(9.80\,\text{m/s}^2\right)t_1^2$$

which yields $t_1 = 3.2\,\text{s}$. The *speed* of the parachutist just before he opens the parachute is given by the positive root $v_1^2 = 2gd_1$, or

$$v_1 = \sqrt{2gh_1} = \sqrt{(2)(9.80\,\text{m/s}^2)(50\,\text{m})} = 31\,\text{m/s}\,.$$

If the final speed is v_2, then the time interval t_2 between the opening of the parachute and the arrival of the parachutist at the ground level is

$$t_2 = \frac{v_1 - v_2}{a} = \frac{31\,\text{m/s} - 3.0\,\text{m/s}}{2\,\text{m/s}^2} = 14\,\text{s}\,.$$

This is a result of Eq. 2-11 where *speeds* are used instead of the (negative-valued) velocities (so that final-velocity minus initial-velocity turns out to equal initial-speed minus final-speed); we also note that the acceleration vector for this part of the motion is positive since it points upward (opposite to the direction of motion – which makes it a deceleration). The total time of flight is therefore $t_1 + t_2 = 17\,\text{s}$.

(b) The distance through which the parachutist falls after the parachute is opened is given by

$$d = \frac{v_1^2 - v_2^2}{2a} = \frac{(31\,\text{m/s})^2 - (3.0\,\text{m/s})^2}{(2)(2.0\,\text{m/s}^2)} \approx 240\,\text{m}\,.$$

In the computation, we have used Eq. 2-16 with both sides multiplied by -1 (which changes the negative-valued Δy into the positive d on the left-hand side, and switches the order of v_1 and v_2 on the right-hand side). Thus the fall begins at a height of $h = 50 + d \approx 290\,\text{m}$.

65. The time t the pot spends passing in front of the window of length $L = 2.0$ m is 0.25 s each way. We use v for its velocity as it passes the top of the window (going up). Then, with $a = -g = -9.8\ \text{m/s}^2$ (taking *down* to be the $-y$ direction), Eq. 2-18 yields

$$L = vt - \frac{1}{2}gt^2 \implies v = \frac{L}{t} - \frac{1}{2}gt\,.$$

The distance H the pot goes above the top of the window is therefore (using Eq. 2-16 with the *final velocity* being zero to indicate the highest point)

$$H = \frac{v^2}{2g} = \frac{(L/t - gt/2)^2}{2g} = \frac{(2.00/0.25 - (9.80)(0.25)/2)^2}{(2)\,(9.80)} = 2.34\,\text{m}\,.$$

66. The time being considered is 6 years and roughly 235 days, which is approximately $\Delta t = 2.1 \times 10^7$ s. Using Eq. 2-3, we find the average speed to be

$$\frac{30600 \times 10^3\ \text{m}}{2.1 \times 10^7\ \text{s}} = 0.15\ \text{m/s}\,.$$

67. We assume constant velocity motion and use Eq. 2-2 (with $v_{\text{avg}} = v > 0$). Therefore,

$$\Delta x = v\Delta t = \left(303\,\frac{\text{km}}{\text{h}}\left(\frac{1000\,\text{m/km}}{3600\,\text{s/h}}\right)\right)\left(100 \times 10^{-3}\,\text{s}\right) = 8.4\,\text{m}\,.$$

68. For each rate, we use distance $d = vt$ and convert to SI using 0.0254 cm = 1 inch (from which we derive the factors appearing in the computations below).

(a) The total distance d comes from summing

$$\begin{aligned} d_1 &= \left(120\,\frac{\text{steps}}{\text{min}}\right)\left(\frac{0.762\,\text{m/step}}{60\,\text{s/min}}\right)(5\,\text{s}) = 7.62\,\text{m} \\ d_2 &= \left(120\,\frac{\text{steps}}{\text{min}}\right)\left(\frac{0.381\,\text{m/step}}{60\,\text{s/min}}\right)(5\,\text{s}) = 3.81\,\text{m} \\ d_3 &= \left(180\,\frac{\text{steps}}{\text{min}}\right)\left(\frac{0.914\,\text{m/step}}{60\,\text{s/min}}\right)(5\,\text{s}) = 13.72\,\text{m} \\ d_4 &= \left(180\,\frac{\text{steps}}{\text{min}}\right)\left(\frac{0.457\,\text{m/step}}{60\,\text{s/min}}\right)(5\,\text{s}) = 6.86\,\text{m} \end{aligned}$$

so that $d = d_1 + d_2 + d_3 + d_4 = 32$ m.

(b) Average velocity is computed using Eq. 2-2: $v_{\text{avg}} = 32/20 = 1.6$ m/s, where we have used the fact that the total time is 20 s.

(c) The total time t comes from summing

$$\begin{aligned} t_1 &= \frac{8\,\text{m}}{\left(120\,\frac{\text{steps}}{\text{min}}\right)\left(\frac{0.762\,\text{m/step}}{60\,\text{s/min}}\right)} = 5.25\,\text{s} \\ t_2 &= \frac{8\,\text{m}}{\left(120\,\frac{\text{steps}}{\text{min}}\right)\left(\frac{0.381\,\text{m/step}}{60\,\text{s/min}}\right)} = 10.5\,\text{s} \\ t_3 &= \frac{8\,\text{m}}{\left(180\,\frac{\text{steps}}{\text{min}}\right)\left(\frac{0.914\,\text{m/step}}{60\,\text{s/min}}\right)} = 2.92\,\text{s} \\ t_4 &= \frac{8\,\text{m}}{\left(180\,\frac{\text{steps}}{\text{min}}\right)\left(\frac{0.457\,\text{m/step}}{60\,\text{s/min}}\right)} = 5.83\,\text{s} \end{aligned}$$

so that $t = t_1 + t_2 + t_3 + t_4 = 24.5$ s.

(d) Average velocity is computed using Eq. 2-2: $v_{\text{avg}} = 32/24.5 = 1.3$ m/s, where we have used the fact that the total distance is $4(8) = 32$ m.

69. The statement that the stoneflies have "constant speed along a straight path" means we are dealing with constant velocity motion (Eq. 2-2 with v_{avg} replaced with v_s or v_{ns}, as the case may be).

(a) We set up the ratio and simplify (using d for the common distance).

$$\frac{v_s}{v_{ns}} = \frac{d/t_s}{d/t_{ns}} = \frac{t_{ns}}{t_s} = \frac{25.0}{7.1} = 3.52$$

(b) We examine Δt and simplify until we are left with an expression having numbers and no variables other than v_s. Distances are understood to be in meters.

$$\begin{aligned} t_{ns} - t_s &= \frac{2}{v_{ns}} - \frac{2}{v_s} \\ &= \frac{2}{(v_s/3.52)} - \frac{2}{v_s} \\ &= \frac{2}{v_s}(3.52 - 1) \\ &\approx \frac{5}{v_s} \end{aligned}$$

70. We orient $+x$ along the direction of motion (so a will be negative-valued, since it is a deceleration), and we use Eq. 2-7 with $a_{avg} = -3400g = -3400(9.8) = -3.33 \times 10^4$ m/s^2 and $v = 0$ (since the recorder finally comes to a stop).

$$a_{avg} = \frac{v - v_0}{\Delta t} \implies v_0 = \left(3.33 \times 10^4 \,\text{m/s}^2\right)\left(6.5 \times 10^{-3}\,\text{s}\right)$$

which leads to $v_0 = 217$ m/s.

71. (a) It is the intent of this problem to treat the $v_0 = 0$ condition rigidly. In other words, we are not fitting the distance to just any second-degree polynomial in t; rather, we requiring $d = At^2$ (which meets the condition that d and its derivative is zero when $t = 0$). If we perform a leastsquares fit with this expression, we obtain $A = 3.587$ (SI units understood). We return to this discussion in part (c). Our expectation based on Eq. 2-15, assuming no error in starting the clock at the moment the acceleration begins, is $d = \frac{1}{2}at^2$ (since he started at the coordinate origin, the location of which presumably is something we can be fairly certain about).

(b) The graph (d on the vertical axis, SI units understood) is shown.

The horizontal axis is t^2 (as indicated by the problem statement) so that we have a straight line instead of a parabola.

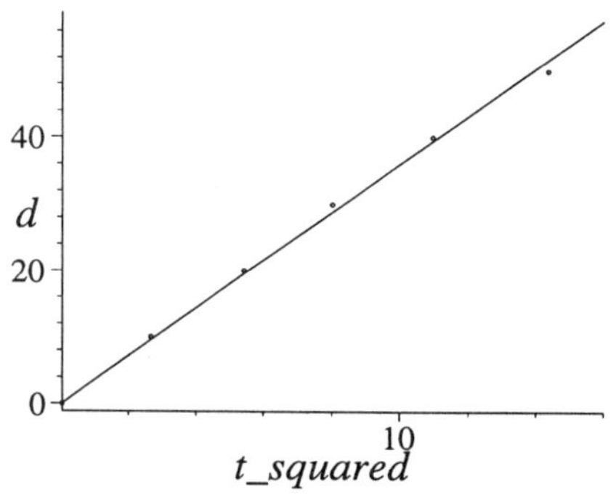

(c) Comparing our two expressions for d, we see the parameter A in our fit should correspond to $\frac{1}{2}a$, so $a = 2(3.587) \approx 7.2\,\text{m/s}^2$. Now, other approaches might be considered (trying to fit the data with $d = Ct^2 + B$ for instance, which leads to $a = 2C = 7.0\,\text{m/s}^2$ and $B \neq 0$), and it might be useful to have the class discuss the assumptions made in each approach.

72. (a) We estimate $x \approx 2$ m at $t = 0.5$ s, and $x \approx 12$ m at $t = 4.5$ s. Hence, using the definition of average velocity Eq. 2-2, we find

$$v_{avg} = \frac{12 - 2}{4.5 - 0.5} = 2.5 \text{ m/s} \ .$$

(b) In the region $4.0 \leq t \leq 5.0$, the graph depicts a straight line, so its slope represents the instantaneous velocity for any point in that interval. Its slope is the average velocity between $t = 4.0$ s and $t = 5.0$ s:

$$v_{avg} = \frac{16.0 - 8.0}{5.0 - 4.0} = 8.0 \text{ m/s} \ .$$

Thus, the instantaneous velocity at $t = 4.5$ s is 8.0 m/s. (Note: similar reasoning leads to a value needed in the next part: the slope of the $0 \leq t \leq 1$ region indicates that the instantaneous velocity at $t = 0.5$ s is 4.0 m/s.)

(c) The average acceleration is defined by Eq. 2-7:

$$a_{avg} = \frac{v_2 - v_1}{t_2 - t_1} = \frac{8.0 - 4.0}{4.5 - 0.5} = 1.0 \text{ m/s}^2 \ .$$

(d) The instantaneous acceleration is the instantaneous rate-of-change of the velocity, and the constant x vs. t slope in the interval $4.0 \leq t \leq 5.0$ indicates that the velocity is constant during that interval. Therefore, $a = 0$ at $t = 4.5$ s.

73. We use the functional notation $x(t)$, $v(t)$ and $a(t)$ and find the latter two quantities by differentiating:

$$v(t) = \frac{dx(t)}{t} = 6.0t^2 \quad \text{and} \quad a(t) = \frac{dv(t)}{dt} = 12t$$

with SI units understood. This expressions are used in the parts that follow.

(a) Using the definition of average velocity, Eq. 2-2, we find

$$v_{\text{avg}} = \frac{x(2) - x(1)}{2.0 - 1.0} = \frac{2(2)^3 - 2(1)^3}{1.0} = 14 \text{ m/s} .$$

(b) The average acceleration is defined by Eq. 2-7:

$$a_{\text{avg}} = \frac{v(2) - v(1)}{2.0 - 1.0} = \frac{6(2)^2 - 6(1)^2}{1.0} = 18 \text{ m/s}^2 .$$

(c) The value of $v(t)$ when $t = 1.0$ s is $v(1) = 6(1)^2 = 6.0$ m/s.

(d) The value of $a(t)$ when $t = 1.0$ s is $a(1) = 12(1) = 12$ m/s^2.

(e) The value of $v(t)$ when $t = 2.0$ s is $v(2) = 6(2)^2 = 24$ m/s.

(f) The value of $a(t)$ when $t = 2.0$ s is $a(2) = 12(2) = 24$ m/s^2.

(g) We don't expect average values of a quantity, say, heights of trees, to equal any particular height for a specific tree, but we are sometimes surprised at the different kinds of averaging that can be performed. Now, the acceleration is a linear function (of time) so its average as defined by Eq. 2-7 is, not surprisingly, equal to the arithmetic average of its $a(1)$ and $a(2)$ values. The velocity is not a linear function so the result of part (a) is not equal to the arithmetic average of parts (c) and (e) (although it is fairly close). This reminds us that the calculus-based definition of the average a function (equivalent to Eq. 2-2 for v_{avg}) is not the same as the simple idea of an arithmetic average of two numbers; in other words,

$$\frac{1}{t' - t} \int_t^{t'} f(\tau)\, d\tau \neq \frac{f(t') + f(t)}{2}$$

except in very special cases (like with linear functions).

(h) The graphs are shown below, $x(t)$ on the left and $v(t)$ on the right. SI units are understood. We do not show the tangent lines (representing instantaneous slope values) at $t = 1$ and $t = 2$, but we do show line segments representing the average quantities computed in parts (a) and (b).

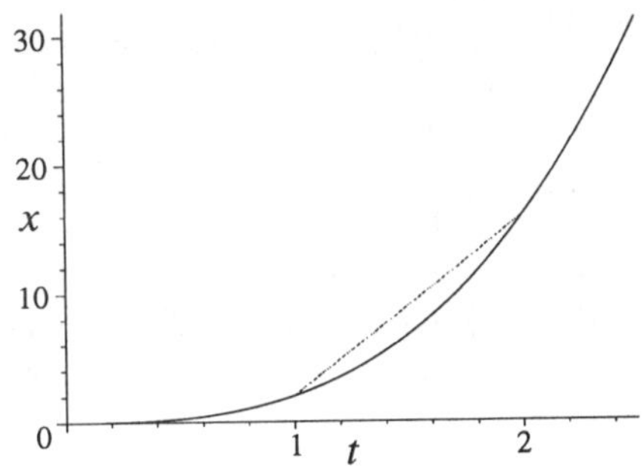

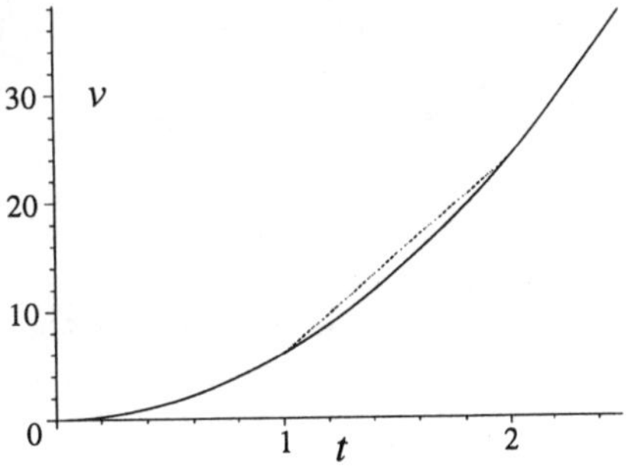

74. We choose *down* as the $+y$ direction and set the coordinate origin at the point where it was dropped (which is when we start the clock). We denote the 1.00 s duration mentioned in the problem as $t - t'$ where t is the value of time when it lands and t' is one second prior to that. The corresponding distance is $y - y' = 0.50h$, where y denotes the location of the ground. In these terms, y is the same as h, so we have $h - y' = 0.50h$ or $0.50h = y'$.

(a) We find t' and t from Eq. 2-15 (with $v_0 = 0$):

$$y' = \frac{1}{2}gt'^2 \implies t' = \sqrt{\frac{2y'}{g}}$$

$$y = \frac{1}{2}gt^2 \implies t = \sqrt{\frac{2y}{g}} .$$

Plugging in $y = h$ and $y' = 0.50h$, and dividing these two equations, we obtain

$$\frac{t'}{t} = \sqrt{\frac{2(0.50h)/g}{2h/g}} = \sqrt{0.50} .$$

Letting $t' = t - 1.00$ (SI units understood) and cross-multiplying, we find

$$t - 1.00 = t\sqrt{0.50} \implies t = \frac{1.00}{1 - \sqrt{0.50}}$$

which yields $t = 3.41$ s.

(b) Plugging this result into $y = \frac{1}{2}gt^2$ we find $h = 57$ m.

(c) In our approach, we did not use the quadratic formula, but we did "choose a root" when we assumed (in the last calculation in part (a)) that $\sqrt{0.50} = +2.236$ instead of -2.236. If we had instead let $\sqrt{0.50} = -2.236$ then our answer for t would have been roughly 0.6 s which would imply that $t' = t - 1$ would equal a negative number (indicating a time *before* it was dropped) which certainly does not fit with the physical situation described in the problem.

75. (a) Let the height of the diving board be h. We choose *down* as the $+y$ direction and set the coordinate origin at the point where it was dropped (which is when we start the clock). Thus, $y = h$ designates the location where the ball strikes the water. Let the depth of the lake be D, and the total time for the ball to descend be T. The speed of the ball as it reaches the surface of the lake is then $v = \sqrt{2gh}$ (from Eq. 2-16), and the time for the ball to fall from the board to the lake surface is $t_1 = \sqrt{2h/g}$ (from Eq. 2-15). Now, the time it spends descending in the lake (at constant velocity v) is

$$t_2 = \frac{D}{v} = \frac{D}{\sqrt{2gh}} .$$

Thus, $T = t_1 + t_2 = \sqrt{\frac{2h}{g}} + \frac{D}{\sqrt{2gh}}$, which gives

$$D = T\sqrt{2gh} - 2h = (4.80)\sqrt{(2)(9.80)(5.20)} - (2)(5.20) = 38.1 \text{ m} .$$

(b) Using Eq. 2-2, the average velocity is

$$v_{avg} = \frac{D + h}{T} = \frac{38.1 + 5.20}{4.80} = 9.02 \text{ m/s}$$

where (recalling our coordinate choices) the positive sign means that the ball is going downward (if, however, upwards had been chosen as the positive direction, then this answer would turn out negative-valued).

(c) We find v_0 from $\Delta y = v_0 t + \frac{1}{2}gt^2$ with $t = T$ and $\Delta y = h + D$. Thus,

$$v_0 = \frac{h + D}{T} - \frac{gT}{2} = \frac{5.20 + 38.1}{4.80} - \frac{(9.8)(4.80)}{2} = -14.5 \text{ m/s}$$

where (recalling our coordinate choices) the negative sign means that the ball is being thrown upward.

76. The time Δt is $2(60) + 41 = 161$ min and the displacement $\Delta x = 370$ cm. Thus, Eq. 2-2 gives

$$v_{\text{avg}} = \frac{\Delta x}{\Delta t} = \frac{370}{161} = 2.3 \text{ cm/min} .$$

77. We use the functional notation $x(t)$, $v(t)$ and $a(t)$ and find the latter two quantities by differentiating:

$$v(t) = \frac{dx(t)}{t} = -15t^2 + 20 \quad \text{and} \quad a(t) = \frac{dv(t)}{dt} = -30t$$

with SI units understood. This expressions are used in the parts that follow.

(a) From $0 = -15t^2 + 20$, we see that the only positive value of t for which the particle is (momentarily) stopped is $t = \sqrt{20/15} = 1.2$ s.

(b) From $0 = -30t$, we find $a(0) = 0$ (that is, it vanishes at $t = 0$).

(c) It is clear that $a(t) = -30t$ is negative for $t > 0$ and positive for $t < 0$.

(d) We show the two of the graphs below (the third graph, $a(t)$, which is a straight line through the origin with slope$= -30$ is omitted in the interest of saving space). SI units are understood.

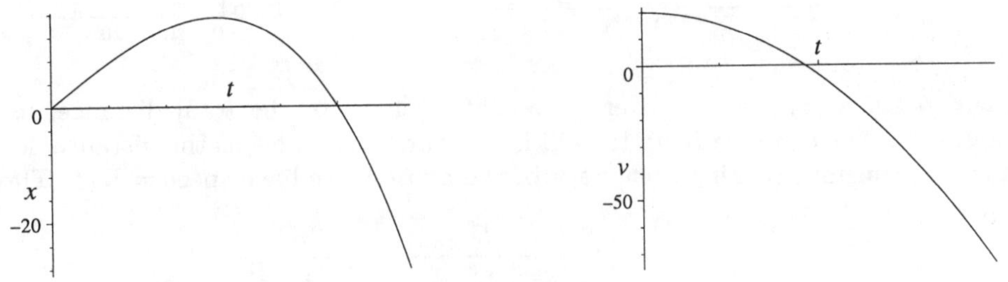

78. (a) It follows from Eq. 2-8 that $v - v_0 = \int a\,dt$, which has the geometric interpretation of being the area under the graph. Thus, with $v_0 = 2.0$ m/s and that area amounting to 3.0 m/s (adding that of a triangle to that of a square, over the interval $0 \le t \le 2$ s), we find $v = 2.0 + 3.0 = 5.0$ m/s (which we will denote as v_2 in the next part). The information given that $x_0 = 4.0$ m is not used in this solution.

(b) During $2 < t \le 4$ s, the graph of a is a straight line with slope 1.0 m/s^3. Extrapolating, we see that the intercept of this line with the a axis is zero. Thus, with SI units understood,

$$v = v_2 + \int_{2.0}^{t} a\,d\tau = 5.0 + \int_{2.0}^{t} (1.0)\tau\,d\tau = 5.0 + \frac{(1.0)t^2 - (1.0)(2.0)^2}{2}$$

which yields $v = 3.0 + 0.50t^2$ in m/s.

79. We assume the train accelerates from rest ($v_0 = 0$ and $x_0 = 0$) at $a_1 = +1.34$ m/s^2 until the midway point and then decelerates at $a_2 = -1.34$ m/s^2 until it comes to a stop ($v_2 = 0$) at the next station. The velocity at the midpoint is v_1 which occurs at $x_1 = 806/2 = 403$ m.

(a) Eq. 2-16 leads to

$$v_1^2 = v_0^2 + 2a_1x_1 \implies v_1 = \sqrt{2(1.34)(403)}$$

which yields $v_1 = 32.9$ m/s.

(b) The time t_1 for the accelerating stage is (using Eq. 2-15)

$$x_1 = v_0t_1 + \frac{1}{2}a_1t_1^2 \implies t_1 = \sqrt{\frac{2(403)}{1.34}}$$

which yields $t_1 = 24.53$ s. Since the time interval for the decelerating stage turns out to be the same, we double this result and obtain $t = 49.1$ s for the travel time between stations.

(c) With a "dead time" of 20 s, we have $T = t + 20 = 69.1$ s for the total time between start-ups. Thus, Eq. 2-2 gives

$$v_{\text{avg}} = \frac{806\,\text{m}}{69.1\,\text{s}} = 11.7 \text{ m/s} .$$

(d) We show the two of the graphs below. The third graph, $a(t)$, is not shown to save space; it consists of three horizontal "steps" – one at 1.34 during $0 < t < 24.53$ and the next at -1.34 during $24.53 < t < 49.1$ and the last at zero during the "dead time" $49.1 < t < 69.1$). SI units are understood.

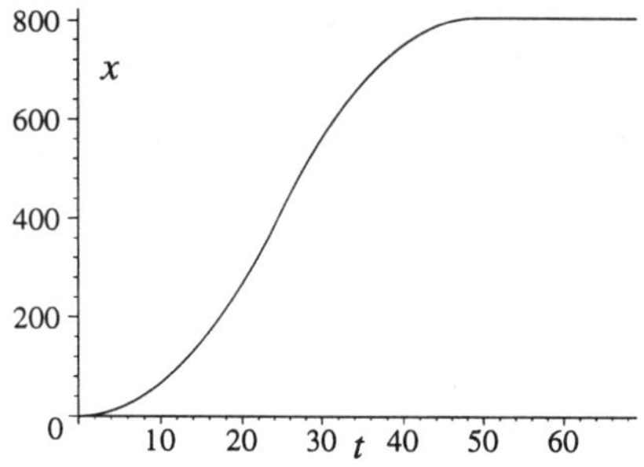

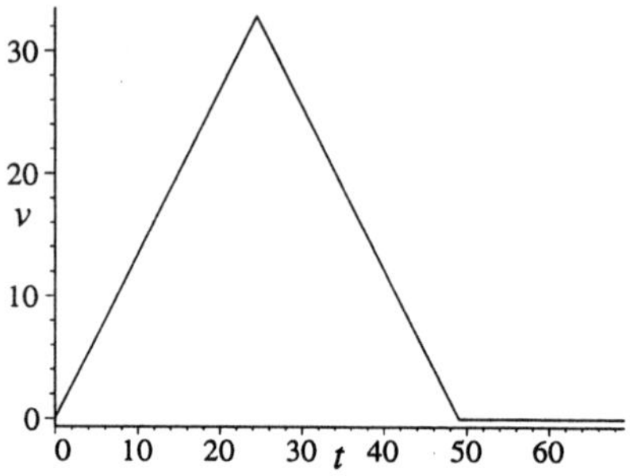

80. Average speed, as opposed to average velocity, relates to the total distance, as opposed to the net displacement. The distance D up the hill is, of course, the same as the distance down the hill, and since the speed is constant (during each stage of the motion) we have speed$= D/t$. Thus, the average speed is

$$\frac{D_{\text{up}} + D_{\text{down}}}{t_{\text{up}} + t_{\text{down}}} = \frac{2D}{\frac{D}{v_{\text{up}}} + \frac{D}{v_{\text{down}}}}$$

which, after canceling D and plugging in $v_{\text{up}} = 40$ km/h and $v_{\text{down}} = 60$ km/h, yields 48 km/h for the average speed.

81. During T_r the velocity v_0 is constant (in the direction we choose as $+x$) and obeys $v_0 = D_r/T_r$ where we note that in SI units the velocity is $v_0 = 200(1000/3600) = 55.6$ m/s. During T_b the acceleration is opposite to the direction of v_0 (hence, for us, $a < 0$) until the car is stopped ($v = 0$).

(a) Using Eq. 2-16 (with $\Delta x_b = 170$ m) we find

$$v^2 = v_0^2 + 2a\Delta x_b \implies a = -\frac{v_0^2}{2\Delta x_b}$$

which yields $|a| = 9.08 \text{ m/s}^2$.

(b) We express this as a multiple of g by setting up a ratio:

$$a = \left(\frac{9.08}{9.8}\right) 9 = 0.926g .$$

(c) We use Eq. 2-17 to obtain the braking time:

$$\Delta x_b = \frac{1}{2}(v_0 + v)\,T_b \implies T_b = \frac{2(170)}{55.6} = 6.12 \text{ s} .$$

(d) We express our result for T_b as a multiple of the reaction time T_r by setting up a ratio:

$$T_b = \left(\frac{6.12}{400 \times 10^{-3}}\right) T_r = 15.3T_r \ .$$

(e) We are only asked what the *increase* in distance D is, due to $\Delta T_r = 0.100$ s, so we simply have

$$\Delta D = v_0 \Delta T_r = (55.6)(0.100) = 5.56 \text{ m} .$$

82. We take $+x$ in the direction of motion. We use subscripts 1 and 2 for the data. Thus, $v_1 = +30$ m/s, $v_2 = +50$ m/s and $x_2 - x_1 = +160$ m.

(a) Using these subscripts, Eq. 2-16 leads to

$$a = \frac{v_2^2 - v_1^2}{2(x_2 - x_1)} = \frac{50^2 - 30^2}{2(160)} = 5.0 \text{ m/s}^2 .$$

(b) We find the time interval corresponding to the displacement $x_2 - x_1$ using Eq. 2-17:

$$t_2 - t_1 = \frac{2(x_2 - x_1)}{v_1 + v_2} = \frac{2(160)}{30 + 50} = 4.0 \text{ s} .$$

(c) Since the train is at rest ($v_0 = 0$) when the clock starts, we find the value of t_1 from Eq. 2-11:

$$v_1 = v_0 + at_1 \implies t_1 = \frac{30}{5.0} = 6.0 \text{ s} .$$

(d) The coordinate origin is taken to be the location at which the train was initially at rest (so $x_0 = 0$). Thus, we are asked to find the value of x_1. Although any of several equations could be used, we choose Eq. 2-17:

$$x_1 = \frac{1}{2}(v_0 + v_1)t_1 = \frac{1}{2}(30)(6.0) = 90 \text{ m} .$$

(e) The graphs are shown below, with SI units assumed.

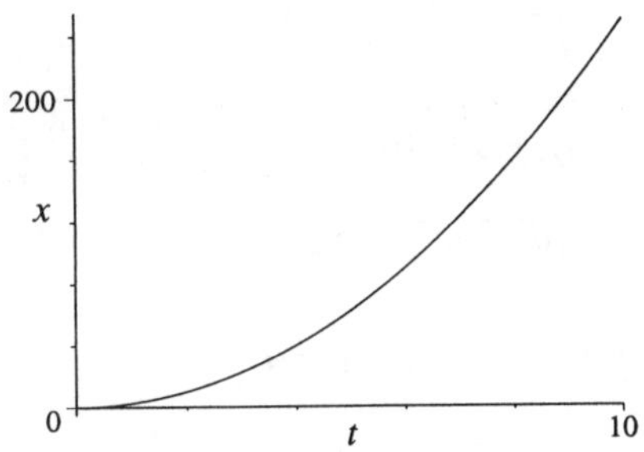

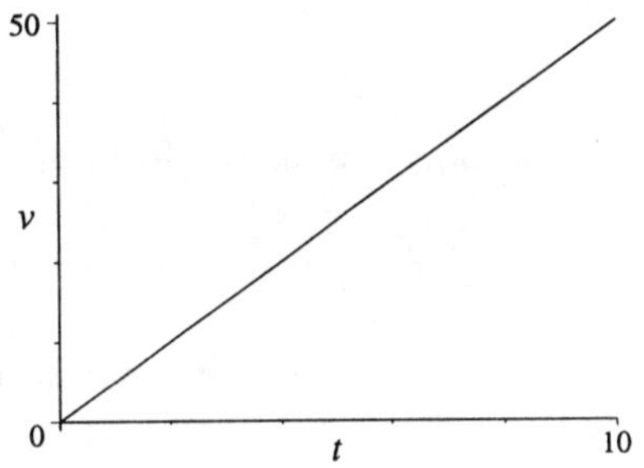

83. Direction of $+x$ is implicit in the problem statement. The initial position (when the clock starts) is $x_0 = 0$ (where $v_0 = 0$), the end of the speeding-up motion occurs at $x_1 = 1100/2 = 550$ m, and the subway comes to a halt ($v_2 = 0$) at $x_2 = 1100$ m.

(a) Using Eq. 2-15, the subway reaches x_1 at

$$t_1 = \sqrt{\frac{2x_1}{a_1}} = \sqrt{\frac{2(550)}{1.2}} = 30.3 \text{ s} .$$

The time interval $t_2 - t_1$ turns out to be the same value (most easily seen using Eq. 2-18 so the total time is $t_2 = 2(30.3) = 60.6$ s.

(b) Its maximum speed occurs at t_1 and equals

$$v_1 = v_0 + a_1 t_1 = 36.3 \text{ m/s} .$$

(c) The graphs are not shown here, in the interest of saving space. They are very similar to those shown in the solution for problem 79, above.

84. We note that the running time for Bill Rodgers is $\Delta t_R = 2(3600) + 10(60) = 7800$ s. We also note that the magnitude of the average velocity (Eq. 2-2) and Eq. 2-3 (for average speed) agree in this exercise (which is not usually the case).

(a) Denoting the Lewis' average velocity as v_L (similarly for Rodgers), we find

$$v_L = \frac{100\,\text{m}}{10\,\text{s}} = 10 \text{ m/s} \quad v_R = \frac{42000\,\text{m}}{7800\,\text{s}} = 5.4 \text{ m/s} .$$

(b) If Lewis continued at this rate, he would covered $D = 42000$ m in

$$\Delta t_L = \frac{D}{v_L} = \frac{42000}{10} = 4200 \text{ s}$$

which is equivalent to 1 h and 10 min.

85. We choose *down* as the $+y$ direction and use the equations of Table 2-1 (replacing x with y) with $a = +g$, $v_0 = 0$ and $y_0 = 0$. We use subscript 2 for the elevator reaching the ground and 1 for the halfway point.

(a) Eq. 2-16, $v_2^2 = v_0^2 + 2a(y_2 - y_0)$, leads to

$$v_2 = \sqrt{2gy_2} = \sqrt{2(9.8)(120)} = 48.5 \text{ m/s} .$$

(b) The time at which it strikes the ground is (using Eq. 2-15)

$$t_2 = \sqrt{\frac{2y_2}{g}} = \sqrt{\frac{2(120)}{9.8}} = 4.95 \text{ s} .$$

(c) Now Eq. 2-16, in the form $v_1^2 = v_0^2 + 2a(y_1 - y_0)$, leads to

$$v_1 = \sqrt{2gy_1} = \sqrt{2(9.8)(60)} = 34.2 \text{ m/s} .$$

(d) The time at which it reaches the halfway point is (using Eq. 2-15)

$$t_1 = \sqrt{\frac{2y_1}{g}} = \sqrt{\frac{2(60)}{9.8}} = 3.50 \text{ s} .$$

86. To find the "launch" velocity of the rock, we apply Eq. 2-11 to the maximum height (where the speed is momentarily zero)

$$v = v_0 - gt \implies 0 = v_0 - (9.8)(2.5)$$

so that $v_0 = 24.5$ m/s (with $+y$ up). Now we use Eq. 2-15 to find the height of the tower (taking $y_0 = 0$ at the ground level)

$$y - y_0 = v_0 t + \frac{1}{2}at^2 \implies y - 0 = (24.5)(1.5) - \frac{1}{2}(9.8)(1.5)^2 .$$

Thus, we obtain $y = 26$ m.

87. We take the direction of motion as $+x$, so $a = -5.18 \text{ m/s}^2$, and we use SI units, so $v_0 = 55(1000/3600) = 15.28$ m/s.

(a) The velocity is constant during the reaction time T, so the distance traveled during it is $d_r = v_0T = (15.28)(0.75) = 11.46$ m. We use Eq. 2-16 (with $v = 0$) to find the distance d_b traveled during braking:

$$v^2 = v_0^2 + 2ad_b \implies d_b = -\frac{15.28^2}{2(-5.18)}$$

which yields $d_b = 22.53$ m. Thus, the total distance is $d_r + d_b = 34.0$ m, which means that the driver *is* able to stop in time. And if the driver were to continue at v_0, the car would enter the intersection in $t = (40\,\mathrm{m})/(15.28\,\mathrm{m/s}) = 2.6$ s which is (barely) enough time to enter the intersection before the light turns, which many people would consider an acceptable situation.

(b) In this case, the total distance to stop (found in part (a) to be 34 m) is greater than the distance to the intersection, so the driver cannot stop without the front end of the car being a couple of meters into the intersection. And the time to reach it at constant speed is $32/15.28 = 2.1$ s, which is too long (the light turns in 1.8 s). The driver is caught between a rock and a hard place.

88. We assume $v_0 = 0$ and integrate the acceleration to find the velocity. In the graphs below (the first is the acceleration, like Fig. 2-35 but with some numbers we adopted, and the second is the velocity) we modeled the curve in the textbook with straight lines and circular arcs for the rounded corners, and literally integrated it. The intent of the textbook was not, however, to go through such an involved procedure, and one should be able to obtain a close approximation to the shape of the velocity graph below (the one on the right) just by applying the idea that constant nonzero acceleration means a linearly changing velocity.

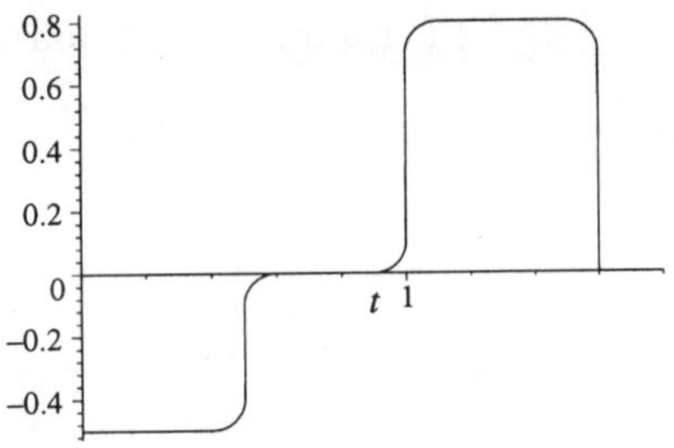

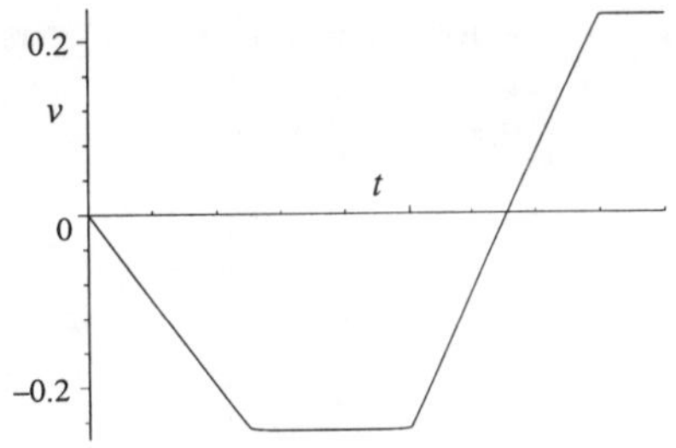

89. We take the direction of motion as $+x$, take $x_0 = 0$ and use SI units, so $v = 1600(1000/3600) = 444$ m/s.

(a) Eq. 2-11 gives $444 = a(1.8)$ or $a = 247\ \mathrm{m/s^2}$. We express this as a multiple of g by setting up a ratio:

$$a = \left(\frac{247}{9.8}\right) g = 25g \; .$$

(b) Eq. 2-17 readily yields

$$x = \frac{1}{2}\,(v_0 + v)\,t = \frac{1}{2}(444)(1.8) = 400 \text{ m} \; .$$

90. The graph is shown below. We assumed each interval described in the problem was one time unit long. A marks where the curve is steepest and B is where it is least steep (where it, in fact, has zero slope).

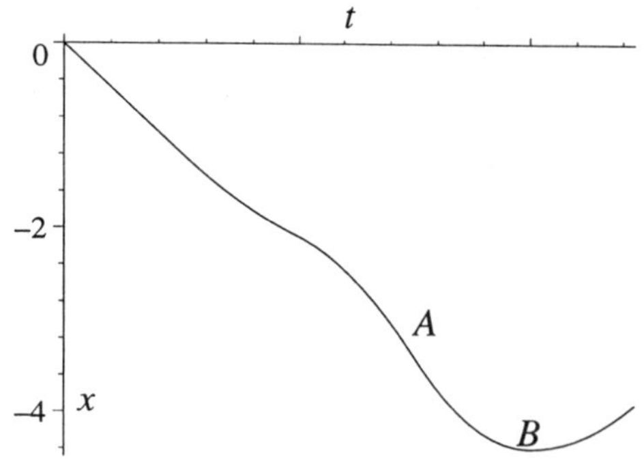

91. We use the functional notation $x(t)$, $v(t)$ and $a(t)$ in this solution, where the latter two quantities are obtained by differentiation:

$$v(t) = \frac{dx(t)}{dt} = -12t \quad \text{and} \quad a(t) = \frac{dv(t)}{dt} = -12$$

with SI units understood.

(a) From $v(t) = 0$ we find it is (momentarily) at rest at $t = 0$.

(b) We obtain $x(0) = 4.0$ m

(c) Requiring $x(t) = 0$ in the expression $x(t) = 4.0 - 6.0t^2$ leads to $t = \pm 0.82$ s for the times when the particle can be found passing through the origin.

(d) We show both the asked-for graph (on the left) as well as the "shifted" graph which is relevant to part (e). In both cases, the time axis is given by $-3 \le t \le 3$ (SI units understood).

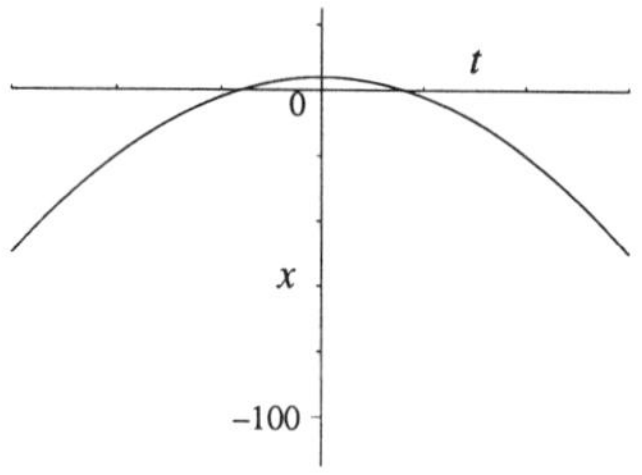

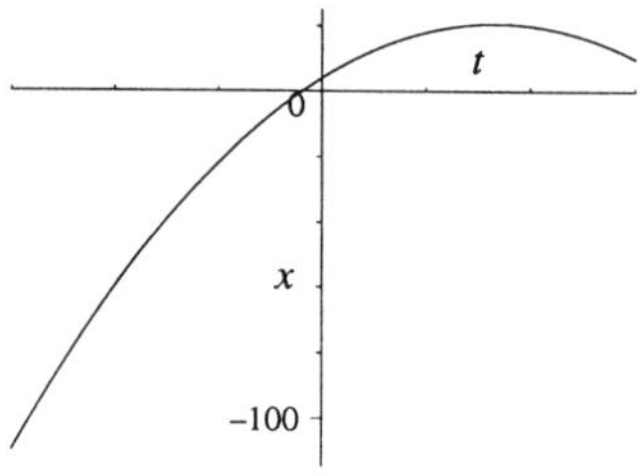

(e) We arrived at the graph on the right (shown above) by adding $20t$ to the $x(t)$ expression.

(f) Examining where the slopes of the graphs become zero, it is clear that the shift causes the $v = 0$ point to correspond to a larger value of x (the top of the second curve shown in part (d) is higher than that of the first).

92. (a) The slope of the graph (at a point) represents the velocity there, and the up-or-down concavity of the curve there indicates the $\pm$ sign of the acceleration. Thus, during AB we have $v > 0$ and $a = 0$ (since it is a straight line). During BC, we still have $v > 0$ but there is some curvature and a downward concavity is indicated (so $a < 0$). The segment CD is horizontal, implying the particle remains at the same position for some time; thus, $v = a = 0$ during CD. Clearly, the slope is negative during DE (so $v < 0$) but whether or not the graph is curved is less clear; we believe it is, with an upward concavity ($a > 0$).

(b) The key word is "obviously." Since it seems plausible to us that the curved portions can be "fit" with parabolic arcs (indications of constant acceleration by Eq. 2-15), then our answer is "no."

(c) Neither signs of slopes nor the sign of the concavity depends on a global shift in one axis or another (or, for that matter, on rescalings of the axes themselves) so the answer again is "no."

93. (a) The slope of the graph (at a point) represents the velocity there, and the up-or-down concavity of the curve there indicates the $\pm$ sign of the acceleration. Thus, during AB we have positive slope ($v > 0$) and $a < 0$ (since it is concave downward). The segment BC is horizontal, implying the particle remains at the same position for some time; thus, $v = a = 0$ during BC. During CD we have $v > 0$ and $a > 0$ (since it is concave upward). Clearly, the slope is positive during DE (so $v > 0$) but whether or not the graph is curved is less clear; we believe it is not, so $a = 0$.

(b) The key word is "obviously." Since it seems plausible to us that the curved portions can be "fit" with parabolic arcs (indications of constant acceleration by Eq. 2-15), then our answer is "no."

(c) Neither signs of slopes nor the sign of the concavity depends on a global shift in one axis or another (or, for that matter, on rescalings of the axes themselves) so the answer again is "no."

94. This problem consists of two parts: part 1 with constant acceleration (so that the equations in Table 2-1 apply), $v_0 = 0$, $v = 11.0$ m/s, $x = 12.0$ m, and $x_0 = 0$ (adopting the starting line as the coordinate origin); and, part 2 with constant velocity (so that $x - x_0 = vt$ applies) with $v = 11.0$ m/s, $x_0 = 12.0$, and $x = 100.0$ m.

(a) We obtain the time for part 1 from Eq. 2-17

$$x - x_0 = \frac{1}{2}(v_0 + v)\,t_1 \implies 12.0 - 0 = \frac{1}{2}(0 + 11.0)t_1$$

so that $t_1 = 2.2$ s, and we find the time for part 2 simply from $88.0 = (11.0)t_2 \rightarrow t_2 = 8.0$ s. Therefore, the total time is $t_1 + t_2 = 10.2$ s.

(b) Here, the total time is required to be 10.0 s, and we are to locate the point x_p where the runner switches from accelerating to proceeding at constant speed. The equations for parts 1 and 2, used above, therefore become

$$\begin{aligned} x_p - 0 &= \frac{1}{2}(0 + 11.0)t_1 \\ 100.0 - x_p &= (11.0)(10.0 - t_1) \end{aligned}$$

where in the latter equation, we use the fact that $t_2 = 10.0 - t_1$. Solving the equations for the two unknowns, we find that $t_1 = 1.8$ s and $x_p = 10.0$ m.

95. We take $+x$ in the direction of motion, so $v_0 = +24.6$ m/s and $a = -4.92$ m/s^2. We also take $x_0 = 0$.

(a) The time to come to a halt is found using Eq. 2-11:

$$0 = v_0 + at \implies t = -\frac{24.6}{-4.92} = 5.00 \text{ s} .$$

(b) Although several of the equations in Table 2-1 will yield the result, we choose Eq. 2-16 (since it does not depend on our answer to part (a)).

$$0 = v_0^2 + 2ax \implies x = -\frac{24.6^2}{2(-4.92)} = 61.5 \text{ m} .$$

(c) Using these results, we plot $v_0 t + \frac{1}{2}at^2$ (the x graph, shown below, on the left) and $v_0 + at$ (the v graph, below right) over $0 \le t \le 5$ s, with SI units understood.

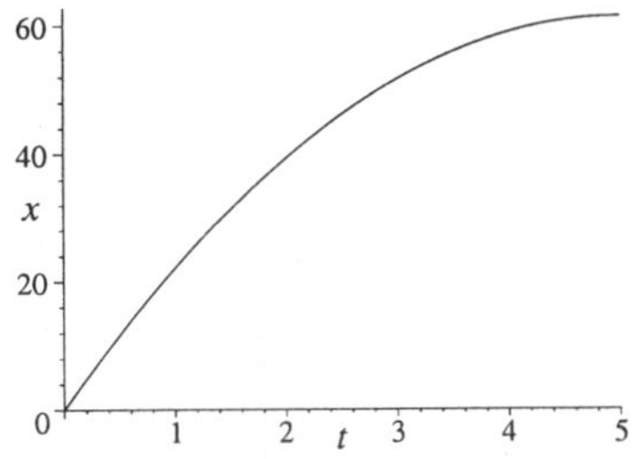

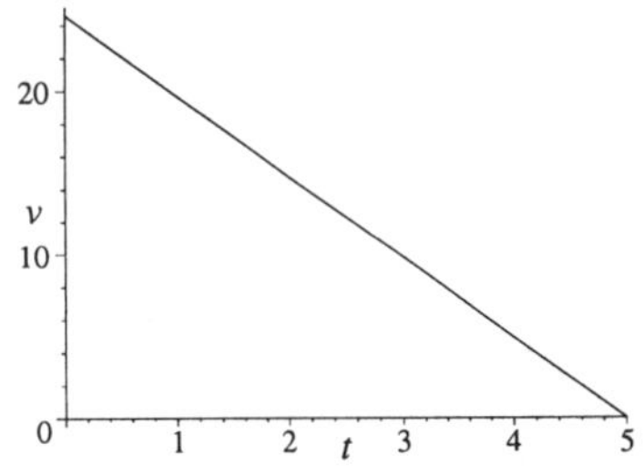

96. We take $+x$ in the direction of motion, so

$$v = (60\,\text{km/h})\left(\frac{1000\,\text{m/km}}{3600\,\text{s/h}}\right) = +16.7\ \text{m/s}$$

and $a > 0$. The location where it starts from rest ($v_0 = 0$) is taken to be $x_0 = 0$.

(a) Eq. 2-7 gives $a_{\text{avg}} = (v - v_0)/t$ where $t = 5.4$ s and the velocities are given above. Thus, $a_{\text{avg}} = 3.1\ \text{m/s}^2$.

(b) The assumption that $a =$ constant permits the use of Table 2-1. From that list, we choose Eq. 2-17:

$$x = \frac{1}{2}(v_0 + v)\,t = \frac{1}{2}(16.7)(5.4) = 45\ \text{m}\ .$$

(c) We use Eq. 2-15, now with $x = 250$ m:

$$x = \frac{1}{2}at^2 \implies t = \sqrt{\frac{2x}{a}} = \sqrt{\frac{2(250)}{3.1}}$$

which yields $t = 13$ s.

97. Converting to SI units, we have $v = 3400(1000/3600) = 944$ m/s (presumed constant) and $\Delta t = 0.10$ s. Thus, $\Delta x = v\Delta t = 94$ m.

98. The (ideal) driving time before the change was $t = \Delta x/v$, and after the change it is $t' = \Delta x/v'$. The time saved by the change is therefore

$$t - t' = \Delta x\left(\frac{1}{v} - \frac{1}{v'}\right) = \Delta x\left(\frac{1}{55} - \frac{1}{65}\right) = \Delta x(0.0028\,\text{h/mi})$$

which becomes, converting $\Delta x = 700/1.61 = 435$ mi (using a conversion found on the inside front cover of the textbook), $t - t' = (435)(0.0028) = 1.2$ h. This is equivalent to 1 h and 13 min.

99. (a) With the understanding that these are good to three significant figures, we write the function (in SI units) as

$$x(t) = -32 + 24t^2e^{-0.03t}$$

and find the velocity and acceleration functions by differentiating (calculus is reviewed Appendix E). We find

$$v(t) = 24t(2 - 0.03t)e^{-0.03t} \quad \text{and} \quad a(t) = 24\left(2 - 0.12t + 0.0009t^2\right)e^{-0.03t}\ .$$

(b) The $v(t)$ and $a(t)$ graphs are shown below (SI units understood). The time axis in both cases runs from $t = 0$ to $t = 100$ s. We include the $x(t)$ graph in the next part, accompanying our discussion of its root (which is, as suggested by the graph, a small positive value of t).

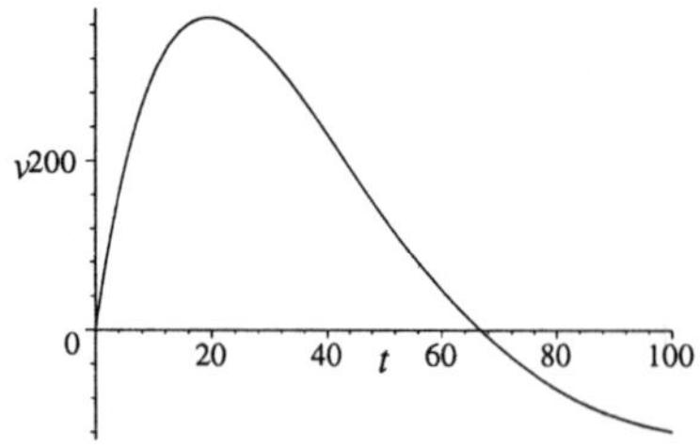

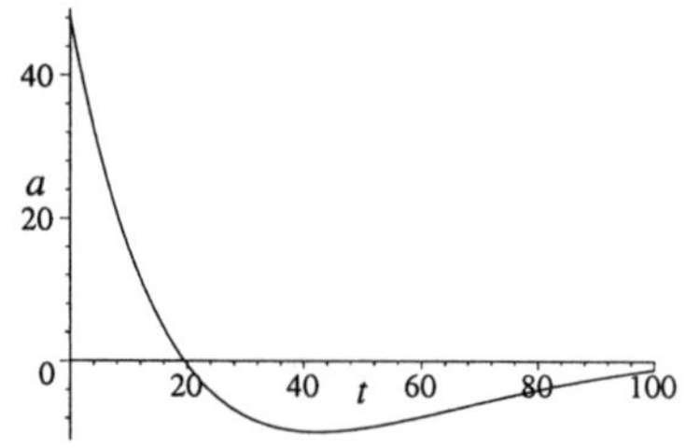

(c) We seek to find a positive value of t for which $24t^2e^{-0.03t} = 32$. We turn to the calculator or to a computer for its (numerical) solution. In this case, we ignore the roots outside the $0 \le t \le 100$ range (such as $t = -1.14$ s and $t = 387.77$ s) and choose $t = 1.175$ s as our answer. All of these are rounded-off values. We find $v = 53.5$ m/s and $a = 43.1$ m/s^2 at this time.

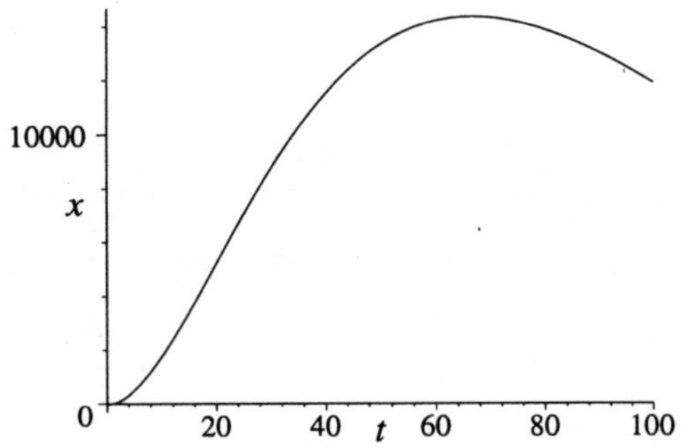

(d) It is much easier to find when $24t(2 - 0.03t)e^{-0.03t} = 0$ since the roots are clearly $t_1 = 0$ and $t_2 = 2/0.03 = 66.7$ s. We find $x(t_1) = -32.0$ m and $a(t_1) = 48.0$ m/s^2 at the first root, and we find $x(t_2) = 1.44 \times 10^4$ m and $a(t_2) = -6.50$ m/s^2 at the second root.

100. We take $+x$ in the direction of motion, so $v_0 = +30$ m/s, $v_1 = +15$ m/s and $a < 0$. The acceleration is found from Eq. 2-11: $a = (v_1 - v_0)/t_1$ where $t_1 = 3.0$ s. This gives $a = -5.0$ m/s^2. The displacement (which in this situation is the same as the distance traveled) to the point it stops ($v_2 = 0$) is, using Eq. 2-16,

$$v_2^2 = v_0^2 + 2a\Delta x \implies \Delta x = -\frac{30^2}{2(-5)} = 90 \text{ m} .$$

101. We choose the direction of motion as the positive direction. We work with the kilometer and hour units, so we write $\Delta x = 0.088$ km.

(a) Eq. 2-16 leads to

$$a = \frac{v^2 - v_0^2}{2\Delta x} = \frac{65^2 - 85^2}{2(0.088)}$$

which yields $a = -1.7 \times 10^4$ km/h^2.

(b) In this case, we obtain

$$a = \frac{60^2 - 80^2}{2(0.088)} = -1.6 \times 10^4 \text{ km/h}^2 .$$

(c) In this final situation, we find

$$a = \frac{40^2 - 50^2}{2(0.088)} = -5.1 \times 10^3 \text{ km/h}^2 .$$

102. Let the vertical distances between Jim's and Clara's feet and the jump-off level be H_J and H_C, respectively. At the instant this photo was taken, Clara has fallen for a time T_C, while Jim has fallen for T_J. Thus (using Eq. 2-15 with $v_0 = 0$) we have

$$H_J = \frac{1}{2}gT_J^2 \quad \text{and} \quad H_C = \frac{1}{2}gT_C^2 .$$

Measuring directly from the photo, we get $H_J \approx 3.6$ m and $H_C \approx 6.3$ m, which yields $T_J \approx 0.86$ s and $T_C \approx 1.13$ s. Jim's waiting time is therefore $T_C - T_J \approx 0.3$ s.

103. We choose *down* as the $+y$ direction and place the coordinate origin at the top of the building (which has height H). During its fall, the ball passes (with velocity v_1) the top of the window (which is at y_1) at time t_1, and passes the bottom (which is at y_2) at time t_2. We are told $y_2 - y_1 = 1.20$ m and $t_2 - t_1 = 0.125$ s. Using Eq. 2-15 we have

$$y_2 - y_1 = v_1(t_2 - t_1) + \frac{1}{2}g(t_2 - t_1)^2$$

which immediately yields

$$v_1 = \frac{1.20 - \frac{1}{2}(9.8)(0.125)^2}{0.125} = 8.99 \text{ m/s} .$$

From this, Eq. 2-16 (with $v_0 = 0$) reveals the value of y_1:

$$v_1^2 = 2gy_1 \implies y_1 = \frac{8.99^2}{2(9.8)} = 4.12 \text{ m} .$$

It reaches the ground ($y_3 = H$) at t_3. Because of the symmetry expressed in the problem ("upward flight is a reverse of the fall") we know that $t_3 - t_2 = 2.00/2 = 1.00$ s. And this means $t_3 - t_1 = 1.00 + 0.125 = 1.125$ s. Now Eq. 2-15 produces

$$\begin{aligned} y_3 - y_1 &= v_1(t_3 - t_1) + \frac{1}{2}g(t_3 - t_1)^2 \\ y_3 - 4.12 &= (8.99)(1.125) + \frac{1}{2}(9.8)(1.125)^2 \end{aligned}$$

which yields $y_3 = H = 20.4$ m.

104. (a) Using the fact that the area of a triangle is $\frac{1}{2}$(base)(height) (and the fact that the integral corresponds to area under the curve) we find, from $t = 0$ through $t = 5$ s, the integral of v with respect to t is 15 m. Since we are told that $x_0 = 0$ then we conclude that $x = 15$ m when $t = 5.0$ s.

(b) We see directly from the graph that $v = 2.0$ m/s when $t = 5.0$ s.

(c) Since $a = \frac{dv}{dt}$ = slope of the graph, we find that the acceleration during the interval $4 < t < 6$ is uniformly equal to $-2.0\,\text{m/s}^2$.

(d) Thinking of $x(t)$ in terms of accumulated area (on the graph), we note that $x(1) = 1$ m; using this and the value found in part (a), Eq. 2-2 produces

$$v_{\text{avg}} = \frac{x(5) - x(1)}{5 - 1} = \frac{15 - 1}{4} = 3.5 \text{ m/s} .$$

(e) From Eq. 2-7 and the values $v(t)$ we read directly from the graph, we find

$$a_{\text{avg}} = \frac{v(5) - v(1)}{5 - 1} = \frac{2 - 2}{4} = 0.$$

105. (First problem of **Cluster 1**)
The two parts of this problem are as follows. Part 1 (motion from A to B) consists of constant acceleration (so Table 2-1 applies) and involves the data $v_0 = 0$, $v = 10.0$ m/s, $x_0 = 0$ and $x = 40.0$ m (taking point A as the coordinate origin and orienting the positive x axis towards B and C). Part 2 (from B to C) consists of constant velocity motion (so the simple equation $\frac{\Delta x}{\Delta t} = v$ applies) with $v = 10.0$ m/s and $\Delta t = 10.0$ s.

(a) Eq. 2-16 is an efficient way of finding the part 1 acceleration:

$$v^2 = v_0^2 + 2a(x - x_0) \implies (10.0)^2 = 0 + 2a(40.0)$$

from which we obtain $a = 1.25\ \text{m/s}^2$.

(b) Using Eq. 2-17 avoids using the result from part (a) and finds the time readily.

$$x - x_0 = \frac{1}{2}(v_0 + v)\,t \implies 40.0 - 0 = \frac{1}{2}(0 + 10.0)t$$

This leads to $t = 8.00$ s, for part 1.

(c) We find the distance traveled in part 2 with $\Delta x = v\Delta t = (10.0)(10.0) = 100$ m.

(d) The average velocity is defined by Eq. 2-2

$$v_{\text{avg}} = \frac{x_C - x_A}{t_C - t_A} = \frac{140 - 0}{18 - 0} = 7.78 \text{ m/s} .$$

106. (Second problem of **Cluster 1**)
The two parts of this problem are as follows. Part 1 (motion from A to B) consists of constant acceleration (so Table 2-1 applies) and involves the data $v_0 = 20.0$ m/s, $v = 30.0$ m/s, $x_0 = 0$ and $t_1 = 10.0$ s (taking point A as the coordinate origin, orienting the positive x axis towards B and C, and starting the clock when it passes point A). Part 2 (from B to C) also involves uniformly accelerated motion but now with the data $v_0 = 30.0$ m/s, $v = 15.0$ m/s, and $\Delta x = x - x_0 = 150$ m.

(a) The distance for part 1 is given by

$$x - x_0 = \frac{1}{2}(v_0 + v)\,t_1 = \frac{1}{2}(20.0 + 30.0)(10.0)$$

which yields $x = 250$ m.

(b) The time t_2 for part 2 is found from the same formula as in part (a).

$$x - x_0 = \frac{1}{2}(v_0 + v)\,t_2 \implies 150 = \frac{1}{2}(30.0 + 15.0)t_2 .$$

This results in $t_2 = 6.67$ s.

(c) The definition of average velocity is given by Eq. 2-2:

$$v_{\text{avg}} = \frac{x_C - x_A}{t_C - t_A} = \frac{400 - 0}{16.7} = 24.0 \text{ m/s} .$$

(d) The definition of average acceleration is given by Eq. 2-7:

$$a_{\text{avg}} = \frac{v_C - v_A}{t_C - t_A} = \frac{15.0 - 20.0}{16.7} = -0.30 \text{ m/s}^2 .$$

107. (Third problem of **Cluster 1**)
The problem consists of two parts (A to B at constant velocity, then B to C with constant acceleration). The constant velocity in part 1 is 20 m/s (taking the positive direction in the direction of motion) and $t_1 = 5.0$ s. In part 2, we have $v_0 = 20$ m/s, $v = 0$, and $t_2 = 10$ s.

(a) We find the distance in part 1 from $x - x_0 = vt_1$, so we obtain $x = 100$ m (taking A to be at the origin). And the position at the end of part 2 is then found using Eq. 2-17.

$$x = x_0 + \frac{1}{2}(v_0 + v)\,t_2 = 100 + \frac{1}{2}(20 + 0)(10) = 200 \text{ m} .$$

(b) The acceleration in part (a) can be found using Eq. 2-11.

$$v = v_0 + at_2 \implies 0 = 20 + a(10) .$$

Thus, we find $a = -2.0$ m/s^2. The negative sign indicates that the acceleration vector points opposite to the chosen positive direction (the direction of motion), which is what we expect of a deceleration.

108. (Fourth problem of **Cluster 1**)
The part 1 motion in this problem is simply that of constant velocity, so $x_B - x_0 = v_1 t_1$ applies with $t_1 = 5.00$ s and $x_0 = x_A = 0$ if we choose point A as the coordinate origin. Next, the part 2 motion consists of constant acceleration (so the equations of Table 2-1, such as Eq. 2-17, apply) with $x_0 = x_B$ (an unknown), $v_0 = v_B$ (also unknown, but equal to the v_1 above), $x_C = 300$ m, $v_C = 10.0$ m/s, and $t_2 = 20.0$ s. The equations describing parts 1 and 2, respectively, are therefore

$$x_B - x_A = v_1 t_1 \quad \Longrightarrow \quad x_B = v_1 (5.00)$$
$$x_C - x_B = \frac{1}{2}(v_B + v_C)\, t_2 \quad \Longrightarrow \quad 300 - x_B = \frac{1}{2}(v_B + 10.0)\,(20.0)$$

(a) We use the fact that $v_A = v_1 = v_B$ in solving this set of simultaneous equations. Adding equations, we obtain the result $v_1 = 13.3$ m/s.

(b) In order to find the acceleration, we use our result from part (a) as the initial velocity in Eq. 2-14 (applied to the part 2 motion):

$$v = v_0 + at_2 \quad \Longrightarrow \quad 10.0 = 13.3 + a(20.0)$$

Thus, $a = -0.167$ m/s^2.

109. (Fifth problem of **Cluster 1**)
The problem consists of two parts, where part 1 (A to B) involves constant velocity motion for $t_1 = 5.00$ s and part 2 (B to C) involves uniformly accelerated motion. Assuming the coordinate origin is at point A and the positive axis is directed towards B and C, then we have $x_C = 250$ m, $a_2 = -0.500$ m/s^2, and $v_C = 0$.

(a) We set up the uniform velocity equation for part 1 ($\Delta x = vt$) and Eq. 2-16 for part 2 ($v^2 = v_0^2 + 2a\Delta x$) as a simultaneous set of equations to be solved:

$$\begin{aligned} x_B - 0 &= v_1 (5.00) \\ 0^2 &= v_B^2 + 2(-0.500)\,(250 - x_B) \ . \end{aligned}$$

Bearing in mind that $v_A = v_1 = v_B$, we can solve the equations by, for instance, substituting the first into the second – eliminating x_B and leading to a quadratic equation for v_1:

$$v_1^2 + 5v_1 - 150 = 0 \ .$$

The positive root gives us $v_1 = 13.5$ m/s.

(b) We obtain the duration t_2 of part 2 from Eq. 2-11:

$$v = v_0 + at_2 \quad \Longrightarrow \quad 0 = 13.5 + (-0.500)t_2$$

which yields the value $t_2 = 27.0$ s. Therefore, the total time is $t_1 + t_2 = 32.0$ s.

110. (Sixth problem of **Cluster 1**)
Both part 1 and part 2 of this problem involve uniformly accelerated motion, but at different rates a_1 and a_2. We take the coordinate origin at point A and direct the positive axis towards B and C. In these terms, we are given $x_A = 0$, $x_C = 1300$ m, $v_A = 0$, and $v_C = 50$ m/s. Further, the time-duration for each part is given: $t_1 = 20$ s and $t_2 = 40$ s.

(a) We have enough information to apply Eq. 2-17 ($\Delta x = \frac{1}{2}(v_0 + v)t$) to parts 1 and 2 and solve the simultaneous set:

$$x_B - x_A = \frac{1}{2}(v_A + v_B)\, t_1 \quad \Longrightarrow \quad x_B = \frac{1}{2} v_B\,(20)$$
$$x_C - x_B = \frac{1}{2}(v_B + v_C)\, t_2 \quad \Longrightarrow \quad 1300 - x_B = \frac{1}{2}(v_B + 50)\,(40)$$

Adding equations, we find $v_B = 10$ m/s.

(b) The other unknown in the above set of equations is now easily found by plugging the result for v_B back in: $x_B = 100$ m.

(c) We can find a_1 a variety of ways, using the just-obtained results. We note that Eq. 2-11 is especially easy to use.

$$v = v_0 + a_1 t_1 \implies 10 = 0 + a_1(20)$$

This leads to $a_1 = 0.50\ \text{m/s}^2$.

(d) To find a_2 we proceed as just as we did in part (c), so that Eq. 2-11 for part 2 becomes $50 = 10 + a_2(40)$. Therefore, the acceleration is $a_2 = 1.0\,\text{m/s}^2$.

Chapter 3

1. The vectors should be parallel to achieve a resultant 7 m long (the unprimed case shown below), antiparallel (in opposite directions) to achieve a resultant 1 m long (primed case shown), and perpendicular to achieve a resultant $\sqrt{3^2 + 4^2} = 5$ m long (the double-primed case shown). In each sketch, the vectors are shown in a "head-to-tail" sketch but the resultant is not shown. The resultant would be a straight line drawn from beginning to end; the beginning is indicated by A (with or without primes, as the case may be) and the end is indicated by B.

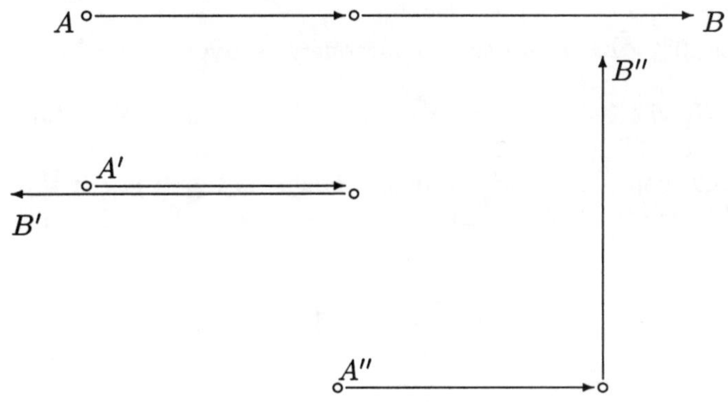

2. A sketch of the displacements is shown. The resultant (not shown) would be a straight line from start (Bank) to finish (Walpole). With a careful drawing, one should find that the resultant vector has length 29.5 km at 35° west of south.

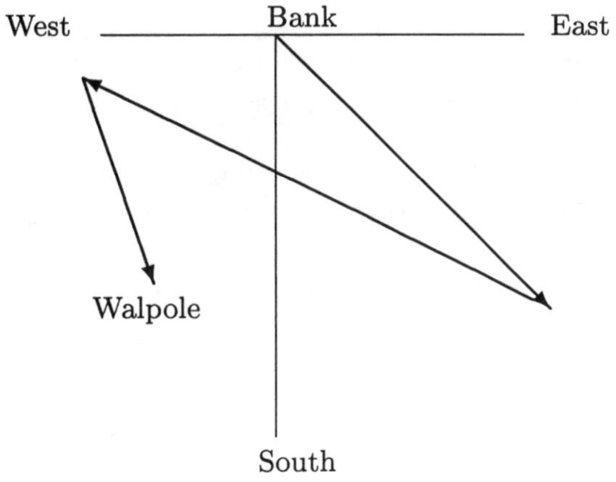

3. The x component of $\vec{a}$ is given by $a_x = 7.3\cos 250° = -2.5$ and the y component is given by $a_y = 7.3\sin 250° = -6.9$. In considering the variety of ways to compute these, we note that the vector is 70° below the $-x$ axis, so the components could also have been found from $a_x = -7.3\cos 70°$ and $a_y = -7.3\sin 70°$. In a similar vein, we note that the vector is 20° from the $-y$ axis, so one could use $a_x = -7.3\sin 20°$ and $a_y = -7.3\cos 20°$ to achieve the same results.

4. The angle described by a full circle is $360° = 2\pi$ rad, which is the basis of our conversion factor. Thus,

$$(20.0°)\,\frac{2\pi\,\text{rad}}{360°} = 0.349\ \text{rad}$$

and (similarly) $50.0° = 0.873$ rad and $100° = 1.75$ rad. Also,

$$(0.330\,\text{rad})\,\frac{360°}{2\pi\,\text{rad}} = 18.9°$$

and (similarly) $2.10\,\text{rad} = 120°$ and $7.70\,\text{rad} = 441°$.

5. The textbook's approach to this sort of problem is through the use of Eq. 3-6, and is illustrated in Sample Problem 3-3. However, most modern graphical calculators can produce the results quite efficiently using rectangular ↔ polar "shortcuts."

   (a) $\sqrt{(-25)^2 + 40^2} = 47.2$ m

   (b) Recalling that $\tan(\theta) = \tan(\theta + 180°)$, we note that the two possibilities for $\tan^{-1}(40/-25)$ are $-58°$ and $122°$. Noting that the vector is in the third quadrant (by the signs of its x and y components) we see that $122°$ is the correct answer. The graphical calculator "shortcuts" mentioned above are designed to correctly choose the right possibility.

6. The x component of $\vec{r}$ is given by $15\cos 30° = 13$ m and the y component is given by $15\sin 30° = 7.5$ m.

7. The point P is displaced vertically by $2R$, where R is the radius of the wheel. It is displaced horizontally by half the circumference of the wheel, or πR. Since $R = 0.450$ m, the horizontal component of the displacement is 1.414 m and the vertical component of the displacement is 0.900 m. If the x axis is horizontal and the y axis is vertical, the vector displacement (in meters) is $\vec{r} = (1.414\,\hat{\imath} + 0.900\,\hat{\jmath})$. The displacement has a magnitude of

$$|\vec{r}| = \sqrt{(\pi R)^2 + (2R)^2} = R\sqrt{\pi^2 + 4} = 1.68\ \text{m}$$

and an angle of

$$\tan^{-1}\left(\frac{2R}{\pi R}\right) = \tan^{-1}\left(\frac{2}{\pi}\right) = 32.5°$$

above the floor. In physics there are no "exact" measurements, yet that angle computation seemed to yield something *exact*. However, there has to be some uncertainty in the observation that the wheel rolled half of a revolution, which introduces some indefiniteness in our result.

8. Although we think of this as a three-dimensional movement, it is rendered effectively two-dimensional by referring measurements to its well-defined plane of the fault.

   (a) The magnitude of the net displacement is

$$|\vec{AB}| = \sqrt{|AD|^2 + |AC|^2} = \sqrt{17^2 + 22^2} = 27.8\ \text{m}\ .$$

   (b) The magnitude of the vertical component of $\vec{AB}$ is $|AD|\sin 52.0° = 13.4$ m.

9. The length unit meter is understood throughout the calculation.

   (a) We compute the distance from one corner to the diametrically opposite corner: $d = \sqrt{3.00^2 + 3.70^2 + 4.30^2}$ 6.42.

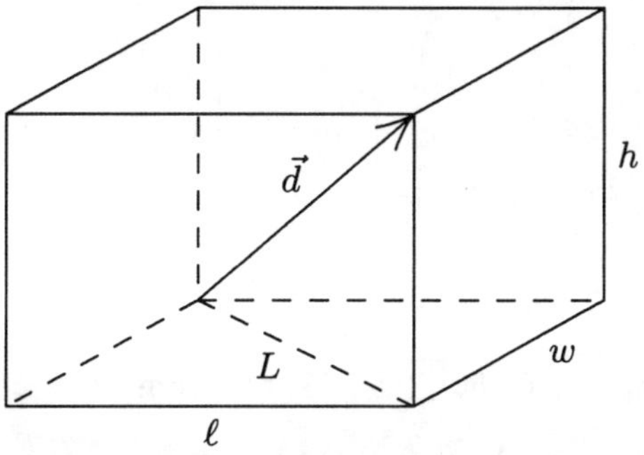

(b) The displacement vector is along the straight line from the beginning to the end point of the trip. Since a straight line is the shortest distance between two points, the length of the path cannot be less than the magnitude of the displacement.

(c) It can be greater, however. The fly might, for example, crawl along the edges of the room. Its displacement would be the same but the path length would be $\ell + w + h$.

(d) The path length is the same as the magnitude of the displacement if the fly flies along the displacement vector.

(e) We take the x axis to be out of the page, the y axis to be to the right, and the z axis to be upward. Then the x component of the displacement is $w = 3.70$, the y component of the displacement is 4.30, and the z component is 3.00. Thus $\vec{d} = 3.70\,\hat{\imath} + 4.30\,\hat{\jmath} + 3.00\,\hat{k}$. An equally correct answer is gotten by interchanging the length, width, and height.

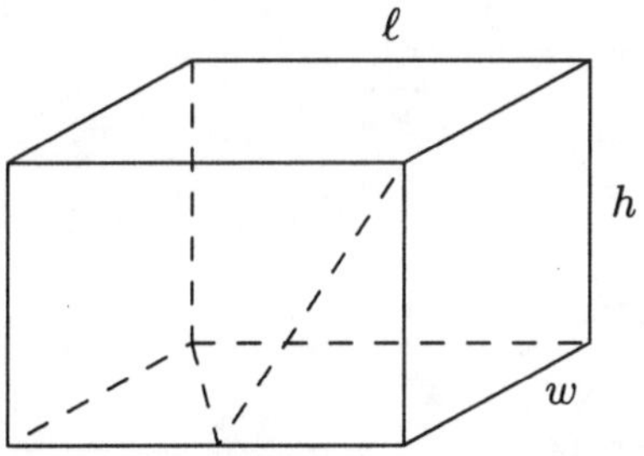

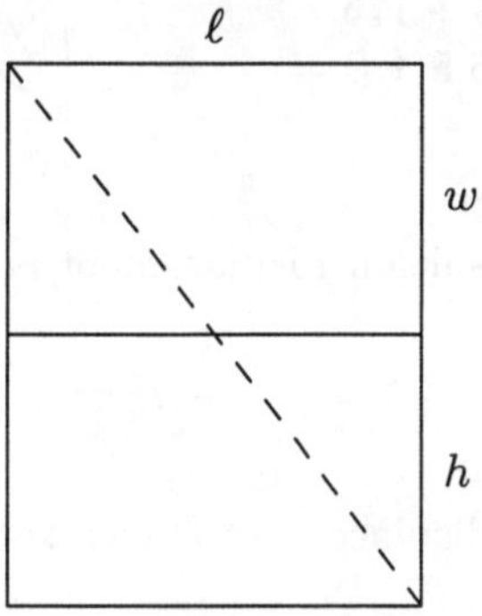

(f) Suppose the path of the fly is as shown by the dotted lines on the upper diagram. Pretend there is a hinge where the front wall of the room joins the floor and lay the wall down as shown on the lower diagram. The shortest walking distance between the lower left back of the room and the upper right front corner is the dotted straight line shown on the diagram. Its length is

$$L_{\text{min}} = \sqrt{(w+h)^2 + \ell^2} = \sqrt{(3.70 + 3.00)^2 + 4.30^2} = 7.96 \text{ m} .$$

10. We label the displacement vectors $\vec{A}$, $\vec{B}$ and $\vec{C}$ (and denote the result of their vector sum as $\vec{r}$). We choose *east* as the $\hat{\imath}$ direction ($+x$ direction) and *north* as the $\hat{\jmath}$ direction ($+y$ direction). All distances are understood to be in kilometers. We note that the angle between $\vec{C}$ and the x axis is 60°. Thus,

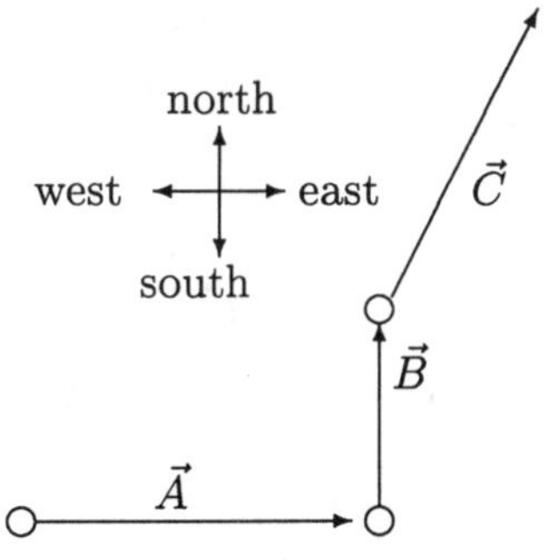

$$\begin{aligned}\vec{A} &= 50\,\hat{\imath}\\ \vec{B} &= 30\,\hat{\jmath}\\ \vec{C} &= 25\,\cos(60°)\,\hat{\imath} + 25\,\sin(60°)\,\hat{\jmath}\\ \vec{r} = \vec{A} + \vec{B} + \vec{C} &= 62.50\,\hat{\imath} + 51.65\,\hat{\jmath}\end{aligned}$$

which means that its magnitude is

$$|\vec{r}| = \sqrt{62.50^2 + 51.65^2} \approx 81 \text{ km} .$$

and its angle (counterclockwise from $+x$ axis) is $\tan^{-1}(51.65/62.50) \approx 40°$, which is to say that it points 40° *north of east*. Although the resultant $\vec{r}$ is shown in our sketch, it would be a direct line from the "tail" of $\vec{A}$ to the "head" of $\vec{C}$.

11. The diagram shows the displacement vectors for the two segments of her walk, labeled $\vec{A}$ and $\vec{B}$, and the total ("final") displacement vector, labeled $\vec{r}$. We take east to be the $+x$ direction and north to be the $+y$ direction. We observe that the angle between $\vec{A}$ and the x axis is 60°. Where the units are not explicitly shown, the distances are understood to be in meters. Thus, the components of $\vec{A}$ are $A_x = 250\cos 60° = 125$ and $A_y = 250\sin 30° = 216.5$. The components of $\vec{B}$ are $B_x = 175$ and $B_y = 0$. The components of the total displacement are $r_x = A_x + B_x = 125 + 175 = 300$ and $r_y = A_y + B_y = 216.5 + 0 = 216.5$.

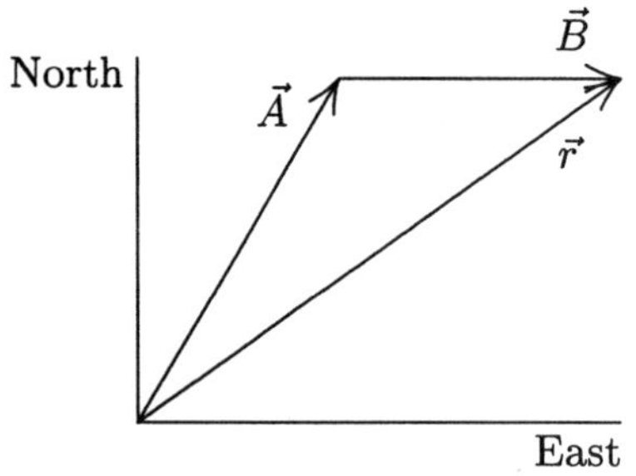

(a) The magnitude of the resultant displacement is

$$|\vec{r}| = \sqrt{r_x^2 + r_y^2} = \sqrt{300^2 + 216.5^2} = 370 \text{ m} .$$

(b) The angle the resultant displacement makes with the $+x$ axis is

$$\tan^{-1}\left(\frac{r_y}{r_x}\right) = \tan^{-1}\left(\frac{216.5}{300}\right) = 36° .$$

(c) The total *distance* walked is $d = 250 + 175 = 425$ m.

(d) The total distance walked is greater than the magnitude of the resultant displacement. The diagram shows why: $\vec{A}$ and $\vec{B}$ are not collinear.

12. We label the displacement vectors $\vec{A}$, $\vec{B}$ and $\vec{C}$ (and denote the result of their vector sum as $\vec{r}$). We choose *east* as the $\hat{\imath}$ direction ($+x$ direction) and *north* as the $\hat{\jmath}$ direction ($+y$ direction). All distances are understood to be in kilometers. Therefore,

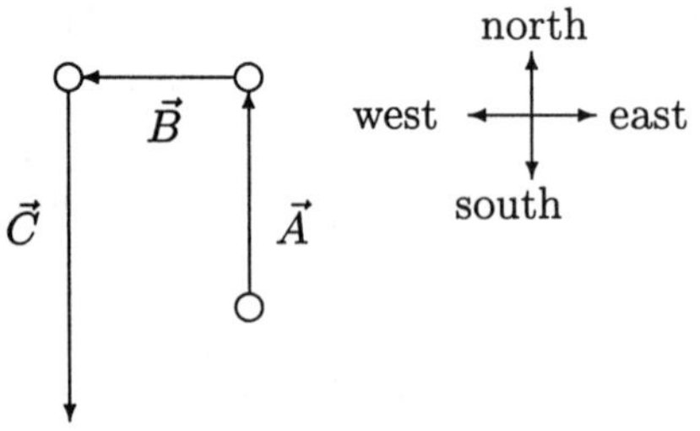

$$\begin{aligned}
\vec{A} &= 3.1\,\hat{\jmath} \\
\vec{B} &= -2.4\,\hat{\imath} \\
\vec{C} &= -5.2\,\hat{\jmath} \\
\vec{r} = \vec{A} + \vec{B} + \vec{C} &= -2.1\,\hat{\imath} - 2.4\,\hat{\jmath}
\end{aligned}$$

which means that its magnitude is

$$|\vec{r}| = \sqrt{(-2.1)^2 + (-2.4)^2} \approx 3.2 \text{ km} .$$

and the two possibilities for its angle are

$$\tan^{-1}\left(\frac{-2.4}{-2.1}\right) = 41^\circ, \quad \text{or} \quad 221^\circ .$$

We choose the latter possibility since $\vec{r}$ is in the third quadrant. It should be noted that many graphical calculators have polar $\leftrightarrow$ rectangular "shortcuts" that automatically produce the correct answer for angle (measured counterclockwise from the $+x$ axis). We may phrase the angle, then, as 221° counterclockwise from East (a phrasing that sounds peculiar, at best) or as 41° south from west or 49° west from south. The resultant $\vec{r}$ is not shown in our sketch; it would be an arrow directed from the "tail" of $\vec{A}$ to the "head" of $\vec{C}$.

13. We write $\vec{r} = \vec{a} + \vec{b}$. When not explicitly displayed, the units here are assumed to be meters. Then $r_x = a_x + b_x = 4.0 - 13 = -9.0$ and $r_y = a_y + b_y = 3.0 + 7.0 = 10$. Thus $\vec{r} = (-9.0\,\text{m})\,\hat{\imath} + (10\,\text{m})\,\hat{\jmath}$. The magnitude of the resultant is

$$r = \sqrt{r_x^2 + r_y^2} = \sqrt{(-9.0)^2 + (10)^2} = 13 \text{ m} .$$

The angle between the resultant and the $+x$ axis is given by $\tan^{-1}(r_y/r_x) = \tan^{-1} 10/(-9.0)$ which is either $-48°$ or 132°. Since the x component of the resultant is negative and the y component is positive, characteristic of the second quadrant, we find the angle is 132° (measured counterclockwise from $+x$ axis).

14. The x, y and z components (with meters understood) of $\vec{r}$ are

$$\begin{aligned}
r_x = c_x + d_x &= 7.4 + 4.4 = 12 \\
r_y = c_y + d_y &= -3.8 - 2.0 = -5.8 \\
r_z = c_z + d_z &= -6.1 + 3.3 = -2.8 .
\end{aligned}$$

15. The vectors are shown on the diagram. The x axis runs from west to east and the y axis run from south to north. Then $a_x = 5.0\,\text{m}$, $a_y = 0$, $b_x = -(4.0\,\text{m})\sin 35° = -2.29\,\text{m}$, and $b_y = (4.0\,\text{m})\cos 35° = 3.28\,\text{m}$.

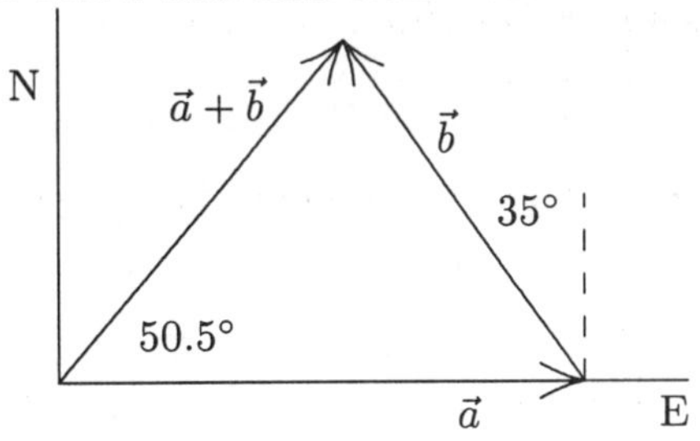

(a) Let $\vec{c} = \vec{a} + \vec{b}$. Then $c_x = a_x + b_x = 5.0\,\text{m} - 2.29\,\text{m} = 2.71\,\text{m}$ and $c_y = a_y + b_y = 0 + 3.28\,\text{m} = 3.28\,\text{m}$. The magnitude of c is

$$c = \sqrt{c_x^2 + c_y^2} = \sqrt{(2.71\,\text{m})^2 + (3.28\,\text{m})^2} = 4.3 \text{ m} .$$

(b) The angle θ that $\vec{c} = \vec{a} + \vec{b}$ makes with the $+x$ axis is

$$\theta = \tan^{-1}\frac{c_y}{c_x} = \tan^{-1}\frac{3.28\,\text{m}}{2.71\,\text{m}} = 50.4^\circ\ .$$

The second possibility ($\theta = 50.4^\circ + 180^\circ = 126^\circ$) is rejected because it would point in a direction opposite to $\vec{c}$.

(c) The vector $\vec{b} - \vec{a}$ is found by adding $-\vec{a}$ to $\vec{b}$. The result is shown on the diagram to the right. Let $\vec{c} = \vec{b} - \vec{a}$. Then $c_x = b_x - a_x = -2.29\,\text{m} - 5.0\,\text{m} = -7.29\,\text{m}$ and $c_y = b_y - a_y = 3.28\,\text{m}$. The magnitude of $\vec{c}$ is $c = \sqrt{c_x^2 + c_y^2} = 8.0\,\text{m}$.

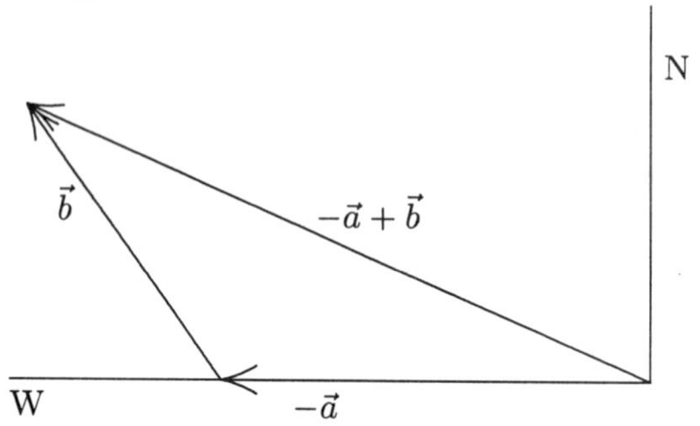

(d) The tangent of the angle θ that $\vec{c}$ makes with the $+x$ axis (east) is

$$\tan\theta = \frac{c_y}{c_x} = \frac{3.28\,\text{m}}{-7.29\,\text{m}} = -4.50,.$$

There are two solutions: -24.2° and 155.8°. As the diagram shows, the second solution is correct. The vector $\vec{c} = -\vec{a} + \vec{b}$ is 24° north of west.

16. All distances in this solution are understood to be in meters.

(a) $\vec{a} + \vec{b} = (3.0\,\hat{\imath} + 4.0\,\hat{\jmath}) + (5.0\,\hat{\imath} - 2.0\,\hat{\jmath}) = 8.0\,\hat{\imath} + 2.0\,\hat{\jmath}$.

(b) The magnitude of $\vec{a} + \vec{b}$ is

$$|\vec{a} + \vec{b}| = \sqrt{8.0^2 + 2.0^2} = 8.2\ \text{m}\ .$$

(c) The angle between this vector and the $+x$ axis is $\tan^{-1}(2.0/8.0) = 14^\circ$.

(d) $\vec{b} - \vec{a} = (5.0\,\hat{\imath} - 2.0\,\hat{\jmath}) - (3.0\,\hat{\imath} + 4.0\,\hat{\jmath}) = 2.0\,\hat{\imath} - 6.0\,\hat{\jmath}$.

(e) The magnitude of the difference vector $\vec{b} - \vec{a}$ is

$$|\vec{b} - \vec{a}| = \sqrt{2.0^2 + (-6.0)^2} = 6.3\ \text{m}\ .$$

(f) The angle between this vector and the $+x$ axis is $\tan^{-1}(-6.0/2.0) = -72^\circ$. The vector is 72° *clockwise* from the axis defined by $\hat{\imath}$.

17. All distances in this solution are understood to be in meters.

(a) $\vec{a} + \vec{b} = (4.0 + (-1.0))\,\hat{\imath} + ((-3.0) + 1.0)\,\hat{\jmath} + (1.0 + 4.0)\,\hat{k} = 3.0\,\hat{\imath} - 2.0\,\hat{\jmath} + 5.0\,\hat{k}$.

(b) $\vec{a} - \vec{b} = (4.0 - (-1.0))\,\hat{\imath} + ((-3.0) - 1.0)\,\hat{\jmath} + (1.0 - 4.0)\,\hat{k} = 5.0\,\hat{\imath} - 4.0\,\hat{\jmath} - 3.0\,\hat{k}$.

(c) The requirement $\vec{a} - \vec{b} + \vec{c} = 0$ leads to $\vec{c} = \vec{b} - \vec{a}$, which we note is the opposite of what we found in part (b). Thus, $\vec{c} = -5.0\,\hat{\imath} + 4.0\,\hat{\jmath} + 3.0\,\hat{k}$.

18. Many of the operations are done efficiently on most modern graphical calculators using their built-in vector manipulation and rectangular $\leftrightarrow$ polar "shortcuts." In this solution, we employ the "traditional" methods (such as Eq. 3-6).

(a) The magnitude of $\vec{a}$ is $\sqrt{4^2 + (-3)^2} = 5.0$ m.

(b) The angle between $\vec{a}$ and the $+x$ axis is $\tan^{-1}(-3/4) = -37°$. The vector is 37° *clockwise* from the axis defined by $\hat{\imath}$.

(c) The magnitude of $\vec{b}$ is $\sqrt{6^2 + 8^2} = 10$ m.

(d) The angle between $\vec{b}$ and the $+x$ axis is $\tan^{-1}(8/6) = 53°$.

(e) $\vec{a} + \vec{b} = (4+6)\,\hat{\imath} + ((-3)+8)\,\hat{\jmath} = 10\,\hat{\imath} + 5\,\hat{\jmath}$, with the unit meter understood. The magnitude of this vector is $\sqrt{10^2 + 5^2} = 11$ m; we rounding to two significant figures in our results.

(f) The angle between the vector described in part (e) and the $+x$ axis is $\tan^{-1}(5/10) = 27°$.

(g) $\vec{b} - \vec{a} = (6-4)\,\hat{\imath} + (8-(-3))\,\hat{\jmath} = 2\,\hat{\imath} + 11\,\hat{\jmath}$, with the unit meter understood. The magnitude of this vector is $\sqrt{2^2 + 11^2} = 11$ m, which is, interestingly, the same result as in part (e) (exactly, not just to 2 significant figures) (this curious coincidence is made possible by the fact that $\vec{a} \perp \vec{b}$).

(h) The angle between the vector described in part (g) and the $+x$ axis is $\tan^{-1}(11/2) = 80°$.

(i) $\vec{a} - \vec{b} = (4-6)\,\hat{\imath} + ((-3)-8)\,\hat{\jmath} = -2\,\hat{\imath} - 11\,\hat{\jmath}$, with the unit meter understood. The magnitude of this vector is $\sqrt{(-2)^2 + (-11)^2} = 11$ m.

(j) The two possibilities presented by a simple calculation for the angle between the vector described in part (i) and the $+x$ direction are $\tan^{-1}(11/2) = 80°$, and $180° + 80° = 260°$. The latter possibility is the correct answer (see part (k) for a further observation related to this result).

(k) Since $\vec{a} - \vec{b} = (-1)(\vec{b} - \vec{a})$, they point in opposite (antiparallel) directions; the angle between them is 180°.

19. Many of the operations are done efficiently on most modern graphical calculators using their built-in vector manipulation and rectangular ↔ polar "shortcuts." In this solution, we employ the "traditional" methods (such as Eq. 3-6). Where the length unit is not displayed, the unit meter should be understood.

(a) Using unit-vector notation,

$$\begin{aligned}
\vec{a} &= 50\cos(30°)\,\hat{\imath} + 50\sin(30°)\,\hat{\jmath} \\
\vec{b} &= 50\cos(195°)\,\hat{\imath} + 50\sin(195°)\,\hat{\jmath} \\
\vec{c} &= 50\cos(315°)\,\hat{\imath} + 50\sin(315°)\,\hat{\jmath} \\
\vec{a} + \vec{b} + \vec{c} &= 30.4\,\hat{\imath} - 23.3\,\hat{\jmath}\,.
\end{aligned}$$

The magnitude of this result is $\sqrt{30.4^2 + (-23.3)^2} = 38$ m.

(b) The two possibilities presented by a simple calculation for the angle between the vector described in part (a) and the $+x$ direction are $\tan^{-1}(-23.2/30.4) = -37.5°$, and $180° + (-37.5°) = 142.5°$. The former possibility is the correct answer since the vector is in the fourth quadrant (indicated by the signs of its components). Thus, the angle is $-37.5°$, which is to say that it is roughly 38° *clockwise* from the $+x$ axis. This is equivalent to 322.5° counterclockwise from $+x$.

(c) We find $\vec{a} - \vec{b} + \vec{c} = (43.3 - (-48.3) + 35.4)\,\hat{\imath} - (25 - (-12.9) + (-35.4))\,\hat{\jmath} = 127\,\hat{\imath} + 2.6\,\hat{\jmath}$ in unit-vector notation. The magnitude of this result is $\sqrt{127^2 + 2.6^2} \approx 1.3 \times 10^2$ m.

(d) The angle between the vector described in part (c) and the $+x$ axis is $\tan^{-1}(2.6/127) \approx 1°$.

(e) Using unit-vector notation, $\vec{d}$ is given by

$$\begin{aligned}
\vec{d} &= \vec{a} + \vec{b} - \vec{c} \\
&= -40.4\,\hat{\imath} + 47.4\,\hat{\jmath}
\end{aligned}$$

which has a magnitude of $\sqrt{(-40.4)^2 + 47.4^2} = 62$ m.

(f) The two possibilities presented by a simple calculation for the angle between the vector described in part (e) and the $+x$ axis are $\tan^{-1}(47.4/(-40.4)) = -50°$, and $180° + (-50°) = 130°$. We choose the latter possibility as the correct one since it indicates that $\vec{d}$ is in the second quadrant (indicated by the signs of its components).

20. Angles are given in 'standard' fashion, so Eq. 3-5 applies directly. We use this to write the vectors in unit-vector notation before adding them. However, a very different-looking approach using the special capabilities of most graphical calculators can be imagined. Where the length unit is not displayed in the solution below, the unit meter should be understood.

(a) Allowing for the different angle units used in the problem statement, we arrive at

$$\begin{aligned}\vec{E} &= 3.73\hat{\imath} + 4.70\hat{\jmath} \\ \vec{F} &= 1.29\hat{\imath} - 4.83\hat{\jmath} \\ \vec{G} &= 1.45\hat{\imath} + 3.73\hat{\jmath} \\ \vec{H} &= -5.20\hat{\imath} + 3.00\hat{\jmath} \\ \vec{E} + \vec{F} + \vec{G} + \vec{H} &= 1.28\hat{\imath} + 6.60\hat{\jmath}\ .\end{aligned}$$

(b) The magnitude of the vector sum found in part (a) is $\sqrt{1.28^2 + 6.60^2} = 6.72$ m. Its angle measured counterclockwise from the $+x$ axis is $\tan^{-1}(6.6/1.28) = 79° = 1.38$ rad.

21. It should be mentioned that an efficient way to work this vector addition problem is with the cosine law for general triangles (and since $\vec{a}$, $\vec{b}$ and $\vec{r}$ form an isosceles triangle, the angles are easy to figure). However, in the interest of reinforcing the usual systematic approach to vector addition, we note that the angle $\vec{b}$ makes with the $+x$ axis is 135° and apply Eq. 3-5 and Eq. 3-6 where appropriate.

(a) The x component of $\vec{r}$ is $10\cos 30° + 10\cos 135° = 1.59$ m.

(b) The y component of $\vec{r}$ is $10\sin 30° + 10\sin 135° = 12.1$ m.

(c) The magnitude of $\vec{r}$ is $\sqrt{1.59^2 + 12.1^2} = 12.2$ m.

(d) The angle between $\vec{r}$ and the $+x$ direction is $\tan^{-1}(12.1/1.59) = 82.5°$.

22. If we wish to use Eq. 3-5 in an unmodified fashion, we should note that the angle between $\vec{C}$ and the $+x$ axis is $180° + 20° = 200°$.

(a) The x component of $\vec{B}$ is given by $C_x - A_x = 15\cos 200° - 12\cos 40° = -23.3$ m, and the y component of $\vec{B}$ is given by $C_y - A_y = 15\sin 200° - 12\sin 40° = -12.8$ m. Consequently, its magnitude is $\sqrt{(-23.3)^2 + (-12.8)^2} = 26.6$ m.

(b) The two possibilities presented by a simple calculation for the angle between $\vec{B}$ and the $+x$ axis are $\tan^{-1}((-12.8)/(-23.3)) = 28.9°$, and $180° + 28.9° = 209°$. We choose the latter possibility as the correct one since it indicates that $\vec{B}$ is in the third quadrant (indicated by the signs of its components). We note, too, that the answer can be equivalently stated as $-151°$.

23. We consider $\vec{A}$ with (x, y) components given by $(A\cos\alpha, A\sin\alpha)$. Similarly, $\vec{B} \to (B\cos\beta, B\sin\beta)$. The angle (measured from the $+x$ direction) for their vector sum must have a slope given by

$$\tan\theta = \frac{A\sin\alpha + B\sin\beta}{A\cos\alpha + B\cos\beta}\ .$$

The problem requires that we now consider the orthogonal direction, where $\tan\theta + 90° = -\cot\theta$. If this (the negative reciprocal of the above expression) is to equal the slope for their vector *difference*, then we must have

$$-\frac{A\cos\alpha + B\cos\beta}{A\sin\alpha + B\sin\beta} = \frac{A\sin\alpha - B\sin\beta}{A\cos\alpha - B\cos\beta}\ .$$

Multiplying both sides by $A\sin\alpha + B\sin\beta$ and doing the same with $A\cos\alpha - B\cos\beta$ yields

$$A^2\cos^2\alpha - B^2\cos^2\beta = A^2\sin^2\alpha - B^2\sin^2\beta\ .$$

Rearranging, using the $\cos^2\phi + \sin^2\phi = 1$ identity, we obtain

$$A^2 = B^2 \implies A = B\ .$$

In a *later* section, the scalar (dot) product of vectors is presented and this result can be revisited with a more compact derivation.

24. If we wish to use Eq. 3-5 directly, we should note that the angles for $\vec{Q}$,$\vec{R}$ and $\vec{S}$ are 100°, 250° and 310°, respectively, if they are measured counterclockwise from the $+x$ axis.

    (a) Using unit-vector notation, with the unit meter understood, we have

$$\begin{aligned}
\vec{P} &= 10\cos(25°)\,\hat{\mathrm{i}} + 10\sin(25°)\,\hat{\mathrm{j}} \\
\vec{Q} &= 12\cos(100°)\,\hat{\mathrm{i}} + 12\sin(100°)\,\hat{\mathrm{j}} \\
\vec{R} &= 8\cos(250°)\,\hat{\mathrm{i}} + 8\sin(250°)\,\hat{\mathrm{j}} \\
\vec{S} &= 9\cos(310°)\,\hat{\mathrm{i}} + 9\sin(310°)\,\hat{\mathrm{j}} \\
\vec{P} + \vec{Q} + \vec{R} + \vec{S} &= 10.0\hat{\mathrm{i}} + 1.6\hat{\mathrm{j}}
\end{aligned}$$

    (b) The magnitude of the vector sum is $\sqrt{10^2 + 1.6^2} = 10.2$ m and its angle is $\tan^{-1}(1.6/10) \approx 9.2°$ measured counterclockwise from the $+x$ axis. The appearance of this solution would be quite different using the vector manipulation capabilities of most modern graphical calculators, although the principle would be basically the same.

25. Without loss of generality, we assume $\vec{a}$ points along the $+x$ axis, and that $\vec{b}$ is at θ measured counterclockwise from $\vec{a}$. We wish to verify that

$$r^2 = a^2 + b^2 + 2ab\cos\theta$$

where $a = |\vec{a}| = a_x$ (we'll call it a for simplicity) and $b = |\vec{b}| = \sqrt{b_x^2 + b_y^2}$. Since $\vec{r} = \vec{a} + \vec{b}$ then $r = |\vec{r}| = \sqrt{(a + b_x)^2 + b_y^2}$. Thus, the above expression becomes

$$\begin{aligned}
\left(\sqrt{(a + b_x)^2 + b_y^2}\right)^2 &= a^2 + \left(\sqrt{b_x^2 + b_y^2}\right)^2 + 2ab\cos\theta \\
a^2 + 2ab_x + b_x^2 + b_y^2 &= a^2 + b_x^2 + b_y^2 + 2ab\cos\theta
\end{aligned}$$

which makes a valid equality since (the last term) $2ab\cos\theta$ is indeed the same as $2ab_x$ (on the left-hand side). In a *later* section, the scalar (dot) product of vectors is presented and this result can be revisited with a somewhat different-looking derivation.

26. The vector equation is $\vec{R} = \vec{A} + \vec{B} + \vec{C} + \vec{D}$. Expressing $\vec{B}$ and $\vec{D}$ in unit-vector notation, we have $1.69\hat{\mathrm{i}} + 3.63\hat{\mathrm{j}}$ and $-2.87\hat{\mathrm{i}} + 4.10\hat{\mathrm{j}}$, respectively. Where the length unit is not displayed in the solution below, the unit meter should be understood.

    (a) Adding corresponding components, we obtain $\vec{R} = -3.18\hat{\mathrm{i}} + 4.72\hat{\mathrm{j}}$.

    (b) and (c) Converting this result to polar coordinates (using Eq. 3-6 or functions on a vector-capable calculator), we obtain

$$(-3.18, 4.72) \longrightarrow (5.69\ \angle\ 124°)$$

    which tells us the magnitude is 5.69 m and the angle (measured counterclockwise from $+x$ axis) is 124°.

27. (a) There are 4 such lines, one from each of the corners on the lower face to the diametrically opposite corner on the upper face. One is shown on the diagram. Using an xyz coordinate system as shown (with the origin at the back lower left corner) The position vector of the "starting point" of the diagonal shown is $a\,\hat{\text{i}}$ and the position vector of the ending point is $a\,\hat{\text{j}} + a\,\hat{\text{k}}$, so the vector along the line is the difference $a\,\hat{\text{j}} + a\,\hat{\text{k}} - a\,\hat{\text{i}}$.

The point diametrically opposite the origin has position vector $a\,\hat{\text{i}} + a\,\hat{\text{j}} + a\,\hat{\text{k}}$ and this is the vector along the diagonal. Another corner of the bottom face is at $a\,\hat{\text{i}} + a\,\hat{\text{j}}$ and the diametrically opposite corner is at $a\,\hat{\text{k}}$, so another cube diagonal is $a\,\hat{\text{k}} - a\,\hat{\text{i}} - a\,\hat{\text{j}}$. The fourth diagonal runs from $a\,\hat{\text{j}}$ to $a\,\hat{\text{i}} + a\,\hat{\text{k}}$, so the vector along the diagonal is $a\,\hat{\text{i}} + a\,\hat{\text{k}} - a\,\hat{\text{j}}$.

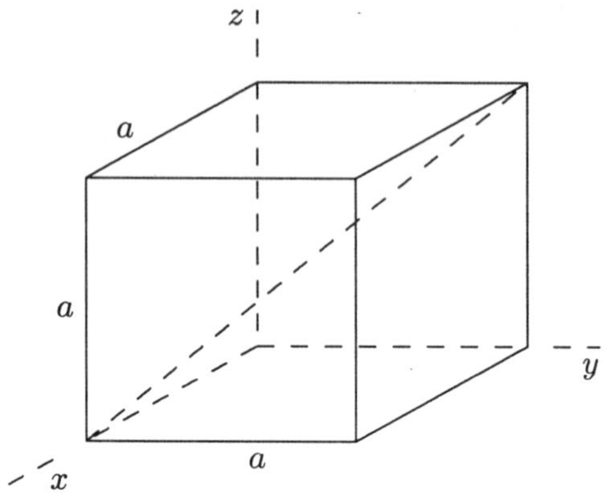

(b) Consider the vector from the back lower left corner to the front upper right corner. It is $a\,\hat{\text{i}} + a\,\hat{\text{j}} + a\,\hat{\text{k}}$. We may think of it as the sum of the vector $a\,\hat{\text{i}}$ parallel to the x axis and the vector $a\,\hat{\text{j}} + a\,\hat{\text{k}}$ perpendicular to the x axis. The tangent of the angle between the vector and the x axis is the perpendicular component divided by the parallel component. Since the magnitude of the perpendicular component is $\sqrt{a^2 + a^2} = a\sqrt{2}$ and the magnitude of the parallel component is a, $\tan\theta = \left(a\sqrt{2}\right)/a = \sqrt{2}$. Thus $\theta = 54.7°$. The angle between the vector and each of the other two adjacent sides (the y and z axes) is the same as is the angle between any of the other diagonal vectors and any of the cube sides adjacent to them.

(c) The length of any of the diagonals is given by $\sqrt{a^2 + a^2 + a^2} = a\sqrt{3}$.

28. Reference to Figure 3-18 (and the accompanying material in that section) is helpful. If we convert $\vec{B}$ to the magnitude-angle notation (as $\vec{A}$ already is) we have $\vec{B} = (14.4\ \angle\ 33.7°)$ (appropriate notation especially if we are using a vector capable calculator in polar mode). Where the length unit is not displayed in the solution, the unit meter should be understood. In the magnitude-angle notation, rotating the axis by $+20°$ amounts to subtracting that angle from the angles previously specified. Thus, $\vec{A} = (12.0\ \angle 40.0°)'$ and $\vec{B} = (14.4\ \angle\ 13.7°)'$, where the 'prime' notation indicates that the description is in terms of the new coordinates. Converting these results to (x, y) representations, we obtain

$$\begin{aligned} \vec{A} &= 9.19\hat{\text{i}}' + 7.71\hat{\text{j}}' \\ \vec{B} &= 14.0\hat{\text{i}}' + 3.41\hat{\text{j}}' \end{aligned}$$

with the unit meter understood, as already mentioned.

29. We apply Eq. 3-20 and Eq. 3-27.

(a) The scalar (dot) product of the two vectors is $\vec{a} \cdot \vec{b} = ab\cos\phi = (10)(6.0)\cos 60° = 30$.

(b) The magnitude of the vector (cross) product of the two vectors is $|\vec{a}\times\vec{b}| = ab\sin\phi = (10)(6.0)\sin 60° = 52$.

30. We consider all possible products and then simplify using relations such as $\hat{\text{i}} \cdot \hat{\text{k}} = 0$ and $\hat{\text{i}} \cdot \hat{\text{i}} = 1$. Thus,

$$\begin{aligned} \vec{a} \cdot \vec{b} &= \left(a_x\,\hat{\text{i}} + a_y\,\hat{\text{j}} + a_z\,\hat{\text{k}}\right) \cdot \left(b_x\,\hat{\text{i}} + b_y\,\hat{\text{j}} + b_z\,\hat{\text{k}}\right) \\ &= a_x b_x\,\hat{\text{i}} \cdot \hat{\text{i}} + a_x b_y\,\hat{\text{i}} \cdot \hat{\text{j}} + a_x b_z\,\hat{\text{i}} \cdot \hat{\text{k}} + a_y b_x\,\hat{\text{j}} \cdot \hat{\text{i}} + a_y b_j\,\hat{\text{j}} \cdot \hat{\text{j}} + \cdots \\ &= a_x b_x(1) + a_x b_y(0) + a_x b_z(0) + a_y b_x(0) + a_y b_j(1) + \cdots \end{aligned}$$

which is seen to reduce to the desired result (one might wish to show this in two dimensions before tackling the additional tedium of working with these three-component vectors).

31. Since $ab\cos\phi = a_x b_x + a_y b_y + a_z b_z$,

$$\cos\phi = \frac{a_x b_x + a_y b_y + a_z b_z}{ab} .$$

The magnitudes of the vectors given in the problem are

$$\begin{aligned} a = |\vec{a}| &= \sqrt{(3.0)^2 + (3.0)^2 + (3.0)^2} = 5.2 \\ b = |\vec{b}| &= \sqrt{(2.0)^2 + (1.0)^2 + (3.0)^2} = 3.7 . \end{aligned}$$

The angle between them is found from

$$\cos\phi = \frac{(3.0)(2.0) + (3.0)(1.0) + (3.0)(3.0)}{(5.2)(3.7)} = 0.926 .$$

The angle is $\phi = 22°$.

32. We consider all possible products and then simplify using relations such as $\hat{\text{i}} \times \hat{\text{i}} = 0$ and the important fundamental products

$$\begin{aligned} \hat{\text{i}} \times \hat{\text{j}} = -\hat{\text{j}} \times \hat{\text{i}} &= \hat{\text{k}} \\ \hat{\text{j}} \times \hat{\text{k}} = -\hat{\text{k}} \times \hat{\text{j}} &= \hat{\text{i}} \\ \hat{\text{k}} \times \hat{\text{i}} = -\hat{\text{i}} \times \hat{\text{k}} &= \hat{\text{j}} . \end{aligned}$$

Thus,

$$\begin{aligned} \vec{a} \times \vec{b} &= \left(a_x\,\hat{\text{i}} + a_y\,\hat{\text{j}} + a_z\,\hat{\text{k}}\right) \times \left(b_x\,\hat{\text{i}} + b_y\,\hat{\text{j}} + b_z\,\hat{\text{k}}\right) \\ &= a_x b_x\,\hat{\text{i}} \times \hat{\text{i}} + a_x b_y\,\hat{\text{i}} \times \hat{\text{j}} + a_x b_z\,\hat{\text{i}} \times \hat{\text{k}} + a_y b_x\,\hat{\text{j}} \times \hat{\text{i}} + a_y b_j\,\hat{\text{j}} \times \hat{\text{j}} + \cdots \\ &= a_x b_x(0) + a_x b_y(\hat{\text{k}}) + a_x b_z(-\hat{\text{j}}) + a_y b_x(-\hat{\text{k}}) + a_y b_j(0) + \cdots \end{aligned}$$

which is seen to simplify to the desired result.

33. The area of a triangle is half the product of its base and altitude. The base is the side formed by vector $\vec{a}$. Then the altitude is $b\sin\phi$ and the area is $A = \frac{1}{2}ab\sin\phi = \frac{1}{2}|\vec{a} \times \vec{b}|$.

34. Applying Eq. 3-23, $\vec{F} = q\vec{v} \times \vec{B}$ (where q is a scalar) becomes

$$F_x\,\hat{\text{i}} + F_y\,\hat{\text{j}} + F_z\,\hat{\text{k}} = q\left(v_y B_z - v_z B_y\right)\hat{\text{i}} + q\left(v_z B_x - v_x B_z\right)\hat{\text{j}} + q\left(v_x B_y - v_y B_x\right)\hat{\text{k}}$$

which – plugging in values – leads to three equalities:

$$\begin{aligned} 4.0 &= 2\,(4.0B_z - 6.0B_y) \\ -20 &= 2\,(6.0B_x - 2.0B_z) \\ 12 &= 2\,(2.0B_y - 4.0B_x) \end{aligned}$$

Since we are told that $B_x = B_y$, the third equation leads to $B_y = -3.0$. Inserting this value into the first equation, we find $B_z = -4.0$. Thus, our answer is

$$\vec{B} = -3.0\,\hat{\text{i}} - 3.0\,\hat{\text{j}} - 4.0\,\hat{\text{k}} .$$

35. Both proofs shown below utilize the fact that the vector (cross) product of $\vec{a}$ and $\vec{b}$ is perpendicular to both $\vec{a}$ and $\vec{b}$. This is mentioned in the book, and is fundamental to its discussion of the right-hand rule.

(a) $(\vec{b} \times \vec{a})$ is a vector that is perpendicular to $\vec{a}$, so the scalar product of $\vec{a}$ with this vector is zero. This can also be verified by using Eq. 3-30, and then (with suitable notation changes) Eq. 3-23.

(b) Let $\vec{c} = \vec{b} \times \vec{a}$. Then the magnitude of $\vec{c}$ is $c = ab\sin\phi$. Since $\vec{c}$ is perpendicular to $\vec{a}$ the magnitude of $\vec{a} \times \vec{c}$ is ac. The magnitude of $\vec{a} \times (\vec{b} \times \vec{a})$ is consequently $|\vec{a} \times (\vec{b} \times \vec{a})| = ac = a^2 b \sin\phi$. This too can be verified by repeated application of Eq. 3-30, although it must be admitted that this is much less intimidating if one is using a math software package such as MAPLE or Mathematica.

36. If a vector capable calculator is used, this makes a good exercise for getting familiar with those features. Here we briefly sketch the method. Eq. 3-30 leads to

$$2\vec{A} \times \vec{B} = 2\left(2\hat{\mathrm{i}} + 3\hat{\mathrm{j}} - 4\hat{\mathrm{k}}\right) \times \left(-3\hat{\mathrm{i}} + 4\hat{\mathrm{j}} + 2\hat{\mathrm{k}}\right) = 44\hat{\mathrm{i}} + 16\hat{\mathrm{j}} + 34\hat{\mathrm{k}}\ .$$

We now apply Eq. 3-23 to evaluate $3\vec{C} \cdot \left(2\vec{A} \times \vec{B}\right)$:

$$3\left(7\hat{\mathrm{i}} - 8\hat{\mathrm{j}}\right) \cdot \left(44\hat{\mathrm{i}} + 16\hat{\mathrm{j}} + 34\hat{\mathrm{k}}\right) = 3\left((7)(44) + (-8)(16) + (0)(34)\right) = 540\ .$$

37. From the figure, we note that $\vec{c} \perp \vec{b}$, which implies that the angle between $\vec{c}$ and the $+x$ axis is 120°.

(a) Direct application of Eq. 3-5 yields the answers for this and the next few parts. $a_x = a\cos 0° = a = 3.00$ m.

(b) $a_y = a\sin 0° = 0$.

(c) $b_x = b\cos 30° = (4.00\,\mathrm{m})\cos 30° = 3.46$ m.

(d) $b_y = b\sin 30° = (4.00\,\mathrm{m})\sin 30° = 2.00$ m.

(e) $c_x = c\cos 120° = (10.0\,\mathrm{m})\cos 120° = -5.00$ m.

(f) $c_y = c\sin 30° = (10.0\,\mathrm{m})\sin 120° = 8.66$ m.

(g) In terms of components (first x and then y), we must have

$$\begin{aligned} -5.00\,\mathrm{m} &= p(3.00\,\mathrm{m}) + q(3.46\,\mathrm{m}) \\ 8.66\,\mathrm{m} &= p(0) + q(2.00\,\mathrm{m})\ . \end{aligned}$$

Solving these equations, we find $p = -6.67$

(h) and $q = 4.33$ (note that it's easiest to solve for q first). The numbers p and q have no units.

38. We apply Eq. 3-20 with Eq. 3-23. Where the length unit is not displayed, the unit meter is understood.

(a) We first note that $a = |\vec{a}| = \sqrt{3.2^2 + 1.6^2} = 3.58$ m and $b = |\vec{b}| = \sqrt{0.5^2 + 4.5^2} = 4.53$ m. Now,

$$\begin{aligned} \vec{a} \cdot \vec{b} = a_x b_x + a_y b_y &= ab\cos\phi \\ (3.2)(0.5) + (1.6)(4.5) &= (3.58)(4.53)\cos\phi \end{aligned}$$

which leads to $\phi = 57°$ (the inverse cosine is double-valued as is the inverse tangent, but we know this is the right solution since both vectors are in the same quadrant).

(b) Since the angle (measured from $+x$) for $\vec{a}$ is $\tan^{-1}(1.6/3.2) = 26.6°$, we know the angle for $\vec{c}$ is $26.6° - 90° = -63.4°$ (the other possibility, $26.6° + 90°$ would lead to a $c_x < 0$). Therefore, $c_x = c\cos -63.4° = (5.0)(0.45) = 2.2$ m.

(c) Also, $c_y = c\sin -63.4° = (5.0)(-0.89) = -4.5$ m.

(d) And we know the angle for $\vec{d}$ to be $26.6° + 90° = 116.6°$, which leads to $d_x = d\cos 116.6° = (5.0)(-0.45) = -2.2$ m.

(e) Finally, $d_y = d\sin 116.6° = (5.0)(0.89) = 4.5$ m.

39. The solution to problem 27 showed that each diagonal has a length given by $a\sqrt{3}$, where a is the length of a cube edge. Vectors along two diagonals are $\vec{b} = a\,\hat{\imath} + a\,\hat{\jmath} + a\,\hat{k}$ and $\vec{c} = -a\,\hat{\imath} + a\,\hat{\jmath} + a\,\hat{k}$. Using Eq. 3-20 with Eq. 3-23, we find the angle between them:

$$\cos\phi = \frac{b_x c_x + b_y c_y + b_z c_z}{bc} = \frac{-a^2 + a^2 + a^2}{3a^2} = \frac{1}{3} .$$

The angle is $\phi = \cos^{-1}(1/3) = 70.5°$.

40. (a) The vector equation $\vec{r} = \vec{a} - \vec{b} - \vec{v}$ is computed as follows: $(5.0 - (-2.0) + 4.0)\hat{\imath} + (4.0 - 2.0 + 3.0)\hat{\jmath} + ((-6.0) - 3.0 + 2.0)\hat{k}$. This leads to $\vec{r} = 11\,\hat{\imath} + 5.0\,\hat{\jmath} - 7.0\,\hat{k}$.

(b) We find the angle from $+z$ by "dotting" (taking the scalar product) $\vec{r}$ with $\hat{k}$. Noting that $r = |\vec{r}| = \sqrt{11^2 + 5^2 + (-7)^2} = 14$, Eq. 3-20 with Eq. 3-23 leads to

$$\vec{r}\cdot\hat{k} = -7.0 = (14)(1)\cos\phi \implies \phi = 120° .$$

(c) To find the component of a vector in a certain direction, it is efficient to "dot" it (take the scalar product of it) with a unit-vector in that direction. In this case, we make the desired unit-vector by

$$\hat{b} = \frac{\vec{b}}{|\vec{b}|} = \frac{-2\,\hat{\imath} + 2\,\hat{\jmath} + 3\,\hat{k}}{\sqrt{(-2)^2 + 2^2 + 3^2}} .$$

We therefore obtain

$$a_b = \vec{a}\cdot\hat{b} = \frac{(5)(-2) + (4)(2) + (-6)(3)}{\sqrt{(-2)^2 + 2^2 + 3^2}} = -4.9 .$$

(d) One approach (if we all we require is the magnitude) is to use the vector cross product, as the problem suggests; another (which supplies more information) is to subtract the result in part (c) (multiplied by $\hat{b}$) from $\vec{a}$. We briefly illustrate both methods. We note that if $a\cos\theta$ (where θ is the angle between $\vec{a}$ and $\vec{b}$) gives a_b (the component along $\hat{b}$) then we expect $a\sin\theta$ to yield the orthogonal component:

$$a\sin\theta = \frac{|\vec{a}\times\vec{b}|}{b} = 7.3$$

(alternatively, one might compute θ form part (c) and proceed more directly). The second method proceeds as follows:

$$\begin{aligned}\vec{a} - a_b\,\hat{b} &= (5.0 - 2.35)\hat{\imath} + (4.0 - (-2.35))\hat{\jmath} + ((-6.0) - (-3.53))\hat{k} \\ &= 2.65\,\hat{\imath} + 6.35\,\hat{\jmath} - 2.47\,\hat{k}\end{aligned}$$

This describes the perpendicular part of $\vec{a}$ completely. To find the magnitude of this part, we compute

$$\sqrt{2.65^2 + 6.35^2 + (-2.47)^2} = 7.3$$

which agrees with the first method.

41. The volume of a parallelepiped is equal to the product of its altitude and the area of its base. Take the base to be the parallelogram formed by the vectors $\vec{b}$ and $\vec{c}$. Its area is $bc\sin\phi$, where ϕ is the angle between $\vec{b}$ and $\vec{c}$. This is just the magnitude of the vector (cross) product $\vec{b}\times\vec{c}$. The altitude of the parallelepiped is $a\cos\theta$, where θ is the angle between $\vec{a}$ and the normal to the plane of $\vec{b}$ and $\vec{c}$. Since $\vec{b}\times\vec{c}$ is normal to that plane, θ is the angle between $\vec{a}$ and $\vec{b}\times\vec{c}$. Thus, the volume of the parallelepiped is $V = a|\vec{b}\times\vec{c}|\cos\theta = \vec{a}\cdot(\vec{b}\times\vec{c})$.

42. We apply Eq. 3-30 and Eq. 3-23.

(a) $\vec{a}\times\vec{b} = (a_x b_y - a_y b_x)\,\hat{k}$ since all other terms vanish, due to the fact that neither $\vec{a}$ nor $\vec{b}$ have any z components. Consequently, we obtain $((3.0)(4.0) - (5.0)(2.0))\,\hat{k} = 2.0\,\hat{k}$.

(b) $\vec{a}\cdot\vec{b} = a_x b_x + a_y b_y$ yields $(3)(2) + (5)(4) = 26$.

(c) $\vec{a}+\vec{b} = (3+2)\,\hat{\mathrm{i}} + (5+4)\,\hat{\mathrm{j}}$, so that $\left(\vec{a}+\vec{b}\right)\cdot\vec{b} = (5)(2) + (9)(4) = 46$.

(d) Several approaches are available. In this solution, we will construct a $\hat{b}$ unit-vector and "dot" it (take the scalar product of it) with $\vec{a}$. In this case, we make the desired unit-vector by

$$\hat{b} = \frac{\vec{b}}{|\vec{b}|} = \frac{2\,\hat{\mathrm{i}} + 4\,\hat{\mathrm{j}}}{\sqrt{2^2+4^2}} \; .$$

We therefore obtain

$$a_b = \vec{a}\cdot\hat{b} = \frac{(3)(2)+(5)(4)}{\sqrt{2^2+4^2}} = 5.8 \; .$$

43. We apply Eq. 3-30 and Eq.3-23. If a vector capable calculator is used, this makes a good exercise for getting familiar with those features. Here we briefly sketch the method.

(a) We note that $\vec{b}\times\vec{c} = -8\,\hat{\mathrm{i}} + 5\,\hat{\mathrm{j}} + 6\,\hat{\mathrm{k}}$. Thus, $\vec{a}\cdot\left(\vec{b}\times\vec{c}\right) = (3)(-8) + (3)(5) + (-2)(6) = -21$.

(b) We note that $\vec{b}+\vec{c} = 1\,\hat{\mathrm{i}} - 2\,\hat{\mathrm{j}} + 3\,\hat{\mathrm{k}}$. Thus, $\vec{a}\cdot\left(\vec{b}+\vec{c}\right) = (3)(1) + (3)(-2) + (-2)(3) = -9$.

(c) Finally, $\vec{a}\times\left(\vec{b}+\vec{c}\right) = ((3)(3)-(-2)(-2))\,\hat{\mathrm{i}} + ((-2)(1)-(3)(3))\,\hat{\mathrm{j}} + ((3)(-2)-(3)(1))\,\hat{\mathrm{k}} = 5\,\hat{\mathrm{i}} - 11\,\hat{\mathrm{j}} - 9\,\hat{\mathrm{k}}$.

44. The components of $\vec{a}$ are $a_x = 0$, $a_y = 3.20\cos 63° = 1.45$, and $a_z = 3.20\sin 63° = 2.85$. The components of $\vec{b}$ are $b_x = 1.40\cos 48° = 0.937$, $b_y = 0$, and $b_z = 1.40\sin 48° = 1.04$.

(a) The scalar (dot) product is therefore

$$\vec{a}\cdot\vec{b} = a_x b_x + a_y b_y + a_z b_z = (0)(0.937) + (1.45)(0) + (2.85)(1.04) = 2.97 \; .$$

(b) The vector (cross) product is

$$\begin{aligned}\vec{a}\times\vec{b} &= (a_y b_z - a_z b_y)\,\hat{\mathrm{i}} + (a_z b_x - a_x b_z)\,\hat{\mathrm{j}} + (a_x b_y - a_y b_x)\,\hat{\mathrm{k}} \\ &= ((1.45)(1.04) - 0))\,\hat{\mathrm{i}} + ((2.85)(0.937) - 0))\,\hat{\mathrm{j}} + (0 - (1.45)(0.94))\,\hat{\mathrm{k}} \\ &= 1.51\,\hat{\mathrm{i}} + 2.67\,\hat{\mathrm{j}} - 1.36\,\hat{\mathrm{k}} \; .\end{aligned}$$

(c) The angle θ between $\vec{a}$ and $\vec{b}$ is given by

$$\theta = \cos^{-1}\left(\frac{\vec{a}\cdot\vec{b}}{ab}\right) = \cos^{-1}\left(\frac{2.96}{(3.30)(1.40)}\right) = 48° \; .$$

45. We observe that $\left|\hat{\mathrm{i}}\times\hat{\mathrm{i}}\right| = \left|\hat{\mathrm{i}}\right|\left|\hat{\mathrm{i}}\right|\sin 0°$ vanishes because $\sin 0° = 0$. Similarly, $\hat{\mathrm{j}}\times\hat{\mathrm{j}} = \hat{\mathrm{k}}\times\hat{\mathrm{k}} = 0$. When the unit vectors are perpendicular, we have to do a little more work to show the cross product results. First, the magnitude of the vector $\hat{\mathrm{i}}\times\hat{\mathrm{j}}$ is

$$\left|\hat{\mathrm{i}}\times\hat{\mathrm{j}}\right| = \left|\hat{\mathrm{i}}\right|\left|\hat{\mathrm{j}}\right|\sin 90°$$

which equals 1 because $\sin 90° = 1$ and these are all units vectors (each has magnitude equal to 1). This is consistent with the claim that $\hat{\mathrm{i}}\times\hat{\mathrm{j}} = \hat{\mathrm{k}}$ since the magnitude of $\hat{\mathrm{k}}$ is certainly 1. Now, we use the right-hand rule to show that $\hat{\mathrm{i}}\times\hat{\mathrm{j}}$ is in the positive z direction. Thus $\hat{\mathrm{i}}\times\hat{\mathrm{j}}$ has the same magnitude and direction as $\hat{\mathrm{k}}$, so it is equal to $\hat{\mathrm{k}}$. Similarly, $\hat{\mathrm{k}}\times\hat{\mathrm{i}} = \hat{\mathrm{j}}$ and $\hat{\mathrm{j}}\times\hat{\mathrm{k}} = \hat{\mathrm{i}}$. If, however, the coordinate system is left-handed, we replace $\hat{\mathrm{k}}\to-\hat{\mathrm{k}}$ in the work we have shown above and get

$$\hat{\mathrm{i}}\times\hat{\mathrm{i}} = \hat{\mathrm{j}}\times\hat{\mathrm{j}} = \hat{\mathrm{k}}\times\hat{\mathrm{k}} = 0 \; .$$

just as before. But the relations that are different are

$$\hat{\mathrm{i}}\times\hat{\mathrm{j}} = -\hat{\mathrm{k}} \quad \hat{\mathrm{k}}\times\hat{\mathrm{i}} = -\hat{\mathrm{j}} \quad \hat{\mathrm{j}}\times\hat{\mathrm{k}} = -\hat{\mathrm{i}} \; .$$

46. (a) By the right-hand rule, $\vec{A} \times \vec{B}$ points upward if $\vec{A}$ points north and $\vec{B}$ points west. If $\vec{A}$ and $\vec{B}$ have magnitude $= 1$ then, by Eq. 3-27, the result also has magnitude equal to 1.
(b) Since $\cos 90° = 0$, the scalar dot product between perpendicular vectors is zero. Thus, $\vec{A} \cdot \vec{B} = 0$ is $\vec{A}$ points down and $\vec{B}$ points south.
(c) By the right-hand rule, $\vec{A} \times \vec{B}$ points south if $\vec{A}$ points east and $\vec{B}$ points up. If $\vec{A}$ and $\vec{B}$ have unit magnitude then, by Eq. 3-27, the result also has unit magnitude.
(d) Since $\cos 0° = 1$, then $\vec{A} \cdot \vec{B} = AB$ (where A is the magnitude of $\vec{A}$ and B is the magnitude of $\vec{B}$). If, additionally, we have $A = B = 1$, then the result is 1.
(e) Since $\sin 0° = 0$, $\vec{A} \times \vec{B} = 0$ if both $\vec{A}$ and $\vec{B}$ point south.

47. Let A denote the magnitude of $\vec{A}$; similarly for the other vectors. The vector equation is $\vec{A} + \vec{B} = \vec{C}$ where $B = 8.0$ m and $C = 2A$. We are also told that the angle (measured in the 'standard' sense) for $\vec{A}$ is 0° and the angle for $\vec{C}$ is 90°, which makes this a right triangle (when drawn in a "head-to-tail" fashion) where B is the size of the hypotenuse. Using the Pythagorean theorem,

$$B = \sqrt{A^2 + C^2} \quad \Longrightarrow \quad 8.0 = \sqrt{A^2 + 4A^2}$$

which leads to $A = 8/\sqrt{5} = 3.6$ m.

48. We choose $+x$ east and $+y$ north and measure all angles in the "standard" way (positive ones are counterclockwise from $+x$). Thus, vector $\vec{d_1}$ has magnitude $d_1 = 4$ (with the unit meter and three significant figures assumed) and direction $\theta_1 = 225°$. Also, $\vec{d_2}$ has magnitude $d_2 = 5$ and direction $\theta_2 = 0°$, and vector $\vec{d_3}$ has magnitude $d_3 = 6$ and direction $\theta_3 = 60°$.

(a) The x-component of $\vec{d_1}$ is $d_1 \cos\theta_1 = -2.83$ m.
(b) The y-component of $\vec{d_1}$ is $d_1 \sin\theta_1 = -2.83$ m.
(c) The x-component of $\vec{d_2}$ is $d_2 \cos\theta_2 = 5.00$ m.
(d) The y-component of $\vec{d_2}$ is $d_2 \sin\theta_2 = 0$.
(e) The x-component of $\vec{d_3}$ is $d_3 \cos\theta_3 = 3.00$ m.
(f) The y-component of $\vec{d_3}$ is $d_3 \sin\theta_3 = 5.20$ m.
(g) The sum of x-components is $-2.83 + 5.00 + 3.00 = 5.17$ m.
(h) The sum of y-components is $-2.83 + 0 + 5.20 = 2.37$ m.
(i) The magnitude of the resultant displacement is $\sqrt{5.17^2 + 2.37^2} = 5.69$ m.
(j) And its angle is $\theta = \tan^{-1}(2.37/5.17) = 24.6°$ which (recalling our coordinate choices) means it points at about 25° north of east.
(k) and (ℓ) This new displacement (the direct line home) when vectorially added to the previous (net) displacement must give zero. Thus, the new displacement is the negative, or opposite, of the previous (net) displacement. That is, it has the same magnitude (5.69 m) but points in the opposite direction (25° south of west).

49. Reading carefully, we see that the (x, y) specifications for each "dart" are to be interpreted as $(\Delta x, \Delta y)$ descriptions of the corresponding displacement vectors. We combine the different parts of this problem into a single exposition. Thus, along the x axis, we have (with the centimeter unit understood)

$$30.0 + b_x - 20.0 - 80.0 = -140.0 \ ,$$

and along y axis we have

$$40.0 - 70.0 + c_y - 70.0 = -20.0 \ .$$

Hence, we find $b_x = -70.0$ cm and $c_y = 80.0$ cm. And we convert the final location $(-140, -20)$ into polar coordinates and obtain $(141 \angle -172°)$, an operation quickly done using a vector capable calculator in polar mode. Thus, the ant is 141 cm from where it started at an angle of $-172°$, which means 172° clockwise from the $+x$ axis or 188° counterclockwise from the $+x$ axis.

50. We find the components and then add them (as scalars, not vectors). With $d = 3.40$ km and $\theta = 35.0°$ we find $d\cos\theta + d\sin\theta = 4.74$ km.

51. We choose $+x$ east and $+y$ north and measure all angles in the "standard" way (positive ones counterclockwise from $+x$, negative ones clockwise). Thus, vector $\vec{d_1}$ has magnitude $d_1 = 3.66$ (with the unit meter and three significant figures assumed) and direction $\theta_1 = 90°$. Also, $\vec{d_2}$ has magnitude $d_2 = 1.83$ and direction $\theta_2 = -45°$, and vector $\vec{d_3}$ has magnitude $d_3 = 0.91$ and direction $\theta_3 = -135°$. We add the x and y components, respectively:

$$x: \quad d_1\cos\theta_1 + d_2\cos\theta_2 + d_3\cos\theta_3 = 0.651 \text{ m}$$
$$y: \quad d_1\sin\theta_1 + d_2\sin\theta_2 + d_3\sin\theta_3 = 1.723 \text{ m} .$$

(a) The magnitude of the direct displacement (the vector sum $\vec{d_1}+\vec{d_2}+\vec{d_3}$) is $\sqrt{0.651^2 + 1.723^2} = 1.84$ m.

(b) The angle (understood in the sense described above) is $\tan^{-1}(1.723/0.651) = 69°$. That is, the first putt must aim in the direction 69° north of east.

52. (a) We write $\vec{b} = b\hat{\mathrm{j}}$ where $b > 0$. We are asked to consider

$$\frac{\vec{b}}{d} = \left(\frac{b}{d}\right)\hat{\mathrm{j}}$$

in the case $d > 0$. Since the coefficient of $\hat{\mathrm{j}}$ is positive, then the vector points in the $+y$ direction.

(b) If, however, $d < 0$, then the coefficient is negative and the vector points in the $-y$ direction.

(c) Since $\cos 90° = 0$, then $\vec{a}\cdot\vec{b} = 0$, using Eq. 3-20.

(d) Since $\vec{b}/d$ is along the y axis, then (by the same reasoning as in the previous part) $\vec{a}\cdot(\vec{b}/d) = 0$.

(e) By the right-hand rule, $\vec{a}\times\vec{b}$ points in the $+z$ direction.

(f) By the same rule, $\vec{b}\times\vec{a}$ points in the $-z$ direction. We note that $\vec{b}\times\vec{a} = -\vec{a}\times\vec{b}$ is true in this case and quite generally.

(g) Since $\sin 90° = 1$, Eq. 3-27 gives $|\vec{a}\times\vec{b}| = ab$ where a is the magnitude of $\vec{a}$. Also, $|\vec{a}\times\vec{b}| = |\vec{b}\times\vec{a}|$ so both results have the same magnitude.

(h) and (i) With $d > 0$, we find that $\vec{a}\times(\vec{b}/d)$ has magnitude ab/d and is pointed in the $+z$ direction.

53. (a) With $a = 17.0$ m and $\theta = 56.0°$ we find $a_x = a\cos\theta = 9.51$ m.

(b) And $a_y = a\sin\theta = 14.1$ m.

(c) The angle relative to the new coordinate system is $\theta' = 56 - 18 = 38°$. Thus, $a'_x = a\cos\theta' = 13.4$ m.

(d) And $a'_y = a\sin\theta' = 10.5$ m.

54. Since $\cos 0° = 1$ and $\sin 0° = 0$, these follows immediately from Eq. 3-20 and Eq. 3-27.

55. (a) The magnitude of the vector $\vec{a} = 4.0\vec{d}$ is $(4.0)(2.5) = 10$ m.

(b) The direction of the vector $\vec{a} = 4.0\vec{d}$ is the same as the direction of $\vec{d}$ (north).

(c) The magnitude of the vector $\vec{c} = -3.0\vec{d}$ is $(3.0)(2.5) = 7.5$ m.

(d) The direction of the vector $\vec{c} = -3.0\vec{d}$ is the opposite of the direction of $\vec{d}$. Thus, the direction of $\vec{c}$ is south.

56. The vector sum of the displacements $\vec{d}_{\text{storm}}$ and $\vec{d}_{\text{new}}$ must give the same result as its originally intended displacement $\vec{d}_o = 120\hat{\mathrm{j}}$ where east is $\hat{\mathrm{i}}$, north is $\hat{\mathrm{j}}$, and the assumed length unit is km. Thus, we write

$$\vec{d}_{\text{storm}} = 100\hat{\mathrm{i}} \quad \text{and} \quad \vec{d}_{\text{new}} = A\hat{\mathrm{i}} + B\hat{\mathrm{j}} .$$

(a) The equation $\vec{d}_{\text{storm}} + \vec{d}_{\text{new}} = \vec{d}_o$ readily yields $A = -100$ km and $B = 120$ km. The magnitude of $\vec{d}_{\text{new}}$ is therefore $\sqrt{A^2 + B^2} = 156$ km.

(b) And its direction is $\tan^{-1}(B/A) = -50.2°$ or $180° + (-50.2°) = 129.8°$. We choose the latter value since it indicates a vector pointing in the second quadrant, which is what we expect here. The answer can be phrased several equivalent ways: 129.8° counterclockwise from east, or 39.8° west from north, or 50.2° north from west.

57. (a) The height is $h = d \sin\theta$, where $d = 12.5$ m and $\theta = 20.0°$. Therefore, $h = 4.28$ m.

(b) The horizontal distance is $d \cos\theta = 11.7$ m.

58. (a) We orient $\hat{\imath}$ eastward, $\hat{\jmath}$ northward, and $\hat{k}$ upward. The displacement in meters is consequently $1000\,\hat{\imath} + 2000\,\hat{\jmath} - 500\,\hat{k}$.

(b) The net displacement is zero since his final position matches his initial position.

59. (a) If we add the equations, we obtain $2\vec{a} = 6\vec{c}$, which leads to $\vec{a} = 3\vec{c} = 9\,\hat{\imath} + 12\,\hat{\jmath}$.

(b) Plugging this result back in, we find $\vec{b} = \vec{c} = 3\,\hat{\imath} + 4\,\hat{\jmath}$.

60. (First problem in **Cluster 1**)
The given angle $\theta = 130°$ is assumed to be measured counterclockwise from the $+x$ axis. Angles (if positive) in our results follow the same convention (but if negative are clockwise from $+x$).

(a) With $A = 4.00$, the x-component of $\vec{A}$ is $A\cos\theta = -2.57$.

(b) The y-component of $\vec{A}$ is $A\sin\theta = 3.06$.

(c) Adding $\vec{A}$ and $\vec{B}$ produces a vector we call R with components $R_x = -6.43$ and $R_y = -1.54$. Using Eq. 3-6 (or special functions on a calculator) we present this in magnitude-angle notation: $\vec{R} = (6.61 \angle -167°)$.

(d) From the discussion in the previous part, it is clear that $\vec{R} = -6.43\,\hat{\imath} - 1.54\,\hat{\jmath}$.

(e) The vector $\vec{C}$ is the difference of $\vec{A}$ and $\vec{B}$. In unit-vector notation, this becomes

$$\vec{C} = \vec{A} - \vec{B} = \left(-2.57\,\hat{\imath} - 3.06\,\hat{\jmath}\right) - \left(-3.86\,\hat{\imath} - 4.60\,\hat{\jmath}\right)$$

which yields $\vec{C} = 1.29\,\hat{\imath} + 7.66\,\hat{\jmath}$.

(f) Using Eq. 3-6 (or special functions on a calculator) we present this in magnitude-angle notation: $\vec{C} = (7.77 \angle 80.5°)$.

(g) We note that $\vec{C}$ is the "constant" in all six pictures. Remembering that the negative of a vector simply reverses it, then we see that in form or another, all six pictures express the relation $\vec{C} = \vec{A} - \vec{B}$.

61. (Second problem in **Cluster 1**)

(a) The dot (scalar) product of $3\vec{A}$ and $\vec{B}$ is found using Eq. 3-23:

$$3\vec{A} \cdot \vec{B} = 3\,(4.00\cos 130°)\,(-3.86) + 3\,(4.00\sin 130°)\,(-4.60) = -12.5\ .$$

(b) We call the result $\vec{D}$ and combine the scalars ((3)(4) = 12). Thus, $\vec{D} = (4\vec{A}) \times (3\vec{B})$ becomes, using Eq. 3-30,

$$12\vec{A} \times \vec{B} = 12\,((4.00\cos 130°)\,(-4.60) - (4.00\sin 130°)\,(-3.86))\,\hat{k}$$

which yields $\vec{D} = 284\,\hat{k}$.

(c) Since $\vec{D}$ has magnitude 284 and points in the $+z$ direction, it has radial coordinate 284 and angle-measured-from-z-axis equal to 0°. The angle measured in the xy plane does not have a well-defined value (since this vector does not have a component in that plane)

(d) Since $\vec{A}$ is in the xy plane, then it is clear that $\vec{A} \perp \vec{D}$. The angle between them is 90°.

(e) Calling this new result $\vec{G}$ we have

$$\vec{G} = (4.00\cos 130°)\,\hat{\imath} + (4.00\sin 130°)\,\hat{\jmath} + (3.00)\hat{k}$$

which yields $\vec{G} = -2.57\hat{\imath} + 3.06\hat{\jmath} + 3.00\,\hat{k}$.

(f) It is straightforward using a vector-capable calculator to convert the above into spherical coordinates. We, however, proceed "the hard way", using the notation in Fig. 3-44 (where θ is in the xy plane and ϕ is measured from the z axis):

$$\begin{aligned} |\vec{G}| = r &= \sqrt{(-2.57)^2 + 3.06^2 + 3.00^2} = 5.00 \\ \phi &= \tan^{-1}(4.00/3.00) = 53.1° \\ \theta &= 130° \quad \text{given in problem 60} . \end{aligned}$$

62. (Third problem in **Cluster 1**)

(a) Looking at the xy plane in Fig. 3-44, it is clear that the angle to $\vec{A}$ (which is the vector lying *in* the plane, not the one rising out of it, which we called $\vec{G}$ in the previous problem) measured counterclockwise from the $-y$ axis is $90° + 130° = 220°$. Had we measured this *clockwise* we would obtain (in absolute value) $360° - 220° = 140°$.

(b) We found in part (b) of the previous problem that $\vec{A}\times\vec{B}$ points along the z axis, so it is perpendicular to the $-y$ direction.

(c) Let $\vec{u} = -\hat{\jmath}$ represent the $-y$ direction, and $\vec{w} = 3\,\hat{k}$ is the vector being added to $\vec{B}$ in this problem. The vector being examined in this problem (we'll call it $\vec{Q}$) is, using Eq. 3-30 (or a vector-capable calculator),

$$\vec{Q} = \vec{A} \times \left(\vec{B} + \vec{w}\right) = 9.19\,\hat{\imath} + 7.71\,\hat{\jmath} + 23.7\,\hat{k}$$

and is clearly in the first octant (since all components are positive); using Pythagorean theorem, its magnitude is $Q = 26.52$. From Eq. 3-23, we immediately find $\vec{u} \cdot \vec{Q} = -7.71$. Since $\vec{u}$ has unit magnitude, Eq. 3-20 leads to

$$\cos^{-1}\left(\frac{\vec{u}\cdot\vec{Q}}{Q}\right) = \cos^{-1}\left(\frac{-7.71}{26.52}\right)$$

which yields a choice of angles 107° or −107°. Since we have already observed that $\vec{Q}$ is in the first octant, the the angle measured counterclockwise (as observed by someone high up on the $+z$ axis) from the $-y$ axis to $\vec{Q}$ is 107°.

Chapter 4

1. Where the length unit is not specified (in this solution), the unit meter should be understood.

   (a) The position vector, according to Eq. 4-1, is $\vec{r} = -5.0\,\hat{\imath} + 8.0\,\hat{\jmath}$ (in meters).

   (b) The magnitude is $|\vec{r}| = \sqrt{x^2 + y^2 + z^2} = 9.4$ m.

   (c) Many calculators have polar $\leftrightarrow$ rectangular conversion capabilities which make this computation more efficient than what is shown below. Noting that the vector lies in the xy plane, we are using Eq. 3-6:
$$\tan^{-1}\left(\frac{8.0}{-5.0}\right) = -58^\circ \quad \text{or} \quad 122^\circ$$
   where we choose the latter possibility (122° measured counterclockwise from the $+x$ direction) since the signs of the components imply the vector is in the second quadrant.

   (d) In the interest of saving space, we omit the sketch. The vector is 32° counterclockwise from the $+y$ direction, where the $+y$ direction is assumed to be (as is standard) $+90^\circ$ counterclockwise from $+x$, and the $+z$ direction would therefore be "out of the paper."

   (e) The displacement is $\Delta\vec{r} = \vec{r}\,' - \vec{r}$ where $\vec{r}$ is given in part (a) and $\vec{r}\,' = 3.0\,\hat{\imath}$. Therefore, $\Delta\vec{r} = 8.0\,\hat{\imath} - 8.0\,\hat{\jmath}$ (in meters).

   (f) The magnitude of the displacement is $|\Delta\vec{r}| = \sqrt{8^2 + (-8)^2} = 11$ m.

   (g) The angle for the displacement, using Eq. 3-6, is found from
$$\tan^{-1}\left(\frac{8.0}{-8.0}\right) = -45^\circ \quad \text{or} \quad 135^\circ$$
   where we choose the former possibility (-45°, which means 45° measured clockwise from $+x$, or 315° counterclockwise from $+x$) since the signs of the components imply the vector is in the fourth quadrant.

2. (a) The magnitude of $\vec{r}$ is $\sqrt{5.0^2 + (-3.0)^2 + 2.0^2} = 6.2$ m.

   (b) A sketch is shown. The coordinate values are in meters.

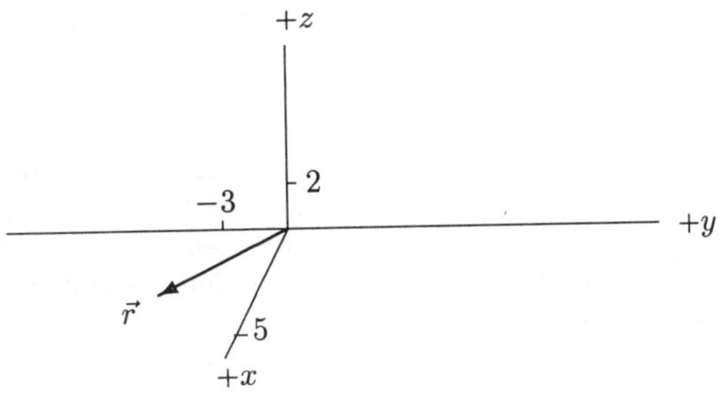

3. Where the unit is not specified, the unit meter is understood. We use Eq. 4-2 and Eq. 4-3.

   (a) With the initial position vector as $\vec{r}_1$ and the later vector as $\vec{r}_2$, Eq. 4-3 yields

   $$\Delta r = ((-2) - 5)\,\hat{\mathrm{i}} + (6 - (-6))\,\hat{\mathrm{j}} + (2 - 2)\,\hat{\mathrm{k}} = -7.0\,\hat{\mathrm{i}} + 12\,\hat{\mathrm{j}}$$

   for the displacement vector in unit-vector notation (in meters).

   (b) Since there is no z component (that is, the coefficient of $\hat{\mathrm{k}}$ is zero), the displacement vector is in the xy plane.

4. We use a coordinate system with $+x$ eastward and $+y$ upward. We note that 123° is the angle between the initial position and later position vectors, so that the angle from $+x$ to the later position vector is $40° + 123° = 163°$. In unit-vector notation, the position vectors are

$$\begin{aligned} \vec{r}_1 &= 360\cos(40°)\,\hat{\mathrm{i}} + 360\sin(40°)\,\hat{\mathrm{j}} = 276\,\hat{\mathrm{i}} + 231\,\hat{\mathrm{j}} \\ \vec{r}_2 &= 790\cos(163°)\,\hat{\mathrm{i}} + 790\sin(163°)\,\hat{\mathrm{j}} = -755\,\hat{\mathrm{i}} + 231\,\hat{\mathrm{j}} \end{aligned}$$

respectively (in meters). Consequently, we plug into Eq. 4-3

$$\Delta r = ((-755) - 276)\,\hat{\mathrm{i}} + (231 - 231)\,\hat{\mathrm{j}}$$

and find the displacement vector is horizontal (westward) with a length of 1.03 km. If unit-vector notation is not a priority in this problem, then the computation can be approached in a variety of ways – particularly in view of the fact that a number of vector capable calculators are on the market which reduce this problem to a very few keystrokes (using magnitude-angle notation throughout).

5. The average velocity is given by Eq. 4-8. The total displacement $\Delta\vec{r}$ is the sum of three displacements, each result of a (constant) velocity during a given time. We use a coordinate system with $+x$ East and $+y$ North. In unit-vector notation, the first displacement is given by

$$\Delta\vec{r}_1 = \left(60\,\frac{\mathrm{km}}{\mathrm{h}}\right)\left(\frac{40\,\mathrm{min}}{60\,\mathrm{min/h}}\right)\hat{\mathrm{i}} = 40\,\hat{\mathrm{i}}$$

in kilometers. The second displacement has a magnitude of $60\,\frac{\mathrm{km}}{\mathrm{h}} \cdot \frac{20\,\mathrm{min}}{60\,\mathrm{min/h}} = 20$ km, and its direction is 40° north of east. Therefore,

$$\Delta\vec{r}_2 = 20\cos(40°)\,\hat{\mathrm{i}} + 20\sin(40°)\,\hat{\mathrm{j}} = 15.3\,\hat{\mathrm{i}} + 12.9\,\hat{\mathrm{j}}$$

in kilometers. And the third displacement is

$$\Delta\vec{r}_3 = -\left(60\,\frac{\mathrm{km}}{\mathrm{h}}\right)\left(\frac{50\,\mathrm{min}}{60\,\mathrm{min/h}}\right)\hat{\mathrm{i}} = -50\,\hat{\mathrm{i}}$$

in kilometers. The total displacement is

$$\begin{aligned} \Delta\vec{r} &= \Delta\vec{r}_1 + \Delta\vec{r}_2 + \Delta\vec{r}_3 \\ &= 40\,\hat{\mathrm{i}} + 15.3\,\hat{\mathrm{i}} + 12.9\,\hat{\mathrm{j}} - 50\,\hat{\mathrm{i}} \\ &= 5.3\,\hat{\mathrm{i}} + 12.9\,\hat{\mathrm{j}} \quad (\mathrm{km})\ . \end{aligned}$$

The time for the trip is $40 + 20 + 50 = 110$ min, which is equivalent to 1.83 h. Eq. 4-8 then yields

$$\vec{v}_{\mathrm{avg}} = \left(\frac{5.3\,\mathrm{km}}{1.83\,\mathrm{h}}\right)\hat{\mathrm{i}} + \left(\frac{12.9\,\mathrm{km}}{1.83\,\mathrm{h}}\right)\hat{\mathrm{j}} = 2.90\,\hat{\mathrm{i}} + 7.01\,\hat{\mathrm{j}}$$

in kilometers-per-hour. If it is desired to express this in magnitude-angle notation, then this is equivalent to a vector of magnitude $\sqrt{2.9^2 + 7.01^2} = 7.59$ km/h, which is inclined 67.5° north of east (or, 22.5° east of north). If unit-vector notation is not a priority in this problem, then the computation can be approached in a variety of ways – particularly in view of the fact that a number of vector capable calculators are on the market which reduce this problem to a very few keystrokes (using magnitude-angle notation throughout).

6. Using Eq. 4-3 and Eq. 4-8, we have

$$\begin{aligned}\vec{v}_{\rm avg} &= \frac{(-2.0\,\hat{\imath}+8.0\,\hat{\jmath}-2.0\,\hat{k})-(5.0\,\hat{\imath}-6.0\,\hat{\jmath}+2.0\,\hat{k})}{10}\\ &= -0.70\,\hat{\imath}+1.40\,\hat{\jmath}-0.40\,\hat{k}\end{aligned}$$

in meters-per-second.

7. To emphasize the fact that the velocity is a function of time, we adopt the notation $v(t)$ for $\frac{dx}{dt}$.

   (a) Eq. 4-10 leads to
   $$v(t)=\frac{d}{dt}\left(3.00t\,\hat{\imath}-4.00t^2\,\hat{\jmath}+2.00\,\hat{k}\right)=3.00\,\hat{\imath}-8.00t\,\hat{\jmath}$$
   in meters-per-second.

   (b) Evaluating this result at $t=2$ s produces $\vec{v}=3.0\,\hat{\imath}-16.0\,\hat{\jmath}$ m/s.

   (c) The speed at $t=2$ s is $v=|\vec{v}|=\sqrt{3^2+(-16)^2}=16.3$ m/s.

   (d) And the angle of $\vec{v}$ at that moment is one of the possibilities
   $$\tan^{-1}\left(\frac{-16}{3}\right)=-79.4^\circ \ \text{ or } \ 101^\circ$$
   where we choose the first possibility (79.4° measured clockwise from the $+x$ direction, or 281° counterclockwise from $+x$) since the signs of the components imply the vector is in the fourth quadrant.

8. On the one hand, we could perform the vector addition of the displacements with a vector capable calculator in polar mode ($(75\ \angle\ 37^\circ)+(65\ \angle-90^\circ)=(63\ \angle-18^\circ)$), but in keeping with Eq. 3-5 and Eq. 3-6 we will show the details in unit-vector notation. We use a 'standard' coordinate system with $+x$ East and $+y$ North. Lengths are in kilometers and times are in hours.

   (a) We perform the vector addition of individual displacements to find the net displacement of the camel.
   $$\begin{aligned}\Delta\vec{r}_1 &= 75\cos(37^\circ)\,\hat{\imath}+75\sin(37^\circ)\,\hat{\jmath}\\ \Delta\vec{r}_2 &= -65\,\hat{\jmath}\\ \Delta\vec{r}_1+\Delta\vec{r}_2 &= 60\,\hat{\imath}-20\,\hat{\jmath}\ \ \text{km}\ .\end{aligned}$$
   If it is desired to express this in magnitude-angle notation, then this is equivalent to a vector of length $\sqrt{60^2+(-20)^2}=63$ km, which is directed at 18° south of east.

   (b) We use the result from part (a) in Eq. 4-8 along with the fact that $\Delta t=90$ h. In unit vector notation, we obtain
   $$\vec{v}_{\rm avg}=\frac{60\,\hat{\imath}-20\,\hat{\jmath}}{90}=0.66\,\hat{\imath}-0.22\,\hat{\jmath}$$
   in kilometers-per-hour. This result in magnitude-angle notation is $\vec{v}_{\rm avg}=0.70$ km/h at 18° south of east.

   (c) Average speed is distinguished from the magnitude of average velocity in that it depends on the total distance as opposed to the net displacement. Since the camel travels 140 km, we obtain $140/90=1.56$ km/h.

   (d) The net displacement is required to be the 90 km East from A to B. The displacement from the resting place to B is denoted $\vec{r}_3$. Thus, we must have (in kilometers)
   $$\vec{r}_1+\vec{r}_2+\vec{r}_3=90\,\hat{\imath}$$

which produces $\vec{r}_3 = 30\,\hat{\text{i}} + 20\,\hat{\text{j}}$ in unit-vector notation, or $(36 \angle 33°)$ in magnitude-angle notation. Therefore, using Eq. 4-8 we obtain

$$|\vec{v}_{\text{avg}}| = \frac{36 \text{ km}}{120 - 90 \text{ h}} = 1.2 \text{ km/h}$$

and the direction of this vector is the same as $\vec{r}_3$ (that is, 33° north of east).

9. We apply Eq. 4-10 and Eq. 4-16.

   (a) Taking the derivative of the position vector with respect to time, we have

   $$\vec{v} = \frac{d}{dt}\left(\hat{\text{i}} + 4t^2\,\hat{\text{j}} + t\,\hat{\text{k}}\right) = 8t\,\hat{\text{j}} + \hat{\text{k}}$$

   in SI units (m/s).

   (b) Taking another derivative with respect to time leads to

   $$\vec{a} = \frac{d}{dt}\left(8t\,\hat{\text{j}} + \hat{\text{k}}\right) = 8\,\hat{\text{j}}$$

   in SI units (m/s^2).

10. We use Eq. 4-15 with $\vec{v}_1$ designating the initial velocity and $\vec{v}_2$ designating the later one.

    (a) The average acceleration during the $\Delta t = 4$ s interval is

    $$\vec{a}_{\text{avg}} = \frac{\left(-2\,\hat{\text{i}} - 2\,\hat{\text{j}} + 5\,\hat{\text{k}}\right) - \left(4\,\hat{\text{i}} - 22\,\hat{\text{j}} + 3\,\hat{\text{k}}\right)}{4} = -1.5\,\hat{\text{i}} + 0.5\,\hat{\text{k}}$$

    in SI units (m/s^2).

    (b) The magnitude of $\vec{a}_{\text{avg}}$ is $\sqrt{(-1.5)^2 + 0.5^2} = 1.6$ m/s^2. Its angle in the xz plane (measured from the $+x$ axis) is one of these possibilities:

    $$\tan^{-1}\left(\frac{0.5}{-1.5}\right) = -18° \quad \text{or} \quad 162°$$

    where we settle on the second choice since the signs of its components imply that it is in the second quadrant.

11. In parts (b) and (c), we use Eq. 4-10 and Eq. 4-16. For part (d), we find the direction of the velocity computed in part (b), since that represents the asked-for tangent line.

    (a) Plugging into the given expression, we obtain

    $$\vec{r}\,\Big|_{t=2} = (2(8) - 5(2))\,\hat{\text{i}} + (6 - 7(16))\,\hat{\text{j}} = 6.00\,\hat{\text{i}} - 106\,\hat{\text{j}}$$

    in meters.

    (b) Taking the derivative of the given expression produces

    $$\vec{v}(t) = \left(6.00t^2 - 5.00\right)\,\hat{\text{i}} + 28.0t^3\,\hat{\text{j}}$$

    where we have written $v(t)$ to emphasize its dependence on time. This becomes, at $t = 2.00$ s, $\vec{v} = 19.0\,\hat{\text{i}} - 224\,\hat{\text{j}}$ m/s.

    (c) Differentiating the $\vec{v}(t)$ found above, with respect to t produces $12.0t\,\hat{\text{i}} - 84.0t^2\,\hat{\text{j}}$, which yields $\vec{a} = 24.0\,\hat{\text{i}} - 336\,\hat{\text{j}}$ m/s^2 at $t = 2.00$ s.

(d) The angle of $\vec{v}$, measured from $+x$, is either

$$\tan^{-1}\left(\frac{-224}{19.0}\right) = -85.2^\circ \quad \text{or} \quad 94.8^\circ$$

where we settle on the first choice (-85.2°, which is equivalent to 275°) since the signs of its components imply that it is in the fourth quadrant.

12. Noting that $\vec{v}_2 = 0$, then, using Eq. 4-15, the average acceleration is

$$\vec{a}_{\text{avg}} = \frac{\Delta\vec{v}}{\Delta t} = \frac{0 - (6.30\,\hat{\imath} - 8.42\,\hat{\jmath})}{3} = -2.1\,\hat{\imath} + 2.8\,\hat{\jmath}$$

in SI units (m/s^2).

13. Constant acceleration in both directions (x and y) allows us to use Table 2-1 for the motion along each direction. This can be handled individually (for Δx and Δy) or together with the unit-vector notation (for Δr). Where units are not shown, SI units are to be understood.

(a) The velocity of the particle at any time t is given by $\vec{v} = \vec{v}_0 + \vec{a}t$, where $\vec{v}_0$ is the initial velocity and $\vec{a}$ is the (constant) acceleration. The x component is $v_x = v_{0x} + a_x t = 3.00 - 1.00t$, and the y component is $v_y = v_{0y} + a_y t = -0.500t$ since $v_{0y} = 0$. When the particle reaches its maximum x coordinate at $t = t_m$, we must have $v_x = 0$. Therefore, $3.00 - 1.00t_m = 0$ or $t_m = 3.00$ s. The y component of the velocity at this time is $v_y = 0 - 0.500(3.00) = -1.50$ m/s; this is the only nonzero component of $\vec{v}$ at t_m.

(b) Since it started at the origin, the coordinates of the particle at any time t are given by $\vec{r} = \vec{v}_0 t + \frac{1}{2}\vec{a}t^2$. At $t = t_m$ this becomes

$$(3.00\,\hat{\imath})(3.00) + \frac{1}{2}(-1.00\,\hat{\imath} - 0.50\,\hat{\jmath})(3.00)^2 = 4.50\,\hat{\imath} - 2.25\,\hat{\jmath}$$

in meters.

14. (a) Using Eq. 4-16, the acceleration as a function of time is

$$\vec{a} = \frac{d\vec{v}}{dt} = \frac{d}{dt}\left((6.0t - 4.0t^2)\hat{\imath} + 8.0\,\hat{\jmath}\right) = (6.0 - 8.0t)\hat{\imath}$$

in SI units. Specifically, we find the acceleration vector at $t = 3.0$ s to be $(6.0 - 8.0(3.0))\hat{\imath} = -18\,\hat{\imath}\ \text{m/s}^2$.

(b) The equation is $\vec{a} = (6.0 - 8.0t)\hat{\imath} = 0$; we find $t = 0.75$ s.

(c) Since the y component of the velocity, $v_y = 8.0$ m/s, is never zero, the velocity cannot vanish.

(d) Since speed is the magnitude of the velocity, we have $v = |\vec{v}| = \sqrt{(6.0t - 4.0t^2)^2 + (8.0)^2} = 10$ in SI units (m/s). We solve for t as follows:

$$\begin{aligned}
\text{squaring} \quad (6t - 4t^2)^2 + 64 &= 100 \\
\text{rearranging} \quad (6t - 4t^2)^2 &= 36 \\
\text{taking square root} \quad 6t - 4t^2 &= \pm 6 \\
\text{rearranging} \quad 4t^2 - 6t \pm 6 &= 0 \\
\text{using quadratic formula} \quad t &= \frac{6 \pm \sqrt{36 - 4(4)(\pm 6)}}{2(8)}
\end{aligned}$$

where the requirement of a real positive result leads to the unique answer: $t = 2.2$ s.

15. Since the x and y components of the acceleration are constants, then we can use Table 2-1 for the motion along both axes. This can be handled individually (for Δx and Δy) or together with the unit-vector notation (for Δr). Where units are not shown, SI units are to be understood.

   (a) Since $\vec{r}_0 = 0$, the position vector of the particle is (adapting Eq. 2-15)

$$\begin{aligned} \vec{r} &= \vec{v}_0 t + \frac{1}{2}\vec{a}t^2 \\ &= (8.0\,\hat{\jmath})t + \frac{1}{2}(4.0\,\hat{\imath} + 2.0\,\hat{\jmath})t^2 \\ &= (2.0t^2)\,\hat{\imath} + (8.0t + 1.0t^2)\,\hat{\jmath}\ . \end{aligned}$$

   Therefore, we find when $x = 29$ m, by solving $2.0t^2 = 29$, which leads to $t = 3.8$ s. The y coordinate at that time is $y = 8.0(3.8) + 1.0(3.8)^2 = 45$ m.

   (b) Adapting Eq. 2-11, the velocity of the particle is given by

$$\vec{v} = \vec{v}_0 + \vec{a}t\ .$$

   Thus, at $t = 3.8$ s, the velocity is

$$\vec{v} = 8.0\,\hat{\jmath} + (4.0\,\hat{\imath} + 2.0\,\hat{\jmath})(3.8) = 15.2\,\hat{\imath} + 15.6\,\hat{\jmath}$$

   which has a magnitude of

$$v = \sqrt{v_x^2 + v_y^2} = \sqrt{15.2^2 + 15.6^2} = 22\ \text{m/s}\ .$$

16. The acceleration is constant so that use of Table 2-1 (for both the x and y motions) is permitted. Where units are not shown, SI units are to be understood. Collision between particles A and B requires two things. First, the y motion of B must satisfy (using Eq. 2-15 and noting that θ is measured from the y axis)

$$y = \frac{1}{2}a_y t^2 \implies 30 = \frac{1}{2}(0.40\cos\theta)\,t^2\ .$$

   Second, the x motions of A and B must coincide:

$$vt = \frac{1}{2}a_x t^2 \implies 3.0t = \frac{1}{2}(0.40\sin\theta)\,t^2\ .$$

   We eliminate a factor of t in the last relationship and formally solve for time:

$$t = \frac{3}{0.2\sin\theta}\ .$$

   This is then plugged into the previous equation to produce

$$30 = \frac{1}{2}(0.40\cos\theta)\left(\frac{3}{0.2\sin\theta}\right)^2$$

   which, with the use of $\sin^2\theta = 1 - \cos^2\theta$, simplifies to

$$30 = \frac{9}{0.2}\,\frac{\cos\theta}{1-\cos^2\theta} \implies 1 - \cos^2\theta = \frac{9}{(0.2)(30)}\cos\theta\ .$$

   We use the quadratic formula (choosing the positive root) to solve for $\cos\theta$:

$$\cos\theta = \frac{-1.5 + \sqrt{1.5^2 - 4(1)(-1)}}{2} = \frac{1}{2}$$

   which yields

$$\theta = \cos^{-1}\left(\frac{1}{2}\right) = 60^\circ\ .$$

17. We adopt the positive direction choices used in the textbook so that equations such as Eq. 4-22 are directly applicable.

   (a) With the origin at the firing point, the y coordinate of the bullet is given by $y = -\frac{1}{2}gt^2$. If t is the time of flight and $y = -0.019$ m indicates where the bullet hits the target, then

$$t = \sqrt{\frac{2(0.019)}{9.8}} = 6.2 \times 10^{-2} \text{ s} .$$

   (b) The muzzle velocity is the initial (horizontal) velocity of the bullet. Since $x = 30$ m is the horizontal position of the target, we have $x = v_0 t$. Thus,

$$v_0 = \frac{x}{t} = \frac{30}{6.3 \times 10^{-2}} = 4.8 \times 10^2 \text{ m/s} .$$

18. We adopt the positive direction choices used in the textbook so that equations such as Eq. 4-22 are directly applicable.

   (a) With the origin at the initial point (edge of table), the y coordinate of the ball is given by $y = -\frac{1}{2}gt^2$. If t is the time of flight and $y = -1.20$ m indicates the level at which the ball hits the floor, then

$$t = \sqrt{\frac{2(1.20)}{9.8}} = 0.495 \text{ s} .$$

   (b) The initial (horizontal) velocity of the ball is $\vec{v} = v_0\,\hat{\mathrm{i}}$. Since $x = 1.52$ m is the horizontal position of its impact point with the floor, we have $x = v_0 t$. Thus,

$$v_0 = \frac{x}{t} = \frac{1.52}{0.495} = 3.07 \text{ m/s} .$$

19. We adopt the positive direction choices used in the textbook so that equations such as Eq. 4-22 are directly applicable. The initial velocity is horizontal so that $v_{0y} = 0$ and $v_{0x} = v_0 = 161$ km/h. Converting to SI units, this is $v_0 = 44.7$ m/s.

   (a) With the origin at the initial point (where the ball leaves the pitcher's hand), the y coordinate of the ball is given by $y = -\frac{1}{2}gt^2$, and the x coordinate is given by $x = v_0 t$. From the latter equation, we have a simple proportionality between horizontal distance and time, which means the time to travel half the total distance is half the total time. Specifically, if $x = 18.3/2$ m, then $t = (18.3/2)/44.7 = 0.205$ s.

   (b) And the time to travel the next 18.3/2 m must also be 0.205 s. It can be useful to write the horizontal equation as $\Delta x = v_0 \Delta t$ in order that this result can be seen more clearly.

   (c) From $y = -\frac{1}{2}gt^2$, we see that the ball has reached the height of $-\frac{1}{2}(9.8)(0.205)^2 = -0.205$ m at the moment the ball is halfway to the batter.

   (d) The ball's height when it reaches the batter is $-\frac{1}{2}(9.8)(0.409)^2 = -0.820$ m, which, when subtracted from the previous result, implies it has fallen another 0.615 m. Since the value of y is not simply proportional to t, we do not expect equal time-intervals to correspond to equal height-changes; in a physical sense, this is due to the fact that the initial y-velocity for the first half of the motion is not the same as the "initial" y-velocity for the second half of the motion.

20. We adopt the positive direction choices used in the textbook so that equations such as Eq. 4-22 are directly applicable. The initial velocity is horizontal so that $v_{0y} = 0$ and $v_{0x} = v_0 = 10$ m/s.

   (a) With the origin at the initial point (where the dart leaves the thrower's hand), the y coordinate of the dart is given by $y = -\frac{1}{2}gt^2$, so that with $y = -PQ$ we have $PQ = \frac{1}{2}(9.8)(0.19)^2 = 0.18$ m.

(b) From $x = v_0 t$ we obtain $x = (10)(0.19) = 1.9$ m.

21. Since this problem involves constant downward acceleration of magnitude a, similar to the projectile motion situation, we use the equations of §4-6 as long as we substitute a for g. We adopt the positive direction choices used in the textbook so that equations such as Eq. 4-22 are directly applicable. The initial velocity is horizontal so that $v_{0y} = 0$ and $v_{0x} = v_0 = 1.0 \times 10^9$ cm/s.

(a) If ℓ is the length of a plate and t is the time an electron is between the plates, then $\ell = v_0 t$, where v_0 is the initial speed. Thus

$$t = \frac{\ell}{v_0} = \frac{2.0\,\text{cm}}{1.0 \times 10^9\,\text{cm/s}} = 2.0 \times 10^{-9}\ \text{s}\,.$$

(b) The vertical displacement of the electron is

$$y = -\frac{1}{2}at^2 = -\frac{1}{2}\left(1.0 \times 10^{17}\,\text{cm/s}^2\right)\left(2.0 \times 10^{-9}\,\text{s}\right)^2 = -0.20\ \text{cm}\,.$$

(c) and (d) The x component of velocity does not change: $v_x = v_0 = 1.0 \times 10^9$ cm/s, and the y component is

$$v_y = a_y t = \left(1.0 \times 10^{17}\,\text{cm/s}^2\right)\left(2.0 \times 10^{-9}\,\text{s}\right) = 2.0 \times 10^8\,\text{cm/s}\,.$$

22. We use Eq. 4-26

$$R_{\text{max}} = \left(\frac{v_0^2}{g}\sin 2\theta_0\right)_{\text{max}} = \frac{v_0^2}{g} = \frac{(9.5\,\text{m/s})^2}{9.80\,\text{m/s}^2} = 9.21\ \text{m}$$

to compare with Powell's long jump; the difference from R_{max} is only $\Delta R = 9.21 - 8.95 = 0.26$ m.

23. We adopt the positive direction choices used in the textbook so that equations such as Eq. 4-22 are directly applicable. The coordinate origin is throwing point (the stone's initial position). The x component of its initial velocity is given by $v_{0x} = v_0 \cos\theta_0$ and the y component is given by $v_{0y} = v_0 \sin\theta_0$, where $v_0 = 20$ m/s is the initial speed and $\theta_0 = 40.0°$ is the launch angle.

(a) At $t = 1.10$ s, its x coordinate is

$$x = v_0 t \cos\theta_0 = (20.0\,\text{m/s})(1.10\,\text{s})\cos 40.0° = 16.9\,\text{m}$$

(b) Its y coordinate at that instant is

$$y = v_0 t \sin\theta_0 - \frac{1}{2}gt^2 = (20.0\,\text{m/s})(1.10\,\text{s})\sin 40° - \frac{1}{2}(9.80\,\text{m/s}^2)(1.10\,\text{s})^2 = 8.21\,\text{m}\,.$$

(c) At $t' = 1.80$ s, its x coordinate is

$$x = (20.0\,\text{m/s})(1.80\,\text{s})\cos 40.0° = 27.6\,\text{m}$$

(d) Its y coordinate at t' is

$$y = (20.0\,\text{m/s})(1.80\,\text{s})\sin 40° - \frac{1}{2}\left(9.80\,\text{m/s}^2\right)(1.80\,\text{s})^2 = 7.26\ \text{m}\,.$$

(e) and (f) The stone hits the ground earlier than $t = 5.0$ s. To find the time when it hits the ground solve $y = v_0 t \sin\theta_0 - \frac{1}{2}gt^2 = 0$ for t. We find

$$t = \frac{2v_0}{g}\sin\theta_0 = \frac{2(20.0\,\text{m/s})}{9.8\,\text{m/s}^2}\sin 40° = 2.62\ \text{s}\,.$$

Its x coordinate on landing is

$$x = v_0 t \cos\theta_0 = (20.0\,\text{m/s})(2.62\,\text{s})\cos 40° = 40.2\ \text{m}$$

(or Eq. 4-26 can be used). Assuming it stays where it lands, its coordinates at $t = 5.00$ s are $x = 40.2$ m and $y = 0$.

24. In this projectile motion problem, we have $v_0 = v_x =$ constant, and what is plotted is $v = \sqrt{v_x^2 + v_y^2}$. We infer from the plot that at $t = 2.5$ s, the ball reaches its maximum height, where $v_y = 0$. Therefore, we infer from the graph that $v_x = 19$ m/s.

  (a) During $t = 5$ s, the horizontal motion is $x - x_0 = v_x t = 95$ m.

  (b) Since $\sqrt{19^2 + v_{0y}^2} = 31$ m/s (the first point on the graph), we find $v_{0y} = 24.5$ m/s. Thus, with $t = 2.5$ s, we can use $y_{\max} - y_0 = v_{0y}t - \frac{1}{2}gt^2$ or $v_y^2 = 0 = v_{0y}^2 - 2g(y_{\max} - y_0)$, or $y_{\max} - y_0 = \frac{1}{2}\left(v_y + v_{0y}\right)t$ to solve. Here we will use the latter:

$$y_{\max} - y_0 = \frac{1}{2}\left(v_y + v_{0y}\right)t \implies y_{\max} = \frac{1}{2}(0 + 24.5)(2.5) = 31 \text{ m}$$

  where we have taken $y_0 = 0$ as the ground level.

25. We adopt the positive direction choices used in the textbook so that equations such as Eq. 4-22 are directly applicable. The coordinate origin is at the end of the rifle (the initial point for the bullet as it begins projectile motion in the sense of §4-5), and we let θ_0 be the firing angle. If the target is a distance d away, then its coordinates are $x = d$, $y = 0$. The projectile motion equations lead to $d = v_0 t \cos\theta_0$ and $0 = v_0 t \sin\theta_0 - \frac{1}{2}gt^2$. Eliminating t leads to $2v_0^2 \sin\theta_0 \cos\theta_0 - gd = 0$. Using $\sin\theta_0 \cos\theta_0 = \frac{1}{2}\sin(2\theta_0)$, we obtain

$$v_0^2 \sin(2\theta_0) = gd \implies \sin(2\theta_0) = \frac{gd}{v_0^2} = \frac{(9.8)(45.7)}{460^2}$$

which yields $\sin(2\theta_0) = 2.12 \times 10^{-3}$ and consequently $\theta_0 = 0.0606°$. If the gun is aimed at a point a distance ℓ above the target, then $\tan\theta_0 = \ell/d$ so that

$$\ell = d\tan\theta_0 = 45.7\tan 0.0606° = 0.0484\,\text{m} = 4.84 \text{ cm} .$$

26. The figure offers many interesting points to analyze, and others are easily inferred (such as the point of maximum height). The focus here, to begin with, will be the final point shown (1.25 s after the ball is released) which is when the ball returns to its original height. In English units, $g = 32$ ft/s^2.

  (a) Using $x - x_0 = v_x t$ we obtain $v_x = (40 \text{ ft})/(1.25 \text{ s}) = 32$ ft/s. And $y - y_0 = 0 = v_{0y}t - \frac{1}{2}gt^2$ yields $v_{0y} = \frac{1}{2}(32)(1.25) = 20$ ft/s. Thus, the initial speed is

$$v_0 = |\vec{v}_0| = \sqrt{32^2 + 20^2} = 38 \text{ ft/s} .$$

  (b) Since $v_y = 0$ at the maximum height and the horizontal velocity stays constant, then the speed at the top is the same as $v_x = 32$ ft/s.

  (c) We can infer from the figure (or compute from $v_y = 0 = v_{0y} - gt$) that the time to reach the top is 0.625 s. With this, we can use $y - y_0 = v_{0y}t - \frac{1}{2}gt^2$ to obtain 9.3 ft (where $y_0 = 3$ ft has been used). An alternative approach is to use $v_y^2 = v_{0y}^2 - 2g(y - y_0)$.

27. Taking the y axis to be upward and placing the origin at the firing point, the y coordinate is given by $y = v_0 t \sin\theta_0 - \frac{1}{2}gt^2$ and the y component of the velocity is given by $v_y = v_0 \sin\theta_0 - gt$. The maximum height occurs when $v_y = 0$. Thus, $t = (v_0/g)\sin\theta_0$ and

$$y = v_0\left(\frac{v_0}{g}\right)\sin\theta_0 \sin\theta_0 - \frac{1}{2}\frac{g(v_0\sin\theta_0)^2}{g^2} = \frac{(v_0\sin\theta_0)^2}{2g} .$$

28. We adopt the positive direction choices used in the textbook so that equations such as Eq. 4-22 are directly applicable. The coordinate origin is at the release point (the initial position for the ball as it begins projectile motion in the sense of §4-5), and we let θ_0 be the angle of throw (shown in the figure). Since the horizontal component of the velocity of the ball is $v_x = v_0 \cos 40.0°$, the time it takes for the ball to hit the wall is

$$t = \frac{\Delta x}{v_x} = \frac{22.0}{25.0\cos 40.0°} = 1.15 \text{ s} .$$

(a) The vertical distance is

$$\begin{aligned}\Delta y &= (v_0 \sin\theta_0)t - \frac{1}{2}gt^2 \\ &= (25.0\sin 40.0^\circ)(1.15) - \frac{1}{2}(9.8)(1.15)^2 = 12.0 \text{ m} .\end{aligned}$$

(b) The horizontal component of the velocity when it strikes the wall does not change from its initial value: $v_x = v_0 \cos 40.0^\circ = 19.2$ m/s, while the vertical component becomes (using Eq. 4-23)

$$v_y = v_0 \sin\theta_0 - gt = 25.0 \sin 40.0^\circ - (9.8)(1.15) = 4.80 \text{ m/s} .$$

(c) Since $v_y > 0$ when the ball hits the wall, it has not reached the highest point yet.

29. We designate the given velocity $\vec{v} = 7.6\hat{\text{i}} + 6.1\hat{\text{j}}$ (SI units understood) as $\vec{v}_1$ – as opposed to the velocity when it reaches the max height $\vec{v}_2$ or the velocity when it returns to the ground $\vec{v}_3$ – and take $\vec{v}_0$ as the launch velocity, as usual. The origin is at its launch point on the ground.

(a) Different approaches are available, but since it will be useful (for the rest of the problem) to first find the initial y velocity, that is how we will proceed. Using Eq. 2-16, we have

$$\begin{aligned}v_{1\,y}^2 &= v_{0\,y}^2 - 2g\Delta y \\ 6.1^2 &= v_{0\,y}^2 - 2(9.8)(9.1)\end{aligned}$$

which yields $v_{0\,y} = 14.7$ m/s. Knowing that $v_{2\,y}$ must equal 0, we use Eq. 2-16 again but now with $\Delta y = h$ for the maximum height:

$$\begin{aligned}v_{2\,y}^2 &= v_{0\,y}^2 - 2gh \\ 0 &= 14.7^2 - 2(9.8)h\end{aligned}$$

which yields $h = 11$ m.

(b) Recalling the derivation of Eq. 4-26, but using $v_{0\,y}$ for $v_0 \sin\theta_0$ and $v_{0\,x}$ for $v_0 \cos\theta_0$, we have

$$\begin{aligned}0 &= v_{0\,y}t - \frac{1}{2}gt^2 \\ R &= v_{0\,x}t\end{aligned}$$

which leads to $R = \frac{2\,v_{0\,x}\,v_{0\,y}}{g}$. Noting that $v_{0\,x} = v_{1\,x} = 7.6$ m/s, we plug in values and obtain $R = 2(7.6)(14.7)/9.8 = 23$ m.

(c) Since $v_{3\,x} = v_{1\,x} = 7.6$ m/s and $v_{3\,y} = -v_{0\,y} = -14.7$ m/s, we have

$$v_3 = \sqrt{v_{3\,x}^2 + v_{3\,y}^2} = \sqrt{(-14.7)^2 + 7.6^2} = 17 \text{ m/s} .$$

(d) The angle (measured from horizontal) for $\vec{v}_3$ is one of these possibilities:

$$\tan^{-1}\left(\frac{-14.7}{7.6}\right) = -63^\circ \quad \text{or} \quad 117^\circ$$

where we settle on the first choice (-63°, which is equivalent to 297°) since the signs of its components imply that it is in the fourth quadrant.

30. We apply Eq. 4-21, Eq. 4-22 and Eq. 4-23.

(a) From $\Delta x = v_{0x}t$, we find $v_{0x} = 40/2 = 20$ m/s.

(b) From $\Delta y = v_{0y}t - \frac{1}{2}gt^2$, we find $v_{0y} = (53 + \frac{1}{2}(9.8)(2)^2)/2 = 36$ m/s.

(c) From $v_y = v_{0y} - gt'$ with $v_y = 0$ as the condition for maximum height, we obtain $t' = 36/9.8 = 3.7$ s. During that time the x-motion is constant, so $x' - x_0 = (20)(3.7) = 74$ m.

31. We adopt the positive direction choices used in the textbook so that equations such as Eq. 4-22 are directly applicable. The coordinate origin is at the the initial position for the football as it begins projectile motion in the sense of §4-5), and we let θ_0 be the angle of its initial velocity measured from the $+x$ axis.

(a) $x = 46$ m and $y = -1.5$ m are the coordinates for the landing point; it lands at time $t = 4.5$ s. Since $x = v_{0x}t$,

$$v_{0\,x} = \frac{x}{t} = \frac{46\,\text{m}}{4.5\,\text{s}} = 10.2 \text{ m/s} .$$

Since $y = v_{0y}t - \frac{1}{2}gt^2$,

$$v_{0\,y} = \frac{y + \frac{1}{2}gt^2}{t} = \frac{(-1.5\,\text{m}) + \frac{1}{2}(9.8\,\text{m/s}^2)(4.5\,\text{s})^2}{4.5\,\text{s}} = 21.7 \text{ m/s} .$$

The magnitude of the initial velocity is

$$v_0 = \sqrt{v_{0\,x}^2 + v_{0\,y}^2} = \sqrt{(10.2\,\text{m/s})^2 + (21.7\,\text{m/s})^2} = 24 \text{ m/s} .$$

(b) The initial angle satisfies $\tan\theta_0 = v_{0\,y}/v_{0\,x}$. Thus, $\theta_0 = \tan^{-1}(21.7/10.2) = 64.8°$.

32. The initial velocity has no vertical component – only an x component equal to $+2.00$ m/s. Also, $y_0 = +10.0$ m if the water surface is established as $y = 0$.

(a) $x - x_0 = v_x t$ readily yields $x - x_0 = 1.60$ m.

(b) Using $y - y_0 = v_{0y}t - \frac{1}{2}gt^2$, we obtain $y = 6.86$ m when $t = 0.800$ s.

(c) With t unknown and $y = 0$, the equation $y - y_0 = v_{0y}t - \frac{1}{2}gt^2$ leads to $t = \sqrt{2(10)/9.8} = 1.43$ s. During this time, the x-displacement of the diver is $x - x_0 = (2.00 \text{ m/s})(1.43 \text{ s}) = 2.86$ m.

33. We adopt the positive direction choices used in the textbook so that equations such as Eq. 4-22 are directly applicable. The coordinate origin is at ground level directly below the release point. We write $\theta_0 = -30°$ since the angle shown in the figure is measured clockwise from horizontal. We note that the initial speed of the decoy is the plane's speed at the moment of release: $v_0 = 290$ km/h, which we convert to SI units: $(290)(1000/3600) = 80.6$ m/s.

(a) We use Eq. 4-12 to solve for the time:

$$\Delta x = (v_0 \cos\theta_0)\, t \implies t = \frac{700}{(80.6)\cos -30°} = 10.0 \text{ s} .$$

(b) And we use Eq. 4-22 to solve for the initial height y_0:

$$\begin{aligned} y - y_0 &= (v_0 \sin\theta_0)\, t - \frac{1}{2}gt^2 \\ 0 - y_0 &= (-40.3)(10.0) - \frac{1}{2}(9.8)(10.0)^2 \end{aligned}$$

which yields $y_0 = 897$ m.

34. We adopt the positive direction choices used in the textbook so that equations such as Eq. 4-22 are directly applicable. The coordinate origin is at its initial position (where it is launched). At maximum height, we observe $v_y = 0$ and denote $v_x = v$ (which is also equal to $v_{0\,x}$). In this notation, we have

$$v_0 = 5v \ .$$

Next, we observe $v_0 \cos\theta_0 = v_{0\,x} = v$, so that we arrive at an equation (where $v \neq 0$ cancels) which can be solved for θ_0:

$$(5v)\cos\theta_0 = v \implies \theta_0 = \cos^{-1}\frac{1}{5} = 78^\circ \ .$$

35. We denote h as the height of a step and w as the width. To hit step n, the ball must fall a distance nh and travel horizontally a distance between $(n-1)w$ and nw. We take the origin of a coordinate system to be at the point where the ball leaves the top of the stairway, and we choose the y axis to be positive in the upward direction. The coordinates of the ball at time t are given by $x = v_{0\,x}t$ and $y = -\frac{1}{2}gt^2$ (since $v_{0\,y} = 0$). We equate y to $-nh$ and solve for the time to reach the level of step n:

$$t = \sqrt{\frac{2nh}{g}} \ .$$

The x coordinate then is

$$x = v_{0\,x}\sqrt{\frac{2nh}{g}} = (1.52\,\text{m/s})\sqrt{\frac{2n(0.203\,\text{m})}{9.8\,\text{m/s}^2}} = (0.309\,\text{m})\sqrt{n} \ .$$

The method is to try values of n until we find one for which x/w is less than n but greater than $n-1$. For $n = 1$, $x = 0.309\,\text{m}$ and $x/w = 1.52$, which is greater than n. For $n = 2$, $x = 0.437\,\text{m}$ and $x/w = 2.15$, which is also greater than n. For $n = 3$, $x = 0.535\,\text{m}$ and $x/w = 2.64$. Now, this is less than n and greater than $n-1$, so the ball hits the third step.

36. Although we could use Eq. 4-26 to find where it lands, we choose instead to work with Eq. 4-21 and Eq. 4-22 (for the soccer ball) since these will give information about where *and when* and these are also considered more fundamental than Eq. 4-26. With $\Delta y = 0$, we have

$$\Delta y = (v_0 \sin\theta_0)\,t - \frac{1}{2}gt^2 \implies t = \frac{(19.5)\sin 45^\circ}{\frac{1}{2}(9.8)} = 2.81 \text{ s} \ .$$

Then Eq. 4-21 yields $\Delta x = (v_0 \cos\theta_0)\,t = 38.3$ m. Thus, using Eq. 4-8 and SI units, the player must have an average velocity of

$$\vec{v}_{\text{avg}} = \frac{\Delta\vec{r}}{\Delta t} = \frac{38.3\,\hat{\imath} - 5\,\hat{\imath}}{2.81} = -5.8\,\hat{\imath}$$

which means his average speed (assuming he ran in only one direction) is 5.8 m/s.

37. We adopt the positive direction choices used in the textbook so that equations such as Eq. 4-22 are directly applicable. The coordinate origin is at ground level directly below the release point. We write $\theta_0 = -37^\circ$ for the angle measured from $+x$, since the angle given in the problem is measured from the $-y$ direction. We note that the initial speed of the projectile is the plane's speed at the moment of release.

(a) We use Eq. 4-22 to find v_0 (SI units are understood).

$$\begin{aligned} y - y_0 &= (v_0 \sin\theta_0)\,t - \frac{1}{2}gt^2 \\ 0 - 730 &= v_0 \sin(-37^\circ)\,(5.00) - \frac{1}{2}(9.8)(5.00)^2 \end{aligned}$$

which yields $v_0 = 202$ m/s.

(b) The horizontal distance traveled is $x = v_0 t \cos\theta_0 = (202)(5.00)\cos -37.0° = 806$ m.

(c) The x component of the velocity (just before impact) is $v_x = v_0 \cos\theta_0 = (202)\cos -37.0° = 161$ m/s.

(d) The y component of the velocity (just before impact) is $v_y = v_0 \sin\theta_0 - gt = (202)\sin(-37°) - (9.80)(5.00) = -171$ m/s.

38. We assume the ball's initial velocity is perpendicular to the plane of the net. We choose coordinates so that $(x_0, y_0) = (0, 3.0)$ m, and $v_x > 0$ (note that $v_{0y} = 0$).

(a) To (barely) clear the net, we have

$$y - y_0 = v_{0y}t - \frac{1}{2}gt^2 \implies 3.0 - 2.24 = 0 - \frac{1}{2}(9.8)t^2$$

which gives $t = 0.39$ s for the time it is passing over the net. This is plugged into the x-equation to yield the (minimum) initial velocity $v_x = (8.0 \text{ m})/(0.39 \text{ s}) = 20.3$ m/s.

(b) We require $y = 0$ and find t from $y - y_0 = v_{0y}t - \frac{1}{2}gt^2$. This value ($t = \sqrt{2(3.0)/9.8} = 0.78$ s) is plugged into the x-equation to yield the (maximum) initial velocity $v_x = (17.0 \text{ m})/(0.78 \text{ s})$ $= 21.7$ m/s.

39. We adopt the positive direction choices used in the textbook so that equations such as Eq. 4-22 are directly applicable. The coordinate origin is at ground level directly below impact point between bat and ball. The *Hint* given in the problem is important, since it provides us with enough information to find v_0 directly from Eq. 4-26.

(a) We want to know how high the ball is from the ground when it is at $x = 97.5$ m, which requires knowing the initial velocity. Using the range information and $\theta_0 = 45°$, we use Eq. 4-26 to solve for v_0:

$$v_0 = \sqrt{\frac{gR}{\sin 2\theta_0}} = \sqrt{\frac{(9.8)(107)}{1}} = 32.4 \text{ m/s} .$$

Thus, Eq. 4-21 tells us the time it is over the fence:

$$t = \frac{x}{v_0 \cos\theta_0} = \frac{97.5}{(32.4)\cos 45°} = 4.26 \text{ s} .$$

At this moment, the ball is at a height (above the ground) of

$$y = y_0 + (v_0 \sin\theta_0)\, t - \frac{1}{2}gt^2 = 9.88 \text{ m}$$

which implies it does indeed clear the 7.32 m high fence.

(b) At $t = 4.26$ s, the center of the ball is $9.88 - 7.32 = 2.56$ m above the fence.

40. We adopt the positive direction choices used in the textbook so that equations such as Eq. 4-22 are directly applicable. The coordinate origin is at ground level directly below the point where the ball was hit by the racquet.

(a) We want to know how high the ball is above the court when it is at $x = 12$ m. First, Eq. 4-21 tells us the time it is over the fence:

$$t = \frac{x}{v_0 \cos\theta_0} = \frac{12}{(23.6)\cos 0°} = 0.508 \text{ s} .$$

At this moment, the ball is at a height (above the court) of

$$y = y_0 + (v_0 \sin\theta_0)\, t - \frac{1}{2}gt^2 = 1.103 \text{ m}$$

which implies it does indeed clear the 0.90 m high fence.

(b) At $t = 0.508$ s, the center of the ball is $1.103 - 0.90 = 0.20$ m above the net.

(c) Repeating the computation in part (a) with $\theta_0 = -5°$ results in $t = 0.510$ s and $y = 0.04$ m, which clearly indicates that it cannot clear the net.

(d) In the situation discussed in part (c), the distance between the top of the net and the center of the ball at $t = 0.510$ s is $0.90 - 0.04 = 0.86$ m.

41. We adopt the positive direction choices used in the textbook so that equations such as Eq. 4-22 are directly applicable. The coordinate origin is at the point where the ball is kicked. Where units are not displayed, SI units are understood. We use x and y to denote the coordinates of ball at the goalpost, and try to find the kicking angle(s) θ_0 so that $y = 3.44$ m when $x = 50$ m. Writing the kinematic equations for projectile motion: $x = v_0 t \cos\theta_0$ and $y = v_0 t \sin\theta_0 - \frac{1}{2}gt^2$, we see the first equation gives $t = x/v_0 \cos\theta_0$, and when this is substituted into the second the result is

$$y = x\tan\theta_0 - \frac{gx^2}{2v_0^2\cos^2\theta_0} .$$

One may solve this by trial and error: systematically trying values of θ_0 until you find the two that satisfy the equation. A little manipulation, however, will give an algebraic solution:
Using the trigonometric identity $1/\cos^2\theta_0 = 1 + \tan^2\theta_0$, we obtain

$$\frac{1}{2}\frac{gx^2}{v_0^2}\tan^2\theta_0 - x\tan\theta_0 + y + \frac{1}{2}\frac{gx^2}{v_0^2} = 0$$

which is a second-order equation for $\tan\theta_0$. To simplify writing the solution, we denote $c = \frac{1}{2}gx^2/v_0^2 = \frac{1}{2}(9.80)(50)^2/(25)^2 = 19.6$ m. Then the second-order equation becomes $c\tan^2\theta_0 - x\tan\theta_0 + y + c = 0$. Using the quadratic formula, we obtain its solution(s).

$$\begin{aligned}\tan\theta_0 &= \frac{x \pm \sqrt{x^2 + 4(y+c)c}}{2c} \\ &= \frac{50 \pm \sqrt{50^2 - 4(3.44 + 19.6)(19.6)}}{2(19.6)} .\end{aligned}$$

The two solutions are given by $\tan\theta_0 = 1.95$ and $\tan\theta_0 = 0.605$. The corresponding (first-quadrant) angles are $\theta_0 = 63°$ and $\theta_0 = 31°$. If kicked at any angle between these two, the ball will travel above the cross bar on the goalposts.

42. The magnitude of the acceleration is

$$a = \frac{v^2}{r} = \frac{(10\,\mathrm{m/s})^2}{25\,\mathrm{m}} = 4.0\,\mathrm{m/s^2} .$$

43. We apply Eq. 4-33 to solve for speed v and Eq. 4-32 to find acceleration a.

(a) Since the radius of Earth is 6.37×10^6 m, the radius of the satellite orbit is $6.37 \times 10^6\,\mathrm{m} + 640 \times 10^3\,\mathrm{m} = 7.01 \times 10^6$ m. Therefore, the speed of the satellite is

$$v = \frac{2\pi r}{T} = \frac{2\pi(7.01 \times 10^6\,\mathrm{m})}{(98.0\,\mathrm{min})(60\,\mathrm{s/min})} = 7.49 \times 10^3\,\mathrm{m/s} .$$

(b) The magnitude of the acceleration is

$$a = \frac{v^2}{r} = \frac{(7.49 \times 10^3\,\mathrm{m/s})^2}{7.01 \times 10^6\,\mathrm{m}} = 8.00\,\mathrm{m/s^2} .$$

44. We note that the period of revolution is $(1200\,\mathrm{rev/min})^{-1} = 8.3 \times 10^{-4}$ min which becomes, in SI units, $T = 0.050$ s.

(a) The circumference is $c = 2\pi r = 2\pi(0.15) = 0.94$ m.

(b) The speed is $v = c/T = (0.94)/(0.050) = 19$ m/s. This is equivalent to using Eq. 4-33.

(c) The magnitude of the acceleration is $a = v^2/r = 19^2/0.15 = 2.4 \times 10^3\ \text{m/s}^2$.

(d) As noted above, $T = 50$ ms.

45. We apply Eq. 4-32 to solve for speed v and Eq. 4-33 to find the period T.

(a) We obtain

$$v = \sqrt{ra} = \sqrt{(5.0\,\text{m})(7.0)(9.8\,\text{m/s}^2)} = 19\ \text{m/s}\ .$$

(b) The time to go around once (the period) is $T = 2\pi r/v = 1.7$ s. Therefore, in one minute ($t = 60$ s), the astronaut executes

$$\frac{t}{T} = \frac{60}{1.7} = 35$$

revolutions. Thus, 35 rev/min is needed to produce a centripetal acceleration of $7g$ when the radius is 5.0 m.

(c) As noted above, $T = 1.7$ s.

46. The magnitude of centripetal acceleration ($a = v^2/r$) and its direction (towards the center of the circle) form the basis of this problem.

(a) If a passenger at this location experiences $\vec{a} = 1.83\ \text{m/s}^2$ east, then the center of the circle is *east* of this location. And the distance is $r = v^2/a = (3.66^2)/(1.83) = 7.32$ m. Thus, relative to the center, the passenger at that moment is located 7.32 m toward the west.

(b) We see the distance is the same, but now the direction of $\vec{a}$ experienced by the passenger is *south* – indicating that the center of the merry-go-round is south of him. Therefore, relative to the center, the passenger at that moment located 7.32 m toward the north.

47. The radius of Earth may be found in Appendix C.

(a) The speed of a person at Earth's equator is $v = 2\pi R/T$, where R is the radius of Earth (6.37×10^6 m) and T is the length of a day (8.64×10^4 s): $v = 2\pi(6.37 \times 10^6\,\text{m})/(8.64 \times 10^4\,\text{s}) = 463\,\text{m/s}$. The magnitude of the acceleration is given by

$$a = \frac{v^2}{R} = \frac{(463\,\text{m/s})^2}{6.37 \times 10^6\,\text{m}} = 0.034\ \text{m/s}^2\ .$$

(b) If T is the period, then $v = 2\pi R/T$ is the speed and $a = v^2/R = 4\pi^2R^2/T^2R = 4\pi^2R/T^2$ is the magnitude of the acceleration. Thus

$$T = 2\pi\sqrt{\frac{R}{a}} = 2\pi\sqrt{\frac{6.37 \times 10^6\,\text{m}}{9.8\,\text{m/s}^2}} = 5.1 \times 10^3\,\text{s} = 84\ \text{min}\ .$$

48. Eq. 4-32 describes an inverse proportionality between r and a, so that a large acceleration results from a small radius. Thus, an upper limit for a corresponds to a lower limit for r.

(a) The minimum turning radius of the train is given by

$$r_{\text{min}} = \frac{v^2}{a_{\text{max}}} = \frac{(216\,\text{km/h})^2}{(0.050)(9.8\,\text{m/s}^2)} = 7.3 \times 10^3\,\text{m}\,.$$

(b) The speed of the train must be reduced to no more than

$$v = \sqrt{a_{\max} r} = \sqrt{0.050(9.8)(1.00 \times 10^3)} = 22 \text{ m/s}$$

which is roughly 80 km/h.

49. (a) Since the wheel completes 5 turns each minute, its period is one-fifth of a minute, or 12 s.

(b) The magnitude of the centripetal acceleration is given by $a = v^2/R$, where R is the radius of the wheel, and v is the speed of the passenger. Since the passenger goes a distance $2\pi R$ for each revolution, his speed is

$$v = \frac{2\pi(15\,\text{m})}{12\,\text{s}} = 7.85\,\text{m/s}$$

and his centripetal acceleration is

$$a = \frac{(7.85\,\text{m/s})^2}{15\,\text{m}} = 4.1\,\text{m/s}^2 .$$

When the passenger is at the highest point, his centripetal acceleration is downward, toward the center of the orbit.

(c) At the lowest point, the centripetal acceleration vector points up, toward the center of the orbit. It has the same magnitude as in part (b).

50. We apply Eq. 4-33 to solve for speed v and Eq. 4-32 to find centripetal acceleration a.

(a) $v = 2\pi r/T = 2\pi(20\,\text{km})/1.0\,\text{s} = 1.3 \times 10^5\,\text{km/s}$.

(b)

$$a = \frac{v^2}{r} = \frac{(126\,\text{km/s})^2}{20\,\text{km}} = 7.9 \times 10^5\,\text{m/s}^2 .$$

(c) Clearly, both v and a will increase if T is reduced.

51. To calculate the centripetal acceleration of the stone, we need to know its speed during its circular motion (this is also its initial speed when it flies off). We use the kinematic equations of projectile motion (discussed in §4-6) to find that speed. Taking the $+y$ direction to be upward and placing the origin at the point where the stone leaves its circular orbit, then the coordinates of the stone during its motion as a projectile are given by $x = v_0 t$ and $y = -\frac{1}{2}gt^2$ (since $v_{0\,y} = 0$). It hits the ground at $x = 10\,\text{m}$ and $y = -2.0\,\text{m}$. Formally solving the second equation for the time, we obtain $t = \sqrt{-2y/g}$, which we substitute into the first equation:

$$v_0 = x\sqrt{-\frac{g}{2y}} = (10\,\text{m})\sqrt{-\frac{9.8\,\text{m/s}^2}{2(-2.0\,\text{m})}} = 15.7 \text{ m/s} .$$

Therefore, the magnitude of the centripetal acceleration is

$$a = \frac{v^2}{r} = \frac{(15.7\,\text{m/s})^2}{1.5\,\text{m}} = 160 \text{ m/s}^2 .$$

52. We write our magnitude-angle results in the form $(R \angle \theta)$ with SI units for the magnitude understood (m for distances, m/s for speeds, m/s^2 for accelerations). All angles θ are measured counterclockwise from $+x$, but we will occasionally refer to angles ϕ which are measured counterclockwise from the vertical line between the circle-center and the coordinate origin and the line drawn from the circle-center to the particle location (see r in the figure). We note that the speed of the particle is $v = 2\pi r/T$ where $r = 3.00$ m and $T = 20.0$ s; thus, $v = 0.942$ m/s. The particle is moving counterclockwise in Fig. 4-37.

(a) At $t = 5.00$ s, the particle has traveled a fraction of

$$\frac{t}{T} = \frac{5.00}{20.0} = \frac{1}{4}$$

of a full revolution around the circle (starting at the origin). Thus, relative to the circle-center, the particle is at

$$\phi = \frac{1}{4}(360°) = 90°$$

measured from vertical (as explained above). Referring to Fig. 4-37, we see that this position (which is the "3 o'clock" position on the circle) corresponds to $x = 3.00$ m and $y = 3.00$ m relative to the coordinate origin. In our magnitude-angle notation, this is expressed as $(R \angle \theta) = (4.24 \angle 45°)$. Although this position is easy to analyze without resorting to trigonometric relations, it is useful (for the computations below) to note that these values of x and y relative to coordinate origin can be gotten from the angle ϕ from the relations $x = r \sin\phi$ and $y = r - r\cos\phi$. Of course, $R = \sqrt{x^2 + y^2}$ and θ comes from choosing the appropriate possibility from $\tan^{-1}(y/x)$ (or by using particular functions of vector capable calculators).

(b) At $t = 7.50$ s, the particle has traveled a fraction of $7.50/20.0 = 3/8$ of a revolution around the circle (starting at the origin). Relative to the circle-center, the particle is therefore at $\phi = 3/8\,(360°) = 135°$ measured from vertical in the manner discussed above. Referring to Fig. 4-37, we compute that this position corresponds to $x = 3.00 \sin 135° = 2.12$ m and $y = 3.00 - 3.00 \cos 135° = 5.12$ m relative to the coordinate origin. In our magnitude-angle notation, this is expressed as $(R \angle \theta) = (5.54 \angle 67.5°)$.

(c) At $t = 10.0$ s, the particle has traveled a fraction of $10.0/20.0 = 1/2$ of a revolution around the circle. Relative to the circle-center, the particle is at $\phi = 180°$ measured from vertical (see explanation, above). Referring to Fig. 4-37, we see that this position corresponds to $x = 0$ and $y = 6.00$ m relative to the coordinate origin. In our magnitude-angle notation, this is expressed as $(R \angle \theta) = (6.00 \angle 90.0°)$.

(d) We subtract the position vector in part (a) from the position vector in part (c): $(6.00 \angle 90.0°) - (4.24 \angle 45°) = (4.24 \angle 135°)$ using magnitude-angle notation (convenient when using vector capable calculators). If we wish instead to use unit-vector notation, we write

$$\Delta\vec{R} = (0 - 3)\,\hat{\imath} + (6 - 3)\,\hat{\jmath} = -3\,\hat{\imath} + 3\,\hat{\jmath}$$

which leads to $|\Delta\vec{R}| = 4.24$ m and $\theta = 135°$.

(e) From Eq. 4-8, we have

$$\vec{v}_{\text{avg}} = \frac{\Delta\vec{R}}{\Delta t} \quad \text{where } \Delta t = 5.00 \text{ s}$$

which produces $-0.6\,\hat{\imath} + 0.6\,\hat{\jmath}$ m/s in unit-vector notation or $(0.849 \angle 135°)$ in magnitude-angle notation.

(f) The speed has already been noted ($v = 0.942$ m/s), but its direction is best seen by referring again to Fig. 4-37. The velocity vector is tangent to the circle at its "3 o'clock position" (see part (a)), which means $\vec{v}$ is vertical. Thus, our result is $(0.942 \angle 90°)$.

(g) Again, the speed has been noted above ($v = 0.942$ m/s), but its direction is best seen by referring to Fig. 4-37. The velocity vector is tangent to the circle at its "12 o'clock position" (see part (c)), which means $\vec{v}$ is horizontal. Thus, our result is $(0.942 \angle 180°)$.

(h) The acceleration has magnitude $v^2/r = 0.296$ m/s^2, and at this instant (see part (a)) it is horizontal (towards the center of the circle). Thus, our result is $(0.296 \angle 180°)$.

(i) Again, $a = v^2/r = 0.296$ m/s^2, but at this instant (see part (c)) it is vertical (towards the center of the circle). Thus, our result is $(0.296 \angle 270°)$.

53. We use Eq. 4-15 first using velocities relative to the truck (subscript t) and then using velocities relative to the ground (subscript g). We work with SI units, so 20 km/h → 5.6 m/s, 30 km/h → 8.3 m/s, and 45 km/h → 12.5 m/s. We choose east as the $+\hat{\imath}$ direction.

(a) The velocity of the cheetah (subscript c) at the end of the 2.0 s interval is (from Eq. 4-42)

$$\vec{v}_{\mathrm{c\,t}} = \vec{v}_{\mathrm{c\,g}} - \vec{v}_{\mathrm{t\,g}} = 12.5\,\hat{\imath} - (-5.6\,\hat{\imath}) = 18.1\,\hat{\imath}\ \mathrm{m/s}$$

relative to the truck. The (average) acceleration vector relative to the cameraman (in the truck) is

$$\vec{a}_{\mathrm{avg}} = \frac{18.1\,\hat{\imath} - (-8.3\,\hat{\imath})}{2.0} = 13\,\hat{\imath}\ \mathrm{m/s^2}\ .$$

(b) The velocity of the cheetah at the start of the 2.0 s interval is (from Eq. 4-42)

$$\vec{v}_{0\,\mathrm{c\,g}} = \vec{v}_{0\,\mathrm{c\,t}} + \vec{v}_{0\,\mathrm{t\,g}} = (-8.3\,\hat{\imath}) + (-5.6\,\hat{\imath}) = -13.9\,\hat{\imath}\ \mathrm{m/s}$$

relative to the ground. The (average) acceleration vector relative to the crew member (on the ground) is

$$\vec{a}_{\mathrm{avg}} = \frac{12.5\,\hat{\imath} - (-13.9\,\hat{\imath})}{2.0} = 13\,\hat{\imath}\ \mathrm{m/s^2}$$

identical to the result of part (a).

54. We choose upstream as the $+\hat{\imath}$ direction, and use Eq. 4-42.

(a) The subscript b is for the boat, w is for the water, and g is for the ground.

$$\vec{v}_{\mathrm{b\,g}} = \vec{v}_{\mathrm{b\,w}} + \vec{v}_{\mathrm{w\,g}} = (14\,\mathrm{km/h})\,\hat{\imath} + (-9\,\mathrm{km/h})\,\hat{\imath} = (5\,\mathrm{km/h})\,\hat{\imath}$$

(b) And we use the subscript c for the child.

$$\vec{v}_{\mathrm{c\,g}} = \vec{v}_{\mathrm{c\,b}} + \vec{v}_{\mathrm{b\,g}} = (-6\,\mathrm{km/h})\,\hat{\imath} + (5\,\mathrm{km/h})\,\hat{\imath} = (-1\,\mathrm{km/h})\,\hat{\imath}$$

55. When the escalator is stalled the speed of the person is $v_p = \ell/t$, where ℓ is the length of the escalator and t is the time the person takes to walk up it. This is $v_p = (15\,\mathrm{m})/(90\,\mathrm{s}) = 0.167\,\mathrm{m/s}$. The escalator moves at $v_e = (15\,\mathrm{m})/(60\,\mathrm{s}) = 0.250\,\mathrm{m/s}$. The speed of the person walking up the moving escalator is $v = v_p + v_e = 0.167\,\mathrm{m/s} + 0.250\,\mathrm{m/s} = 0.417\,\mathrm{m/s}$ and the time taken to move the length of the escalator is

$$t = \ell/v = (15\,\mathrm{m})/(0.417\,\mathrm{m/s}) = 36\,\mathrm{s}\ .$$

If the various times given are independent of the escalator length, then the answer does not depend on that length either. In terms of ℓ (in meters) the speed (in meters per second) of the person walking on the stalled escalator is $\ell/90$, the speed of the moving escalator is $\ell/60$, and the speed of the person walking on the moving escalator is $v = (\ell/90) + (\ell/60) = 0.0278\ell$. The time taken is $t = \ell/v = \ell/0.0278\ell = 36\,\mathrm{s}$ and is independent of ℓ.

56. We denote the velocity of the player with $\vec{v}_1$ and the relative velocity between the player and the ball be $\vec{v}_2$. Then the velocity $\vec{v}$ of the ball relative to the field is given by $\vec{v} = \vec{v}_1 + \vec{v}_2$. The smallest angle θ_{min} corresponds to the case when
$\vec{v} \perp \vec{v}_1$. Hence,

$$\begin{aligned}\theta_{\mathrm{min}} &= 180^\circ - \cos^{-1}\left(\frac{|\vec{v}_1|}{|\vec{v}_2|}\right) \\ &= 180^\circ - \cos^{-1}\left(\frac{4.0\,\mathrm{m/s}}{6.0\,\mathrm{m/s}}\right) \\ &\approx 130^\circ\ .\end{aligned}$$

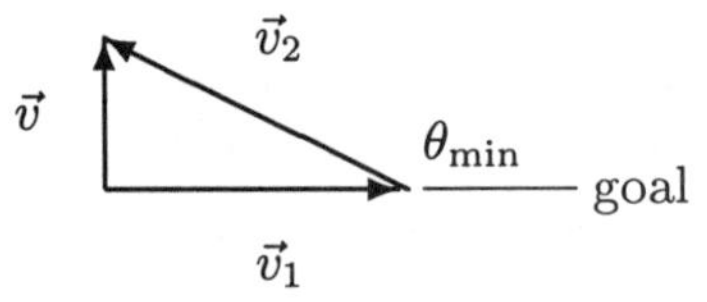

57. Relative to the car the velocity of the snowflakes has a vertical component of 8.0 m/s and a horizontal component of 50 km/h = 13.9 m/s. The angle θ from the vertical is found from

$$\tan\theta = v_h/v_v = (13.9\,\text{m/s})/(8.0\,\text{m/s}) = 1.74$$

which yields $\theta = 60°$.

58. We denote the police and the motorist with subscripts p and m, respectively. The coordinate system is indicated in Fig. 4-38.

(a) The velocity of the motorist with respect to the police car is

$$\vec{v}_{m\,p} = \vec{v}_m - \vec{v}_p = -60\hat{\jmath} - (-80\hat{\imath}) = 80\hat{\imath} - 60\hat{\jmath}\ (\text{km/h})\ .$$

(b) $\vec{v}_{m\,p}$ does happen to be along the line of sight. Referring to Fig. 4-38, we find the vector pointing from car to another is $\vec{r} = 800\hat{\imath} - 600\hat{\jmath}$ m (from M to P). Since the ratio of components in $\vec{r}$ is the same as in $\vec{v}_{m\,p}$, they must point the same direction.

(c) No, they remain unchanged.

59. Since the raindrops fall vertically relative to the train, the horizontal component of the velocity of a raindrop is $v_h = 30\,\text{m/s}$, the same as the speed of the train. If v_v is the vertical component of the velocity and θ is the angle between the direction of motion and the vertical, then $\tan\theta = v_h/v_v$. Thus $v_v = v_h/\tan\theta = (30\,\text{m/s})/\tan 70° = 10.9\,\text{m/s}$. The speed of a raindrop is $v = \sqrt{v_h^2 + v_v^2} = \sqrt{(30\,\text{m/s})^2 + (10.9\,\text{m/s})^2} = 32\,\text{m/s}$.

60. Here, the subscript W refers to the water. Our coordinates are chosen with $+x$ being *east* and $+y$ being *north*. In these terms, the angle specifying *east* would be 0° and the angle specifying *south* would be −90° or 270°. Where the length unit is not displayed, km is to be understood.

(a) We have $\vec{v}_{A\,W} = \vec{v}_{A\,B} + \vec{v}_{B\,W}$, so that $\vec{v}_{A\,B} = (22\ \angle -90°) - (40\ \angle\ 37°) = (56\ \angle -125°)$ in the magnitude-angle notation (conveniently done with a vector capable calculator in polar mode). Converting to rectangular components, we obtain

$$\vec{v}_{A\,B} = -32\hat{\imath} - 46\hat{\jmath}\ \ \text{km/h}\ .$$

Of course, this could have been done in unit-vector notation from the outset.

(b) Since the velocity-components are constant, integrating them to obtain the position is straightforward ($\vec{r} - \vec{r}_0 = \int \vec{v}\,dt$)

$$\vec{r} = (2.5 - 32t)\,\hat{\imath} + (4.0 - 46t)\,\hat{\jmath}$$

with lengths in kilometers and time in hours.

(c) The magnitude of this $\vec{r}$ is

$$r = \sqrt{(2.5 - 32t)^2 + (4.0 - 46t)^2}$$

We minimize this by taking a derivative and requiring it to equal zero – which leaves us with an equation for t

$$\frac{dr}{dt} = \frac{1}{2}\frac{6286t - 528}{\sqrt{(2.5 - 32t)^2 + (4.0 - 46t)^2}} = 0$$

which yields $t = 0.084$ h.

(d) Plugging this value of t back into the expression for the distance between the ships (r), we obtain $r = 0.2$ km. Of course, the calculator offers more digits ($r = 0.225...$), but they are not significant; in fact, the uncertainties implicit in the given data, here, should make the ship captains worry.

61. The velocity vector (relative to the shore) for ships A and B are given by

$$\vec{v}_A = -\left(v_A \cos 45°\right)\hat{\text{i}} + \left(v_A \sin 45°\right)\hat{\text{j}}$$

and

$$\vec{v}_B = -\left(v_B \sin 40°\right)\hat{\text{i}} - \left(v_B \cos 40°\right)\hat{\text{j}}$$

respectively (where $v_A = 24$ knots and $v_B = 28$ knots). We are taking East as $+\hat{\text{i}}$ and North as $\hat{\text{j}}$.

(a) Their relative velocity is

$$\vec{v}_{A\,B} = \vec{v}_A - \vec{v}_B = \left(v_B \sin 40° - v_A \cos 45°\right)\hat{\text{i}} + \left(v_B \cos 40° + v_A \sin 45°\right)\hat{\text{j}}$$

the magnitude of which is $|\vec{v}_{A\,B}| = \sqrt{1.0^2 + 38.4^2} \approx 38$ knots. The angle θ which $\vec{v}_{A\,B}$ makes with North is given by

$$\theta = \tan^{-1}\left(\frac{v_{AB,x}}{v_{AB,y}}\right) = \tan^{-1}\left(\frac{1.0}{38.4}\right) = 1.5°$$

which is to say that $\vec{v}_{A\,B}$ points 1.5° east of north.

(b) Since they started at the same time, their relative velocity describes at what rate the distance between them is increasing. Because the rate is steady, we have

$$t = \frac{|\Delta r_{A\,B}|}{|\vec{v}_{A\,B}|} = \frac{160}{38} = 4.2 \text{ h} .$$

(c) The velocity $\vec{v}_{A\,B}$ does not change with time in this problem, and $\vec{r}_{A\,B}$ is in the same direction as $\vec{v}_{A\,B}$ since they started at the same time. Reversing the points of view, we have $\vec{v}_{A\,B} = -\vec{v}_{B\,A}$ so that $\vec{r}_{A\,B} = -\vec{r}_{B\,A}$ (i.e., they are 180° opposite to each other). Hence, we conclude that B stays at a bearing of 1.5° west of south relative to A during the journey (neglecting the curvature of Earth).

62. The (box)car has velocity $\vec{v}_{c\,g} = v_1\,\hat{\text{i}}$ relative to the ground, and the bullet has velocity

$$\vec{v}_{0\,b\,g} = v_2 \cos\theta\,\hat{\text{i}} + v_2 \sin\theta\,\hat{\text{j}}$$

relative to the ground before entering the car (we are neglecting the effects of gravity on the bullet). While in the car, its velocity relative to the outside ground is $\vec{v}_{b\,g} = 0.8v_2 \cos\theta\,\hat{\text{i}} + 0.8v_2 \sin\theta\,\hat{\text{j}}$ (due to the 20% reduction mentioned in the problem). The problem indicates that the velocity of the bullet in the car *relative to the car* is (with v_3 unspecified) $\vec{v}_{b\,c} = v_3\,\hat{\text{j}}$. Now, Eq. 4-42 provides the condition

$$\begin{aligned} \vec{v}_{b\,g} &= \vec{v}_{b\,c} + \vec{v}_{c\,g} \\ 0.8v_2 \cos\theta\,\hat{\text{i}} + 0.8v_2 \sin\theta\,\hat{\text{j}} &= v_3\,\hat{\text{j}} + v_1\,\hat{\text{i}} \end{aligned}$$

so that equating x components allows us to find θ. If one wished to find v_3 one could also equate the y components, and from this, if the car width were given, one could find the time spent by the bullet in the car, but this information is not asked for (which is why the width is irrelevant). Therefore, examining the x components in SI units leads to

$$\theta = \cos^{-1}\left(\frac{v_1}{0.8v_2}\right) = \cos^{-1}\left(\frac{85\left(\frac{1000}{3600}\right)}{0.8(650)}\right)$$

which yields 87° for the direction of $\vec{v}_{b\,g}$ (measured from $\hat{\text{i}}$, which is the direction of motion of the car). The problem asks, "from what direction was it fired?" – which means the answer is not 87° but rather its supplement 93° (measured from the direction of motion). Stating this more carefully, in the coordinate system we have adopted in our solution, the bullet velocity vector is in the first quadrant, at 87° measured counterclockwise from the $+x$ direction (the direction of train motion), which means that the direction from which the bullet came (where the sniper is) is in the third quadrant, at $-93°$ (that is, 93° measured clockwise from $+x$).

63. We construct a right triangle starting from the clearing on the south bank, drawing a line (200 m long) due north (*upward* in our sketch) across the river, and then a line due west (upstream, leftward in our sketch) along the north bank for a distance $(82\,\text{m}) + (1.1\,\text{m/s})t$, where the t-dependent contribution is the distance that the river will carry the boat downstream during time t.

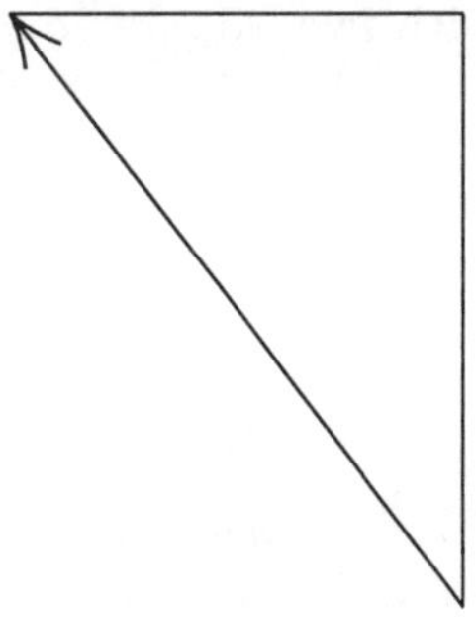

The hypotenuse of this right triangle (the arrow in our sketch) also depends on t and on the boat's speed (relative to the water), and we set it equal to the Pythagorean "sum" of the triangle's sides:

$$(4.0)t = \sqrt{200^2 + (82 + 1.1t)^2}$$

which leads to a quadratic equation for t

$$46724 + 180.4t - 14.8t^2 = 0\ .$$

We solve this and find a positive value: $t = 62.6$ s. The angle between the northward (200 m) leg of the triangle and the hypotenuse (which is measured "west of north") is then given by

$$\theta = \tan^{-1}\left(\frac{82 + 1.1t}{200}\right) = \tan^{-1}\left(\frac{151}{200}\right) = 37°\ .$$

64. (a) We compute the coordinate pairs (x, y) from $x = v_0 \cos\theta t$ and $x = v_0 \sin\theta t - \frac{1}{2}gt^2$ for $t = 20$ s and the speeds and angles given in the problem. We obtain (in kilometers)

$$(x_A\,,\,y_A) = (10.1, 0.56) \qquad (x_B\,,\,y_B) = (12.1, 1.51)$$
$$(x_C\,,\,y_C) = (14.3, 2.68) \qquad (x_D\,,\,y_D) = (16.4, 3.99)$$

and $(x_E\,,\,y_E) = (18.5, 5.53)$ which we plot in the next part.

(b) The vertical (y) and horizontal (x) axes are in kilometers. The graph does not start at the origin. The curve to "fit" the data is not shown, but is easily imagined (forming the "curtain of death").

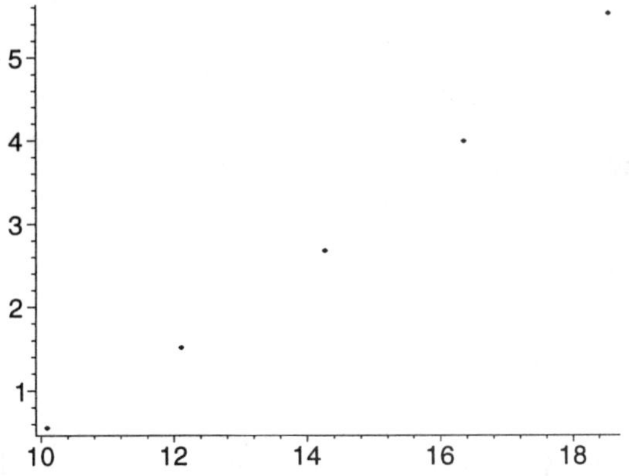

65. We denote $\vec{v}_{\text{PG}}$ as the velocity of the plane relative to the ground, $\vec{v}_{\text{AG}}$ as the velocity of the air relative to the ground, and $\vec{v}_{\text{PA}}$ be the velocity of the plane relative to the air.

(a) The vector diagram is shown below. $\vec{v}_{PG} = \vec{v}_{PA} + \vec{v}_{AG}$. Since the magnitudes v_{PG} and v_{PA} are equal the triangle is isosceles, with two sides of equal length. Consider either of the right triangles formed when the bisector of θ is drawn (the dashed line). It bisects $\vec{v}_{AG}$, so

$$\sin(\theta/2) = \frac{v_{AG}}{2v_{PG}} = \frac{70.0\,\text{mi/h}}{2(135\,\text{mi/h})}$$

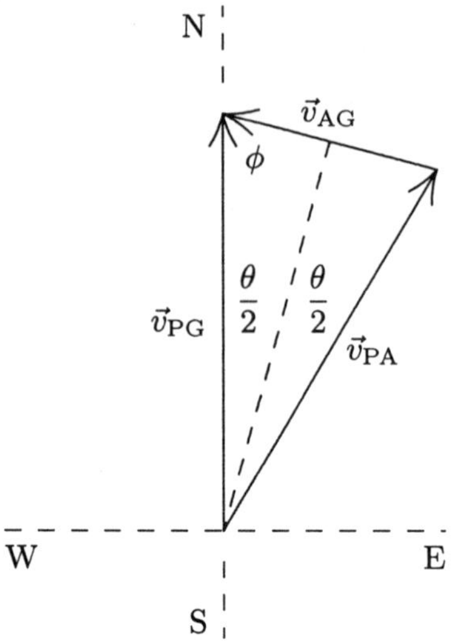

which leads to $\theta = 30.1°$. Now $\vec{v}_{AG}$ makes the same angle with the E-W line as the dashed line does with the N-S line. The wind is blowing in the direction 15° north of west. Thus, it is blowing *from* 75° east of south.

(b) The plane is headed along $\vec{v}_{PA}$, in the direction 30° east of north. There is another solution, with the plane headed 30° west of north and the wind blowing 15° north of east (that is, from 75° west of south).

66. (a) The ball must increase in height by $\Delta y = 0.193$ m and cover a horizontal distance $\Delta x = 0.910$ m during a very short time $t_0 = 1.65 \times 10^{-2}$ s. The statement that the "initial curvature of the ball's path can be neglected" is essentially the same as saying the average velocity for $0 \le t \le t_0$ may be taken equal to the instantaneous initial velocity $\vec{v}_0$. Thus, using Eq. 4-8 to figure its two components, we have

$$\tan\theta_0 = \frac{v_{0y}}{v_{0x}} = \frac{\frac{\Delta y}{t_0}}{\frac{\Delta x}{t_0}} = \frac{\Delta y}{\Delta x}$$

so that $\theta_0 = \tan^{-1}(0.193/0.910) = 12°$.

(b) The magnitude of $\vec{v}_0$ is

$$\sqrt{v_{0x}^2 + v_{0y}^2} = \sqrt{\left(\frac{\Delta x}{t_0}\right)^2 + \left(\frac{\Delta y}{t_0}\right)^2} = \frac{\sqrt{\Delta x^2 + \Delta y^2}}{t_0}$$

which yields $v_0 = 56.4$ m/s.

(c) The range is given by Eq. 4-26:

$$R = \frac{v_0^2}{g}\sin 2\theta_0 = 132 \text{ m} .$$

(d) Partly because of its dimpled surface (but other air-flow related effects are important here) the golf ball travels farther than one would expect based on the simple projectile-motion analysis done here.

67. (a) Since the performer returns to the original level, Eq. 4-26 applies. With $R = 4.0$ m and $\theta_0 = 30°$, the initial speed (for the projectile motion) is consequently

$$v_0 = \sqrt{\frac{gR}{\sin 2\theta_0}} = 6.7 \text{ m/s} .$$

This is, of course, the final speed v for the Air Ramp's acceleration process (for which the initial speed is taken to be zero) Then, for that process, Eq. 2-11 leads to

$$a = \frac{v}{t} = \frac{6.7}{0.25} = 27 \text{ m/s}^2 .$$

We express this as a multiple of g by setting up a ratio: $a = (27/9.8)g = 2.7g$.

(b) Repeating the above steps for $R = 12$ m, $t = 0.29$ s and $\theta_0 = 45°$ gives $a = 3.8g$.

68. The initial position vector $\vec{r}_o$ satisfies $\vec{r} - \vec{r}_o = \Delta\vec{r}$, which results in

$$\vec{r}_o = \vec{r} - \Delta\vec{r} = (3\,\hat{j} - 4\,\hat{k}) - (2\,\hat{i} - 3\,\hat{j} + 6\,\hat{k}) = -2.0\,\hat{i} + 6.0\,\hat{j} - 10\,\hat{k}$$

where the understood unit is meters.

69. We adopt a coordinate system with $\hat{i}$ pointed east and $\hat{j}$ pointed north; the coordinate origin is the flagpole. With SI units understood, we "translate" the given information into unit-vector notation as follows:

$$\vec{r}_o = 40\,\hat{i} \quad \text{and} \quad \vec{v}_o = -10\,\hat{j}$$
$$\vec{r} = 40\,\hat{j} \quad \text{and} \quad \vec{v} = 10\,\hat{i}\,.$$

(a) Using Eq. 4-2, the displacement $\Delta\vec{r}$ is

$$\vec{r} - \vec{r}_o = (56.6 \angle\ 135°)$$

where we have expressed the result in magnitude-angle notation. The displacement has magnitude $40\sqrt{2} = 56.6$ m and points due Northwest.

(b) Eq. 4-8 shows that $\vec{v}_{\rm avg}$ points in the same direction as $\Delta\vec{r}$, and that its magnitude is simply the magnitude of the displacement divided by the time ($\Delta t = 30$ s). Thus, the average velocity has magnitude $56.6/30 = 1.89\,\text{m/s}$ and points due Northwest.

(c) Using Eq. 4-15, we have

$$\vec{a}_{\rm avg} = \frac{\vec{v} - \vec{v}_o}{\Delta t} = 0.333\,\hat{i} + 0.333\,\hat{j}$$

in SI units. The magnitude of the average acceleration vector is therefore $0.333\sqrt{2} = 0.471\,\text{m/s}^2$, and it points due Northeast.

70. The velocity of Larry is v_1 and that of Curly is v_2. Also, we denote the length of the corridor by L. Now, Larry's time of passage is $t_1 = 150\,$s (which must equal L/v_1), and Curly's time of passage is $t_2 = 70\,$s (which must equal L/v_2). The time Moe takes is therefore

$$t = \frac{L}{v_1 + v_2} = \frac{1}{v_1/L + v_2/L} = \frac{1}{\frac{1}{150} + \frac{1}{70}} = 48 \text{ s}\,.$$

71. We choose horizontal x and vertical y axes such that both components of $\vec{v}_0$ are positive. Positive angles are counterclockwise from $+x$ and negative angles are clockwise from it. In unit-vector notation, the velocity at each instant during the projectile motion is

$$\vec{v} = v_0 \cos\theta_0\,\hat{i} + (v_0 \sin\theta_0 - gt)\,\hat{j}\,.$$

(a) With $v_0 = 30$ m/s and $\theta_0 = 60°$, we obtain $\vec{v} = 15\,\hat{i} + \hat{j}$ in m/s, for $t = 2.0$ s. Converting to magnitude-angle notation, this is $\vec{v} = (16 \angle\ 23°)$ with the magnitude in m/s.

(b) Now with $t = 5.0$ s, we find $\vec{v} = (27 \angle -57°)$.

72. (a) The helicopter's speed is $v' = 6.2$ m/s. From the discussions in §4-9 we see that the speed of the package is $v_0 = 12 - v' = 5.8$ m/s, relative to the ground.

(b) Letting $+x$ be in the direction of $\vec{v}_0$ for the package and $+y$ be downward, we have (for the motion of the package)

$$\Delta x = v_0 t \quad \text{and} \quad \Delta y = \frac{1}{2}gt^2$$

where $\Delta y = 9.5$ m. From these, we find $t = 1.39$ s and $\Delta x = 8.08$ m for the package, while $\Delta x'$ (for the helicopter, which is moving in the opposite direction) is $-v'\,t = -8.63$ m. Thus, the horizontal separation between them is $8.08 - (-8.63) = 16.7$ m.

(c) The components of $\vec{v}$ at the moment of impact are $(v_x, v_y) = (5.8, 13.6)$ in SI units. The vertical component has been computed using Eq. 2-11. The angle (which is below horizontal) for this vector is $\tan^{-1}(13.6/5.8) = 67°$.

73. (a) By symmetry, $y = H$ occurs at $x = R/2$ (taking the coordinate origin to be at the launch point). Substituting this into Eq. 4-25 gives

$$H = \frac{R}{2}\tan\theta_0 - \frac{gR^2/4}{2v_0^2\cos^2\theta_0}$$

which leads immediately to

$$\frac{H}{R} = \frac{1}{2}\left(\tan\theta_0 - \frac{gR}{4v_0^2\cos^2\theta_0}\right) .$$

In the far right term, we substitute from Eq. 4-26 for the range:

$$\frac{H}{R} = \frac{1}{2}\left(\tan\theta_0 - \frac{g\left(v_0^2\sin(2\theta_0)/g\right)}{4v_0^2\cos^2\theta_0}\right)$$

which, upon setting $\sin 2\theta_0 = 2\sin\theta_0\cos\theta_0$ and simplifying that last term, yields

$$\frac{H}{R} = \frac{1}{2}\left(\tan\theta_0 - \frac{\sin\theta_0}{2\cos\theta_0}\right)$$

which clearly leads to the relation we wish to prove.

(b) Setting $H/R = 1$ in that relation, we have $\theta_0 = \tan^{-1}(4) = 76°$.

74. (a) The tangent of the angle ϕ is found from the ratio of y to x coordinates of the highest point (taking the coordinate origin to be at the launch point). Using the same notation as in problem 73, we have

$$\phi = \tan^{-1}\left(\frac{H}{\frac{1}{2}R}\right)\tan^{-1}\left(2\frac{H}{R}\right) .$$

Substituting $H/R = \frac{1}{4}\tan\theta_0$ from problem 73, we obtain the relation

$$\tan^{-1}\left(\frac{1}{2}\tan\theta_0\right) .$$

(b) Since $\tan 45° = 1$, then $\phi = \tan^{-1}\left(\frac{1}{2}\right) = 27°$.

75. The initial velocity has magnitude v_0 and because it is horizontal, it is equal to v_x the horizontal component of velocity at impact. Thus, the speed at impact is

$$\sqrt{v_0^2 + v_y^2} = 3v_0 \quad \text{where} \quad v_y = \sqrt{2gh}$$

where we use Eq. 2-16 with Δx replaced with the $h = 20$ m to obtain that second equality. Squaring both sides of the first equality and substituting from the second, we find

$$v_0^2 + 2gh = (3v_0)^2$$

which leads to $gh = 4v_0^2$ and therefore to $v_0 = \sqrt{(9.8)(20)}/2 = 7.0$ m/s.

76. (a) The magnitude of the displacement vector $\Delta\vec{r}$ is given by

$$|\Delta\vec{r}| = \sqrt{21.5^2 + 9.7^2 + 2.88^2} = 23.8 \text{ km} .$$

Thus,

$$|\vec{v}_{\text{avg}}| = \frac{|\Delta\vec{r}|}{\Delta t} = \frac{23.8}{3.50} = 6.79 \text{ km/h} .$$

(b) The angle θ in question is given by

$$\theta = \tan^{-1}\left(\frac{2.88}{\sqrt{21.5^2 + 9.7^2}}\right) = 6.96^\circ \ .$$

77. With no acceleration in the x direction yet a constant acceleration of 1.4 m/s^2 in the y direction, the position (in meters) as a function of time (in seconds) must be

$$\vec{r} = (6.0t)\hat{\imath} + \left(\frac{1}{2}(1.4)t^2\right)\hat{\jmath}$$

and $\vec{v}$ is its derivative with respect to t.

(a) At $t = 3.0$ s, therefore, $\vec{v} = 6.0\hat{\imath} + 4.2\hat{\jmath}$ m/s.

(b) At $t = 3.0$ s, the position is $\vec{r} = 18\hat{\imath} + 6.3\hat{\jmath}$ m.

78. We choose a coordinate system with origin at the clock center and $+x$ rightward (towards the "3:00" position) and $+y$ upward (towards "12:00").

(a) In unit-vector notation, we have (in centimeters) $\vec{r}_1 = 10\hat{\imath}$ and $\vec{r}_2 = -10\hat{\jmath}$. Thus, Eq. 4-2 gives

$$\Delta\vec{r} = \vec{r}_2 - \vec{r}_1 = -10\hat{\imath} - 10\hat{\jmath} \longrightarrow (14 \angle - 135^\circ)$$

where we have switched to magnitude-angle notation in the last step.

(b) In this case, $\vec{r}_1 = -10\hat{\jmath}$ and $\vec{r}_2 = 10\hat{\jmath}$, and $\Delta\vec{r} = 20\hat{\jmath}$ cm.

(c) In a full-hour sweep, the hand returns to its starting position, and the displacement is zero.

79. We let g_p denote the magnitude of the gravitational acceleration on the planet. A number of the points on the graph (including some "inferred" points – such as the max height point at $x = 12.5$ m and $t = 1.25$ s) can be analyzed profitably; for future reference, we label (with subscripts) the first ($(x_0, y_0) = (0, 2)$ at $t_0 = 0$) and last ("final") points ($(x_f, y_f) = (25, 2)$ at $t_f = 2.5$), with lengths in meters and time in seconds.

(a) The x-component of the initial velocity is found from $x_f - x_0 = v_{0x}t_f$. Therefore, $v_{0x} = 25/2.5 =$ 10 m/s. And we try to obtain the y-component from $y_f - y_0 = 0 = v_{0y}t_f - \frac{1}{2}g_p t_f^2$. This gives us $v_{0y} = 1.25g_p$, and we see we need another equation (by analyzing another point, say, the next-to-last one) $y - y_0 = v_{0y}t - \frac{1}{2}g_p t^2$ with $y = 6$ and $t = 2$; this produces our second equation $v_{0y} = 2 + g_p$. Simultaneous solution of these two equations produces results for v_{0y} and g_p (relevant to part (b)). Thus, our complete answer for the initial velocity is $\vec{v} = 10\hat{\imath} + 10\hat{\jmath}$ m/s.

(b) As a by-product of the part (a) computations, we have $g_p = 8.0\,\text{m/s}^2$.

(c) Solving for t_g (the time to reach the ground) in $y_g = 0 = y_0 + v_{0y}t_g - \frac{1}{2}g_p t_g^2$ leads to a positive answer: $t_g = 2.7$ s.

(d) With $g = 9.8\,\text{m/s}^2$, the method employed in part (c) would produce the quadratic equation $-4.9t_g^2 + 10t_g + 2 = 0$ and then the positive result $t_g = 2.2$ s.

80. At maximum height, the y-component of a projectile's velocity vanishes, so the given 10 m/s is the (constant) x-component of velocity.

(a) Using v_{0y} to denote the y-velocity 1.0 s before reaching the maximum height, then (with $v_y = 0$) the equation $v_y = v_{0y} - gt$ leads to $v_{0y} = 9.8$ m/s. The magnitude of the velocity vector at that moment (also known as the *speed*) is therefore

$$\sqrt{v_x^2 + {v_{0y}}^2} = \sqrt{10^2 + 9.8^2} = 14 \text{ m/s} \ .$$

(b) It is clear from the symmetry of the problem that the speed is the same 1.0 s after reaching the top, as it was 1.0 s before (14 m/s again). This may be verified by using $v_y = v_{0y} - gt$ again but now "starting the clock" at the highest point so that $v_{0y} = 0$ (and $t = 1.0$ s). This leads to $v_y = -9.8$ m/s and ultimately to $\sqrt{10^2 + (-18)^2} = 14$ m/s.

(c) With v_{0y} denoting the y-component of velocity one second before the top of the trajectory – as in part (a) – then we have $y = 0 = y_0 + v_{0y}t - \frac{1}{2}gt^2$ where $t = 1.0$ s. This yields $y_0 = -4.9$ m. Alternatively, Eq. 2-18 could have been used, with $v_y = 0$ to the same end. The x_0 value more simply results from $x = 0 = x_0 + (10\,\text{m/s})(1.0\,\text{s})$. Thus, the coordinates (in meters) of the projectile one second before reaching maximum height is $(-10, -4.9)$.

(d) It is clear from symmetry that the coordinate one second after the maximum height is reached is $(10, -4.9)$ (in meters). But this can be verified by considering $t = 0$ at the top and using $y - y_0 = v_{0y}t - \frac{1}{2}gt^2$ where $y_0 = v_{0y} = 0$ and $t = 1$ s. And by using $x - x_0 = (10\,\text{m/s})(1.0\,\text{s})$ where $x_0 = 0$. Thus, $x = 10$ m and $y = -4.9$ m is obtained.

81. With $g_B = 9.8128$ m/s^2 and $g_M = 9.7999$ m/s^2, we apply Eq. 4-26:

$$R_M - R_B = \frac{v_0^2 \sin 2\theta_0}{g_M} - \frac{v_0^2 \sin 2\theta_0}{g_B} = \frac{v_0^2 \sin 2\theta_0}{g_B}\left(\frac{g_B}{g_M} - 1\right)$$

which becomes

$$R_M - R_B = R_B\left(\frac{9.8128}{9.7999} - 1\right)$$

and yields (upon substituting $R_B = 8.09$ m) $R_M - R_B = 0.01$ m.

82. (a) Using the same coordinate system assumed in Eq. 4-25, we rearrange that equation to solve for the initial speed:

$$v_0 = \frac{x}{\cos\theta_0}\sqrt{\frac{g}{2\,(x\tan\theta_0 - y)}}$$

which yields $v_0 = 255.5 \approx 2.6 \times 10^2$ m/s for $x = 9400$ m, $y = -3300$ m, and $\theta_0 = 35°$.

(b) From Eq. 4-21, we obtain the time of flight:

$$t = \frac{x}{v_0\cos\theta_0} = \frac{9400}{255.5\cos 35°} = 45 \text{ s} .$$

(c) We expect the air to provide resistance but no appreciable lift to the rock, so we would need a greater launching speed to reach the same target.

83. (a) Using the same coordinate system assumed in Eq. 4-22, we solve for $y = h$:

$$h = y_0 + v_0\sin\theta_0 - \frac{1}{2}gt^2$$

which yields $h = 51.8$ m for $y_0 = 0$, $v_0 = 42$ m/s, $\theta_0 = 60°$ and $t = 5.5$ s.

(b) The horizontal motion is steady, so $v_x = v_{0x} = v_0\cos\theta_0$, but the vertical component of velocity varies according to Eq. 4-23. Thus, the speed at impact is

$$v = \sqrt{(v_0\cos\theta_0)^2 + (v_0\sin\theta_0 - gt)^2} = 27 \text{ m/s} .$$

(c) We use Eq. 4-24 with $v_y = 0$ and $y = H$:

$$H = \frac{(v_0\sin\theta_0)^2}{2g} = 67.5 \text{ m} .$$

84. (a) Using the same coordinate system assumed in Eq. 4-25, we find

$$y = x\tan\theta_0 - \frac{gx^2}{2\left(v_0\cos\theta_0\right)^2} = -\frac{gx^2}{2v_0^2} \quad \text{if } \theta_0 = 0 \ .$$

Thus, with $v_0 = 3 \times 10^6$ m/s and $x = 1$ m, we obtain $y = -5.4 \times 10^{-13}$ m which is not practical to measure (and suggests why gravitational processes play such a small role in the fields of atomic and subatomic physics).

(b) It is clear from the above expression that $|y|$ decreases as v_0 is reduced.

85. (a) Using the same coordinate system assumed in Eq. 4-21, we obtain the time of flight

$$t = \frac{\Delta x}{v_0\cos\theta_0} = \frac{20}{15\cos 35^\circ} = 1.63 \text{ s} \ .$$

(b) At that moment, its height of above the ground (taking $y_0 = 0$) is

$$y = \left(v_0\sin\theta_0\right)t - \frac{1}{2}gt^2 = 1.02 \text{ m} \ .$$

Thus, the ball is 18 cm below the center of the circle; since the circle radius is 15 cm, we see that it misses it altogether.

(c) The horizontal component of velocity (at $t = 1.63$ s) is the same as initially:

$$v_x = v_{0x} = v_0\cos\theta_0 = 15\cos 35^\circ = 12.3 \text{ m/s} \ .$$

The vertical component is given by Eq. 4-23:

$$v_y = v_0\sin\theta_0 - gt = 15\sin 35^\circ - (9.8)(1.63) = -7.3 \text{ m/s} \ .$$

Thus, the magnitude of its speed at impact is $\sqrt{v_x^2 + v_y^2} = 14.3$ m/s.

(d) As we saw in the previous part, the sign of v_y is negative, implying that it is now heading down (after reaching its max height).

86. (a) From Eq. 4-22 (with $\theta_0 = 0$), the time of flight is

$$t = \sqrt{\frac{2h}{g}} = \sqrt{\frac{2(45)}{9.8}} = 3.03 \text{ s} \ .$$

(b) The horizontal distance traveled is given by Eq. 4-21:

$$\Delta x = v_0 t = (250)(3.03) = 758 \text{ m} \ .$$

(c) And from Eq. 4-23, we find

$$|v_y| = gt = (9.80)(3.03) = 29.7 \text{ m/s} \ .$$

87. Using the same coordinate system assumed in Eq. 4-25, we find x for the elevated cannon from

$$y = x\tan\theta_0 - \frac{gx^2}{2\left(v_0\cos\theta_0\right)^2} \quad \text{where } y = -30 \text{ m.}$$

Using the quadratic formula (choosing the positive root), we find

$$x = v_0\cos\theta_0\left(\frac{v_0\sin\theta_0 + \sqrt{\left(v_0\sin\theta_0\right)^2 - 2gy}}{g}\right)$$

which yields $x = 715$ m for $v_0 = 82$ m/s (from Sample Problem 4-7) and $\theta_0 = 45°$. This is 29 m longer than the 686 m found in that Sample Problem. The "9" in 29 m is not reliable, considering the low level of precision in the given data.

88. (a) With $r = 0.15$ m and $a = 3.0 \times 10614$ m/s^2, Eq. 4-32 gives

$$v = \sqrt{ra} = 6.7 \times 10^6 \text{ m/s} .$$

(b) The period is given by Eq. 4-33:

$$T = \frac{2\pi r}{v} = 1.4 \times 10^{-7} \text{ s} .$$

89. The type of acceleration involved in steady-speed circular motion is the centripetal acceleration $a = v^2/r$ which is at each moment directed towards the center of the circle. The radius of the circle is $r = 12^2/3 = 48$ m. Thus, the car is at the present moment 48 m west of the center of its circular path; this is equally true in part (a) and part (b).

90. (a) With $v = c/10 = 3 \times 10^7$ m/s and $a = 20g = 196$ m/s^2, Eq. 4-32 gives

$$r = \frac{v^2}{a} = 4.6 \times 10^{12} \text{ m} .$$

(b) The period is given by Eq. 4-33:

$$T = \frac{2\pi r}{v} = 9.6 \times 10^5 \text{ s} .$$

Thus, the time to make a quarter-turn is $T/4 = 2.4 \times 10^5$ s or about 2.8 days.

91. (a) Using the same coordinate system assumed in Eq. 4-21 and Eq. 4-22 (so that $\theta_0 = -20.0°$), we use $v_0 = 15.0$ m/s and find the horizontal displacement of the ball at $t = 2.30$ s:

$$\Delta x = (v_0 \cos\theta_0)\, t = 32.4 \text{ m} .$$

(b) And we find the vertical displacement:

$$\Delta y = (v_0 \sin\theta_0)\, t - \frac{1}{2} g t^2 = -37.7 \text{ m} .$$

92. This is a classic problem involving two-dimensional relative motion; see §4-9. The steps in Sample Problem 4-11 in the textbook are similar to those used here. We align our coordinates so that *east* corresponds to $+x$ and *north* corresponds to $+y$. We write the vector addition equation as $\vec{v}_{BG} = \vec{v}_{BW} + \vec{v}_{WG}$. We have $\vec{v}_{WG} = (2.0 \angle 0°)$ in the magnitude-angle notation (with the unit m/s understood), or $\vec{v}_{WG} = 2.0\hat{\imath}$ in unit-vector notation. We also have $\vec{v}_{BW} = (8.0 \angle 120°)$ where we have been careful to phrase the angle in the 'standard' way (measured counterclockwise from the $+x$ axis), or $\vec{v}_{BW} = -4.0\hat{\imath} + 6.9\hat{\jmath}$.

(a) We can solve the vector addition equation for $\vec{v}_{BG}$:

$$\vec{v}_{BG} = \vec{v}_{BW} + \vec{v}_{WG} = (2.0 \angle 0°) + (8.0 \angle 120°) = (7.2 \angle 106°)$$

which is very efficiently done using a vector capable calculator in polar mode. Thus $|\vec{v}_{BG}| = 7.2$ m/s, and its direction is 16° west of north, or 74° north of west.

(b) The velocity is constant, and we apply $y - y_0 = v_y t$ in a reference frame. Thus, in the *ground* reference frame, we have $200 = 7.2 \sin(106°) t \rightarrow t = 29$ s. Note: if a student obtains "28 s", then the student has probably neglected to take the y component properly (a common mistake).

93. The topic of relative motion (with constant velocity motion) in a two-dimensional setting is covered in §4-9. We note that

$$\vec{v}_{PG} = \vec{v}_{PA} + \vec{v}_{AG}$$

describes a right triangle, with one leg being $\vec{v}_{PG}$ (east), another leg being $\vec{v}_{AG}$ (magnitude = 20, direction = south), and the hypotenuse being $\vec{v}_{PA}$ (magnitude = 70). Lengths are in kilometers and time is in hours. Using the Pythagorean theorem, we have

$$|\vec{v}_{PA}| = \sqrt{|\vec{v}_{PG}|^2 + |\vec{v}_{AG}|^2} \implies 70 = \sqrt{|\vec{v}_{PG}|^2 + 20^2}$$

which is easily solved for the ground speed: $|\vec{v}_{PG}| = 67$ km/h.

94. Our coordinate system has $\hat{\imath}$ pointed east and $\hat{\jmath}$ pointed north. All distances are in kilometers, times in hours, and speeds in km/h. The first displacement is $\vec{r}_{AB} = 483\hat{\imath}$ and the second is $\vec{r}_{BC} = -966\hat{\jmath}$.

(a) The net displacement is

$$\vec{r}_{AC} = \vec{r}_{AB} + \vec{r}_{BC} = 483\hat{\imath} - 966\hat{\jmath} \longrightarrow (1080 \angle -63.4°)$$

where we have expressed the result in magnitude-angle notation in the last step. We observe that the angle can be alternatively expressed as 63.4° south of east, or 26.6° east of south.

(b) Dividing the magnitude of $\vec{r}_{AC}$ by the total time (2.25 h) gives the magnitude of $\vec{v}_{\text{avg}}$ and its direction is the same as in part (a). Thus, $\vec{v}_{\text{avg}} = (480 \angle -63.4°)$ in magnitude-angle notation (with km/h understood).

(c) Assuming the AB trip was a straight one, and similarly for the BC trip, then $|\vec{r}_{AB}|$ is the distance traveled during the AB trip, and $|\vec{r}_{BC}|$ is the distance traveled during the BC trip. Since the average speed is the total distance divided by the total time, it equals

$$\frac{483 + 966}{2.25} = 644 \text{ km/h} .$$

95. We take the initial (x, y) specification to be $(0.000, 0.762)$ m, and the positive x direction to be towards the "green monster." The components of the initial velocity are $(33.53 \angle 55°) \rightarrow (19.23, 27.47)$ m/s.

(a) With $t = 5.00$ s, we have $x = x_0 + v_x t = 96.2$ m.

(b) At that time, $y = y_0 + v_{0y}t - \frac{1}{2}gt^2 = 15.59$ m, which is 4.31 m above the wall.

(c) The moment in question is specified by $t = 4.50$ s. At that time, $x - x_0 = (19.23)(4.5) = 86.5$ m, and $y = y_0 + v_{0y}t - \frac{1}{2}gt^2 = 25.1$ m.

96. The displacement of the one-way trip is the same as the displacement, which has magnitude $D = 4350$ km for the flight (we are in a frame of reference that rotates with the earth). The velocity of the flight relative to the earth is

$$\vec{v}_{fe} = \vec{va} + \vec{ae}$$

where $\vec{ae}$ is the velocity of the (eastward) jet stream (with magnitude $v > 0$), and $\vec{ae}$ is the velocity of the plane relative to the air (with magnitude $u = 966$ m/s). And the magnitudes of the eastward flight velocity (relative to earth) and of the westward flight velocity (primed) are, respectively,

$$|\vec{v}_{fe}| = \frac{D}{t} \quad \text{and} \quad |\vec{v}\,'_{fe}| = \frac{D}{t'} .$$

The time difference (5/6 of an hour) is therefore

$$\begin{aligned} t' - t &= \frac{D}{|\vec{v}\,'_{fe}|} - \frac{D}{|\vec{v}_{fe}|} \\ \Delta t &= \frac{D}{u - v} - \frac{D}{u + v} . \end{aligned}$$

Using the quadratic formula to solve for v, we obtain

$$v = \frac{-D + \sqrt{D^2 + u^2(\Delta t)^2}}{\Delta t} = 89 \text{ km/h} .$$

97. Using the same coordinate system assumed in Eq. 4-25, we rearrange that equation to solve for the initial speed:

$$v_0 = \frac{x}{\cos\theta_0}\sqrt{\frac{g}{2\,(x\tan\theta_0 - y)}}$$

which yields $v_0 = 23$ ft/s for $g = 32$ ft/s^2, $x = 13$ ft, $y = 3$ ft and $\theta_0 = 55°$.

98. We establish coordinates with $\hat{\imath}$ pointing to the far side of the river (perpendicular to the current) and $\hat{\jmath}$ pointing in the direction of the current. We are told that the magnitude (presumed constant) of the velocity of the boat relative to the water is $|\vec{v}_{bw}| = u = 6.4$ km/h. Its angle, relative to the x axis is θ. With km and h as the understood units, the velocity of the water (relative to the ground) is $\vec{v}_{wg} = 3.2\hat{\jmath}$.

(a) To reach a point "directly opposite" means that the velocity of her boat relative to ground must be $\vec{bg} = v\,\hat{\imath}$ where $v > 0$ is unknown. Thus, all $\hat{\jmath}$ components must cancel in the vector sum

$$\vec{v}_{bw} + \vec{v}_{wg} = \vec{v}_{bg}$$

which means the $u\sin\theta = -3.2$, so $\theta = \sin^{-1}(-3.2/6.4) = -30°$.

(b) Using the result from part (a), we find $v = u\cos\theta = 5.5$ km/h. Thus, traveling a distance of $\ell = 6.4$ km requires a time of $6.4/5.5 = 1.15$ h or 69 min.

(c) If her motion is completely along the y axis (as the problem implies) then with $v_w = 3.2$ km/h (the water speed) we have

$$t_{\text{total}} = \frac{D}{u + v_w} + \frac{D}{u - v_w} = 1.33 \text{ h}$$

where $D = 3.2$ km. This is equivalent to 80 min.

(d) Since

$$\frac{D}{u + v_w} + \frac{D}{u - v_w} = \frac{D}{u - v_w} + \frac{D}{u + v_w}$$

the answer is the same as in the previous part.

(e) The case of general θ leads to

$$\vec{v}_{bg} = \vec{v}_{bw} + \vec{v}_{wg} = u\cos\theta\,\hat{\imath} + (u\sin\theta + v_w)\,\hat{\jmath}$$

where the x component of $\vec{v}_{bg}$ must equal ell/t. Thus,

$$t = \frac{\ell}{u\cos\theta}$$

which can be minimized using $dt/d\theta = 0$ (though, of course, an easier way is to appeal to either physical or mathematical intuition – concluding that the shortest-time path should have $\theta = 0$). Then $t = 6.4/6.4 = 1.0$ h, or 60 min.

99. With $v_0 = 30$ m/s and $R = 20$ m, Eq. 4-26 gives

$$\sin 2\theta_0 = \frac{gR}{v_0^2} = 0.218\ .$$

Because $\sin(\phi) = \sin(180° - \phi)$, there are two roots of the above equation:

$$2\theta_0 = \sin^{-1}(0.218) = 12.6° \quad \text{and} \quad 167.4°\ .$$

Therefore, the two possible launch angles that will hit the target (in the absence of air friction and related effects) are $\theta_0 = 6.3°$ and $\theta_0 = 83.7°$. An alternative approach to this problem in terms of Eq. 4-25 (with $y = 0$ and $1/\cos^2 = 1 + \tan^2$) is possible – and leads to a quadratic equation for $\tan\theta_0$ with the roots providing these two possible θ_0 values.

100. (a) The time available before the train arrives at the impact spot is

$$t_{\text{train}} = \frac{40\,\text{m}}{30\,\text{m/s}} = 1.33 \text{ s}$$

(the train does not reduce its speed). We interpret the phrase "distance between the car and the center of the crossing" to refer to the distance from the front bumper of the car to that point. In which case, the car needs to travel a total distance of $\Delta x = 40 + 5 + 1.5 = 46.5$ m in order for its rear bumper and the edge of the train not to collide (the distance from the center of the train to either edge of the train is 1.5 m). With a starting velocity of $v_0 = 30$ m/s and an acceleration of $a = 1.5$ m/s^2, Eq. 2-15 leads to

$$\Delta x = v_0 t + \frac{1}{2}at^2 \implies t = \frac{-v_0 \pm \sqrt{v_0^2 + 2a\Delta x}}{a}$$

which yields (upon taking the positive root) a time $t_{\text{car}} = 1.49$ s needed for the car to make it. Recalling our result for t_{train} we see the car doesn't have enough time available to make it across.

(b) The difference is $t_{\text{car}} - t_{\text{train}} = 0.16$ s. We note that at $t = t_{\text{train}}$ the front bumper of the car is $v_0 t + \frac{1}{2}at^2 = 41.33$ m from where it started, which means it is 1.33 m past the center of the track (but the edge of the track is 1.5 m from the center). If the car was coming from the south, then the point P on the car impacted by the southern-most corner of the front of the train is 2.83 m behind the front bumper (or 2.17 m in front of the rear bumper). The motion of P is what is plotted below (the top graph – looking like a line instead of a parabola because the final speed of the car is not much different than its initial speed). Since the position of the train is on an entirely different axis than that of the car, we plot the distance (in meters) from P to "south" rail of the tracks (the top curve shown), and the distance of the "south" front corner of the train to the line-of-motion of the car (the bottom line shown).

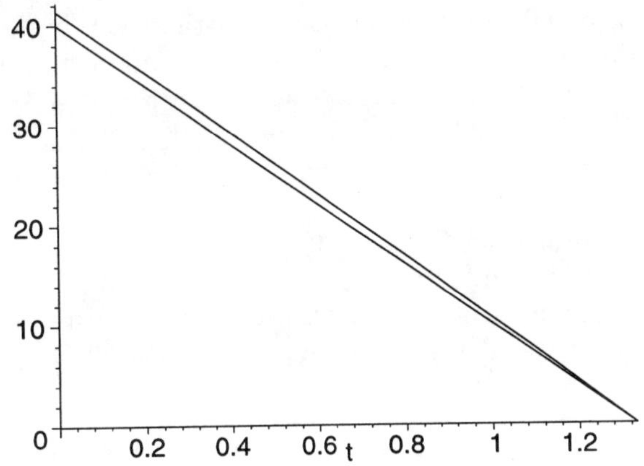

101. (a) With $v_0 = 6.3$ m/s and $R = 0.40$ m, Eq. 4-26 gives

$$\sin 2\theta_0 = \frac{gR}{v_0^2} = 0.0988 .$$

Because $\sin(\phi) = \sin(180° - \phi)$, there are two roots of the above equation:

$$2\theta_0 = \sin^{-1}(0.0988) = 5.7° \quad \text{and} \quad 174.3° .$$

Therefore, the two possible launch angles that will hit the target (in the absence of air friction and related effects) are $\theta_0 = 2.8°$ and $\theta_0 = 87.1°$. But the juggler is trying to achieve a visual effect by having a relatively high trajectory for the balls, so $\theta_0 = 87.1°$ is the result he should choose.

(b) We do not show the graph here. It would be very much like the higher parabola shown in Fig. 4-51.

(c) , (d) and (e) The problem requests that the student work with his graphs, here, but we – for doublechecking purposes – use Eq. 4-26 to calculate $R - 0.40$ m for $\theta_0 - 87.1° = -2°, -1°, 1°$, and $2°$. We obtain the respective values (in meters) $0.28, 0.14, -0.14$, and -0.28.

102. (First problem in **Cluster 1**)
Using the coordinate system employed in §4-5 and §4-6, we have $v_{0x} = v_x > 0$ and $v_{0y} = 0$. Also, $y_0 = h > 0$, $x_0 = 0$, $y = 0$ (when it hits the ground at $t = 3.00$)), and $x = 150$, with lengths in meters and time in seconds.

   (a) The equation $y - y_0 = v_{0y}t - \frac{1}{2}gt^2$ becomes $-h = -\frac{1}{2}(9.8)(3.00)^2$, so that $h = 44.1$ m.

   (b) The equation $v_y = v_{0y} - gt$ gives the y-component of the "final" velocity as $v_y = -(9.8)(3.00) = 29.4$ m/s. The x-component of velocity (which is constant) is computed from $v_x = (x - x_0)/t = 150/3.00 = 50.0$ m/s. Therefore,

$$|\vec{v}| = \sqrt{v_x^2 + v_y^2} = \sqrt{50^2 + 29.4^2} = 58.0 \text{ m/s} .$$

103. (Second problem in **Cluster 1**)
Using the coordinate system employed in §4-5 and §4-6, we have $v_{0x} = v_0 \cos 30° > 0$ and $v_{0y} = v_0 \sin 30° > 0$. Also, $y_0 = 0$ (corresponding to the dashed line in the figure), $x_0 = 0$, $y = h > 0$ (where it lands at $t = 3.00$), and $x = 100$, with lengths in meters and time in seconds.

   (a) The x-equation determines v_0

$$x - x_0 = v_0 \cos(30)t \implies 100 = v_0(0.866)(3.00)$$

   which leads to $v_0 = 38.5$ m/s. The y-equation $y - y_0 = v_{0y}t - \frac{1}{2}gt^2$ becomes $h = (38.5)(\sin 30)(3.00) - \frac{1}{2}(9.8)(3.00)^2 = 13.6$ m.

   (b) As a byproduct of part (a)'s computation, we found $v_0 = 38.5$ m/s.

   (c) Although a somewhat easier method will be found in the energy chapter (especially Chapter 8), we will find the "final" velocity components with the methods of §4-6. We have $v_x = v_{0x} = 38.5 \cos 30 = 33.3$ m/s. And $v_y = v_{0y} - gt = 38.5 \sin 30 - (9.8)(3.00) = -10.2$ m/s. Therefore,

$$|\vec{v}| = \sqrt{v_x^2 + v_y^2} = \sqrt{(33.3)^2 + (-10.2)^2} = 34.8 \text{ m/s} .$$

104. (Third problem in **Cluster 1**)
Following the hint, we have the time-reversed problem with the ball thrown from the roof, towards the left, at 60° measured clockwise from a leftward axis. We see in this time-reversed situation that it is convenient to take $+x$ as *leftward* with positive angles measured clockwise. Lengths are in meters and time is in seconds.

   (a) With $y_0 = 20.0$, and $y = 0$ at $t = 4.00$, we have $y - y_0 = v_{0y}t - \frac{1}{2}gt^2$ where $v_{0y} = v_0 \sin 60°$. This leads to $v_0 = 16.9$ m/s. This plugs into the x-equation (with $x_0 = 0$ and $x = d$) to produce $d = (16.9 \cos 60°)(4.00) = 33.7$ m.

   (b) Although a somewhat easier method will be found in the energy chapter (especially Chapter 8), we will find the "final" velocity components with the methods of §4-6. Note that we're still working the time-reversed problem; this "final" $\vec{v}$ is actually the velocity with which it was thrown. We have $v_x = v_{0x} = 16.9 \cos 60° = 8.43$ m/s. And $v_y = v_{0y} - gt = 16.9 \sin 60° - (9.8)(4.00) = -24.6$ m/s. We convert from rectangular components to polar (that is, magnitude-angle) representation:

$$\vec{v} = (8.43, -24.6) \longrightarrow (26.0 \angle -71.1°) .$$

   and we now interpret our result ("undoing" the time reversal) as an initial velocity of magnitude 26 m/s with angle (up from rightward) of 71°.

105. (Fourth problem in **Cluster 1**)
Following the hint, we have the time-reversed problem with the ball thrown from the ground, towards the right, at 60° measured counterclockwise from a rightward axis. We see in this time-reversed situation that it is convenient to use the familiar coordinate system with $+x$ as *rightward* and with positive angles measured counterclockwise. Lengths are in meters and time is in seconds.

(a) The x-equation (with $x_0 = 0$ and $x = 25.0$) leads to $25 = (v_0 \cos 60°)(1.50)$, so that $v_0 = 33.3$ m/s. And with $y_0 = 0$, and $y = h > 0$ at $t = 1.50$, we have $y - y_0 = v_{0y}t - \frac{1}{2}gt^2$ where $v_{0y} = v_0 \sin 60°$. This leads to $h = 32.3$ m.

(b) Although a somewhat easier method will be found in the energy chapter (especially Chapter 8), we will find the "final" velocity components with the methods of §4-6. Note that we're still working the time-reversed problem; this "final" $\vec{v}$ is actually the velocity with which it was thrown. We have $v_x = v_{0x} = 33.3 \cos 60° = 16.7$ m/s. And $v_y = v_{0y} - gt = 33.3 \sin 60° - (9.8)(1.50) = 14.2$ m/s. We convert from rectangular to polar in terms of the magnitude-angle notation:

$$\vec{v} = (16.7, 14.2) \longrightarrow (21.9 \angle\ 40.4°)\ .$$

We now interpret this result ("undoing" the time reversal) as an initial velocity (from the edge of the building) of magnitude 22 m/s with angle (down from leftward) of 40°.

106. (Fifth problem in **Cluster 1**)
Let $y_0 = 1.0$ m at $x_0 = 0$ when the ball is hit. Let $y_1 = h$ (the height of the wall) and x_1 describe the point where it first rises above the wall one second after being hit; similarly, $y_2 = h$ and x_2 describe the point where it passes back down behind the wall four seconds later. And $y_f = 1.0$ m at $x_f = R$ is where it is caught. Lengths are in meters and time is in seconds.

(a) Keeping in mind that v_x is constant, we have $x_2 - x_1 = 50.0 = v_{1x}(4.00)$, which leads to $v_{1x} = 12.5$ m/s. Thus, applied to the full six seconds of motion: $x_f - x_0 = R = v_x(6.00) = 75.0$ m.

(b) We apply $y - y_0 = v_{0y}t - \frac{1}{2}gt^2$ to the motion above the wall.

$$y_2 - y_1 = 0 = v_{1y}(4.00) - \frac{1}{2}g(4.00)^2$$

leads to $v_{1y} = 19.6$ m/s. One second earlier, using $v_{1y} = v_{0y} - g(1.00)$, we find $v_{0y} = 29.4$ m/s. We convert from (x, y) to magnitude-angle (polar) representation:

$$\vec{v}_0 = (16.7, 14.2) \longrightarrow (31.9 \angle\ 66.9°)\ .$$

We interpret this result as a velocity of magnitude 32 m/s, with angle (up from rightward) of 67°.

(c) During the first 1.00 s of motion, $y = y_0 + v_{0y}t - \frac{1}{2}gt^2$ yields $h = 1.0 + (29.4)(1.00) - \frac{1}{2}(9.8)(1.00)^2 = 25.5$ m.

107. (First problem in **Cluster 2**)

(a) Since $v_y^2 = v_{0y}^2 - 2g\Delta y$, and $v_y = 0$ at the target, we obtain $v_{0y} = \sqrt{2(9.8)(5.00)} = 9.90$ m/s. Since $v_0 \sin\theta_0 = v_{0y}$, with $v_0 = 12$ m/s, we find $\theta_0 = 55.6°$.

(b) Now, $v_y = v_{0y} - gt$ gives $t = 9.90/9.8 = 1.01$ s. Thus, $\Delta x = (v_0 \cos\theta_0)t = 6.85$ m.

(c) The velocity at the target has only the v_x component, which is equal to $v_{0x} = v_0 \cos\theta_0 = 6.78$ m/s.

108. (Second problem in **Cluster 2**)

(a) The magnitudes of the components are equal at point A, but in terms of the coordinate system usually employed in projectile motion problems, we have $v_x > 0$ and $v_y = -v_x$. The problem gives v_0 which is related to its components by $v_0^2 = v_{0x}^2 + v_{0y}^2$ which suggests that we look at the pair of equations

$$\begin{aligned} v_y^2 &= v_{0y}^2 - 2g\Delta y \\ v_x^2 &= v_{0x}^2 \end{aligned}$$

which we can add to obtain $2v_x^2 = v_0^2 - 2g\Delta y$ (this is closely related to the type of reasoning that will be employed in some Chapter 8 problems). Therefore, we find $v_x = -v_y = 6.53$ m/s. Therefore, $\Delta y = v_y t + \frac{1}{2} g t^2$ (Eq. 2-16) can be used to find t.

$$3.00 = (-6.53)t + \frac{1}{2}(9.8)t^2 \implies t = 1.69 \text{ or } -0.36$$

from the quadratic formula or or with a polynomial solver available with some calculators. We choose the positive root: $t = 1.69$ s. Finally, we obtain

$$\Delta x = v_x t = 11.1 \text{ m} \ .$$

(b) The speed is $v = \sqrt{v_x^2 + v_y^2} = 9.23$ m/s.

109. (Third problem in **Cluster 2**)

(a) Eq. 4-25, which assumes $(x_0, y_0) = (0, 0)$, gives

$$y = 5.00 = (\tan\theta_0)x - \frac{gx^2}{2\left(v_0 \cos\theta_0\right)^2}$$

where $x = 30.0$ (lengths are in meters and time is in seconds). Using the trig identity suggested in the problem and letting u stand for $\tan\theta_0$, we have a second-degree equation for u (its two roots leading to the values $\theta_{0\,\text{min}}$, and $\theta_{0\,\text{max}}$) parameterized by the initial speed v_0.

$$\frac{4410}{v_0^2}u^2 - 30.0u + \left(\frac{4410}{v_0^2} + 5.00\right) = 0$$

where numerical simplifications have already been made. To see these steps written with the *variables* x, y, v_0 and g made explicit, see the solution to problem 111, below. Now, we solve for u using the quadratic formula, and then find the angles:

$$\theta_0 = \tan^{-1}\left(\frac{1}{294}v_0^2 \pm \frac{1}{294}\sqrt{v_0^4 - 98\,v_0^2 - 86436}\right)$$

where the plus is chosen for $\theta_{0\text{max}}$ and the negative is chosen for $\theta_{0\text{min}}$.

(b) These angles are plotted (in degrees) versus v_0 (in m/s) as follows. There are no (real) solutions of the above equations for $18.0 \le v_0 \le 18.6$ m/s (this is further discussed in the next problem).

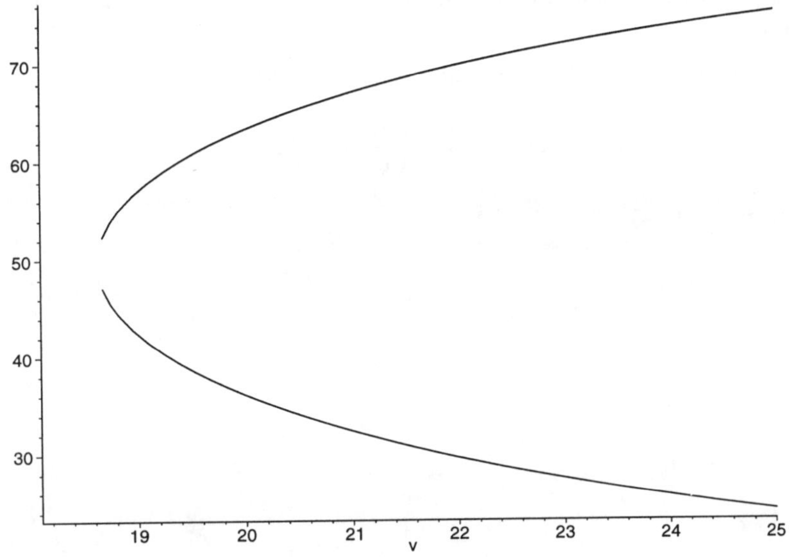

110. (Fourth problem in **Cluster 2**)
Following the hint in the problem (regarding *analytic* solution), we equate the square root expression, above, to zero:

$$\sqrt{v_0^4 - 98\,v_0^2 - 86436} = 0 \implies v_0 = 18.6 \text{ m/s} .$$

That solution can be obtained either with the quadratic formula (by writing the equation, first, in terms of $w = v_0^2$) or with a polynomial solver built into many calculators; in the latter approach, this is straightforwardly handled as a fourth degree polynomial. Note that the other root ($v_0 = 15.8$ m/s) is dismissed since we are finding where the *real* solutions for angle disappear as one decreases the initial speed from roughly 20 m/s. In case this problem was assigned without assigning Problem 109 first, then this (the choice of root) might be a confusing point. Plugging $v_0 = 18.6$ m/s into

$$\theta_0 = \tan^{-1}\left(\frac{1}{294}\,v_0^2 \pm \frac{1}{294}\sqrt{v_0^4 - 98\,v_0^2 - 86436}\right)$$

(which is unambiguous since the square root factor is zero) provides the launch angle: $\theta_0 = 49.7°$ in this "critical" case.

111. (Fifth problem in **Cluster 2**)

(a) This builds directly on the solutions of the previous two problems. If we return to the solution of problem 109 without plugging in the data for x, y, and g, we obtain the following expression for the θ_0 roots.

$$\theta_0 = \tan^{-1}\left(\frac{v_0^2}{gx}\left(1 \pm \sqrt{1 - \frac{g}{v_0^2}\left(2y + \frac{gx^2}{v_0^2}\right)}\right)\right)$$

And for the "critical case" of maximum distance for a given launch-speed, we set the square root expression to zero (as in the previous problem) and solve for x_{max}.

$$x_{\text{max}} = \frac{v_0^2}{g}\sqrt{1 - \frac{2gy}{v_0^2}}$$

which one might wish to check for the "straight-up" case (where $x = 0$, and the familiar result $y_{\rm max} = \frac{1}{2}v_0^2/g$ is obtained) and for the "range" case (where $y = 0$ and this then agrees with Eq. 4-26 where $\theta_0 = 45°$). In the problem at hand, we have $y = 5.00$ m, and $v_0 = 15.0$ m/s. This leads to $x_{\rm max} = 17.2$ m.

(b) When the square root term vanishes, the expression for θ_0 becomes

$$\theta_0 = \tan^{-1}\left(\frac{v_0^2}{gx}\right) = 53.1°$$

using $x = x_{\rm max}$ from part (a).

Chapter 5

1. We apply Newton's second law (specifically, Eq. 5-2).

   (a) We find the x component of the force is

   $$F_x = ma_x = ma\cos 20° = (1.00\,\text{kg})(2.00\,\text{m/s}^2)\cos 20° = 1.88\text{ N}\ .$$

   (b) The y component of the force is

   $$F_y = ma_y = ma\sin 20° = (1.0\,\text{kg})(2.00\,\text{m/s}^2)\sin 20° = 0.684\text{ N}\ .$$

   (c) In unit-vector notation, the force vector (in Newtons) is

   $$\vec{F} = F_x\,\hat{\imath} + F_y\,\hat{\jmath} = 1.88\hat{\imath} + 0.684\hat{\jmath}\ .$$

2. We apply Newton's second law (Eq. 5-1 or, equivalently, Eq. 5-2). The net force applied on the chopping block is $\vec{F}_{\text{net}} = \vec{F}_1 + \vec{F}_2$, where the vector addition is done using unit-vector notation. The acceleration of the block is given by $\vec{a} = \left(\vec{F}_1 + \vec{F}_2\right)/m$.

   (a) In the first case

   $$\vec{F}_1 + \vec{F}_2 = \left((3.0\,\text{N})\hat{\imath} + (4.0\,\text{N})\hat{\jmath}\right) + \left((-3.0\,\text{N})\hat{\imath} + (-4.0\,\text{N})\hat{\jmath}\right) = 0$$

   so $\vec{a} = 0$.

   (b) In the second case, the acceleration $\vec{a}$ equals

   $$\frac{\vec{F}_1 + \vec{F}_2}{m} = \frac{\left((3.0\,\text{N})\,\hat{\imath} + (4.0\,\text{N})\hat{\jmath}\right) + \left((-3.0\,\text{N})\,\hat{\imath} + (4.0\,\text{N})\hat{\jmath}\right)}{2.0\,\text{kg}} = 4.0\hat{\jmath}\ \text{m/s}^2\ .$$

   (c) In this final situation, $\vec{a}$ is

   $$\frac{\vec{F}_1 + \vec{F}_2}{m} = \frac{\left((3.0\,\text{N})\,\hat{\imath} + (4.0\,\text{N})\hat{\jmath}\right) + \left((3.0\,\text{N})\hat{\imath} + (-4.0\,\text{N})\hat{\jmath}\right)}{2.0\,\text{kg}} = 3.0\hat{\imath}\ \text{m/s}^2\ .$$

3. We are only concerned with horizontal forces in this problem (gravity plays no direct role). We take East as the $+x$ direction and North as $+y$. This calculation is efficiently implemented on a vector capable calculator, using magnitude-angle notation (with SI units understood).

   $$\vec{a} = \frac{\vec{F}}{m} = \frac{(9.0\ \angle\ 0°) + (8.0\ \angle\ 118°)}{3.0} = (2.9\ \angle\ 53°)$$

   Therefore, the acceleration has a magnitude of $2.9\ \text{m/s}^2$.

4. Since $\vec{v} = \text{constant}$, we have $\vec{a} = 0$, which implies

$$\vec{F}_{\text{net}} = \vec{F}_1 + \vec{F}_2 = m\vec{a} = 0 \ .$$

Thus, the other force must be

$$\vec{F}_2 = -\vec{F}_1 = -2\hat{\imath} + 6\hat{\jmath} \ \text{N} \ .$$

5. Since the velocity of the particle does not change, it undergoes no acceleration and must therefore be subject to zero net force. Therefore,

$$\vec{F}_{\text{net}} = \vec{F}_1 + \vec{F}_2 + \vec{F}_3 = 0 \ .$$

Thus, the third force $\vec{F}_3$ is given by

$$\begin{aligned} \vec{F}_3 &= -\vec{F}_1 - \vec{F}_2 \\ &= -\left(2\hat{\imath} + 3\hat{\jmath} - 2\hat{k}\right) - \left(-5\hat{\imath} + 8\hat{\jmath} - 2\hat{k}\right) \\ &= 3\hat{\imath} - 11\hat{\jmath} + 4\hat{k} \end{aligned}$$

in Newtons. The specific value of the velocity is not used in the computation.

6. The net force applied on the chopping block is $\vec{F}_{\text{net}} = \vec{F}_1 + \vec{F}_2 + \vec{F}_3$, where the vector addition is done using unit-vector notation. The acceleration of the block is given by $\vec{a} = \left(\vec{F}_1 + \vec{F}_2 + \vec{F}_3\right)/m$.

(a) The forces exerted by the three astronauts can be expressed in unit-vector notation as follows:

$$\begin{aligned} \vec{F}_1 &= 32(\cos 30^\circ \,\hat{\imath} + \sin 30^\circ \,\hat{\jmath}) \\ &= 27.7\hat{\imath} + 16\hat{\jmath} \\ \vec{F}_2 &= 55(\cos 0^\circ \,\hat{\imath} + \sin 0^\circ \,\hat{\jmath}) \\ &= 55\hat{\imath} \end{aligned}$$

in Newtons, and

$$\vec{F}_3 = 41\left(\cos(-60^\circ)\hat{\imath} + \sin(-60^\circ)\hat{\jmath}\right) = 20.5\hat{\imath} - 35.5\hat{\jmath}$$

in Newtons. The resultant acceleration of the asteroid of mass $m = 120$ kg is therefore

$$\begin{aligned} \vec{a} &= \frac{(27.7\hat{\imath} + 16\hat{\jmath}) + (55\hat{\imath}) + (20.5\hat{\imath} - 35.5\hat{\jmath})}{120} \\ &= 0.86\hat{\imath} - 0.16\hat{\jmath} \ \ \text{m/s}^2 \ . \end{aligned}$$

(b) The magnitude of the acceleration vector is

$$|\vec{a}| = \sqrt{a_x^2 + a_y^2} = \sqrt{0.86^2 + (-0.16)^2} = 0.88 \ \text{m/s}^2 \ .$$

(c) The vector $\vec{a}$ makes an angle θ with the $+x$ axis, where

$$\theta = \tan^{-1}\left(\frac{a_y}{a_x}\right) = \tan^{-1}\left(\frac{-0.16}{0.86}\right) = -11^\circ \ .$$

7. We denote the two forces $\vec{F}_1$ and $\vec{F}_2$. According to Newton's second law, $\vec{F}_1 + \vec{F}_2 = m\vec{a}$, so $\vec{F}_2 = m\vec{a} - \vec{F}_1$.

(a) In unit vector notation $\vec{F}_1 = (20.0\,\text{N})\hat{\imath}$ and

$$\vec{a} = -(12\sin 30^\circ \ \text{m/s}^2)\hat{\imath} - (12\cos 30^\circ \ \text{m/s}^2)\hat{\jmath} = -(6.0\,\text{m/s}^2)\hat{\imath} - (10.4\,\text{m/s}^2)\hat{\jmath} \ .$$

Therefore,

$$\begin{aligned} \vec{F}_2 &= (2.0\,\text{kg})\left(-6.0\,\text{m/s}^2\right)\hat{\imath} + (2.0\,\text{kg})\left(-10.4\,\text{m/s}^2\right)\hat{\jmath} - (20.0\,\text{N})\hat{\imath} \\ &= (-32\,\text{N})\hat{\imath} - (21\,\text{N})\hat{\jmath} \ . \end{aligned}$$

(b) The magnitude of $\vec{F}_2$ is

$$\left|\vec{F}_2\right| = \sqrt{F_{2x}^2 + F_{2y}^2} = \sqrt{(-32)^2 + (-21)^2} = 38 \text{ N} .$$

(c) The angle that $\vec{F}_2$ makes with the positive x axis is found from $\tan\theta = F_{2y}/F_{2x} = 21/32 = 0.656$. Consequently, the angle is either 33° or 33° + 180° = 213°. Since both the x and y components are negative, the correct result is 213°.

8. The goal is to arrive at the least magnitude of $\vec{F}_{\text{net}}$, and as long as the magnitudes of $\vec{F}_2$ and $\vec{F}_3$ are (in total) less than or equal to $|\vec{F}_1|$ then we should orient them opposite to the direction of $\vec{F}_1$ (which is the $+x$ direction).

   (a) We orient both $\vec{F}_2$ and $\vec{F}_3$ in the $-x$ direction. Then, the magnitude of the net force is $50 - 30 - 20 = 0$, resulting in zero acceleration for the tire.

   (b) We again orient $\vec{F}_2$ and $\vec{F}_3$ in the negative x direction. We obtain an acceleration along the $+x$ axis with magnitude

$$a = \frac{F_1 - F_2 - F_3}{m} = \frac{50\,\text{N} - 30\,\text{N} - 10\,\text{N}}{12\,\text{kg}} = 0.83 \text{ m/s}^2 .$$

   (c) In this case, the forces $\vec{F}_2$ and $\vec{F}_3$ are collectively strong enough to have y components (one positive and one negative) which cancel each other and still have enough x contributions (in the $-x$ direction) to cancel $\vec{F}_1$. Since $|\vec{F}_2| = |\vec{F}_3|$, we see that the angle above the $-x$ axis to one of them should equal the angle below the $-x$ axis to the other one (we denote this angle θ). We require

$$\begin{aligned} -50\,\text{N} &= \vec{F}_{2x} + \vec{F}_{3x} \\ &= -(30\,\text{N})\cos\theta - (30\,\text{N})\cos\theta \end{aligned}$$

   which leads to

$$\theta = \cos^{-1}\left(\frac{50\,\text{N}}{60\,\text{N}}\right) = 34° .$$

9. In all three cases the scale is not accelerating, which means that the two cords exert forces of equal magnitude on it. The scale reads the magnitude of either of these forces. In each case the tension force of the cord attached to the salami must be the same in magnitude as the weight of the salami because the salami is not accelerating. Thus the scale reading is mg, where m is the mass of the salami. Its value is $(11.0\,\text{kg})(9.8\,\text{m/s}^2) = 108\,\text{N}$.

10. Three vertical forces are acting on the block: the earth pulls down on the block with gravitational force 3.0 N; a spring pulls up on the block with elastic force 1.0 N; and, the surface pushes up on the block with normal force N. There is no acceleration, so

$$\sum F_y = 0 = N + (1.0 \text{ N}) + (-3.0 \text{ N})$$

   yields $N = 2.0$ N. By Newton's third law, the force exerted by the block on the surface has that same magnitude but opposite direction: 2.0 N down.

11. We apply Eq. 5-12.

   (a) The mass is $m = W/g = (22\,\text{N})/(9.8\,\text{m/s}^2) = 2.2\,\text{kg}$. At a place where $g = 4.9\,\text{m/s}^2$, the mass is still 2.2 kg but the gravitational force is $F_g = mg = (2.2\,\text{kg})(4.9\,\text{m/s}^2) = 11\,\text{N}$.

   (b) As noted, $m = 2.2$ kg.

   (c) At a place where $g = 0$ the gravitational force is zero.

(d) The mass is still 2.2 kg.

12. We use $W_p = mg_p$, where W_p is the weight of an object of mass m on the surface of a certain planet p, and g_p is the acceleration of gravity on that planet.

    (a) The weight of the space ranger on Earth is $W_e = mg_e$ which we compute to be $(75\,\text{kg})\left(9.8\,\text{m/s}^2\right) = 7.4 \times 10^2\,\text{N}$.

    (b) The weight of the space ranger on Mars is $W_m = mg_m$ which we compute to be $(75\,\text{kg})(3.8\,\text{m/s}^2) = 2.9 \times 10^2\,\text{N}$.

    (c) The weight of the space ranger in interplanetary space is zero, where the effects of gravity are negligible.

    (d) The mass of the space ranger remains the same (75 kg) at all the locations.

13. According to Newton's second law, the magnitude of the force is given by $F = ma$, where a is the magnitude of the acceleration of the neutron. We use kinematics (Table 2-1) to find the acceleration that brings the neutron to rest in a distance d. Assuming the acceleration is constant, then $v^2 = v_0^2 + 2ad$ produces the value of a:

$$a = \frac{(v^2 - v_0^2)}{2d} = \frac{-(1.4 \times 10^7\,\text{m/s})^2}{2(1.0 \times 10^{-14}\,\text{m})} = -9.8 \times 10^{27}\,\text{m/s}^2 .$$

The magnitude of the force is consequently

$$F = m|a| = (1.67 \times 10^{-27}\,\text{kg})(9.8 \times 10^{27}\,\text{m/s}^2) = 16\ \text{N} .$$

14. The child-backpack is in static equilibrium while he waits, so Newton's second law applies with $\sum \vec{F} = 0$. Since students sometimes confuse this with Newton's third law, we phrase our results carefully.

    (a) The magnitude of the normal force $\vec{N}$ exerted upward by the sidewalk is equal, in this situation, to the total weight of the child-backpack, as a result of $\sum \vec{F} = 0$. Thus, $\vec{N} = (33.5\ \text{kg})(9.8\ \text{m/s}^2) = 328\ \text{N}$ and is directed up; this is $\vec{F}_{sc}$ – the force of the sidewalk exerted up on the child's feet. By Newton's third law, the force exerted down (at the child's feet) on the sidewalk is $\vec{F}_{cs} = 328\ \text{N}$ downward.

    (b) Except for an entirely negligible gravitation attraction between the child and the concrete, there is no force exerted on the sidewalk by the child when the child is not in contact with it.

    (c) Earth pulls gravitationally on the child, and the child pulls equally in the opposite direction on Earth. This force is the previously computed weight $(29.0)(9.8) = 284$ N. The gravitational force on Earth exerted by the child is 284 N up. But the contact force exerted by the child on the sidewalk (hence, on Earth) is (see part (a)) 328 N downward. Thus, the *net* force exerted by the child on Earth is zero.

    (d) Here the answer is simply the gravitational interaction: 284 N up.

15. We note that the free-body diagram is shown in Fig. 5-18 of the text.

    (a) Since the acceleration of the block is zero, the components of the Newton's second law equation yield $T - mg\sin\theta = 0$ and $N - mg\cos\theta = 0$. Solving the first equation for the tension in the string, we find

$$T = mg\sin\theta = (8.5\,\text{kg})(9.8\,\text{m/s}^2)\sin 30^\circ = 42\ \text{N} .$$

    (b) We solve the second equation in part (a) for the normal force N:

$$N = mg\cos\theta = (8.5\,\text{kg})(9.8\,\text{m/s}^2)\cos 30^\circ = 72\ \text{N} .$$

(c) When the string is cut, it no longer exerts a force on the block and the block accelerates. The x component of the second law becomes $-mg\sin\theta = ma$, so the acceleration becomes

$$a = -g\sin\theta = -9.8\sin 30° = -4.9$$

in SI units. The negative sign indicates the acceleration is down the plane. The magnitude of the acceleration is 4.9 m/s^2.

16. An excellent analysis of the accelerating elevator is given in Sample Problem 5-8 in the textbook.

(a) From Newton's second law

$$N - mg = ma \quad \text{where } a = a_{\text{max}} = 2.0 \text{ m/s}^2$$

we obtain $N = 590$ N upward, for $m = 50$ kg.

(b) Again, we use Newton's second law

$$N - mg = ma \quad \text{where } a = a_{\text{max}} = -3.0 \text{ m/s}^2 \ .$$

Now, we obtain $N = 340$ N upward.

(c) Returning to part (a), we use Newton's third law, and conclude that the force exerted by the passenger on the floor is $\vec{F}_{PF} = 590$ N downward.

17. (a) The acceleration is

$$a = \frac{F}{m} = \frac{20\,\text{N}}{900\,\text{kg}} = 0.022 \text{ m/s}^2 \ .$$

(b) The distance traveled in 1 day (= 86400 s) is

$$s = \frac{1}{2}at^2 = \frac{1}{2}\left(0.0222\,\text{m/s}^2\right)(86400\,\text{s})^2 = 8.3 \times 10^7 \text{ m} \ .$$

(c) The speed it will be traveling is given by

$$v = at = (0.0222\,\text{m/s}^2)(86400\,\text{s}) = 1.9 \times 10^3 \text{ m/s} \ .$$

18. Some assumptions (not so much for realism but rather in the interest of using the given information efficiently) are needed in this calculation: we assume the fishing line and the path of the salmon are horizontal. Thus, the weight of the fish contributes only (via Eq. 5-12) to information about its mass ($m = W/g = 8.7$ kg). Our $+x$ axis is in the direction of the salmon's velocity (away from the fisherman), so that its acceleration ("deceleration") is negative-valued and the force of tension is in the $-x$ direction: $\vec{T} = -T$. We use Eq. 2-16 and SI units (noting that $v = 0$).

$$v^2 = v_0^2 + 2a\Delta x \implies a = -\frac{v_0^2}{2\Delta x} = -\frac{2.8^2}{2(0.11)}$$

which yields $a = -36$ m/s^2. Assuming there are no significant horizontal forces other than the tension, Eq. 5-1 leads to

$$\vec{T} = m\vec{a} \implies -T = (8.7\,\text{kg})\left(-36\,\text{m/s}^2\right)$$

which results in $T = 3.1 \times 10^2$ N.

19. In terms of magnitudes, Newton's second law is $F = ma$, where F represents $|\vec{F}_{\rm net}|$, a represents $|\vec{a}|$ (which it does not always do; note the use of a in the previous solution), and m is the (always positive) mass. The magnitude of the acceleration can be found using constant acceleration kinematics (Table 2-1). Solving $v = v_0 + at$ for the case where it starts from rest, we have $a = v/t$ (which we interpret in terms of magnitudes, making specification of coordinate directions unnecessary). The velocity is $v = (1600\,{\rm km/h})(1000\,{\rm m/km})/(3600\,{\rm s/h}) = 444\,{\rm m/s}$, so

$$F = (500\,{\rm kg})\frac{444\,{\rm m/s}}{1.8\,{\rm s}} = 1.2 \times 10^5\,{\rm N}\ .$$

20. The stopping force $\vec{F}$ and the path of the car are horizontal. Thus, the weight of the car contributes only (via Eq. 5-12) to information about its mass ($m = W/g = 1327$ kg). Our $+x$ axis is in the direction of the car's velocity, so that its acceleration ("deceleration") is negative-valued and the stopping force is in the $-x$ direction: $\vec{F} = -F$.

(a) We use Eq. 2-16 and SI units (noting that $v = 0$ and $v_0 = 40(1000/3600) = 11.1$ m/s).

$$v^2 = v_0^2 + 2a\Delta x \implies a = -\frac{v_0^2}{2\Delta x} = -\frac{11.1^2}{2(15)}$$

which yields $a = -4.12$ m/s^2. Assuming there are no significant horizontal forces other than the stopping force, Eq. 5-1 leads to

$$\vec{F} = m\vec{a} \implies -F = (1327\,{\rm kg})\left(-4.12\,{\rm m/s^2}\right)$$

which results in $F = 5.5 \times 10^3$ N.

(b) Eq. 2-11 readily yields $t = -v_0/a = 2.7$ s.

(c) Keeping F the same means keeping a the same, in which case (since $v = 0$) Eq. 2-16 expresses a direct proportionality between Δx and v_0^2. Therefore, doubling v_0 means quadrupling Δx. That is, the new over the old stopping distances is a factor of 4.0.

(d) Eq. 2-11 illustrates a direct proportionality between t and v_0 so that doubling one means doubling the other. That is, the new time of stopping is a factor of 2.0 greater than the one found in part (c).

21. The acceleration of the electron is vertical and for all practical purposes the only force acting on it is the electric force. The force of gravity is negligible. We take the $+x$ axis to be in the direction of the initial velocity and the $+y$ axis to be in the direction of the electrical force, and place the origin at the initial position of the electron. Since the force and acceleration are constant, we use the equations from Table 2-1: $x = v_0 t$ and

$$y = \frac{1}{2}at^2 = \frac{1}{2}\left(\frac{F}{m}\right)t^2\ .$$

The time taken by the electron to travel a distance x (= 30 mm) horizontally is $t = x/v_0$ and its deflection in the direction of the force is

$$y = \frac{1}{2}\frac{F}{m}\left(\frac{x}{v_0}\right)^2 = \frac{1}{2}\left(\frac{4.5 \times 10^{-16}}{9.11 \times 10^{-31}}\right)\left(\frac{30 \times 10^{-3}}{1.2 \times 10^7}\right)^2 = 1.5 \times 10^{-3}\ {\rm m}\ .$$

22. The stopping force $\vec{F}$ and the path of the passenger are horizontal. Our $+x$ axis is in the direction of the passenger's motion, so that the passenger's acceleration ("deceleration") is negative-valued and the stopping force is in the $-x$ direction: $\vec{F} = -F$. We use Eq. 2-16 and SI units (noting that $v = 0$ and $v_0 = 53(1000/3600) = 14.7$ m/s).

$$v^2 = v_0^2 + 2a\Delta x \implies a = -\frac{v_0^2}{2\Delta x} = -\frac{14.7^2}{2(0.65)}$$

which yields $a = -167\ \text{m/s}^2$. Assuming there are no significant horizontal forces other than the stopping force, Eq. 5-1 leads to

$$\vec{F} = m\vec{a} \implies -F = (41\,\text{kg})\left(-167\,\text{m/s}^2\right)$$

which results in $F = 6.8 \times 10^3$ N.

23. We note that The rope is 22° from vertical – and therefore 68° from horizontal.

(a) With $T = 760$ N, then its components are

$$\vec{T} = T\cos 68°\,\hat{\text{i}} + T\sin 68°\,\hat{\text{j}} = 285\,\hat{\text{i}} + 705\,\hat{\text{j}}$$

understood to be in newtons.

(b) No longer in contact with the cliff, the only other force on Tarzan is due to earth's gravity (his weight). Thus,

$$\vec{F}_{\text{net}} = \vec{T} + \vec{W} = 285\,\hat{\text{i}} + 705\,\hat{\text{j}} - 820\,\hat{\text{j}} = 285\,\hat{\text{i}} - 115\,\hat{\text{j}}$$

again understood to be in newtons.

(c) In a manner that is efficiently implemented on a vector capable calculator, we convert from rectangular (x, y) components to magnitude-angle notation:

$$\vec{F}_{\text{net}} = (285, -115) \longrightarrow (307\ \angle -22°)$$

so that the net force has a magnitude of 307 N.

(d) The angle (see part (c)) has been found to be 22° below horizontal (away from cliff)

(e) Since $\vec{a} = \vec{F}_{\text{net}} \,/\, m$ where $m = W/g = 84$ kg, we obtain $\vec{a} = 3.67\ \text{m/s}^2$

(f) Eq. 5-1 requires that $\vec{a} \parallel \vec{F}_{\text{net}}$ so that it is also directed at 22° below horizontal (away from cliff).

24. The analysis of coordinates and forces (the free-body diagram) is exactly as in the textbook in Sample Problem 5-7 (see Fig. 5-18(b) and (c)).

(a) Constant velocity implies zero acceleration, so the "uphill" force must equal (in magnitude) the "downhill" force: $T = mg\sin\theta$. Thus, with $m = 50$ kg and $\theta = 8.0°$, the tension in the rope equals 68 N.

(b) With an uphill acceleration of $0.10\ \text{m/s}^2$, Newton's second law (applied to the x axis shown in Fig. 5-18(b)) yields

$$T - mg\sin\theta = ma \implies T - (50)(9.8)\sin 8.0° = (50)(0.10)$$

which leads to $T = 73$ N.

25. (a) Since friction is negligible the force of the girl is the only horizontal force on the sled. The vertical forces (the force of gravity and the normal force of the ice) sum to zero. The acceleration of the sled is

$$a_s = \frac{F}{m_s} = \frac{5.2\,\text{N}}{8.4\,\text{kg}} = 0.62\ \text{m/s}^2\ .$$

(b) According to Newton's third law, the force of the sled on the girl is also 5.2 N. Her acceleration is

$$a_g = \frac{F}{m_g} = \frac{5.2\,\text{N}}{40\,\text{kg}} = 0.13\ \text{m/s}^2\ .$$

(c) The accelerations of the sled and girl are in opposite directions. Assuming the girl starts at the origin and moves in the $+x$ direction, her coordinate is given by $x_g = \frac{1}{2}a_g t^2$. The sled starts at $x_0 = 1.5\,\text{m}$ and moves in the $-x$ direction. Its coordinate is given by $x_s = x_0 - \frac{1}{2}a_s t^2$. They meet when

$$\begin{aligned} x_g &= x_s \\ \frac{1}{2}a_g t^2 &= x_0 - \frac{1}{2}a_s t^2 \,. \end{aligned}$$

This occurs at time

$$t = \sqrt{\frac{2x_0}{a_g + a_s}} \,.$$

By then, the girl has gone the distance

$$x_g = \frac{1}{2}a_g t^2 = \frac{x_0 a_g}{a_g + a_s} = \frac{(15)(0.13)}{0.13 + 0.62} = 2.6 \text{ m} \,.$$

26. We assume the direction of motion is $+x$ and assume the refrigerator starts from rest (so that the speed being discussed is the velocity v which results from the process). The only force along the x axis is the x component of the applied force $\vec{F}$.

(a) Since $v_0 = 0$, the combination of Eq. 2-11 and Eq. 5-2 leads simply to

$$F_x = m\left(\frac{v}{t}\right) \implies v_i = \left(\frac{F\cos\theta_i}{m}\right) t$$

for $i = 1$ or 2 (where we denote $\theta_1 = 0$ and $\theta_2 = \theta$ for the two cases). Hence, we see that the ratio v_2 over v_1 is equal to $\cos\theta$.

(b) Since $v_0 = 0$, the combination of Eq. 2-16 and Eq. 5-2 leads to

$$F_x = m\left(\frac{v^2}{2\Delta x}\right) \implies v_i = \sqrt{2\left(\frac{F\cos\theta_i}{m}\right)\Delta x}$$

for $i = 1$ or 2 (again, $\theta_1 = 0$ and $\theta_2 = \theta$ is used for the two cases). In this scenario, we see that the ratio v_2 over v_1 is equal to $\sqrt{\cos\theta}$.

27. We choose up as the $+y$ direction, so $\vec{a} = -3.00 \text{ m/s}^2\,\hat{\text{j}}$ (which, without the unit-vector, we denote as a since this is a 1-dimensional problem in which Table 2-1 applies). From Eq. 5-12, we obtain the firefighter's mass: $m = W/g = 72.7$ kg.

(a) We denote the force exerted by the pole on the firefighter $\vec{F}_{\text{f p}} = F\,\hat{\text{j}}$ and apply Eq. 5-1 (using SI units).

$$\begin{aligned} \vec{F}_{\text{net}} &= m\vec{a} \\ F - F_g &= ma \\ F - 712 &= (72.7)(-3.00) \end{aligned}$$

which yields $F = 494$ N. The fact that the result is positive means $\vec{F}_{\text{f p}}$ points up.

(b) Newton's third law indicates $\vec{F}_{\text{f p}} = -\vec{F}_{\text{p f}}$, which leads to the conclusion that $\vec{F}_{\text{p f}} = 494$ N down.

28. The coordinate choices are made in the problem statement.

(a) We write the velocity of the armadillo as $\vec{v} = v_x\,\hat{\imath} + v_y\,\hat{\jmath}$. Since there is no net force exerted on it in the x direction, the x component of the velocity of the armadillo is a constant: $v_x = 5.0\,\text{m/s}$. In the y direction at $t = 3.0\,\text{s}$, we have (using Eq. 2-11 with $v_{0\,y} = 0$)

$$v_y = v_{0\,y} + a_y t = v_{0\,y} + \left(\frac{F_y}{m}\right) t = \left(\frac{17}{12}\right)(3.0) = 4.3$$

in SI units. Thus

$$\vec{v} = 5.0\hat{\imath} + 4.3\hat{\jmath} \quad \text{m/s} .$$

(b) We write the position vector of the armadillo as $\vec{r} = r_x\,\hat{\imath} + r_y\,\hat{\jmath}$. At $t = 3.0\,\text{s}$ we have $r_x = (5.0)(3.0) = 15$ and (using Eq. 2-15 with $v_{0\,y} = 0$)

$$r_y = v_{0\,y}\,t + \frac{1}{2}a_y\,t^2 = \frac{1}{2}\left(\frac{F_y}{m}\right)t^2 = \frac{1}{2}\left(\frac{17}{12}\right)(3.0)^2 = 6.4$$

in SI units. The position vector at $t = 3.0\,\text{s}$ is therefore

$$\vec{r} = 15\hat{\imath} + 6.4\hat{\jmath} \quad \text{m} .$$

29. The solutions to parts (a) and (b) have been combined here. The free-body diagram is shown below, with the tension of the string $\vec{T}$, the force of gravity $m\vec{g}$, and the force of the air $\vec{F}$. Our coordinate system is shown. The x component of the net force is $T\sin\theta - F$ and the y component is $T\cos\theta - mg$, where $\theta = 37°$.

Since the sphere is motionless the net force on it is zero. We answer the questions in the reverse order. Solving $T\cos\theta - mg = 0$ for the tension, we obtain $T = mg/\cos\theta = (3.0 \times 10^{-4})(9.8)/\cos 37° = 3.7 \times 10^{-3}\,\text{N}$. Solving $T\sin\theta - F = 0$ for the force of the air: $F = T\sin\theta = (3.7 \times 10^{-3})\sin 37° = 2.2 \times 10^{-3}\,\text{N}$.

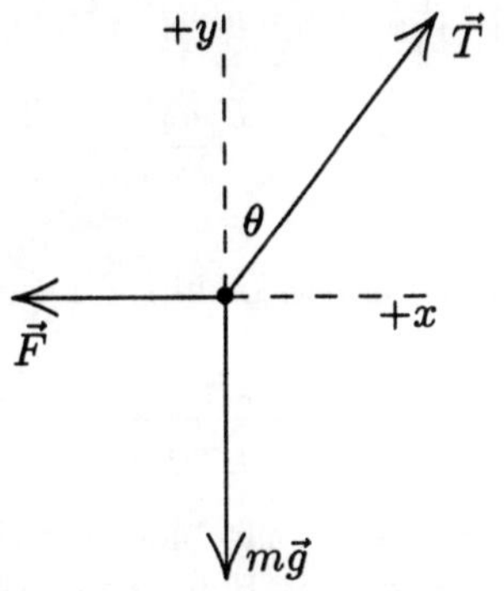

30. We label the 40 kg skier "m" which is represented as a block in the figure shown. The force of the wind is denoted $\vec{F}_w$ and might be either "uphill" or "downhill" (it is shown uphill in our sketch). The incline angle θ is 10°. The $+x$ direction is downhill.

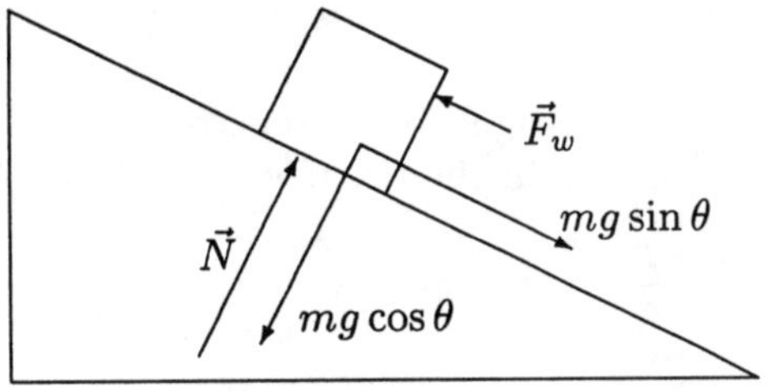

(a) Constant velocity implies zero acceleration; thus, application of Newton's second law along the x axis leads to

$$mg\sin\theta - F_w = 0 \ .$$

This yields $F_w = 68$ N (uphill).

(b) Given our coordinate choice, we have $a = +1.0\ \text{m/s}^2$. Newton's second law

$$mg\sin\theta - F_w = ma$$

now leads to $F_w = 28$ N (uphill).

(c) Continuing with the forces as shown in our figure, the equation

$$mg\sin\theta - F_w = ma$$

will lead to $F_w = -12$ N when $a = +2.0$ m/s^2. This simply tells us that the wind is opposite to the direction shown in our sketch; in other words, $\vec{F}_w = 12$ N *downhill.*

31. The free-body diagrams for part (a) are shown below. $\vec{F}$ is the applied force and $\vec{f}$ is the force exerted by block 1 on block 2. We note that $\vec{F}$ is applied directly to block 1 and that block 2 exerts the force $-\vec{f}$ on block 1 (taking Newton's third law into account).

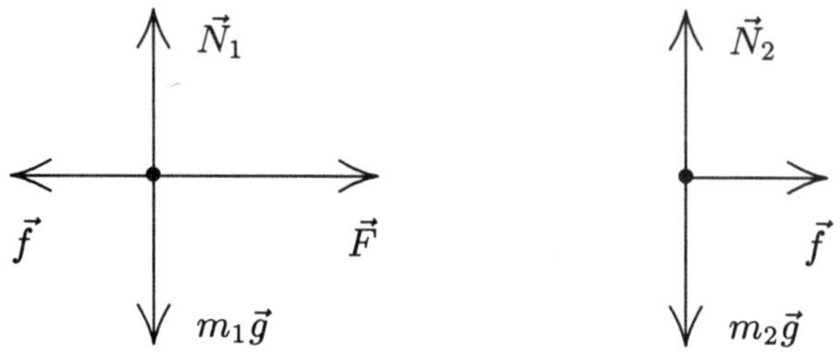

(a) Newton's second law for block 1 is $F - f = m_1 a$, where a is the acceleration. The second law for block 2 is $f = m_2 a$. Since the blocks move together they have the same acceleration and the same symbol is used in both equations. From the second equation we obtain the expression $a = f/m_2$, which we substitute into the first equation to get $F - f = m_1 f/m_2$. Therefore,

$$f = \frac{Fm_2}{m_1 + m_2} = \frac{(3.2\,\text{N})(1.2\,\text{kg})}{2.3\,\text{kg} + 1.2\,\text{kg}} = 1.1\,\text{N}\ .$$

(b) If $\vec{F}$ is applied to block 2 instead of block 1 (and in the opposite direction), the force of contact between the blocks is

$$f = \frac{Fm_1}{m_1 + m_2} = \frac{(3.2\,\text{N})(2.3\,\text{kg})}{2.3\,\text{kg} + 1.2\,\text{kg}} = 2.1\,\text{N}\ .$$

(c) We note that the acceleration of the blocks is the same in the two cases. In part (a), the force f is the only horizontal force on the block of mass m_2 and in part (b) f is the only horizontal force on the block with $m_1 > m_2$. Since $f = m_2 a$ in part (a) and $f = m_1 a$ in part (b), then for the accelerations to be the same, f must be larger in part (b).

32. The additional "apparent weight" experienced during upward acceleration is well treated in Sample Problem 5-8. The discussion in the textbook surrounding Eq. 5-13 is also relevant to this.

(a) When $\vec{F}_{\text{net}} = 3F - mg = 0$, we have

$$F = \frac{1}{3}mg = \frac{1}{3}(1400\,\text{kg})\left(9.8\,\text{m/s}^2\right) = 4.6 \times 10^3\ \text{N}$$

for the force exerted by each bolt on the engine.

(b) The force on each bolt now satisfies $3F - mg = ma$, which yields

$$F = \frac{1}{3}m(g+a) = \frac{1}{3}(1400)(9.8 + 2.6) = 5.8 \times 10^3\ \text{N}\ .$$

33. The free-body diagram is shown below. $\vec{T}$ is the tension of the cable and $m\vec{g}$ is the force of gravity. If the upward direction is positive, then Newton's second law is $T - mg = ma$, where a is the acceleration.

Thus, the tension is $T = m(g + a)$. We use constant acceleration kinematics (Table 2-1) to find the acceleration (where $v = 0$ is the final velocity, $v_0 = -12\,\text{m/s}$ is the initial velocity, and $y = -42\,\text{m}$ is the coordinate at the stopping point). Consequently, $v^2 = v_0^2 + 2ay$ leads to $a = -v_0^2/2y = -(-12)^2/2(-42) = 1.71\,\text{m/s}^2$. We now return to calculate the tension:

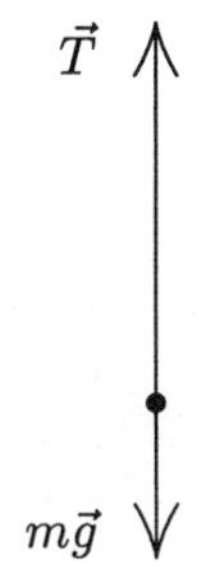

$$\begin{aligned} T &= m(g+a) \\ &= (1600\,\text{kg})(9.8\,\text{m/s}^2 + 1.71\,\text{m/s}^2) \\ &= 1.8 \times 10^4\ \text{N}\ . \end{aligned}$$

34. First, we consider all the penguins (1 through 4, counting left to right) as one system, to which we apply Newton's second law:

$$\begin{aligned} F_{\text{net}} &= (m_1 + m_2 + m_3 + m_4)a \\ 222\,\text{N} &= (20\,\text{kg} + 15\,\text{kg} + m_3 + 12\,\text{kg})a\ . \end{aligned}$$

Second, we consider penguins 3 and 4 as one system, for which we have

$$\begin{aligned} F'_{\text{net}} &= (m_3 + m_4)a \\ 111\,\text{N} &= (m_3 + 12\,\text{kg})a\ . \end{aligned}$$

We solve these two equations for m_3 to obtain $m_3 = 23\,\text{kg}$. The solution step can be made a little easier, though, by noting that the net force on penguins 1 and 2 is also 111 N and applying Newton's law to them as a single system to solve first for a.

35. We take the down to be the $+y$ direction.

(a) The first diagram (below) is the free-body diagram for the person and parachute, considered as a single object with a mass of $80\,\text{kg} + 5\,\text{kg} = 85\,\text{kg}$. $\vec{F}_a$ is the force of the air on the parachute and $m\vec{g}$ is the force of gravity. Application of Newton's second law produces $mg - F_a = ma$, where a is the acceleration. Solving for F_a we find

$$F_a = m(g-a) = (85\,\text{kg})(9.8\,\text{m/s}^2 - 2.5\,\text{m/s}^2) = 620\ \text{N}\ .$$

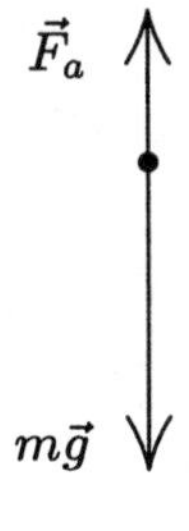

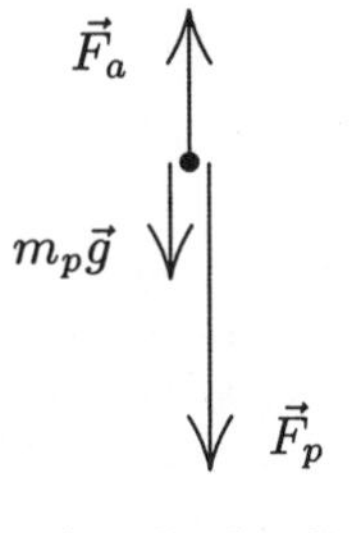

(b) The second diagram (above) is the free-body diagram for the parachute alone. $\vec{F}_a$ is the force of the air, $m_p\vec{g}$ is the force of gravity, and $\vec{F}_p$ is the force of the person. Now, Newton's second law leads to $m_p g + F_p - F_a = m_p a$. Solving for F_p, we obtain

$$F_p = m_p(a-g) + F_a = (5.0)(2.5 - 9.8) + 620 = 580\ \text{N}\ .$$

36. We apply Newton's second law first to the three blocks as a single system and then to the individual blocks. The $+x$ direction is to the right in Fig. 5-37.

(a) With $m_{\text{sys}} = m_1 + m_2 + m_3 = 67.0$ kg, we apply Eq. 5-2 to the x motion of the system – in which case, there is only one force $\vec{T}_3 = +T_3\,\hat{\imath}$.

$$\begin{aligned} T_3 &= m_{\text{sys}}\, a \\ 65.0\,\text{N} &= (67.0\,\text{kg})a \end{aligned}$$

which yields $a = 0.970$ m/s^2 for the system (and for each of the blocks individually).

(b) Applying Eq. 5-2 to block 1, we find

$$T_1 = m_1 a = (12.0\,\text{kg})\left(0.970\ \text{m/s}^2\right) = 11.6\ \text{N}\ .$$

(c) In order to find T_2, we can either analyze the forces on block 3 or we can treat blocks 1 and 2 as a system and examine its forces. We choose the latter.

$$T_2 = (m_1 + m_2)\,a = (12.0 + 24.0)(0.970) = 34.9\ \text{N}\ .$$

37. We use the notation g as the acceleration due to gravity near the surface of Callisto, m as the mass of the landing craft, a as the acceleration of the landing craft, and F as the rocket thrust. We take down to be the positive direction. Thus, Newton's second law takes the form $mg - F = ma$. If the thrust is F_1 $(= 3260\,\text{N})$, then the acceleration is zero, so $mg - F_1 = 0$. If the thrust is F_2 $(= 2200\,\text{N})$, then the acceleration is a_2 $(= 0.39\,\text{m/s}^2)$, so $mg - F_2 = ma_2$.

(a) The first equation gives the weight of the landing craft: $mg = F_1 = 3260\,\text{N}$.

(b) The second equation gives the mass:

$$m = \frac{mg - F_2}{a_2} = \frac{3260\,\text{N} - 2200\,\text{N}}{0.39\,\text{m/s}^2} = 2.7 \times 10^3\ \text{kg}\ .$$

(c) The weight divided by the mass gives the acceleration due to gravity: $g = (3260\,\text{N})/(2.7 \times 10^3\,\text{kg}) = 1.2\,\text{m/s}^2$.

38. Although the full specification of $\vec{F}_{\text{net}} = m\vec{a}$ in this situation involves both x and y axes, only the x-application is needed to find what this particular problem asks for. We note that $a_y = 0$ so that there is no ambiguity denoting a_x simply as a. We choose $+x$ to the right and $+y$ up, in Fig. 5-38. We also note that the x component of the rope's tension (acting on the crate) is $T_x = +450\cos 38° = 355$ N, and the resistive force (pointing in the $-x$ direction) has magnitude $f = 125$ N.

(a) Newton's second law leads to

$$T_x - f = ma \implies a = \frac{355 - 125}{310} = 0.74\ \text{m/s}^2\ .$$

(b) In this case, we use Eq. 5-12 to find the mass: $m = W/g = 31.6$ kg. Now, Newton's second law leads to

$$T_x - f = ma \implies a = \frac{355 - 125}{31.6} = 7.3\ \text{m/s}^2\ .$$

39. The force diagrams in Fig. 5-18 are helpful to refer to. In adapting Fig. 5-18(b) to this problem, the normal force $\vec{N}$ and the tension $\vec{T}$ should be labeled $F_{\text{m,r}_y}$ and $F_{\text{m,r}_x}$, respectively, and thought of as the y and x components of the force $\vec{F}_{\text{m,r}}$ exerted by the motorcycle on the rider. We adopt the coordinates used in Fig. 5-18 and note that they are not the usual horizontal and vertical axes.

(a) Since the net force equals ma, then the magnitude of the net force on the rider is $(60.0\,\text{kg})(3.0\,\text{m/s}^2) = 1.8 \times 10^2$ N.

(b) We apply Newton's second law to the x axis:

$$F_{m,r_x} - mg\sin\theta = ma$$

where $m = 60.0$ kg, $a = 3.0$ m/s^2, and $\theta = 10°$. Thus, $F_{m,r_x} = 282$ N. Applying it to the y axis (where there is no acceleration), we have

$$F_{m,r_y} - mg\cos\theta = 0$$

which produces $F_{m,r_y} = 579$ N. Using the Pythagorean theorem, we find

$$\sqrt{F_{m,r_x}^2 + F_{m,r_y}^2} = 644 \text{ N} .$$

Now, the magnitude of the force exerted on the rider by the motorcycle is the same magnitude of force exerted by the rider on the motorcycle, so the answer is 6.4×10^2 N.

40. Referring to Fig. 5-10(c) is helpful. In this case, viewing the man-rope-sandbag as a system means that we should be careful to choose a consistent positive direction of motion (though there are other ways to proceed – say, starting with individual application of Newton's law to each mass). We take *down* as positive for the man's motion and *up* as positive for the sandbag's motion and, without ambiguity, denote their acceleration as a. The net force on the system is the difference between the weight of the man and that of the sandbag. The system mass is $m_{sys} = 85 + 65 = 150$ kg. Thus, Eq. 5-1 leads to

$$(85)(9.8) - (65)(9.8) = m_{sys}\, a$$

which yields $a = 1.3$ m/s^2. Since the system starts from rest, Eq. 2-16 determines the speed (after traveling $\Delta y = 10$ m) as follows:

$$v = \sqrt{2a\Delta y} = \sqrt{2(1.3)(10)} = 5.1 \text{ m/s} .$$

41. (a) The links are numbered from bottom to top. The forces on the bottom link are the force of gravity $m\vec{g}$, downward, and the force $\vec{F}_{2on1}$ of link 2, upward. Take the positive direction to be upward. Then Newton's second law for this link is $F_{2on1} - mg = ma$. Thus $F_{2on1} = m(a+g) = (0.100\,\text{kg})(2.50\,\text{m/s}^2 + 9.8\,\text{m/s}^2) = 1.23\,\text{N}$.

(b) The forces on the second link are the force of gravity $m\vec{g}$, downward, the force $\vec{F}_{1on2}$ of link 1, downward, and the force $\vec{F}_{3on2}$ of link 3, upward. According to Newton's third law $\vec{F}_{1on2}$ has the same magnitude as $\vec{F}_{2on1}$. Newton's second law for the second link is $F_{3on2} - F_{1on2} - mg = ma$, so $F_{3on2} = m(a+g) + F_{1on2} = (0.100\,\text{kg})(2.50\,\text{m/s}^2 + 9.8\,\text{m/s}^2) + 1.23\,\text{N} = 2.46\,\text{N}$.

(c) Newton's second for link 3 is $F_{4on3} - F_{2on3} - mg = ma$, so $F_{4on3} = m(a+g) + F_{2on3} = (0.100\,\text{N})(2.50\,\text{m/s}^2 + 9.8\,\text{m/s}^2) + 2.46\,\text{N} = 3.69\,\text{N}$, where Newton's third law implies $F_{2on3} = F_{3on2}$ (since these are magnitudes of the force vectors).

(d) Newton's second law for link 4 is $F_{5on4} - F_{3on4} - mg = ma$, so $F_{5on4} = m(a+g) + F_{3on4} = (0.100\,\text{kg})(2.50\,\text{m/s}^2 + 9.8\,\text{m/s}^2) + 3.69\,\text{N} = 4.92\,\text{N}$, where Newton's third law implies $F_{3on4} = F_{4on3}$.

(e) Newton's second law for the top link is $F - F_{4on5} - mg = ma$, so $F = m(a+g) + F_{4on5} = (0.100\,\text{kg})(2.50\,\text{m/s}^2 + 9.8\,\text{m/s}^2) + 4.92\,\text{N} = 6.15\,\text{N}$, where $F_{4on5} = F_{5on4}$ by Newton's third law.

(f) Each link has the same mass and the same acceleration, so the same net force acts on each of them: $F_{net} = ma = (0.100\,\text{kg})(2.50\,\text{m/s}^2) = 0.25\,\text{N}$.

42. The mass of the jet is $m = W/g = 2.36 \times 10^4$ kg. Its acceleration is found from Eq. 2-16:

$$v^2 = v_0^2 + 2a\Delta x \implies a = \frac{85^2}{2(90)} = 40 \text{ m/s}^2 .$$

Thus, Newton's second law provides the needed force F from the catapult.

$$F + F_{thrust} = ma \implies F = (2.36 \times 10^4)(40) - 107 \times 10^3$$

which yields $F = 8.4 \times 10^5$ N.

43. The free-body diagram for each block is shown below. T is the tension in the cord and $\theta = 30°$ is the angle of the incline. For block 1, we take the $+x$ direction to be up the incline and the $+y$ direction to be in the direction of the normal force $\vec{N}$ that the plane exerts on the block. For block 2, we take the $+y$ direction to be down. In this way, the accelerations of the two blocks can be represented by the same symbol a, without ambiguity. Applying Newton's second law to the x and y axes for block 1 and to the y axis of block 2, we obtain

$$\begin{aligned} T - m_1 g \sin\theta &= m_1 a \\ N - m_1 g \cos\theta &= 0 \\ m_2 g - T &= m_2 a \end{aligned}$$

respectively. The first and third of these equations provide a simultaneous set for obtaining values of a and T. The second equation is not needed in this problem, since the normal force is neither asked for nor is it needed as part of some further computation (such as can occur in formulas for friction).

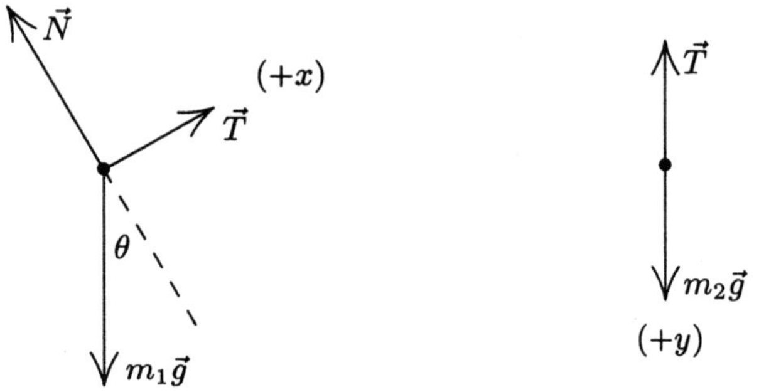

(a) We add the first and third equations above: $m_2 g - m_1 g \sin\theta = m_1 a + m_2 a$. Consequently, we find

$$a = \frac{(m_2 - m_1 \sin\theta)g}{m_1 + m_2} = \frac{(2.30\,\text{kg}) - 3.70 \sin 30.0°)\,(9.8)}{3.70 + 2.30} = 0.735\ \text{m/s}^2 .$$

(b) The result for a is positive, indicating that the acceleration of block 1 is indeed up the incline and that the acceleration of block 2 is vertically down.

(c) The tension in the cord is

$$T = m_1 a + m_1 g \sin\theta = (3.70)(0.735) + (3.70)(9.8) \sin 30° = 20.8\ \text{N} .$$

44. For convenience, we have labeled the 2.0 kg mass m and the 3.0 kg mass M. The $+x$ direction for m is "downhill" and the $+x$ direction for M is rightward; thus, they accelerate with the same sign.

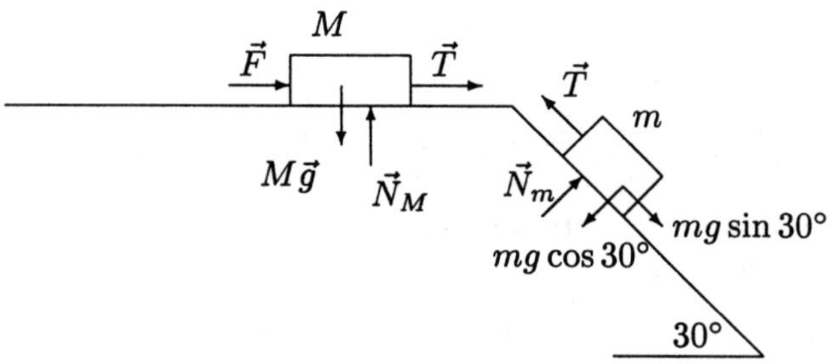

(a) We apply Newton's second law to each block's x axis:

$$\begin{aligned} mg \sin 30° - T &= ma \\ F + T &= Ma \end{aligned}$$

Adding the two equations allows us to solve for the acceleration. With $F = 2.3$ N, we have $a = 1.8\ \text{m/s}^2$. We plug back in to find the tension $T = 3.1$ N.

(b) We consider the "critical" case where the F has reached the *max* value, causing the tension to vanish. The first of the equations in part (a) shows that $a = g\sin 30°$ in this case; thus, $a = 4.9\ \text{m/s}^2$. This implies (along with $T = 0$ in the second equation in part (a)) that $F = (3.0)(4.9) = 14.7$ N in the critical case.

45. The free-body diagram is shown below. $\vec{N}$ is the normal force of the plane on the block and $m\vec{g}$ is the force of gravity on the block. We take the $+x$ direction to be down the incline, in the direction of the acceleration, and the $+y$ direction to be in the direction of the normal force exerted by the incline on the block. The x component of Newton's second law is then $mg\sin\theta = ma$; thus, the acceleration is $a = g\sin\theta$.

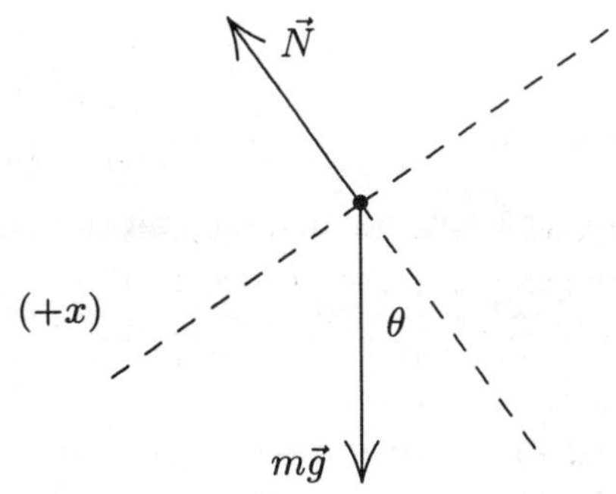

(a) Placing the origin at the bottom of the plane, the kinematic equations (Table 2-1) for motion along the x axis which we will use are $v^2 = v_0^2 + 2ax$ and $v = v_0 + at$. The block momentarily stops at its highest point, where $v = 0$; according to the second equation, this occurs at time $t = -v_0/a$. The position where it stops is

$$\begin{aligned} x &= -\frac{1}{2}\frac{v_0^2}{a} \\ &= -\frac{1}{2}\left(\frac{(-3.50\ \text{m/s})^2}{(9.8\ \text{m/s}^2)\sin 32.0°}\right) \\ &= -1.18\ \text{m}\ . \end{aligned}$$

(b) The time is

$$t = -\frac{v_0}{a} = -\frac{v_0}{g\sin\theta} = -\frac{-3.50\ \text{m/s}}{(9.8\ \text{m/s}^2)\sin 32.0°} = 0.674\ \text{s}\ .$$

(c) That the return-speed is identical to the initial speed is to be expected since there are no dissipative forces in this problem. In order to prove this, one approach is to set $x = 0$ and solve $x = v_0 t + \frac{1}{2}at^2$ for the total time (up and back down) t. The result is

$$t = -\frac{2v_0}{a} = -\frac{2v_0}{g\sin\theta} = -\frac{2(-3.50\ \text{m/s})}{(9.8\ \text{m/s}^2)\sin 32.0°} = 1.35\ \text{s}\ .$$

The velocity when it returns is therefore

$$v = v_0 + at = v_0 + gt\sin\theta = -3.50 + (9.8)(1.35)\sin 32° = 3.50\ \text{m/s}\ .$$

46. We write the length unit light-month as c·month in this solution.

(a) The magnitude of the required acceleration is given by

$$a = \frac{\Delta v}{\Delta t} = \frac{(0.10)(3.0\times 10^8\ \text{m/s})}{(3.0\ \text{days})(86400\ \text{s/day})} = 1.2\times 10^2\ \text{m/s}^2\ .$$

(b) The acceleration in terms of g is

$$a = \left(\frac{a}{g}\right)g = \left(\frac{1.2\times 10^2\ \text{m/s}^2}{9.8\ \text{m/s}^2}\right)g = 12g\ .$$

(c) The force needed is

$$F = ma = \left(1.20 \times 10^6\right)\left(1.2 \times 10^2\right) = 1.4 \times 10^8 \text{ N} .$$

(d) The spaceship will travel a distance $d = 0.1$ c·month during one month. The time it takes for the spaceship to travel at constant speed for 5.0 light-months is

$$t = \frac{d}{v} = \frac{5.0\,\text{c} \cdot \text{months}}{0.1c} = 50 \text{ months}$$

which is about 4.2 years.

47. We take $+y$ to be up for both the monkey and the package.

(a) The force the monkey pulls downward on the rope has magnitude F. According to Newton's third law, the rope pulls upward on the monkey with a force of the same magnitude, so Newton's second law for forces acting on the monkey leads to $F - m_m g = m_m a_m$, where m_m is the mass of the monkey and a_m is its acceleration. Since the rope is massless $F = T$ is the tension in the rope. The rope pulls upward on the package with a force of magnitude F, so Newton's second law for the package is $F + N - m_p g = m_p a_p$, where m_p is the mass of the package, a_p is its acceleration, and N is the normal force exerted by the ground on it. Now, if F is the minimum force required to lift the package, then $N = 0$ and $a_p = 0$. According to the second law equation for the package, this means $F = m_p g$. Substituting $m_p g$ for F in the equation for the monkey, we solve for a_m:

$$a_m = \frac{F - m_m g}{m_m} = \frac{(m_p - m_m)\,g}{m_m} = \frac{(15 - 10)(9.8)}{10} = 4.9 \text{ m/s}^2 .$$

(b) As discussed, Newton's second law leads to $F - m_p g = m_p a_p$ for the package and $F - m_m g = m_m a_m$ for the monkey. If the acceleration of the package is downward, then the acceleration of the monkey is upward, so $a_m = -a_p$. Solving the first equation for F

$$F = m_p(g + a_p) = m_p(g - a_m)$$

and substituting this result into the second equation, we solve for a_m:

$$a_m = \frac{(m_p - m_m)\,g}{m_p + m_m} = \frac{(15 - 10)(9.8)}{15 + 10} = 2.0 \text{ m/s}^2 .$$

(c) The result is positive, indicating that the acceleration of the monkey is upward.

(d) Solving the second law equation for the package, we obtain

$$F = m_p\,(g - a_m) = (15)(9.8 - 2.0) = 120 \text{ N} .$$

48. The direction of motion (the direction of the barge's acceleration) is $+\hat{\imath}$, and $+\hat{\jmath}$ is chosen so that the pull $\vec{F}_h$ from the horse is in the first quadrant. The components of the unknown force of the water are denoted simply F_x and F_y.

(a) Newton's second law applied to the barge, in the x and y directions, leads to

$$\begin{aligned}(7900\,\text{N})\cos 18° + F_x &= ma \\ (7900\,\text{N})\sin 18° + F_y &= 0\end{aligned}$$

respectively. Plugging in $a = 0.12$ m/s^2 and $m = 9500$ kg, we obtain $F_x = 6.4 \times 10^3$ N and $F_y = -2.4 \times 10^3$ N. The magnitude of the force of the water is therefore

$$F_{\text{water}} = \sqrt{F_x^2 + F_y^2} = 6.8 \times 10^3 \text{ N} .$$

(b) Its angle measured from $+\hat{\text{i}}$ is either

$$\tan^{-1}\left(\frac{F_y}{F_x}\right) = -21° \quad \text{or} \quad 159° .$$

The signs of the components indicate the former is correct, so $\vec{F}_{\text{water}}$ is at 21° measured clockwise from the line of motion.

49. The force diagram (not to scale) for the block is shown below. $\vec{N}$ is the normal force exerted by the floor and $m\vec{g}$ is the force of gravity.

(a) The x component of Newton's second law is $F\cos\theta = ma$, where m is the mass of block and a is the x component of its acceleration. We obtain

$$a = \frac{F\cos\theta}{m} = \frac{(12.0\,\text{N})\cos 25.0°}{5.00\,\text{kg}} = 2.18\,\text{m/s}^2 .$$

This is its acceleration provided it remains in contact with the floor. Assuming it does, we find the value of N (and if N is positive, then the assumption is true but if N is negative then the block leaves the floor). The y component of Newton's second law becomes $N + F\sin\theta - mg = 0$, so $N = mg - F\sin\theta = (5.00)(9.8) - (12.0)\sin 25.0° = 43.9\,\text{N}$. Hence the block remains on the floor and its acceleration is $a = 2.18\,\text{m/s}^2$.

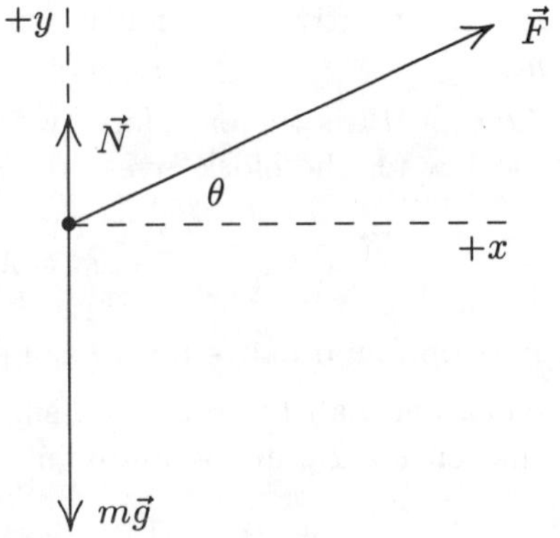

(b) If F is the minimum force for which the block leaves the floor, then $N = 0$ and the y component of the acceleration vanishes. The y component of the second law becomes $F\sin\theta - mg = 0$, so

$$F = \frac{mg}{\sin\theta} = \frac{(5.00)(9.8)}{\sin 25.0°} = 116\ \text{N} .$$

(c) The acceleration is still in the x direction and is still given by the equation developed in part (a):

$$a = \frac{F\cos\theta}{m} = \frac{116\cos 25°}{5.00} = 21.0\ \text{m/s}^2 .$$

50. The motion of the man-and-chair is positive if upward.

(a) When the man is grasping the rope, pulling with a force equal to the tension T in the rope, the total upward force on the man-and-chair due its two contact points with the rope is $2T$. Thus, Newton's second law leads to

$$2T - mg = ma$$

so that when $a = 0$, the tension is $T = 466$ N.

(b) When $a = +1.3\ \text{m/s}^2$ the equation in part (a) predicts that the tension will be $T = 527$ N.

(c) When the man is not holding the rope (instead, the co-worker attached to the ground is pulling on the rope with a force equal to the tension T in it), there is only one contact point between the rope and the man-and-chair, and Newton's second law now leads to

$$T - mg = ma$$

so that when $a = 0$, the tension is $T = 931$ N.

(d) When $a = +1.3\ \text{m/s}^2$ the equation in part (c) predicts that the tension will be $T = 1.05 \times 10^3$ N.

(e) The rope comes into contact (pulling down in each case) at the left edge and the right edge of the pulley, producing a total downward force of magnitude $2T$ on the ceiling. Thus, in part (a) this gives $2T = 931$ N.

(f) In part (b) the downward force on the ceiling has magnitude $2T = 1.05 \times 10^3$ N.

(g) In part (c) the downward force on the ceiling has magnitude $2T = 1.86 \times 10^3$ N.

(h) In part (d) the downward force on the ceiling has magnitude $2T = 2.11 \times 10^3$ N.

51. (a) A small segment of the rope has mass and is pulled down by the gravitational force of the Earth. Equilibrium is reached because neighboring portions of the rope pull up sufficiently on it. Since tension is a force *along* the rope, at least one of the neighboring portions must slope up away from the segment we are considering. Then, the tension has an upward component which means the rope sags.

(b) The only force acting with a horizontal component is the applied force $\vec{F}$. Treating the block and rope as a single object, we write Newton's second law for it: $F = (M + m)a$, where a is the acceleration and the positive direction is taken to be to the right. The acceleration is given by $a = F/(M + m)$.

(c) The force of the rope F_r is the only force with a horizontal component acting on the block. Then Newton's second law for the block gives

$$F_r = Ma = \frac{MF}{M + m}$$

where the expression found above for a has been used.

(d) Treating the block and half the rope as a single object, with mass $M + \frac{1}{2}m$, where the horizontal force on it is the tension T_m at the midpoint of the rope, we use Newton's second law:

$$T_m = (M + \frac{1}{2}m)a = \frac{(M + \frac{1}{2}m)F}{(M + m)} = \frac{(2M + m)F}{2(M + m)} .$$

52. The coordinate system we wish to use is shown in Fig. 5-18(c) in the textbook, so we resolve this horizontal force into appropriate components.

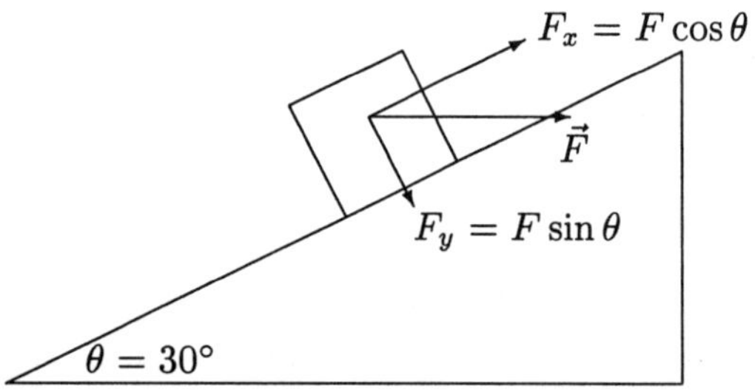

(a) Referring to Fig. 5-18 in the textbook, we see that Newton's second law applied to the x axis produces

$$F\cos\theta - mg\sin\theta = ma .$$

For $a = 0$, this yields $F = 566$ N.

(b) Applying Newton's second law to the y axis (where there is no acceleration), we have

$$N - F\sin\theta - mg\cos\theta = 0$$

which yields the normal force $N = 1.13 \times 10^3$ N.

53. The forces on the balloon are the force of gravity $m\vec{g}$ (down) and the force of the air $\vec{F}_a$ (up). We take the $+y$ to be up, and use a to mean the *magnitude* of the acceleration (which is not its usual use in this chapter). When the mass is M (before the ballast is thrown out) the acceleration is downward and Newton's second law is $F_a - Mg = -Ma$. After the ballast is thrown out, the mass is $M - m$ (where m is the mass of the ballast) and the acceleration is upward. Newton's second law leads to $F_a - (M-m)g = (M-m)a$. The earlier equation gives $F_a = M(g-a)$, and this plugs into the new equation to give

$$M(g-a) - (M-m)g = (M-m)a \implies m = \frac{2Ma}{g+a} .$$

54. The free-body diagram is shown below. Newton's second law for the mass m for the x direction leads to

$$T_1 - T_2 - mg\sin\theta = ma$$

which gives the difference in the tension in the pull cable:

$$\begin{aligned} T_1 - T_2 &= m(g\sin\theta + a) \\ &= (2800)\,(9.8\sin 35^\circ + 0.81) \\ &= 1.8 \times 10^4 \text{ N} . \end{aligned}$$

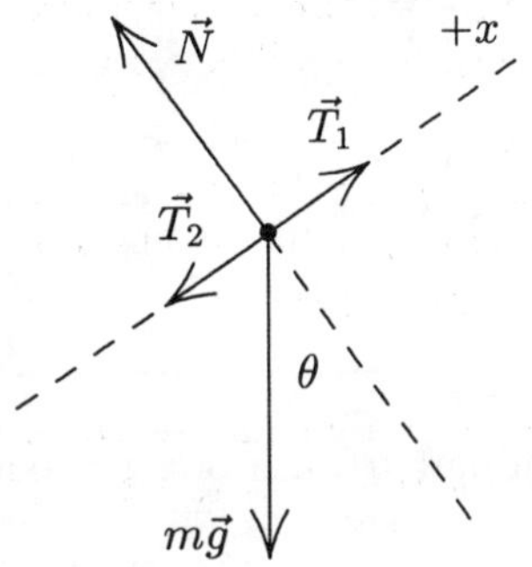

55. (a) The mass of the elevator is $m = 27800/9.8 = 2837$ kg and (with $+y$ upward) the acceleration is $a = +1.22$ m/s^2. Newton's second law leads to

$$T - mg = ma \implies T = m(g+a)$$

which yields $T = 3.13 \times 10^4$ N for the tension.

(b) The term "deceleration" means the acceleration vector is in the direction opposite to the velocity vector (which the problem tells us is upward). Thus (with $+y$ upward) the acceleration is now $a = -1.22$ m/s^2, so that the tension $T = m(g+a)$ turns out to be $T = 2.43 \times 10^4$ N in this case.

56. (a) The term "deceleration" means the acceleration vector is in the direction opposite to the velocity vector (which the problem tells us is downward). Thus (with $+y$ upward) the acceleration is $a = +2.4$ m/s^2. Newton's second law leads to

$$T - mg = ma \implies m = \frac{T}{g+a}$$

which yields $m = 7.3$ kg for the mass.

(b) Repeating the above computation (now to solve for the tension) with $a = +2.4$ m/s^2 will, of course, leads us right back to $T = 89$ N. Since the direction of the velocity did not enter our computation, this is to be expected.

57. The mass of the bundle is $m = 449/9.8 = 45.8$ kg and we choose $+y$ upward.

(a) Newton's second law, applied to the bundle, leads to

$$T - mg = ma \implies a = \frac{387 - 449}{45.8}$$

which yields $a = -1.35$ m/s^2 for the acceleration. The minus sign in the result indicates the acceleration vector points down. Any downward acceleration of magnitude greater than this is also acceptable (since that would lead to even smaller values of tension).

(b) We use Eq. 2-16 (with Δx replaced by $\Delta y = -6.1$ m). We assume $v_0 = 0$.

$$|v| = \sqrt{2a\Delta y} = \sqrt{2(-1.35)(-6.1)} = 4.1 \text{ m/s} .$$

For downward accelerations greater than 1.35 m/s^2, the speeds at impact will be larger than 4.1 m/s.

58. For convenience, we have labeled the 2.0 kg box m_1 and the 3.0 kg box m_2 – and their weights w' and w, respectively. The $+x$ axis is "downhill" for m_1 and "uphill" for m_2 (so they both accelerate with the same sign).

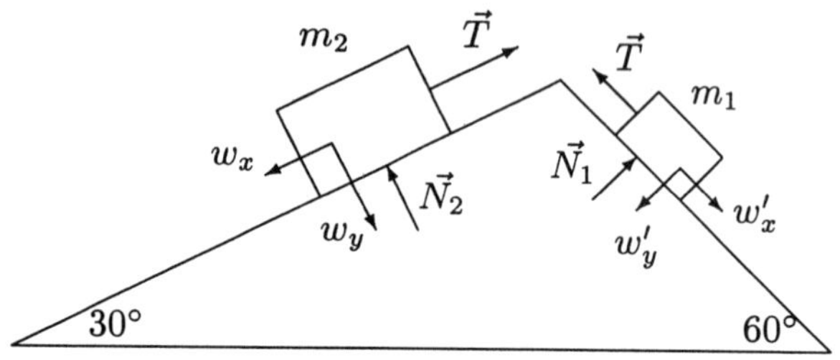

We apply Newton's second law to each box's x axis:

$$\begin{aligned} m_1 g \sin 60° - T &= m_1 a \\ T - m_2 g \sin 30° &= m_2 a \end{aligned}$$

Adding the two equations allows us to solve for the acceleration $a = 0.45$ m/s^2. This value is plugged back into either of the two equations to yield the tension $T = 16$ N.

59. (a) There are six legs, and the vertical component of the tension force in each leg is $T \sin\theta$ where $\theta = 40°$. For vertical equilibrium (zero acceleration in the y direction) then Newton's second law leads to

$$6T \sin\theta = mg \implies T = \frac{mg}{6 \sin\theta}$$

which (expressed as a multiple of the bug's weight mg) gives roughly $0.26mg$ for the tension.

(b) The angle θ is measured from horizontal, so as the insect "straightens out the legs" θ will increase (getting closer to 90°), which causes $\sin\theta$ to increase (getting closer to 1) and consequently (since $\sin\theta$ is in the denominator) causes T to decrease.

60. (a) Choosing the direction of motion as $+x$, Eq. 2-11 gives

$$a = \frac{88.5\,\text{km/h} - 0}{6.0\,\text{s}} = 15 \text{ km/h/s} .$$

Converting to SI, this is $a = 4.1$ m/s^2.

(b) With mass $m = 2000/9.8 = 204$ kg, Newton's second law gives $\vec{F} = m\vec{a} = 836$ N in the $+x$ direction.

61. (a) Intuition readily leads to the conclusion (that the heavier block should be the hanging one, for largest acceleration). The force that "drives" the system into motion is the weight of the hanging block (gravity acting on the block on the table has no effect on the dynamics, so long as we ignore friction).

(b) In Sample Problem 5-5 (where it was assumed the m is the hanging block) Eq. 5-21 gave the acceleration. Now that we have switched $m \leftrightarrow M$ (so that now M is the hanging block) our new version of Eq. 5-21 is

$$a = \frac{M}{m + M} g = 6.5 \text{ m/s}^2 .$$

(c) Switching $m \leftrightarrow M$ has no effect on Eq. 5-22, which yields

$$T = \frac{mM}{m+M}\, g = 13 \text{ N} .$$

62. Making separate free-body diagrams for the helicopter and the truck, one finds there are two forces on the truck ($\vec{T}$ upward, caused by the tension, which we'll think of as that of a single cable, and $m\vec{g}$ downward, where $m = 4500$ kg) and three forces on the helicopter ($\vec{T}$ downward, $\vec{F}_{\text{lift}}$ upward, and $M\vec{g}$ downward, where $M = 15000$ kg). With $+y$ upward, then $a = +1.4$ m/s^2 for both the helicopter and the truck.

(a) Newton's law applied to the helicopter and truck separately gives

$$\begin{aligned} F_{\text{lift}} - T - Mg &= Ma \\ T - mg &= ma \end{aligned}$$

which we add together to obtain

$$F_{\text{lift}} - (M+m)g = (M+m)a .$$

From this equation, we find $F_{\text{lift}} = 2.2 \times 10^5$ N.

(b) From the truck equation $T - mg = ma$ we obtain $T = 5.0 \times 10^4$ N.

63. (a) With SI units understood, the net force is

$$\vec{F}_{\text{net}} = \vec{F}_1 + \vec{F}_2 = (3.0 + (-2.0))\,\hat{\imath} + (4.0 + (-6.0))\,\hat{\jmath}$$

which yields $\vec{F}_{\text{net}} = 1.0\,\hat{\imath} - 2.0\,\hat{\jmath}$ in Newtons.

(b) Using magnitude-angle notation (especially convenient on a vector-capable calculator), the answer to part (a) becomes
$\vec{F}_{\text{net}} = (2.2\,\text{N} \angle -63°)$.

(c) Since $\vec{F}_{\text{net}}$ is equal to $\vec{a}$ multiplied by a positive scalar (which cannot affect the direction of the vector it multiplies), then the acceleration has the same angle as the net force. The magnitude of $\vec{a}$ comes from dividing the magnitude of $\vec{F}_{\text{net}}$ by the mass ($m = 1.0$ kg). Thus, in magnitude-angle notation, the answer is $\vec{a} = (2.2\,\text{m/s}^2 \angle -63°)$.

64. We take rightwards as the $+x$ direction. Thus, $\vec{F}_1 = 20\,\hat{\imath}$ in Newtons. In each case, we use Newton's second law $\vec{F}_1 + \vec{F}_2 = m\vec{a}$ where $m = 2.0$ kg.

(a) If $\vec{a} = +10\,\hat{\imath}$ in SI units, then the equation above gives $\vec{F}_2 = 0$.

(b) If $\vec{a} = +20\,\hat{\imath}$ m/s^2, then that equation gives $\vec{F}_2 = 20\,\hat{\imath}$ N.

(c) If $\vec{a} = 0$, then the equation gives $\vec{F}_2 = -20\,\hat{\imath}$ N.

(d) If $\vec{a} = -10\,\hat{\imath}$ m/s^2, the equation gives $\vec{F}_2 = -40\,\hat{\imath}$ N.

(e) If $\vec{a} = -20\,\hat{\imath}$ m/s^2, the equation gives $\vec{F}_2 = -60\,\hat{\imath}$ N.

65. (a) Since the performer's weight is $(52)(9.8) = 510$ N, the rope breaks.

(b) Setting $T = 425$ N in Newton's second law (with $+y$ upward) leads to

$$T - mg = ma \implies a = \frac{T}{m} - g$$

which yields $|a| = 1.6$ m/s^2.

66. The mass of the pilot is $m = 735/9.8 = 75$ kg. Denoting the upward force exerted by the spaceship (his seat, presumably) on the pilot as $\vec{F}$ and choosing upward the $+y$ direction, then Newton's second law leads to

$$F - mg_{\text{moon}} = ma \implies F = (75)(1.6 + 1.0) = 195 \text{ N} .$$

67. With SI units understood, the net force on the box is

$$\vec{F}_{\text{net}} = (3.0 + 14\cos 30° - 11)\,\hat{\imath} + (14\sin 30° + 5.0 - 17)\,\hat{\jmath}$$

which yields $\vec{F}_{\text{net}} = 4.1\,\hat{\imath} - 5.0\,\hat{\jmath}$ in Newtons.

(a) Newton's second law applied to the $m = 4.0$ kg box leads to

$$\vec{a} = \frac{\vec{F}_{\text{net}}}{m} = 1.0\hat{\imath} - 1.3\hat{\jmath} \text{ m/s}^2 .$$

(b) The magnitude of $\vec{a}$ is $\sqrt{1.0^2 + (-1.3)^2} = 1.6$ m/s^2. Its angle is $\tan^{-1}(-1.3/1.0) = -50°$ (that is, 50° measured clockwise from the rightward axis).

68. The net force is in the y direction, so the unknown force must have an x component that cancels the $(8.0\,\text{N})\hat{\imath}$ value of the known force, and it must also have enough y component to give the 3.0 kg object an acceleration of $(3.0\,\text{m/s}^2)\hat{\jmath}$. Thus, the magnitude of the unknown force is

$$\left|\vec{F}\right| = \sqrt{F_x^2 + F_y^2} = \sqrt{(-8.0)^2 + 9.0^2} = 12 \text{ N} .$$

69. We are only concerned with horizontal forces in this problem (gravity plays no direct role). Thus, $\sum \vec{F} = m\vec{a}$ reduces to $\vec{F}_{\text{avg}} = m\vec{a}$, and we see that the magnitude of the force is ma, where $m = 0.20$ kg and

$$a = |\vec{a}| = \sqrt{a_x^2 + a_y^2}$$

and the direction of the force is the same as that of $\vec{a}$. We take *east* as the $+x$ direction and *north* as $+y$. The acceleration is the *average* acceleration in the sense of Eq. 4-15.

(a) We find the (average) acceleration to be

$$\vec{a} = \frac{\vec{v} - \vec{v}_0}{\Delta t} = \frac{(-5.0\,\hat{\imath}) - (2.0\,\hat{\imath})}{0.50} = -14\hat{\imath} \text{ m/s}^2 .$$

Thus, the magnitude of the force is $(0.20\,\text{kg})(14\,\text{m/s}^2) = 2.8\,\text{N}$ and its direction is $-\hat{\imath}$ which means *west* in this context.

(b) A computation similar to the one in part (a) yields the (average) acceleration with two components, which can be expressed various ways:

$$\vec{a} = -4.0\hat{\imath} - 10.0\hat{\jmath} \;\rightarrow\; (-4.0, -10.0) \;\rightarrow\; (10.8\;\angle\; -112°)$$

Therefore, the magnitude of the force is $(0.20 \text{ kg})(10.8 \text{ m/s}^2) = 2.2$ N and its direction is 112° clockwise from east – which means it is 22° west of south, stated more conventionally.

70. The "certain force" is denoted F is assumed to be the net force on the object when it gives m_1 an acceleration $a_1 = 12$ m/s^2 and when it gives m_2 an acceleration $a_2 = 3.3$ m/s^2. Thus, we substitute $m_1 = F/a_1$ and $m_2 = F/a_2$ in appropriate places during the following manipulations.

(a) Now we seek the acceleration a of an object of mass $m_2 - m_1$ when F is the net force on it. Thus,

$$a = \frac{F}{m_2 - m_1} = \frac{F}{(F/a_2) - (F/a_1)} = \frac{a_1 a_2}{a_1 - a_2}$$

which yields $a = 4.6$ m/s^2.

(b) Similarly for an object of mass $m_2 + m_1$:

$$a = \frac{F}{m_2 + m_1} = \frac{F}{(F/a_2) + (F/a_1)} = \frac{a_1 a_2}{a_1 + a_2}$$

which yields $a = 2.6$ m/s^2.

71. We mention that the textbook treats this particular arrangement of blocks and pulleys in extensive detail in Sample Problem 5-5. Using the usual coordinate system (*right* $= +x$ and *up* $= +y$) for both blocks has the important consequence that for the 3.0 kg block to have a positive acceleration ($a > 0$), block M must have a negative acceleration of the same magnitude ($-a$). Thus, applying Newton's second law to the two blocks, we have

$$\begin{aligned} T &= (3.0 \text{ kg}) \left(1.0 \text{ m/s}^2\right) && \text{along } x \text{ axis} \\ T - Mg &= M\left(-1.0 \text{ m/s}^2\right) && \text{along } y \text{ axis} . \end{aligned}$$

(a) The first equation yields the tension $T = 3.0$ N.

(b) The second equation yields the mass $M = 3.0/8.8 = 0.34$ kg.

72. We take $+x$ uphill for the $m = 1.0$ kg box and $+x$ rightward for the $M = 3.0$ kg box (so the accelerations of the two boxes have the same magnitude and the same sign). The uphill force on m is F and the downhill forces on it are T and $mg \sin\theta$, where $\theta = 37°$. The only horizontal force on M is the rightward-pointed tension. Applying Newton's second law to each box, we find

$$\begin{aligned} F - T - mg\sin\theta &= ma \\ T &= Ma \end{aligned}$$

which are added to obtain $F - mg \sin\theta = (m + M)a$. This yields the acceleration

$$a = \frac{12 - (1.0)(9.8)\sin 37°}{1.0 + 3.0} = 1.53 \text{ m/s}^2 .$$

Thus, the tension is $T = Ma = (3.0)(1.53) = 4.6$ N.

73. (a) With $v_0 = 0$, Eq. 2-16 leads to

$$a = \frac{v^2}{2\Delta x} = \frac{\left(6.0 \times 10^6 \text{ m/s}\right)^2}{2(0.015 \text{ m})}$$

which yields 1.2×10^{15} m/s^2 for the acceleration. The force responsible for producing this acceleration is

$$F = ma = \left(9.11 \times 10^{-31} \text{ kg}\right)\left(1.2 \times 10^{15} \text{ m/s}^2\right) = 1.1 \times 10^{-15} \text{ N} .$$

(b) The weight is $mg = 8.9 \times 10^{-30}$ N, many orders of magnitude smaller than the result of part (a). As a result, gravity plays a negligible role in most atomic and subatomic processes.

74. We denote the thrust as T and choose $+y$ upward. Newton's second law leads to

$$T - Mg = Ma \implies a = \frac{2.6 \times 10^5}{1.3 \times 10^4} - 9.8$$

which yields $a = 10$ m/s^2.

75. (a) The reaction force to $\vec{F}_{MW} = 180$ N west is, by Newton's third law, $\vec{F}_{WM} = 180$ N east.

(b) Applying $\vec{F} = m\vec{a}$ to the woman gives an acceleration $a = 180/45 = 4.0\,\text{m/s}^2$, directed west.

(c) Applying $\vec{F} = m\vec{a}$ to the man gives an acceleration $a = 180/90 = 2.0\,\mathrm{m/s^2}$, directed east.

76. We note that $mg = (15)(9.8) = 147$ N.

   (a) The penguin's weight is $W = 147$ N.

   (b) The normal force exerted upward on the penguin by the scale is equal to the gravitational pull W on the penguin because the penguin is not accelerating. So, by Newton's second law, $N = W = 147$ N.

   (c) The reading on the scale, by Newton's third law, is the reaction force to that found in part (b). Its magnitude is therefore the same: 147 N.

77. Sample Problem 5-8 has a good treatment of the forces in an elevator. We apply Newton's second law (with $+y$ up)

$$N - mg = ma$$

where $m = 100$ kg and a must be estimated from the graph (it is the instantaneous slope at the various moments).

   (a) At $t = 1.8$ s, we estimate the slope to be $+1.0\,\mathrm{m/s^2}$. Thus, Newton's law yields $N \approx 1100$ N (up).

   (b) At $t = 4.4$ s, the slope is zero, so $N = 980$ N (up).

   (c) At $t = 6.8$ s, we estimate the slope to be $-1.7\,\mathrm{m/s^2}$. Thus, Newton's law yields $N = 810$ N (up).

78. From the reading when the elevator was at rest, we know the mass of the object is $m = 65/9.8 = 6.6$ kg. We choose $+y$ upward and note there are two forces on the object: mg downward and T upward (in the cord that connects it to the balance; T is the reading on the scale by Newton's third law).

   (a) "Upward at constant speed" means constant velocity, which means no acceleration. Thus, the situation is just as it was at rest: $T = 65$ N.

   (b) The term "deceleration" is used when the acceleration vector points in the direction opposite to the velocity vector. We're told the velocity is upward, so the acceleration vector points downward ($a = -2.4$ m/s^2). Newton's second law gives

$$T - mg = ma \implies T = (6.6)(9.8 - 2.4) = 49 \text{ N} .$$

79. Since $(x_0, y_0) = (0, 0)$ and $\vec{v}_0 = 6.0\,\hat{\imath}$, we have from Eq. 2-15

$$\begin{aligned} x &= (6.0)t + \frac{1}{2}a_x t^2 \\ y &= \frac{1}{2}a_y t^2 . \end{aligned}$$

These equations express uniform acceleration along each axis; the x axis points east and the y axis presumably points north (the assumption is that the figure shown in the problem is a view *from above*). Lengths are in meters, time is in seconds, and force is in newtons.

Examination of any non-zero (x, y) point will suffice, though it is certainly a good idea to check results by examining more than one. Here we will look at the $t = 4.0$ s point, at $(8.0, 8.0)$. The x equation becomes $8.0 = (6.0)(4.0) + \frac{1}{2}a_x(4.0)^2$. Therefore, $a_x = -2.0$ m/s^2. The y equation becomes $8.0 = \frac{1}{2}a_y(4.0)^2$. Thus, $a_y = 1.0$ m/s^2. The force, then, is

$$\vec{F} = m\vec{a} = -24\,\hat{\imath} + 12\,\hat{\jmath} \longrightarrow (27 \angle\ 153^\circ)$$

where the vector has been expressed in unit-vector and then magnitude-angle notation. Thus, the force has magnitude 27 N and is directed 63° west of north (or, equivalently, 27° north of west).

80. We label the 1.0 kg mass m and label the 2.0 kg mass M. We first analyze the forces on m.

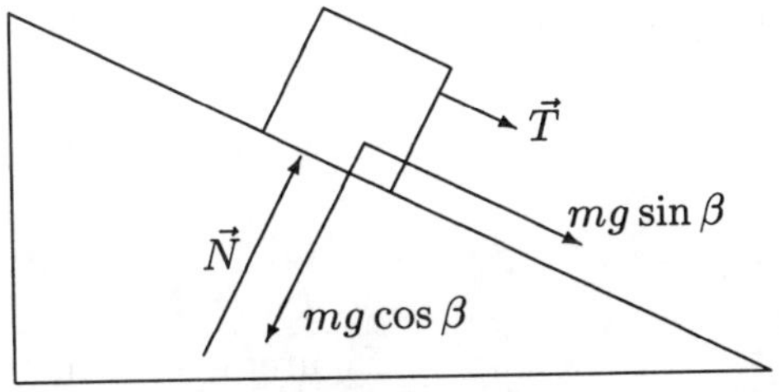

The $+x$ direction is "downhill" (parallel to $\vec{T}$).

With the acceleration $(5.5\,\mathrm{m/s^2})$ in the positive x direction for m, then Newton's second law, applied to the x axis, becomes

$$T + mg \sin \beta = m(5.5\,\mathrm{m/s^2})$$

But for M, using the more familiar vertical y axis (with *up* as the positive direction), we have the acceleration in the negative direction:

$$F + T - Mg = M(-5.5\,\mathrm{m/s^2})$$

where the tension comes in as an upward force (the cord can pull, not push).

(a) From the equation for M, with $F = 6.0\,\mathrm{N}$, we find the tension $T = 2.6\,\mathrm{N}$.

(b) From the equation for m, using the result from part (a), we obtain the angle $\beta = 17°$.

81. (a) The bottom cord is only supporting a mass of 4.5 kg against gravity, so its tension is $(4.5)(9.8) = 44$ N.

(b) The top cord is supporting a total mass of 8.0 kg against gravity, so the tension there is $(8.0)(9.8) = 78$ N.

(c) In the second picture, the lowest cord supports a mass of 5.5 kg against gravity and consequently has a tension of $(5.5)(9.8) = 54$ N.

(d) The top cord, we are told, has tension 199 N which supports a total of $199/9.8 = 20.3$ kg, 10.3 of which is accounted for in the figure. Thus, the unknown mass in the middle must be $20.3 - 10.3 = 10.0$ kg, and the tension in the cord above it must be enough to support $10.0 + 5.5 = 15.5$ kg, so $T = (15.5)(9.8) = 152$ N. Another way to analyze this is to examine the forces on the 4.8 kg piece; one of the downward forces on it is this T.

82. The mass of the automobile is $17000/9.8 = 1735$ kg, so the net force has magnitude $F = (1735)(3.66) = 6.35 \times 10^2$ N.

83. (First problem in **Cluster 1**)

(a) Using the coordinate system and force resolution shown in the textbook Figure 5-18(c), we apply Newton's second law along the x axis

$$-mg \sin \theta = ma$$

where $\theta = 30.0°$. Thus, $a = -4.9\,\mathrm{m/s^2}$. The magnitude of the acceleration, then, is $4.9\,\mathrm{m/s^2}$.

(b) Applying Newton's second law along the y axis (where there is no acceleration), we have

$$N - mg \cos \theta = 0 \,.$$

Thus, with $m = 10.0$ kg, we obtain $N = 84.9$ N.

84. (Second problem in **Cluster 1**)

(a) Newton's second law applied to the x axis yields $F - mg\sin\theta = ma$. Thus, with $F = 40.0$ N, we find $a = -0.90\ \mathrm{m/s^2}$. The interpretation is that the magnitude of the acceleration is $0.90\ \mathrm{m/s^2}$ and its direction is downhill.

(b) Substituting $F = 60.0$ N into $F - mg\sin\theta = ma$, we find $a = 1.1\ \mathrm{m/s^2}$. Thus, the acceleration is $1.1\ \mathrm{m/s^2}$ uphill.

85. (Third problem in **Cluster 1**)
The coordinate system we wish to use is shown in Figure 5-18(c) in the textbook, so we resolve this vertical force into appropriate components.

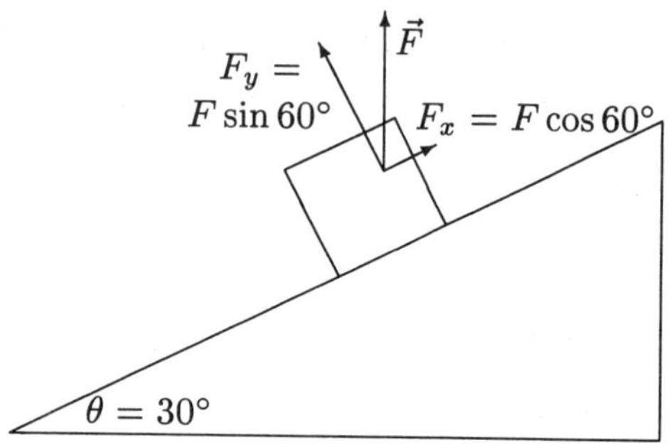

(a) Assuming the block is not pulled entirely off the incline, Newton's second law applied to the x axis yields

$$F_x - mg\sin\theta = ma \ .$$

This leads to $a = -1.9\,\mathrm{m/s^2}$, which we interpret as a acceleration of $1.9\,\mathrm{m/s^2}$ directed *downhill.*

(b) The assumption stated in part (a) implies there is no acceleration in the y direction. Newton's second law along the y axis gives

$$N + F_y - mg\cos\theta = 0 \ .$$

Therefore, $N = 32.9$ N. We note that a negative value of N would have been a sure sign that our assumption was incorrect.

(c) The equation in part (a) can be used to solve for the equilibrium ($a = 0$) value of F:

$$F\cos 60° = mg\sin 30° = 49 \text{ N} \ .$$

Therefore, $F = 98$ N.

(d) There are three forces acting on the block: $\vec{N}, \vec{F}$, and $m\vec{g}$. Equilibrium generally suggests that the "vector triangle" formed by three such vectors closes on itself. In this case, however, two sides of that "triangle" are vertical! $\vec{F}$ is *up* and $m\vec{g}$ is *down*! The insight behind this "squashed triangle" is that $\vec{N}$ (the only vector that is not vertical) has zero magnitude. Thus, the block is not "bearing down" on the incline surface. In fact, in this circumstance, the incline is not needed at all for support; the value $F = 98.0$ N is just what is needed to hold the block (which weighs 98.0 N) aloft.

86. (Fourth problem in **Cluster 1**)
The coordinate system we wish to use is shown in Fig. 5-18 in the textbook, so we resolve this horizontal force into appropriate components.

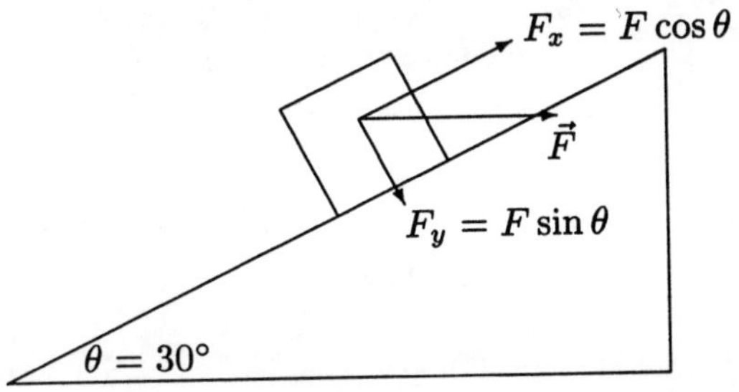

(a) We apply Newton's second law to the x axis:

$$F_x - mg\sin\theta = ma$$

This yields $a = -1.44\,\text{m/s}^2$, which is interpreted as an acceleration of $1.44\,\text{m/s}^2$ downhill.

(b) Applying Newton's second law to the y axis (where there is no acceleration), we have

$$N - F_y - mg\cos\theta = 0 .$$

This yields the normal force $N = 105$ N.

(c) When we set $a = 0$ in the part (a) equation, we obtain

$$F\cos 30° = mg\sin 30° .$$

Therefore, $F = 56.6$ N. Alternatively, we can use a "vector triangle" approach, referred to in the previous problem solution. We form a closed triangle.

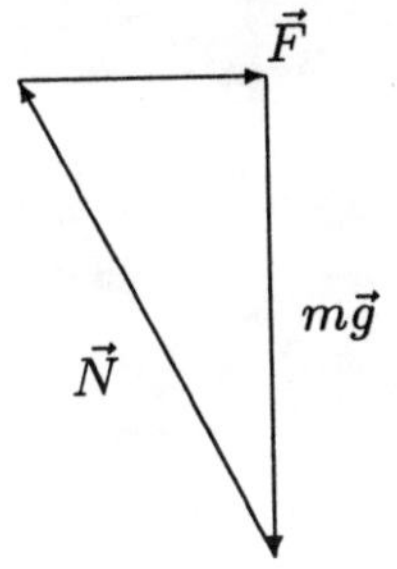

We note that the angle between the weight vector and the normal force is θ. Thus, we see $mg\tan\theta = F$, which gives $F = 56.6$ N.

Chapter 6

1. We do not consider the possibility that the bureau might tip, and treat this as a purely horizontal motion problem (with the person's push $\vec{F}$ in the $+x$ direction). Applying Newton's second law to the x and y axes, we obtain

$$\begin{aligned} F - f_{s,\text{max}} &= ma \\ N - mg &= 0 \end{aligned}$$

respectively. The second equation yields the normal force $N = mg$, whereupon the maximum static friction is found to be (from Eq. 6-1) $f_{s,\text{max}} = \mu_s mg$. Thus, the first equation becomes

$$F - \mu_s mg = ma = 0$$

where we have set $a = 0$ to be consistent with the fact that the static friction is still (just barely) able to prevent the bureau from moving.

(a) With $\mu_s = 0.45$ and $m = 45$ kg, the equation above leads to $F = 198$ N. To bring the bureau into a state of motion, the person should push with any force greater than this value. Rounding to two significant figures, we can therefore say the minimum required push is $F = 2.0 \times 10^2$ N.

(b) Replacing $m = 45$ kg with $m = 28$ kg, the reasoning above leads to roughly $F = 1.2 \times 10^2$ N.

2. An excellent discussion and equation development related to this problem is given in Sample Problem 6-3. We merely quote (and apply) their main result (Eq. 6-13)

$$\theta = \tan^{-1} \mu_s = \tan^{-1} 0.04 \approx 2^\circ \ .$$

3. The free-body diagram for the player is shown below. $\vec{N}$ is the normal force of the ground on the player, $m\vec{g}$ is the force of gravity, and $\vec{f}$ is the force of friction. The force of friction is related to the normal force by $f = \mu_k N$. We use Newton's second law applied to the vertical axis to find the normal force. The vertical component of the acceleration is zero, so we obtain $N - mg = 0$; thus, $N = mg$. Consequently,

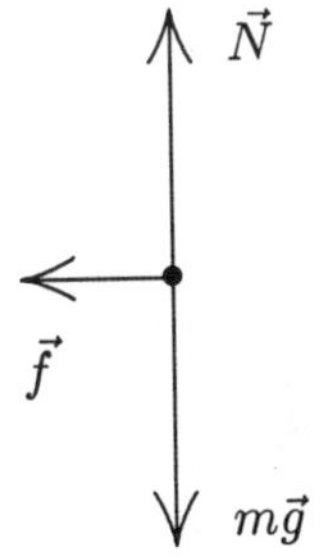

$$\begin{aligned} \mu_k &= \frac{f}{N} \\ &= \frac{470\,\text{N}}{(79\,\text{kg})\left(9.8\,\text{m/s}^2\right)} \\ &= 0.61 \ . \end{aligned}$$

4. To maintain the stone's motion, a horizontal force (in the $+x$ direction) is needed that cancels the retarding effect due to kinetic friction. Applying Newtons' second to the x and y axes, we obtain

$$\begin{aligned} F - f_k &= ma \\ N - mg &= 0 \end{aligned}$$

respectively. The second equation yields the normal force $N = mg$, so that (using Eq. 6-2) the kinetic friction becomes $f_k = \mu_k mg$. Thus, the first equation becomes

$$F - \mu_k mg = ma = 0$$

where we have set $a = 0$ to be consistent with the idea that the horizontal velocity of the stone should remain constant. With $m = 20$ kg and $\mu_k = 0.80$, we find $F = 1.6 \times 10^2$ N.

5. We denote $\vec{F}$ as the horizontal force of the person exerted on the crate (in the $+x$ direction), $\vec{f_k}$ is the force of kinetic friction (in the $-x$ direction), $\vec{N}$ is the vertical normal force exerted by the floor (in the $+y$ direction), and $m\vec{g}$ is the force of gravity. The magnitude of the force of friction is given by $f_k = \mu_k N$ (Eq. 6-2). Applying Newtons' second to the x and y axes, we obtain

$$\begin{aligned} F - f_k &= ma \\ N - mg &= 0 \end{aligned}$$

respectively.

(a) The second equation yields the normal force $N = mg$, so that the friction is

$$f_k = \mu_k mg = (0.35)(55\,\text{kg})\left(9.8\,\text{m/s}^2\right) = 1.9 \times 10^2\ \text{N}\ .$$

(b) The first equation becomes

$$F - \mu_k mg = ma$$

which (with $F = 220$ N) we solve to find

$$a = \frac{F}{m} - \mu_k g = 0.56\ \text{m/s}^2\ .$$

6. An excellent discussion and equation development related to this problem is given in Sample Problem 6-3. We merely quote (and apply) their main result (Eq. 6-13)

$$\theta = \tan^{-1} \mu_s = \tan^{-1} 0.5 = 27°$$

which implies that the angle through which the slope should be *reduced* is $\phi = 45° - 27° \approx 20°$.

7. The free-body diagram for the puck is shown below. $\vec{N}$ is the normal force of the ice on the puck, $\vec{f}$ is the force of friction (in the $-x$ direction), and $m\vec{g}$ is the force of gravity.

(a) The horizontal component of Newton's second law gives $-f = ma$, and constant acceleration kinematics (Table 2-1) can be used to find the acceleration.

Since the final velocity is zero, $v^2 = v_0^2 + 2ax$ leads to $a = -v_0^2/2x$. This is substituted into the Newton's law equation to obtain

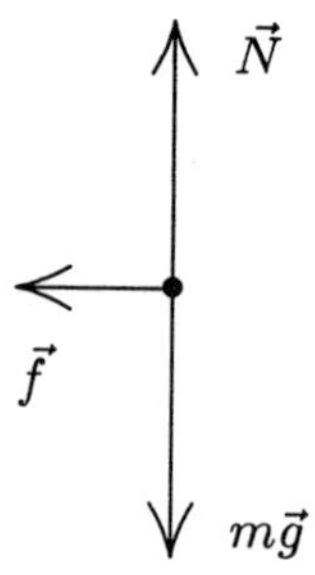

$$\begin{aligned} f &= \frac{mv_0^2}{2x} \\ &= \frac{(0.110\,\text{kg})(6.0\,\text{m/s})^2}{2(15\,\text{m})} \\ &= 0.13\ \text{N}\ . \end{aligned}$$

(b) The vertical component of Newton's second law gives $N - mg = 0$, so $N = mg$ which implies (using Eq. 6-2) $f = \mu_k mg$. We solve for the coefficient:

$$\mu_k = \frac{f}{mg} = \frac{0.13\,\mathrm{N}}{(0.110\,\mathrm{kg})\left(9.8\,\mathrm{m/s^2}\right)} = 0.12\ .$$

8. (a) The free-body diagram for the person (shown as an L-shaped block) is shown below. The force that she exerts on the rock slabs is not directly shown (since the diagram should only show forces exerted on her), but it is related by Newton's third law) to the normal forces $\vec{N}_1$ and $\vec{N}_2$ exerted horizontally by the slabs onto her shoes and back, respectively. We will show in part (b) that $N_1 = N_2$ so that we there is no ambiguity in saying that the magnitude of her push is N_2. The total upward force due to (maximum) static friction is $\vec{f} = \vec{f}_1 + \vec{f}_2$ where (using Eq. 6-1) $f_1 = \mu_{s\,1} N_1$ and $f_2 = \mu_{s\,2} N_2$. The problem gives the values $\mu_{s\,1} = 1.2$ and $\mu_{s\,2} = 0.8$.

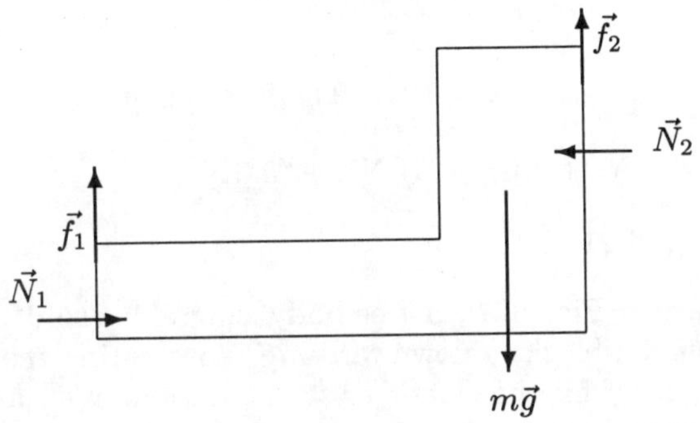

(b) We apply Newton's second law to the x and y axes (with $+x$ rightward and $+y$ upward and there is no acceleration in either direction).

$$\begin{aligned} N_1 - N_2 &= 0 \\ f_1 + f_2 - mg &= 0 \end{aligned}$$

The first equation tells us that the normal forces are equal $N_1 = N_2 = N$. Consequently, from Eq. 6-1

$$\begin{aligned} f_1 &= \mu_{s\,1} N \\ f_2 &= \mu_{s\,2} N \end{aligned}$$

we conclude that

$$f_1 = \left(\frac{\mu_{s\,1}}{\mu_{s\,2}}\right) f_2\ .$$

Therefore, $f_1 + f_2 - mg = 0$ leads to

$$\left(\frac{\mu_{s\,1}}{\mu_{s\,2}} + 1\right) f_2 = mg$$

which (with $m = 49$ kg) yields $f_2 = 192$ N. From this we find $N = f_2/\mu_{s\,2} = 240$ N. This is equal to the magnitude of the push exerted by the rock climber.

(c) From the above calculation, we find $f_1 = \mu_{s\,1} N = 288$ N which amounts to a fraction

$$\frac{f_1}{W} = \frac{288}{(49)(9.8)} = 0.60$$

or 60% of her weight.

9. (a) The free-body diagram for the block is shown below. $\vec{F}$ is the applied force, $\vec{N}$ is the normal force of the wall on the block, $\vec{f}$ is the force of friction, and $m\vec{g}$ is the force of gravity. To determine if the block falls, we find the magnitude f of the force of friction required to hold it without accelerating and also find the normal force of the wall on the block. We compare f and $\mu_s N$. If $f < \mu_s N$, the block does not slide on the wall but if $f > \mu_s N$, the block does slide. The horizontal component of Newton's second law is $F - N = 0$, so $N = F = 12\,\text{N}$ and $\mu_s N = (0.60)(12\,\text{N}) = 7.2\,\text{N}$. The vertical component is $f - mg = 0$, so $f = mg = 5.0\,\text{N}$. Since $f < \mu_s N$ the block does not slide.

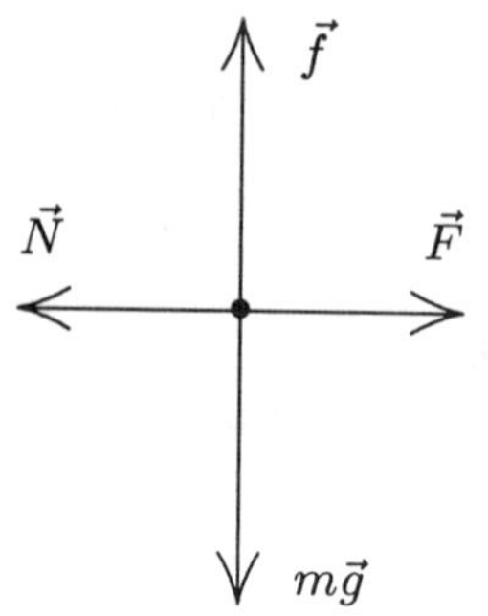

(b) Since the block does not move $f = 5.0\,\text{N}$ and $N = 12\,\text{N}$. The force of the wall on the block is

$$\vec{F}_w = -N\,\hat{\text{i}} + f\,\hat{\text{j}} = -(12\,\text{N})\,\hat{\text{i}} + (5.0\,\text{N})\,\hat{\text{j}}$$

where the axes are as shown on Fig. 6-21 of the text.

10. In addition to the forces already shown in Fig. 6-22, a free-body diagram would include an upward normal force $\vec{N}$ exerted by the floor on the block, a downward $m\vec{g}$ representing the gravitational pull exerted by Earth, and an assumed-leftward $\vec{f}$ for the kinetic or static friction. We choose $+x$ rightwards and $+y$ upwards. We apply Newton's second law to these axes:

$$\begin{aligned}(6.0\,\text{N}) - f &= ma \\ P + N - mg &= 0\end{aligned}$$

where $m = 2.5$ kg is the mass of the block.

(a) In this case, $P = 8.0$ N leads to $N = (2.5)(9.8) - 8.0$ so that the normal force is $N = 16.5$ N. Using Eq. 6-1, this implies $f_{s,\text{max}} = \mu_s N = 6.6$ N, which is larger than the 6.0 N rightward force – so the block (which was initially at rest) does not move. Putting $a = 0$ into the first of our equations above yields a static friction force of $f = P = 6.0$ N. Since its value is positive, then our assumption for the direction of $\vec{f}$ (leftward) is correct.

(b) In this case, $P = 10$ N leads to $N = (2.5)(9.8) - 10$ so that the normal force is $N = 14.5$ N. Using Eq. 6-1, this implies $f_{s,\text{max}} = \mu_s N = 5.8$ N, which is less than the 6.0 N rightward force – so the block does move. Hence, we are dealing not with static but with kinetic friction, which Eq. 6-2 reveals to be $f_k = \mu_k N = 3.6$ N. Again, its value is positive, so our assumption for the direction of $\vec{f}$ (leftward) is correct.

(c) In this last case, $P = 12$ N leads to $N = 12.5$ N and thus to $f_{s,\text{max}} = \mu_s N = 5.0$ N, which (as expected) is less than the 6.0 N rightward force – so the block moves. The kinetic friction force, then, is $f_k = \mu_k N = 3.1$ N. Once again, its value is positive, so our assumption for the direction of $\vec{f}$ (leftward) is correct.

11. A cross section of the cone of sand is shown below. To pile the most sand without extending the radius, sand is added to make the height h as great as possible. Eventually, however, the sides become so steep that sand at the surface begins to slide. The goal is to find the greatest height (corresponding to greatest slope) for which the sand does not slide. A grain of sand is shown on the diagram and the forces on it are labeled. $\vec{N}$ is the normal force of the surface, $m\vec{g}$ is the force of gravity, and $\vec{f}$ is the force of (static) friction. We take the x axis to be down the plane and the y axis to be in the direction of the normal

force. We assume the grain does not slide, so its acceleration is zero. Then the x component of Newton's second law is $mg\sin\theta - f = 0$ and the y component is $N - mg\cos\theta = 0$.

The first equation gives $f = mg\sin\theta$ and the second gives $N = mg\cos\theta$. If the grain does not slide, the condition $f < \mu_s N$ must hold. This means $mg\sin\theta < \mu_s mg\cos\theta$ or $\tan\theta < \mu_s$. The surface of the cone has the greatest slope (and the height of the cone is the greatest) if $\tan\theta = \mu_s$.

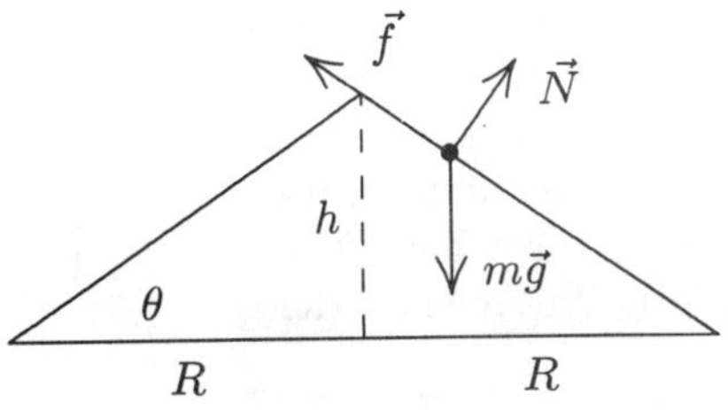

Since R and h are two sides of a right triangle, $h = R\tan\theta$. Replacing $\tan\theta$ with μ_s we obtain $h = \mu_s R$. We substitute this into the volume equation $V = \pi R^2 h/3$ to obtain the result $V = \pi\mu_s R^3/3$.

12. We denote the magnitude of 110 N force exerted by the worker on the crate as F. The magnitude of the static frictional force can vary between zero and $f_{s,\,\mathrm{max}} = \mu_s N$.

(a) In this case, application of Newton's second law in the vertical direction yields $N = mg$. Thus,

$$\begin{aligned} f_{s,\,\mathrm{max}} &= \mu_s N = \mu_s mg \\ &= (0.37)(35\,\mathrm{kg})\left(9.8\,\mathrm{m/s^2}\right) = 126.9\ \mathrm{N} \end{aligned}$$

which is greater than F. The block, which is initially at rest, stays at rest. This implies, by applying Newton's second law to the horizontal direction, that the magnitude of the frictional force exerted on the crate is $f_s = F = 110\,\mathrm{N}$.

(b) As calculated in part (a), $f_{s,\,\mathrm{max}} = 1.3 \times 10^2\,\mathrm{N}$.

(c) As remarked above, the crate does not move (since $F < f_{s,\,\mathrm{max}}$).

(d) Denoting the upward force exerted by the second worker as F_2, then application of Newton's second law in the vertical direction yields $N = mg - F_2$. Therefore, in this case, $f_{s,\,\mathrm{max}} = \mu_s N = \mu_s(mg - F_2)$. In order to move the crate, F must satisfy $F > f_{s,\,\mathrm{max}} = \mu_s(mg - F_2)$, i.e.,

$$110\,\mathrm{N} > (0.37)\left((35\,\mathrm{kg})\left(9.8\,\mathrm{m/s^2}\right) - F_2\right).$$

The minimum value of F_2 that satisfies this inequality is a value slightly bigger than 45.7 N, so we express our answer as $F_{2,\,\mathrm{min}} = 46\,\mathrm{N}$.

(e) In this final case, moving the crate requires a greater horizontal push from the worker than static friction (as computed in part (a)) can resist. Thus, Newton's law in the horizontal direction leads to

$$\begin{aligned} F + F_2 &> f_{s,\,\mathrm{max}} \\ 110\,\mathrm{N} + F_2 &> 126.9\,\mathrm{N} \end{aligned}$$

which leads (after appropriate rounding) to $F_{2,\,\mathrm{min}} = 17\,\mathrm{N}$.

13. (a) The free-body diagram for the crate is shown below. $\vec{T}$ is the tension force of the rope on the crate, $\vec{N}$ is the normal force of the floor on the crate, $m\vec{g}$ is the force of gravity, and $\vec{f}$ is the force of friction. We take the $+x$ direction to be horizontal to the right and the $+y$ direction to be up. We assume the crate is motionless. The x component of Newton's second law leads to $T\cos\theta - f = 0$ and the y component becomes $T\sin\theta + N - mg = 0$, where $\theta = 15°$ is the angle between the rope

and the horizontal.
The first equation gives $f = T\cos\theta$ and the second gives $N = mg - T\sin\theta$. If the crate is to remain at rest, f must be less than $\mu_s N$, or $T\cos\theta < \mu_s(mg - T\sin\theta)$. When the tension force is sufficient to just start the crate moving, we must have $T\cos\theta = \mu_s(mg - T\sin\theta)$. We solve for the tension:

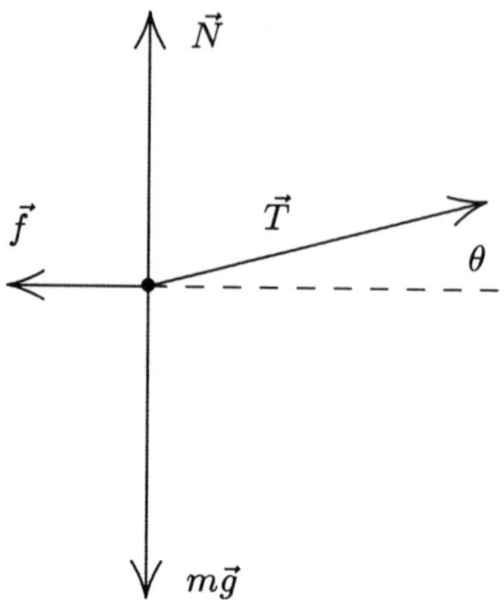

$$\begin{aligned} T &= \frac{\mu_s mg}{\cos\theta + \mu_s\sin\theta} \\ &= \frac{(0.50)(68)(9.8)}{\cos 15^\circ + 0.50\sin 15^\circ} \\ &= 304 \approx 300 \text{ N} . \end{aligned}$$

(b) The second law equations for the moving crate are $T\cos\theta - f = ma$ and $N + T\sin\theta - mg = 0$. Now $f = \mu_k N$. The second equation gives $N = mg - T\sin\theta$, as before, so $f = \mu_k(mg - T\sin\theta)$. This expression is substituted for f in the first equation to obtain $T\cos\theta - \mu_k(mg - T\sin\theta) = ma$, so the acceleration is

$$a = \frac{T(\cos\theta + \mu_k\sin\theta)}{m} - \mu_k g$$

which we evaluate:

$$a = \frac{(304\,\text{N})(\cos 15^\circ + 0.35\sin 15^\circ)}{68\,\text{kg}} - (0.35)(9.8\,\text{m/s}^2) = 1.3\,\text{m/s}^2 .$$

14. We first analyze the forces on the pig of mass m. The incline angle is θ.

The $+x$ direction is "downhill."

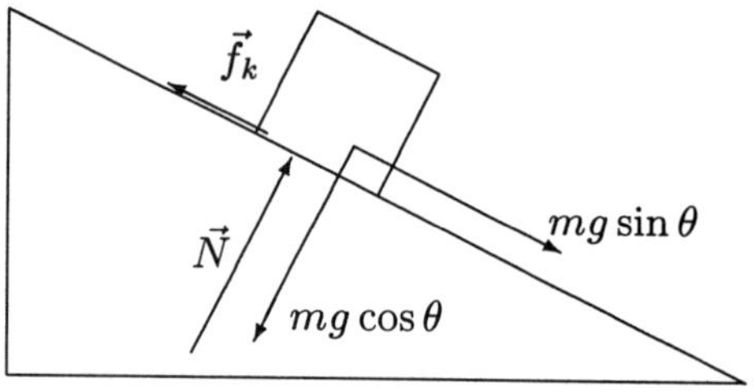

Application of Newton's second law to the x and y axes leads to

$$\begin{aligned} mg\sin\theta - f_k &= ma \\ N - mg\cos\theta &= 0 . \end{aligned}$$

Solving these along with Eq. 6-2 ($f_k = \mu_k N$) produces the following result for the pig's downhill acceleration:

$$a = g(\sin\theta - \mu_k\cos\theta) .$$

To compute the time to slide from rest through a downhill distance ℓ, we use Eq. 2-15:

$$\ell = v_0 t + \frac{1}{2}at^2 \implies t = \sqrt{\frac{2\ell}{a}} .$$

We denote the frictionless ($\mu_k = 0$) case with a prime and set up a ratio:

$$\frac{t}{t'} = \frac{\sqrt{2\ell/a}}{\sqrt{2\ell/a'}} = \sqrt{\frac{a'}{a}}$$

which leads us to conclude that if $t/t' = 2$ then $a' = 4a$. Putting in what we found out above about the accelerations, we have

$$g\sin\theta = 4g\left(\sin\theta - \mu_k\cos\theta\right) \ .$$

Using $\theta = 35°$, we obtain $\mu_k = 0.53$.

15. (a) Free-body diagrams for the blocks A and C, considered as a single object, and for the block B are shown below. T is the magnitude of the tension force of the rope, N is the magnitude of the normal force of the table on block A, f is the magnitude of the force of friction, W_{AC} is the combined weight of blocks A and C (the magnitude of force $\vec{F}_{g\,AC}$ shown in the figure), and W_B is the weight of block B (the magnitude of force $\vec{F}_{g\,B}$ shown). Assume the blocks are not moving. For the blocks on the table we take the x axis to be to the right and the y axis to be upward. The x component of Newton's second law is then $T - f = 0$ and the y component is $N - W_{AC} = 0$. For block B take the downward direction to be positive. Then Newton's second law for that block is $W_B - T = 0$. The third equation gives $T = W_B$ and the first gives $f = T = W_B$. The second equation gives $N = W_{AC}$. If sliding is not to occur, f must be less than $\mu_s N$, or $W_B < \mu_s W_{AC}$. The smallest that W_{AC} can be with the blocks still at rest is $W_{AC} = W_B/\mu_s = (22\,\text{N})/(0.20) = 110\,\text{N}$. Since the weight of block A is 44 N, the least weight for C is $110 - 44 = 66$ N.

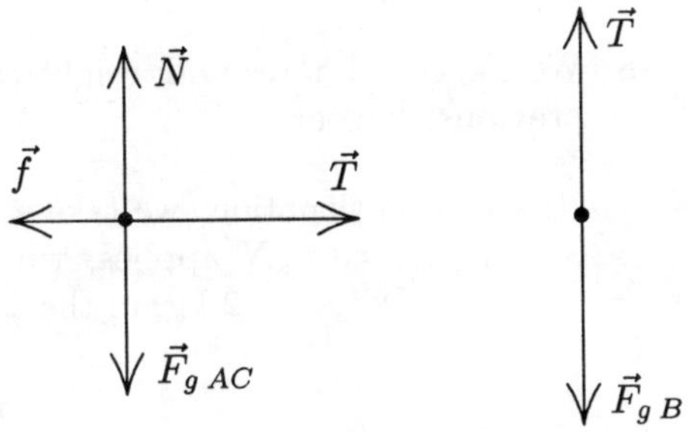

(b) The second law equations become $T - f = (W_A/g)a$, $N - W_A = 0$, and $W_B - T = (W_B/g)a$. In addition, $f = \mu_k N$. The second equation gives $N = W_A$, so $f = \mu_k W_A$. The third gives $T = W_B - (W_B/g)a$. Substituting these two expressions into the first equation, we obtain $W_B - (W_B/g)a - \mu_k W_A = (W_A/g)a$. Therefore,

$$a = \frac{g(W_B - \mu_k W_A)}{W_A + W_B} = \frac{(9.8\,\text{m/s}^2)\,(22\,\text{N} - (0.15)(44\,\text{N}))}{44\,\text{N} + 22\,\text{N}} = 2.3\,\text{m/s}^2 \ .$$

16. We choose $+x$ horizontally rightwards and $+y$ upwards and observe that the 15 N force has components $F_x = F\cos\theta$ and $F_y = -F\sin\theta$.

(a) We apply Newton's second law to the y axis:

$$N - F\sin\theta - mg = 0 \implies N = (15)\sin 40° + (3.5)(9.8) = 44$$

in SI units. With $\mu_k = 0.25$, Eq. 6-2 leads to $f_k = 11$ N.

(b) We apply Newton's second law to the x axis:

$$F\cos\theta - f_k = ma \implies a = \frac{(15)\cos 40° - 11}{3.5} = 0.14$$

in SI units (m/s^2). Since the result is positive-valued, then the block is accelerating in the $+x$ (rightward) direction.

17. (a) Although details in Fig. 6-27 might suggest otherwise, we assume (as the problem states) that only static friction holds block B in place. An excellent discussion and equation development related to this topic is given in Sample Problem 6-3. We merely quote (and apply) their main result (Eq. 6-13) for the maximum angle for which static friction applies (in the absence of additional forces such as the $\vec{F}$ of part (b) of this problem).

$$\theta_{\max} = \tan^{-1}\mu_s = \tan^{-1} 0.63 \approx 32^\circ \ .$$

This is greater than the dip angle in the problem, so the block does not slide.

(b) We analyze forces in a manner similar to that shown in Sample Problem 6-3, but with the addition of a downhill force F.

$$\begin{aligned} F + mg\sin\theta - f_{s,\max} &= ma = 0 \\ N - mg\cos\theta &= 0 \ . \end{aligned}$$

Along with Eq. 6-1 ($f_{s,\max} = \mu_s N$) we have enough information to solve for F. With $\theta = 24^\circ$ and $m = 1.8 \times 10^7$ kg, we find

$$F = mg\,(\mu_s\cos\theta - \sin\theta) = 3.0 \times 10^7 \text{ N} \ .$$

18. We use coordinates and weight-components as indicated in Fig. 5-18 (see Sample Problem 5-7 from the previous chapter).

(a) In this situation, we take $\vec{f_s}$ to point uphill and to be equal to its maximum value, in which case $f_{s,\max} = \mu_s N$ applies, where $\mu_s = 0.25$. Applying Newton's second law to the block of mass $m = W/g = 8.2$ kg, in the x and y directions, produces

$$\begin{aligned} F_{\min 1} - mg\sin\theta + f_{s,\max} &= ma = 0 \\ N - mg\cos\theta &= 0 \end{aligned}$$

which (with $\theta = 20^\circ$) leads to

$$F_{\min 1} = mg\,(\sin\theta - \mu_s\cos\theta) = 8.6 \text{ N} \ .$$

(b) Now we take $\vec{f_s}$ to point downhill and to be equal to its maximum value, in which case $f_{s,\max} = \mu_s N$ applies, where $\mu_s = 0.25$. Applying Newton's second law to the block of mass $m = W/g = 8.2$ kg, in the x and y directions, produces

$$\begin{aligned} F_{\min 2} - mg\sin\theta - f_{s,\max} &= ma = 0 \\ N - mg\cos\theta &= 0 \end{aligned}$$

which (with $\theta = 20^\circ$) leads to

$$F_{\min 2} = mg\,(\sin\theta + \mu_s\cos\theta) = 46 \text{ N} \ .$$

A value slightly larger than the "exact" result of this calculation is required to make it accelerate up hill, but since we quote our results here to two significant figures, 46 N is a "good enough" answer.

(c) Finally, we are dealing with kinetic friction (pointing downhill), so that

$$\begin{aligned} F - mg\sin\theta - f_k &= ma = 0 \\ N - mg\cos\theta &= 0 \end{aligned}$$

along with $f_k = \mu_k N$ (where $\mu_k = 0.15$) brings us to

$$F = mg\,(\sin\theta + \mu_k\cos\theta) = 39 \text{ N} \ .$$

19. The free-body diagrams for block B and for the knot just above block A are shown below. $\vec{T}_1$ is the tension force of the rope pulling on block B or pulling on the knot (as the case may be), $\vec{T}_2$ is the tension force exerted by the second rope (at angle $\theta = 30°$) on the knot, $\vec{f}$ is the force of static friction exerted by the horizontal surface on block B, $\vec{N}$ is normal force exerted by the surface on block B, W_A is the weight of block A (W_A is the magnitude of $m_A\vec{g}$), and W_B is the weight of block B ($W_B = 711$ N is the magnitude of $m_B\vec{g}$).

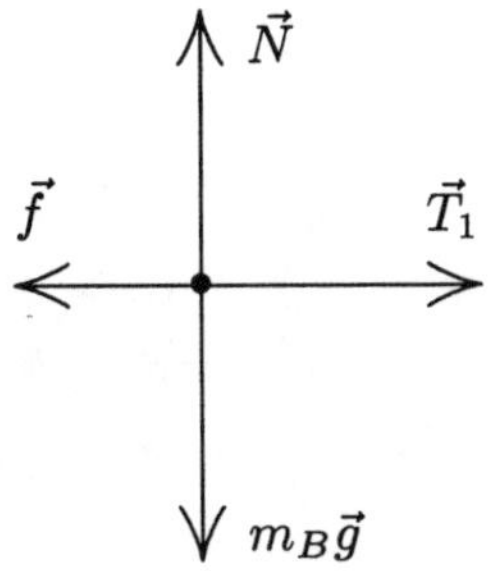

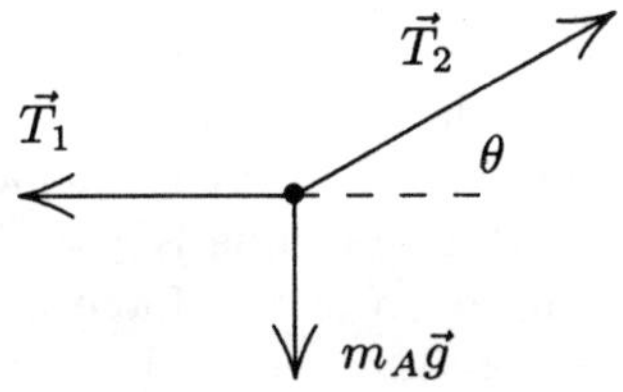

For each object we take $+x$ horizontally rightward and $+y$ upward. Applying Newton's second law in the x and y directions for block B and then doing the same for the knot results in four equations:

$$\begin{aligned} T_1 - f_{s,\max} &= 0 \\ N - W_B &= 0 \\ T_2\cos\theta - T_1 &= 0 \\ T_2\sin\theta - W_A &= 0 \end{aligned}$$

where we assume the static friction to be at its maximum value (permitting us to use Eq. 6-1). Solving these equations with $\mu_s = 0.25$, we obtain $W_A = 103 \approx 100$ N.

20. If the block is sliding then we compute the kinetic friction from Eq. 6-2; if it is not sliding, then we determine the extent of static friction from applying Newton's law, with zero acceleration, to the x axis (which is parallel to the incline surface). The question of whether or not it is sliding is therefore crucial, and depends on the maximum static friction force, as calculated from Eq. 6-1. The forces are resolved in the incline plane coordinate system in Figure 6-5 in the textbook. The acceleration, if there is any, is along the x axis, and we are taking uphill as $+x$. The net force along the y axis, then, is certainly zero, which provides the following relationship:

$$\sum \vec{F}_y = 0 \quad \Longrightarrow \quad N = W\cos\theta$$

where $W = 45$ N is the weight of the block, and $\theta = 15°$ is the incline angle. Thus, $N = 43.5$ N, which implies that the maximum static friction force should be $f_{s,\max} = (0.50)(43.5) = 21.7$ N.

(a) For $\vec{P} = 5.0$ N downhill, Newton's second law, applied to the x axis becomes

$$f - P - W\sin\theta = ma \quad \text{where } m = \frac{W}{g}\ .$$

Here we are assuming $\vec{f}$ is pointing uphill, as shown in Figure 6-5, and if it turns out that it points downhill (which *is* a possibility), then the result for f_s will be negative. If $f = f_s$ then $a = 0$, we obtain $f_s = 17$ N, which is clearly allowed since it is less than $f_{s,\max}$.

(b) For $\vec{P} = 8.0$ N downhill, we obtain (from the same equation) $f_s = 20$ N, which is still allowed since it is less than $f_{s,\max}$.

(c) But for $\vec{P} = 15$ N downhill, we obtain (from the same equation) $f_s = 27$ N, which is not allowed since it is larger than $f_{s,\max}$. Thus, we conclude that it is the kinetic friction, not the static friction, that is relevant in this case. We compute the result $f_k = (0.34)(43.5) = 15$ N. Here, as in the other parts of this problem, the friction is directed uphill.

21. First, we check to see if the bodies start to move. We assume they remain at rest and compute the force of (static) friction which holds them there, and compare its magnitude with the maximum value $\mu_s N$. The free-body diagrams are shown below. T is the magnitude of the tension force of the string, f is the magnitude of the force of friction on body A, N is the magnitude of the normal force of the plane on body A, $m_A\vec{g}$ is the force of gravity on body A (with magnitude $W_A = 102$ N), and $m_B\vec{g}$ is the force of gravity on body B (with magnitude $W_B = 32$ N). $\theta = 40°$ is the angle of incline. We are not told the direction of $\vec{f}$ but we assume it is downhill. If we obtain a negative result for f, then we know the force is actually up the plane.

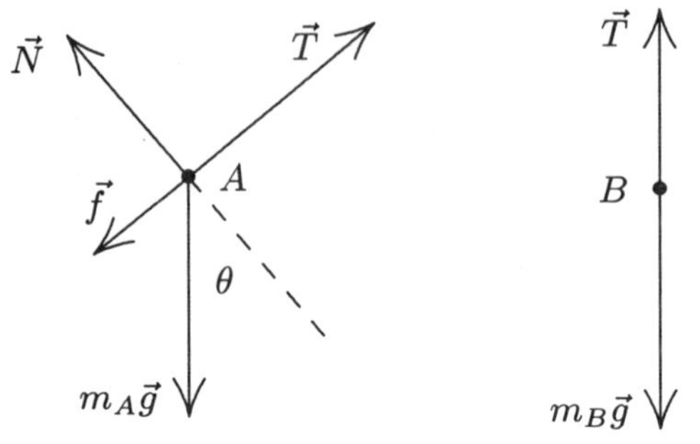

(a) For A we take the $+x$ to be uphill and $+y$ to be in the direction of the normal force. The x and y components of Newton's second law become

$$\begin{aligned} T - f - W_A \sin\theta &= 0 \\ N - W_A \cos\theta &= 0\,. \end{aligned}$$

Taking the positive direction to be *downward* for body B, Newton's second law leads to

$$W_B - T = 0\,.$$

Solving these three equations leads to

$$f = W_B - W_A \sin\theta = 32 - 102\sin 40° = -34 \text{ N}$$

(indicating that the force of friction is *uphill*) and to

$$N = W_A \cos\theta = 102\cos 40° = 78 \text{ N}$$

which means that $f_{s,\text{max}} = \mu_s N = (0.56)(78) = 44$ N. Since the magnitude f of the force of friction that holds the bodies motionless is less than $f_{s,\text{max}}$ the bodies remain at rest. The acceleration is zero.

(b) Since A is moving up the incline, the force of friction is downhill with magnitude $f_k = \mu_k N$. Newton's second law, using the same coordinates as in part (a), leads to

$$\begin{aligned} T - f_k - W_A \sin\theta &= m_A\, a \\ N - W_A \cos\theta &= 0 \\ W_B - T &= m_B\, a \end{aligned}$$

for the two bodies. We solve for the acceleration:

$$\begin{aligned} a &= \frac{W_B - W_A\sin\theta - \mu_k W_A \cos\theta}{m_B + m_A} \\ &= \frac{32\text{ N} - (102\text{ N})\sin 40° - (0.25)(102\text{ N})\cos 40°}{(32\text{ N} + 102\text{ N})\,/\left(9.8\text{ m/s}^2\right)} \\ &= -3.9\text{ m/s}^2\,. \end{aligned}$$

The acceleration is down the plane, which is to say (since the initial velocity was uphill) that the objects are slowing down. We note that $m = W/g$ has been used to calculate the masses in the calculation above.

(c) Now body A is initially moving down the plane, so the force of friction is uphill with magnitude $f_k = \mu_k N$. The force equations become

$$\begin{aligned} T + f_k - W_A \sin\theta &= m_A\, a \\ N - W_A \cos\theta &= 0 \\ W_B - T &= m_B\, a \end{aligned}$$

which we solve to obtain

$$\begin{aligned} a &= \frac{W_B - W_A \sin\theta + \mu_k W_A \cos\theta}{m_B + m_A} \\ &= \frac{32\,\text{N} - (102\,\text{N})\sin 40^\circ + (0.25)(102\,\text{N})\cos 40^\circ}{(32\,\text{N} + 102\,\text{N}) / \left(9.8\,\text{m/s}^2\right)} \\ &= -1.0\ \text{m/s}^2\,. \end{aligned}$$

The acceleration is again downhill the plane. In this case, the objects are speeding up.

22. The free-body diagrams are shown below. T is the magnitude of the tension force of the string, f is the magnitude of the force of friction on block A, N is the magnitude of the normal force of the plane on block A, $m_A\vec{g}$ is the force of gravity on body A (where $m_A = 10$ kg), and $m_B\vec{g}$ is the force of gravity on block B. $\theta = 30^\circ$ is the angle of incline. For A we take the $+x$ to be uphill and $+y$ to be in the direction of the normal force; the positive direction is chosen *downward* for block B.

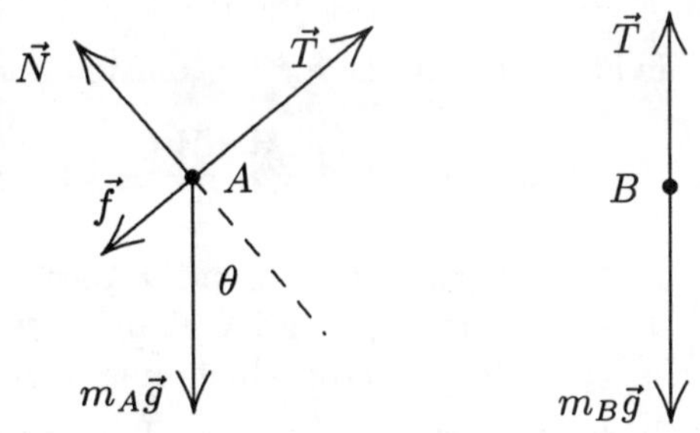

Since A is moving down the incline, the force of friction is uphill with magnitude $f_k = \mu_k N$ (where $\mu_k = 0.20$). Newton's second law leads to

$$\begin{aligned} T - f_k + m_A g \sin\theta &= m_A\, a = 0 \\ N - m_A g \cos\theta &= 0 \\ m_B g - T &= m_B\, a = 0 \end{aligned}$$

for the two bodies (where $a = 0$ is a consequence of the velocity being constant). We solve these for the mass of block B.

$$m_B = m_A\,(\sin\theta - \mu_k \cos\theta) = 3.3\ \text{kg}\,.$$

23. The free-body diagrams for the two blocks are shown below. T is the magnitude of the tension force of the string, $\vec{N}_A$ is the normal force on block A (the leading block), $\vec{N}_B$ is the the normal force on block B, $\vec{f}_A$ is kinetic friction force on block A, $\vec{f}_B$ is kinetic friction force on block B. Also, m_A is the mass of block A (where $m_A = W_A/g$ and $W_A = 3.6$ N), and m_B is the mass of block B (where $m_B = W_B/g$ and $W_B = 7.2$ N). The angle of the incline is $\theta = 30^\circ$.

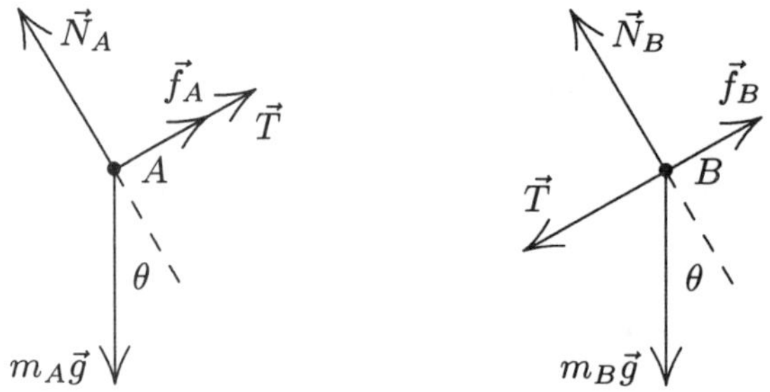

For each block we take $+x$ downhill (which is toward the lower-left in these diagrams) and $+y$ in the direction of the normal force. Applying Newton's second law to the x and y directions of first block A and next block B, we arrive at four equations:

$$\begin{aligned} W_A \sin\theta - f_A - T &= m_A\, a \\ N_A - W_A \cos\theta &= 0 \\ W_B \sin\theta - f_B + T &= m_B\, a \\ N_B - W_B \cos\theta &= 0\,. \end{aligned}$$

which, when combined with Eq. 6-2 ($f_A = \mu_{k\,A} N_A$ where $\mu_{k\,A} = 0.10$ and $f_B = \mu_{k\,B} N_B$ where $\mu_{k\,B} = 0.20$), fully describe the dynamics of the system so long as the blocks have the same acceleration and $T > 0$.

(a) These equations lead to an acceleration equal to

$$a = g\left(\sin\theta - \left(\frac{\mu_{k\,A} W_A + \mu_{k\,B} W_B}{W_A + W_B}\right)\cos\theta\right) = 3.5 \text{ m/s}^2\,.$$

(b) We solve the above equations for the tension and obtain

$$T = \left(\frac{W_A\, W_B}{W_A + W_B}\right)(\mu_{k\,B} - \mu_{k\,A})\cos\theta = 0.21 \text{ N}\,.$$

Simply returning the value for a found in part (a) into one of the above equations is certainly fine, and probably easier than solving for T algebraically as we have done, but the algebraic form does illustrate the $\mu_{k\,B} - \mu_{k\,A}$ factor which aids in the understanding of the next part.

(c) Reversing the blocks is equivalent to switching the labels (so A is now the block of weight 7.2 N and $\mu_{k\,A}$ is now the 0.20 value). We see from our algebraic result in part (b) that this gives a negative value for T, which is impossible. We conclude that the above set of four equations are not valid in this circumstance (specifically, a for one block is not equal to a for the other block). The blocks move independently of each other.

24. Treating the two boxes as a single system of total mass $1.0 + 3.0 = 4.0$ kg, subject to a total (leftward) friction of magnitude $2.0 + 4.0 = 6.0$ N, we apply Newton's second law (with $+x$ rightward):

$$\begin{aligned} F - f_{\text{total}} &= m_{\text{total}}\, a \\ 12.0 - 6.0 &= (4.0)a \end{aligned}$$

which yields the acceleration $a = 1.5$ m/s^2. We have treated F as if it were known to the nearest tenth of a Newton so that our acceleration is "good" to two significant figures. Turning our attention to the larger box (the Wheaties box of mass 3.0 kg) we apply Newton's second law to find the contact force F' exerted by the smaller box on it.

$$\begin{aligned} F' - f_{\text{W}} &= m_{\text{W}}\, a \\ F' - 4.0 &= (3.0)(1.5) \end{aligned}$$

This yields the contact force $F' = 8.5$ N.

25. The free-body diagrams for the two blocks, treated individually, are shown below (first m and then M). F' is the contact force between the two blocks, and the static friction force $\vec{f}_s$ is at its maximum value (so Eq. 6-1 leads to $f_s = f_{s,\max} = \mu_s F'$ where $\mu_s = 0.38$).

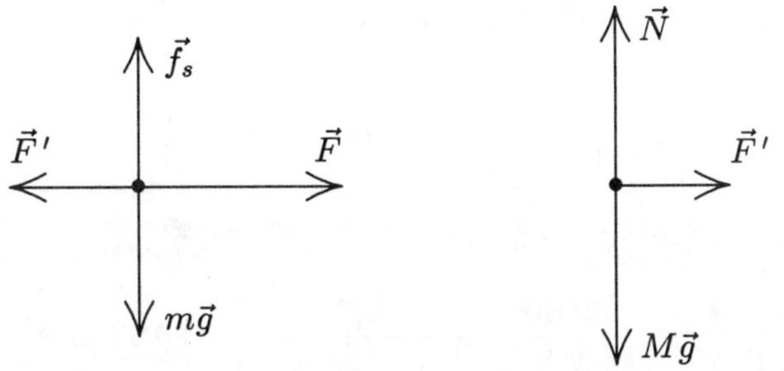

Treating the two blocks together as a single system (sliding across a frictionless floor), we apply Newton's second law (with $+x$ rightward) to find an expression for the acceleration.

$$F = m_{\text{total}}\, a \quad \Longrightarrow a = \frac{F}{m+M}$$

This is equivalent to having analyzed the two blocks individually and then combined their equations. Now, when we analyze the small block individually, we apply Newton's second law to the x and y axes, substitute in the above expression for a, and use Eq. 6-1.

$$F - F' = ma \quad \Longrightarrow \quad F' = F - m\left(\frac{F}{m+M}\right)$$

$$f_s - mg = 0 \quad \Longrightarrow \quad \mu_s F' - mg = 0$$

These expressions are combined (to eliminate F') and we arrive at

$$F = \frac{mg}{\mu_s\left(1 - \frac{m}{m+M}\right)}$$

which we find to be $F = 4.9 \times 10^2$ N.

26. The free-body diagrams for the two boxes are shown below. T is the magnitude of the force in the rod (when $T > 0$ the rod is said to be in tension and when $T < 0$ the rod is under compression), $\vec{N}_2$ is the normal force on box 2 (the uncle box), $\vec{N}_1$ is the the normal force on the aunt box (box 1), $\vec{f}_1$ is kinetic friction force on the aunt box, and $\vec{f}_2$ is kinetic friction force on the uncle box. Also, $m_1 = 1.65$ kg is the mass of the aunt box and $m_2 = 3.30$ kg is the mass of the uncle box (which is a lot of ants!).

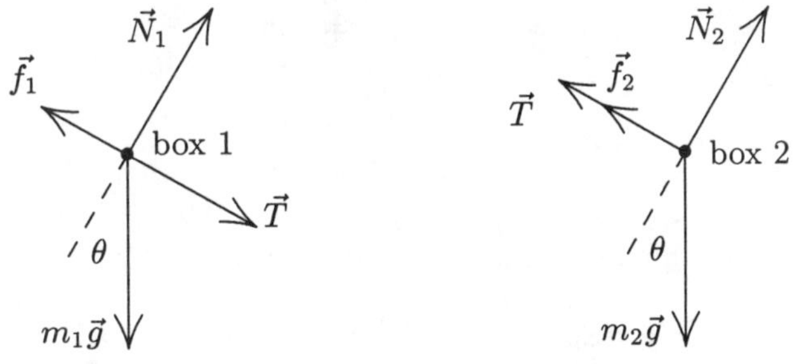

For each block we take $+x$ downhill (which is toward the lower-right in these diagrams) and $+y$ in the direction of the normal force. Applying Newton's second law to the x and y directions of first box 2 and next box 1, we arrive at four equations:

$$\begin{aligned} m_2 g \sin\theta - f_2 - T &= m_2\, a \\ N_2 - m_2 g \cos\theta &= 0 \\ m_1 g \sin\theta - f_1 + T &= m_1\, a \\ N_1 - m_1 g \cos\theta &= 0\,. \end{aligned}$$

which, when combined with Eq. 6-2 ($f_1 = \mu_1 N_1$ where $\mu_1 = 0.226$ and $f_2 = \mu_2 N_2$ where $\mu_2 = 0.113$), fully describe the dynamics of the system.

(a) We solve the above equations for the tension and obtain

$$T = \left(\frac{m_2\, m_1\, g}{m_2 + m_1}\right)(\mu_1 - \mu_2)\cos\theta = 1.05 \text{ N} .$$

(b) These equations lead to an acceleration equal to

$$a = g\left(\sin\theta - \left(\frac{\mu_2 m_2 + \mu_1 m_1}{m_2 + m_1}\right)\cos\theta\right) = 3.62 \text{ m/s}^2 .$$

(c) Reversing the blocks is equivalent to switching the labels. We see from our algebraic result in part (a) that this gives a negative value for T (equal in magnitude to the result we got before). Thus, the situation is as it was before except that the rod is now in a state of compression.

27. The free-body diagrams for the slab and block are shown below. $\vec{F}$ is the 100 N force applied to the block, $\vec{N}_s$ is the normal force of the floor on the slab, N_b is the magnitude of the normal force between the slab and the block, $\vec{f}$ is the force of friction between the slab and the block, m_s is the mass of the slab, and m_b is the mass of the block. For both objects, we take the $+x$ direction to be to the left and the $+y$ direction to be up.

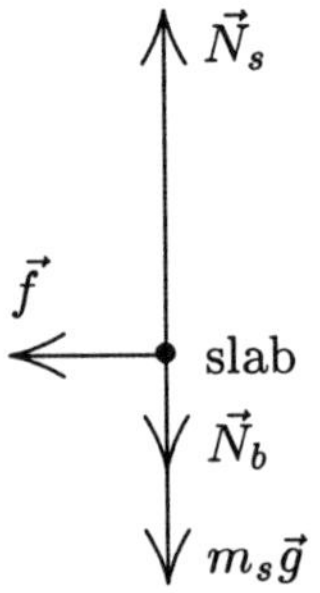

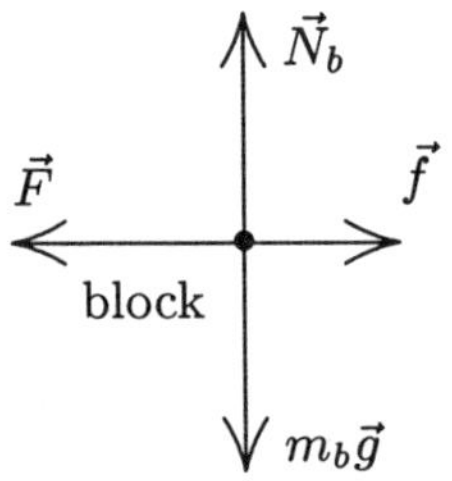

Applying Newton's second law for the x and y axes for (first) the slab and (second) the block results in four equations:

$$\begin{aligned} f &= m_s a_s \\ N_s - N_b - m_s g &= 0 \\ F - f &= m_b a_b \\ N_b - m_b g &= 0 \end{aligned}$$

from which we note that the maximum possible static friction magnitude would be

$$\mu_s N_b = \mu_s m_b g = (0.60)(10\,\text{kg})(9.8\,\text{m/s}^2) = 59 \text{ N} .$$

We check to see if the block slides on the slab. Assuming it does not, then $a_s = a_b$ (which we denote simply as a) and we solve for f:

$$f = \frac{m_s F}{m_s + m_b} = \frac{(40\,\text{kg})(100\,\text{N})}{40\,\text{kg} + 10\,\text{kg}} = 80 \text{ N}$$

which is greater than $f_{s,\max}$ so that we conclude the block is sliding across the slab (their accelerations are different).

(a) Using $f = \mu_k N_b$ the above equations yield

$$a_b = \frac{F - \mu_k m_b g}{m_b} = \frac{100\,\text{N} - (0.40)(10\,\text{kg})(9.8\,\text{m/s}^2)}{10\,\text{kg}} = 6.1\ \text{m/s}^2\ .$$

The result is positive which means (recalling our choice of $+x$ direction) that it accelerates leftward.

(b) We also obtain

$$a_s = \frac{\mu_k m_b g}{m_s} = \frac{(0.40)(10\,\text{kg})(9.8\,\text{m/s}^2)}{40\,\text{kg}} = 0.98\ \text{m/s}^2\ .$$

As mentioned above, this means it accelerates to the left.

28. We may treat all 25 cars as a single object of mass $m = 25 \times 5.0 \times 10^4$ kg and (when the speed is 30 km/h = 8.3 m/s) subject to a friction force equal to $f = 25 \times 250 \times 8.3 = 5.2 \times 10^4$ N.

(a) Along the level track, this object experiences a "forward" force T exerted by the locomotive, so that Newton's second law leads to

$$T - f = ma \implies T = 5.2 \times 10^4 + \left(1.25 \times 10^6\right)(0.20)$$

which yields $T = 3.0 \times 10^5$ N.

(b) The free-body diagram is shown below, with θ as the angle of the incline. The $+x$ direction (which is the only direction to which we will be applying Newton's second law) is uphill (to the upper right in our sketch).

Thus, we obtain

$$T - f - mg\sin\theta = ma$$

where we set $a = 0$ (implied by the problem statement) and solve for the angle. We obtain $\theta = 1.2°$.

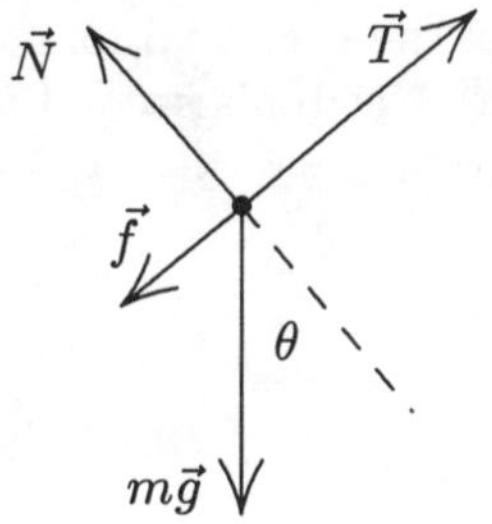

29. Each side of the trough exerts a normal force on the crate. The first diagram shows the view looking in toward a cross section. The net force is along the dashed line. Since each of the normal forces makes an angle of 45° with the dashed line, the magnitude of the resultant normal force is given by $N_r = 2N\cos 45° = \sqrt{2}N$. The second diagram is the free-body diagram for the crate (from a "side" view, similar to that shown in the first picture in Fig. 6-36). The force of gravity has magnitude mg, where m is the mass of the crate, and the magnitude of the force of friction is denoted by f. We take the $+x$ direction to be down the incline and $+y$ to be in the direction of $\vec{N}_r$. Then the x component of Newton's second law is $mg\sin\theta - f = ma$ and the y component is $N_r - mg\cos\theta = 0$. Since the crate is moving, each side of the trough exerts a force of kinetic friction, so the total frictional force has magnitude $f = 2\mu_k N = 2\mu_k N_r/\sqrt{2} = \sqrt{2}\mu_k N_r$. Combining this expression with $N_r = mg\cos\theta$ and substituting into the x component equation, we obtain $mg\sin\theta - \sqrt{2}mg\cos\theta = ma$. Therefore $a = g(\sin\theta - \sqrt{2}\mu_k\cos\theta)$.

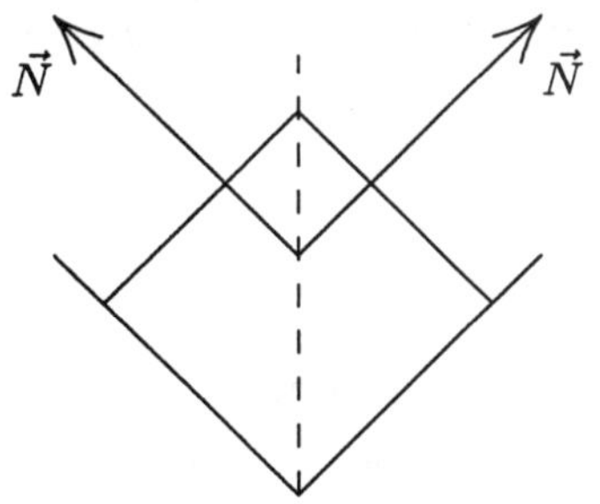

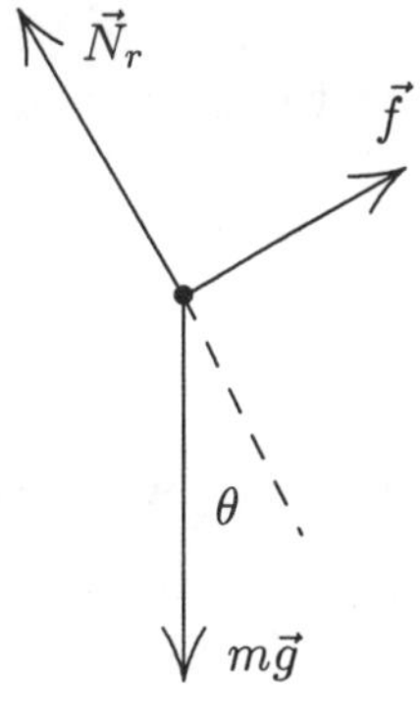

30. Fig. 6-4 in the textbook shows a similar situation (using ϕ for the unknown angle) along with a free-body diagram. We use the same coordinate system as in that figure.

(a) Thus, Newton's second law leads to

$$\begin{aligned} T\cos\phi - f &= ma \qquad \text{along } x \text{ axis} \\ T\sin\phi + N - mg &= 0 \qquad \text{along } y \text{ axis} \end{aligned}$$

Setting $a = 0$ and $f = f_{s,\max} = \mu_s N$, we solve for the mass of the box-and-sand (as a function of angle):

$$m = \frac{T}{g}\left(\sin\phi + \frac{\cos\phi}{\mu_s}\right)$$

which we will solve with calculus techniques (to find the angle ϕ_m corresponding to the maximum mass that can be pulled).

$$\frac{dm}{dt} = \frac{T}{g}\left(\cos\phi_m - \frac{\sin\phi_m}{\mu_s}\right) = 0$$

This leads to $\tan\phi_m = \mu_s$ which (for $\mu_s = 0.35$) yields $\phi_m = 19°$.

(b) Plugging our value for ϕ_m into the equation we found for the mass of the box-and-sand yields $m = 340$ kg. This corresponds to a weight of $mg = 3.3 \times 10^3$ N.

31. We denote the magnitude of the frictional force αv, where $\alpha = 70\,\mathrm{N\cdot s/m}$. We take the direction of the boat's motion to be positive. Newton's second law gives

$$-\alpha v = m\frac{dv}{dt} .$$

Thus,

$$\int_{v_0}^{v} \frac{dv}{v} = -\frac{\alpha}{m}\int_0^t dt$$

where v_0 is the velocity at time zero and v is the velocity at time t. The integrals are evaluated with the result

$$\ln\frac{v}{v_0} = -\frac{\alpha t}{m} .$$

We take $v = v_0/2$ and solve for time:

$$t = \frac{m}{\alpha}\ln 2 = \frac{1000\,\mathrm{kg}}{70\,\mathrm{N\cdot s/m}}\ln 2 = 9.9\ \mathrm{s} .$$

32. In the solution to exercise 4, we found that the force provided by the wind needed to equal $F = 157$ N (where that last figure is not "significant").

(a) Setting $F = D$ (for Drag force) we use Eq. 6-14 to find the wind speed V along the ground (which actually is relative to the moving stone, but we assume the stone is moving slowly enough that this does not invalidate the result):

$$V = \sqrt{\frac{2F}{C\rho A}} = \sqrt{\frac{2(157)}{(0.80)(1.21)(0.040)}}$$

which yields $V = 90$ m/s which converts to $V = 3.2 \times 10^2$ km/h.

(b) Doubling our previous result, we find the reported speed to be 6.5×10^2 km/h, which is not reasonable for a terrestrial storm. (A category 5 hurricane has speeds on the order of 2.6×10^2 m/s.)

33. We use Eq. 6-14, $D = \frac{1}{2}C\rho A v^2$, where ρ is the air density, A is the cross-sectional area of the missile, v is the speed of the missile, and C is the drag coefficient. The area is given by $A = \pi R^2$, where $R = 0.265\,\text{m}$ is the radius of the missile. Thus

$$D = \frac{1}{2}(0.75)(1.2\,\text{kg/m}^3)\pi(0.265\,\text{m})^2(250\,\text{m/s})^2 = 6.2 \times 10^3\,\text{N} .$$

34. Using Eq. 6-16, we solve for the area

$$A = \frac{2mg}{C\rho v_t^2}$$

which illustrates the inverse proportionality between the area and the speed-squared. Thus, when we set up a ratio of areas – of the slower case to the faster case – we obtain

$$\frac{A_{\text{slow}}}{A_{\text{fast}}} = \left(\frac{310\,\text{km/h}}{160\,\text{km/h}}\right)^2 = 3.75 .$$

35. For the passenger jet $D_j = \frac{1}{2}C\rho_1 A v_j^2$, and for the prop-driven transport $D_t = \frac{1}{2}C\rho_2 A v_t^2$, where ρ_1 and ρ_2 represent the air density at 10 km and 5.0 km, respectively. Thus the ratio in question is

$$\frac{D_j}{D_t} = \frac{\rho_1 v_j^2}{\rho_2 v_t^2} = \frac{(0.38\,\text{kg/m}^3)(1000\,\text{km/h})^2}{(0.67\,\text{kg/m}^3)(500\,\text{km/h})^2} = 2.3 .$$

36. With $v = 96.6$ km/h $= 26.8$ m/s, Eq. 6-17 readily yields

$$a = \frac{v^2}{R} = \frac{26.8^2}{7.6} = 94.7\ \text{m/s}^2$$

which we express as a multiple of g:

$$a = \left(\frac{a}{g}\right)g = \left(\frac{94.7}{9.8}\right)g = 9.7g .$$

37. The magnitude of the acceleration of the car as it rounds the curve is given by v^2/R, where v is the speed of the car and R is the radius of the curve. Since the road is horizontal, only the frictional force of the road on the tires makes this acceleration possible. The horizontal component of Newton's second law is $f = mv^2/R$. If N is the normal force of the road on the car and m is the mass of the car, the vertical component of Newton's second law leads to $N = mg$. Thus, using Eq. 6-1, the maximum value of static friction is $f_{s,\text{max}} = \mu_s N = \mu_s mg$. If the car does not slip, $f \le \mu_s mg$. This means

$$\frac{v^2}{R} \le \mu_s g \implies v \le \sqrt{\mu_s R g} .$$

Consequently, the maximum speed with which the car can round the curve without slipping is

$$v_{\text{max}} = \sqrt{\mu_s R g} = \sqrt{(0.60)(30.5)(9.8)} = 13\ \text{m/s} .$$

38. We will start by assuming that the normal force (on the car from the rail) points up. Note that gravity points down, and the y axis is chosen positive upwards. Also, the direction to the center of the circle (the direction of centripetal acceleration) is down. Thus, Newton's second law leads to

$$N - mg = m\left(-\frac{v^2}{r}\right) .$$

(a) When $v = 11$ m/s, we obtain $N = 3.7 \times 10^3$ N. The fact that this answer is positive means that $\vec{N}$ does indeed point upward as we had assumed.

(b) When $v = 14$ m/s, we obtain $N = -1.3 \times 10^3$ N. The fact that this answer is negative means that $\vec{N}$ points opposite to what we had assumed. Thus, the magnitude of $\vec{N}$ is 1.3 kN and its direction is *down*.

39. The magnitude of the acceleration of the cyclist as it rounds the curve is given by v^2/R, where v is the speed of the cyclist and R is the radius of the curve. Since the road is horizontal, only the frictional force of the road on the tires makes this acceleration possible. The horizontal component of Newton's second law is $f = mv^2/R$. If N is the normal force of the road on the bicycle and m is the mass of the bicycle and rider, the vertical component of Newton's second law leads to $N = mg$. Thus, using Eq. 6-1, the maximum value of static friction is $f_{s,\text{max}} = \mu_s N = \mu_s mg$. If the bicycle does not slip, $f \le \mu_s mg$. This means

$$\frac{v^2}{R} \le \mu_s g \implies R \ge \frac{v^2}{\mu_s g} .$$

Consequently, the minimum radius with which a cyclist moving at 29 km/h = 8.1 m/s can round the curve without slipping is

$$R_{\text{min}} = \frac{v^2}{\mu_s g} = \frac{8.1^2}{(0.32)(9.8)} = 21 \text{ m} .$$

40. The situation is somewhat similar to that shown in the "loop-the-loop" example done in the textbook (see Figure 6-10) except that, instead of a downward normal force, we are dealing with the force of the boom $\vec{F}_B$ on the car – which is capable of pointing any direction. We will assume it to be upward as we apply Newton's second law to the car (of total weight 5000 N):

$$F_B - W = ma \text{ where } m = \frac{W}{g} , \text{ and } a = -\frac{v^2}{r}$$

Note that the centripetal acceleration is downward (our choice for negative direction) for a body at the top of its circular trajectory.

(a) If $r = 10$ m and $v = 5.0$ m/s, we obtain $F_B = 3.7 \times 10^3$ N = 3.7 kN (up).

(b) If $r = 10$ m and $v = 12$ m/s, we obtain $F_B = -2.3 \times 10^3$ N = -2.3 kN where the minus sign indicates that $\vec{F}_B$ points downward.

41. For the puck to remain at rest the magnitude of the tension force T of the cord must equal the gravitational force Mg on the cylinder. The tension force supplies the centripetal force that keeps the puck in its circular orbit, so $T = mv^2/r$. Thus $Mg = mv^2/r$. We solve for the speed: $v = \sqrt{Mgr/m}$.

42. The magnitude of the acceleration of the cyclist as it moves along the horizontal circular path is given by v^2/R, where v is the speed of the cyclist and R is the radius of the curve.

(a) The horizontal component of Newton's second law is $f = mv^2/R$, where f is the static friction exerted horizontally by the ground on the tires. Thus,

$$f = \frac{(85.0)(9.00)^2}{25.0} = 275 \text{ N} .$$

(b) If N is the vertical force of the ground on the bicycle and m is the mass of the bicycle and rider, the vertical component of Newton's second law leads to $N = mg = 833$ N. The magnitude of the force exerted by the ground on the bicycle is therefore

$$\sqrt{f^2 + N^2} = \sqrt{275^2 + 833^2} = 877 \text{ N} .$$

43. (a) At the top (the highest point in the circular motion) the seat pushes up on the student with a force of magnitude $N = 556$ N. Earth pulls down with a force of magnitude $W = 667$ N. The seat is pushing up with a force that is smaller than the student's weight, and we say the student experiences a decrease in his "apparent weight" at the highest point.

(b) When the student is at the highest point, the net force toward the center of the circular orbit is $W - F_t$ (note that we are choosing downward as the positive direction). According to Newton's second law, this must equal mv^2/R, where v is the speed of the student and R is the radius of the orbit. Thus

$$mv^2/R = W - N = 667\,\text{N} - 556\,\text{N} = 111\ \text{N}\ .$$

(c) Now N is the magnitude of the upward force exerted by the seat when the student is at the lowest point. The net force toward the center of the circle is $F_b - W = mv^2/R$ (note that we are now choosing upward as the positive direction). The Ferris wheel is "steadily rotating" so the value mv^2/R is the same as in part (a). Thus,

$$N = \frac{mv^2}{R} + W = 111\,\text{N} + 667\,\text{N} = 778\ \text{N}\ .$$

(d) If the speed is doubled, mv^2/R increases by a factor of 4, to 444 N. Therefore, at the highest point we have $W - N = mv^2/R$, which leads to

$$N = 667\,\text{N} - 444\,\text{N} = 223\ \text{N}\ .$$

Similarly, the normal force at the lowest point is now found to be $N = 667 + 444 \approx 1.1$ kN.

44. The free-body diagram (for the hand straps of mass m) is the view that a passenger might see if she was looking forward and the streetcar was curving towards the right (so $\vec{a}$ points rightwards in the figure) We note that $|\vec{a}| = v^2/R$ where $v = 16$ km/h $= 4.4$ m/s.
Applying Newton's law to the axes of the problem ($+x$ is rightward and $+y$ is upward) we obtain

$$\begin{aligned} T\sin\theta &= m\frac{v^2}{R} \\ T\cos\theta &= mg\ . \end{aligned}$$

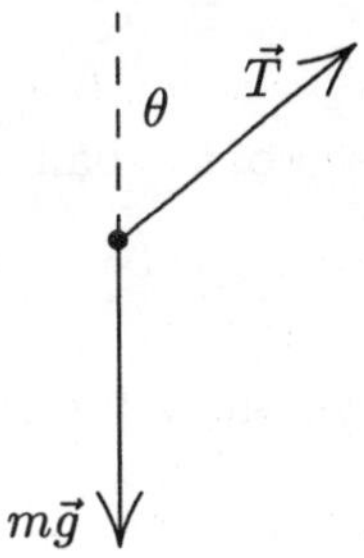

We solve these equations for the angle:

$$\theta = \tan^{-1}\left(\frac{v^2}{Rg}\right)$$

which yields $\theta = 12°$.

45. The free-body diagram (for the airplane of mass m) is shown below. We note that $\vec{F}_\ell$ is the force of aerodynamic lift and $\vec{a}$ points rightwards in the figure. We also note that $|\vec{a}| = v^2/R$ where $v = 480$ km/h $= 133$ m/s.
Applying Newton's law to the axes of the problem ($+x$ rightward and $+y$ upward) we obtain

$$\begin{aligned} \vec{F}_\ell\sin\theta &= m\frac{v^2}{R} \\ \vec{F}_\ell\cos\theta &= mg \end{aligned}$$

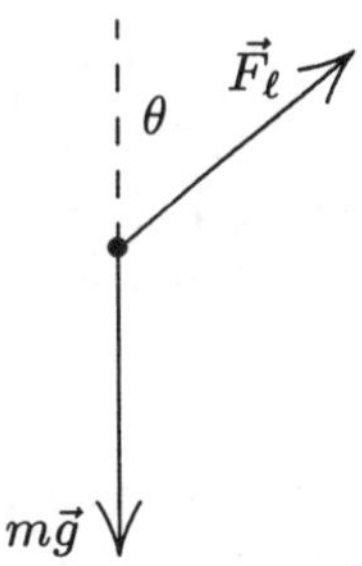

where $\theta = 40°$. Eliminating mass from these equations leads to

$$\tan\theta = \frac{v^2}{gR}$$

which yields $R = v^2/g\tan\theta = 2.2 \times 10^3$ m.

46. (a) The upward force exerted by the car on the passenger is equal to the downward force of gravity ($W = 500$ N) on the passenger. So the *net* force does not have a vertical contribution; it only has the contribution from the horizontal force (which is necessary for maintaining the circular motion). Thus $\left|\vec{F}_{\text{net}}\right| = F = 210$ N.

(b) Using Eq. 6-18, we have

$$v = \sqrt{\frac{FR}{m}} = \sqrt{\frac{(210)(470)}{51.0}} = 44.0 \text{ m/s} .$$

47. (a) The free-body diagram for the ball is shown below. $\vec{T}_u$ is the tension exerted by the upper string on the ball, $\vec{T}_\ell$ is the tension force of the lower string, and m is the mass of the ball. Note that the tension in the upper string is greater than the tension in the lower string. It must balance the downward pull of gravity and the force of the lower string.

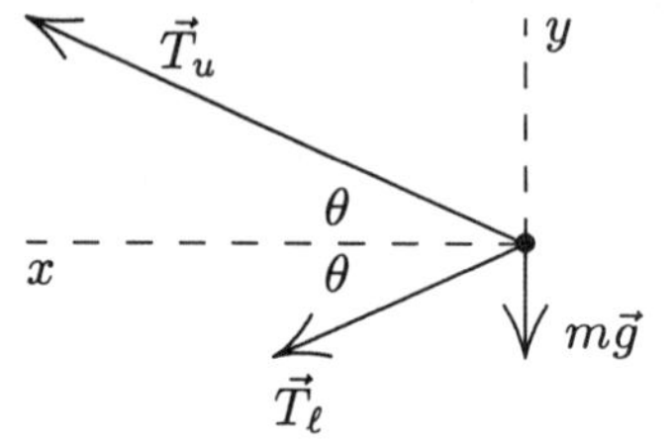

(b) We take the $+x$ direction to be leftward (toward the center of the circular orbit) and $+y$ upward. Since the magnitude of the acceleration is $a = v^2/R$, the x component of Newton's second law is

$$T_u \cos\theta + T_\ell \cos\theta = \frac{mv^2}{R} ,$$

where v is the speed of the ball and R is the radius of its orbit. The y component is

$$T_u \sin\theta - T_\ell \sin\theta - mg = 0 .$$

The second equation gives the tension in the lower string: $T_\ell = T_u - mg/\sin\theta$. Since the triangle is equilateral $\theta = 30°$. Thus

$$T_\ell = 35 - \frac{(1.34)(9.8)}{\sin 30°} = 8.74 \text{ N} .$$

(c) The net force is leftward ("radially inward") and has magnitude

$$F_{\text{net}} = (T_u + T_\ell)\cos\theta = (35 + 8.74)\cos 30° = 37.9 \text{ N} .$$

(d) The radius of the path is $[(1.70\,\text{m})/2)]\tan 30° = 1.47\,\text{m}$. Using $F_{\text{net}} = mv^2/R$, we find that the speed of the ball is

$$v = \sqrt{\frac{RF_{\text{net}}}{m}} = \sqrt{\frac{(1.47\,\text{m})(37.9\,\text{N})}{1.34\,\text{kg}}} = 6.45 \text{ m/s} .$$

48. In the solution to exercise 4, we found that the force provided by the wind needed to equal $F = \mu_k mg$. In this situation, we have a much smaller value of μ_k (0.10) and a much larger mass (one hundred stones and the layer of ice). The layer of ice has a mass of

$$m_{\text{ice}} = \left(917\,\text{kg/m}^3\right)(400\,\text{m} \times 500\,\text{m} \times 0.0040\,\text{m})$$

which yields $m_{\text{ice}} = 7.34 \times 10^5$ kg. This added to the mass of the hundred stones (at 20 kg each) comes to $m = 7.36 \times 10^5$ kg.

(a) Setting $F = D$ (for Drag force) we use Eq. 6-14 to find the wind speed v along the ground (which actually is relative to the moving stone, but we assume the stone is moving slowly enough that this does not invalidate the result):

$$v = \sqrt{\frac{\mu_k mg}{4C_{\text{ice}}\rho A_{\text{ice}}}} = \sqrt{\frac{(0.10)\,(7.36\times10^5)\,(9.8)}{4(0.002)(1.21)(400\times500)}}$$

which yields $v = 19$ m/s which converts to $v = 69$ km/h.

(b) and (c) Doubling our previous result, we find the reported speed to be 139 km/h, which is a reasonable for a storm winds. (A category 5 hurricane has speeds on the order of 2.6×10^2 m/s.)

49. (a) The distance traveled by the coin in 3.14 s is $3(2\pi r) = 6\pi(0.050) = 0.94$ m. Thus, its speed is $v = 0.94/3.14 = 0.30$ m/s.

(b) The acceleration vector (at any instant) is horizontal and points from the coin towards the center of the turntable. This centripetal acceleration is given by Eq. 6-17:

$$a = \frac{v^2}{r} = \frac{0.30^2}{0.050} = 1.8\ \text{m/s}^2\ .$$

(c) The only horizontal force acting on the coin is static friction f_s and must be large enough to supply the acceleration of part (b) for the $m = 0.0020$ kg coin. Using Newton's second law,

$$f_s = ma = (0.0020)(1.8) = 3.6\times10^{-3}\ \text{N}$$

which must point in the same direction as the acceleration (towards the center of the turntable).

(d) We note that the normal force exerted upward on the coin by the turntable must equal the coin's weight (since there is no vertical acceleration in the problem). We also note that if we repeat the computations in parts (a) and (b) for $r' = 0.10$ m, then we obtain $v' = 0.60$ m/s and $a' = 3.6$ m/s^2. Now, if friction is at its maximum at $r = r'$, then, by Eq. 6-1, we obtain

$$\mu_s = \frac{f_{s,\text{max}}}{mg} = \frac{ma'}{mg} = 0.37\ .$$

50. (a) The angle made by the cord with the vertical axis is given by $\theta = \cos^{-1}(18/30) = 53°$. This means the radius of the plane's circular path is $r = 30\sin\theta = 24$ m (we also could have arrived at this using the Pythagorean theorem). The speed of the plane is

$$v = \frac{4.4(2\pi r)}{1\,\text{min}} = \frac{8.8\pi(24\,\text{m})}{60\,\text{s}}$$

which yields $v = 11$ m/s. Eq. 6-17 then gives the acceleration (which at any instant is horizontally directed from the plane to the center of its circular path)

$$a = \frac{v^2}{r} = \frac{11^2}{24} = 5.1\ \text{m/s}^2\ .$$

(b) The only horizontal force on the airplane is that component of tension, so Newton's second law gives

$$T\sin\theta = \frac{mv^2}{r} \implies T = \frac{(0.75)(11)^2}{24\sin 53°}$$

which yields $T = 4.8$ N.

(c) The net vertical force on the airplane is zero (since its only acceleration is horizontal), so

$$F_{\text{lift}} = T\cos\theta + mg = 4.8\cos 53° + (0.75)(9.8) = 10\ \text{N}\ .$$

51. (a) The centripetal force is given by Eq. 6-18:

$$F = \frac{mv^2}{R} = \frac{(1)(465)^2}{6.4 \times 10^6} = 0.034 \text{ N} .$$

(b) Calling downward (towards the center of Earth) the positive direction, Newton's second law leads to

$$mg - T = ma$$

where $mg = 9.80$ N and $ma = 0.034$ N, calculated in part (a). Thus, the tension in the cord by which the body hangs from the balance is $T = 9.80 - 0.03 = 9.77$ N. Thus, this is the reading for a standard kilogram mass, of the scale at the equator of the spinning Earth.

52. There is no acceleration, so the (upward) static friction forces (there are four of them, one for each thumb and one for each set of opposing fingers) equals the magnitude of the (downward) pull of gravity. Using Eq. 6-1, we have

$$4\mu_s N = mg = (79\,\text{kg})\left(9.8\,\text{m/s}^2\right)$$

which, with $\mu_s = 0.70$, yields $N = 2.8 \times 10^2$ N.

53. (a) From Table 6-1 and Eq. 6-16, we have

$$v_t = \sqrt{\frac{2F_g}{C\rho A}} \implies C\rho A = 2\frac{mg}{v_t^2}$$

where $v_t = 60$ m/s. We estimate the pilot's mass at about $m = 70$ kg. Now, we convert $v = 1300(1000/3600) \approx 360$ m/s and plug into Eq. 6-14:

$$D = \frac{1}{2}C\rho A v^2 = \frac{1}{2}\left(2\frac{mg}{v_t^2}\right)v^2 = mg\left(\frac{v}{v_t}\right)^2$$

which yields $D = (690)(360/60)^2 \approx 2 \times 10^4$ N.

(b) We assume the mass of the ejection seat is roughly equal to the mass of the pilot. Thus, Newton's second law (in the horizontal direction) applied to this system of mass $2m$ gives the magnitude of acceleration:

$$|a| = \frac{D}{2m} = \frac{g}{2}\left(\frac{v}{v_t}\right)^2 = 18g .$$

54. Although the object in question is a sphere, the area A in Eq. 6-16 is the cross sectional area presented by the object as it moves through the air (the cross section is perpendicular to $\vec{v}$). Thus, A is that of a circle: $A = \pi R^2$. We also note that 16 lb equates to an SI weight of 71 N. Thus,

$$v_t = \sqrt{\frac{2F_g}{C\rho\pi R^2}} \implies R = \frac{1}{145}\sqrt{\frac{2(71)}{(0.49)(1.2)\pi}}$$

which yields a diameter of $2R = 0.12$ m.

55. In the following sketch, T and T' are the tensions in the left and right strings, respectively. Also, $m_1 = M = 2.0$ kg, $m_2 = 2M = 4.0$ kg, and $m_3 = 2M = 4.0$ kg. Since it does, in fact, slide (presumably rightward), the type of friction that is acting upon m_2 is *kinetic* friction.

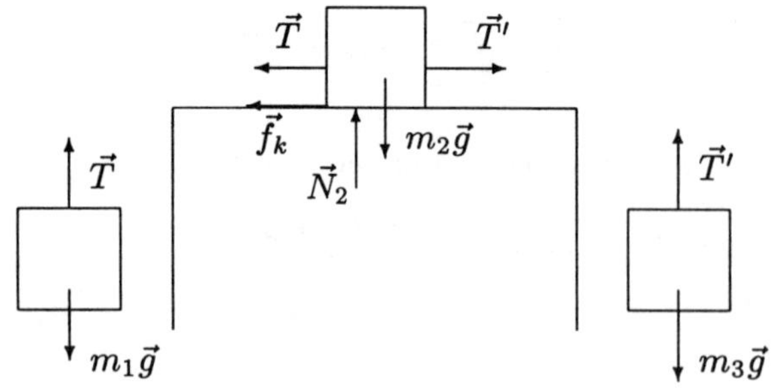

We use the familiar axes with $+x$ rightward and $+y$ upward for each block. This has the consequence that m_1 and m_2 accelerate with the same sign, but the acceleration of m_3 has the opposite sign. We take this into account as we apply Newton's second law to the three blocks.

$$\begin{aligned} T - m_1 g &= m_1(+a) \\ T' - T - f_k &= m_2(+a) \\ T' - m_3 g &= m_3(-a) \end{aligned}$$

Adding the first two equations, and subtracting the last, we obtain

$$(m_3 - m_1)\,g - f_k = (m_1 + m_2 + m_3)\,a$$

or (using M as in the problem statement)

$$Mg - f_k = 5Ma\ .$$

With $a = 1.5\ \text{m/s}^2$, we find $f_k = 4.6$ N.

56. (a) The component of the weight along the incline (with downhill understood as the positive direction) is $mg\sin\theta$ where $m = 630$ kg and $\theta = 10.2°$. With $f = 62.0$ N, Newton's second law leads to

$$mg\sin\theta - f = ma$$

which yields $a = 1.64\ \text{m/s}^2$. Using Eq. 2-15, we have

$$80.0\ \text{m} = \left(6.20\ \frac{\text{m}}{\text{s}}\right)t + \frac{1}{2}\left(1.64\ \frac{\text{m}}{\text{s}^2}\right)t^2\ .$$

This is solved using the quadratic formula. The positive root is $t = 6.80$ s.

(b) Running through the calculation of part (a) with $f = 42.0$ N instead of $f = 62$ N results in $t = 6.76$ s.

57. We convert to SI units: $v = 94(1000/3600) = 26$ m/s. Eq. 6-18 yields

$$F = \frac{mv^2}{R} = \frac{(85)(26)^2}{220} = 263\ \text{N}$$

for the horizontal force exerted on the passenger by the seat. But the seat also exerts an upward force equal to $mg = 833$ N. The magnitude of force is therefore $\sqrt{263^2 + 833^2} = 874$ N.

58. (a) Comparing the $t = 2.0$ s photo with the $t = 0$ photo, we see that the distance traveled by the box is

$$d = \sqrt{4.0^2 + 2.0^2} = 4.5\ \text{m}\ .$$

Thus (from Table 2-1, with *downhill* positive) $d = v_0 t + \frac{1}{2}at^2$, we obtain $a = 2.2\ \text{m/s}^2$; note that the boxes are assumed to start from rest.

(b) For the axis along the incline surface, we have

$$mg\sin\theta - f_k = ma\ .$$

We compute mass m from the weight $m = 240/9.8 = 24$ kg, and θ is figured from the absolute value of the slope of the graph: $\theta = \tan^{-1} 2.5/5.0 = 27°$. Therefore, we find $f_k = 53$ N.

59. (a) If the skier covers a distance L during time t with zero initial speed and a constant acceleration a, then $L = at^2/2$, which gives the acceleration a_1 for the first (old) pair of skis:

$$a_1 = \frac{2L}{t_1^2} = \frac{2(200\,\text{m})}{(61\,\text{s})^2} = 0.11\,\text{m/s}^2$$

and the acceleration a_2 for the second (new) pair:

$$a_2 = \frac{2L}{t_2^2} = \frac{2(200\,\text{m})}{(42\,\text{s})^2} = 0.23\,\text{m/s}^2\ .$$

(b) The net force along the slope acting on the skier of mass m is

$$F_{\text{net}} = mg\sin\theta - f_k = mg(\sin\theta - \mu_k\cos\theta) = ma$$

which we solve for μ_{k1} for the first pair of skis:

$$\mu_{k1} = \tan\theta - \frac{a_1}{g\cos\theta} = \tan 3.0° - \frac{0.11}{9.8\cos 3.0°} = 0.041$$

and μ_{k2} for the second pair:

$$\mu_{k2} = \tan\theta - \frac{a_2}{g\cos\theta} = \tan 3.0° - \frac{0.23}{9.8\cos 3.0°} = 0.029 \ .$$

60. (a) The box doesn't move until $t = 2.8$ s, which is when the applied force $\vec{F}$ reaches a magnitude of $F = (1.8)(2.8) = 5.0$ N, implying therefore that $f_{s,\max} = 5.0$ N. Analysis of the vertical forces on the block leads to the observation that the normal force magnitude equals the weight $N = mg = 15$ N. Thus, $\mu_s = f_{s,\max}/N = 0.34$.

(b) We apply Newton's second law to the horizontal x axis (positive in the direction of motion).

$$F - f_k = ma \implies 1.8t - f_k = (1.5)(1.2t - 2.4)$$

Thus, we find $f_k = 3.6$ N. Therefore, $\mu_k = f_k/N = 0.24$.

61. In both cases (highest point and lowest point), the normal force (on the child from the seat) points up, gravity points down, and the y axis is chosen positive upwards. At the high point, the direction to the center of the circle (the direction of centripetal acceleration) is down, and at the low point that direction is up.

(a) Newton's second law (using Eq. 6-17 for the magnitude of the acceleration) leads to

$$N - mg = m\left(-\frac{v^2}{R}\right) \ .$$

With $m = 26$ kg, $v = 5.5$ m/s and $R = 12$ m, this yields $N = 189$ N which we round off to $N \approx 190$ N.

(b) Now, Newton's second law leads to

$$N - mg = m\left(\frac{v^2}{r}\right)$$

which yields $N = 320$ N. As already mentioned, the direction of $\vec{N}$ is *up* in both cases.

62. The mass of the car is $m = 10700/9.8 = 1.09 \times 10^3$ kg. We choose "inward" (horizontally towards the center of the circular path) as the positive direction.

(a) With $v = 13.4$ m/s and $R = 61$ m, Newton's second law (using Eq. 6-18) leads to

$$f_s = \frac{mv^2}{R} = 3.21 \times 10^3 \text{ N} \ .$$

(b) Noting that $N = mg$ in this situation, the maximum possible static friction is found to be

$$f_{s,\max} = \mu_s mg = (0.35)(10700) = 3.75 \times 10^3 \text{ N}$$

using Eq. 6-1. We see that the static friction found in part (a) is less than this, so the car rolls (no skidding) and successfully negotiates the curve.

63. (a) The distance traveled in one revolution is $2\pi R = 2\pi(4.6) = 29$ m. The (constant) speed is consequently $v = 29/30 = 0.96$ m/s.

(b) Newton's second law (using Eq. 6-17 for the magnitude of the acceleration) leads to

$$f_s = m\left(\frac{v^2}{R}\right) = m(0.20)$$

in SI units. Noting that $N = mg$ in this situation, the maximum possible static friction is $f_{s,\max} = \mu_s mg$ using Eq. 6-1. Equating this with $f_s = m(0.20)$ we find the mass m cancels and we obtain $\mu_s = 0.20/9.8 = 0.021$.

64. At the top of the hill the vertical forces on the car are the upward normal force exerted by the ground and the downward pull of gravity. Designating $+y$ downward, we have

$$mg - N = \frac{mv^2}{R}$$

from Newton's second law. To find the greatest speed without leaving the hill, we set $N = 0$ and solve for v:

$$v = \sqrt{gR} = \sqrt{(9.8)(250)} = 49.5 \text{ m/s}$$

which converts to $49.5(3600/1000) = 178$ km/h.

65. For simplicity, we denote the 70° angle as θ and the magnitude of the push (80 N) as P. The vertical forces on the block are the downward normal force exerted on it by the ceiling, the downward pull of gravity (of magnitude mg) and the vertical component of $\vec{P}$ (which is upward with magnitude $P\sin\theta$). Since there is no acceleration in the vertical direction, we must have

$$N = P\sin\theta - mg$$

in which case the leftward-pointed kinetic friction has magnitude

$$f_k = \mu_k\left(P\sin\theta - mg\right) .$$

Choosing $+x$ rightward, Newton's second law leads to

$$P\cos\theta - f_k = ma \implies a = \frac{P\cos\theta - \mu_k\left(P\sin\theta - mg\right)}{m}$$

which yields $a = 3.4$ m/s^2 when $\mu_k = 0.40$ and $m = 5.0$ kg.

66. Probably the most appropriate picture in the textbook to represent the situation in this problem is in the previous chapter: Fig. 5-9. We adopt the familiar axes with $+x$ rightward and $+y$ upward, and refer to the 85 N horizontal push of the worker as P (and assume it to be rightward). Applying Newton's second law to the x axis and y axis, respectively, produces

$$\begin{aligned} P - f_k &= ma \\ N - mg &= 0 \ . \end{aligned}$$

Using $v^2 = v_0^2 + 2a\Delta x$ we find $a = 0.36$ m/s^2. Consequently, we obtain $f_k = 71$ N and $N = 392$ N. Therefore, $\mu_k = f_k/N = 0.18$.

67. In the figure below, $m = 140/9.8 = 14.3$ kg is the mass of the child. We use $\vec{w}_x$ and $\vec{w}_y$ as the components of the gravitational pull of Earth on the block; their magnitudes are $w_x = mg\sin\theta$ and $w_y = mg\cos\theta$.

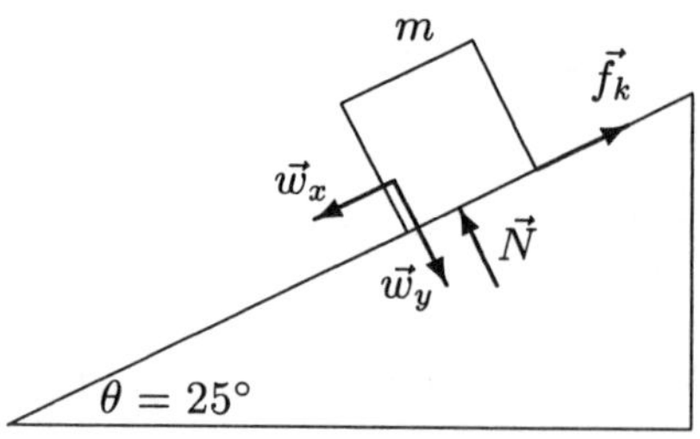

(a) With the x axis directed up along the incline (so that $a = -0.86\ \text{m/s}^2$), Newton's second law leads to

$$f_k - 140 \sin 25° = m\,(-0.86)$$

which yields $f_k = 47$ N. We also apply Newton's second law to the y axis (perpendicular to the incline surface), where the acceleration-component is zero:

$$N - 140 \cos 25° = 0 \implies N = 127\ \text{N}\ .$$

Therefore, $\mu_k = f_k/N = 0.37$.

(b) Returning to our first equation in part (a), we see that if the downhill component of the weight force were insufficient to overcome static friction, the child would not slide at all. Therefore, we require $140 \sin 25° > f_{s,\max} = \mu_s N$, which leads to $\tan 25° = 0.47 > \mu_s$. The minimum value of μ_s equals μ_k and is more subtle; reference to §6-1 is recommended. If μ_k exceeded μ_s then when static friction were overcome (as the incline is raised) then it should start to move – which is impossible if f_k is large enough to cause deceleration! The bounds on μ_s are therefore given by $\tan 25° > \mu_s > \mu_k$.

68. (a) The intuitive conclusion, that the tension is greatest at the bottom of the swing, is certainly supported by application of Newton's second law there:

$$T - mg = \frac{mv^2}{R} \implies T = m\left(g + \frac{v^2}{R}\right)$$

where Eq. 6-18 has been used. Increasing the speed eventually leads to the tension at the bottom of the circle reaching that breaking value of 40 N.

(b) Solving the above equation for the speed, we find

$$v = \sqrt{R\left(\frac{T}{m} - g\right)} = \sqrt{(0.91)\left(\frac{40}{0.37} - 9.8\right)}$$

which yields $v = 9.5$ m/s.

69. (a) We denote the apparent weight of the crew member of mass m on the spaceship as $W_a = 300\,\text{N}$, his weight on Earth as $W_e = mg = 600\,\text{N}$, and the radius of the spaceship as $R = 500\,\text{m}$. Since $mv_s^2/R = W_a$, we get

$$v_s = \sqrt{\frac{W_a R}{m}} = \sqrt{\left(\frac{W_a}{W_e}\right) gR}$$

where we substituted $m = W_e/g$. Thus,

$$v_s = \sqrt{\left(\frac{300\,\text{N}}{600\,\text{N}}\right)(9.8\,\text{m/s}^2)\,(500\,\text{m})} = 49.5\ \text{m/s}\ .$$

(b) For any object of mass m on the spaceship $W_a = mv^2/R \propto v^2$, where v is the speed of the circular motion of the object relative to the center of the circle. In the previous case $v = v_s = 49.5\,\mathrm{m/s}$, and in the present case $v = 10\,\mathrm{m/s} + 49.5\,\mathrm{m/s} = 59.5\,\mathrm{m/s} \equiv v'$. Thus the apparent weight of the running crew member is

$$W_a' = W_a \left(\frac{v'}{v}\right)^2 = (300\,\mathrm{N}) \left(\frac{59.5\,\mathrm{m/s}}{49.5\,\mathrm{m/s}}\right)^2 = 4.3 \times 10^2\,\mathrm{N}\ .$$

70. We refer the reader to Sample Problem 6-11, and use the result Eq. 6-29:

$$\theta = \tan^{-1}\left(\frac{v^2}{gR}\right)$$

with $v = 60(1000/3600) = 17$ m/s and $R = 200$ m. The banking angle is therefore $\theta = 8.1°$. Now we consider a vehicle taking this banked curve at $v' = 40(1000/3600) = 11$ m/s. Its (horizontal) acceleration is $a' = v'^2/R$, which has components parallel the incline and perpendicular to it.

$$a_{\|} = a' \cos\theta = \frac{v'^2 \cos\theta}{R} \quad \text{and} \quad a_{\perp} = a' \sin\theta = \frac{v'^2 \sin\theta}{R}$$

These enter Newton's second law as follows (choosing downhill as the $+x$ direction and away-from-incline as $+y$):

$$mg\sin\theta - f_s = ma_{\|} \quad \text{and} \quad N - mg\cos\theta = ma_{\perp}$$

and we are led to

$$\frac{f_s}{N} = \frac{mg\sin\theta - mv'^2\cos\theta/R}{mg\cos\theta + mv'^2\sin\theta/R}\ .$$

We cancel the mass and plug in, obtaining $f_s/N = 0.078$. The problem implies we should set $f_s = f_{s,\max}$ so that, by Eq. 6-1, we have $\mu_s = 0.078$.

71. (a) The force which provides the horizontal acceleration v^2/R necessary for the circular motion of radius $R = 0.25$ m is $T\sin\theta$, where T in the tension in the $L = 1.2$ m string and θ is the angle of the string measured from vertical. The other component of tension must equal the bob's weight so that there is no vertical acceleration: $T\cos\theta = mg$. Combining these observations leads to

$$\frac{v^2}{R} = g\tan\theta \quad \text{where} \quad \sin\theta = \frac{R}{L}$$

so that $\theta = \sin^{-1}(0.25/1.2) = 12°$ and $v = \sqrt{gR\tan\theta} = 0.72\,\mathrm{m/s}$. It should be mentioned that Sample Problem 6-11 discusses the conical pendulum.

(b) Thus, $a = v^2/R = 2.1\ \mathrm{m/s^2}$.

(c) The tension is

$$T = \frac{mg}{\cos\theta} = \frac{(0.050)(9.8)}{\cos 12°} = 0.50\ \mathrm{N}\ .$$

72. (a) Our $+x$ direction is horizontal and is chosen (as we also do with $+y$) so that the components of the 100 N force $\vec{F}$ are non-negative. Thus, $F_x = F\cos\theta = 100$ N, which the textbook denotes F_h in this problem.

(b) Since there is no vertical acceleration, application of Newton's second law in the y direction gives

$$N + F_y = mg \implies N = mg - F\sin\theta$$

where $m = 25$ kg. This yields $N = 245$ N in this case ($\theta = 0°$).

(c) Now, $F_x = F_h = F\cos\theta = 86.6$ N for $\theta = 30°$.

(d) And $N = mg - F\sin\theta = 195$ N.

(e) We find $F_x = F_h = F\cos\theta = 50$ N for $\theta = 60°$.

(f) And $N = mg - F\sin\theta = 158$ N.

(g) The condition for the chair to slide is

$$F_x > f_{s,\max} = \mu_s N \quad \text{where} \quad \mu_s = 0.42 .$$

For $\theta = 0°$, we have

$$F_x = 100\,\text{N} < f_{s,\max} = (0.42)(245) = 103\,\text{N}$$

so the crate remains at rest.

(h) For $\theta = 30.0°$, we find

$$F_x = 86.6\,\text{N} > f_{s,\max} = (0.42)(195) = 81.9\,\text{N}$$

so the crate slides.

(i) For $\theta = 60°$, we get

$$F_x = 50.0\,\text{N} < f_{s,\max} = (0.42)(158) = 66.4\,\text{N}$$

which means the crate must remain at rest.

73. We note that $N = mg$ in this situation, so $f_k = \mu_k mg = (0.32)(220) = 70.4\,\text{N}$ and $f_{s,\max} = \mu_s mg = (0.41)(220) = 90.2\,\text{N}$.

(a) The person needs to push at least as hard as the static friction maximum if he hopes to start it moving. Denoting his force as P, this means a value of P slightly larger than 90.2 N is sufficient. Rounding to two figures, we obtain $P = 90$ N.

(b) Constant velocity (zero acceleration) implies the push equals the kinetic friction, so $P = 70$ N.

(c) Applying Newton's second law, we have

$$P - f_k = ma \implies a = \frac{\mu_s mg - \mu_k mg}{m}$$

which simplifies to $a = g(\mu_s - \mu_k) = 0.88\ \text{m/s}^2$.

74. Except for replacing f_s with f_k, Fig. 6-5 in the textbook is appropriate. With that figure in mind, we choose uphill as the $+x$ direction. Applying Newton's second law to the x axis, we have

$$f_k - W\sin\theta = ma \quad \text{where} \quad m = \frac{W}{g} ,$$

and where $W = 40$ N, $a = +0.80\ \text{m/s}^2$ and $\theta = 25°$. Thus, we find $f_k = 20$ N. Along the y axis, we have

$$\sum \vec{F}_y = 0 \implies N = W\cos\theta$$

so that $\mu_k = f_k/N = 0.56$.

75. We use the familiar horizontal and vertical axes for x and y directions, with rightward and upward positive, respectively. The rope is assumed massless so that the force exerted by the child $\vec{F}$ is identical to the tension uniformly through the rope. The x and y components of $\vec{F}$ are $F\cos\theta$ and $F\sin\theta$, respectively. The static friction force points leftward.

(a) Newton's Law applied to the y axis, where there is presumed to be no acceleration, leads to

$$N + F\sin\theta - mg = 0$$

which implies that the maximum static friction is $\mu_s(mg - F\sin\theta)$. If $f_s = f_{s,\,\max}$ is assumed, then Newton's second law applied to the x axis (which also has $a = 0$ even though it is "verging" on moving) yields

$$\begin{aligned} F\cos\theta - f_s &= ma \ , \quad \text{or} \\ F\cos\theta - \mu_s(mg - F\sin\theta) &= 0 \end{aligned}$$

which we solve, for $\theta = 42°$ and $\mu_s = 0.42$, to obtain $F = 74$ N.

(b) Solving the above equation algebraically for F, with W denoting the weight, we obtain

$$F = \frac{\mu_s W}{\cos\theta + \mu_s \sin\theta} \ .$$

(c) We minimize the above expression for F by working through the $\frac{dF}{d\theta} = 0$ condition:

$$\frac{dF}{d\theta} = \frac{\mu_s W\,(\sin\theta - \mu_s\cos\theta)}{(\cos\theta + \mu_s\sin\theta)^2} = 0$$

which leads to the result $\theta = \tan^{-1}\mu_s = 23°$.

(d) Plugging $\theta = 23°$ into the above result for F, with $\mu_s = 0.42$ and $W = 180$ N, yields $F = 70$ N.

76. (a) We note that $N = mg$ in this situation, so $f_{s,\max} = \mu_s mg = (0.52)(11)(9.8) = 56\,\text{N}$. Consequently, the horizontal force $\vec{F}$ needed to initiate motion must be (at minimum) slightly more than 56 N.

(b) Analyzing vertical forces when $\vec{F}$ is at nonzero θ yields

$$F\sin\theta + N = mg \implies f_{s,\max} = \mu_s\,(mg - F\sin\theta) \ .$$

Now, the horizontal component of $\vec{F}$ needed to initiate motion must be (at minimum) slightly more than this, so

$$F\cos\theta = \mu_s\,(mg - F\sin\theta) \implies F = \frac{\mu_s mg}{\cos\theta + \mu_s\sin\theta}$$

which yields $F = 59$ N when $\theta = 60°$.

(c) We now set $\theta = -60°$ and obtain

$$F = \frac{(0.52)(11)(9.8)}{\cos(-60°) + (0.52)\sin(-60°)} = 1.1 \times 10^3\,\text{N} \ .$$

77. The coordinate system we wish to use is shown in Fig. 5-18 in the textbook, so we resolve this horizontal force into appropriate components.

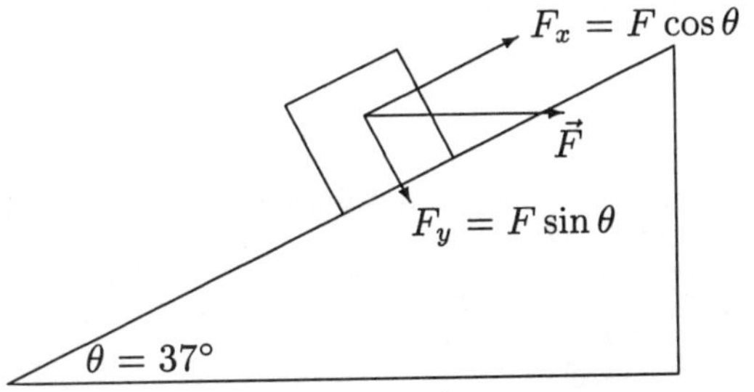

(a) Applying Newton's second law to the x (directed uphill) and y (directed away from the incline surface) axes, we obtain

$$\begin{aligned} F\cos\theta - f_k - mg\sin\theta &= ma \\ N - F\sin\theta - mg\cos\theta &= 0 . \end{aligned}$$

Using $f_k = \mu_k N$, these equations lead to

$$a = \frac{F}{m}(\cos\theta - \mu_k\sin\theta) - g(\sin\theta + \mu_k\cos\theta)$$

which yields $a = -2.1$ m/s^2 for $\mu_k = 0.30$, $F = 50$ N and $m = 5.0$ kg.

(b) With $v_0 = +4.0$ m/s and $v = 0$, Eq. 2-16 gives

$$\Delta x = -\frac{4.0^2}{2(-2.1)} = 3.9 \text{ m} .$$

(c) We expect $\mu_s \geq \mu_k$; otherwise, an object started into motion would immediately start decelerating (before it gained any speed)! In the minimal expectation case, where $\mu_s = 0.30$, the maximum possible (downhill) static friction is, using Eq. 6-1,

$$f_{s,\max} = \mu_s N = \mu_s(F\sin\theta + mg\cos\theta)$$

which turns out to be 21 N. But in order to have no acceleration along the x axis, we must have

$$f_s = F\cos\theta - mg\sin\theta = 10 \text{ N}$$

(the fact that this is positive reinforces our suspicion that $\vec{f}_s$ points downhill). Since the f_s needed to remain at rest is less than $f_{s,\max}$ then it stays at that location.

78. Since the problem is allowing for student creativity and research here, we only present a problem and solution for part (a).

(a) We show below two blocks M and m, the first on a horizontal surface with $\mu_k = 0.25$ and the second on a frictionless incline. They are connected by a rope (not shown) in which the tension is T. The goal is to find T given $M = 2.0$ kg and $m = 3.0$ kg. We assume f_s is not relevant to this computation.

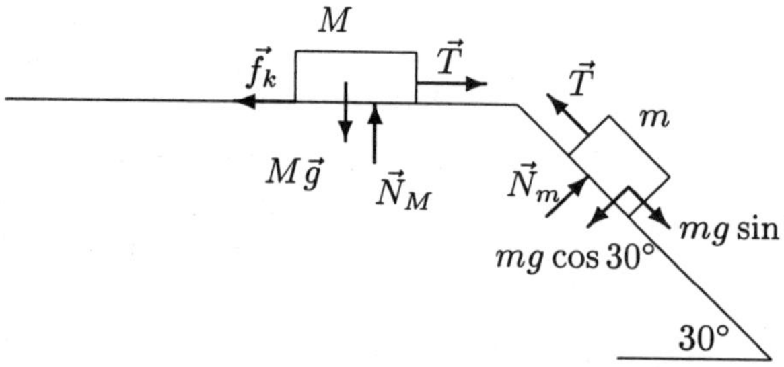

Solution: We apply Newton's second law to each block's x axis, which for M is positive rightward and for m is positive downhill:

$$\begin{aligned} T - f_k &= Ma \\ mg\sin 30° - T &= ma \end{aligned}$$

Adding the equations, we obtain the acceleration.

$$a = \frac{mg\sin 30° - f_k}{m + M}$$

For $f_k = \mu_k N_M = \mu_k Mg$, we obtain $a = 1.96 \text{ m/s}^2$. Returning this value to either of the above equations, we find $T = 8.8$ N.

79. (First problem in **Cluster 1**)
Since the block remains stationary, then $\sum \vec{F} = 0$, and we have (along the horizontal x axis) $f_s = 25$ N, where $\vec{f_s}$ points left.

80. (Second problem in **Cluster 1**)
To keep the block stationary, we require $\sum \vec{F} = 0$ (equilibrium of forces), which leads (along the horizontal x axis) to $f_s = 50$ N. Now, we take $f_s = f_{s,\max} = \mu_s N$ and find that N must equal $50/0.4 = 125$ N. Equilibrium of forces along the y axis implies $N - mg - F = 0$, so that (with $mg = 98$ N) we must have $F = 27$ N.

81. (Third problem in **Cluster 1**)
A useful diagram (where some of these forces are analyzed) is Fig. 6-5 in the textbook. Using that figure for this problem, W is the weight (equal to $mg = 98$ N), and $\theta = 25°$.

   (a) The maximum static friction is given by Eq. 6-1:

$$f_{s,\max} = \mu_s N = (0.60) W \cos\theta = 53 \text{ N} \ .$$

   (b) $W \sin\theta = 41$ N.

   (c) If there is *no* motion, then $\sum \vec{F} = 0$ along the incline, so $f_s - W\sin\theta - F = 0$ (if uphill is positive). And if the system verges on motion, then $f_s = f_{s,\max}$. Therefore, in that case we find $F = 53 - 41 = 12$ N.

   (d) With the block sliding, with no applied force F, then Newton's second law yields $f_k - W\sin\theta = ma$ (if uphill is positive) where $f_k = \mu_k N = (0.20) W\cos\theta = 18$ N. We thus obtain $a = -2.4 \text{ m/s}^2$. Therefore, the magnitude of $\vec{a}$ is 2.4 m/s^2 and the direction is downhill.

82. (Fourth problem in **Cluster 1**)
A useful diagram (where some of these forces are analyzed) is Fig. 6-5 in the textbook; however, since the block is about to move uphill, one must imagine $\vec{f_s}$ turned around (so that it points downhill). Using that figure for this problem, W is the weight (equal to $mg = 98$ N), and $\theta = 25°$.

   (a) If there is *no* motion, then $\sum \vec{F} = 0$ along the incline, so $F - f_s - W\sin\theta = 0$ (if uphill is positive). And if the system verges on motion, then $f_s = f_{s,\max} = \mu_s W\cos\theta = 53$ N. Therefore, in that case we find $F = 95$ N.

   (b) With the block sliding, and the applied force F still equal to the value found in part (a), then Newton's second law yields $F - f_k - W\sin\theta = ma$ (if uphill is positive) where $f_k = \mu_k N = (0.20) W\cos\theta = 18$ N. We thus obtain $a = 3.6 \text{ m/s}^2$. Therefore, the magnitude of $\vec{a}$ is 3.6 m/s^2 and the direction is uphill.

   (c) With the block sliding uphill, but with no applied force F, then Newton's second law yields $-f_k - W\sin\theta = ma$ (if uphill is positive) where $f_k = 18$ N. We thus obtain $a = -5.9 \text{ m/s}^2$. Therefore, the magnitude of $\vec{a}$ is 5.9 m/s^2 and the direction is downhill. It is decelerating and will ultimately come to a stop and remain at there at equilibrium.

83. (Fifth problem in **Cluster 1**)
A useful diagram (where these forces are analyzed) is Fig. 6-5 in the textbook. In that figure, W is the weight (equal to $mg = 98$ N).

(a) Since there is no motion, then $\sum \vec{F} = 0$ along the incline, so $f_s - W \sin\theta = 0$ (if uphill is positive, which is the direction assumed for $\vec{f_s}$). We therefore obtain $f_s = 25$ N. Our result is positive, so it indeed points uphill as we had assumed. One can check that this value of f_s does not exceed the maximum possible value $f_{s,\,\text{max}}$ (see next part).

(b) As in part (a), we have $f_s - W \sin\theta = 0$, but since the system is on the verge of motion we also have $f_s = f_{s,\,\text{max}} = \mu_s W \cos\theta$. Therefore,

$$\mu_s W \cos\theta - W \sin\theta = 0 \quad \Longrightarrow \quad \mu_s = \tan\theta$$

which leads to $\theta_s = \tan^{-1}\mu_s = 31°$ (this is often called "the angle of repose").

(c) If the block slides with no acceleration then we have $f_k - W \sin\theta = 0$ from Newton's second law applied along the incline surface. With $f_k = \mu_k W \cos\theta$ we are led to $\theta_k = \tan^{-1}\mu_k$ as the condition for this constant velocity sliding downhill. Since $\mu_k < \mu_s$ then we see that $\theta_k < \theta_s$ from part (b).

(d) We find $\theta_k = \tan^{-1}\mu_k = 11°$.

Chapter 7

1. The kinetic energy is given by $K = \frac{1}{2}mv^2$, where m is the mass and v is the speed of the electron. The speed is therefore

$$v = \sqrt{\frac{2K}{m}} = \sqrt{\frac{2(6.7 \times 10^{-19}\,\mathrm{J})}{9.11 \times 10^{-31}\,\mathrm{kg}}} = 1.2 \times 10^6 \text{ m/s} .$$

2. (a) The change in kinetic energy for the meteorite would be

$$\begin{aligned} \Delta K &= K_f - K_i = -K_i = -\frac{1}{2}m_i v_i^2 \\ &= -\frac{1}{2}\left(4 \times 10^6\,\mathrm{kg}\right)\left(15 \times 10^3\,\mathrm{m/s}\right)^2 \\ &= -5 \times 10^{14}\ \mathrm{J} \end{aligned}$$

where the negative sign indicates that kinetic energy is lost.

(b) The energy loss in units of megatons of TNT would be

$$-\Delta K = \left(5 \times 10^{14}\,\mathrm{J}\right)\left(\frac{1\text{ megaton TNT}}{4.2 \times 10^{15}\,\mathrm{J}}\right) = 0.1 \text{ megaton TNT} .$$

(c) The number of bombs N that the meteorite impact would correspond to is found by noting that megaton = 1000 kilotons and setting up the ratio:

$$N = \frac{0.1 \times 1000\text{ kiloton TNT}}{13\text{ kiloton TNT}} = 8 .$$

3. We convert to SI units (where necessary) and use $K = \frac{1}{2}mv^2$.

(a) $K = \frac{1}{2}(110)(8.1)^2 = 3.6 \times 10^3$ J.

(b) Since 1000 g = kg, we find

$$K = \frac{1}{2}\left(4.2 \times 10^{-3}\,\mathrm{kg}\right)\left(950\,\mathrm{m/s}\right)^2 = 1.9 \times 10^3\ \mathrm{J} .$$

(c) We note that the conversion from knots to m/s can be obtained from the information in Appendix D (knot = 1.688 ft/s where ft = 0.3048 m), which is also where the ton → kilogram conversion can be found. Therefore,

$$K = \frac{1}{2}\left(91400\text{ tons})\,\frac{907.2\,\mathrm{kg}}{\text{ton}}\right)\left((32\text{ knots})\,\frac{0.515\,\mathrm{m/s}}{\text{knot}}\right)^2 = 1.1 \times 10^{10}\ \mathrm{J} .$$

4. We denote the mass of the father as m and his initial speed v_i. The initial kinetic energy of the father is

$$K_i = \frac{1}{2} K_{\text{son}}$$

and his final kinetic energy (when his speed is $v_f = v_i + 1.0$ m/s) is

$$K_f = K_{\text{son}} \ .$$

We use these relations along with Eq. 7-1 in our solution.

(a) We see from the above that $K_i = \frac{1}{2} K_f$ which (with SI units understood) leads to

$$\frac{1}{2} m v_i^2 = \frac{1}{2}\left(\frac{1}{2} m (v_i + 1.0)^2\right) \ .$$

The mass cancels and we find a second-degree equation for v_i:

$$\frac{1}{2} v_i^2 - v_i - \frac{1}{2} = 0 \ .$$

The positive root (from the quadratic formula) yields $v_i = 2.4$ m/s.

(b) From the first relation above ($K_i = \frac{1}{2} K_{\text{son}}$), we have

$$\frac{1}{2} m v_i^2 = \frac{1}{2}\left(\frac{1}{2}\left(\frac{m}{2}\right) v_{\text{son}}^2\right)$$

and (after canceling m and one factor of 1/2) are led to $v_{\text{son}} = 2v_i = 4.8$ m/s.

5. (a) From Table 2-1, we have $v^2 = v_0^2 + 2a\Delta x$. Thus,

$$v = \sqrt{v_0^2 + 2a\Delta x} = \sqrt{(2.4 \times 10^7)^2 + 2\,(3.6 \times 10^{15})\,(0.035)} = 2.9 \times 10^7 \text{ m/s} \ .$$

(b) The initial kinetic energy is

$$K_i = \frac{1}{2} m v_0^2 = \frac{1}{2}\left(1.67 \times 10^{-27}\text{ kg}\right)\left(2.4 \times 10^7\text{ m/s}\right)^2 = 4.8 \times 10^{-13}\text{ J} \ .$$

The final kinetic energy is

$$K_f = \frac{1}{2} m v^2 = \frac{1}{2}\left(1.67 \times 10^{-27}\text{ kg}\right)\left(2.9 \times 10^7\text{ m/s}\right)^2 = 6.9 \times 10^{-13}\text{ J} \ .$$

The change in kinetic energy is $\Delta K = 6.9 \times 10^{-13} - 4.8 \times 10^{-13} = 2.1 \times 10^{-13}$ J.

6. Using Eq. 7-8 (and Eq. 3-23), we find the work done by the water on the ice block:

$$\begin{aligned} W &= \vec{F} \cdot \vec{d} \\ &= \left(210\,\hat{\imath} - 150\,\hat{\jmath}\right) \cdot \left(15\,\hat{\imath} - 12\,\hat{\jmath}\right) \\ &= (210)(15) + (-150)(-12) \\ &= 5.0 \times 10^3 \text{ J} \ . \end{aligned}$$

7. (a) The force of the worker on the crate is constant, so the work it does is given by $W_F = \vec{F} \cdot \vec{d} = Fd\cos\phi$, where $\vec{F}$ is the force, $\vec{d}$ is the displacement of the crate, and ϕ is the angle between the force and the displacement. Here $F = 210$ N, $d = 3.0$ m, and $\phi = 20°$. Thus $W_F = (210\text{ N})(3.0\text{ m})\cos 20° = 590$ J.

(b) The force of gravity is downward, perpendicular to the displacement of the crate. The angle between this force and the displacement is 90° and $\cos 90° = 0$, so the work done by the force of gravity is zero.

(c) The normal force of the floor on the crate is also perpendicular to the displacement, so the work done by this force is also zero.

(d) These are the only forces acting on the crate, so the total work done on it is 590 J.

8. Since this involves constant-acceleration motion, we can apply the equations of Table 2-1, such as $x = v_0t + \frac{1}{2}at^2$ (where $x_0 = 0$). We choose to analyze the third and fifth points, obtaining

$$\begin{aligned} 0.2\,\text{m} &= v_0(1.0\,\text{s}) + \frac{1}{2}a\,(1.0\,\text{s})^2 \\ 0.8\,\text{m} &= v_0(2.0\,\text{s}) + \frac{1}{2}a\,(2.0\,\text{s})^2 \end{aligned}$$

Simultaneous solution of the equations leads to $v_0 = 0$ and $a = 0.40\,\text{m/s}^2$. We now have two ways to finish the problem. One is to compute force from $F = ma$ and then obtain the work from Eq. 7-7. The other is to find ΔK as a way of computing W (in accordance with Eq. 7-10). In this latter approach, we find the velocity at $t = 2.0$ s from $v = v_0 + at$ (so $v = 0.80$ m/s). Thus,

$$W = \Delta K = \frac{1}{2}(1.0\,\text{kg})(0.80\,\text{m/s})^2 = 0.32\ \text{J}\ .$$

9. We choose $+x$ as the direction of motion (so $\vec{a}$ and $\vec{F}$ are negative-valued).

(a) Newton's second law readily yields $\vec{F} = (85\,\text{kg})(-2.0\,\text{m/s}^2)$ so that $F = |\vec{F}| = 170$ N.

(b) From Eq. 2-16 (with $v = 0$) we have

$$0 = v_0^2 + 2a\Delta x \implies \Delta x = -\frac{(37\,\text{m/s})^2}{2(-2.0\,\text{m/s}^2)}$$

which gives $\Delta x = 3.4 \times 10^2$ m. Alternatively, this can be worked using the work-energy theorem.

(c) Since $\vec{F}$ is opposite to the direction of motion (so the angle ϕ between $\vec{F}$ and $\vec{d} = \Delta x$ is 180°) then Eq. 7-7 gives the work done as $W = -F\Delta x = -5.8 \times 10^4$ J.

(d) In this case, Newton's second law yields $\vec{F} = (85\,\text{kg})(-4.0\,\text{m/s}^2)$ so that $F = |\vec{F}| = 340$ N.

(e) From Eq. 2-16, we now have

$$\Delta x = -\frac{(37\,\text{m/s})^2}{2(-4.0\,\text{m/s}^2)} = 1.7 \times 10^2\ \text{m}\ .$$

(f) The force $\vec{F}$ is again opposite to the direction of motion (so the angle ϕ is again 180°) so that Eq. 7-7 leads to $W = -F\Delta x = -5.8 \times 10^4$ J. The fact that this agrees with the result of part (c) provides insight into the concept of work.

10. We choose to work this using Eq. 7-10 (the work-kinetic energy theorem). To find the initial and final kinetic energies, we need the speeds, so

$$v = \frac{dx}{dt} = 3.0 - 8.0t + 3.0t^2$$

in SI units. Thus, the initial speed is $v_i = 3.0$ m/s and the speed at $t = 4$ s is $v_f = 19$ m/s. The change in kinetic energy for the object of mass $m = 3.0$ kg is therefore

$$\Delta K = \frac{1}{2}m\left(v_f^2 - v_i^2\right) = 528\ \text{J}$$

which we round off to two figures and (using the work-kinetic energy theorem) conclude that the work done is $W = 5.3 \times 10^2$ J.

11. (a) The forces are constant, so the work done by any one of them is given by $W = \vec{F} \cdot \vec{d}$, where $\vec{d}$ is the displacement. Force $\vec{F_1}$ is in the direction of the displacement, so

$$W_1 = F_1 d \cos\phi_1 = (5.00\,\text{N})(3.00\,\text{m})\cos 0° = 15.0\ \text{J}\ .$$

Force $\vec{F_2}$ makes an angle of 120° with the displacement, so

$$W_2 = F_2 d \cos\phi_2 = (9.00\,\text{N})(3.00\,\text{m})\cos 120° = -13.5\ \text{J}\ .$$

Force $\vec{F_3}$ is perpendicular to the displacement, so $W_3 = F_3 d \cos\phi_3 = 0$ since $\cos 90° = 0$. The net work done by the three forces is

$$W = W_1 + W_2 + W_3 = 15.0\,\text{J} - 13.5\,\text{J} + 0 = +1.5\ \text{J}\ .$$

(b) If no other forces do work on the box, its kinetic energy increases by 1.5 J during the displacement.

12. By the work-kinetic energy theorem,

$$\begin{aligned} W &= \Delta K \\ &= \frac{1}{2}mv_f^2 - \frac{1}{2}mv_i^2 \\ &= \frac{1}{2}(2.0\,\text{kg})\left((6.0\,\text{m/s})^2 - (4.0\,\text{m/s})^2\right) \\ &= 20\ \text{J}\ . \end{aligned}$$

We note that the *directions* of $\vec{v}_f$ and $\vec{v}_i$ play no role in the calculation.

13. The forces are all constant, so the total work done by them is given by $W = F_{\text{net}}\,\Delta x$, where F_{net} is the magnitude of the net force and Δx is the magnitude of the displacement. We add the three vectors, finding the x and y components of the net force:

$$\begin{aligned} F_{\text{net}\,x} &= -F_1 - F_2\sin 50° + F_3\cos 35° \\ &= -3.00\,\text{N} - (4.00\,\text{N})\sin 35° + (10.0\,\text{N})\cos 35° \\ &= 2.127\ \text{N} \\ F_{\text{net}\,y} &= -F_2\cos 50° + F_3\sin 35° \\ &= -(4.00\,\text{N})\cos 50° + (10.0\,\text{N})\sin 35° \\ &= 3.165\ \text{N}\ . \end{aligned}$$

The magnitude of the net force is

$$F_{\text{net}} = \sqrt{F_{\text{net}\,x}^2 + F_{\text{net}\,y}^2} = \sqrt{2.127^2 + 3.165^2} = 3.813\ \text{N}\ .$$

The work done by the net force is

$$W = F_{\text{net}}\, d = (3.813\,\text{N})(4.00\,\text{m}) = 15.3\ \text{J}$$

where we have used the fact that $\vec{d} \parallel \vec{F}_{\text{net}}$ (which follows from the fact that the canister started from rest and moved horizontally under the action of horizontal forces – the resultant effect of which is expressed by $\vec{F}_{\text{net}}$).

14. In both cases, there is no acceleration, so the lifting force is equal to the weight of the object.

(a) Eq. 7-8 leads to $W = \vec{F} \cdot \vec{d} = (360\,\text{kN})(0.10\,\text{m}) = 36$ kJ.

(b) In this case, we find $W = (4000\,\text{N})(0.050\,\text{m}) = 200$ J.

15. There is no acceleration, so the lifting force is equal to the weight of the object. We note that the person's pull $\vec{F}$ is equal (in magnitude) to the tension in the cord.

(a) As indicated in the *hint*, tension contributes twice to the lifting of the canister: $2T = mg$. Since, $|\vec{F}| = T$, we find $|\vec{F}| = 98$ N.

(b) To rise 0.020 m, two segments of the cord (see Fig. 7-28) must shorten by that amount. Thus, the amount of string pulled down at the left end (this is the magnitude of $\vec{d}$, the downward displacement of the hand) is $d = 0.040$ m.

(c) Since (at the left end) both $\vec{F}$ and $\vec{d}$ are downward, then Eq. 7-7 leads to $W = \vec{F} \cdot \vec{d} = (98)(0.040) = 3.9$ J.

(d) Since the force of gravity $\vec{F}_g$ (with magnitude mg) is opposite to the displacement $\vec{d}_c = 0.020$ m (up) of the canister, Eq. 7-7 leads to $W = \vec{F}_g \cdot \vec{d}_c = -(196)(0.020) = -3.9$ J. This is consistent with Eq. 7-15 since there is no change in kinetic energy.

16. (a) The component of the force of gravity exerted on the ice block (of mass m) along the incline is $mg \sin\theta$, where $\theta = \sin^{-1}(0.91/1.5)$ gives the angle of inclination for the inclined plane. Since the ice block slides down with uniform velocity, the worker must exert a force $\vec{F}$ "uphill" with a magnitude equal to $mg \sin\theta$. Consequently,

$$F = mg \sin\theta = (45\,\text{kg})\left(9.8\,\text{m/s}^2\right)\left(\frac{0.91\,\text{m}}{1.5\,\text{m}}\right) = 2.7 \times 10^2 \text{ N} .$$

(b) Since the "downhill" displacement is opposite to $\vec{F}$, the work done by the worker is

$$W_1 = -\left(2.7 \times 10^2\,\text{N}\right)(1.5\,\text{m}) = -4.0 \times 10^2 \text{ J} .$$

(c) Since the displacement has a vertically downward component of magnitude 0.91 m (in the same direction as the force of gravity), we find the work done by gravity to be

$$W_2 = (45\,\text{kg})\left(9.8\,\text{m/s}^2\right)(0.91\,\text{m}) = 4.0 \times 10^2 \text{ J} .$$

(d) Since $\vec{N}$ is perpendicular to the direction of motion of the block, and $\cos 90° = 0$, work done by the normal force is $W_3 = 0$ by Eq. 7-7.

(e) The resultant force $\vec{F}_{\text{net}}$ is zero since there is no acceleration. Thus, it's work is zero, as can be checked by adding the above results $W_1 + W_2 + W_3 = 0$.

17. (a) We use $\vec{F}$ to denote the upward force exerted by the cable on the astronaut. The force of the cable is upward and the force of gravity is mg downward. Furthermore, the acceleration of the astronaut is $g/10$ upward. According to Newton's second law, $F - mg = mg/10$, so $F = 11mg/10$. Since the force $\vec{F}$ and the displacement $\vec{d}$ are in the same direction, the work done by $\vec{F}$ is

$$W_F = Fd = \frac{11mgd}{10} = \frac{11(72\,\text{kg})\left(9.8\,\text{m/s}^2\right)(15\,\text{m})}{10} = 1.164 \times 10^4 \text{ J}$$

which (with respect to significant figures) should be quoted as 1.2×10^4 J.

(b) The force of gravity has magnitude mg and is opposite in direction to the displacement. Thus, using Eq. 7-7, the work done by gravity is

$$W_g = -mgd = -(72\,\text{kg})\left(9.8\,\text{m/s}^2\right)(15\,\text{m}) = -1.058 \times 10^4 \text{ J}$$

which should be quoted as -1.1×10^4 J.

(c) The total work done is $W = 1.164 \times 10^4\,\text{J} - 1.058 \times 10^4\,\text{J} = 1.06 \times 10^3\,\text{J}$. Since the astronaut started from rest, the work-kinetic energy theorem tells us that this (which we round to 1.1×10^3 J) is her final kinetic energy.

(d) Since $K = \frac{1}{2}mv^2$, her final speed is

$$v = \sqrt{\frac{2K}{m}} = \sqrt{\frac{2(1.06 \times 10^3\,\text{J})}{72\,\text{kg}}} = 5.4\ \text{m/s}\ .$$

18. We use d to denote the magnitude of the spelunker's displacement during each stage. The mass of the spelunker is $m = 80.0$ kg. The work done by the lifting force is denoted W_i where $i = 1, 2, 3$ for the three stages. We apply the work-energy theorem, Eq. 17-15.

(a) For stage 1, $W_1 - mgd = \Delta K_1 = \frac{1}{2}mv_1^2$, where $v_1 = 5.00\,\text{m/s}$. This gives

$$W_1 = mgd + \frac{1}{2}mv_1^2 = (80.0)(9.8)(10.0) + \frac{1}{2}(80.0)(5.00)^2 = 8.84 \times 10^3\ \text{J}\ .$$

(b) For stage 2, $W_2 - mgd = \Delta K_2 = 0$, which leads to

$$W_2 = mgd = (80.0\,\text{kg})\left(9.8\,\text{m/s}^2\right)(10.0\,\text{m}) = 7.84 \times 10^3\ \text{J}\ .$$

(c) For stage 3, $W_3 - mgd = \Delta K_3 = -\frac{1}{2}mv_1^2$. We obtain

$$W_3 = mgd - \frac{1}{2}mv_1^2 = (80.0)(9.8)(10.0) - \frac{1}{2}(80.0)(5.00)^2 = 6.84 \times 10^3\ \text{J}\ .$$

19. (a) We use F to denote the magnitude of the force of the cord on the block. This force is upward, opposite to the force of gravity (which has magnitude Mg). The acceleration is $\vec{a} = g/4$ downward. Taking the downward direction to be positive, then Newton's second law yields

$$\vec{F}_{\text{net}} = m\vec{a} \implies Mg - F = M\left(\frac{g}{4}\right)$$

so $F = 3Mg/4$. The displacement is downward, so the work done by the cord's force is $W_F = -Fd = -3Mgd/4$, using Eq. 7-7.

(b) The force of gravity is in the same direction as the displacement, so it does work $W_g = Mgd$.

(c) The total work done on the block is $-3Mgd/4 + Mgd = Mgd/4$. Since the block starts from rest, we use Eq. 7-15 to conclude that this $(Mgd/4)$ is the block's kinetic energy K at the moment it has descended the distance d.

(d) Since $K = \frac{1}{2}Mv^2$, the speed is

$$v = \sqrt{\frac{2K}{M}} = \sqrt{\frac{2\,(Mgd/4)}{M}} = \sqrt{\frac{gd}{2}}$$

at the moment the block has descended the distance d.

20. The spring constant is $k = 100$ N/m and the maximum elongation is $x_i = 5.00$ m. Using Eq. 7-25 with $x_f = 0$, the work is found to be

$$W = \frac{1}{2}kx_i^2 = \frac{1}{2}(100)(5.00)^2 = 1.25 \times 10^3\ \text{J}\ .$$

21. (a) The spring constant is $k = 1500$ N/m and the elongation is $x = 0.0076$ m. Our $+x$ direction is rightward. Using Eq. 7-26, the work is found to be

$$W = -\frac{1}{2}kx^2 = -\frac{1}{2}(1500)(0.0076)^2 = -0.043\ \text{J}\ .$$

(b) We use Eq. 7-25 with $x_i = x = 0.0076$ m and $x_f = 2x = 0.0152$ m to find the additional work:

$$\begin{aligned} W &= \frac{1}{2}k\left(x_i^2 - x_f^2\right) \\ &= \frac{1}{2}k\left(x^2 - 4x^2\right) \\ &= -\frac{3}{2}kx^2 \\ &= -\frac{3}{2}(1500)(0.0076)^2 = -0.13 \text{ J} . \end{aligned}$$

We note that this is greater (in magnitude) than the work done in the first interval (even though the displacements have the same magnitude), due to the fact that the force is larger throughout the second interval.

22. (a) The compression of the spring is $d = 0.12$ m. The work done by the force of gravity (acting on the block) is, by Eq. 7-12,

$$W_1 = mgd = (0.25\,\text{kg})\left(9.8\,\text{m/s}^2\right)(0.12\,\text{m}) = 0.29 \text{ J} .$$

(b) The work done by the spring is, by Eq. 7-26,

$$W_2 = -\frac{1}{2}kd^2 = -\frac{1}{2}(250\,\text{N/m})(0.12\,\text{m})^2 = -1.8 \text{ J} .$$

(c) The speed v_i of the block just before it hits the spring is found from the work-kinetic energy theorem (Eq. 7-15).

$$\Delta K = 0 - \frac{1}{2}mv_i^2 = W_1 + W_2$$

which yields

$$v_i = \sqrt{\frac{(-2)(W_1 + W_2)}{m}} = \sqrt{\frac{(-2)(0.29 - 1.8)}{0.25}} = 3.5 \text{ m/s} .$$

(d) If we instead had $v_i' = 7$ m/s, we reverse the above steps and solve for d'. Recalling the theorem used in part (c), we have

$$\begin{aligned} 0 - \frac{1}{2}mv_i'^2 &= W_1' + W_2' \\ &= mgd' - \frac{1}{2}kd'^2 \end{aligned}$$

which (choosing the positive root) leads to

$$d' = \frac{mg + \sqrt{m^2g^2 + mkv_i'^2}}{k}$$

which yields $d' = 0.23$ m. In order to obtain this, we have used more digits in our intermediate results than are shown above (so $v_i = \sqrt{12.048} = 3.471$ m/s and $v_i' = 6.942$ m/s).

23. (a) As the body moves along the x axis from $x_i = 3.0\,\text{m}$ to $x_f = 4.0\,\text{m}$ the work done by the force is

$$\begin{aligned} W &= \int_{x_i}^{x_f} F_x\,dx \\ &= \int_{x_i}^{x_f} -6x\,dx = -3(x_f^2 - x_i^2) \\ &= -3\left(4.0^2 - 3.0^2\right) = -21 \text{ J} . \end{aligned}$$

According to the work-kinetic energy theorem, this gives the change in the kinetic energy:

$$W = \Delta K = \frac{1}{2} m \left(v_f^2 - v_i^2\right)$$

where v_i is the initial velocity (at x_i) and v_f is the final velocity (at x_f). The theorem yields

$$v_f = \sqrt{\frac{2W}{m} + v_i^2} = \sqrt{\frac{2(-21)}{2.0} + 8.0^2} = 6.6 \text{ m/s} .$$

(b) The velocity of the particle is $v_f = 5.0\,\text{m/s}$ when it is at $x = x_f$. The work-kinetic energy theorem is used to solve for x_f. The net work done on the particle is $W = -3(x_f^2 - x_i^2)$, so the theorem leads to

$$-3\left(x_f^2 - x_i^2\right) = \frac{1}{2} m \left(v_f^2 - v_i^2\right) .$$

Thus,

$$\begin{aligned} x_f &= \sqrt{-\frac{m}{6}\left(v_f^2 - v_i^2\right) + x_i^2} \\ &= \sqrt{-\frac{2.0\,\text{kg}}{6\,\text{N/m}}\left((5.0\,\text{m/s})^2 - (8.0\,\text{m/s})^2\right) + (3.0\,\text{m})^2} \\ &= 4.7 \text{ m} . \end{aligned}$$

24. From Eq. 7-32, we see that the "area" in the graph is equivalent to the work done. Finding that area (in terms of rectangular [length×width] and triangular [$\frac{1}{2}$base×height] areas) we obtain

$$\begin{aligned} W &= W_{0<x<2} + W_{2<x<4} + W_{4<x<6} + W_{6<x<8} \\ &= 20 + 10 + 0 - 5 = 25 \text{ J} . \end{aligned}$$

25. According to the graph the acceleration a varies linearly with the coordinate x. We may write $a = \alpha x$, where α is the slope of the graph. Numerically,

$$\alpha = \frac{20\,\text{m/s}^2}{8.0\,\text{m}} = 2.5 \text{ s}^{-2} .$$

The force on the brick is in the positive x direction and, according to Newton's second law, its magnitude is given by $F = a/m = (\alpha/m)x$. If x_f is the final coordinate, the work done by the force is

$$W = \int_0^{x_f} F\,dx = \frac{\alpha}{m}\int_0^{x_f} x\,dx = \frac{\alpha}{2m} x_f^2 = \frac{2.5}{2(10)}(8.0)^2 = 800 \text{ J} .$$

26. From Eq. 7-32, we see that the"area" in the graph is equivalent to the work done. We find the area in terms of rectangular [length×width] and triangular [$\frac{1}{2}$base×height] areas and use the work-kinetic energy theorem appropriately. The initial point is taken to be $x = 0$, where $v_0 = 4.0$ m/s.

(a) With $K_i = \frac{1}{2}mv_0^2 = 16$ J, we have

$$K_3 - K_0 = W_{0<x<1} + W_{1<x<2} + W_{2<x<3} = -4 \text{ J}$$

so that K_3 (the kinetic energy when $x = 3.0$ m) is found to equal 12 J.

(b) With SI units understood, we write $W_{3<x<x_f}$ as $F_x \Delta x = (-4)(x_f - 3.0)$ and apply the work-kinetic energy theorem:

$$\begin{aligned} K_{x_f} - K_3 &= W_{3<x<x_f} \\ K_{x_f} - 12 &= (-4)(x_f - 3.0) \end{aligned}$$

so that the requirement $K_{x_f} = 8$ J leads to $x_f = 4.0$ m.

(c) As long as the work is positive, the kinetic energy grows. The graph shows this situation to hold until $x = 1.0$ m. At that location, the kinetic energy is

$$\begin{aligned} K_1 &= K_0 + W_{0<x<1} \\ &= 16 + 2 = 18 \text{ J} . \end{aligned}$$

27. (a) The graph shows F as a function of x assuming x_0 is positive. The work is negative as the object moves from $x = 0$ to $x = x_0$ and positive as it moves from $x = x_0$ to $x = 2x_0$. Since the area of a triangle is $\frac{1}{2}$(base)(altitude), the work done from $x = 0$ to $x = x_0$ is $-\frac{1}{2}(x_0)(F_0)$ and the work done from $x = x_0$ to $x = 2x_0$ is $\frac{1}{2}(2x_0 - x_0)(F_0) = \frac{1}{2}(x_0)(F_0)$. The total work is the sum, which is zero.

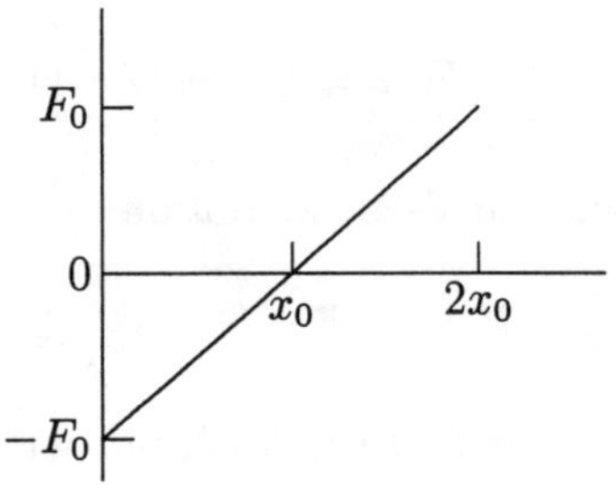

(b) The integral for the work is

$$W = \int_0^{2x_0} F_0 \left(\frac{x}{x_0} - 1 \right) dx = F_0 \left(\frac{x^2}{2x_0} - x \right) \Bigg|_0^{2x_0} = 0 .$$

28. (a) Using the work-kinetic energy theorem

$$K_f = K_i + \int_0^2 (2.5 - x^2)\, dx = 0 + (2.5)(2) - \frac{1}{3}(2)^3$$

we obtain $K_f = 2.3$ J.

(b) For a variable end-point, we have K_f as a function of x, which could be differentiated to find the extremum value, but we recognize that this is equivalent to solving $F = 0$ for x:

$$F = 0 \implies 2.5 - x^2 = 0$$

Thus, K is extremized at $x = \sqrt{2.5}$ and we compute

$$K_f = K_i + \int_0^{\sqrt{2.5}} (2.5 - x^2)\, dx = 0 + (2.5)(\sqrt{2.5}) - \frac{1}{3}(\sqrt{2.5})^3 .$$

Therefore, $K = 2.6$ J at $x = \sqrt{2.5} = 1.6$ m. Recalling our answer for part (a), it is clear that this extreme value is a maximum.

29. One approach is to assume a "path" from $\vec{r}_i$ to $\vec{r}_f$ and do the line-integral accordingly. Another approach is to simply use Eq. 7-36, which we demonstrate:

$$\begin{aligned} W &= \int_{x_i}^{x_f} F_x\, dx + \int_{y_i}^{y_f} F_y\, dy \\ &= \int_2^{-4} (2x)\, dx + \int_3^{-3} (3)\, dy \end{aligned}$$

with SI units understood. Thus, we obtain $W = 12 - 18 = -6$ J.

30. Recognizing that the force in the cable must equal the total weight (since there is no acceleration), we employ Eq. 7-47:

$$P = Fv\cos\theta = mg\left(\frac{\Delta x}{\Delta t}\right)$$

where we have used the fact that $\theta = 0°$ (both the force of the cable and the elevator's motion are upward). Thus,

$$P = \left(3.0 \times 10^3\,\mathrm{kg}\right)\left(9.8\,\mathrm{m/s^2}\right)\left(\frac{210\,\mathrm{m}}{23\,\mathrm{s}}\right) = 2.7 \times 10^5\ \mathrm{W}\ .$$

31. The power associated with force $\vec{F}$ is given by $P = \vec{F} \cdot \vec{v}$, where $\vec{v}$ is the velocity of the object on which the force acts. Thus,

$$P = \vec{F} \cdot \vec{v} = Fv\cos\phi = (122\,\mathrm{N})(5.0\,\mathrm{m/s})\cos 37° = 490\ \mathrm{W}\ .$$

32. (a) Using Eq.7-48 and Eq. 3-23, we obtain

$$P = \vec{F} \cdot \vec{v} = (4.0\,\mathrm{N})(-2.0\,\mathrm{m/s}) + (9.0\,\mathrm{N})(4.0\,\mathrm{m/s}) = 28\ \mathrm{W}\ .$$

(b) We again use Eq.7-48 and Eq. 3-23, but with a one-component velocity: $\vec{v} = v\hat{\mathrm{j}}$.

$$\begin{aligned} P &= \vec{F} \cdot \vec{v} \\ -12\,\mathrm{W} &= (-2.0\,\mathrm{N})v \end{aligned}$$

which yields $v = 6$ m/s.

33. (a) The power is given by $P = Fv$ and the work done by $\vec{F}$ from time t_1 to time t_2 is given by

$$W = \int_{t_1}^{t_2} P\,\mathrm{d}t = \int_{t_1}^{t_2} Fv\,\mathrm{d}t\,.$$

Since $\vec{F}$ is the net force, the magnitude of the acceleration is $a = F/m$, and, since the initial velocity is $v_0 = 0$, the velocity as a function of time is given by $v = v_0 + at = (F/m)t$. Thus

$$W = \int_{t_1}^{t_2} (F^2/m)t\,\mathrm{d}t = \frac{1}{2}(F^2/m)(t_2^2 - t_1^2)\,.$$

For $t_1 = 0$ and $t_2 = 1.0\,\mathrm{s}$,

$$W = \frac{1}{2}\left(\frac{(5.0\,\mathrm{N})^2}{15\,\mathrm{kg}}\right)(1.0\,\mathrm{s})^2 = 0.83\ \mathrm{J}\ .$$

(b) For $t_1 = 1.0\,\mathrm{s}$ and $t_2 = 2.0\,\mathrm{s}$,

$$W = \frac{1}{2}\left(\frac{(5.0\,\mathrm{N})^2}{15\,\mathrm{kg}}\right)\left((2.0\,\mathrm{s})^2 - (1.0\,\mathrm{s})^2\right) = 2.5\ \mathrm{J}\ .$$

(c) For $t_1 = 2.0\,\mathrm{s}$ and $t_2 = 3.0\,\mathrm{s}$,

$$W = \frac{1}{2}\left(\frac{(5.0\,\mathrm{N})^2}{15\,\mathrm{kg}}\right)\left((3.0\,\mathrm{s})^2 - (2.0\,\mathrm{s})^2\right) = 4.2\ \mathrm{J}\ .$$

(d) Substituting $v = (F/m)t$ into $P = Fv$ we obtain $P = F^2t/m$ for the power at any time t. At the end of the third second

$$P = \frac{(5.0\,\mathrm{N})^2(3.0\,\mathrm{s})}{15\,\mathrm{kg}} = 5.0\ \mathrm{W}\ .$$

34. (a) Since constant speed implies $\Delta K = 0$, we require $W_a = -W_g$, by Eq. 7-15. Since W_g is the same in both cases (same weight and same path), then $W_a = 900$ J just as it was in the first case.

(b) Since the speed of 1.0 m/s is constant, then 8.0 meters is traveled in 8.0 seconds. Using Eq. 7-42, and noting that average power is *the* power when the work is being done at a steady rate, we have

$$P = \frac{W}{\Delta t} = \frac{900\ \text{J}}{8.0\ \text{s}}$$

which results in $P = 113$ W.

(c) Since the speed of 2.0 m/s is constant, 8.0 meters is traveled in 4.0 seconds. Using Eq. 7-42, with *average power* replaced by *power*, we have

$$P = \frac{W}{\Delta t} = \frac{900\ \text{J}}{4.0\ \text{s}}$$

from which we obtain $P = 225$ W.

35. The total work is the sum of the work done by gravity on the elevator, the work done by gravity on the counterweight, and the work done by the motor on the system: $W_T = W_e + W_c + W_s$. Since the elevator moves at constant velocity, its kinetic energy does not change and according to the work-kinetic energy theorem the total work done is zero. This means $W_e + W_c + W_s = 0$. The elevator moves upward through 54 m, so the work done by gravity on it is

$$W_e = -m_e g d = -(1200\,\text{kg})(9.8\,\text{m/s}^2)(54\,\text{m}) = -6.35 \times 10^5\,\text{J}\,.$$

The counterweight moves downward the same distance, so the work done by gravity on it is

$$W_c = m_c g d = (950\,\text{kg})\left(9.8\,\text{m/s}^2\right)(54\,\text{m}) = 5.03 \times 10^5\ \text{J}\,.$$

Since $W_T = 0$, the work done by the motor on the system is

$$W_s = -W_e - W_c = 6.35 \times 10^5\ \text{J} - 5.03 \times 10^5\ \text{J} = 1.32 \times 10^5\ \text{J}\,.$$

This work is done in a time interval of $\Delta t = 3.0\,\text{min} = 180\,\text{s}$, so the power supplied by the motor to lift the elevator is

$$P = \frac{W_s}{\Delta t} = \frac{1.32 \times 10^5\ \text{J}}{180\,\text{s}} = 7.4 \times 10^2\ \text{W}\,.$$

36. (a) Since the force exerted by the spring on the mass is zero when the mass passes through the equilibrium position of the spring, the rate at which the spring is doing work on the mass at this instant is also zero.

(b) The rate is given by $P = \vec{F}\cdot\vec{v} = -Fv$, where the minus sign corresponds to the fact that $\vec{F}$ and $\vec{v}$ are antiparallel to each other. The magnitude of the force is given by $F = kx = (500\,\text{N/m})(0.10\,\text{m}) = 50\,\text{N}$, while v is obtained from conservation of energy for the spring-mass system:

$$E = K + U = 10\,\text{J} = \frac{1}{2}mv^2 + \frac{1}{2}kx^2 = \frac{1}{2}(0.30\,\text{kg})v^2 + \frac{1}{2}(500\,\text{N/m})(0.10\,\text{m})^2$$

which gives $v = 7.1\,\text{m/s}$. Thus

$$P = -Fv = -(50\,\text{N})(7.1\,\text{m/s}) = -3.5 \times 10^2\ \text{W}\,.$$

37. We write the force as $F = \alpha v$, where v is the speed and α is a constant of proportionality. The power required is $P = Fv = \alpha v^2$. Let P_1 be the power required for speed v_1 and P_2 be the power required for speed v_2. Dividing $P_2 = \alpha v_2^2$ by $P_1 = \alpha v_1^2$, we find

$$P_2 = \left(\frac{v_2}{v_1}\right)^2 P_1\,.$$

Since $P_1 = 7.5\,\text{kW}$ and $v_2 = 3v_1$,

$$P_2 = (3)^2(7.5\,\text{kW}) = 68\ \text{kW}\,.$$

38. (a) The force $\vec{F}$ of the incline is a combination of normal and friction force which is serving to "cancel" the tendency of the box to fall downward (due to its 19.6 N weight). Thus, $\vec{F} = mg$ upward. In this part of the problem, the angle ϕ between the belt and $\vec{F}$ is 80°. From Eq. 7-47, we have

$$P = Fv\cos\phi = (19.6)(0.50)\cos 80°$$

which leads to $P = 1.7$ W.

(b) Now the angle between the belt and $\vec{F}$ is 90°, so that $P = 0$.

(c) In this part, the angle between the belt and $\vec{F}$ is 100°, so that $P = (19.6)(0.50)\cos 100° = -1.7$ W.

39. (a) In 10 min the cart moves

$$\left(6.0\,\frac{\text{mi}}{\text{h}}\right)\left(\frac{5280\,\text{ft/mi}}{60\,\text{min/h}}\right)(10\,\text{min}) = 5280\ \text{ft}$$

so that Eq. 7-7 yields

$$W = Fd\cos\phi = (40\,\text{lb})(5280\,\text{ft})\cos 30° = 1.8\times 10^5\ \text{ft}\cdot\text{lb}\ .$$

(b) The average power is given by Eq. 7-42, and the conversion to horsepower (hp) can be found on the inside back cover. We note that 10 min is equivalent to 600 s.

$$P_{\text{avg}} = \frac{1.8\times 10^5\ \text{ft}\cdot\text{lb}}{600\,\text{s}} = 305\ \text{ft}\cdot\text{lb/s}$$

which (upon dividing by 550) converts to $P_{\text{avg}} = 0.55$ hp.

40. The acceleration is constant, so we may use the equations in Table 2-1. We choose the direction of motion as $+x$ and note that the displacement is the same as the distance traveled, in this problem. We designate the force (assumed singular) along the x direction acting on the $m = 2.0$ kg object as F.

(a) With $v_0 = 0$, Eq. 2-11 leads to $a = v/t$. And Eq. 2-17 gives $\Delta x = \frac{1}{2}vt$ Newton's second law yields the force $F = ma$. Eq. 7-8, then, gives the work:

$$W = F\Delta x = m\left(\frac{v}{t}\right)\left(\frac{1}{2}vt\right) = \frac{1}{2}mv^2$$

as we expect from the work-kinetic energy theorem. With $v = 10$ m/s, this yields $W = 100$ J.

(b) Instantaneous power is defined in Eq. 7-48. With $t = 3.0$ s, we find

$$P = Fv = m\left(\frac{v}{t}\right)v = 67\ \text{W}\ .$$

(c) The velocity at $t' = 1.5$ s is $v' = at' = 5.0$ m/s. Thus,

$$P' = Fv' = 33\ \text{W}\ .$$

41. The total weight is $(100)(660) = 6.6\times 10^4$ N, and the words "raises ... at constant speed" imply zero acceleration, so the lift-force is equal to the total weight. Thus $P = Fv = (6.6\times 10^4)(150/60) = 1.65\times 10^5$ W.

42. Using Eq. 7-32, we find

$$W = \int_{0.25}^{1.25} e^{-4x^2}\,dx = 0.21\ \text{J}$$

where the result has been obtained numerically. Many modern calculators have that capability, as well as most math software packages that a great many students have access to.

43. (a) and (b) Hooke's law and the work done by a spring is discussed in the chapter. We apply Work-kinetic energy theorem, in the form of $\Delta K = W_a + W_s$, to the points in Figure 7-48 at $x = 1.0$ m and $x = 2.0$ m, respectively. The "applied" work W_a is that due to the constant force $\vec{P}$.

$$\begin{aligned} 4 &= P(1.0) - \frac{1}{2}k(1.0)^2 \\ 0 &= P(2.0) - \frac{1}{2}k(2.0)^2 \end{aligned}$$

Simultaneous solution leads to $P = 8.0$ N and $k = 8.0$ N/m.

44. Using Eq. 7-8, we find

$$W = \vec{F}\cdot\vec{d} = \left(F\cos\theta\,\hat{\imath} + F\sin\theta\,\hat{\jmath}\right)\cdot\left(x\hat{\imath} + y\hat{\jmath}\right) = Fx\cos\theta + Fy\sin\theta$$

where $x = 2.0$ m, $y = -4.0$ m, $F = 10$ N, and $\theta = 150°$. Thus, we obtain $W = -37$ J. Note that the given mass value (2.0 kg) is not used in the computation.

45. (a) Estimating the initial speed from the slope of the graph near the origin is somewhat difficult, and it may be simpler to determine it from the constant-acceleration equations from chapter 2: $v = v_0 + at$ and $x = v_0 + \frac{1}{2}at^2$, where $x_0 = 0$ has been used. Applying these to the last point on the graph (where the slope is apparently zero) or applying just the x equation to any two points on the graph, leads to a pair of simultaneous equations from which $a = -2$ m/s^2 and $v_0 = 10$ m/s can be found. Then,

$$K_0 = \frac{1}{2}mv_0^2 = 2.5\times10^3 \text{ J} = 2.5 \text{ kJ} .$$

(b) The speed at $t = 3.0$ s is obtained by

$$v = v_0 + at = 10 + (-2)(3) = 4 \text{ m/s}$$

or by estimating the slope from the graph (not recommended). Then the work-kinetic energy theorem yields

$$W = \Delta K = \frac{1}{2}(50 \text{ kg})(4 \text{ m/s})^2 - 2.5\times10^3 \text{ J} = -2.1 \text{ kJ} .$$

46. (a) Using Eq. 7-8 and SI units, we find

$$W = \vec{F}\cdot\vec{d} = \left(2\hat{\imath} - 4\hat{\jmath}\right)\cdot\left(8\hat{\imath} + c\hat{\jmath}\right) = 16 - 4c$$

which, if equal zero, implies $c = 16/4 = 4$ m.

(b) If $W > 0$ then $16 > 4c$, which implies $c < 4$ m.

(c) If $W < 0$ then $16 < 4c$, which implies $c > 4$ m.

47. With speed $v = 11200$ m/s, we find

$$K = \frac{1}{2}mv^2 = \frac{1}{2}\left(2.9\times10^5\right)(11200)^2 = 1.8\times10^{13} \text{ J} .$$

48. (a) Hooke's law and the work done by a spring is discussed in the chapter. Taking absolute values, and writing that law in terms of differences ΔF and Δx, we analyze the first two pictures as follows:

$$\begin{aligned} |\Delta F| &= k\,|\Delta x| \\ 240\,\text{N} - 110\,\text{N} &= k(60\,\text{mm} - 40\,\text{mm}) \end{aligned}$$

which yields $k = 6.5$ N/mm. Designating the relaxed position (as read by that scale) as x_o we look again at the first picture:

$$110\,\text{N} = k\,(40\,\text{mm} - x_o)$$

which (upon using the above result for k) yields $x_o = 23$ mm.

(b) Using the results from part (a) to analyze that last picture, we find

$$W = k\,(30\,\text{mm} - x_o\,) = 45\ \text{N}\ .$$

49. (a) To hold the crate at equilibrium in the final situation, $\vec{F}$ must have the same magnitude as the horizontal component of the rope's tension $T\sin\theta$, where θ is the angle between the rope (in the final position) and vertical:

$$\theta = \sin^{-1}\left(\frac{4.00}{12.0}\right) = 19.5^\circ\ .$$

But the vertical component of the tension supports against the weight: $T\cos\theta = mg$. Thus, the tension is $T = (230)(9.8)/\cos 19.5^\circ = 2391\,\text{N}$ and $F = (2391)\sin 19.5^\circ = 797\,\text{N}$. An alternative approach based on drawing a vector triangle (of forces) in the final situation provides a quick solution.

(b) Since there is no change in kinetic energy, the net work on it is zero.

(c) The work done by gravity is $W_g = \vec{F}_g \cdot \vec{d} = -mgh$, where $h = L(1-\cos\theta)$ is the vertical component of the displacement. With $L = 12.0$ m, we obtain $W_g = -1547$ J which should be rounded to three figures: -1.55 kJ.

(d) The tension vector is everywhere perpendicular to the direction of motion, so its work is zero (since $\cos 90^\circ = 0$).

(e) The implication of the previous three parts is that the work due to $\vec{F}$ is $-W_g$ (so the net work turns out to be zero). Thus, $W_F = -W_g = 1.55$ kJ.

(f) Since $\vec{F}$ does not have constant magnitude, we cannot expect Eq. 7-8 to apply.

50. (a) In the work-kinetic energy theorem, we include both the work due to an applied force W_a and work done by gravity W_g in order to find the latter quantity.

$$\Delta K = W_a + W_g \quad\Longrightarrow\quad 30 = (100)(1.8)\cos 180^\circ + W_g$$

leading to $W_g = 210$ J.

(b) The value of W_g obtained in part (a) still applies since the weight and the path of the child remain the same, so $\Delta K = W_g = 210$ J.

51. Using Eq. 7-7, we have $W = Fd\cos\phi = 1504$ J. Then, by the work-kinetic energy theorem, we find the kinetic energy $K_f = K_i + W = 0 + 1504$ J. The answer is therefore 1.5 kJ.

52. (a) Before the cord is cut, each spring (which might be described as being "in series" in this case) is stretched by the force $F = 100$ N. Thus, each spring is stretched by $x = 100/500 = 0.20\,\text{m}$ in the initial configuration. Since the relaxed length of each spring is 0.50 m, then the full length of each spring in the initial configuration is $0.20 + 0.50 = 0.70$ m. Therefore (including that 0.10 m length of string) the distance from the box to the ceiling is $2(0.70) + 0.10 = 1.50$ m, before the string is cut. In the moments after the short string is cut, there is some "transient motion" that is difficult to analyze, but after it has settled down again (in its new equilibrium position) the springs (which now might be described as being "in parallel") are sharing half the weight, so the force stretching each one is $F/2 = 50$ N. This means the elongation of each is $x/2 = 0.10$ m. The total distance (recalling that the longer cords are each of length 0.85 m) of the box to the ceiling is now $0.85 + 0.10 + 0.50 = 1.45$ m. Thus, the box is closer to the ceiling now than it was before. It has moved up.

(b) The distance moved up by the box is $d = 1.50 - 1.45 = 0.05$ m.

(c) To avoid worrying about friction-related (dissipative) processes which are involved in making the "transient motion" ultimately disappear, we consider that the person who cut the cord (and has predicted the new equilibrium position) very carefully and gradually moves it up to that new position, in which case the work being done on the system is due to the person. In this variation of the problem, it is easy to see that the work done by the person *against gravity* is $-W_g = mgd = 5.0$ J (though this is not the full work done by the person, since Eq. 7-25 hasn't been used). Returning to the problem in its original form, we can say that the work done on the block in raising it the distance d is 5.0 J, regardless of the agent doing the work (and in its original form, that agent is the pair of springs, and this represents part of the full work they do).

53. (a) We set up the ratio

$$\frac{50\,\text{km}}{1\,\text{km}} = \left(\frac{E}{1\,\text{megaton}}\right)^{1/3}$$

and find $E = 50^3 \approx 1 \times 10^5$ megatons of TNT.

(b) We note that 15 kilotons is equivalent to 0.015 megatons. Dividing the result from part (a) by 0.013 yields about ten million bombs.

54. (a) With SI units (and three significant figures) understood, the object's displacement is

$$\vec{d} = \vec{d}_f - \vec{d}_i = -8\,\hat{\text{i}} + 6\,\hat{\text{j}} + 2\,\hat{\text{k}} \ .$$

Thus, Eq. 7-8 gives

$$W = \vec{F} \cdot \vec{d} = (3)(-8) + (7)(6) + (7)(2) = 32.0 \text{ J} \ .$$

(b) The average power is given by Eq. 7-42:

$$P_{\text{avg}} = \frac{W}{t} = \frac{32}{4} = 8.00 \text{ W} \ .$$

(c) The distance from the coordinate origin to the initial position is $d_i = \sqrt{3^2 + (-2)^2 + 5^2} = 6.16$ m, and the magnitude of the distance from the coordinate origin to the final position is $d_f = \sqrt{(-5)^2 + 4^2 + 7^2} = 9.49$ m. Their scalar (dot) product is

$$\vec{d}_i \cdot \vec{d}_f = (3)(-5) + (-2)(4) + (5)(7) = 12.0 \text{ m}^2 \ .$$

Thus, the angle between the two vectors is

$$\phi = \cos^{-1}\left(\frac{\vec{d}_i \cdot \vec{d}_f}{d_i\, d_f}\right) = \cos^{-1}\left(\frac{12.0}{(6.16)(9.49)}\right)$$

which yields $\phi = 78°$.

Chapter 8

1. The potential energy stored by the spring is given by $U = \frac{1}{2}kx^2$, where k is the spring constant and x is the displacement of the end of the spring from its position when the spring is in equilibrium. Thus

$$k = \frac{2U}{x^2} = \frac{2(25\,\mathrm{J})}{(0.075\,\mathrm{m})^2} = 8.9 \times 10^3\ \mathrm{N/m}\ .$$

2. (a) Noting that the vertical displacement is $10.0 - 1.5 = 8.5$ m downward (same direction as $\vec{F}_g$), Eq. 7-12 yields

$$W_g = mgd\cos\phi = (2.00)(9.8)(8.5)\cos 0° = 167\ \mathrm{J}\ .$$

(b) One approach (which is fairly trivial) is to use Eq. 8-1, but we feel it is instructive to instead calculate this as ΔU where $U = mgy$ (with upwards understood to be the $+y$ direction).

$$\Delta U = mgy_f - mgy_i = (2.00)(9.8)(1.5) - (2.00)(9.8)(10.0) = -167\ \mathrm{J}\ .$$

(c) In part (b) we used the fact that $U_i = mgy_i = 196$ J.

(d) In part (b), we also used the fact $U_f = mgy_f = 29$ J.

(e) The computation of W_g does not use the new information (that $U = 100$ J at the ground), so we again obtain $W_g = 167$ J.

(f) As a result of Eq. 8-1, we must again find $\Delta U = -W_g = -167$ J.

(g) With this new information (that $U_0 = 100$ J where $y = 0$) we have $U_i = mgy_i + U_0 = 296$ J.

(h) With this new information (that $U_0 = 100$ J where $y = 0$) we have $U_f = mgy_f + U_0 = 129$ J. We can check part (f) by subtracting the new U_i from this result.

3. (a) The force of gravity is constant, so the work it does is given by $W = \vec{F} \cdot \vec{d}$, where $\vec{F}$ is the force and $\vec{d}$ is the displacement. The force is vertically downward and has magnitude mg, where m is the mass of the flake, so this reduces to $W = mgh$, where h is the height from which the flake falls. This is equal to the radius r of the bowl. Thus

$$W = mgr = (2.00 \times 10^{-3}\,\mathrm{kg})(9.8\,\mathrm{m/s^2})(22.0 \times 10^{-2}\,\mathrm{m}) = 4.31 \times 10^{-3}\ \mathrm{J}\ .$$

(b) The force of gravity is conservative, so the change in gravitational potential energy of the flake-Earth system is the negative of the work done: $\Delta U = -W = -4.31 \times 10^{-3}$ J.

(c) The potential energy when the flake is at the top is greater than when it is at the bottom by $|\Delta U|$. If $U = 0$ at the bottom, then $U = +4.31 \times 10^{-3}$ J at the top.

(d) If $U = 0$ at the top, then $U = -4.31 \times 10^{-3}$ J at the bottom.

(e) All the answers are proportional to the mass of the flake. If the mass is doubled, all answers are doubled.

4. We use Eq. 7-12 for W_g and Eq. 8-9 for U.

(a) The displacement between the initial point and A is horizontal, so $\phi = 90°$ and $W_g = 0$ (since $\cos 90° = 0$).

(b) The displacement between the initial point and B has a vertical component of $h/2$ downward (same direction as $\vec{F}_g$), so we obtain $W_g = \vec{F}_g \cdot \vec{d} = mgh/2$.

(c) The displacement between the initial point and C has a vertical component of h downward (same direction as $\vec{F}_g$), so we obtain $W_g = \vec{F}_g \cdot \vec{d} = mgh$.

(d) With the reference position at C, we obtain $U_B = mgh/2$.

(e) Similarly, we find $U_A = mgh$.

(f) All the answers are proportional to the mass of the object. If the mass is doubled, all answers are doubled.

5. (a) The only force that does work on the ball is the force of gravity; the force of the rod is perpendicular to the path of the ball and so does no work. In going from its initial position to the lowest point on its path, the ball moves vertically through a distance equal to the length L of the rod, so the work done by the force of gravity is $W = mgL$.

(b) In going from its initial position to the highest point on its path, the ball moves vertically through a distance equal to L, but this time the displacement is upward, opposite the direction of the force of gravity. The work done by the force of gravity is $W = -mgL$.

(c) The final position of the ball is at the same height as its initial position. The displacement is horizontal, perpendicular to the force of gravity. The force of gravity does no work during this displacement.

(d) The force of gravity is conservative. The change in the gravitational potential energy of the ball-Earth system is the negative of the work done by gravity: $\Delta U = -mgL$ as the ball goes to the lowest point

(e) Continuing this line of reasoning, we find $\Delta U = mgL$ as it goes to the highest point.

(f) Continuing this line of reasoning, we have $\Delta U = 0$ as it goes to the point at the same height.

(g) The change in the gravitational potential energy depends only on the initial and final positions of the ball, not on its speed anywhere. The change in the potential energy is the *same* since the initial and final positions are the same.

6. We use Eq. 7-12 for W_g and Eq. 8-9 for U.

(a) The displacement between the initial point and Q has a vertical component of $h - R$ downward (same direction as $\vec{F}_g$), so (with $h = 5R$) we obtain $W_g = \vec{F}_g \cdot \vec{d} = 4mgR$.

(b) The displacement between the initial point and the top of the loop has a vertical component of $h - 2R$ downward (same direction as $\vec{F}_g$), so (with $h = 5R$) we obtain $W_g = \vec{F}_g \cdot \vec{d} = 3mgR$.

(c) With $y = h = 5R$, we find $U = 5mgR$ at P.

(d) With $y = R$, we find $U = mgR$ at Q.

(e) With $y = 2R$, we find $U = 2mgR$ at the top of the loop.

(f) The new information ($v_i \neq 0$) is not involved in any of the preceding computations; the above results are unchanged.

7. (a) The force of gravity is constant, so the work it does is given by $W = \vec{F} \cdot \vec{d}$, where $\vec{F}$ is the force and $\vec{d}$ is the displacement. The force is vertically downward and has magnitude mg, where m is the mass of the snowball. The expression for the work reduces to $W = mgh$, where h is the height through which the snowball drops. Thus

$$W = mgh = (1.50\,\text{kg})(9.8\,\text{m/s}^2)(12.5\,\text{m}) = 184\ \text{J}\ .$$

(b) The force of gravity is conservative, so the change in the potential energy of the snowball-Earth system is the negative of the work it does: $\Delta U = -W = -184\,\text{J}$.

(c) The potential energy when it reaches the ground is less than the potential energy when it is fired by $|\Delta U|$, so $U = -184\,\text{J}$ when the snowball hits the ground.

8. The main challenge for students in this type of problem seems to be working out the trigonometry in order to obtain the height of the ball (relative to the low point of the swing) $h = L - L\cos\theta$ (for angle θ measured from vertical as shown in Fig. 8-29). Once this relation (which we will not derive here since we have found this to be most easily illustrated at the blackboard) is established, then the principal results of this problem follow from Eq. 7-12 (for W_g) and Eq. 8-9 (for U).

(a) The vertical component of the displacement vector is downward with magnitude h, so we obtain

$$W_g = \vec{F}_g \cdot \vec{d} = mgh = mgL\,(1 - \cos\theta) \ .$$

(b) From Eq. 8-1, we have $\Delta U = -W_g = -mgL(1 - \cos\theta)$.

(c) With $y = h$, Eq. 8-9 yields $U = mgL(1 - \cos\theta)$.

(d) As the angle increases, we intuitively see that the height h increases (and, less obviously, from the mathematics, we see that $\cos\theta$ decreases so that $1 - \cos\theta$ increases), so the answers to parts (a) and (c) increase, and the absolute value of the answer to part (b) also increases.

9. (a) If K_i is the kinetic energy of the flake at the edge of the bowl, K_f is its kinetic energy at the bottom, U_i is the gravitational potential energy of the flake-Earth system with the flake at the top, and U_f is the gravitational potential energy with it at the bottom, then $K_f + U_f = K_i + U_i$. Taking the potential energy to be zero at the bottom of the bowl, then the potential energy at the top is $U_i = mgr$ where $r = 0.220$ m is the radius of the bowl and m is the mass of the flake. $K_i = 0$ since the flake starts from rest. Since the problem asks for the speed at the bottom, we write $\frac{1}{2}mv^2$ for K_f. Energy conservation leads to

$$mgr = \frac{1}{2}mv^2 \implies v = \sqrt{2gr} = \sqrt{2(9.8)(0.220)} = 2.08 \text{ m/s} \ .$$

(b) We note that the expression for the speed ($v = \sqrt{2gr}$) does not contain the mass of the flake. The speed would be the same, 2.08 m/s, regardless of the mass of the flake.

(c) The final kinetic energy is given by $K_f = K_i + U_i - U_f$. Since K_i is greater than before, K_f is greater. This means the final speed of the flake is greater.

10. We use Eq. 8-17, representing the conservation of mechanical energy (which neglects friction and other dissipative effects).

(a) In the solution to exercise 2 (to which this problem refers), we found $U_i = mgy_i = 196$ J and $U_f = mgy_f = 29$ J (assuming the reference position is at the ground). Since $K_i = 0$ in this case, we have

$$\begin{aligned} K_i + U_i &= K_f + U_f \\ 0 + 196 &= K_f + 29 \end{aligned}$$

which gives $K_f = 167$ J and thus leads to

$$v = \sqrt{\frac{2K_f}{m}} = \sqrt{\frac{2(167)}{2.00}} = 12.9 \text{ m/s} \ .$$

(b) If we proceed algebraically through the calculation in part (a), we find $K_f = -\Delta U = mgh$ where $h = y_i - y_f$ and is positive-valued. Thus,

$$v = \sqrt{\frac{2K_f}{m}} = \sqrt{2gh}$$

as we might also have derived from the equations of Table 2-1 (particularly Eq. 2-16). The fact that the answer is independent of mass means that the answer to part (b) is identical to that of part (a).

(c) If $K_i \neq 0$, then we find $K_f = mgh + K_i$ (where K_i is necessarily positive-valued). This represents a larger value for K_f than in the previous parts, and thus leads to a larger value for v.

11. We use Eq. 8-18, representing the conservation of mechanical energy (which neglects friction and other dissipative effects).

(a) In the solution to exercise 5 (to which this problem refers), we found $\Delta U = mgL$ as it goes to the highest point. Thus, we have

$$\begin{aligned} \Delta K + \Delta U &= 0 \\ K_{\text{top}} - K_0 + mgL &= 0 \end{aligned}$$

which, upon requiring $K_{\text{top}} = 0$, gives $K_0 = mgL$ and thus leads to

$$v_0 = \sqrt{\frac{2K_0}{m}} = \sqrt{2gL} \; .$$

(b) We also found in the solution to exercise 5 that the potential energy change is $\Delta U = -mgL$ in going from the initial point to the lowest point (the bottom). Thus,

$$\begin{aligned} \Delta K + \Delta U &= 0 \\ K_{\text{bottom}} - K_0 - mgL &= 0 \end{aligned}$$

which, with $K_0 = mgL$, leads to $K_{\text{bottom}} = 2mgL$. Therefore,

$$v_{\text{bottom}} = \sqrt{\frac{2K_{\text{bottom}}}{m}} = \sqrt{4gL}$$

which simplifies to $2\sqrt{gL}$.

(c) Since there is no change in height (going from initial point to the rightmost point), then $\Delta U = 0$, which implies $\Delta K = 0$. Consequently, the speed is the same as what it was initially ($\sqrt{2gL}$).

(d) It is evident from the above manipulations that the results do not depend on mass. Thus, a different mass for the ball must lead to the same results.

12. We use Eq. 8-17, representing the conservation of mechanical energy (which neglects friction and other dissipative effects).

(a) In the solution to exercise 4, we found $U_A = mgh$ (with the reference position at C). Referring again to Fig. 8-26, we see that this is the same as U_0 which implies that $K_A = K_0$ and thus that $v_A = v_0$.

(b) In the solution to exercise 4, we also found $U_B = mgh/2$. In this case, we have

$$\begin{aligned} K_0 + U_0 &= K_B + U_B \\ \frac{1}{2}mv_0^2 + mgh &= \frac{1}{2}mv_B^2 + mg\left(\frac{h}{2}\right) \end{aligned}$$

which leads to $v_B = \sqrt{v_0^2 + gh}$.

(c) Similarly, $v_C = \sqrt{v_0^2 + 2gh}$.

(d) To find the "final" height, we set $K_f = 0$. In this case, we have

$$\begin{aligned} K_0 + U_0 &= K_f + U_f \\ \frac{1}{2}mv_0^2 + mgh &= 0 + mgh_f \end{aligned}$$

which leads to $h_f = h + v_0^2/2g$.

(e) It is evident that the above results do not depend on mass. Thus, a different mass for the coaster must lead to the same results.

13. We neglect any work done by friction. We work with SI units, so the speed is converted: $v = 130(1000/3600) = 36.1$ m/s.

(a) We use Eq. 8-17: $K_f + U_f = K_i + U_i$ with $U_i = 0$, $U_f = mgh$ and $K_f = 0$. Since $K_i = \frac{1}{2}mv^2$, where v is the initial speed of the truck, we obtain

$$\frac{1}{2}mv^2 = mgh \implies h = \frac{v^2}{2g} = \frac{36.1^2}{2(9.8)} = 66.5 \text{ m} .$$

If L is the length of the ramp, then $L \sin 15° = 66.5$ m so that $L = 66.5/\sin 15° = 257$ m. Therefore, the ramp must be about 260 m long if friction is negligible.

(b) The answers do not depend on the mass of the truck. They remain the same if the mass is reduced.

(c) If the speed is decreased, h and L both decrease (note that h is proportional to the square of the speed and that L is proportional to h).

14. We use Eq. 8-18, representing the conservation of mechanical energy. The reference position for computing U is the lowest point of the swing; it is also regarded as the "final" position in our calculations.

(a) In the solution to problem 8 (to which this problem refers), we found $U = mgL(1 - \cos\theta)$ at the position shown in Fig. 8-29 (which we consider to be the initial position). Thus, we have

$$\begin{aligned} K_i + U_i &= K_f + U_f \\ 0 + mgL(1 - \cos\theta) &= \frac{1}{2}mv^2 + 0 \end{aligned}$$

which leads to

$$v = \sqrt{\frac{2mgL(1 - \cos\theta)}{m}} = \sqrt{2gL(1 - \cos\theta)} .$$

Plugging in $L = 2.00$ m and $\theta = 30.0°$ we find $v = 2.29$ m/s.

(b) It is evident that the result for v does not depend on mass. Thus, a different mass for the ball must not change the result.

15. We use Eq. 8-18, representing the conservation of mechanical energy. We choose the reference position for computing U to be at the ground below the cliff; it is also regarded as the "final" position in our calculations.

(a) Using Eq. 8-9, the initial potential energy is $U_i = mgh$ where $h = 12.5$ m and $m = 1.50$ kg. Thus, we have

$$\begin{aligned} K_i + U_i &= K_f + U_f \\ \frac{1}{2}mv_i^2 + mgh &= \frac{1}{2}mv^2 + 0 \end{aligned}$$

which leads to the speed of the snowball at the instant before striking the ground:

$$v = \sqrt{\frac{2}{m}\left(\frac{1}{2}mv_i^2 + mgh\right)} = \sqrt{v_i^2 + 2gh}$$

where $v_i = 14.0$ m/s is the magnitude of its initial velocity (not just one component of it). Thus we find $v = 21.0$ m/s.

(b) As noted above, v_i is the magnitude of its initial velocity and not just one component of it; therefore, there is no dependence on launch angle. The answer is again 21.0 m/s.

(c) It is evident that the result for v in part (a) does not depend on mass. Thus, changing the mass of the snowball does not change the result for v.

16. We convert to SI units and choose upward as the $+y$ direction. Also, the relaxed position of the top end of the spring is the origin, so the initial compression of the spring (defining an equilibrium situation between the spring force and the force of gravity) is $y_0 = -0.100$ m and the additional compression brings it to the position $y_1 = -0.400$ m.

(a) When the stone is in the equilibrium ($a = 0$) position, Newton's second law becomes

$$\begin{aligned} \vec{F}_{\text{net}} &= ma \\ F_{\text{spring}} - mg &= 0 \\ -k(-0.100) - (8.00)(9.8) &= 0 \end{aligned}$$

where Hooke's law (Eq. 7-21) has been used. This leads to a spring constant equal to $k = 784$ N/m.

(b) With the additional compression (and release) the acceleration is no longer zero, and the stone will start moving upwards, turning some of its elastic potential energy (stored in the spring) into kinetic energy. The amount of elastic potential energy at the moment of release is, using Eq. 8-11,

$$U = \frac{1}{2}ky_1^2 = \frac{1}{2}(784)(-0.400)^2 = 62.7 \text{ J} .$$

(c) Its maximum height y_2 is beyond the point that the stone separates from the spring (entering free-fall motion). As usual, it is characterized by having (momentarily) zero speed. If we choose the y_1 position as the reference position in computing the gravitational potential energy, then

$$\begin{aligned} K_1 + U_1 &= K_2 + U_2 \\ 0 + \frac{1}{2}ky_1^2 &= 0 + mgh \end{aligned}$$

where $h = y_2 - y_1$ is the height above the release point. Thus, mgh (the gravitational potential energy) is seen to be equal to the previous answer, 62.7 J, and we proceed with the solution in the next part.

(d) We find $h = ky_1^2/2mg = 0.800$ m, or 80.0 cm.

17. We take the reference point for gravitational potential energy at the position of the marble when the spring is compressed.

(a) The gravitational potential energy when the marble is at the top of its motion is $U_g = mgh$, where $h = 20$ m is the height of the highest point. Thus,

$$U_g = (5.0 \times 10^{-3}\,\text{kg})\left(9.8\,\text{m/s}^2\right)(20\,\text{m}) = 0.98 \text{ J} .$$

(b) Since the kinetic energy is zero at the release point and at the highest point, then conservation of mechanical energy implies $\Delta U_g + \Delta U_s = 0$, where ΔU_s is the change in the spring's elastic potential energy. Therefore, $\Delta U_s = -\Delta U_g = -0.98$ J.

(c) We take the spring potential energy to be zero when the spring is relaxed. Then, our result in the previous part implies that its initial potential energy is $U_s = 0.98$ J. This must be $\frac{1}{2}kx^2$, where k is the spring constant and x is the initial compression. Consequently,

$$k = \frac{2U_s}{x^2} = \frac{2(0.98\,\text{J})}{(0.080\,\text{m})^2} = 3.1 \times 10^2\ \text{N/m} = 3.1\ \text{N/cm}\ .$$

18. We use Eq. 8-18, representing the conservation of mechanical energy (which neglects friction and other dissipative effects). The reference position for computing U (and height h) is the lowest point of the swing; it is also regarded as the "final" position in our calculations.

(a) Careful examination of the figure leads to the trigonometric relation $h = L - L\cos\theta$ when the angle is measured from vertical as shown. Thus, the gravitational potential energy is $U = mgL(1 - \cos\theta_0)$ at the position shown in Fig. 8-32 (the initial position). Thus, we have

$$\begin{aligned} K_0 + U_0 &= K_f + U_f \\ \frac{1}{2}mv_0^2 + mgL(1 - \cos\theta_0) &= \frac{1}{2}mv^2 + 0 \end{aligned}$$

which leads to

$$v = \sqrt{\frac{2}{m}\left(\frac{1}{2}mv_0^2 + mgL(1 - \cos\theta_0)\right)} = \sqrt{v_0^2 + 2gL(1 - \cos\theta_0)}\ .$$

(b) We look for the initial speed required to barely reach the horizontal position – described by $v_h = 0$ and $\theta = 90°$ (or $\theta = -90°$, if one prefers, but since $\cos(-\phi) = \cos\phi$, the sign of the angle is not a concern).

$$\begin{aligned} K_0 + U_0 &= K_h + U_h \\ \frac{1}{2}mv_0^2 + mgL(1 - \cos\theta_0) &= 0 + mgL \end{aligned}$$

which leads to $v_0 = \sqrt{2gL\cos\theta_0}$.

(c) For the cord to remain straight, then the centripetal force (at the top) must be (at least) equal to gravitational force:

$$\frac{mv_t^2}{r} = mg \implies mv_t^2 = mgL$$

where we recognize that $r = L$. We plug this into the expression for the kinetic energy (at the top, where $\theta = 180°$).

$$\begin{aligned} K_0 + U_0 &= K_t + U_t \\ \frac{1}{2}mv_0^2 + mgL(1 - \cos\theta_0) &= \frac{1}{2}mv_t^2 + mg\,(1 - \cos 180°) \\ \frac{1}{2}mv_0^2 + mgL(1 - \cos\theta_0) &= \frac{1}{2}(mgL) + mg(2L) \end{aligned}$$

which leads to $v_0 = \sqrt{gL(3 + 2\cos\theta_0)}$.

(d) The more initial potential energy there is, the less initial kinetic energy there needs to be, in order to reach the positions described in parts (b) and (c). Increasing θ_0 amounts to increasing U_0, so we see that a greater value of θ_0 leads to smaller results for v_0 in parts (b) and (c).

19. The reference point for the gravitational potential energy U_g (and height h) is at the block when the spring is maximally compressed. When the block is moving to its highest point, it is first accelerated by the spring; later, it separates from the spring and finally reaches a point where its speed v_f is (momentarily) zero. The x axis is along the incline, pointing uphill (so x_0 for the initial compression is negative-valued); its origin is at the relaxed position of the spring. We use SI units, so $k = 1960$ N/m and $x_0 = -0.200$ m.

(a) The elastic potential energy is $\frac{1}{2}kx_0^2 = 39.2$ J.

(b) Since initially $U_g = 0$, the change in U_g is the same as its final value mgh where $m = 2.00$ kg. That this must equal the result in part (a) is made clear in the steps shown in the next part. Thus, $\Delta U_g = U_g = 39.2$ J.

(c) The principle of mechanical energy conservation leads to

$$\begin{aligned} K_0 + U_0 &= K_f + U_f \\ 0 + \frac{1}{2}kx_0^2 &= 0 + mgh \end{aligned}$$

which yields $h = 2.00$ m. The problem asks for the distance *along the incline*, so we have $d = h/\sin 30° = 4.00$ m.

20. (a) At Q the block (which is in circular motion at that point) experiences a centripetal acceleration v^2/R leftward. We find v^2 from energy conservation:

$$\begin{aligned} K_P + U_P &= K_Q + U_Q \\ 0 + mgh &= \frac{1}{2}mv^2 + mgR \end{aligned}$$

Using the fact (mentioned in problem 6) that $h = 5R$, we find $mv^2 = 8mgR$. Thus, the horizontal component of the net force on the block at Q is $mv^2/R = 8mg$ and points left (in the same direction as $\vec{a}$).

(b) The downward component of the net force on the block at Q is the force of gravity mg downward.

(c) To barely make the top of the loop, the centripetal force there must equal the force of gravity:

$$\frac{mv_t^2}{R} = mg \implies mv_t^2 = mgR$$

This requires a different value of h than was used above.

$$\begin{aligned} K_P + U_P &= K_t + U_t \\ 0 + mgh &= \frac{1}{2}mv_t^2 + mgh_t \\ mgh &= \frac{1}{2}(mgR) + mg(2R) \end{aligned}$$

Consequently, $h = 2.5R$.

(d) The normal force N, for speeds v_t greater than $\sqrt{gR}$ (which are the only possibilities for non-zero N – see the solution in the previous part), obeys

$$N = \frac{mv_t^2}{R} - mg$$

from Newton's second law. Since v_t^2 is related to h by energy conservation

$$K_P + U_P = K_t + U_t \implies gh = \frac{1}{2}v_t^2 + 2gR$$

then the normal force, as a function for h (so long as $h \geq 2.5R$ – see solution in previous part), becomes

$$N = \frac{2mg}{R}h - 5mg\ .$$

Thus, the graph for $h \geq 2.5R$ consists of a straight line of positive slope $2mg/R$ (which can be set to some convenient values for graphing purposes). For $h \leq 2.5R$, the normal force is zero. In the interest of saving space, we do not show the graph here.

21. We refer to its starting point as A, the point where it first comes into contact with the spring as B, and the point where the spring is compressed $|x| = 0.055\,\text{m}$ as C. Point C is our reference point for computing gravitational potential energy. Elastic potential energy (of the spring) is zero when the spring is relaxed. Information given in the second sentence allows us to compute the spring constant. From Hooke's law, we find

$$k = \frac{F}{x} = \frac{270\,\text{N}}{0.02\,\text{m}} = 1.35 \times 10^4\ \text{N/m} .$$

(a) The distance between points A and B is ℓ, and we note that the total sliding distance $\ell + |x|$ is related to the initial height h of the block (measured relative to C) by

$$\frac{h}{\ell + |x|} = \sin\theta$$

where the incline angle θ is 30°. Mechanical energy conservation leads to

$$\begin{aligned} K_A + U_A &= K_C + U_C \\ 0 + mgh &= 0 + \frac{1}{2}kx^2 \end{aligned}$$

which yields

$$h = \frac{kx^2}{2mg} = \frac{(1.35 \times 10^4\,\text{N/m})(0.055\,\text{m})^2}{2(12\,\text{kg})\left(9.8\,\text{m/s}^2\right)} = 0.174\ \text{m} .$$

Therefore,

$$\ell + |x| = \frac{h}{\sin 30°} = \frac{0.174\,\text{m}}{\sin 30°} = 0.35\ \text{m} .$$

(b) From this result, we find $\ell = 0.35 - 0.055 = 0.29$ m, which means $\Delta y = -\ell\sin\theta = -0.15$ m in sliding from point A to point B. Thus, Eq. 8-18 gives

$$\begin{aligned} \Delta K + \Delta U &= 0 \\ \frac{1}{2}mv_B^2 + mg\Delta h &= 0 \end{aligned}$$

which yields

$$v_B = \sqrt{-2g\Delta h} = \sqrt{-(9.8)(-0.15)} = 1.7\ \text{m/s} .$$

22. Since time does not directly enter into the energy formulations, we return to Chapter 4 (or Table 2-1 in Chapter 2) to find the change of height during this $t = 6$ s flight.

$$\Delta y = v_{0_y}t - \frac{1}{2}gt^2$$

This leads to $\Delta y = -32$ m. Therefore $\Delta U = mg\Delta y = -318 \approx -320$ J.

23. (a) As the string reaches its lowest point, its original potential energy $U = mgL$ (measured relative to the lowest point) is converted into kinetic energy. Thus,

$$mgL = \frac{1}{2}mv^2 \implies v = \sqrt{2gL} .$$

With $L = 1.20$ m we obtain $v = 4.85$ m/s.

(b) In this case, the total mechanical energy is shared between kinetic $\frac{1}{2}mv_b^2$ and potential mgy_b. We note that $y_b = 2r$ where $r = L - d = 0.450$ m. Energy conservation leads to

$$mgL = \frac{1}{2}mv_b^2 + mgy_b$$

which yields

$$v_b = \sqrt{2gL - 2g(2r)} = 2.42\ \text{m/s} .$$

24. From Chapter 4, we know the height h of the skier's jump can be found from $v_y^2 = 0 = v_{0\,y}^2 - 2gh$ where $v_{0\,y} = v_0 \sin 28°$ is the upward component of the skier's "launch velocity." To find v_0 we use energy conservation.

    (a) The skier starts at rest $y = 20$ m above the point of "launch" so energy conservation leads to

    $$mgy = \frac{1}{2}mv^2 \implies v = \sqrt{2gy} = 20 \text{ m/s}$$

    which becomes the initial speed v_0 for the launch. Hence, the above equation relating h to v_0 yields

    $$h = \frac{(v_0 \sin 28°)^2}{2g} = 4.4 \text{ m} .$$

    (b) We see that all reference to mass cancels from the above computations, so a new value for the mass will yield the same result as before.

25. We denote m as the mass of the block, $h = 0.40$ m as the height from which it dropped (measured from the relaxed position of the spring), and x the compression of the spring (measured downward so that it yields a positive value). Our reference point for the gravitational potential energy is the initial position of the block. The block drops a total distance $h + x$, and the final gravitational potential energy is $-mg(h + x)$. The spring potential energy is $\frac{1}{2}kx^2$ in the final situation, and the kinetic energy is zero both at the beginning and end. Since energy is conserved

    $$\begin{aligned} K_i + U_i &= K_f + U_f \\ 0 &= -mg(h + x) + \frac{1}{2}kx^2 \end{aligned}$$

    which is a second degree equation in x. Using the quadratic formula, its solution is

    $$x = \frac{mg \pm \sqrt{(mg)^2 + 2mghk}}{k} .$$

    Now $mg = 19.6$ N, $h = 0.40$ m, and $k = 1960$ N/m, and we choose the positive root so that $x > 0$.

    $$x = \frac{19.6 + \sqrt{19.6^2 + 2(19.6)(0.40)(1960)}}{1960} = 0.10 \text{ m} .$$

26. To find out whether or not the vine breaks, it is sufficient to examine it at the moment Tarzan swings through the lowest point, which is when the vine – if it didn't break – would have the greatest tension. Choosing upward positive, Newton'second law leads to

    $$T - mg = m\frac{v^2}{r}$$

    where $r = 18$ m and $m = W/g = 688/9.8 = 70.2$ kg. We find the v^2 from energy conservation (where the reference position for the potential energy is at the lowest point).

    $$mgh = \frac{1}{2}mv^2 \implies v^2 = 2gh$$

    where $h = 3.2$ m. Combining these results, we have

    $$T = mg + m\frac{2gh}{r} = mg\left(1 + \frac{2h}{r}\right)$$

    which yields 933 N. Thus, the vine does not break. And rounding to an appropriate number of significant figures, we see the maximum tension is roughly 930 N.

27. The distance the marble travels is determined by its initial speed (and the methods of Chapter 4), and the initial speed is determined (using energy conservation) by the original compression of the spring. We denote h as the height of the table, and x as the horizontal distance to the point where the marble lands. Then $x = v_0 t$ and $h = \frac{1}{2}gt^2$ (since the vertical component of the marble's "launch velocity" is zero). From these we find $x = v_0\sqrt{2h/g}$. We note from this that the distance to the landing point is directly proportional to the initial speed. We denote $v_{0\,1}$ be the initial speed of the first shot and $x_1 = 1.93$ m be the horizontal distance to its landing point; similarly, $v_{0\,2}$ is the initial speed of the second shot and $x_2 = 2.20$ m is the horizontal distance to its landing spot. Then

$$\frac{v_{0\,2}}{v_{0\,1}} = \frac{x_2}{x_1} \implies v_{0\,2} = \frac{x_2}{x_1} v_{0\,1} \ .$$

When the spring is compressed an amount ℓ, the elastic potential energy is $\frac{1}{2}k\ell^2$. When the marble leaves the spring its kinetic energy is $\frac{1}{2}mv_0^2$. Mechanical energy is conserved: $\frac{1}{2}mv_0^2 = \frac{1}{2}k\ell^2$, and we see that the initial speed of the marble is directly proportional to the original compression of the spring. If ℓ_1 is the compression for the first shot and ℓ_2 is the compression for the second, then $v_{02} = (\ell_2/\ell_1)v_{01}$. Relating this to the previous result, we obtain

$$\ell_2 = \frac{x_2}{x_1}\ell_1 = \left(\frac{2.20\,\text{m}}{1.93\,\text{m}}\right)(1.10\,\text{cm}) = 1.25\ \text{cm} \ .$$

28. We place the reference position for evaluating gravitational potential energy at the relaxed position of the spring. We use x for the spring's compression, measured positively downwards (so $x > 0$ means it is compressed).

(a) With $x = 0.190$ m, Eq. 7-26 gives $W_s = -\frac{1}{2}kx^2 = -7.22$ J for the work done by the spring force. Using Newton's third law, we conclude the work done on the spring is 7.22 J.

(b) As noted above, $W_s = -7.22$ J.

(c) Energy conservation leads to

$$\begin{aligned} K_i + U_i &= K_f + U_f \\ mgh_0 &= -mgx + \frac{1}{2}kx^2 \end{aligned}$$

which (with $m = 0.700$ kg) yields $h_0 = 0.862$ m.

(d) With a new value for the height $h_0' = 2h_0 = 1.72$ m, we solve for a new value of x using the quadratic formula (taking its positive root so that $x > 0$).

$$mgh_0' = -mgx + \frac{1}{2}kx^2 \implies x = \frac{mg + \sqrt{(mg)^2 + 2mgkh_0'}}{k}$$

which yields $x = 0.261$ m.

29. We use conservation of mechanical energy: the mechanical energy must be the same at the top of the swing as it is initially. Newton's second law is used to find the speed, and hence the kinetic energy, at the top. There the tension force T of the string and the force of gravity are both downward, toward the center of the circle. We notice that the radius of the circle is $r = L - d$, so the law can be written $T + mg = mv^2/(L - d)$, where v is the speed and m is the mass of the ball. When the ball passes the highest point with the least possible speed, the tension is zero. Then

$$mg = m\frac{v^2}{L - d} \implies v = \sqrt{g(L - d)} \ .$$

We take the gravitational potential energy of the ball-Earth system to be zero when the ball is at the bottom of its swing. Then the initial potential energy is mgL. The initial kinetic energy is zero since the ball starts from rest. The final potential energy, at the top of the swing, is $2mg(L-d)$ and the final kinetic energy is $\frac{1}{2}mv^2 = \frac{1}{2}mg(L-d)$ using the above result for v. Conservation of energy yields

$$mgL = 2mg(L-d) + \frac{1}{2}mg(L-d) \implies d = 3L/5 \; .$$

If d is greater than this value, so the highest point is lower, then the speed of the ball is greater as it reaches that point and the ball passes the point. If d is less, the ball cannot go around. Thus the value we found for d is a lower limit.

30. The connection between angle θ (measured from vertical – see Fig. 8-29) and height h (measured from the lowest point, which is our choice of reference position in computing the gravitational potential energy) is given by $h = L(1 - \cos\theta)$ where L is the length of the pendulum.

(a) We use energy conservation in the form of Eq. 8-17.

$$\begin{aligned} K_1 + U_1 &= K_2 + U_2 \\ 0 + mgL\,(1-\cos\theta_1) &= \frac{1}{2}mv_2^2 + mgL\,(1-\cos\theta_2) \end{aligned}$$

This leads to

$$v_2 = \sqrt{2gL\,(\cos\theta_2 - \cos\theta_1)} = 1.4 \text{ m/s}$$

since $L = 1.4$ m, $\theta_1 = 30°$, and $\theta_2 = 20°$.

(b) The maximum speed v_3 is at the lowest point. Our formula for h gives $h_3 = 0$ when $\theta_3 = 0°$, as expected.

$$\begin{aligned} K_1 + U_1 &= K_3 + U_3 \\ 0 + mgL\,(1-\cos\theta_1) &= \frac{1}{2}mv_3^2 + 0 \end{aligned}$$

This yields $v_3 = 1.9$ m/s.

(c) We look for an angle θ_4 such that the speed there is $v_4 = v_3/3$. To be as accurate as possible, we proceed algebraically (substituting $v_3^2 = 2gL\,(1-\cos\theta_1)$ at the appropriate place) and plug numbers in at the end. Energy conservation leads to

$$\begin{aligned} K_1 + U_1 &= K_4 + U_4 \\ 0 + mgL\,(1-\cos\theta_1) &= \frac{1}{2}mv_4^2 + mgL\,(1-\cos\theta_4) \\ mgL\,(1-\cos\theta_1) &= \frac{1}{2}m\frac{v_3^2}{9} + mgL\,(1-\cos\theta_4) \\ -gL\cos\theta_1 &= \frac{1}{2}\frac{2gL\,(1-\cos\theta_1)}{9} - gL\cos\theta_4 \end{aligned}$$

where in the last step we have subtracted out mgL and then divided by m. Thus, we obtain

$$\theta_4 = \cos^{-1}\left(\frac{1}{9} + \frac{8}{9}\cos\theta_1\right) = 28.2°$$

where we have quoted the answer to three significant figures since the problem gives θ_1 to three figures.

31. The connection between angle θ (measured from vertical – see Fig. 8-29) and height h (measured from the lowest point, which is our choice of reference position in computing the gravitational potential energy mgh) is given by $h = L(1-\cos\theta)$ where L is the length of the pendulum.

(a) Using this formula (or simply using intuition) we see the initial height is $h_1 = 2L$, and of course $h_2 = 0$. We use energy conservation in the form of Eq. 8-17.

$$\begin{aligned} K_1 + U_1 &= K_2 + U_2 \\ 0 + mg\,(2L) &= \frac{1}{2}mv^2 + 0 \end{aligned}$$

This leads to $v = 2\sqrt{gL}$.

(b) The ball is in circular motion with the center of the circle above it, so $\vec{a} = v^2/r$ upward, where $r = L$. Newton's second law leads to

$$T - mg = m\,\frac{v^2}{r} \implies T = m\left(g + \frac{4gL}{L}\right) = 5mg\ .$$

(c) The pendulum is now started (with zero speed) at $\theta_i = 90°$ (that is, $h_i = L$), and we look for an angle θ such that $T = mg$. When the ball is moving through a point at angle θ, then Newton's second law applied to the axis along the rod yields

$$T - mg\cos\theta = m\,\frac{v^2}{r}$$

which (since $r = L$) implies $v^2 = gL(1 - \cos\theta)$ at the position we are looking for. Energy conservation leads to

$$\begin{aligned} K_i + U_i &= K + U \\ 0 + mgL &= \frac{1}{2}mv^2 + mgL\,(1 - \cos\theta) \\ gL &= \frac{1}{2}\,(gL(1 - \cos\theta)) + gL\,(1 - \cos\theta) \end{aligned}$$

where we have divided by mass in the last step. Simplifying, we obtain

$$\theta = \cos^{-1}\left(\frac{1}{3}\right) = 70.5°\ .$$

32. All heights h are measured from the lower end of the incline (which is our reference position for computing gravitational potential energy mgh). Our x axis is along the incline, with $+x$ being uphill (so spring compression corresponds to $x > 0$) and its origin being at the relaxed end of the spring. The 1.00 m distance indicated in Fig. 8-40 will be referred to as ℓ, and the 37.0° angle will be referred to as θ. Thus, the height that corresponds to the canister's initial position (with spring compressed amount $x = 0.200$ m) is given by $h_1 = (\ell + x)\sin\theta$.

(a) Energy conservation leads to

$$\begin{aligned} K_1 + U_1 &= K_2 + U_2 \\ 0 + mg(\ell + x)\sin\theta + \frac{1}{2}kx^2 &= \frac{1}{2}mv_2^2 + mg\ell\sin\theta \end{aligned}$$

which yields $v_2 = \sqrt{2gx\sin\theta + kx^2/m} = 2.40$ m/s using the data $m = 2.00$ kg and $k = 170$ N/m.

(b) In this case, energy conservation leads to

$$\begin{aligned} K_1 + U_1 &= K_3 + U_3 \\ 0 + mg(\ell + x)\sin\theta + \frac{1}{2}kx^2 &= \frac{1}{2}mv_3^2 + 0 \end{aligned}$$

which yields $v_3 = \sqrt{2g(\ell + x)\sin\theta + kx^2/m} = 4.19$ m/s.

33. The work required is the change in the gravitational potential energy as a result of the chain being pulled onto the table. Dividing the hanging chain into a large number of infinitesimal segments, each of length dy, we note that the mass of a segment is $(m/L)\,dy$ and the change in potential energy of a segment when it is a distance $|y|$ below the table top is $dU = (m/L)g|y|\,dy = -(m/L)gy\,dy$ since y is negative-valued (we have $+y$ upward and the origin is at the tabletop). The total potential energy change is

$$\Delta U = -\frac{mg}{L}\int_{-L/4}^{0} y\,dy = \frac{1}{2}\frac{mg}{L}(L/4)^2 = mgL/32\ .$$

The work required to pull the chain onto the table is therefore $W = \Delta U = mgL/32$.

34. If the description of the scenario seems confusing, reference to Figure 8-31 in the textbook is helpful. We note that the block being unattached means that for $y > 0.25$ m, the elastic potential energy vanishes. With $k = 400$ N/m, $m = 40.0/9.8 = 4.08$ kg and length in meters, the energy equation is

$$E = \begin{cases} \frac{1}{2}k\left(\frac{1}{4}\right)^2 & y = 0 \\ K + mgy + \frac{1}{2}k\left(\frac{1}{4} - y\right)^2 & 0 \le y \le \frac{1}{4} \\ K + mgy & \frac{1}{4} \le y \end{cases}$$

In this way, the kinetic energy K for each region is related to E – which by conservation of energy is always equal to the value 12.5 J that it had at $y = 0$. We arrange our results in a table (with energies in Joules) where it is clear that the sum of each column (of energies) is 12.5 J:

part	(a)	(b)	(c)	(d)	(e)	(f)	(g)
position y	0	0.05	0.10	0.15	0.20	0.25	0.30
U_g	0	2.0	4.0	6.0	8.0	10.0	12.0
U_e	12.5	8.0	4.5	2.0	0.5	0	0
K	0	2.5	4.0	4.5	4.0	2.5	0.5

Finally (for part (h)), where $y \ge 0.25$ m, we have $K = E - mgy$, so that $K = 0$ occurs when $y = (12.5\text{ J})/(40\text{ N}) = 0.313$ m.

35. The free-body diagram for the boy is shown below. $\vec{N}$ is the normal force of the ice on him and m is his mass. The net inward force is $mg\cos\theta - N$ and, according to Newton's second law, this must be equal to mv^2/R, where v is the speed of the boy. At the point where the boy leaves the ice $N = 0$, so $g\cos\theta = v^2/R$. We wish to find his speed. If the gravitational potential energy is taken to be zero when he is at the top of the ice mound, then his potential energy at the time shown is $U = -mgR(1 - \cos\theta)$. He starts from rest and his kinetic energy at the time shown is $\frac{1}{2}mv^2$. Thus conservation of energy gives $0 = \frac{1}{2}mv^2 - mgR(1 - \cos\theta)$, or $v^2 = 2gR(1 - \cos\theta)$. We substitute this expression into the equation developed from the second law to obtain $g\cos\theta = 2g(1 - \cos\theta)$. This gives $\cos\theta = 2/3$. The height of the boy above the bottom of the mound is $R\cos\theta = 2R/3$.

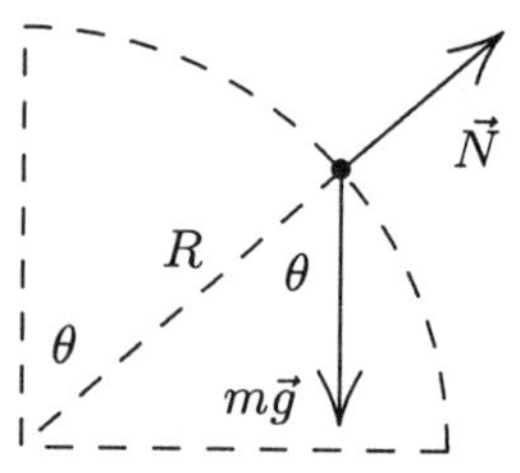

36. We use Eq. 8-20 and various observations made in §8-5.

(a) The force at $x = 2.0$ m is

$$F = -\frac{dU}{dx} \approx -\frac{(-17.5) - (-2.8)}{4.0 - 1.0} = 4.9\text{ N}$$

in the $+x$ direction (but there is some uncertainty in reading the graph which makes the last digit not very significant).

(b) The total mechanical energy at $x = 2.0$ m is

$$E = \frac{1}{2}mv^2 + U \approx \frac{1}{2}(2.0)(-1.5)^2 - 7.7 = -5.5$$

in SI units (Joules). Again, there is some uncertainty in reading the graph which makes the last digit not very significant. At that level (−5.5 J) on the graph, we find two points where the potential energy curve has that value – at $x \approx 1.5$ m and $x \approx 13.5$ m. Therefore, the particle remains in the region $1.5 < x < 13.5$ m.

(c) At $x = 7.0$ m, we read $U \approx -17.5$ J. Thus, if its total energy (calculated in the previous part) is $E \approx -5.5$ J, then we find

$$\frac{1}{2}mv^2 = E - U \approx 12\,\mathrm{J} \implies v = \sqrt{\frac{2}{m}(E-U)} \approx 3.5\ \mathrm{m/s}$$

where there is certainly room for disagreement on that last digit for the reasons cited above.

37. We use Eq. 8-20 and various observations made in §8-5.

(a) The force at the equilibrium position $r = r_{\rm eq}$ is

$$\begin{aligned} F = -\left.\frac{dU}{dr}\right|_{r=r_{\rm eq}} &= 0 \\ -\frac{12A}{r_{\rm eq}^{13}} + \frac{6B}{r_{\rm eq}^{7}} &= 0 \end{aligned}$$

which leads to the result

$$r_{\rm eq} = \left(\frac{2A}{B}\right)^{\frac{1}{6}} = 1.12\left(\frac{A}{B}\right)^{\frac{1}{6}} .$$

(b) This defines a minimum in the potential energy curve (as can be verified either by a graph or by taking another derivative and verifying that it is concave upward at this point), which means that for values of r slightly smaller than $r_{\rm eq}$ the slope of the curve is negative (so the force is positive, repulsive).

(c) And for values of r slightly larger than $r_{\rm eq}$ the slope of the curve must be positive (so the force is negative, attractive).

38. (a) The energy at $x = 5.0$ m is $E = K + U = 2.0 - 5.7 = -3.7$ J.

(b) A plot of the potential energy curve (SI units understood) and the energy E (the horizontal line) is shown for $0 \le x \le 10$ m.

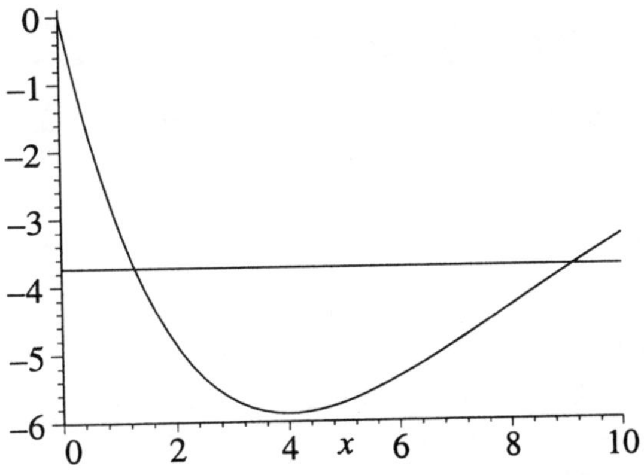

(c) The problem asks for a graphical determination of the turning points, which are the points on the curve corresponding to the total energy computed in part (a). The result for the smallest turning point (determined, to be honest, by more careful means) is $x = 1.29$ m

(d) And the result for the largest turning point is $x = 9.12$ m.

(e) Since $K = E - U$, then maximizing K involves finding the minimum of U. A graphical determination suggests that this occurs at $x = 4.0$ m, which plugs into the expression $E - U = -3.7 - \left(-4xe^{-x/4}\right)$ to give $K = 2.16$ J. Alternatively, one can measure from the graph from the minimum of the U curve up to the level representing the total energy E and thereby obtain an estimate of K at that point.

(f) As mentioned in the previous part, the minimum of the U curve occurs at $x = 4.0$ m.

(g) The force (understood to be in Newtons) follows from the potential energy, using Eq. 8-20 (and Appendix E if students are unfamiliar with such derivatives).

$$F = \frac{dU}{dx} = (4 - x)\, e^{-x/4}$$

(h) This revisits the considerations of parts (d) and (e) (since we are returning to the minimum of $U(x)$) – but now with the advantage of having the analytic result of part (g). We see that the location which produces $F = 0$ is exactly $x = 4$ m.

39. (a) Using Eq. 7-8, we have

$$W_{\text{applied}} = (8.0\,\text{N})(0.70\,\text{m}) = 5.6\ \text{J}\ .$$

(b) Using Eq. 8-29, the thermal energy generated is

$$\Delta E_{\text{th}} = f_k d = (5.0\,\text{N})(0.70\,\text{m}) = 3.5\ \text{J}\ .$$

40. Since the speed is constant $\Delta K = 0$ and Eq. 8-31 (an application of the energy conservation concept) implies

$$W_{\text{applied}} = \Delta E_{\text{th}} = \Delta E_{\text{th (cube)}} + \Delta E_{\text{th (floor)}}\ .$$

Thus, if $W_{\text{applied}} = (15)(3.0) = 45$ J, and we are told that $\Delta E_{\text{th (cube)}} = 20$ J, then we conclude that $\Delta E_{\text{th (floor)}} = 25$ J.

41. (a) The work done on the block by the force in the rope is, using Eq. 7-7,

$$W = Fd\cos\theta = (7.68\,\text{N})(4.06\,\text{m})\cos 15.0^\circ = 30.1\ \text{J}\ .$$

(b) Using f for the magnitude of the kinetic friction force, Eq. 8-29 reveals that the increase in thermal energy is

$$\Delta E_{\text{th}} = fd = (7.42\,\text{N})(4.06\,\text{m}) = 30.1\ \text{J}\ .$$

(c) We can use Newton's second law of motion to obtain the frictional and normal forces, then use $\mu_k = f/N$ to obtain the coefficient of friction. Place the x axis along the path of the block and the y axis normal to the floor. The x component of Newton's second law is $F\cos\theta - f = 0$ and the y component is $N + F\sin\theta - mg = 0$, where m is the mass of the block, F is the force exerted by the rope, and θ is the angle between that force and the horizontal. The first equation gives $f = F\cos\theta = (7.68)\cos 15.0^\circ = 7.42\,\text{N}$ and the second gives $N = mg - F\sin\theta = (3.57)(9.8) - (7.68)\sin 15.0^\circ = 33.0\,\text{N}$. Thus

$$\mu_k = \frac{f}{N} = \frac{7.42\,\text{N}}{33.0\,\text{N}} = 0.22\ .$$

42. Since the velocity is constant, $\vec{a} = 0$ and the horizontal component of the worker's push $F\cos\theta$ (where $\theta = 32°$) must equal the friction force magnitude $f_k = \mu_k N$. Also, the vertical forces must cancel, implying

$$N = mg + F\sin\theta \implies F\cos\theta = \mu_k\,(mg + F\sin\theta)$$

which is solved to find $F = 71$ N.

(a) The work done on the block by the worker is, using Eq. 7-7,

$$W = Fd\cos\theta = (71\,\text{N})(9.2\,\text{m})\cos 32° = 5.6 \times 10^2 \text{ J} .$$

(b) Since $f_k = \mu_k\,(mg + F\sin\theta)$, we find

$$\Delta E_{\text{th}} = f_k d = (60\,\text{N})(9.2\,\text{m}) = 5.6 \times 10^2 \text{ J} .$$

43. (a) We take the initial gravitational potential energy to be $U_i = 0$. Then the final gravitational potential energy is $U_f = -mgL$, where L is the length of the tree. The change is

$$U_f - U_i = -mgL = -(25\,\text{kg})\left(9.8\,\text{m/s}^2\right)(12\,\text{m}) = -2.9 \times 10^3 \text{ J} .$$

(b) The kinetic energy is

$$K = \frac{1}{2}mv^2 = \frac{1}{2}(25\,\text{kg})(5.6\,\text{m/s})^2 = 3.9 \times 10^2 \text{ J} .$$

(c) The changes in the mechanical and thermal energies must sum to zero. The change in thermal energy is $\Delta E_{\text{th}} = fL$, where f is the magnitude of the average frictional force; therefore,

$$f = -\frac{\Delta K + \Delta U}{L} = -\frac{3.92 \times 10^2\text{ J} - 2.94 \times 10^3\text{ J}}{12\,\text{m}} = 210 \text{ N} .$$

44. We use SI units so $m = 0.030$ kg and $d = 0.12$ m.

(a) Since there is no change in height (and we assume no changes in elastic potential energy), then $\Delta U = 0$ and we have

$$\Delta E_{\text{mech}} = \Delta K = -\frac{1}{2}mv_0^2 = -3.8 \times 10^3 \text{ J}$$

where $v_0 = 500$ m/s and the final speed is zero.

(b) By Eq. 8-31 (with $W = 0$) we have $\Delta E_{\text{th}} = 3.8 \times 10^3$ J, which implies

$$f = \frac{\Delta E_{\text{th}}}{d} = 3.1 \times 10^4 \text{ N}$$

using Eq. 8-29 with f_k replaced by f (effectively generalizing that equation to include a greater variety of dissipative forces than just those obeying Eq. 6-2).

45. Equation 8-31 provides $\Delta E_{\text{th}} = -\Delta E_{\text{mec}}$ for the energy "lost" in the sense of this problem. Thus,

$$\begin{aligned}\Delta E_{\text{th}} &= \frac{1}{2}m\left(v_i^2 - v_f^2\right) + mg\left(y_i - y_f\right)\\ &= \frac{1}{2}(60)(24^2 - 22^2) + (60)(9.8)(14)\\ &= 1.1 \times 10^4 \text{ J} .\end{aligned}$$

That the angle of 25° is nowhere used in this calculation is indicative of the fact that energy is a scalar quantity.

46. We use SI units so $m = 0.075$ kg. Equation 8-30 provides $\Delta E_{\text{th}} = -\Delta E_{\text{mec}}$ for the energy "lost" in the sense of this problem. Thus,

$$\begin{aligned}\Delta E_{\text{th}} &= \frac{1}{2}m\left(v_i^2 - v_f^2\right) + mg\left(y_i - y_f\right)\\ &= \frac{1}{2}(0.075)(12^2 - 10.5^2) + (0.075)(9.8)(1.1 - 2.1)\\ &= 0.53 \text{ J} .\end{aligned}$$

47. We work this using the English units (with $g = 32$ ft/s), but for consistency we convert the weight to pounds

$$mg = (9.0\,\text{oz})\left(\frac{1\,\text{lb}}{16\,\text{oz}}\right) = 0.56 \text{ lb}$$

which implies $m = 0.018$ lb·s^2/ft (which can be phrased as 0.018 slug as explained in Appendix D). And we convert the initial speed to feet-per-second

$$v_i = (81.8\,\text{mi/h})\left(\frac{5280\,\text{ft/mi}}{3600\,\text{s/h}}\right) = 120 \text{ ft/s}$$

or a more "direct" conversion from Appendix D can be used. Equation 8-30 provides $\Delta E_{\text{th}} = -\Delta E_{\text{mec}}$ for the energy "lost" in the sense of this problem. Thus,

$$\begin{aligned}\Delta E_{\text{th}} &= \frac{1}{2}m\left(v_i^2 - v_f^2\right) + mg\left(y_i - y_f\right)\\ &= \frac{1}{2}(0.018)(120^2 - 110^2) + 0\\ &= 20 \text{ ft} \cdot \text{lb} .\end{aligned}$$

48. (a) During one second, the decrease is potential energy is

$$-\Delta U = mg(-\Delta y) = (5.5 \times 10^6\,\text{kg})\left(9.8\,\text{m/s}^2\right)(50\,\text{m}) = 2.7 \times 10^9 \text{ J}$$

where $+y$ is upward and $\Delta y = y_f - y_i$.

(b) The information relating mass to volume is not needed in the computation. By Eq. 8-36 (and the SI relation W = J/s), the result follows: $(2.7 \times 10^9 \text{ J})/(1\,\text{s}) = 2.7 \times 10^9$ W.

(c) One year is equivalent to $24 \times 365.25 = 8766$ h which we write as 8.77 kh. Thus, the energy supply rate multiplied by the cost and by the time is

$$\left(2.7 \times 10^9\,\text{W}\right)(8.77\,\text{kh})\left(\frac{1\,\text{cent}}{1\,\text{kWh}}\right) = 2.4 \times 10^{10}\,\text{cents}$$

which equals $\$2.4 \times 10^8$.

49. (a) The initial potential energy is

$$U_i = mgy_i = (520\,\text{kg})\left(9.8\,\text{m/s}^2\right)(300\,\text{m}) = 1.53 \times 10^6 \text{ J}$$

where $+y$ is upward and $y = 0$ at the bottom (so that $U_f = 0$).

(b) Since $f_k = \mu_k N = \mu_k mg\cos\theta$ we have

$$\Delta E_{\text{th}} = f_k d = \mu_k mgd\cos\theta$$

from Eq. 8-29. Now, the hillside surface (of length $d = 500$ m) is treated as an hypotenuse of a 3-4-5 triangle, so $\cos\theta = x/d$ where $x = 400$ m. Therefore,

$$\Delta E_{\text{th}} = \mu_k mgd\frac{x}{d} = \mu_k mgx = (0.25)(520)(9.8)(400) = 5.1 \times 10^5 \text{ J} .$$

(c) Using Eq. 8-31 (with $W = 0$) we find

$$\begin{aligned} K_f &= K_i + U_i - U_f - \Delta E_{\text{th}} \\ &= 0 + 1.53 \times 10^6 - 0 - 5.1 \times 10^5 \\ &= 0 + 1.02 \times 10^6 \text{ J} . \end{aligned}$$

(d) From $K_f = \frac{1}{2}mv^2$ we obtain $v = 62.6$ m/s.

50. Energy conservation, as expressed by Eq. 8-31 (with $W = 0$) leads to

$$\begin{aligned} \Delta E_{\text{th}} &= K_i - K_f + U_i - U_f \\ f_k d &= 0 - 0 + \frac{1}{2}kx^2 - 0 \\ \mu_k mgd &= \frac{1}{2}(200\,\text{N/m})(0.15\,\text{m})^2 \\ \mu_k(2.0\,\text{kg})\left(9.8\,\text{m/s}^2\right)(0.75\,\text{m}) &= 2.25 \text{ J} \end{aligned}$$

which yields $\mu_k = 0.15$ as the coefficient of kinetic friction.

51. (a) The vertical forces acting on the block are the normal force, upward, and the force of gravity, downward. Since the vertical component of the block's acceleration is zero, Newton's second law requires $N = mg$, where m is the mass of the block. Thus $f = \mu_k N = \mu_k mg$. The increase in thermal energy is given by $\Delta E_{\text{th}} = fd = \mu_k mgd$, where d is the distance the block moves before coming to rest. Using Eq. 8-29, we have

$$\Delta E_{\text{th}} = (0.25)(3.5\,\text{kg})\left(9.8\,\text{m/s}^2\right)(7.8\,\text{m}) = 67 \text{ J} .$$

(b) The block has its maximum kinetic energy K_{max} just as it leaves the spring and enters the region where friction acts. Therefore, the maximum kinetic energy equals the thermal energy generated in bringing the block back to rest, 67 J.

(c) The energy that appears as kinetic energy is originally in the form of potential energy in the compressed spring. Thus $K_{\text{max}} = U_i = \frac{1}{2}kx^2$, where k is the spring constant and x is the compression. Thus,

$$x = \sqrt{\frac{2K_{\text{max}}}{k}} = \sqrt{\frac{2(67\,\text{J})}{640\,\text{N/m}}} = 0.46 \text{ m} .$$

52. We use Eq. 8-29

$$\Delta E_{\text{th}} = f_k d = (10\,\text{N})(5.0\,\text{m}) = 50 \text{ J}$$

and Eq. 7-8

$$W = Fd = (2.0\,\text{N})(5.0\,\text{m}) = 10 \text{ J}$$

and Eq. 8-31

$$\begin{aligned} W &= \Delta K + \Delta U + \Delta E_{\text{th}} \\ 10 &= 35 + \Delta U + 50 \end{aligned}$$

which yields $\Delta U = -75$ J. By Eq. 8-1, then, the work done by gravity is $W = -\Delta U = 75$ J.

53. (a) To stretch the spring an external force, equal in magnitude to the force of the spring but opposite to its direction, is applied. Since a spring stretched in the positive x direction exerts a force in the negative x direction, the applied force must be $F = 52.8x + 38.4x^2$, in the $+x$ direction. The work it does is

$$W = \int_{0.50}^{1.00} (52.8x + 38.4x^2)\,dx = \left[\frac{52.8}{2}x^2 + \frac{38.4}{3}x^3\right]_{0.50}^{1.00} = 31.0 \text{ J} .$$

(b) The spring does 31.0 J of work and this must be the increase in the kinetic energy of the particle. Its speed is then

$$v = \sqrt{\frac{2K}{m}} = \sqrt{\frac{2(31.0\,\mathrm{J})}{2.17\,\mathrm{kg}}} = 5.35\ \mathrm{m/s}\ .$$

(c) The force is conservative since the work it does as the particle goes from any point x_1 to any other point x_2 depends only on x_1 and x_2, not on details of the motion between x_1 and x_2.

54. We look for the distance along the incline d which is related to the height ascended by $\Delta h = d\sin\theta$. By a force analysis of the style done in Ch. 6, we find the normal force has magnitude $N = mg\cos\theta$ which means $f_k = \mu_k mg\cos\theta$. Thus, Eq. 8-31 (with $W = 0$) leads to

$$\begin{aligned} 0 &= K_f - K_i + \Delta U + \Delta E_{\text{th}} \\ &= 0 - K_i + mgd\sin\theta + \mu_k mgd\cos\theta \end{aligned}$$

which leads to

$$d = \frac{K_i}{mg(\sin\theta + \mu_k\cos\theta)} = \frac{128}{(4.0)(9.8)(\sin 30^\circ + 0.30\cos 30^\circ)} = 4.3\ \mathrm{m}\ .$$

55. (a) We take the gravitational potential energy of the skier-Earth system to be zero when the skier is at the bottom of the peaks. The initial potential energy is $U_i = mgh_i$, where m is the mass of the skier, and h_i is the height of the higher peak. The final potential energy is $U_f = mgh_f$, where h_f is the height of the lower peak. The skier initially has a kinetic energy of $K_i = 0$, and the final kinetic energy is $K_f = \frac{1}{2}mv^2$, where v is the speed of the skier at the top of the lower peak. The normal force of the slope on the skier does no work and friction is negligible, so mechanical energy is conserved.

$$\begin{aligned} U_i + K_i &= U_f + K_f \\ mgh_i &= mgh_f + \frac{1}{2}mv^2 \end{aligned}$$

Thus,

$$v = \sqrt{2g\,(h_i - h_f)} = \sqrt{2(9.8)(850 - 750)} = 44\ \mathrm{m/s}\ .$$

(b) We recall from analyzing objects sliding down inclined planes that the normal force of the slope on the skier is given by $N = mg\cos\theta$, where θ is the angle of the slope from the horizontal, 30° for each of the slopes shown. The magnitude of the force of friction is given by $f = \mu_k N = \mu_k mg\cos\theta$. The thermal energy generated by the force of friction is $fd = \mu_k mgd\cos\theta$, where d is the total distance along the path. Since the skier gets to the top of the lower peak with no kinetic energy, the increase in thermal energy is equal to the decrease in potential energy. That is, $\mu_k mgd\cos\theta = mg(h_i - h_f)$. Consequently,

$$\mu_k = \frac{(h_i - h_f)}{d\cos\theta} = \frac{(850 - 750)}{(3.2\times 10^3)\cos 30^\circ} = 0.036\ .$$

56. (a) By a force analysis in the style of Chapter 6, we find the normal force $N = mg\cos\theta$ (where $mg = 267$ N) which means $f_k = \mu_k mg\cos\theta$. Thus, Eq. 8-29 yields

$$\Delta E_{\text{th}} = f_k d = \mu_k mgd\cos\theta = (0.10)(267)(6.1)\cos 20^\circ = 1.5\times 10^2\ \mathrm{J}\ .$$

(b) The potential energy change is $\Delta U = mg(-d\sin\theta) = (267)(-6.1\sin 20^\circ) = -5.6\times 10^2$ J. The initial kinetic energy is

$$K_i = \frac{1}{2}mv_i^2 = \frac{1}{2}\left(\frac{267\,\mathrm{N}}{9.8\,\mathrm{m/s^2}}\right)(0.457\,\mathrm{m/s})^2 = 2.8\ \mathrm{J}\ .$$

Therefore, using Eq. 8-31 (with $W = 0$), the final kinetic energy is

$$K_f = K_i - \Delta U - \Delta E_{\text{th}} = 2.8 - \left(-5.6 \times 10^2\right) - 1.5 \times 10^2 = 4.1 \times 10^2 \text{ J} .$$

Consequently, the final speed is $v_f = \sqrt{2K_f/m} = 5.5$ m/s

57. (a) With $x = 0.075$ m and $k = 320$ N/m, Eq. 7-26 yields $W_s = -\frac{1}{2}kx^2 = -0.90$ J. For later reference, this is equal to the negative of ΔU.

(b) Analyzing forces, we find $N = mg$ which means $f_k = \mu_k mg$. With $d = x$, Eq. 8-29 yields

$$\Delta E_{\text{th}} = f_k d = \mu_k mgx = (0.25)(2.5)(9.8)(0.075) = 0.46 \text{ J} .$$

(c) Eq. 8-31 (with $W = 0$) indicates that the initial kinetic energy is

$$K_i = \Delta U + \Delta E_{\text{th}} = 0.90 + 0.46 = 1.36 \text{ J}$$

which leads to $v_i = \sqrt{2K_i/m} = 1.0$ m/s.

58. This can be worked entirely by the methods of Chapters 2-6, but we will use energy methods in as many steps as possible.

(a) By a force analysis in the style of Chapter 6, we find the normal force has magnitude $N = mg\cos\theta$ (where $\theta = 39°$) which means $f_k = \mu_k mg\cos\theta$ where $\mu_k = 0.28$. Thus, Eq. 8-29 yields $\Delta E_{\text{th}} = f_k d = \mu_k mgd\cos\theta$. Also, elementary trigonometry leads us to conclude that $\Delta U = -mgd\sin\theta$ where $d = 3.7$ m. Since $K_i = 0$, Eq. 8-31 (with $W = 0$) indicates that the final kinetic energy is

$$K_f = -\Delta U - \Delta E_{\text{th}} = mgd\,(\sin\theta - \mu_k\cos\theta)$$

which leads to the speed at the bottom of the ramp

$$v = \sqrt{\frac{2K_f}{m}} = \sqrt{2gd\,(\sin\theta - \mu_k\cos\theta)} = 5.5 \text{ m/s} .$$

(b) This speed begins its horizontal motion, where $f_k = \mu_k mg$ and $\Delta U = 0$. It slides a distance d' before it stops. According to Eq. 8-31 (with $W = 0$),

$$\begin{aligned} 0 &= \Delta K + \Delta U + \Delta E_{\text{th}} \\ &= 0 - \frac{1}{2}mv^2 + 0 + \mu_k mgd' \\ &= -\frac{1}{2}\left(2gd\,(\sin\theta - \mu_k\cos\theta)\right) + \mu_k gd' \end{aligned}$$

where we have divided by mass and substituted from part (a) in the last step. Therefore,

$$d' = \frac{d\,(\sin\theta - \mu_k\cos\theta)}{\mu_k} = 5.4 \text{ m} .$$

(c) We see from the algebraic form of the results, above, that the answers do not depend on mass. A 90 kg crate should have the same speed at the bottom and sliding distance across the floor, to the extent that the friction relations in Ch. 6 are accurate. Interestingly, since g does not appear in the relation for d', the sliding distance would seem to be the same if the experiment were performed on Mars!

59. Since the valley is frictionless, the only reason for the speed being less when it reaches the higher level is the gain in potential energy $\Delta U = mgh$ where $h = 1.1$ m. Sliding along the rough surface of the higher level, the block finally stops since its remaining kinetic energy has turned to thermal energy

$$\Delta E_{\text{th}} = f_k d = \mu mgd$$

where $\mu = 0.60$. Thus, Eq. 8-31 (with $W = 0$) provides us with an equation to solve for the distance d:

$$K_i = \Delta U + \Delta E_{\text{th}} = mg\,(h + \mu d)$$

where $K_i = \frac{1}{2}mv_i^2$ and $v_i = 6.0$ m/s. Dividing by mass and rearranging, we obtain

$$d = \frac{v_i^2}{2\mu g} - \frac{h}{\mu} = 1.2 \text{ m} .$$

60. This can be worked entirely by the methods of Chapters 2-6, but we will use energy methods in as many steps as possible.

(a) By a force analysis of the style done in Ch. 6, we find the normal force has magnitude $N = mg\cos\theta$ (where $\theta = 40°$) which means $f_k = \mu_k mg\cos\theta$ where $\mu_k = 0.15$. Thus, Eq. 8-29 yields $\Delta E_{\text{th}} = f_k d = \mu_k mgd\cos\theta$. Also, elementary trigonometry leads us to conclude that $\Delta U = mgd\sin\theta$. Eq. 8-31 (with $W = 0$ and $K_f = 0$) provides an equation for determining d:

$$\begin{aligned} K_i &= \Delta U + \Delta E_{\text{th}} \\ \frac{1}{2}mv_i^2 &= mgd\,(\sin\theta + \mu_k\cos\theta) \end{aligned}$$

where $v_i = 1.4$ m/s. Dividing by mass and rearranging, we obtain

$$d = \frac{v_i^2}{2g\,(\sin\theta + \mu_k\cos\theta)} = 0.13 \text{ m} .$$

(b) Now that we know where on the incline it stops ($d' = 0.13 + 0.55 = 0.68$ m from the bottom), we can use Eq. 8-31 again (with $W = 0$ and now with $K_i = 0$,) to describe the final kinetic energy (at the bottom):

$$\begin{aligned} K_f &= -\Delta U - \Delta E_{\text{th}} \\ \frac{1}{2}mv^2 &= mgd'\,(\sin\theta - \mu_k\cos\theta) \end{aligned}$$

which – after dividing by the mass and rearranging – yields

$$v = \sqrt{2gd'\,(\sin\theta - \mu_k\cos\theta)} = 2.7 \text{ m/s} .$$

(c) In part (a) it is clear that d increases if μ_k decreases – both mathematically (since it is a positive term in the denominator) and intuitively (less friction – less energy "lost"). In part (b), there are two terms in the expression for v which imply that it should increase if μ_k were smaller: the increased value of $d' = d_0 + d$ and that last factor $\sin\theta - \mu_k\cos\theta$ which indicates that less is being subtracted from $\sin\theta$ when μ_k is less (so the factor itself increases in value).

61. (a) The maximum height reached is h. The thermal energy generated by air resistance as the stone rises to this height is $\Delta E_{\text{th}} = fh$ by Eq. 8-29. We use energy conservation in the form of Eq. 8-31 (with $W = 0$):

$$K_f + U_f + \Delta E_{\text{th}} = K_i + U_i$$

and we take the potential energy to be zero at the throwing point (ground level). The initial kinetic energy is $K_i = \frac{1}{2}mv_0^2$, the initial potential energy is $U_i = 0$, the final kinetic energy is $K_f = 0$, and the final potential energy is $U_f = wh$. Thus $wh + fh = \frac{1}{2}mv_0^2$, and we solve for the height:

$$h = \frac{mv_0^2}{2(w+f)} = \frac{wv_0^2}{2g(w+f)} = \frac{v_0^2}{2g(1+f/w)} .$$

(b) We notice that the force of the air is downward on the trip up and upward on the trip down, since it is opposite to the direction of motion. Over the entire trip the increase in thermal energy is $\Delta E_{\text{th}} = 2fh$. The final kinetic energy is $K_f = \frac{1}{2}mv^2$, where v is the speed of the stone just before it hits the ground. The final potential energy is $U_f = 0$. Thus, using Eq. 8-31 (with $W = 0$), we find

$$\frac{1}{2}mv^2 + 2fh = \frac{1}{2}mv_0^2 .$$

We substitute the expression found for h to obtain

$$-\frac{2fv_0^2}{2g(1+f/w)} = \frac{1}{2}mv^2 - \frac{1}{2}mv_0^2$$

which leads to

$$v^2 = v_0^2 - \frac{2fv_0^2}{mg(1+f/w)} = v_0^2 - \frac{2fv_0^2}{w(1+f/w)} = v_0^2\left(1 - \frac{2f}{w+f}\right) = v_0^2\,\frac{w-f}{w+f}$$

where w was substituted for mg and some algebraic manipulations were carried out. Therefore,

$$v = v_0\sqrt{\frac{w-f}{w+f}} .$$

62. (a) The assumption is that the slope of the bottom of the slide is horizontal, like the ground. A useful analogy is that of the pendulum of length $R = 12$ m that is pulled leftward to an angle θ (corresponding to being at the top of the slide at height $h = 4.0$ m) and released so that the pendulum swings to the lowest point (zero height) gaining speed $v = 6.2$ m/s. Exactly as we would analyze the trigonometric relations in the pendulum problem, we find

$$h = R(1 - \cos\theta) \implies \theta = \cos^{-1}\left(1 - \frac{h}{R}\right) = 48°$$

or 0.84 radians. The slide, representing a circular arc of length $s = R\theta$, is therefore $(12)(0.84) = 10$ m long.

(b) To find the magnitude f of the frictional force, we use Eq. 8-31 (with $W = 0$):

$$\begin{aligned} 0 &= \Delta K + \Delta U + \Delta E_{\text{th}} \\ &= \frac{1}{2}mv^2 - mgh + fs \end{aligned}$$

so that (with $m = 25$ kg) we obtain $f = 49$ N.

(c) The assumption is no longer that the slope of the bottom of the slide is horizontal, but rather that the slope of the top of the slide is vertical (and 12 m to the left of the center of curvature). Returning to the pendulum analogy, this corresponds to releasing the pendulum from horizontal (at $\theta_1 = 90°$ measured from vertical) and taking a snapshot of its motion a few moments later when it is at angle θ_2 with speed $v = 6.2$ m/s. The difference in height between these two positions is (just as we would figure for the pendulum of length R)

$$\Delta h = R(1 - \cos\theta_2) - R(1 - \cos\theta_1) = -R\cos\theta_2$$

where we have used the fact that $\cos\theta_1 = 0$. Thus, with $\Delta h = -4.0$ m, we obtain $\theta_2 = 70.5°$ which means the arc subtends an angle of $|\Delta\theta| = 19.5°$ or 0.34 radians. Multiplying this by the radius gives a slide length of $s' = 4.1$ m.

(d) We again find the magnitude f' of the frictional force by using Eq. 8-31 (with $W = 0$):

$$\begin{aligned} 0 &= \Delta K + \Delta U + \Delta E_{\text{th}} \\ &= \frac{1}{2}mv^2 - mgh + f's' \end{aligned}$$

so that we obtain $f' = 1.2 \times 10^2$ N.

63. The initial and final kinetic energies are zero, and we set up energy conservation in the form of Eq. 8-31 (with $W = 0$) according to our assumptions. Certainly, it can only come to a permanent stop somewhere in the flat part, but the question is whether this occurs during its first pass through (going rightward) or its second pass through (going leftward) or its third pass through (going rightward again), and so on. If it occurs during its first pass through, then the thermal energy generated is $\Delta E_{\text{th}} = f_k d$ where $d \leq L$ and $f_k = \mu_k mg$. If it occurs during its second pass through, then the total thermal energy is $\Delta E_{\text{th}} = \mu_k mg(L + d)$ where we again use the symbol d for how far through the level area it goes during that last pass (so $0 \leq d \leq L$). Generalizing to the nth pass through, we see that $\Delta E_{\text{th}} = \mu_k mg((n-1)L + d)$. In this way, Eq. 8-39 leads to

$$mgh = \mu_k mg\left((n-1)L + d\right)$$

which simplifies (when h = L/2 is inserted) to

$$\frac{d}{L} = 1 + \frac{1}{2\mu_k} - n \; .$$

The first two terms give $1 + 1/2\mu_k = 3.5$, so that the requirement $0 \leq d/L \leq 1$ demands that $n = 3$. We arrive at the conclusion that $d/L = \frac{1}{2}$ and that this occurs on its third pass through the flat region.

64. We observe that the last line of the problem indicates that static friction is not to be considered a factor in this problem. The friction force of magnitude $f = 4400$ N mentioned in the problem is kinetic friction and (as mentioned) is constant (and directed upward), and the thermal energy change associated with it is $\Delta E_{\text{th}} = fd$ (Eq. 8-29) where $d = 3.7$ m in part (a) (but will be replaced by x, the spring compression, in part (b)).

(a) With $W = 0$ and the reference level for computing $U = mgy$ set at the top of the (relaxed) spring, Eq. 8-31 leads to

$$U_i = K + \Delta E_{\text{th}} \quad \Longrightarrow \quad v = \sqrt{2d\left(g - \frac{f}{m}\right)}$$

which yields $v = 7.4$ m/s for $m = 1800$ kg.

(b) We again utilize Eq. 8-31 (with $W = 0$), now relating its kinetic energy at the moment it makes contact with the spring to the system energy at the bottom-most point. Using the same reference level for computing $U = mgy$ as we did in part (a), we end up with gravitational potential energy equal to $mg(-x)$ at that bottom-most point, where the spring (with spring constant $k = 1.5 \times 10^5$ N/m) is fully compressed.

$$K = mg(-x) + \frac{1}{2}kx^2 + fx$$

where $K = \frac{1}{2}mv^2 = 4.9 \times 10^4$ J using the speed found in part (a). Using the abbreviation $\xi = mg - f = 1.3 \times 10^4$ N, the quadratic formula yields

$$x = \frac{\xi \pm \sqrt{\xi^2 + 2kK}}{k} = 0.90 \text{ m}$$

where we have taken the positive root.

(c) We relate the energy at the bottom-most point to that of the highest point of rebound (a distance d' above the relaxed position of the spring). We assume $d' > x$. We now use the bottom-most point as the reference level for computing gravitational potential energy.

$$\frac{1}{2}kx^2 = mgd' + fd' \quad \Longrightarrow \quad d' = \frac{kx^2}{2(mg + d)} = 2.8 \text{ m} \; .$$

(d) The non-conservative force (§8-1) is friction, and the energy term associated with it is the one that keeps track of the total distance traveled (whereas the potential energy terms, coming as they do from conservative forces, depend on positions – but not on the paths that led to them). We assume the elevator comes to final rest at the equilibrium position of the spring, with the spring compressed an amount d_{eq} given by

$$mg = kd_{\text{eq}} \implies d_{\text{eq}} = \frac{mg}{k} = 0.12 \text{ m} .$$

In this part, we use that final-rest point as the reference level for computing gravitational potential energy, so the original $U = mgy$ becomes $mg(d_{\text{eq}} + d)$. In that final position, then, the gravitational energy is zero and the spring energy is $\frac{1}{2}kd_{\text{eq}}^2$. Thus, Eq. 8-31 becomes

$$\begin{aligned} mg\,(d_{\text{eq}} + d) &= \frac{1}{2}kd_{\text{eq}}^2 + fd_{\text{total}} \\ (1800)(9.8)(0.12 + 3.7) &= \frac{1}{2}\left(1.5 \times 10^5\right)(0.12)^2 + (4400)d_{\text{total}} \end{aligned}$$

which yields $d_{\text{total}} = 15$ m.

65. (a) Since the speed of the crate of mass m increases from 0 to 1.20 m/s relative to the factory ground, the kinetic energy supplied to it is

$$K = \frac{1}{2}mv^2 = \frac{1}{2}(300\,\text{kg})(1.20\,\text{m/s})^2 = 216 \text{ J}.$$

(b) The magnitude of the kinetic frictional force is

$$f = \mu N = \mu mg = (0.400)(300\,\text{kg})\left(9.8\,\text{m/s}^2\right) = 1.18 \times 10^3 \text{ N} .$$

(c) and (d) The energy supplied by the motor is the work W it does on the system, and must be greater than the kinetic energy gained by the crate computed in part (b). This is due to the fact that part of the energy supplied by the motor is being used to compensate for the energy dissipated ΔE_{th} while it was slipping. Let the distance the crate moved relative to the conveyor belt before it stops slipping be d, then from Eq. 2-16 ($v^2 = 2ad = 2(f/m)d$) we find

$$\Delta E_{\text{th}} = fd = \frac{1}{2}mv^2 = K .$$

Thus, the total energy that must be supplied by the motor is

$$W = K + \Delta E_{\text{th}} = 2K = (2)(216\,\text{J}) = 432 \text{ J} .$$

66. (a) The compression is "spring-like" so the maximum force relates to the distance x by Hooke's law:

$$F_x = kx \implies x = \frac{750}{2.5 \times 10^5} = 0.0030 \text{ m} .$$

(b) The work is what produces the "spring-like" potential energy associated with the compression. Thus, using Eq. 8-11,

$$W = \frac{1}{2}kx^2 = \frac{1}{2}\left(2.5 \times 10^5\right)(0.0030)^2 = 1.1 \text{ J} .$$

(c) By Newton's third law, the force F exerted by the tooth is equal and opposite to the "spring-like" force exerted by the licorice, so the graph of F is a straight line of slope k. We plot F (in Newtons) versus x (in millimeters); both are taken as positive.

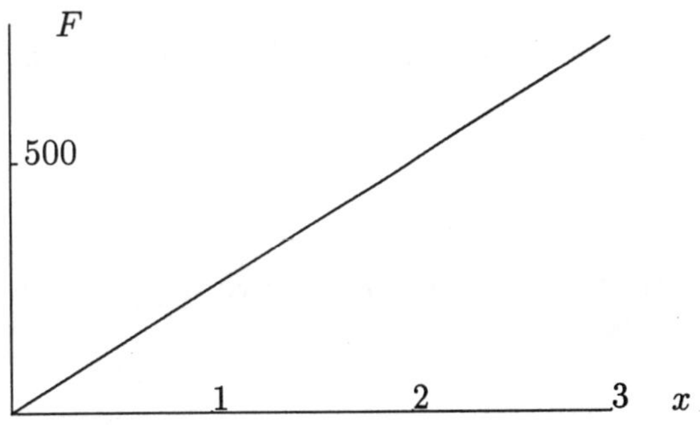

(d) As mentioned in part (b), the spring potential energy expression is relevant. Now, whether or not we can ignore dissipative processes is a deeper question. In other words, it seems unlikely that – if the tooth at any moment were to reverse its motion – that the licorice could "spring back" to its original shape. Still, to the extent that $U = \frac{1}{2}kx^2$ applies, the graph is a parabola (not shown here) which has its vertex at the origin and is either concave upward or concave downward depending on how one wishes to define the sign of F (the connection being $F = -dU/dx$).

(e) As a crude estimate, the area under the curve is roughly half the area of the entire plotting-area (8000 N by 12 mm). This leads to an approximate work of $\frac{1}{2}(8000)(0.012) \approx 50$ J. Estimates in the range $40 \le W \le 50$ J are acceptable.

(f) Certainly dissipative effects dominate this process, and we cannot assign it a meaningful potential energy.

67. (a) The drawings in the Figure (especially pictures (b) and (c)) show this geometric relationship very clearly. But we can work out the details, if need be. If ℓ is the length we are to compute (that of the still moving upper section) and ℓ' is the length of the lower (motionless) section, then clearly $\ell + \ell' = L$. Also, (as is especially easy to see in picture (c)) $x + \ell$ must equal ℓ'. These two equations, then, lead to the conclusion $\ell = \frac{1}{2}(L - x)$.

(b) The mass of the still moving upper section is

$$m = \rho\ell = \frac{\rho}{2}(L - x) \ .$$

(c) The assumptions stated in the problem lead to

$$\frac{1}{2}(\rho L + m_f)\, v_0^2 = \frac{1}{2}\left(\frac{\rho}{2}(L - x) + m_f\right) v^2$$

which yields the speed of the still moving upper section:

$$v = v_0\sqrt{\frac{\rho L + m_f}{\rho(L - x)/2 + m_f}} \ .$$

(d) As x approaches L, we obtain

$$v_f = v_0\sqrt{\frac{\rho L + m_f}{m_f}} = (6.0)\sqrt{\frac{(1.3)(20) + 0.8}{0.8}}$$

which yields $v_f = 35$ m/s.

68. (a) The effect of a (sliding) friction is described in terms of energy dissipated as shown in Eq. 8-29. We have

$$\Delta E = K + \frac{1}{2}k(0.08)^2 - \frac{1}{2}k(0.10)^2 = -f_k(0.02)$$

where distances are in meters and energies are in Joules. With $k = 4000$ N/m and $f_k = 80$ N, we obtain $K = 5.6$ J.

(b) In this case, we have $d = 0.10$ m. Thus,

$$\Delta E = K + 0 - \frac{1}{2}k(0.10)^2 = -f_k(0.10)$$

which leads to $K = 12$ J.

(c) We can approach this two ways. One way is to examine the dependence of energy on the variable d:

$$\Delta E = K + \frac{1}{2}k\,(d_0 - d)^2 - \frac{1}{2}kd_0^2 = -f_k d$$

where $d_0 = 0.10$ m, and solving for K as a function of d:

$$K = -\frac{1}{2}kd^2 + (kd_0)\,d - f_k d \ .$$

In this first approach, we could work through the $\frac{dK}{dd} = 0$ condition (or with the special capabilities of a graphing calculator) to obtain the answer $K_{\text{max}} = \frac{1}{2k}(kd_0 - f_k)^2$. In the second (and perhaps easier) approach, we note that K is maximum where v is maximum – which is where $a = 0 \implies$ equilibrium of forces. Thus, the second approach simply solves for the equilibrium position

$$|F_{\text{spring}}| = f_k \implies kx = 80 \ .$$

Thus, with $k = 4000$ N/m we obtain $x = 0.02$ m. But $x = d_0 - d$ so this corresponds to $d = 0.08$ m. Then the methods of part (a) lead to the answer $K_{\text{max}} = 12.8 \approx 13$ J.

69. Let the amount of stretch of the spring be x. For the object to be in equilibrium

$$kx - mg = 0 \implies x = mg/k \ .$$

Thus the gain in elastic potential energy for the spring is

$$\Delta U_e = \frac{1}{2}kx^2 = \frac{1}{2}k\left(\frac{mg}{k}\right)^2 = \frac{m^2g^2}{2k}$$

while the loss in the gravitational potential energy of the system is

$$-\Delta U_g = mgx = mg\left(\frac{mg}{k}\right) = \frac{m^2g^2}{k}$$

which we see (by comparing with the previous expression) is equal to $2\Delta U_e$. The reason why $|\Delta U_g| \neq \Delta U_e$ is that, since the object is slowly lowered, an upward external force (e.g., due to the hand) must have been exerted on the object during the lowering process, preventing it from accelerating downward. This force does *negative* work on the object, reducing the total mechanical energy of the system.

70. (a) The rate of change of the gravitational potential energy is

$$\frac{dU}{dt} = mg\frac{dy}{dt} = -mg|v| = -(68)(9.8)(59) = -3.9 \times 10^4 \text{ J/s} \ .$$

Thus, the gravitational energy is being reduced at the rate of 3.9×10^4 W.

(b) Since the velocity is constant, the rate of change of the kinetic energy is zero. Thus the rate at which the mechanical energy is being dissipated is the same as that of the gravitational potential energy (3.9×10^4 W).

71. The power generation (assumed constant, so average power is the same as instantaneous power) is

$$P = \frac{mgh}{t} = \frac{(3/4)\,(1200\,\text{m}^3)\left(10^3\,\text{kg/m}^3\right)(9.8\,\text{m/s}^2)\,(100\,\text{m})}{1.0\,\text{s}}$$

which yields $P = 8.8 \times 10^8$ W.

72. The free-body diagram for the trunk is shown.

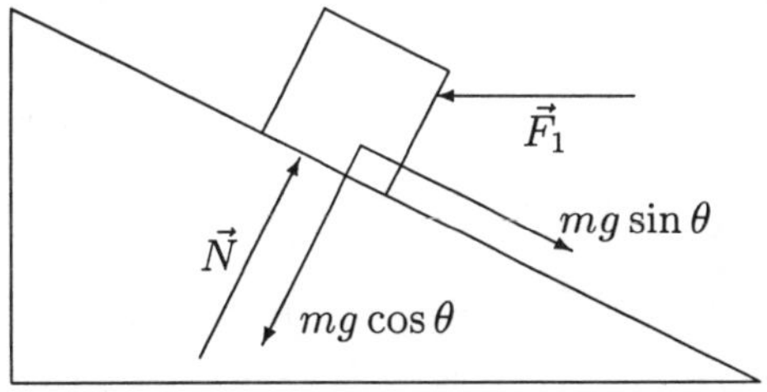

The x and y applications of Newton's second law provide two equations:

$$\begin{aligned} F_1\cos\theta - f_k - mg\sin\theta &= ma \\ N - F_1\sin\theta - mg\cos\theta &= 0 \ . \end{aligned}$$

(a) The trunk is moving up the incline at constant velocity, so $a = 0$. Using $f_k = \mu_k N$, we solve for the push-force F_1 and obtain

$$F_1 = \frac{mg(\sin\theta + \mu_k \cos\theta)}{\cos\theta - \mu_k \sin\theta} .$$

The work done by the push-force $\vec{F_1}$ as the trunk is pushed through a distance ℓ up the inclined plane is therefore

$$\begin{aligned} W_1 &= F_1 \ell \cos\theta = \frac{(mg\ell\cos\theta)(\sin\theta + \mu_k \cos\theta)}{\cos\theta - \mu_k \sin\theta} \\ &= \frac{(50\,\text{kg})(9.8\,\text{m/s}^2)(6.0\,\text{m})(\cos 30^\circ)\,(\sin 30^\circ + (0.20)\cos 30^\circ)}{\cos 30^\circ - (0.20)\sin 30^\circ} \\ &= 2.2 \times 10^3\ \text{J} . \end{aligned}$$

(b) The increase in the gravitational potential energy of the trunk is

$$\Delta U = mg\ell \sin\theta = (50\,\text{kg})\,(9.8\,\text{m/s}^2)\,(6.0\,\text{m}) \sin 30^\circ = 1.5 \times 10^3\ \text{J} .$$

Since the speed (and, therefore, the kinetic energy) of the trunk is unchanged, Eq. 8-31 leads to

$$W_1 = \Delta U + \Delta E_{\text{th}} .$$

Thus, using more precise numbers than are shown above, the increase in thermal energy (generated by the kinetic friction) is $2.24 \times 10^3 - 1.47 \times 10^3 = 7.7 \times 10^2$ J. An alternate way to this result is to use $\Delta E_{\text{th}} = f_k \ell$ (Eq. 8-29).

73. The style of reasoning used here is presented in §8-5.

(a) The horizontal line representing E_1 intersects the potential energy curve at a value of $r \approx 0.07$ nm and seems not to intersect the curve at larger r (though this is somewhat unclear since $U(r)$ is graphed only up to $r = 0.4$ nm). Thus, if m were propelled towards M from large r with energy E_1 it would "turn around" at 0.07 nm and head back in the direction from which it came.

(b) The line representing E_2 has two intersections points $r_1 \approx 0.16$ nm and $r_2 \approx 0.28$ nm with the $U(r)$ plot. Thus, if m starts in the region $r_1 < r < r_2$ with energy E_2 it will bounce back and forth between these two points, presumably forever.

(c) At $r = 0.3$ nm, the potential energy is roughly $U = -1.1 \times 10^{-19}$ J.

(d) With $M >> m$, the kinetic energy is essentially just that of m. Since $E = 1 \times 10^{-19}$ J, its kinetic energy is $K = E - U \approx 2.1 \times 10^{-19}$ J.

(e) Since force is related to the slope of the curve, we must (crudely) estimate $|F| \approx 1 \times 10^{-9}$ N at this point. The sign of the slope is positive, so by Eq. 8-20, the force is negative-valued. This is interpreted to mean that the atoms are attracted to each other.

(f) Recalling our remarks in the previous part, we see that the sign of F is positive (meaning it's repulsive) for $r < 0.2$ nm.

(g) And the sign of F is negative (attractive) for $r > 0.2$ nm.

(h) At $r = 0.2$ nm, the slope (hence, F) vanishes.

74. We take her original elevation to be the $y = 0$ reference level and observe that the top of the hill must consequently have $y_A = R(1 - \cos 20^\circ) = 1.2$ m, where R is the radius of the hill. The mass of the skier is $600/9.8 = 61$ kg.

(a) Applying energy conservation, Eq. 8-17, we have

$$K_B + U_B = K_A + U_A \implies K_B + 0 = K_A + m\ .gy_A$$

Using $K_B = \frac{1}{2}(61\,\text{kg})(8.0\,\text{m/s})^2$, we obtain $K_A = 1.2 \times 10^3$ J. Thus, we find the speed at the hilltop is $v = \sqrt{2K/m} = 6.4$ m/s. (Note: one might wish to check that the skier stays in contact with the hill – which is indeed the case, here. For instance, at A we find $v^2/r \approx 2$ m/s^2 which is considerably less than g.)

(b) With $K_A = 0$, we have

$$K_B + U_B = K_A + U_A \implies K_B + 0 = 0 + mgy_A$$

which yields $K_B = 724$ J, and the corresponding speed is $v = \sqrt{2K/m} = 4.9$ m/s.

(c) Expressed in terms of mass, we have

$$\begin{aligned} K_B + U_B &= K_A + U_A \implies \\ \frac{1}{2}mv_B^2 + mgy_B &= \frac{1}{2}mv_A^2 + mgy_A\ . \end{aligned}$$

Thus, the mass m cancels, and we observe that solving for speed does not depend on the value of mass (or weight).

75. The spring is relaxed at $y = 0$, so the elastic potential energy (Eq. 8-11) is $U_{\text{el}} = \frac{1}{2}ky^2$. The total energy is conserved, and is zero (determined by evaluating it at its initial position). We note that U is the same as ΔU in these manipulations. Thus, we have

$$0 = K + U_g + U_e \implies K = -U_g - U_e$$

where $U_g = mgy = (20\,\text{N})y$ with y in meters (so that the energies are in Joules). We arrange the results in a table:

position y	-0.05	-0.10	-0.15	-0.20
U_g	-1.0	-2.0	-3.0	-4.0
U_e	0.25	1.0	2.25	4.0
K	0.75	1.0	0.75	0

76. From Eq. 8-6, we find (with SI units understood)

$$U(\xi) = -\int_0^\xi \left(-3x - 5x^2\right) dx = \frac{3}{2}\xi^2 + \frac{5}{3}\xi^3\ .$$

(a) Using the above formula, we obtain $U(2) \approx 19$ J.

(b) When its speed is $v = 4$ m/s, its mechanical energy is $\frac{1}{2}mv^2 + U(5)$. This must equal the energy at the origin:

$$\frac{1}{2}mv^2 + U(5) = \frac{1}{2}mv_o^2 + U(0)$$

so that the speed at the origin is

$$v_o = \sqrt{v^2 + \frac{2}{m}\left(U(5) - U(0)\right)}\ .$$

Thus, with $U(5) = 246$ J, $U(0) = 0$ and $m = 20$ kg, we obtain $v_o = 6.4$ m/s.

(c) Our original formula for U is changed to $U(x) = -8 + \frac{3}{2}x^2 + \frac{5}{3}x^3$ in this case. Therefore, $U(2) = 11$ J. But we still have $v_o = 6.4$ m/s since that calculation only depended on the difference of potential energy values (specifically, $U(5) - U(0)$).

77. (a) At the top of its flight, the vertical component of the velocity vanishes, and the horizontal component (neglecting air friction) is the same as it was when it was thrown. Thus,

$$K_{\text{top}} = \frac{1}{2}mv_x^2 = \frac{1}{2}(0.050\,\text{kg})\left((8.0\,\text{m/s})\cos 30°\right)^2$$

which yields $K_{\text{top}} = 1.2$ J.

(b) We choose the point 3.0 m below the window as the reference level for computing the potential energy. Thus, equating the mechanical energy when it was thrown to when it is at this reference level, we have (with SI units understood)

$$\begin{aligned} mgy_0 + K_0 &= K \\ m(9.8)(3.0) + \frac{1}{2}m(8.0)^2 &= \frac{1}{2}mv^2 \end{aligned}$$

which yields (after canceling m and simplifying) $v = 11$ m/s.

(c) As mentioned, m cancels – and is therefore not relevant to that computation.

(d) The v in the kinetic energy formula is the magnitude of the velocity vector; it does not depend on the direction.

78. From the slope of the graph, we find the spring constant

$$k = \frac{\Delta F}{\Delta x} = 0.10\ \text{N/cm} = 10\ \text{N/m}\ .$$

(a) Equating the potential energy of the compressed spring to the kinetic energy of the cork at the moment of release, we have

$$\frac{1}{2}kx^2 = \frac{1}{2}mv^2 \implies v = x\sqrt{\frac{k}{m}}$$

which yields $v = 2.8$ m/s for $m = 0.0038$ kg and $x = 0.055$ m.

(b) The new scenario involves some potential energy at the moment of release. With $d = 0.015$ m, energy conservation becomes

$$\frac{1}{2}kx^2 = \frac{1}{2}mv^2 + \frac{1}{2}kd^2 \implies v = \sqrt{\frac{k}{m}\left(x^2 - d^2\right)}$$

which yields $v = 2.7$ m/s.

79. We assume his initial kinetic energy (when he jumps) is negligible. Then, his initial gravitational potential energy measured relative to where he momentarily stops is what becomes the elastic potential energy of the stretched net (neglecting air friction). Thus,

$$U_{\text{net}} = U_{\text{grav}} = mgh$$

where $h = 11.0 + 1.5 = 12.5$ m. Consequently, with $m = 70$ kg, we obtain $U_{\text{net}} = 8.6 \times 10^3$ J.

80. The work done by $\vec{F}$ is the negative of its potential energy change (see Eq. 8-6), so $U_B = U_A - 25 = 15$ J.

81. (a) During the final $d = 12$ m of motion, we use

$$K_1 + U_1 = K_2 + U_2 + f_k d \implies \frac{1}{2}mv^2 + 0 = 0 + 0 + f_k d$$

where $v = 4.2$ m/s. This gives $f_k = 0.31$ N. Therefore, the thermal energy change is $f_k d = 3.7$ J.

(b) Using $f_k = 0.31$ N we obtain $f_k d_{\text{total}} = 4.3$ J for the thermal energy generated by friction; here, $d_{\text{total}} = 14$ m.

(c) During the initial $d' = 2$ m of motion, we have

$$K_0 + U_0 + W_{\text{app}} = K_1 + U_1 + f_k d' \implies 0 + 0 + W_{\text{app}} = \frac{1}{2}mv^2 + 0 + f_k d'$$

which essentially combines Eq. 8-31 and Eq. 8-29. This leads to the result $W_{\text{app}} = 4.3$ J, and – reasonably enough – is the same as our answer in part (b).

82. (a) This part is essentially a free-fall problem, which can be easily done with Chapter 2 methods. Instead, choosing energy methods, we take $y = 0$ to be the ground level.

$$K_i + U_i = K + U \implies 0 + mgy_i = \frac{1}{2}mv^2 + 0$$

Therefore $v = \sqrt{2gy_i} = 9.2$ m/s, where $y_i = 4.3$ m.

(b) Eq. 8-29 provides $\Delta E_{\text{th}} = f_k d$ for thermal energy generated by the kinetic friction force. We apply Eq. 8-31:

$$K_i + U_i = K + U \implies 0 + mgy_i = \frac{1}{2}mv^2 + 0 + f_k d$$

With $d = y_i$, $m = 70$ kg and $f_k = 500$ N, this yields $v = 4.8$ m/s.

83. We want to convert (at least in theory) the water that falls through $h = 500$ m into electrical energy. The problem indicates that in one year, a volume of water equal to $A\Delta z$ lands in the form of rain on the country, where $A = 8 \times 10^{12}$ m^2 and $\Delta z = 0.75$ m. Multiplying this volume by the density $\rho = 1000\,\text{kg/m}^3$ leads to

$$m_{\text{total}} = \rho A \Delta z = (1000)\left(8 \times 10^{12}\right)(0.75) = 6 \times 10^{15}\,\text{kg}$$

for the mass of rainwater. One-third of this "falls" to the ocean, so it is $m = 2 \times 10^{15}$ kg that we want to use in computing the gravitational potential energy mgh (which will turn into electrical energy during the year). Since a year is equivalent to 3.2×10^7 s, we obtain

$$P_{\text{avg}} = \frac{\left(2 \times 10^{15}\right)(9.8)(500)}{3.2 \times 10^7} = 3.1 \times 10^{11}\,\text{W} .$$

84. With the potential energy reference level set at the point of throwing, we have (with SI units understood)

$$\Delta E = mgh - \frac{1}{2}mv_0^2 = m\left((9.8)(8.1) - \frac{1}{2}(14)^2\right)$$

which yields $\Delta E = -12$ J for $m = 0.63$ kg. This "loss" of mechanical energy is presumably due to air friction.

85. We note that in one second, the block slides $d = 1.34$ m up the incline, which means its height increase is $h = d \sin\theta$ where

$$\theta = \tan^{-1}\left(\frac{30}{40}\right) = 37^\circ .$$

We also note that the force of kinetic friction in this inclined plane problem is $f_k = \mu_k mg\cos\theta$ where $\mu_k = 0.40$ and $m = 1400$ kg. Thus, using Eq. 8-31 and Eq. 8-29, we find

$$W = mgh + f_k d = mgd\,(\sin\theta + \mu_k \cos\theta)$$

or $W = 1.69 \times 10^4$ J for this one-second interval. Thus, the power associated with this is

$$P = \frac{1.69 \times 10^4\,\text{J}}{1\,\text{s}} = 1.69 \times 10^4\,\text{W} .$$

86. We take the original height of the box to be the $y = 0$ reference level and observe that, in general, the height of the box (when the box has moved a distance d downhill) is $y = -d\sin 40°$.

 (a) Using the conservation of energy, we have

$$K_i + U_i = K + U \implies 0 + 0 = \frac{1}{2}mv^2 + mgy + \frac{1}{2}kd^2 \ .$$

 Therefore, with $d = 0.10$ m, we obtain $v = 0.81$ m/s.

 (b) We look for a value of $d \neq 0$ such that $K = 0$.

$$K_i + U_i = K + U \implies 0 + 0 = 0 + mgy + \frac{1}{2}kd^2$$

 Thus, we obtain $mgd\sin 40° = \frac{1}{2}kd^2$ and find $d = 0.21$ m.

 (c) The uphill force is caused by the spring (Hooke's law) and has magnitude $kd = 25.2$ N. The downhill force is the component of gravity $mg\sin 40° = 12.6$ N. Thus, the net force on the box is $25.2 - 12.6 = 12.6$ N uphill, and the acceleration is uphill with magnitude $12.6/2 = 6.3$ m/s^2.

87. Equating the mechanical energy at his initial position (as he emerges from the canon, where we set the reference level for computing potential energy) to his energy as he lands, we obtain

$$\begin{aligned} K_i &= K_f + U_f \\ \frac{1}{2}(60\,\text{kg})(16\,\text{m/s})^2 &= K_f + (60\,\text{kg})\left(9.8\,\text{m/s}^2\right)(3.9\,\text{m}) \end{aligned}$$

which leads to $K_f = 5.4 \times 10^3$ J.

88. (a) The initial kinetic energy is $K_i = \frac{1}{2}(1.5)(20)^2 = 300$ J.

 (b) At the point of maximum height, the vertical component of velocity vanishes but the horizontal component remains what it was when it was "shot" (if we neglect air friction). Its kinetic energy at that moment is

$$K = \frac{1}{2}(1.5)\,(20\cos 34°)^2 = 206 \text{ J} \ .$$

 Thus, $\Delta U = K_i - K = 300 - 206 = 94$ J.

 (c) Since $\Delta U = mg\Delta y$, we obtain

$$\Delta y = \frac{94\,\text{J}}{(1.5\,\text{kg})\,(9.8\,\text{m/s}^2)} = 6.4 \text{ m} \ .$$

89. We note that if the larger mass ($M = 2$ kg) falls $d = 0.25$ m, then the smaller mass ($m = 1$ kg) must increase its height by $h = d\sin 30°$. Thus, by mechanical energy conservation, the kinetic energy of the system is

$$K_{\text{total}} = Mgd - mgh = 3.7 \text{ J} \ .$$

90. (a) At the point of maximum height, where $y = 140$ m, the vertical component of velocity vanishes but the horizontal component remains what it was when it was launched (if we neglect air friction). Its kinetic energy at that moment is

$$K = \frac{1}{2}(0.55\,\text{kg})v_x^2 \ .$$

 Also, its potential energy (with the reference level chosen at the level of the cliff edge) at that moment is $U = mgy = 755$ J. Thus, by mechanical energy conservation,

$$K = K_i - U = 1550 - 755 \implies v_x = \sqrt{\frac{2(1550 - 755)}{0.55}}$$

 which yields $v_x = 54$ m/s.

(b) As mentioned $v_x = v_{i\,x}$ so that the initial kinetic energy

$$K_i = \frac{1}{2}m\left(v_{i\,x}^2 + v_{i\,y}^2\right)$$

can be used to find $v_{i\,y}$. We obtain $v_{i\,y} = 52$ m/s.

(c) Applying Eq. 2-16 to the vertical direction (with $+y$ upward), we have

$$\begin{aligned} v_y^2 &= v_{i\,y}^2 - 2g\Delta y \\ 65^2 &= 52^2 - 2(9.8)\Delta y \end{aligned}$$

which yields $\Delta y = -76$ m. The minus sign tells us it is below its launch point.

91. (a) The initial kinetic energy is $K_i = \frac{1}{2}(1.5)(3)^2 = 6.75$ J.

(b) The work of gravity is the negative of its change in potential energy. At the highest point, all of K_i has converted into U (if we neglect air friction) so we conclude the work of gravity is -6.75 J.

(c) And we conclude that $\Delta U = 6.75$ J.

(d) The potential energy there is $U_f = U_i + \Delta U = 6.75$ J.

(e) If $U_f = 0$, then $U_i = U_f - \Delta U = -6.75$ J.

(f) Since $mg\Delta y = \Delta U$, we obtain $\Delta y = 0.46$ m.

92. (a) With energy in Joules and length in meters, we have

$$\Delta U = U(x) - U(0) = -\int_0^x (6x' - 12)\, dx' \ .$$

Therefore, with $U(0) = 27$ J, we obtain $U(x)$ (written simply as U) by integrating and rearranging:

$$U = 27 + 12x - 3x^2 \ .$$

(b) We can maximize the above function by working through the $\frac{dU}{dx} = 0$ condition, or we can treat this as a force equilibrium situation – which is the approach we show.

$$F = 0 \implies 6x_{eq} - 12 = 0$$

Thus, $x_{eq} = 2.0$ m, and the above expression for the potential energy becomes $U = 39$ J.

(c) Using the quadratic formula or using the polynomial solver on an appropriate calculator, we find the values of x for which $U = 0$ to be 5.6 m and -1.6 m.

93. Since the aim of this problem is to invite student creativity (and possibly some research), we "invent a problem" (and give its solution) somewhat along the lines of part (b) (in fact, the student might consider running our example "in reverse"). Consider a block of mass M that falls from rest a distance H to a vertical spring of spring constant k. The spring compresses by x_c in order to halt the block, but on the rebound (due to the fact that the block is stuck on the end of the spring) the spring stretches (relative to its original relaxed length) an amount x_s before the block is momentarily at rest again. Take both values of x to be positive. Find x_c and x_s and their difference.

Solution: The height to which the spring reaches when it is relaxed is our $y = 0$ reference level. We relate the initial situation (when the block is dropped) to the situation of maximum compression using energy conservation.

$$K_0 + U_0 = K_c + U_c \implies 0 + MgH = 0 + Mg(-x_c) + \frac{1}{2}kx_c^2$$

The positive root stemming from a quadratic formula solution for x_c yields

$$x_c = \frac{Mg}{k}\left(1+\sqrt{1+\frac{2kH}{Mg}}\right) .$$

Next, we relate the initial situation to the final situation (of maximal stretch) using energy conservation.

$$K_0 + U_0 = K_s + U_s \implies 0 + MgH = 0 + Mgx_s + \frac{1}{2}kx_s^2$$

The positive root stemming from a quadratic formula solution for x_s yields

$$x_s = \frac{Mg}{k}\left(-1+\sqrt{1+\frac{2kH}{Mg}}\right) .$$

Finally, we note that $x_c > x_s$ with the difference being $x_c - x_s = 2Mg/k$.

94. (First problem in **Cluster 1**)
We take the bottom of the incline to be the $y = 0$ reference level. The incline angle is $\theta = 30°$. The distance along the incline d (measured from the bottom) is related to height y by the relation $y = d\sin\theta$.

(a) Using the conservation of energy, we have

$$K_0 + U_0 = K_{\text{top}} + U_{\text{top}} \implies \frac{1}{2}mv_0^2 + 0 = 0 + mgy$$

with $v_0 = 5.0$ m/s. This yields $y = 1.3$ m, from which we obtain $d = 2.6$ m.

(b) An analysis of forces in the manner of Chapter 6 reveals that the magnitude of the friction force is $f_k = \mu_k mg\cos\theta$. Now, we write Eq. 8-31 as

$$\begin{aligned} K_0 + U_0 &= K_{\text{top}} + U_{\text{top}} + f_k d \\ \frac{1}{2}mv_0^2 + 0 &= 0 + mgy + f_k d \\ \frac{1}{2}mv_0^2 &= mgd\sin\theta + \mu_k mgd\cos\theta \end{aligned}$$

which – upon cancelling the mass and rearranging – provides the result for d:

$$d = \frac{v_0^2}{2g\,(\mu_k\cos\theta + \sin\theta)} = 1.5 \text{ m} .$$

(c) The thermal energy generated by friction is $f_k d = \mu_k mgd\cos\theta = 26$ J.

(d) The slide back down, from the height $y = 1.5\sin 30°$ is also described by Eq. 8-31. With ΔE_{th} again equal to 26 J, we have

$$K_{\text{top}} + U_{\text{top}} = K_{\text{bot}} + U_{\text{bot}} + f_k d \implies 0 + mgy = \frac{1}{2}mv_{\text{bot}}^2 + 0 + 26$$

from which we find $v_{\text{bot}} = 2.1$ m/s.

95. (Second problem in **Cluster 1**)
Converting to SI units, $v_0 = 8.3$ m/s and $v = 11.1$ m/s. The incline angle is $\theta = 5.0°$. The height difference between the car's highest and lowest points is $(50\text{ m})\sin\theta = 4.4$ m. We take the lowest point (the car's final reported location) to correspond to the $y = 0$ reference level.

(a) Using Eq. 8-31 and Eq. 8-29, we find

$$f_k d = -\Delta K - \Delta U \quad \Longrightarrow \quad f_k d = \frac{1}{2} m \left(v_0^2 - v^2\right) + mgy_0 \ .$$

Therefore, the mechanical energy reduction (due to friction) is $f_k d = 2.4 \times 10^4$ J.

(b) With $d = 50$ m, we solve for f_k and obtain 471 N, which can be rounded to 470 N.

96. (Third problem in **Cluster 1**)

(a) When there is no change in potential energy, Eq. 8-24 leads to

$$W_{\text{app}} = \Delta K = \frac{1}{2} m \left(v^2 - v_0^2\right) \ .$$

Therefore, $\Delta E = 6.0 \times 10^3$ J.

(b) From the above manipulation, we see $W_{\text{app}} = 6.0 \times 10^3$ J. Also, from Chapter 2, we know that $\Delta t = \Delta v / a = 10$ s. Thus, using Eq. 7-42,

$$P_{\text{avg}} = \frac{W}{\Delta t} = \frac{6.0 \times 10^3}{10} = 600 \text{ W} \ .$$

(c) and (d) The constant applied force is $ma = 30$ N and clearly in the direction of motion, so Eq. 7-48 provides the results for instantaneous power

$$P = \vec{F} \cdot \vec{v} = \begin{cases} 300 \text{ W} & \text{for } v = 10 \text{ m/s} \\ 900 \text{ W} & \text{for } v = 30 \text{ m/s} \end{cases}$$

We note that the average of these two values agrees with the result in part (b).

97. (Fourth problem in **Cluster 1**)
The distance traveled up the incline can be figured with Chapter 2 techniques: $v^2 = v_0^2 + 2a\Delta x \longrightarrow \Delta x = 200\,\text{m}$. This corresponds to an increase in height equal to $y = 200 \sin\theta = 17$ m, where $\theta = 5.0°$. We take its initial height to be $y = 0$.

(a) Eq. 8-24 leads to

$$W_{\text{app}} = \Delta E = \frac{1}{2} m \left(v^2 - v_0^2\right) + mgy \ .$$

Therefore, $\Delta E = 8.6 \times 10^3$ J.

(b) From the above manipulation, we see $W_{\text{app}} = 8.6 \times 10^3$ J. Also, from Chapter 2, we know that $\Delta t = \Delta v / a = 10$ s. Thus, using Eq. 7-42,

$$P_{\text{avg}} = \frac{W}{\Delta t} = \frac{8.6 \times 10^3}{10} = 860 \text{ W}$$

where the answer has been rounded off (from the 856 value that is provided by the calculator).

(c) and (d) Taking into account the component of gravity along the incline surface, the applied force is $ma + mg \sin\theta = 43$ N and clearly in the direction of motion, so Eq. 7-48 provides the results for instantaneous power

$$P = \vec{F} \cdot \vec{v} = \begin{cases} 430 \text{ W} & \text{for } v = 10 \text{ m/s} \\ 1300 \text{ W} & \text{for } v = 30 \text{ m/s} \end{cases}$$

where these answers have been rounded off (from 428 and 1284, respectively). We note that the average of these two values agrees with the result in part (b).

Chapter 9

1. (a) We locate the coordinate origin at the center of Earth. Then the distance r_{com} of the center of mass of the Earth-Moon system is given by

$$r_{\text{com}} = \frac{m_M r_M}{m_M + m_E}$$

where m_M is the mass of the Moon, m_E is the mass of Earth, and r_M is their separation. These values are given in Appendix C. The numerical result is

$$r_{\text{com}} = \frac{(7.36 \times 10^{22}\,\text{kg})(3.82 \times 10^{8}\,\text{m})}{7.36 \times 10^{22}\,\text{kg} + 5.98 \times 10^{24}\,\text{kg}} = 4.64 \times 10^{6}\ \text{m}\ .$$

(b) The radius of Earth is $R_E = 6.37 \times 10^6$ m, so $r_{\text{com}} = 0.73R_E$.

2. We locate the coordinate origin at the center of the carbon atom, and we consider both atoms to be "point particles." We will use the non-SI units for mass found in Appendix F; since they will cancel they will not prevent the answer from being in SI units.

$$r_{\text{com}} = \frac{(15.9994\,\text{grams/mole})(1.131 \times 10^{-10}\,\text{m})}{12.01115\,\text{grams/mole} + 15.9994\,\text{grams/mole}} = 6.46 \times 10^{-11}\ \text{m}\ .$$

3. Our notation is as follows: $x_1 = 0$ and $y_1 = 0$ are the coordinates of the $m_1 = 3.0\,\text{kg}$ particle; $x_2 = 1.0\,\text{m}$ and $y_2 = 2.0\,\text{m}$ are the coordinates of the $m_2 = 8.0\,\text{kg}$ particle; and, $x_3 = 2.0\,\text{m}$ and $y_3 = 1.0\,\text{m}$ are the coordinates of the $m_3 = 4.0\,\text{kg}$ particle.

(a) The x coordinate of the center of mass is

$$\begin{aligned} x_{\text{com}} &= \frac{m_1 x_1 + m_2 x_2 + m_3 x_3}{m_1 + m_2 + m_2} \\ &= \frac{0 + (8.0\,\text{kg})(1.0\,\text{m}) + (4.0\,\text{kg})(2.0\,\text{m})}{3.0\,\text{kg} + 8.0\,\text{kg} + 4.0\,\text{kg}} \\ &= 1.1\,\text{m}\ . \end{aligned}$$

(b) The y coordinate of the center of mass is

$$\begin{aligned} y_{\text{com}} &= \frac{m_1 y_1 + m_2 y_2 + m_3 y_3}{m_1 + m_2 + m_3} \\ &= \frac{0 + (8.0\,\text{kg})(2.0\,\text{m}) + (4.0\,\text{kg})(1.0\,\text{m})}{3.0\,\text{kg} + 8.0\,\text{kg} + 4.0\,\text{kg}} \\ &= 1.3\ \text{m}\ . \end{aligned}$$

(c) As the mass of the topmost particle is increased, the center of mass shifts toward that particle. As we approach the limit as the topmost particle is infinitely more massive than the others, the center of mass becomes infinitesimally close to the position of that particle.

4. We will refer to the arrangement as a "table." We locate the coordinate origin at the center of the tabletop and note that the center of mass of each "leg" is a distance $L/2$ below the top. With $+x$ rightward and $+y$ upward, then the center of mass of the right leg is at $(x, y) = (+L/2, -L/2)$ and the center of mass of the left leg is at $(x, y) = (-L/2, -L/2)$. Thus, the x coordinate of the (whole table) center of mass is

$$x_{\text{com}} = \frac{M(+L/2) + M(-L/2)}{M + M + 3M} = 0$$

as expected. And the y coordinate of the (whole table) center of mass is

$$y_{\text{com}} = \frac{M(-L/2) + M(-L/2)}{M + M + 3M} = -\frac{L}{5}$$

so that the whole table center of mass is a small distance $(0.2L)$ directly below the middle of the tabletop.

5. First, we imagine that the small square piece (of mass m) that was cut from the large plate is returned to it so that the large plate is again a complete $6\,\text{m} \times 6\,\text{m}$ square plate (which has its center of mass at the origin). Then we "add" a square piece of "negative mass" $(-m)$ at the appropriate location to obtain what is shown in Fig. 9-24. If the mass of the whole plate is M, then the mass of the small square piece cut from it is obtained from a simple ratio of areas:

$$m = \left(\frac{2.0\,\text{m}}{6.0\,\text{m}}\right)^2 M \implies M = 9m .$$

(a) The x coordinate of the small square piece is $x = 2.0\,\text{m}$ (the middle of that square "gap" in the figure). Thus the x coordinate of the center of mass of the remaining piece is

$$x_{\text{com}} = \frac{(-m)x}{M + (-m)} = \frac{-m(2.0\,\text{m})}{9m - m} = -0.25\ \text{m} .$$

(b) Since the y coordinate of the small square piece is zero, we have $y_{\text{com}} = 0$.

6. We locate the coordinate origin at the lower left corner of the iron side of the composite slab. We orient the x axis along the length of the slab (the 22.0-cm side); the y axis along the width of the slab (the 13.0-cm side); and, the z axis along the height of the slab (the 2.80-cm side). The coordinates for the opposite corner on the aluminum side are then $x = 22.0\,\text{cm}$, $y = 13.0\,\text{cm}$, and $z = 2.80\,\text{cm}$. By symmetry $y_{\text{com}} = 13.0\,\text{cm}/2 = 6.50\,\text{cm}$ and $z_{\text{com}} = 2.80\,\text{cm}/2 = 1.40\,\text{cm}$. We use Eq. 9-5 to find x_{com}:

$$\begin{aligned} x_{\text{com}} &= \frac{m_i x_{\text{com},i} + m_a x_{\text{com},a}}{m_i + m_a} = \frac{\rho_i V_i x_{\text{com},i} + \rho_a V_a x_{\text{cm},a}}{\rho_i V_i + \rho_a V_a} \\ &= \frac{(11.0\,\text{cm}/2)\left(7.85\,\text{g/cm}^3\right) + 3(11.0\,\text{cm}/2)\left(2.70\,\text{g/cm}^3\right)}{7.85\,\text{g/cm}^3 + 2.70\,\text{g/cm}^3} = 8.30\ \text{cm} . \end{aligned}$$

Therefore, the center of mass is at $11.0\,\text{cm} - 8.3\,\text{cm} = 2.7\,\text{cm}$ from the midpoint of the slab.

7. By symmetry the center of mass is located on the axis of symmetry of the molecule. We denote the distance between the nitrogen atom and the center of mass of NH_3 as x. Then $m_N x = 3m_H(d - x)$, where d is the distance from the nitrogen atom to the plane containing the three hydrogen atoms:

$$d = \sqrt{(10.14 \times 10^{-11}\,\text{m})^2 - (9.4 \times 10^{-11}\text{m})^2} = 3.803 \times 10^{-11}\ \text{m} .$$

Thus,

$$x = \frac{3m_H d}{m_N + 3m_H} = \frac{3(1.00797)(3.803 \times 10^{-11}\text{m})}{14.0067 + 3(1.00797)} = 6.8 \times 10^{-12}\,\text{m}$$

where Appendix F has been used to find the masses.

8. The centers of mass (with centimeters understood) for each of the five sides are as follows:

$(x_1, y_1, z_1) = (0, 20, 20)$	for the side in the yz plane
$(x_2, y_2, z_2) = (20, 0, 20)$	for the side in the xz plane
$(x_3, y_3, z_3) = (20, 20, 0)$	for the side in the xy plane
$(x_4, y_4, z_4) = (40, 20, 20)$	for the remaining side parallel to side 1
$(x_5, y_5, z_5) = (20, 40, 20)$	for the remaining side parallel to side 2

Recognizing that all sides have the same mass m, we plug these into Eq. 9-5 to obtain the results (the first two being expected based on the symmetry of the problem).

(a)

$$x_{\text{com}} = \frac{mx_1 + mx_2 + mx_3 + mx_4 + mx_5}{5m} = \frac{0 + 20 + 20 + 40 + 20}{5} = 20\text{ cm}$$

(b)

$$y_{\text{com}} = \frac{my_1 + my_2 + my_3 + my_4 + my_5}{5m} = \frac{20 + 0 + 20 + 20 + 40}{5} = 20\text{ cm}$$

(c)

$$z_{\text{com}} = \frac{mz_1 + mz_2 + mz_3 + mz_4 + mz_5}{5m} = \frac{20 + 20 + 0 + 20 + 20}{5} = 16\text{ cm}$$

9. (a) Since the can is uniform, its center of mass is at its geometrical center, a distance $H/2$ above its base. The center of mass of the soda alone is at its geometrical center, a distance $x/2$ above the base of the can. When the can is full this is $H/2$. Thus the center of mass of the can and the soda it contains is a distance

$$h = \frac{M(H/2) + m(H/2)}{M + m} = \frac{H}{2}$$

above the base, on the cylinder axis.

(b) We now consider the can alone. The center of mass is $H/2$ above the base, on the cylinder axis.

(c) As x decreases the center of mass of the soda in the can at first drops, then rises to $H/2$ again.

(d) When the top surface of the soda is a distance x above the base of the can, the mass of the soda in the can is $m_p = m(x/H)$, where m is the mass when the can is full ($x = H$). The center of mass of the soda alone is a distance $x/2$ above the base of the can. Hence

$$h = \frac{M(H/2) + m_p(x/2)}{M + m_p} = \frac{M(H/2) + m(x/H)(x/2)}{M + (mx/H)} = \frac{MH^2 + mx^2}{2(MH + mx)} .$$

We find the lowest position of the center of mass of the can and soda by setting the derivative of h with respect to x equal to 0 and solving for x. The derivative is

$$\frac{dh}{dx} = \frac{2mx}{2(MH + mx)} - \frac{(MH^2 + mx^2)m}{2(MH + mx)^2} = \frac{m^2x^2 + 2MmHx - MmH^2}{2(MH + mx)^2} .$$

The solution to $m^2x^2 + 2MmHx - MmH^2 = 0$ is

$$x = \frac{MH}{m}\left(-1 + \sqrt{1 + \frac{m}{M}}\right) .$$

The positive root is used since x must be positive. Next, we substitute the expression found for x into $h = (MH^2 + mx^2)/2(MH + mx)$. After some algebraic manipulation we obtain

$$h = \frac{HM}{m}\left(\sqrt{1 + \frac{m}{M}} - 1\right) .$$

10. Since the center of mass of the two-skater system does not move, both skaters will end up at the center of mass of the system. Let the center of mass be a distance x from the 40-kg skater, then

$$(65\,\text{kg})(10\,\text{m} - x) = (40\,\text{kg})x \implies x = 6.2\,\text{m} .$$

Thus the 40-kg skater will move by 6.2 m.

11. Let m_c be the mass of the Chrysler and v_c be its velocity. Let m_f be the mass of the Ford and v_f be its velocity. Then the velocity of the center of mass is

$$v_{\text{com}} = \frac{m_c v_c + m_f v_f}{m_c + m_f} = \frac{(2400\,\text{kg})(80\,\text{km/h}) + (1600\,\text{kg})(60\,\text{km/h})}{2400\,\text{kg} + 1600\,\text{kg}} = 72\,\text{km/h} .$$

We note that the two velocities are in the same direction, so the two terms in the numerator have the same sign.

12. (a) Since the center of mass of the man-balloon system does not move, the balloon will move downward with a certain speed u relative to the ground as the man climbs up the ladder. The speed of the man relative to the ground is $v_g = v - u$. Thus, the speed of the center of mass of the system is

$$v_{\text{com}} = \frac{mv_g - Mu}{M+m} = \frac{m(v-u) - Mu}{M+m} = 0 .$$

This yields $u = mv/(M+m)$.

(b) Now that there is no relative motion within the system, the speed of both the balloon and the man is equal to v_{com}, which is zero. So the balloon will again be stationary.

13. We use the constant-acceleration equations of Table 2-1 (with $+y$ downward and the origin at the release point), Eq. 9-5 for y_{com} and Eq. 9-17 for $\vec{v}_{\text{com}}$.

(a) The location of the first stone (of mass m_1) at $t = 300\times10^{-3}$ s is $y_1 = (1/2)gt^2 = (1/2)(9.8)\left(300 \times 10^{-3}\right)^2 =$ 0.44 m, and the location of the second stone (of mass $m_2 = 2m_1$) at $t = 300 \times 10^{-3}$ s is $y_2 = (1/2)gt^2 = (1/2)(9.8)(300 \times 10^{-3} - 100 \times 10^{-3})^2 = 0.20$ m. Thus, the center of mass is at

$$y_{\text{com}} = \frac{m_1 y_1 + m_2 y_2}{m_1 + m_2} = \frac{m_1(0.44\,\text{m}) + 2m_1(0.20\,\text{m})}{m_1 + 2m_2} = 0.28\ \text{m} .$$

(b) The speed of the first stone at time t is $v_1 = gt$, while that of the second stone is $v_2 = g(t - 100 \times 10^{-3}\,\text{s})$. Thus, the center-of-mass speed at $t = 300 \times 10^{-3}$ s is

$$\begin{aligned} v_{\text{com}} &= \frac{m_1 v_1 + m_2 v_2}{m_1 + m_2} \\ &= \frac{m_1(9.8)\left(300 \times 10^{-3}\right) + 2m_1(9.8)\left(300 \times 10^{-3} - 100 \times 10^{-3}\right)}{m_1 + 2m_1} \\ &= 2.3\ \text{m/s} . \end{aligned}$$

14. We use the constant-acceleration equations of Table 2-1 (with the origin at the traffic light), Eq. 9-5 for x_{com} and Eq. 9-17 for $\vec{v}_{\text{com}}$. At $t = 3.0$ s, the location of the automobile (of mass m_1) is $x_1 = \frac{1}{2}at^2 = \frac{1}{2}(4.0\,\text{m/s}^2)(3.0\,\text{s})^2 = 18$ m, while that of the truck (of mass m_2) is $x_2 = vt = (8.0\,\text{m/s})(3.0\,\text{s}) = 24$ m. The speed of the automobile then is $v_1 = at = \left(4.0\,\text{m/s}^2\right)(3.0\,\text{s}) = 12$ m/s, while the speed of the truck remains $v_2 = 8.0$ m/s.

(a) The location of their center of mass is

$$x_{\text{com}} = \frac{m_1 x_1 + m_2 x_2}{m_1 + m_2} = \frac{(1000\,\text{kg})(18\,\text{m}) + (2000\,\text{kg})(24\,\text{m})}{1000\,\text{kg} + 2000\,\text{kg}} = 22\ \text{m} .$$

(b) The speed of the center of mass is

$$v_{\text{com}} = \frac{m_1 v_1 + m_2 v_2}{m_1 + m_2} = \frac{(1000\,\text{kg})(12\,\text{m/s}) + (2000\,\text{kg})(8.0\,\text{m/s})}{1000\,\text{kg} + 2000\,\text{kg}} = 9.3\,\text{m/s}\,.$$

15. We need to find the coordinates of the point where the shell explodes and the velocity of the fragment that does not fall straight down. The coordinate origin is at the firing point, the $+x$ axis is rightward, and the $+y$ direction is upward. The y component of the velocity is given by $v = v_{0\,y} - gt$ and this is zero at time $t = v_{0\,y}/g = (v_0/g)\sin\theta_0$, where v_0 is the initial speed and θ_0 is the firing angle. The coordinates of the highest point on the trajectory are

$$x = v_{0x}t = v_0 t\cos\theta_0 = \frac{v_0^2}{g}\sin\theta_0\cos\theta_0 = \frac{(20\,\text{m/s})^2}{9.8\,\text{m/s}^2}\sin 60^\circ\cos 60^\circ = 17.7\text{ m}$$

and

$$y = v_{0\,y}t - \frac{1}{2}gt^2 = \frac{1}{2}\frac{v_0^2}{g}\sin^2\theta_0 = \frac{1}{2}\frac{(20\,\text{m/s})^2}{9.8\,\text{m/s}^2}\sin^2 60^\circ = 15.3\text{ m}\ .$$

Since no horizontal forces act, the horizontal component of the momentum is conserved. Since one fragment has a velocity of zero after the explosion, the momentum of the other equals the momentum of the shell before the explosion. At the highest point the velocity of the shell is $v_0\cos\theta_0$, in the positive x direction. Let M be the mass of the shell and let V_0 be the velocity of the fragment. Then $Mv_0\cos\theta_0 = MV_0/2$, since the mass of the fragment is $M/2$. This means

$$V_0 = 2v_0\cos\theta_0 = 2(20\,\text{m/s})\cos 60^\circ = 20\text{ m/s}\ .$$

This information is used in the form of initial conditions for a projectile motion problem to determine where the fragment lands. Resetting our clock, we now analyze a projectile launched horizontally at time $t = 0$ with a speed of $20\,\text{m/s}$ from a location having coordinates $x_0 = 17.7\,\text{m}$, $y_0 = 15.3\,\text{m}$. Its y coordinate is given by $y = y_0 - \frac{1}{2}gt^2$, and when it lands this is zero. The time of landing is $t = \sqrt{2y_0/g}$ and the x coordinate of the landing point is

$$x = x_0 + V_0 t = x_0 + V_0\sqrt{\frac{2y_0}{g}} = 17.7\,\text{m} + (20\,\text{m/s})\sqrt{\frac{2(15.3\,\text{m})}{9.8\,\text{m/s}^2}} = 53\text{ m}\ .$$

16. The implication in the problem regarding $\vec{v}_0$ is that the olive and the nut start at rest. Although we could proceed by analyzing the forces on each object, we prefer to approach this using Eq. 9-14. The total force on the nut-olive system is $\vec{F}_o + \vec{F}_n = -\hat{\text{i}} + \hat{\text{j}}$ with the unit newton understood. Thus, Eq. 9-14 becomes

$$-\hat{\text{i}} + \hat{\text{j}} = M\vec{a}_{\text{com}}$$

where $M = 2.0$ kg. Thus, $\vec{a}_{\text{com}} = -\frac{1}{2}\hat{\text{i}} + \frac{1}{2}\hat{\text{j}}$ in SI units. Each component is constant, so we apply the equations discussed in Chapters 2 and 4.

$$\Delta\vec{r}_{\text{com}} = \frac{1}{2}\vec{a}_{\text{com}}\,t^2 = -4.0\hat{\text{i}} + 4.0\hat{\text{j}}$$

(in meters) when $t = 4.0$ s. It is perhaps instructive to work through this problem the *long way* (separate analysis for the olive and the nut and then application of Eq. 9-5) since it helps to point out the computational advantage of Eq. 9-14.

17. (a) We place the origin of a coordinate system at the center of the pulley, with the x axis horizontal and to the right and with the y axis downward. The center of mass is halfway between the containers, at $x = 0$ and $y = \ell$, where ℓ is the vertical distance from the pulley center to either of the containers. Since the diameter of the pulley is 50 mm, the center of mass is 25 mm from each container.

(b) Suppose 20 g is transferred from the container on the left to the container on the right. The container on the left has mass $m_1 = 480\,\text{g}$ and is at $x_1 = -25\,\text{mm}$. The container on the right has mass $m_2 = 520\,\text{g}$ and is at $x_2 = +25\,\text{mm}$. The x coordinate of the center of mass is then

$$x_{\text{com}} = \frac{m_1x_1 + m_2x_2}{m_1 + m_2} = \frac{(480\,\text{g})(-25\,\text{mm}) + (520\,\text{g})(25\,\text{mm})}{480\,\text{g} + 520\,\text{g}} = 1.0\,\text{mm}\,.$$

The y coordinate is still ℓ. The center of mass is 26 mm from the lighter container, along the line that joins the bodies.

(c) When they are released the heavier container moves downward and the lighter container moves upward, so the center of mass, which must remain closer to the heavier container, moves downward.

(d) Because the containers are connected by the string, which runs over the pulley, their accelerations have the same magnitude but are in opposite directions. If a is the acceleration of m_2, then $-a$ is the acceleration of m_1. The acceleration of the center of mass is

$$a_{\text{com}} = \frac{m_1(-a) + m_2a}{m_1 + m_2} = a\,\frac{m_2 - m_1}{m_1 + m_2}\,.$$

We must resort to Newton's second law to find the acceleration of each container. The force of gravity m_1g, down, and the tension force of the string T, up, act on the lighter container. The second law for it is $m_1g - T = -m_1a$. The negative sign appears because a is the acceleration of the heavier container. The same forces act on the heavier container and for it the second law is $m_2g - T = m_2a$. The first equation gives $T = m_1g + m_1a$. This is substituted into the second equation to obtain $m_2g - m_1g - m_1a = m_2a$, so $a = (m_2 - m_1)g/(m_1 + m_2)$. Thus

$$a_{\text{com}} = \frac{g(m_2 - m_1)^2}{(m_1 + m_2)^2} = \frac{(9.8\,\text{m/s}^2)(520\,\text{g} - 480\,\text{g})^2}{(480\,\text{g} + 520\,\text{g})^2} = 1.6 \times 10^{-2}\,\text{m/s}^2\,.$$

The acceleration is downward.

18. We denote the mass of Ricardo as M_R and that of Carmelita as M_C. Let the center of mass of the two-person system (assumed to be closer to Ricardo) be a distance x from the middle of the canoe of length L and mass m. Then $M_R(L/2 - x) = mx + M_C(L/2 + x)$. Now, after they switch positions, the center of the canoe has moved a distance $2x$ from its initial position. Therefore, $x = 40\,\text{cm}/2 = 0.20\,\text{m}$, which we substitute into the above equation to solve for M_C:

$$M_C = \frac{M_R(L/2 - x) - mx}{L/2 + x} = \frac{(80)\left(\frac{3.0}{2} - 0.20\right) - (30)(0.20)}{(3.0/2) + 0.20} = 58\ \text{kg}\,.$$

19. There is no net horizontal force on the dog-boat system, so their center of mass does not move. Therefore by Eq. 9-16,

$$M\Delta x_{\text{com}} = 0 = m_b\Delta x_b + m_d\Delta x_d$$

which implies

$$|\Delta x_b| = \frac{m_d}{m_b}\,|\Delta x_d|\,.$$

Now we express the geometrical condition that *relative to the boat* the dog has moved a distance $d = 2.4$ m:

$$|\Delta x_b| + |\Delta x_d| = d$$

which accounts for the fact that the dog moves one way and the boat moves the other. We substitute for $|\Delta x_b|$ from above:

$$\frac{m_d}{m_b}\,|\Delta x_d| + |\Delta x_d| = d$$

which leads to

$$|\Delta x_d| = \frac{d}{1 + \frac{m_d}{m_b}} = \frac{2.4}{1 + \frac{4.5}{18}} = 1.92\ \text{m}\,.$$

The dog is therefore 1.9 m closer to the shore than initially (where it was 6.1 m from it). Thus, it is now 4.2 m from the shore.

20. We apply Eq. 9-22 ($p = mv$) and Eq. 7-1 ($K = \frac{1}{2}mv^2$).

(a) The speed of the VW Beetle of mass m is

$$v = \frac{p}{m} = \frac{(2650\,\text{kg})(16\,\text{km/h})}{816\,\text{kg}} = 52\,\text{km/h} .$$

(b) In this case, the speed of the VW Beetle must be

$$v = \sqrt{\frac{2K}{m}} = \sqrt{\frac{2(2650\,\text{kg})(16\,\text{km/h})^2/2}{816\,\text{kg}}} = 29\,\text{km/h} .$$

21. Using Eq. 9-22, the necessary speed v is

$$v = \frac{p}{m} = \frac{(1600\,\text{kg})(1.2\,\text{km/h})}{80\,\text{kg}} = 24\,\text{km/h} .$$

22. The magnitude of the ball's momentum change is

$$\Delta p = |mv_i - mv_f| = (0.70\,\text{kg})\,|5.0\,\text{m/s} - (-2.0\,\text{m/s})| = 4.9\ \text{kg}\cdot\text{m/s} .$$

23. (a) The change in kinetic energy is

$$\begin{aligned} \Delta K &= \frac{1}{2}mv_f^2 - \frac{1}{2}mv_i^2 \\ &= \frac{1}{2}(2100\,\text{kg})\left((51\,\text{km/h})^2 - (41\,\text{km/h})^2\right) \\ &= 9.66 \times 10^4\,\text{kg}\cdot(\text{km/h})^2\left((10^3\,\text{m/km})(1\,\text{h}/3600\,\text{s})\right)^2 \\ &= 7.5 \times 10^4\ \text{J} . \end{aligned}$$

(b) The magnitude of the change in velocity is

$$\begin{aligned} |\Delta\vec{v}| &= \sqrt{(-v_i)^2 + (v_f)^2} \\ &= \sqrt{(-41\,\text{km/h})^2 + (51\,\text{km/h})^2} \\ &= 65.4\ \text{km/h} \end{aligned}$$

so the magnitude of the change in momentum is

$$|\Delta\vec{p}| = m\,|\Delta\vec{v}| = (2100\,\text{kg})(65.4\,\text{km/h})\left(\frac{1000\,\text{m/km}}{3600\,\text{s/h}}\right) = 3.8 \times 10^4\,\text{kg}\cdot\text{m/s} .$$

(c) The vector $\Delta\vec{p}$ points at an angle θ south of east, where

$$\theta = \tan^{-1}\left(\frac{v_i}{v_f}\right) = \tan^{-1}\left(\frac{41\,\text{km/h}}{51\,\text{km/h}}\right) = 39^\circ .$$

24. (a) Since the force of impact on the ball is in the y direction, p_x is conserved: $p_{x\,i} = mv_i \sin 30^\circ = p_{x\,f} = mv_i \sin\theta$. Thus $\theta = 30^\circ$.

(b) The momentum change is

$$\begin{aligned} \Delta\vec{p} &= mv_i\cos\theta\,(-\hat{\jmath}) - mv_i\cos\theta\,(+\hat{\jmath}) \\ &= -2(0.165\,\text{kg})(2.00\,\text{m/s})(\cos 30^\circ)\,\hat{\jmath} \\ &= -0.572\hat{\jmath}\ \ \text{kg}\cdot\text{m/s} . \end{aligned}$$

25. The velocity of the object is

$$\vec{v} = \frac{d\vec{r}}{dt} = \frac{d}{dt}\left((3500 - 160t)\,\hat{\imath} + 2700\,\hat{\jmath} + 300\,\hat{k}\right) = -160\,\hat{\imath}\ \text{m/s} .$$

(a) The linear momentum is

$$\vec{p} = m\vec{v} = (250)(-160\,\hat{\imath}) = -4.0 \times 10^4\,\hat{\imath}\ \ \text{kg·m/s} .$$

(b) The object is moving west (our $-\hat{\imath}$ direction).

(c) Since the value of $\vec{p}$ does not change with time, the net force exerted on the object is zero, by Eq. 9-23.

26. We use coordinates with $+x$ horizontally toward the pitcher and $+y$ upward. Angles are measured counterclockwise from the $+x$ axis. Mass, velocity and momentum units are SI. Thus, the initial momentum can be written $\vec{p}_0 = (4.5 \angle\ 215°)$ in magnitude-angle notation.

(a) In magnitude-angle notation, the momentum change is $(6.0 \angle -90°) - (4.5 \angle\ 215°) = (5.0 \angle -43°)$ (efficiently done with a vector capable calculator in polar mode). The magnitude of the momentum change is therefore 5.0 kg·m/s.

(b) The momentum change is $(6.0 \angle\ 0°) - (4.5 \angle\ 215°) = (10 \angle\ 15°)$. Thus, the magnitude of the momentum change is 10 kg·m/s.

27. No external forces with horizontal components act on the man-stone system and the vertical forces sum to zero, so the total momentum of the system is conserved. Since the man and the stone are initially at rest, the total momentum is zero both before and after the stone is kicked. Let m_s be the mass of the stone and v_s be its velocity after it is kicked; let m_m be the mass of the man and v_m be his velocity after he kicks the stone. Then $m_s v_s + m_m v_m = 0 \rightarrow v_m = -m_s v_s / m_m$. We take the axis to be positive in the direction of motion of the stone. Then

$$v_m = -\frac{(0.068\,\text{kg})(4.0\,\text{m/s})}{91\,\text{kg}} = -3.0 \times 10^{-3}\ \text{m/s} .$$

The negative sign indicates that the man moves in the direction opposite to the direction of motion of the stone.

28. The fact that they are connected by a spring is not used in the solution. We use Eq. 9-17 for $\vec{v}_{\text{com}}$:

$$\begin{aligned} M\vec{v}_{\text{com}} &= m_1\vec{v}_1 + m_2\vec{v}_2 \\ 0 &= (1.0)(1.7) + (3.0)\vec{v}_2 \end{aligned}$$

which yields $|\vec{v}_2| = 0.57$ m/s. The direction of $\vec{v}_2$ is opposite that of $\vec{v}_1$ (that is, they are both headed towards the center of mass, but from opposite directions).

29. No external forces with horizontal components act on the cart-man system and the vertical forces sum to zero, so the total momentum of the system is conserved. Let m_c be the mass of the cart, v be its initial velocity, and v_c be its final velocity (after the man jumps off). Let m_m be the mass of the man. His initial velocity is the same as that of the cart and his final velocity is zero. Conservation of momentum yields $(m_m + m_c)v = m_c v_c$. Consequently, the final speed of the cart is

$$v_c = \frac{v(m_m + m_c)}{m_c} = \frac{(2.3\,\text{m/s})(75\,\text{kg} + 39\,\text{kg})}{39\,\text{kg}} = 6.7\,\text{m/s} .$$

The cart speeds up by $6.7 - 2.3 = 4.4\,\text{m/s}$. In order to slow himself, the man gets the cart to push backward on him by pushing forward on it, so the cart speeds up.

30. We apply Eq. 9-17, with $M = \sum m = 1.3$ kg,

$$\begin{aligned} M\vec{v}_{\text{com}} &= m_A\vec{v}_A + m_B\vec{v}_B + m_C\vec{v}_C \\ (1.3)\left(-0.40\hat{\mathrm{i}}\right) &= (0.50)\vec{v}_A + (0.60)\left(0.20\hat{\mathrm{i}}\right) + (0.20)\left(0.30\hat{\mathrm{i}}\right) \end{aligned}$$

which leads to $\vec{v}_A = -1.4\hat{\mathrm{i}}$ in SI units (m/s).

31. Our notation is as follows: the mass of the motor is M; the mass of the module is m; the initial speed of the system is v_0; the relative speed between the motor and the module is v_r; and, the speed of the module relative to the Earth is v after the separation. Conservation of linear momentum requires $(M+m)v_0 = mv + M(v - v_r)$. Therefore,

$$v = v_0 + \frac{Mv_r}{M+m} = 4300\,\text{km/h} + \frac{(4m)(82\,\text{km/h})}{4m+m} = 4.4\times10^3\,\text{km/h}\ .$$

32. Denoting the new speed of the car as v, then the new speed of the man relative to the ground is $v - v_{\text{rel}}$. Conservation of momentum requires

$$\left(\frac{W}{g} + \frac{w}{g}\right)v_0 = \left(\frac{W}{g}\right)v + \left(\frac{w}{g}\right)(v - v_{\text{rel}})\ .$$

Consequently, the change of velocity is

$$\Delta\vec{v} = v - v_0 = \frac{w\,v_{\text{rel}}}{W+w}\ .$$

33. We assume no external forces act on the system composed of the two parts of the last stage. Hence, the total momentum of the system is conserved. Let m_c be the mass of the rocket case and m_p be the mass of the payload. At first they are traveling together with velocity v. After the clamp is released m_c has velocity v_c and m_p has velocity v_p. Conservation of momentum yields $(m_c + m_p)v = m_cv_c + m_pv_p$.

(a) After the clamp is released the payload, having the lesser mass, will be traveling at the greater speed. We write $v_p = v_c + v_{\text{rel}}$, where v_{rel} is the relative velocity. When this expression is substituted into the conservation of momentum condition, the result is

$$(m_c + m_p)\,v = m_cv_c + m_pv_c + m_pv_{\text{rel}}\ .$$

Therefore,

$$\begin{aligned} v_c &= \frac{(m_c+m_p)\,v - m_pv_{\text{rel}}}{m_c+m_p} \\ &= \frac{(290.0\,\text{kg} + 150.0\,\text{kg})(7600\,\text{m/s}) - (150.0\,\text{kg})(910.0\,\text{m/s})}{290.0\,\text{kg} + 150.0\,\text{kg}} \\ &= 7290\ \text{m/s}\ . \end{aligned}$$

(b) The final speed of the payload is $v_p = v_c + v_{\text{rel}} = 7290\,\text{m/s} + 910.0\,\text{m/s} = 8200\,\text{m/s}$.

(c) The total kinetic energy before the clamp is released is

$$K_i = \frac{1}{2}(m_c + m_p)v^2 = \frac{1}{2}(290.0\,\text{kg} + 150.0\,\text{kg})(7600\,\text{m/s})^2 = 1.271\times10^{10}\ \text{J}\ .$$

(d) The total kinetic energy after the clamp is released is

$$\begin{aligned} K_f &= \frac{1}{2}m_cv_c^2 + \frac{1}{2}m_pv_p^2 \\ &= \frac{1}{2}(290.0\,\text{kg})(7290\,\text{m/s})^2 + \frac{1}{2}(150.0\,\text{kg})(8200\,\text{m/s})^2 \\ &= 1.275\times10^{10}\ \text{J}\ . \end{aligned}$$

The total kinetic energy increased slightly. Energy originally stored in the spring is converted to kinetic energy of the rocket parts.

34. Our $+x$ direction is east and $+y$ direction is north. The linear momenta for the two $m = 2.0\,\text{kg}$ parts are then

$$\vec{p}_1 = m\vec{v}_1 = mv_1\,\hat{\text{j}}$$

where $v_1 = 3.0\,\text{m/s}$, and

$$\vec{p}_2 = m\vec{v}_2 = m(v_{2x}\,\hat{\text{i}} + v_{2y}\,\hat{\text{j}}) = mv_2(\cos\theta\,\hat{\text{i}} + \sin\theta\,\hat{\text{j}})$$

where $v_2 = 5.0\,\text{m/s}$ and $\theta = 30°$. The combined linear momentum of both parts is then

$$\begin{aligned}\vec{P} &= \vec{p}_1 + \vec{p}_2 \\ &= mv_1\,\hat{\text{j}} + mv_2\left(\cos\theta\,\hat{\text{i}} + \sin\theta\,\hat{\text{j}}\right) = (mv_2\cos\theta)\,\hat{\text{i}} + (mv_1 + mv_2\sin\theta)\,\hat{\text{j}} \\ &= (2.0\,\text{kg})(5.0\,\text{m/s})(\cos 30°)\hat{\text{i}} + (2.0\,\text{kg})\,(3.0\,\text{m/s} + (5.0\,\text{m/s})(\sin 30°))\,\hat{\text{j}} \\ &= \left(8.66\,\hat{\text{i}} + 11\,\hat{\text{j}}\right)\ \text{kg}\cdot\text{m/s}\ .\end{aligned}$$

From conservation of linear momentum we know that this is also the linear momentum of the whole kit before it splits. Thus the speed of the 4.0-kg kit is

$$v = \frac{P}{M} = \frac{\sqrt{P_x^2 + P_y^2}}{M} = \frac{\sqrt{(8.66\,\text{kg}\cdot\text{m/s})^2 + (11\,\text{kg}\cdot\text{m/s})^2}}{4.0\,\text{kg}} = 3.5\ \text{m/s}\ .$$

35. We establish a coordinate system with the origin at the position of initial nucleus of mass $m_{m\,i}$ (which was stationary), with the electron momentum $\vec{p}_e$ in the $-x$ direction and the neutrino momentum $\vec{p}_\nu$ in the $-y$ direction. We will use unit-vector notation, although the problem does not specifically request it.

(a) We find the momentum $\vec{p}_{n\,r}$ of the residual nucleus from momentum conservation.

$$\begin{aligned}\vec{p}_{n\,i} &= \vec{p}_e + \vec{p}_\nu + \vec{p}_{n\,r} \\ 0 &= -1.2\times10^{-22}\,\hat{\text{i}} - 6.4\times10^{-23}\,\hat{\text{j}} + \vec{p}_{n\,r}\end{aligned}$$

Thus, $\vec{p}_{n\,r} = 1.2\times10^{-22}\,\hat{\text{i}} + 6.4\times10^{-23}\,\hat{\text{j}}$ in SI units (kg·m/s). Its magnitude is

$$|\vec{p}_{n\,r}| = \sqrt{(1.2\times10^{-22})^2 + (6.4\times10^{-23})^2} = 1.4\times10^{-22}\ \text{kg}\cdot\text{m/s}\ .$$

(b) The angle measured from the $+x$ axis to $\vec{p}_{n\,r}$ is

$$\theta = \tan^{-1}\left(\frac{6.4\times10^{-23}}{1.2\times10^{-22}}\right) = 28°\ .$$

Therefore, the angle between $\vec{p}_e$ (which is in the $-x$ direction) and $\vec{p}_{n\,r}$ is $180° - 28° \approx 150°$.

(c) Measuring clockwise (but not using the "traditional" minus sign with that sense) we find the angle between $\vec{p}_{n\,r}$ and $\vec{p}_\nu$ (which points in the $-y$ direction) is $90° + 28° \approx 120°$.

(d) Combining the two equations $p = mv$ and $K = \frac{1}{2}mv^2$, we obtain (with $p = p_{n\,r}$ and $m = m_{n\,r}$)

$$K = \frac{p^2}{2m} = \frac{(1.4\times10^{-22})^2}{2\,(5.8\times10^{-26})} = 1.6\times10^{-19}\ \text{J}\ .$$

36. This problem involves both mechanical energy conservation

$$U_i = K_1 + K_2$$

where $U_i = 60$ J, and momentum conservation

$$0 = m_1\vec{v}_1 + m_2\vec{v}_2$$

where $m_2 = 2m_1$. From the second equation, we find $|\vec{v}_1| = 2\,|\vec{v}_2|$ which in turn implies (since $v_1 = |\vec{v}_1|$ and likewise for v_2)

$$K_1 = \frac{1}{2}m_1v_1^2 = \frac{1}{2}\left(\frac{1}{2}m_2\right)(2v_2)^2 = 2\left(\frac{1}{2}m_2v_2^2\right) = 2K_2 .$$

(a) We substitute $K_1 = 2K_2$ into the energy conservation relation and find

$$U_i = 2K_2 + K_2 \implies K_2 = \frac{1}{3}U_i = 20 \text{ J} .$$

(b) And we obtain $K_1 = 2(20) = 40$ J.

37. Our notation is as follows: the mass of the original body is $M = 20.0$ kg; its initial velocity is $\vec{v}_0 = 200\hat{\text{i}}$ in SI units (m/s); the mass of one fragment is $m_1 = 10.0$ kg; ; its velocity is $\vec{v}_1 = 100\hat{\text{j}}$ in SI units; the mass of the second fragment is $m_2 = 4.0$ kg; ; its velocity is $\vec{v}_2 = -500\hat{\text{i}}$ in SI units; and, the mass of the third fragment is $m_3 = 6.00$ kg.

(a) Conservation of linear momentum requires

$$M\vec{v}_0 = m_1\vec{v}_1 + m_2\vec{v}_2 + m_3\vec{v}_3$$

which (using the above information) leads to

$$\vec{v}_3 = 1000\,\hat{\text{i}} - 167\,\hat{\text{j}}$$

in SI units. The magnitude of $\vec{v}_3$ is $v_3 = \sqrt{1000^2 + (-167)^2} = 1.01 \times 10^3$ m/s. It points at $\tan^{-1}(-167/1000) = -9.48°$ (that is, at 9.5° measured clockwise from the $+x$ axis).

(b) We are asked to calculate ΔK or

$$\left(\frac{1}{2}m_1v_1^2 + \frac{1}{2}m_2v_2^2 + \frac{1}{2}m_3v_3^2\right) - \frac{1}{2}Mv_0^2 = 3.23 \times 10^6 \text{ J} .$$

38. Our notation (and, implicitly, our choice of coordinate system) is as follows: the mass of the original body is m; its initial velocity is $\vec{v}_0 = v\,\hat{\text{i}}$; the mass of the less massive piece is m_1; ; its velocity is $\vec{v}_1 = 0$; and, the mass of the more massive piece is m_2. We note that the conditions $m_2 = 3m_1$ (specified in the problem) and $m_1 + m_2 = m$ generally assumed in classical physics (before Einstein) lead us to conclude

$$m_1 = \frac{1}{4}m \quad \text{and} \quad m_2 = \frac{3}{4}m .$$

Conservation of linear momentum requires

$$\begin{aligned} m\vec{v}_0 &= m_1\vec{v}_1 + m_2\vec{v}_2 \\ mv\,\hat{\text{i}} &= 0 + \frac{3}{4}m\,\vec{v}_2 \end{aligned}$$

which leads to

$$\vec{v}_2 = \frac{4}{3}v\,\hat{\text{i}} .$$

The increase in the system's kinetic energy is therefore

$$\begin{aligned} \Delta K &= \frac{1}{2}m_1v_1^2 + \frac{1}{2}m_2v_2^2 - \frac{1}{2}mv_0^2 \\ &= 0 + \frac{1}{2}\left(\frac{3}{4}m\right)\left(\frac{4}{3}v\right)^2 - \frac{1}{2}mv^2 \\ &= \frac{1}{6}m\,v^2 . \end{aligned}$$

39. Our notation (and, implicitly, our choice of coordinate system) is as follows: the mass of one piece is $m_1 = m$; ; its velocity is $\vec{v}_1 = -30\,\hat{\mathrm{i}}$ in SI units (m/s); the mass of the second piece is $m_2 = m$; ; its velocity is $\vec{v}_2 = -30\,\hat{\mathrm{j}}$ in SI units; and, the mass of the third piece is $m_3 = 3m$. Conservation of linear momentum requires

$$\begin{aligned} m\vec{v}_0 &= m_1\vec{v}_1 + m_2\vec{v}_2 + m_3\vec{v}_3 \\ 0 &= m\left(-30\,\hat{\mathrm{i}}\right) + m\left(-30\,\hat{\mathrm{j}}\right) + 3m\vec{v}_3 \end{aligned}$$

which leads to

$$\vec{v}_3 = 10\,\hat{\mathrm{i}} + 10\,\hat{\mathrm{j}}$$

in SI units. Its magnitude is $v_3 = 10\sqrt{2} \approx 14$ m/s and its angle is 45° counterclockwise from $+x$ (in this system where we have m_1 flying off in the $-x$ direction and m_2 flying off in the $-y$ direction).

40. One approach is to choose a *moving* coordinate system which travels the center of mass of the body, and another is to do a little extra algebra analyzing it in the original coordinate system (in which the speed of the $m = 8.0$ kg mass is $v_0 = 2$ m/s, as given). Our solution is in terms of the latter approach since we are assuming that this is the approach most students would take. Conservation of linear momentum (along the direction of motion) requires

$$\begin{aligned} mv_0 &= m_1v_1 + m_2v_2 \\ (8.0)(2.0) &= (4.0)v_1 + (4.0)v_2 \end{aligned}$$

which leads to

$$v_2 = 4 - v_1$$

in SI units (m/s). We require

$$\begin{aligned} \Delta K &= \left(\frac{1}{2}m_1v_1^2 + \frac{1}{2}m_2v_2^2\right) - \frac{1}{2}mv_0^2 \\ 16 &= \left(\frac{1}{2}(4.0)v_1^2 + \frac{1}{2}(4.0)v_2^2\right) - \frac{1}{2}(8.0)(2.0)^2 \end{aligned}$$

which simplifies to

$$v_2^2 = 16 - v_1^2$$

in SI units. If we substitute for v_2 from above, we find

$$(4 - v_1)^2 = 16 - v_1^2$$

which simplifies to

$$2v_1^2 - 8v_1 = 0$$

and yields either $v_1 = 0$ or $v_1 = 4$ m/s. If $v_1 = 0$ then $v_2 = 4 - v_1 = 4$ m/s, and if $v_1 = 4$ then $v_2 = 0$. Stated more simply, one of the chunks has zero speed and the other has a velocity of 4.0 m/s (along the original direction of motion).

41. We use Eq. 9-43. Then

$$\begin{aligned} v_f &= v_i + v_{\text{rel}} \ln \frac{M_i}{M_f} \\ &= 105\,\text{m/s} + (253\,\text{m/s}) \ln \frac{6090\,\text{kg}}{6010\,\text{kg}} \\ &= 108 \text{ m/s} . \end{aligned}$$

42. (a) We use Eq. 9-42. The thrust is

$$\begin{aligned} R\,v_{\rm rel} &= Ma \\ &= \left(4.0\times10^4\,{\rm kg}\right)\left(2.0\,{\rm m/s^2}\right) \\ &= 8.0\times10^4\ {\rm N}\ . \end{aligned}$$

(b) Since $v_{\rm rel} = 3000$ m/s, we see from part (a) that $R \approx 27$ kg/s.

43. (a) The thrust of the rocket is given by $T = Rv_{\rm rel}$, where R is the rate of fuel consumption and $v_{\rm rel}$ is the speed of the exhaust gas relative to the rocket. For this problem $R = 480\,{\rm kg/s}$ and $v_{\rm rel} = 3.27\times10^3\,{\rm m/s}$, so

$$T = (480\,{\rm kg/s})(3.27\times10^3\,{\rm m/s}) = 1.57\times10^6\ {\rm N}\ .$$

(b) The mass of fuel ejected is given by $M_{\rm fuel} = R\Delta t$, where Δt is the time interval of the burn. Thus, $M_{\rm fuel} = (480\,{\rm kg/s})(250\,{\rm s}) = 1.20\times10^5\,{\rm kg}$. The mass of the rocket after the burn is $M_f = M_i - M_{\rm fuel} = 2.55\times10^5\,{\rm kg} - 1.20\times10^5\,{\rm kg} = 1.35\times10^5\,{\rm kg}$.

(c) Since the initial speed is zero, the final speed is given by

$$v_f = v_{\rm rel}\ln\frac{M_i}{M_f} = (3.27\times10^3)\ln\left(\frac{2.55\times10^5}{1.35\times10^5}\right) = 2.08\times10^3\ {\rm m/s}\ .$$

44. We use Eq. 9-43 and simplify with $v_i = 0$, $v_f = v$, and $v_{\rm rel} = u$.

$$v_f - v_i = v_{\rm rel}\ln\frac{M_i}{M_f} \implies \frac{M_i}{M_f} = e^{v/u}$$

(a) If $v = u$, we obtain $\frac{M_i}{M_f} = e^1 \approx 2.7$.

(b) If $v = 2u$, we obtain $\frac{M_i}{M_f} = e^2 \approx 7.4$.

45. We use Eq. 9-43 and simplify with $v_f - v_i = \Delta v$, and $v_{\rm rel} = u$.

$$v_f - v_i = v_{\rm rel}\ln\frac{M_i}{M_f} \implies \frac{M_f}{M_i} = e^{-\Delta v/u}$$

If $\Delta v = 2.2$ m/s and $u = 1000$ m/s, we obtain $\frac{M_i - M_f}{M_i} = 1 - e^{-0.0022} \approx 0.0022$.

46. We convert mass rate to SI units: $R = 540/60 = 9.00$ kg/s. In the absence of the asked-for additional force, the car would decelerate with a magnitude given by Eq. 9-42:

$$R\,v_{\rm rel} = M\,|a|$$

so that if $a = 0$ is desired then the additional force must have a magnitude equal to $R\,v_{\rm rel}$ (so as to cancel that effect).

$$F = R\,v_{\rm rel} = (9.00)(3.20) = 28.8\ {\rm N}\ .$$

47. (a) We consider what must happen to the coal that lands on the faster barge during one minute ($\Delta t = 60\,{\rm s}$). In that time, a total of $m = 1000\,{\rm kg}$ of coal must experience a change of velocity

$$\Delta v = 20\,{\rm km/h} - 10\,{\rm km/h} = 10\,{\rm km/h} = 2.8\ {\rm m/s}$$

where rightwards is considered the positive direction. The rate of change in momentum for the coal is therefore

$$\frac{\Delta\vec{p}}{\Delta t} = \frac{m\Delta\vec{v}}{\Delta t} = \frac{(1000)(2.8)}{60} = 46\ {\rm N}$$

which, by Eq. 9-23, must equal the force exerted by the (faster) barge on the coal. The processes (the shoveling, the barge motions) are constant, so there is no ambiguity in equating $\frac{\Delta p}{\Delta t}$ with $\frac{dp}{dt}$.

(b) The problem states that the frictional forces acting on the barges does not depend on mass, so the loss of mass from the slower barge does not affect its motion (so no extra force is required as a result of the shoveling).

48. (a) The thrust is $Rv_{\rm rel}$ where $v_{\rm rel} = 1200$ m/s. For this to equal the weight Mg where $M = 6100$ kg, we must have $R = (6100)(9.8)/1200 \approx 50$ kg/s.

(b) Using Eq. 9-42 with the additional effect due to gravity, we have

$$Rv_{\rm rel} - Mg = Ma$$

so that requiring $a = 21$ m/s^2 leads to $R = (6100)(9.8 + 21)/1200 = 1.6 \times 10^2$ kg/s.

49. (a) We assume his mass is between $m_1 = 50$ kg and $m_2 = 70$ kg (corresponding to a weight between 110 lb and 154 lb). His increase in gravitational potential energy is therefore in the range

$$\begin{aligned} m_1gh \leq &\ \Delta U \leq m_2gh \\ 2 \times 10^5 \leq &\ \Delta U \leq 3 \times 10^5 \end{aligned}$$

in SI units (J), where $h = 443$ m.

(b) The problem only asks for the amount of internal energy which converts into gravitational potential energy, so this result is the same as in part (a). But if we were to consider his *total* internal energy "output" (much of which converts to heat) we can expect that external climb is quite different from taking the stairs.

50. (a) The (internal) energy the climber must convert to gravitational potential energy is

$$\Delta U = mgh = (90)(9.8)(8850) = 7.8 \times 10^6 \text{ J} .$$

(b) The number of candy bars this corresponds to is

$$N = \frac{7.8 \times 10^6 \text{ J}}{1.25 \times 10^6 \text{ J/bar}} \approx 6 \text{ bars} .$$

51. (a) The acceleration of the sprinter is (using Eq. 2-15)

$$a = \frac{2\Delta x}{t^2} = \frac{(2)(7.0\,\text{m})}{(1.6\,\text{s})^2} = 5.47 \text{ m/s}^2 .$$

Consequently, the speed at $t = 1.6$ s is

$$v = at = \left(5.47\,\text{m/s}^2\right)(1.6\,\text{s}) = 8.8 \text{ m/s} .$$

Alternatively, Eq. 2-17 could be used.

(b) The kinetic energy of the sprinter (of weight w and mass $m = w/g$) is

$$K = \frac{1}{2}mv^2 = \frac{1}{2}\left(\frac{w}{g}\right)v^2 = \frac{(670)(8.8)^2}{2(9.8)} = 2.6 \times 10^3 \text{ J} .$$

(c) The average power is

$$P_{\rm avg} = \frac{\Delta K}{\Delta t} = \frac{2.6 \times 10^3 \text{ J}}{1.6\,\text{s}} = 1.6 \times 10^3 \text{ W} .$$

52. We use $P = Fv$ (Eq. 7-48) to compute the force:

$$F = \frac{P}{v} = \frac{92 \times 10^6 \text{ W}}{(32.5\,\text{knot})\left(1.852\,\frac{\text{km/h}}{\text{knot}}\right)\left(\frac{1000\,\text{m/km}}{3600\,\text{s/h}}\right)} = 5.5 \times 10^6 \text{ N} .$$

53. To swim at constant velocity the swimmer must push back against the water with a force of 110 N. Relative to him the water is going at 0.22 m/s toward his rear, in the same direction as his force. Using Eq. 7-48, his power output is obtained:

$$P = \vec{F} \cdot \vec{v} = Fv = (110\,\text{N})(0.22\,\text{m/s}) = 24\ \text{W}\ .$$

54. The initial kinetic energy of the automobile of mass m moving at speed v_i is $K_i = \frac{1}{2}mv_i^2$, where $m = 16400/9.8 = 1673$ kg. Using Eq. 8-29 and Eq. 8-31, this relates to the effect of friction force f in stopping the auto over a distance d by

$$K_i = fd$$

where the road is assumed level (so $\Delta U = 0$). Thus,

$$d = \frac{K_i}{f} = \frac{mv_i^2}{2f} = \frac{(1673\,\text{kg})\left((113\,\text{km/h})\left(\frac{1000\,\text{m/km}}{3600\,\text{s/h}}\right)\right)^2}{2(8230\,\text{N})} = 100\ \text{m}\ .$$

55. (a) By combining Newton's second law $F - mg = ma$ (where F is the force exerted up on her by the floor) and Eq. 2-16 $v^2 = 2ad_1$ (where $d_1 = 0.90 - 0.40 = 0.50$ m is the distance her center of mass moves while her feet are on the floor) it is straightforward to derive the equation

$$K_{\text{launch}} = (F - mg)d_1$$

where $K_{\text{launch}} = \frac{1}{2}mv^2$ is her kinetic energy as her feet leave the floor. We mention this method of deriving that equation (which also follows from the work-kinetic energy theorem Eq. 7-10, or – suitably interpreted – from energy conservation as expressed by Eq. 8-31) since the energy approaches might seem paradoxical (one might sink into the quagmire of questions such as "how can the floor possibly provide energy to the person?"); the Newton's law approach leads to no such quandaries. Next, her feet leave the floor and this kinetic energy is converted to gravitational potential energy. Then mechanical energy conservation leads straightforwardly to

$$K_{\text{launch}} = mgd_2$$

where $d_2 = 1.20 - 0.90 = 0.30$ m is the distance her center of mass rises from the time her feet leave the floor to the time she reaches the top of her leap. Now we combine these two equations and solve $(F - mg)d_1 = mgd_2$ for the force:

$$F = \frac{mg(d_1 + d_2)}{d_1} = \frac{(55\,\text{kg})(9.8\,\text{m/s}^2)(0.50\,\text{m} + 0.30\,\text{m})}{0.50\,\text{m}} = 860\ \text{N}\ .$$

(b) She has her maximum speed at the time her feet leave the floor (this is her "launch" speed). Consequently, the equation derived above becomes

$$\frac{1}{2}mv^2 = (F - mg)d_1$$

from which we obtain

$$\begin{aligned} v &= \sqrt{\frac{2(F - mg)d_1}{m}} \\ &= \sqrt{\frac{2\left(860\,\text{N} - (55\,\text{kg})\left(9.8\,\text{m/s}^2\right)\right)(0.50\,\text{m})}{55\,\text{kg}}} \\ &= 2.4\ \text{m/s}\ . \end{aligned}$$

56. (a) The kinetic energy K of the automobile of mass m at $t = 30\,\text{s}$ is

$$K = \frac{1}{2}mv^2 = \frac{1}{2}(1500\,\text{kg})\left((72\,\text{km/h})\left(\frac{1000\,\text{m/km}}{3600\,\text{s/h}}\right)\right)^2 = 3.0\times10^5\ \text{J}\ .$$

(b) The average power required is

$$P_{\text{avg}} = \frac{\Delta K}{\Delta t} = \frac{3.0\times10^5\,\text{J}}{30\,\text{s}} = 1.0\times10^4\ \text{W}\ .$$

(c) We use Eq. 7-48 ($P = Fv$) for the instantaneous power delivered at t. Since the acceleration a is constant, the power is $P = Fv = mav = ma(at) = ma^2t$, using Eq. 2-11. By contrast, from part (b), the average power is $P_{\text{avg}} = \frac{mv^2}{2t}$ which becomes $\frac{1}{2}ma^2t$ when $v = at$ is again utilized. Thus, the instantaneous power at the end of the interval is twice the average power during it:

$$P = 2P_{\text{avg}} = (2)\left(1.0\times10^4\,\text{W}\right) = 2.0\times10^4\ \text{W}\ .$$

57. (a) With $P = 1.5\,\text{MW} = 1.5\times10^6\,\text{W}$ (assumed constant) and $t = 6.0\,\text{min} = 360\,\text{s}$, the work-kinetic energy theorem (along with Eq. 7-48) becomes

$$W = Pt = \Delta K = \frac{1}{2}m\left(v_f^2 - v_i^2\right)\ .$$

The mass of the locomotive is then

$$m = \frac{2Pt}{v_f^2 - v_i^2} = \frac{(2)(1.5\times10^6\,\text{W})(360\,\text{s})}{(25\,\text{m/s})^2 - (10\,\text{m/s})^2} = 2.1\times10^6\,\text{kg}\,.$$

(b) With t arbitrary, we use $Pt = \frac{1}{2}m\left(v^2 - v_i^2\right)$ to solve for the speed $v = v(t)$ as a function of time and obtain

$$v(t) = \sqrt{v_i^2 + \frac{2Pt}{m}} = \sqrt{(10)^2 + \frac{(2)(1.5\times10^6)t}{2.1\times10^6}} = \sqrt{100 + 1.5t}$$

in SI units (v in m/s and t in s).

(c) Using Eq. 7-48, the force $F(t)$ as a function of time is

$$F(t) = \frac{P}{v(t)} = \frac{1.5\times10^6}{\sqrt{100+1.5t}}$$

in SI units (F in N and t in s).

(d) The distance d the train moved is given by

$$d = \int_0^t v(t')\,dt' = \int_0^{360}\left(100 + \frac{3}{2}t\right)^{\frac{1}{2}}\,dt = \frac{4}{9}\left(100 + \frac{3}{2}t\right)^{\frac{3}{2}}\Bigg|_0^{360}$$

which yields 6.7×10^3 m.

58. We work this in SI units and convert to horsepower in the last step. Thus,

$$v = (80\,\text{km/h})\left(\frac{1000\,\text{m/km}}{3600\,\text{s/h}}\right) = 22.2\ \text{m/s}\ .$$

The force F_P needed to propel the car (of weight w and mass $m = w/g$) is found from Newton's second law:

$$F_{\text{net}} = F_P - F = ma = \frac{wa}{g}$$

where $F = 300 + 1.8v^2$ in SI units. Therefore, the power required is

$$\begin{aligned} P &= \vec{F}_P \cdot \vec{v} \\ &= \left(F + \frac{wa}{g}\right) v \\ &= \left(300 + 1.8(22.2)^2 + \frac{(12000)(0.92)}{9.8}\right)(22.2) \\ &= 5.14 \times 10^4 \text{ W} \\ &= (5.14 \times 10^4 \text{ W}) \left(\frac{1 \text{ hp}}{746 \text{ W}}\right) = 69 \text{ hp} . \end{aligned}$$

59. The third-to-last statement in the problem about the peeling-off rate of the top layer and the thickening rate of the bottom layer is best interpreted, we feel, in the rest frame of the layer. Thus, imagining that we are in a reference frame moving up at v_t, then it is clear from the uniform nature of the described peeling-off of the top and thickening of the bottom that *in this moving reference frame* the center of mass of the layer must move downward with a speed $2v_f$ (if the rates were denoted R and were different then this would be $R_{\text{bottom}} + R_{\text{top}}$). Returning to the original reference frame, where we see the trapped bubbles rising at v_t, we find (with $+y$ upward) the center of mass velocity is

$$v_{\text{com}} = v_t - 2v_f = -1.5 \text{ cm/s} .$$

60. (a) Since the initial momentum is zero, then the final momenta must add (in the vector sense) to 0. Therefore, with SI units understood, we have

$$\begin{aligned} \vec{p}_3 &= -\vec{p}_1 - \vec{p}_2 \\ &= -m_1\vec{v}_1 - m_2\vec{v}_2 \\ &= -\left(16.7 \times 10^{-27}\right)\left(6.00 \times 10^6\,\hat{\text{i}}\right) - \left(8.35 \times 10^{-27}\right)\left(-8.00 \times 10^6\,\hat{\text{j}}\right) \\ &= -1.00 \times 10^{-19}\,\hat{\text{i}} + 0.67 \times 10^{-19}\,\hat{\text{j}} \text{ kg}\cdot\text{m/s} . \end{aligned}$$

(b) Dividing by $m_3 = 11.7 \times 10^{-27}$ kg and using Pythagorean's theorem we find the speed of the third particle to be $v_3 = 1.03 \times 10^7$ m/s. The total amount of kinetic energy is

$$\frac{1}{2}m_1v_1^2 + \frac{1}{2}m_2v_2^2 + \frac{1}{2}m_3v_3^2 = 1.19 \times 10^{-12} \text{ J} .$$

61. By conservation of momentum, the final speed v of the sled satisfies

$$(2900 \text{ kg})(250 \text{ m/s}) = (2900 \text{ kg} + 920 \text{ kg})v$$

which gives $v = 190$ m/s.

62. We denote the mass of the car as M and that of the sumo wrestler as m. Let the initial velocity of the sumo wrestler be $v_0 > 0$ and the final velocity of the car be v. We apply the momentum conservation law.

(a) From $mv_0 = (M + m)v$ we get

$$v = \frac{mv_0}{M + m} = \frac{(242 \text{ kg})(5.3 \text{ m/s})}{2140 \text{ kg} + 242 \text{ kg}} = 0.54 \text{ m/s} .$$

(b) Since $v_{\text{rel}} = v_0$, we have

$$mv_0 = Mv + m\left(v + v_{\text{rel}}\right) = mv_0 + (M + m)v$$

and obtain $v = 0$ for the final speed of the flatcar.

(c) Now $mv_0 = Mv + m(v - v_{\rm rel})$, which leads to

$$v = \frac{m(v_0 + v_{\rm rel})}{m + M} = \frac{(242\,\text{kg})(5.3\,\text{m/s} + 5.3\,\text{m/s})}{242\,\text{kg} + 2140\,\text{kg}} = 1.1 \text{ m/s} .$$

63. (a) We use coordinates with $+x$ eastward and $+y$ northward, and employ magnitude-angle notation which is well suited for computations with vector-capable calculators. Positive angles are measured counterclockwise from the $+x$ axis (negative angles are clockwise). Length is in meters and time is in seconds. The mass of each piece is designated m. Thus, the conservation of momentum becomes

$$\begin{aligned}\vec{p}_0 &= \vec{p}_1 + \vec{p}_2 + \vec{p}_3 \\ \vec{p}_0 &= m(7.0 \angle\ 90^\circ) + m(4.0 \angle\ 210^\circ) + m(4.0 \angle -30^\circ) \\ \vec{p}_0 &= m(3.0 \angle\ 90^\circ)\end{aligned}$$

which implies that the velocity of the package had magnitude $|\vec{p}|/(3m) = 1.0$ m/s and was directed north.

(b) The center of mass proceeds at 1.0 m/s unaffected by the explosion. Its displacement during the 3.0 s interval is $(1.0\,\text{m/s})(3.0\,\text{s}) = 3.0$ m. The displacement is directed north, in accordance with its velocity.

64. The width ℓ of the pyramid measured at variable height z is seen to decrease from L at the base (where $z = 0$) to zero at the top (where $z = H$). This is a linear decrease, so we must have

$$\ell = L\left(1 - \frac{z}{H}\right) .$$

If we imagine the pyramid layered into a large number N of horizontal (square) slabs (each of thickness Δz) then the volume of each slab is $V' = \ell^2 \Delta z$ and the mass of each slab is $m' = \rho V' = \rho \ell^2 \Delta z$. If we make the continuum approximation ($N \to \infty$ while $\Delta z \to dz$) and substitute from above for ℓ, the mass element becomes

$$dm = \rho L^2 \left(1 - \frac{z}{H}\right)^2 dz .$$

We note, for later use, that the total mass M is given by $\rho L^2 H/3$ using the volume relation mentioned in the problem, but this can also be derived by integrating the above expression for dm.

(a) Using Eq. 9-9 we find

$$z_{\rm com} = \frac{1}{M} \int z\, dm = \frac{3}{\rho L^2 H} \int_0^H z \rho L^2 \left(1 - \frac{z}{H}\right)^2 dz$$

where ρ and L^2 are constants (and, in fact, cancel) so we obtain

$$z_{\rm com} = \frac{3}{H} \int_0^H \left(z - \frac{2z^2}{H} + \frac{z^3}{H}\right) dz = \frac{H}{4} = 36.8 \text{ m} .$$

(b) Although we could do the integral $\int dU = \int gz\, dm$ to find the work done against gravity, it is easier to use the conclusion drawn in the book that this should be equivalent to lifting a point mass M to height $z_{\rm com}$.

$$W = \Delta U = Mgz_{\rm com} = \left(\frac{\rho L^2 H}{3}\right) g \frac{H}{4} = 1.7 \times 10^{12} \text{ J} .$$

65. Although it is expected that the boat will have a slight downward recoil (of brief duration) from the upward component of the father's leap, the problem's intent is to concentrate only on the horizontal components, since – if the effects of friction are small – the boat can continue moving horizontally for a significant time. Mass, velocity and momentum units are SI. We use coordinates with $+x$ eastward and

$+y$ northward. Angles are positive if measured counterclockwise from the $+x$ axis. Using magnitude-angle notation, momentum conservation is expressed as

$$\begin{aligned} \vec{p}_0 &= \vec{p}_c + \vec{p}_f + \vec{p}_b \\ (0 \angle 0°) &= (80 \angle 0°) + (90 \angle -90°) + \vec{p}_b \end{aligned}$$

where it must be stressed that the relevant component of the father's momentum is $\vec{p}_f = (75)(1.5)\cos 37°$ *south* (represented as $(90 \angle -90°)$ in the expression above). Thus, we obtain $\vec{p}_b = (120 \angle 132°)$, which implies that the boat's (horizontal) velocity is $|\vec{p}|/m = 120/100 = 1.2$ m/s at an angle of 132° counterclockwise from east; this can also be expressed as 48° north of west.

66. (a) Ignoring air friction amounts to assuming that the ball has the same speed v when it returns to its original height.

$$K_i = K_f = \frac{1}{2}mv^2 = \frac{1}{2}(0.050\,\text{kg})(16\,\text{m/s})^2 = 6.4\text{ J}\,.$$

(b) The momentum at the moment it is thrown (taking $+y$ upward) is

$$|\vec{p}_i| = |\vec{p}_f| = mv = (0.050\,\text{kg})(16\,\text{m/s}) = 0.80\text{ kg}\cdot\text{m/s}\,.$$

The vector $\vec{p}_i$ is $\theta = 30°$ above the horizontal, while $\vec{p}_f$ is 30° below the horizontal (since the vertical component is now downward). We note for later reference that the magnitude of the change in momentum is

$$|\Delta\vec{p}| = |\vec{p}_f - \vec{p}_i| = 2mv\sin\theta = 0.80\text{ kg}\cdot\text{m/s}$$

and $\Delta\vec{p}$ points vertically downward.

(c) The time of flight for the ball is $t = 2v_i\sin\theta/g$, thus

$$mgt = mg\left(\frac{2v\sin\theta}{g}\right) = 2mv\sin\theta = 2p_i\sin\theta = 0.80\text{ kg}\cdot\text{m/s}$$

which (recalling our result in part (b)) illustrates the relation $|\Delta p| = Ft$ where $F = mg$.

67. Choosing downward as the $+y$ direction and placing the coordinate origin at the top of the building, we apply the equations from Table 2-1 to this two-block system:

$$\begin{aligned} y_1 &= \frac{1}{2}gt^2 && \text{for} \quad 0 \le t \le 5 \\ y_2 &= \frac{1}{2}g(t-1)^2 && \text{for} \quad 1 \le t \le 6 \\ v_1 &= gt && \text{for} \quad 0 \le t \le 5 \\ v_2 &= g(t-1) && \text{for} \quad 1 \le t \le 6 \end{aligned}$$

with SI units understood.

(a) With $m_1 = 2.00$ kg and $m_2 = 3.00$ kg, Eq. 9-5 provides

$$y_{\text{com}} = \frac{m_1y_1 + m_2y_2}{m_1 + m_2} = \frac{1}{2}gt^2 - \frac{3}{5}gt + \frac{3}{10}g$$

while they are both in free fall ($1 \le t \le 5$). But during the interval when m_2 is "waiting" at the top of the building, we have

$$y_{\text{com}} = \frac{m_1y_1 + m_2(0)}{m_1 + m_2} = \frac{1}{5}gt^2 \quad \text{for } 0 \le t \le 1$$

and during the interval where m_1 is sitting on the ground (at $y = \frac{1}{2}(9.8)(5)^2$) we have

$$y_{\text{com}} = \frac{m_1\left(\frac{25g}{2}\right) + m_2 y_2}{m_1 + m_2} = \frac{3}{10}gt^2 - \frac{3}{5}gt + \frac{53}{10}g$$

for $5 \le t \le 6$. This behavior is plotted below, with y_{com} in meters and t in seconds.

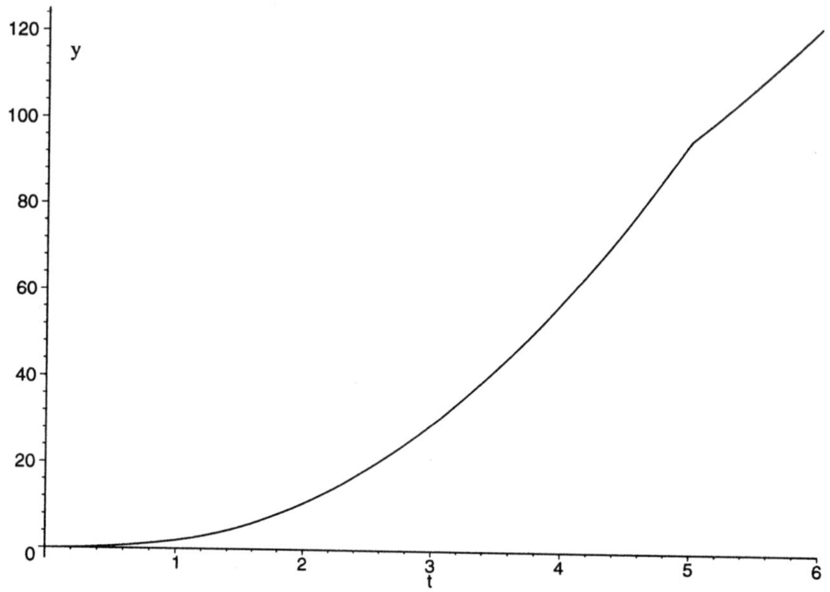

(b) We turn now to Eq. 9-17 which gives

$$v_{\text{com}} = \frac{m_1 v_1 + m_2 v_2}{m_1 + m_2} = gt - \frac{3}{5}g$$

while they are both in free fall ($1 \le t \le 5$). We note that we could have easily gotten this by taking the derivative of the corresponding y_{com} expression in part (a). During the interval when m_2 sits on the top of the building, we have

$$v_{\text{com}} = \frac{m_1 v_1 + m_2(0)}{m_1 + m_2} = \frac{2}{5}gt \quad \text{for } 0 \le t \le 1$$

and during the interval where m_1 sits on the ground we have

$$v_{\text{com}} = \frac{m_1(0) + m_2 v_2}{m_1 + m_2} = \frac{3}{5}gt - \frac{3}{5}g$$

for $5 \le t \le 6$. This behavior is plotted below, with v_{com} in m/s and t in s. The sudden drop at $t = 5$ s is understandable, since m_1 stops, but it should be noted that we are ignoring the dynamics of how the ground decelerates that block – the effects of which might be to (slightly) smooth out that transition.

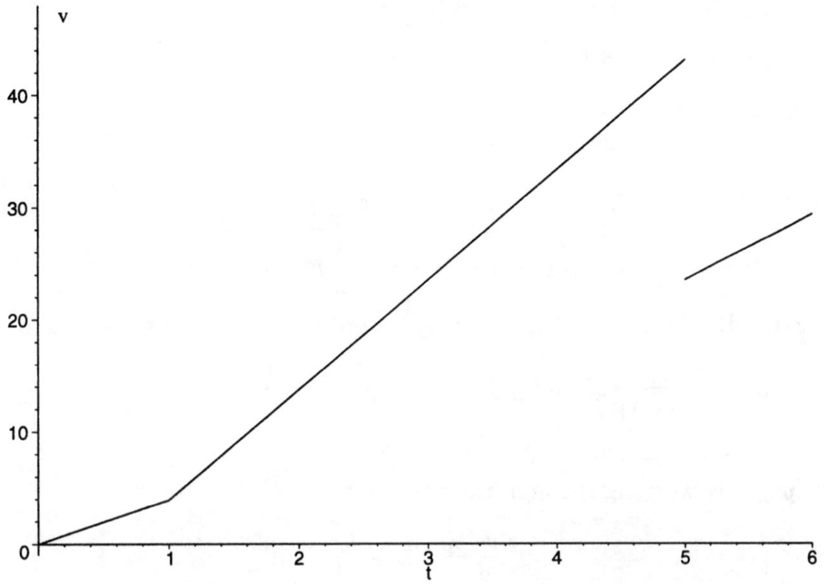

68. The velocity of the first particle (of mass $m_1 = 3.0$ kg) is $\vec{v}_1 = -6.0\hat{\jmath}$ m/s while that of the second one (of mass $m_2 = 4.0$ kg) is $\vec{v}_2 = 7.0\hat{\imath}$ m/s. The center-of-mass velocity is then

$$\vec{v}_{\text{com}} = \frac{m_1\vec{v}_1 + m_2\vec{v}_2}{m_1 + m_2} = \frac{(3.0)(-6.0\hat{\jmath}) + (4.0)(7.0\hat{\imath})}{3.0 + 4.0} = -2.6\hat{\imath} + 4.0\hat{\jmath}$$

in SI units. The corresponding speed is

$$v_{\text{com}} = \sqrt{v_x^2 + v_y^2} = \sqrt{(-2.6)^2 + (4.0)^2} = 4.8 \text{ m/s} .$$

69. We use Eq. 9-17, or – equivalently – we differentiate Eq. 9-5.

$$\begin{aligned} v_{\text{com}\,x} &= \frac{1}{M}\left((1500\,\text{kg})(0 \text{ m/s}) + (4000\,\text{kg})v_{\text{truck}}\right) \\ v_{\text{com}\,y} &= \frac{1}{M}\left((1500\,\text{kg})v_{\text{car}} + (4000\,\text{kg})(0\,\text{m/s})\right) \end{aligned}$$

where $M = 5500$ kg. From $v_{\text{com}\,x} = (11)\cos 55° = 6.3$ m/s and $v_{\text{com}\,y} = (11)\sin 55° = 9.0$ m/s, we get the following results for v_{truck} and v_{car} from the above formulas.

(a) $v_{\text{car}} = 33$ m/s.

(b) $v_{\text{truck}} = 8.7$ m/s.

70. (a) We use Eq. 7-48:

$$P = Fv \implies F = \frac{16.0\,\text{kW}}{15.0\,\text{m/s}} = 1.07 \text{ kN} .$$

(b) We add to our previous result the downhill pull of gravity $mg\sin\theta$ where $\theta = \tan^{-1}(8/100)$.

$$F' = 1.07 \times 10^3 + (1710)(9.8)\sin 4.57° = 2.40 \times 10^3$$

in SI units (N). Therefore,

$$P' = F'v = (2.40\,\text{kN}))(15.0\,\text{m/s}) = 36 \text{ kW} .$$

(c) For the engine to be off but the (downhill) velocity to remain constant, the downhill component of gravity must equal the magnitude of the retarding forces:

$$mg\sin\theta = F \ .$$

Using F from part (a), we find $\theta = 3.65°$ which corresponds to $\tan\theta = 0.0638 \approx 6.4\%$.

71. Using Eq. 2-15 for both object j (the jelly jar) and object p (the peanut butter), with $y = 0$ designating the base of the building in both cases, we have

$$\begin{aligned} y_j &= 40t - \frac{1}{2}gt^2 \\ y_p - 50 &= 0 - \frac{1}{2}gt^2 \end{aligned}$$

with SI units understood. Thus, using Eq. 9-5, the center of mass of this system is at

$$y_{\rm com} = \frac{1}{3.0\ {\rm kg}}\left((1.0\ {\rm kg})y_j + (2.0\ {\rm kg})y_p\right) = \frac{100}{3} + \frac{40}{3}t - \frac{1}{2}gt^2 \ .$$

(a) With $t = 3.0$ s, the above equation gives $y_{\rm com} = 29$ m.

(b) We maximize $y_{\rm com}$ by working through the condition

$$\frac{dy_{\rm com}}{dt} = 0 = \frac{40}{3} - gt \ .$$

Thus, we find $t = 1.4$ s, which produces $y_{\rm com} = 42$ m as its highest value.

72. (a) We denote the mass of the car (and cannon) as M (excluding that of the cannonballs) and the mass of all the cannonballs as m. For concreteness, we assume that before firing all the cannonballs are at the front (left side of Fig. 9-52) of the car, which we choose to be the origin of the x axis; we choose $+x$ rightward. The coordinate of the center of mass of the car-cannonball system is

$$x_{\rm com} = \frac{(0)m + \left(\frac{L}{2}\right)M}{M+m} = \frac{LM}{2(M+m)} \ .$$

After the firing, we assume all the cannonballs are at the other end of the car; the train will have moved (in the negative x direction) by a distance d, at which time

$$x_{\rm com} = \frac{\left(\frac{L}{2} - d\right)M + (L-d)m}{M+m} \ .$$

Equating the two expressions, we obtain $d = \frac{mL}{M+m} < L$. If $m \gg M$, the distance d can be very close to (but can never exceed) L. Thus $d_{\rm max} = L$.

(b) After each impact, there is no relative motion in the system; thus, the final speed of the car is equal to that of the center of mass of the system, which is zero.

73. Let the velocity of the shell (of mass m_s) relative to the ground be $\vec{v}_s$, the recoiling velocity of the cannon (of mass m_c) be $\vec{v}_c$ (pointed in our $-x$ direction), and the velocity of the shell relative to the muzzle be $\vec{v}_s'$, where $\vec{v}_s\,' + \vec{v}_c = \vec{v}_s$. In component form, this becomes

$$\begin{aligned} v_s'\cos 39.0° - v_c &= v_{sx} \\ v_s'\sin 39.0° &= v_{sy} \end{aligned}$$

where $v_c = |\vec{v}_c|$. Conservation of linear momentum in the horizontal direction provides us with the additional relation $m_s v_{sx} = m_c v_c$. We solve these equations for the components of $\vec{v}_s$:

$$\begin{aligned} v_{sx} &= \frac{m_c v_s'\cos 39.0°}{m_s + m_c} = \frac{(1400\,{\rm kg})(556\,{\rm m/s})\cos 39.0°}{1400\,{\rm kg} + 70.0\,{\rm kg}} = 412\,{\rm m/s} \\ v_{sy} &= v_s'\sin 39.0° = (556\,{\rm m/s})(\sin 39.0°) = 350\,{\rm m/s} \ . \end{aligned}$$

(a) The speed of the shell relative to the Earth is then

$$v_s = \sqrt{v_{sx}^2 + v_{sy}^2} = \sqrt{412^2 + 350^2} = 540 \text{ m/s} .$$

(b) The angle (relative to a stationary observer) at which the shell is fired is given by

$$\theta = \tan^{-1}\left(\frac{v_{sy}}{v_{sx}}\right) = \tan^{-1}\left(\frac{350}{412}\right) = 40.4^\circ .$$

74. The value 0.368 comes from rounding off e^{-1}. We will use e^{-1} in our solution. The speed of the rocket v as a function of the instantaneous rocket mass M' is given by $v = v_{\text{rel}} \ln(M/M')$ (Eq. 9-43 with $v_i = 0$). Thus, when $M' = e^{-1} M$, the speed of the fuel as measured by observers in the initial reference frame (defined when the rocket was at rest with $M' = M$) is

$$v_{\text{fuel}} = v - v_{\text{rel}} = v_{\text{rel}}\left(\ln\frac{M}{M'} - 1\right) = v(\ln\left(\frac{1}{e^{-1}}\right) - 1) = 0 .$$

75. We use momentum conservation choosing $+x$ forward and recognizing that the initial momentum is zero. We analyze this from the point of view of an observer at rest on the ice.

(a) If $v_{1\,\text{and}\,2}$ is the speed of the stones, then the speeds are related by $v_{1\,\text{and}\,2} + v_{\text{boat}} = v_{\text{rel}}$. Thus, with $m_1 = 2m_2$ and $M = 12m_2$, we obtain

$$\begin{aligned} 0 &= (m_1 + m_2)(-v_{1\,\text{and}\,2}) + M v_{\text{boat}} \\ &= (2m_2 + m_2)(-v_{\text{rel}} + v_{\text{boat}}) + 12 m_2 v_{\text{boat}} \\ &= -3m_2 v_{\text{rel}} + 15 m_2 v_{\text{boat}} \end{aligned}$$

which yields $v_{\text{boat}} = \frac{1}{5} v_{\text{rel}} = 0.2000 v_{\text{rel}}$.

(b) Using $v_1 + v'_{\text{boat}} = v_{\text{rel}}$, we find – as a result of the first throw – the boat's speed:

$$\begin{aligned} 0 &= m_1(-v_1) + (M + m_2) v'_{\text{boat}} \\ &= 2m_2(-v_{\text{rel}} + v'_{\text{boat}}) + (12m_2 + m_2) v'_{\text{boat}} \\ &= -2m_2 v_{\text{rel}} + 15 m_2 v'_{\text{boat}} \end{aligned}$$

which yields $v'_{\text{boat}} = \frac{2}{15} v_{\text{rel}} \approx 0.133 v_{\text{rel}}$. Then, using $v_2 + v_{\text{boat}} = v_{\text{rel}}$, we consider the second throw:

$$\begin{aligned} (M + m_2) v'_{\text{boat}} &= m_2(-v_2) + M v_{\text{boat}} \\ (12m_2 + m_2)\left(\frac{2}{15} v_{\text{rel}}\right) &= m_2(-v_{\text{rel}} + v_{\text{boat}}) + 12 m_2 v_{\text{boat}} \\ \frac{26}{15} m_2 v_{\text{rel}} &= -m_2 v_{\text{rel}} + 13 m_2 v_{\text{boat}} \end{aligned}$$

which yields $v_{\text{boat}} = \frac{41}{195} v_{\text{rel}} \approx 0.2103 v_{\text{rel}}$.

(c) Finally, using $v_2 + v'_{\text{boat}} = v_{\text{rel}}$, we find – as a result of the first throw – the boat's speed:

$$\begin{aligned} 0 &= m_2(-v_2) + (M + m_1) v'_{\text{boat}} \\ &= m_2(-v_{\text{rel}} + v'_{\text{boat}}) + (12m_2 + 2m_2) v'_{\text{boat}} \\ &= -m_2 v_{\text{rel}} + 15 m_2 v'_{\text{boat}} \end{aligned}$$

which yields $v'_{\text{boat}} = \frac{1}{15} v_{\text{rel}} \approx 0.0673 v_{\text{rel}}$. Then, using $v_1 + v_{\text{boat}} = v_{\text{rel}}$, we consider the second throw:

$$\begin{aligned} (M + m_1) v'_{\text{boat}} &= m_1(-v_1) + M v_{\text{boat}} \\ (12m_2 + 2m_2)\left(\frac{1}{15} v_{\text{rel}}\right) &= 2m_2(-v_{\text{rel}} + v_{\text{boat}}) + 12 m_2 v_{\text{boat}} \\ \frac{14}{15} m_2 v_{\text{rel}} &= -2m_2 v_{\text{rel}} + 14 m_2 v_{\text{boat}} \end{aligned}$$

which yields $v_{\text{boat}} = \frac{22}{105}v_{\text{rel}} \approx 0.2095 v_{\text{rel}}$.

76. (a) It is clear from the problem that $\vec{v}_{\text{air,plane}} = -180\,\hat{\imath}$ m/s where $+\hat{\imath}$ is the plane's direction of motion (relative to the ground).

(b) Let ΔM_a be the mass of air taken in and ejected and let ΔM_f be the mass of fuel ejected in time Δt. From the viewpoint of ground-based observers, the initial velocity of the air is zero and its final velocity is $v - u$, where u is the exhaust speed (labeled v_{rel} in the textbook) and v is the velocity of the plane. The initial velocity of the fuel is v and its final velocity is $v - u$. The velocity of the plane changes from v to $v + \Delta v$ over this time interval. The change in the total momentum of the plane-fuel-air system is $\Delta P = M_p \Delta v + \Delta M_f(u) + \Delta M_a(u - v)$ so the net external force is

$$\frac{\Delta P}{\Delta t} = M_p\frac{\Delta v}{\Delta t} - u\frac{\Delta M_f}{\Delta t} + (v-u)\frac{\Delta M_a}{\Delta t}\ .$$

We examine some of these terms individually. The $v\,\Delta M_a/\Delta t$ term gives the magnitude of the force on the plane due to air intake (most easily seen from the point of view of observers on the plane) and is equal to $(180)(70) = 1.3 \times 10^4$ N.

(c) We interpret the question as asking for the force due to ejection of both the air and the combustion products due to consuming the fuel. This corresponds then to the $u\,\Delta M_a/\Delta t$ and $u\,\Delta M_f/\Delta t$ terms above, and is equal to $(490)(70 + 2.9) = 3.6 \times 10^4$ N.

(d) We require $\Delta P/\Delta t = 0$ since this (the air, plane and fuel) forms an isolated system (Eq. 9-29). Therefore, our equation above leads to

$$M_p\frac{\Delta v}{\Delta t} = u\frac{\Delta M_f}{\Delta t} + (u-v)\frac{\Delta M_a}{\Delta t}$$

with all the terms on the right hand side constituting the net thrust (compare Eq. 9-42). These are the values (with appropriate signs) found in parts (b) and (c), so we obtain $3.6 \times 10^4 - 1.3 \times 10^4 = 2.3 \times 10^4$ N.

(e) Using Eq. 7-48, we multiply the net thrust by the plane speed and obtain $\left(2.3 \times 10^4\right)(180) = 4.2 \times 10^6$ W.

77. Using Eq. 9-5, we have

$$\begin{aligned} x_{\text{com}} &= 0 = \frac{1}{M}\left((4.0\text{ kg})(0\text{ m}) + (3.0\text{ kg})(3.0\text{ m}) + (2.0\text{ kg})x\right) \\ y_{\text{com}} &= 0 = \frac{1}{M}\left((4.0\text{ kg})(2.0\text{ m}) + (3.0\text{ kg})(1.0\text{ m}) + (2.0\text{ kg})y\right) \end{aligned}$$

where $M = 9.0$ kg.

(a) Evaluating the above, we find $x = -4.5$ m.

(b) And we find $y = -5.5$ m.

78. (a) We use Eq. 9-5 to compute the center of mass coordinates.

$$\begin{aligned} x_{\text{com}} &= \frac{(4\,\text{kg})(0) + (3\,\text{kg})(7\,\text{m}) + (5\,\text{kg})(3\,\text{m})}{4\,\text{kg} + 3\,\text{kg} + 5\,\text{kg}} = 3.00\text{ m} \\ y_{\text{com}} &= \frac{(4\,\text{kg})(0) + (3\,\text{kg})(3\,\text{m}) + (5\,\text{kg})(2\,\text{m})}{4\,\text{kg} + 3\,\text{kg} + 5\,\text{kg}} = 1.58\text{ m} \end{aligned}$$

(b) Using Eq. 9-17 and SI units, we obtain

$$\begin{aligned} \vec{v} &= \frac{(4\,\text{kg})\left(1.5\hat{\imath} - 2.5\hat{\jmath}\right) + (3\,\text{kg})(0) + (5\,\text{kg})\left(2.0\hat{\imath} - 1.0\hat{\jmath}\right)}{4\,\text{kg} + 3\,\text{kg} + 5\,\text{kg}} \\ &= 1.33\hat{\imath} - 1.25\hat{\jmath}\ \text{m/s}\ . \end{aligned}$$

(c) Multiplying the previous result by the total mass yields $\vec{P} = 16.0\hat{\text{i}} - 15.0\hat{\text{j}}$ in SI units (kg·m/s). This can also be gotten by adding up the individual momenta.

79. Although we do not show graphs here, we do jot down down an idea for each part.

(a) Find the center of mass of a rod in which the density is not uniform. If the rod extends along the x axis from the origin to $x = 5$ m, then with mass-per-unit-length (as a function of x) equal to e^{-x} in SI units, use Eq. 9-9 to find x_{com}. *A sketch of the solution is*

$$x_{\text{com}} = \frac{1}{M}\int_0^5 xe^{-x}\,dx \approx 37 \text{ m}$$

where M was figured from $\int_0^5 e^{-x}dx \approx 1$ kg.

(b) A firecracker is dropped from a height of 20 m. Halfway down it explodes into two identical pieces. As a result of the explosion, one of the pieces is (momentarily) at rest. What is the speed of the other piece immediately after the explosion? *A sketch of the solution is*

$$v_{\text{firecrack}} = \sqrt{2g(10\,\text{m})} = 14 \text{ m/s}$$

and we use momentum conservation:

$$mv_{\text{firecrack}} = \frac{m}{2}v_{\text{piece}} \implies v_{\text{piece}} = 28 \text{ m/s}\ .$$

(c) An 80 kg person is climbing a ladder at a steady rate of 25 cm/s. If we assume his total power output P is three times his rate of gaining gravitational potential energy, then compute P. *The solution is*

$$P = 3\frac{\Delta U}{\Delta t} = 3\frac{mg\Delta y}{\Delta t} \approx 590 \text{ J}$$

where $\Delta t = 1$ s and $\Delta y = 0.25$ m.

(d) Unlike the ideal physics of point particles moving through a vacuum, a runner cannot continue at a constant velocity effortlessly. If a runner's total power output is 650 W while running at 5.8 m/s, then what is the force retarding him (which includes several friction-related effects)? *The solution is*

$$P = Fv = \implies F = \frac{P}{v} \approx 110 \text{ N}\ .$$

80. (First problem in **Cluster**)

(a) The length of each of the tall sides is $\ell = \sqrt{H^2 + (B/2)^2}$, so that the total length of the wire is $L = 2\sqrt{H^2 + (B/2)^2} + B$. If A is the cross-section area and ρ is the density, then the total mass of the wire is $M = \rho AL$ and the mass of each of the tall sides is

$$m_\ell = \rho A\sqrt{H^2 + (B/2)^2} = M\frac{\sqrt{H^2 + (B/2)^2}}{2\sqrt{H^2 + (B/2)^2} + B}\ .$$

It is clear by symmetry that $x_{\text{com}} = B/2$ for the system, but the value of y_{com} is not obvious. Note that the base does not contribute to this computation:

$$y_{\text{com}} = \frac{1}{M}\left(m_\ell\frac{H}{2} + m_\ell\frac{H}{2}\right)$$

which can be 'simplified' to the following form.

$$y_{\text{com}} = \frac{H}{2 + \frac{B}{\sqrt{H^2 + (B/2)^2}}}$$

(b) The element of mass on the left-hand tall side is related to $d\ell = \sqrt{dx^2 + dy^2}$ and ultimately to the individual coordinate elements (since $dy = (2H/B)dx$):

$$dm_\ell = \rho A d\ell = \begin{cases} \rho A\sqrt{1+(2H/B)^2}\,dx \\ \rho A\sqrt{(B/2H)^2+1}\,dy \end{cases}$$

where $\rho A = m_\ell/\sqrt{H^2 + (B/2)^2}$ (see part (a)). Therefore, using Eq. 9-9, we have

$$\begin{aligned} x_{\text{com}} &= \frac{1}{m_\ell}\int_0^{B/2} x\,\frac{m_\ell}{\sqrt{H^2+(B/2)^2}}\sqrt{1+(2H/B)^2}\,dx \\ &= \frac{\sqrt{1+(2H/B)^2}}{\sqrt{H^2+(B/2)^2}}\int_0^{B/2} x\,dx \\ &= \frac{2}{B}\frac{\sqrt{(B/2)^2+H^2}}{\sqrt{H^2+(B/2)^2}}\left(\frac{B^2}{8}-0\right) \\ &= \frac{B}{4} \quad \text{and} \\ y_{\text{com}} &= \frac{1}{m_\ell}\int_0^{H} y\,\frac{m_\ell}{\sqrt{H^2+(B/2)^2}}\sqrt{(B/2H)^2+1}\,dy \\ &= \frac{\sqrt{(B/2H)^2+1}}{\sqrt{H^2+(B/2)^2}}\int_0^{H} y\,dy \\ &= \frac{1}{H}\frac{\sqrt{(B/2)^2+H^2}}{\sqrt{H^2+(B/2)^2}}\left(\frac{H^2}{2}-0\right) \\ &= \frac{H}{2}\,. \end{aligned}$$

81. (Second problem in **Cluster**)
It is clear by symmetry that $x_{\text{com}} = B/2$ for the system, but the value of y_{com} is not obvious. If the thickness is Δz and the density is ρ, then the relation between the mass element dm and a height element dy is

$$dm = \rho\Delta z \ell_y\,dy = \frac{M}{A_\triangle}\ell_y\,dy$$

where the area of the triangle is $A_\triangle = \frac{1}{2}BH$ and the length of each horizontal "strip" at height y is $\ell_y = B(1 - y/H)$. Therefore, using Eq. 9-9, we have

$$\begin{aligned} y_{\text{com}} &= \frac{1}{M}\int_0^H y\,\frac{M}{A_\triangle}B\left(1-\frac{y}{H}\right)dy \\ &= \frac{B}{\frac{1}{2}BH}\int_0^H y\left(1-\frac{y}{H}\right)dy \\ &= \frac{2}{H}\left(\frac{H^2}{2}-\frac{H^3}{3H}\right) \\ &= \frac{H}{3}\,. \end{aligned}$$

82. (Third problem in **Cluster**)
It is clear by symmetry that $x_{\text{com}} = B/2$ for the system, but the value of y_{com} is not obvious. If the cross-section area of the wire is A and the density is ρ, then in one quadrant the relation between the mass element dm and height element dy is

$$dm = \rho A\frac{R}{\sqrt{R^2-y^2}}\,dy = \frac{M}{\ell_\cap}\frac{R}{\sqrt{R^2-y^2}}\,dy$$

where the length of the semicircle is $\ell_\cap = \pi R$. To include the contributions from both quadrants shown, we multiply by 2, and Eq. 9-9 becomes

$$\begin{aligned} y_{\rm com} &= \frac{2}{M}\int_0^R y\,\frac{M}{\ell_\cap}\,\frac{R}{\sqrt{R^2-y^2}}\,dy \\ &= \frac{2}{\pi}\int_0^R \frac{y}{\sqrt{R^2-y^2}}\,dy \\ &= \frac{2}{\pi}\left[-\sqrt{R^2-y^2}\right]_0^R \\ &= \frac{2R}{\pi}\,. \end{aligned}$$

83. (Fourth problem in **Cluster**)
It is clear by symmetry that $x_{\rm com} = B/2$ for the system. The value of $y_{\rm com}$ is found as follows. If the thickness is Δz and the density is ρ, then the relation between the mass element dm and a height element dy is

$$dm = \rho\Delta z \ell_y\,dy = \frac{M}{A}\ell_y\,dy$$

where the area of the semicircle is $A = \frac{1}{2}\pi R^2$ and the length of each horizontal "strip" at height y is $\ell_y = 2\sqrt{R^2-y^2}$. Therefore, using Eq. 9-9, we find

$$\begin{aligned} y_{\rm com} &= \frac{1}{M}\int_0^R y\,\frac{M}{A}\,2\sqrt{R^2-y^2}\,dy \\ &= \frac{2}{\frac{1}{2}\pi R^2}\int_0^R y\sqrt{R^2-y^2}\,dy \\ &= \frac{4}{\pi R^2}\left[-\frac{1}{3}\left(R^2-y^2\right)^{3/2}\right]_0^R \\ &= \frac{4R}{3\pi}\,. \end{aligned}$$

Chapter 10

1. If F_{avg} is the magnitude of the average force, then the magnitude of the impulse is $J = F_{avg}\Delta t$, where Δt is the time interval over which the force is exerted (see Eq. 10-8). This equals the magnitude of the change in the momentum of the ball. Since the ball is initially at rest, J is equal to the magnitude of the final momentum mv. When $F_{avg}\Delta t = mv$ is solved for the speed, the result is

$$v = \frac{F_{avg}\Delta t}{m} = \frac{(50\,\text{N})\,(10 \times 10^{-3}\,\text{s})}{0.20\,\text{kg}} = 2.5 \text{ m/s} .$$

2. The magnitude of the average force is

$$\left|\vec{F}_{avg}\right| = \frac{|\Delta\vec{p}|}{m} = \frac{m|\Delta\vec{v}|}{\Delta t} = \frac{(2300\,\text{kg})(15\,\text{m/s})}{0.56\,\text{s}} = 6.2 \times 10^4 \text{ N} .$$

3. We take the final direction of motion to be the $+\hat{\imath}$ direction (when it is headed back to the pitcher) so that $\vec{v}_f = +60\hat{\imath}$ and $\vec{v}_i = -40\hat{\imath}$ in SI units. Therefore, $\Delta\vec{v} = 60 - (-40) = 100\hat{\imath}$ m/s. The magnitude of the average force is

$$\left|\vec{F}_{avg}\right| = \frac{|\Delta\vec{p}|}{\Delta t} = \frac{m|\Delta\vec{v}|}{\Delta t} = \frac{(0.150\,\text{kg})(100\,\text{m/s})}{5.0 \times 10^{-3}\,\text{s}} = 3.0 \times 10^3 \text{ N} .$$

4. We estimate his mass in the neighborhood of 70 kg and compute the upward force F of the water from Newton's second law.

$$F - mg = ma$$

where we have chosen $+y$ upward, so that $a > 0$ (the acceleration is upward since it represents a deceleration of his downward motion through the water). His speed when he arrives at the surface of the water is found either from Eq. 2-16 or from energy conservation:

$$v = \sqrt{2gh}$$

where $h = 12$ m, and since the deceleration a reduces the speed to zero over a distance $d = 0.30$ m we also obtain $v = \sqrt{2ad}$. We use these observations in the following.

(a) Equating our two expressions for v leads to $a = gh/d$. Our force equation, then, leads to

$$F = mg + m\left(g\frac{h}{d}\right) = mg\left(1 + \frac{h}{d}\right)$$

which yields $F \approx 2.8 \times 10^4$ kg. Since we are not at all certain of his mass, we express this as a guessed-at range (in kN) $25 < F < 30$.

(b) Since $F \gg mg$, the impulse $\vec{J}$ due to the net force (while he is in contact with the water) is overwhelmingly caused by the upward force of the water: $\int F\,dt = \vec{J}$ to a good approximation. Thus, by Eq. 10-2,

$$\int F\,dt = \vec{p}_f - \vec{p}_i = 0 - m\left(-\sqrt{2gh}\right)$$

(the minus sign with the initial velocity is due to the fact that downward is the negative direction) which yields $(70)\sqrt{2(9.8)(12)} = 1.1 \times 10^3$ kg·m/s. Expressing this as a range (in kN·s) we estimate $1.0 < \int F\,dt < 1.2$.

5. We take the initial direction of motion to be positive and use F_{avg} to denote the magnitude of the average force, Δt as the duration of the force, m as the mass of the ball, v_i as the initial velocity of the ball, and v_f as the final velocity of the ball. The force is in the negative direction and the impulse-momentum theorem (Eq. 10-4 with Eq. 10-8) yields $-F_{avg}\Delta t = mv_f - mv_i$. Thus,

$$v_f = \frac{mv_i - F_{avg}\Delta t}{m} = \frac{(0.40\,\text{kg})(14\,\text{m/s}) - (1200\,\text{N})(27\times10^{-3}\,\text{s})}{0.40\,\text{kg}} = -67\,\text{m/s}\ .$$

The final speed of the ball is 67 m/s. The negative sign indicates that the velocity is opposite to the initial direction of travel.

6. We choose $+y$ upward, which implies $a > 0$ (the acceleration is upward since it represents a deceleration of his downward motion through the snow).

(a) The maximum deceleration a_{max} of the paratrooper (of mass m and initial speed $v = 56$ m/s) is found from Newton's second law

$$F_{snow} - mg = ma_{max}$$

where we require $F_{snow} = 1.2 \times 10^5$ N. Using Eq. 2-15 $v^2 = 2a_{max}d$, we find the minimum depth of snow for the man to survive:

$$d = \frac{v^2}{2a_{max}} = \frac{mv^2}{2\left(F_{snow} - mg\right)} \approx \frac{(85\,\text{kg})(56\,\text{m/s})^2}{2(1.2\times10^5\,\text{N})} = 1.1\,\text{m}\ .$$

(b) His short trip through the snow involves a change in momentum

$$\vec{p}_f - \vec{p}_i = 0 - (85\,\text{kg})(-56\,\text{m/s})$$

(the negative value of the initial velocity is due to the fact that downward is the negative direction) which yields 4.8×10^3 kg·m/s. By the impulse-momentum theorem, this equals the impulse due to the net force $F_{snow} - mg$, but since $F_{snow} \gg mg$ we can approximate this as the impulse on him just from the snow.

7. We choose $+y$ upward, which means $\vec{v}_i = -25$ m/s and $\vec{v}_f = +10$ m/s. During the collision, we make the reasonable approximation that the net force on the ball is equal to F_{avg} – the average force exerted by the floor up on the ball.

(a) Using the impulse momentum theorem (Eq. 10-4) we find

$$\vec{J} = m\vec{v}_f - m\vec{v}_i = (1.2)(10) - (1.2)(-25) = 42\ \text{kg}\cdot\text{m/s}\ .$$

(b) From Eq. 10-8, we obtain

$$\vec{F}_{avg} = \frac{\vec{J}}{\Delta t} = \frac{42}{0.020} = 2.1 \times 10^3\ \text{N}\ .$$

8. We choose the positive direction in the direction of rebound so that $\vec{v}_f > 0$ and $\vec{v}_i < 0$. Since they have the same speed v, we write this as $\vec{v}_f = v$ and $\vec{v}_i = -v$. Therefore, the change in momentum for each bullet of mass m is $\Delta\vec{p} = m\Delta v = 2mv$. Consequently, the total change in momentum for the 100 bullets (each minute) $\Delta\vec{P} = 100\Delta\vec{p} = 200mv$. The average force is then

$$\vec{F}_{\text{avg}} = \frac{\Delta\vec{P}}{\Delta t} = \frac{(200)(3 \times 10^{-3}\,\text{kg})(500\,\text{m/s})}{(1\,\text{min})(60\,\text{s/min})} \approx 5\,\text{N} .$$

9. (a) The initial momentum of the car is $\vec{p}_i = m\vec{v}_i = (1400\,\text{kg})(5.3\,\text{m/s})\,\hat{\text{j}} = (7400\,\text{kg}\cdot\text{m/s})\,\hat{\text{j}}$ and the final momentum is $\vec{p}_f = (7400\,\text{kg}\cdot\text{m/s})\,\hat{\text{i}}$. The impulse on it equals the change in its momentum:

$$\vec{J} = \vec{p}_f - \vec{p}_i = (7400\,\text{N}\cdot\text{s})(\hat{\text{i}} - \hat{\text{j}}) .$$

(b) The initial momentum of the car is $\vec{p}_i = (7400\,\text{kg}\cdot\text{m/s})\,\hat{\text{i}}$ and the final momentum is $\vec{p}_f = 0$. The impulse acting on it is

$$\vec{J} = \vec{p}_f - \vec{p}_i = -7400\,\hat{\text{i}}\ \text{N}\cdot\text{s} .$$

(c) The average force on the car is

$$\vec{F}_{\text{avg}} = \frac{\Delta\vec{p}}{\Delta t} = \frac{\vec{J}}{\Delta t} = \frac{(7400\,\text{kg}\cdot\text{m/s})(\hat{\text{i}} - \hat{\text{j}})}{4.6\,\text{s}} = (1600\,\text{N})(\hat{\text{i}} - \hat{\text{j}})$$

and its magnitude is $F_{\text{avg}} = (1600\,\text{N})\sqrt{2} = 2300\,\text{N}$.

(d) The average force is

$$\vec{F}_{\text{avg}} = \frac{\vec{J}}{\Delta t} = \frac{(-7400\,\text{kg}\cdot\text{m/s})\,\hat{\text{i}}}{350 \times 10^{-3}\,\text{s}} = (-2.1 \times 10^4\,\text{N})\,\hat{\text{i}}$$

and its magnitude is $F_{\text{avg}} = 2.1 \times 10^4\,\text{N}$.

(e) The average force is given above in unit vector notation. Its x and y components have equal magnitudes. The x component is positive and the y component is negative, so the force is 45° below the positive x axis.

10. We use coordinates with $+x$ rightward and $+y$ upward, with the usual conventions for measuring the angles (so that the initial angle becomes $180 + 35 = 215°$). Using SI units and magnitude-angle notation (efficient to work with when using a vector capable calculator), the change in momentum is

$$\vec{p}_f - \vec{p}_i = (3.0 \angle 90°) - (3.6 \angle 215°) = (5.9 \angle 60°) .$$

This equals the impulse delivered to the ball (by the bat). Then, Eq. 10-8 leads to

$$F_{\text{avg}}\Delta t = 5.9 \implies F_{\text{avg}} = \frac{5.9}{2.0 \times 10^{-3}} \approx 2.9 \times 10^3\ \text{N} .$$

We note that this force is very much larger than the weight of the ball, which justifies our (implicit) assumption that gravity played no significant role in the collision.

11. We take the magnitude of the force to be $F = At$, where A is a constant of proportionality. The condition that $F = 50\,\text{N}$ when $t = 4.0\,\text{s}$ leads to $A = (50\,\text{N})/(4.0\,\text{s}) = 12.5\,\text{N/s}$. The magnitude of the impulse exerted on the object is

$$J = \int_0^{4.0} F\,dt = \int_0^{4.0} At\,dt = \frac{1}{2}At^2\bigg|_0^{4.0} = \frac{1}{2}(12.5)(4.0)^2 = 100\ \text{N}\cdot\text{s} .$$

This equals the magnitude of the change in the momentum of the object (by the impulse-momentum theorem), and since the ball started from rest, we have $J = mv_f$. Therefore, $v_f = J/m = (100\,\text{N}\cdot\text{s})/(10\,\text{kg}) = 10\,\text{m/s}$.

12. (a) The mass of each spherical hailstone of radius $r = 0.5$ cm and density $\rho = 0.92$ g/cm^3 is

$$m = \rho\left(\frac{4\pi r^3}{3}\right) = 0.48 \text{ g} = 4.8 \times 10^{-4} \text{ kg} .$$

(b) If the final speed is zero, then Eq. 10-4 and Eq. 10-8 (with $+y$ upward) lead to

$$\vec{F}_{\text{avg}}\Delta t = -m\vec{v}_i = -\left(4.8 \times 10^{-4}\right)(-25) = 0.012$$

in SI units (N·s). This gives the impulse imparted to a single hailstone by the roof (and is equal to the magnitude of the force on the roof by the hailstone, by Newton's third law). An imagined "cube" of falling air, $\ell = 1$ m on each side (falling with the hail at $v = 25$ m/s), takes a time of

$$\Delta t = \frac{\ell}{v} = \frac{1\,\text{m}}{25\,\text{m/s}} = 0.04 \text{ s}$$

to fully "collapse" onto a square meter of roof top (delivering its load of 120 hailstones). We can cover an area of $10\,\text{m} \times 20\,\text{m}$ with 200 of these "collapsing cubes" of air. Therefore, in this time, the total impulse is of magnitude

$$\vec{F}_{\text{avg,total}}\Delta t = 200(120)(0.012\,\text{N}\cdot\text{s}) \approx 290\,\text{N}\cdot\text{s}$$

which leads to $\vec{F}_{\text{avg,total}} = 290/0.04 = 7.2 \times 10^3$ N.

13. (a) If m is the mass of a pellet and v is its velocity as it hits the wall, then its momentum is $p = mv = (2.0 \times 10^{-3}\,\text{kg})(500\,\text{m/s}) = 1.0\,\text{kg}\cdot\text{m/s}$, toward the wall.

(b) The kinetic energy of a pellet is

$$K = \frac{1}{2}mv^2 = \frac{1}{2}\left(2.0 \times 10^{-3}\,\text{kg}\right)\left(500\,\text{m/s}\right)^2 = 2.5 \times 10^2 \text{ J} .$$

(c) The force on the wall is given by the rate at which momentum is transferred from the pellets to the wall. Since the pellets do not rebound, each pellet that hits transfers $p = 1.0\,\text{kg}\cdot\text{m/s}$. If ΔN pellets hit in time Δt, then the average rate at which momentum is transferred is

$$F_{\text{avg}} = \frac{p\Delta N}{\Delta t} = (1.0\,\text{kg}\cdot\text{m/s})\left(10\,\text{s}^{-1}\right) = 10 \text{ N} .$$

The force on the wall is in the direction of the initial velocity of the pellets.

(d) If Δt is the time interval for a pellet to be brought to rest by the wall, then the average force exerted on the wall by a pellet is

$$F_{\text{avg}} = \frac{p}{\Delta t} = \frac{1.0\,\text{kg}\cdot\text{m/s}}{0.6 \times 10^{-3}\,\text{s}} = 1.7 \times 10^3\,\text{N} .$$

The force is in the direction of the initial velocity of the pellet.

(e) In part (d) the force is averaged over the time a pellet is in contact with the wall, while in part (c) it is averaged over the time for many pellets to hit the wall. During the majority of this time, no pellet is in contact with the wall, so the average force in part (c) is much less than the average force in part (d).

14. We choose our positive direction in the direction of the rebound (so the ball's initial velocity is negative-valued). We evaluate the integral $J = \int F\,dt$ by adding the appropriate areas (of a triangle, a rectangle,

and another triangle) shown in the graph (but with the t converted to seconds). With $m = 0.058\,\text{kg}$ and $v = 34\,\text{m/s}$, we apply the impulse-momentum theorem:

$$\begin{aligned}
\int F_{\text{wall}}\,dt &= m\vec{v}_f - m\vec{v}_i \\
\int_0^{0.002} F\,dt + \int_{0.002}^{0.004} F\,dt + \int_{0.004}^{0.006} F\,dt &= m(+v) - m(-v) \\
\frac{1}{2}F_{\text{max}}\,(0.002\,\text{s}) + F_{\text{max}}\,(0.002\,\text{s}) + \frac{1}{2}F_{\text{max}}\,(0.002\,\text{s}) &= 2mv \\
F_{\text{max}}\,(0.004\,\text{s}) &= 2\,(0.058\,\text{kg})\,(34\,\text{m/s})
\end{aligned}$$

which yields $F_{\text{max}} = 9.9 \times 10^2$ N.

15. We first consider the 1200 kg part. The impulse has magnitude J and is (by our choice of coordinates) in the positive direction. Let m_1 be the mass of the part and v_1 be its velocity after the bolts are exploded. We assume both parts are at rest before the explosion. Then $J = m_1 v_1$, so

$$v_1 = \frac{J}{m_1} = \frac{300\,\text{N}\cdot\text{s}}{1200\,\text{kg}} = 0.25\ \text{m/s}\ .$$

The impulse on the 1800 kg part has the same magnitude but is in the opposite direction, so $-J = m_2 v_2$, where m_2 is the mass and v_2 is the velocity of the part. Therefore,

$$v_2 = -\frac{J}{m_2} = -\frac{300\,\text{N}\cdot\text{s}}{1800\,\text{kg}} = -0.167\ \text{m/s}\ .$$

Consequently, the relative speed of the parts after the explosion is $0.25\,\text{m/s} - (-0.167\,\text{m/s}) = 0.417\,\text{m/s}$.

16. We choose our positive direction in the direction of the rebound (so the ball's initial velocity is negative-valued: $\vec{v}_i = -5.2$ m/s).

(a) The speed of the ball right after the collision is

$$\begin{aligned}
v_f &= \sqrt{\frac{2K_f}{m}} \\
&= \sqrt{\frac{2\left(\frac{1}{2}K_i\right)}{m}} \\
&= \sqrt{\frac{\frac{1}{2}mv_i^2}{m}} \\
&= \frac{v_i}{\sqrt{2}} \approx 3.7\ \text{m/s}\ .
\end{aligned}$$

(b) With $m = 0.15$ kg, the impulse-momentum theorem (Eq. 10-4) yields

$$\vec{J} = m\vec{v}_f - m\vec{v}_i = (0.15)(3.7) - (0.15)(-5.2) = 1.3$$

in SI units (N·s).

(c) Eq. 10-8 leads to $F_{\text{avg}} = J/\Delta t = 1.3/0.0076 = 1.8 \times 10^2$ N.

17. We choose $+y$ in the direction of the rebound (directly away from the wall) and $+x$ towards the right in the figure (parallel to the wall; see Fig. 10-30). Using unit-vector notation, the the ball's initial and final velocities are

$$\begin{aligned}
\vec{v}_i &= v\cos\theta\,\hat{\text{i}} - v\sin\theta\,\hat{\text{j}} = 5.2\,\hat{\text{i}} - 3.0\,\hat{\text{j}} \\
\vec{v}_f &= v\cos\theta\,\hat{\text{i}} + v\sin\theta\,\hat{\text{j}} = 5.2\,\hat{\text{i}} + 3.0\,\hat{\text{j}}
\end{aligned}$$

respectively (with SI units understood).

(a) With $m = 0.30$ kg, the impulse-momentum theorem (Eq. 10.4) yields

$$\vec{J} = m\vec{v}_f - m\vec{v}_i = 2(0.30)(3.0\hat{\jmath})$$

so that the magnitude of the impulse delivered on the ball by the wall is 1.8 N·s and its direction is directly away from the wall (which, in terms of Fig. 10-30, is "up").

(b) Using Eq. 10-8, the force on the ball by the wall is $\vec{J}/\Delta t = 1.8\hat{\jmath}/0.010 = 180\hat{\jmath}$ N. By Newton's third law, the force on the wall by the ball is $-180\hat{\jmath}$ N (that is, its magnitude is 180 N and its direction is directly into the wall, or "down" in the view provided by Fig. 10-30).

18. (a) Regardless of the direction of the thrust, the change in linear momentum of the space probe is given by the impulse-momentum theorem (also using Eq. 10-8):

$$\Delta p = (3000\,\text{N})(65.0\,\text{s}) = 1.95 \times 10^5\,\text{kg·m/s} .$$

(b) The change in speed for the probe of mass m is

$$\Delta v = \frac{\Delta p}{m} = \frac{1.95 \times 10^5\,\text{kg·m/s}}{2500\,\text{kg}} = 78.0\,\text{m/s} .$$

Let the initial and final speeds of the probe be v_i and v_f, respectively. Then, the change in its kinetic energy is $\Delta K = \frac{1}{2}m(v_f^2 - v_i^2)$. If the thrust is backward then $v_f = v_i - \Delta v$, and

$$\begin{aligned}\Delta K &= \frac{1}{2}m\left((v_i - \Delta v)^2 - v_i^2\right)\\ &= \frac{1}{2}(2500\,\text{kg})\left((300\,\text{m/s} - 78.0\,\text{m/s})^2 - (300\,\text{m/s})^2\right)\\ &= -5.09 \times 10^7\ \text{J}\end{aligned}$$

If the thrust is forward then $v_f = v_i + \Delta v$, and

$$\begin{aligned}\Delta K &= \frac{1}{2}m\left((v_i + \Delta v)^2 - v_i^2\right)\\ &= \frac{1}{2}(2500\,\text{kg})\left((300\,\text{m/s} + 78.0\,\text{m/s})^2 - (300\,\text{m/s})^2\right)\\ &= 6.61 \times 10^7\ \text{J} .\end{aligned}$$

If the thrust is sideways then $v_f = \sqrt{(\Delta v)^2 + v_i^2}$, and

$$\Delta K = \frac{1}{2}m\left((\Delta v)^2 + v_i^2 - v_i^2\right) = \frac{1}{2}(2500\,\text{kg})(78.0\,\text{m/s})^2 = 7.61 \times 10^6\ \text{J} .$$

19. (a) We take the force to be in the positive direction, at least for earlier times. Then the impulse is

$$\begin{aligned}J &= \int_0^{3.0\times10^{-3}} F\,dt\\ &= \int_0^{3.0\times10^{-3}} (6.0\times10^6)\,t - (2.0\times10^9)\,t^2\ dt\\ &= \left[\frac{1}{2}(6.0\times10^6)t^2 - \frac{1}{3}(2.0\times10^9)t^3\right]_0^{3.0\times10^{-3}}\\ &= 9.0\ \text{N·s} .\end{aligned}$$

(b) Since $J = F_{\text{avg}}\,\Delta t$, we find

$$F_{\text{avg}} = \frac{J}{\Delta t} = \frac{9.0\,\text{N·s}}{3.0\times10^{-3}\,\text{s}} = 3.0\times10^3\ \text{N} .$$

(c) To find the time at which the maximum force occurs, we set the derivative of F with respect to time equal to zero – and solve for t. The result is $t = 1.5 \times 10^{-3}$ s. At that time the force is

$$F_{\max} = (6.0 \times 10^6)(1.5 \times 10^{-3}) - (2.0 \times 10^9)(1.5 \times 10^{-3})^2 = 4.5 \times 10^3\,\mathrm{N}\ .$$

(d) Since it starts from rest, the ball acquires momentum equal to the impulse from the kick. Let m be the mass of the ball and v be its speed as it leaves the foot. Then,

$$v = \frac{p}{m} = \frac{J}{m} = \frac{9.0\,\mathrm{N\cdot s}}{0.45\,\mathrm{kg}} = 20\ \mathrm{m/s}\ .$$

20. (a) We choose $+x$ along the initial direction of motion and apply momentum conservation:

$$\begin{aligned} m_{\text{bullet}}\vec{v}_i &= m_{\text{bullet}}\vec{v}_1 + m_{\text{block}}\vec{v}_2 \\ (5.2\,\mathrm{g})(672\,\mathrm{m/s}) &= (5.2\,\mathrm{g})(428\,\mathrm{m/s}) + (700\,\mathrm{g})\vec{v}_2 \end{aligned}$$

which yields $v_2 = 1.81$ m/s.

(b) It is a consequence of momentum conservation that the velocity of the center of mass is unchanged by the collision. We choose to evaluate it before the collision:

$$\vec{v}_{\text{com}} = \frac{m_{\text{bullet}}\vec{v}_i}{m_{\text{bullet}} + m_{\text{block}}} = \frac{(5.2\,\mathrm{g})(672\,\mathrm{m/s})}{5.2\,\mathrm{g} + 700\,\mathrm{g}}$$

which gives the result $\vec{v}_{\text{com}} = 4.96$ m/s.

21. We examine the horizontal components of the momenta of the package and sled. Let m_s be the mass of the sled and v_s be its initial velocity. Let m_p be the mass of the package and let v be the final velocity of the sled and package together. The horizontal component of the total momentum is conserved, so $m_s v_s = (m_s + m_p)v$ and

$$v = \frac{v_s m_s}{m_s + m_p} = \frac{(9.0\,\mathrm{m/s})(6.0\,\mathrm{kg})}{6.0\,\mathrm{kg} + 12\,\mathrm{kg}} = 3.0\ \mathrm{m/s}\ .$$

22. We refer to the discussion in the textbook (see Sample Problem 10-2, which uses the same notation that we use here) for many of the important details in the reasoning. Here we only present the primary computational step (using SI units):

$$v = \frac{m+M}{m}\sqrt{2gh} = \frac{2.010}{0.010}\sqrt{2(9.8)(0.12)} = 3.1 \times 10^2\ \mathrm{m/s}\ .$$

23. Let m_m be the mass of the meteor and m_e be the mass of Earth. Let v_m be the velocity of the meteor just before the collision and let v be the velocity of Earth (with the meteor) just after the collision. The momentum of the Earth-meteor system is conserved during the collision. Thus, in the reference frame of Earth before the collision, $m_m v_m = (m_m + m_e)v$, so

$$v = \frac{v_m m_m}{m_m + m_e} = \frac{(7200\,\mathrm{m/s})(5 \times 10^{10}\,\mathrm{kg})}{5.98 \times 10^{24}\,\mathrm{kg} + 5 \times 10^{10}\,\mathrm{kg}} = 6 \times 10^{-11}\,\mathrm{m/s}\ .$$

We convert this as follows:

$$\left(6 \times 10^{-11}\,\frac{\mathrm{m}}{\mathrm{s}}\right)\left(\frac{1000\,\mathrm{mm}}{\mathrm{m}}\right)\left(\frac{3.2 \times 10^7\,\mathrm{s}}{\mathrm{y}}\right) = 2\ \mathrm{mm/y}\ .$$

24. (a) To relate the sliding distance to the speed V of the bullet-plus-block at the instant it has finished embedding itself in the block, we can either use Eq. 2-16 and $\vec{F}_{\text{net}} = m\vec{a}$, or energy conservation as expressed by Eq. 8-31 (with $W = 0$ and $f_k = \mu_k(m + M)g$ using Eq. 6-2). We choose the latter approach:

$$\begin{aligned} K_{\text{bullet plus block}} &= \Delta E_{\text{th}} \\ \frac{1}{2}(m + M)V^2 &= \mu_k(m + M)gd \end{aligned}$$

which yields $V = \sqrt{2\mu_k gd} = 2.7$ m/s.

(b) For the collision itself, we use momentum conservation (with the direction of motion being positive).

$$\begin{aligned} m_{\text{bullet}}v_i &= (m + M)V \\ (0.0045\,\text{kg})v_i &= (2.4045\,\text{kg})(2.7\text{m/s}) \end{aligned}$$

which gives the result $v_i = 1.4 \times 10^3$ m/s.

25. (a) The magnitude of the deceleration of each of the cars is $a = f/m = \mu_k mg/m = \mu_k g$. If a car stops in distance d, then its speed v just after impact is obtained from Eq. 2-16:

$$v^2 = v_0^2 + 2ad \implies v = \sqrt{2ad} = \sqrt{2\mu_k gd}$$

since $v_0 = 0$ (this could alternatively have been derived using Eq. 8-31). Thus,

$$v_A = \sqrt{2(0.13)(9.8)(8.2)} = 4.6 \text{ m/s} \quad , \quad \text{and}$$

(b) $v_B = \sqrt{2(0.13)(9.8)(6.1)} = 3.9$ m/s.

(c) Let the speed of car B be v just before the impact. Conservation of linear momentum gives $m_B v = m_A v_A + m_B v_B$, or

$$v = \frac{(m_A v_A + m_B v_B)}{m_B} = \frac{(1100)(4.6) + (1400)(3.9)}{1400} = 7.5 \text{ m/s} .$$

The conservation of linear momentum during the impact depends on the fact that the only significant force (during impact of duration Δt) is the force of contact between the bodies. In this case, that implies that the force of friction exerted by the road on the cars is neglected during the brief Δt. This neglect would introduce some error in the analysis. Related to this is the assumption we are making that the transfer of momentum occurs at one location – that the cars do not slide appreciably during Δt – which is certainly an approximation (though probably a good one). Another source of error is the application of the friction relation Eq. 6-2 for the sliding portion of the problem (after the impact); friction is a complex force that Eq. 6-2 only partially describes.

26. We note that the "(a)" and "(b)" in Fig. 10-32 do not correspond to parts (a) and (b) (in fact, it's somewhat the reverse). Our $+x$ direction is to the right (so all velocities are positive-valued).

(a) We apply momentum conservation to relate the situation just before the bullet strikes the second block to the situation where the bullet is embedded within the block.

$$(0.0035\,\text{kg})v = (1.8035\,\text{kg})(1.4\,\text{m/s}) \implies v = 721 \text{ m/s} .$$

(b) We apply momentum conservation to relate the situation just before the bullet strikes the first block to the instant it has passed through it (having speed v found in part (a)).

$$(0.0035\,\text{kg})v_0 = (1.2\,\text{kg})(0.63\,\text{m/s}) + (0.0035\,\text{kg})(721\,\text{m/s})$$

which yields $v_0 = 937$ m/s.

27. (a) We want to calculate the force that the scale exerts on the marbles. This is the sum of the force that holds the marbles already on the scale against the downward force of gravity and the force that brings the falling marbles to rest. At the end of time t, the number of marbles on the scale is Rt. At this moment, the gravitational force on them is $Rtmg$ and the upward force of the scale that holds them is $F_1 = Rtmg$. Just before striking the scale, a marble that fell from height h has speed $v = \sqrt{2gh}$ and momentum $p = m\sqrt{2gh}$. To stop the falling marbles, the scale must exert an upward force $F_2 = Rp = Rm\sqrt{2gh}$. The total force of the scale on the marbles is

$$F = F_1 + F_2 = Rtmg + Rm\sqrt{2gh} = Rm\left(gt + \sqrt{2gh}\right) .$$

(b) For the given data (using SI units, so $m = 0.0045$ kg), we find

$$F = (100)(0.0045)\left((9.8)(10.0) + \sqrt{2(9.8)(7.60)}\right)$$

which yields $F = 49.6$ N. Assuming the scale is calibrated to read in terms of an equivalent mass, its reading is $F/g = 49.6/9.8 = 5.06$ kg.

28. We choose $+x$ in the direction of (initial) motion of the blocks, which have masses $m_1 = 5$ kg and $m_2 = 10$ kg. Where units are not shown in the following, SI units are to be understood.

(a) Momentum conservation leads to

$$\begin{aligned} m_1\vec{v}_{1i} + m_2\vec{v}_{2i} &= m_1\vec{v}_{1f} + m_2\vec{v}_{2f} \\ (5)(3) + (10)(2) &= 5\vec{v}_{1f} + (10)(2.5) \end{aligned}$$

which yields $\vec{v}_{1f} = 2$. Thus, the speed of the 5 kg block immediately after the collision is 2.0 m/s.

(b) We find the reduction in total kinetic energy:

$$K_i - K_f = \frac{1}{2}(5)(3)^2 + \frac{1}{2}(10)(2)^2 - \frac{1}{2}(5)(2)^2 - \frac{1}{2}(10)(2.5)^2$$

which gives the result 1.25 J. Rounding to two figures and recalling that $\Delta K = K_f - K_i$ then our answer is $\Delta K = -1.3$ J.

(c) In this new scenario where $\vec{v}_{2f} = 4.0$ m/s, momentum conservation leads to $\vec{v}_{1f} = -1.0$ m/s and we obtain $\Delta K = +40$ J.

(d) The creation of additional kinetic energy is possible if, say, some gunpowder were on the surface where the impact occurred (initially stored chemical energy would then be contributing to the result).

29. Let m_F be the mass of the freight car and v_F be its initial velocity. Let m_C be the mass of the caboose and v be the common final velocity of the two when they are coupled. Conservation of the total momentum of the two-car system leads to $m_F v_F = (m_F + m_C)v$, so $v = v_F m_F/(m_F + m_C)$. The initial kinetic energy of the system is

$$K_i = \frac{1}{2}m_F v_F^2$$

and the final kinetic energy is

$$K_f = \frac{1}{2}(m_F + m_C)v^2 = \frac{1}{2}(m_F + m_C)\frac{m_F^2 v_F^2}{(m_F + m_C)^2} = \frac{1}{2}\frac{m_F^2 v_F^2}{(m_F + m_C)} .$$

Since 27% of the original kinetic energy is lost, we have $K_f = 0.73 K_i$. Thus,

$$\frac{1}{2}\frac{m_F^2 v_F^2}{(m_F + m_C)} = (0.73)\left(\frac{1}{2}m_F v_F^2\right) .$$

Simplifying, we obtain $m_F/(m_F + m_C) = 0.73$, which we use in solving for the mass of the caboose:

$$m_C = \frac{0.27}{0.73}m_F = 0.37\,m_F = (0.37)\left(3.18 \times 10^4 \text{ kg}\right) = 1.18 \times 10^4 \text{ kg} .$$

30. We think of this as having two parts: the first is the collision itself – where the bullet passes through the block so quickly that the block has not had time to move through any distance yet – and then the subsequent "leap" of the block into the air (up to height h measured from its initial position). The first part involves momentum conservation (with $+y$ upward):

$$(0.01\,\text{kg})(1000\,\text{m/s}) = (5.0\,\text{kg})\vec{v} + (0.01\,\text{kg})(400\,\text{m/s})$$

which yields $\vec{v} = 1.2$ m/s. The second part involves either the free-fall equations from Ch. 2 (since we are ignoring air friction) or simple energy conservation from Ch. 8. Choosing the latter approach, we have

$$\frac{1}{2}(5.0\,\text{kg})(1.2\,\text{m/s})^2 = (5.0\,\text{kg})\left(9.8\,\text{m/s}^2\right)h$$

which gives the result $h = 0.073$ m.

31. (a) Let v be the final velocity of the ball-gun system. Since the total momentum of the system is conserved $mv_i = (m+M)v$. Therefore, $v = mv_i/(m+M)$.

    (b) The initial kinetic energy is $K_i = \frac{1}{2}mv_i^2$ and the final kinetic energy is $K_f = \frac{1}{2}(m+M)v^2 = \frac{1}{2}m^2v_i^2/(m+M)$. The problem indicates $\Delta E_{\text{th}} = 0$, so the difference $K_i - K_f$ must equal the energy U_s stored in the spring:

$$U_s = \frac{1}{2}mv_i^2 - \frac{1}{2}\frac{m^2v_i^2}{(m+M)} = \frac{1}{2}mv_i^2\left(1 - \frac{m}{m+M}\right) = \frac{1}{2}mv_i^2\frac{M}{m+M}\,.$$

    Consequently, the fraction of the initial kinetic energy that becomes stored in the spring is $U_s/K_i = M/(m+M)$.

32. For a picture of this one-dimensional example of an "explosion" involving two objects ($m_1 = 4.0$ kg and $m_2 = 6.0$ kg), see Fig. 9-40 (but reverse the velocity arrows). Since the system was initially at rest, momentum conservation leads to

$$0 = m_2\vec{v}_2 + m_1\vec{v}_1 \implies |\vec{v}_1| = \frac{m_2}{m_1}\,|\vec{v}_2|$$

which yields 6.0 m/s for the speed of the physics book. Mechanical energy conservation tells us that the initial potential energy is

$$U_i = K_{f\,\text{total}} = \frac{1}{2}m_1v_1^2 + \frac{1}{2}m_2v_2^2$$

which gives the result $U_i = 120$ J.

33. As hinted in the problem statement, the velocity v of the system as a whole – when the spring reaches the maximum compression x_{m} – satisfies $m_1v_{1i} + m_2v_{2i} = (m_1+m_2)v$. The change in kinetic energy of the system is therefore

$$\begin{aligned}\Delta K &= \frac{1}{2}(m_1+m_2)v^2 - \frac{1}{2}m_1v_{1i}^2 - \frac{1}{2}m_2v_{2i}^2 \\ &= \frac{(m_1v_{1i}+m_2v_{2i})^2}{2\,(m_1+m_2)} - \frac{1}{2}m_1v_{1i}^2 - \frac{1}{2}m_2v_{2i}^2\end{aligned}$$

which yields $\Delta K = -35$ J. (Although it is not necessary to do so, still it is worth noting that algebraic manipulation of the above expression leads to $|\Delta K| = \frac{1}{2}\left(\frac{m_1m_2}{m_1+m_2}\right)v_{\text{rel}}^2$ where $v_{\text{rel}} = v_1 - v_2$). Conservation of energy then requires

$$\frac{1}{2}kx_{\text{m}}^2 = -\Delta K \implies x_{\text{m}} = \sqrt{\frac{-2\Delta K}{k}} = \sqrt{\frac{-2(-35)}{1120}}$$

which gives the result $x_{\text{m}} = 0.25$ m.

34. We think of this as having two parts: the first is the collision itself – where the blocks "join" so quickly that the 1.0-kg block has not had time to move through any distance yet – and then the subsequent motion of the 3.0 kg system as it compresses the spring to the maximum amount x_m. The first part involves momentum conservation (with $+x$ rightward):

$$(2.0\,\text{kg})(4.0\,\text{m/s}) = (3.0\,\text{kg})\vec{v}$$

which yields $\vec{v} = 2.7$ m/s. The second part involves mechanical energy conservation:

$$\frac{1}{2}(3.0\,\text{kg})(2.7\,\text{m/s})^2 = \frac{1}{2}(200\,\text{N/m})x_m^2$$

which gives the result $x_m = 0.33$ m.

35. (a) Let m_1 be the mass of the block on the left, v_{1i} be its initial velocity, and v_{1f} be its final velocity. Let m_2 be the mass of the block on the right, v_{2i} be its initial velocity, and v_{2f} be its final velocity. The momentum of the two-block system is conserved, so $m_1v_{1i} + m_2v_{2i} = m_1v_{1f} + m_2v_{2f}$ and

$$v_{1f} = \frac{m_1v_{1i} + m_2v_{2i} - m_2v_{2f}}{m_1} = \frac{(1.6)(5.5) + (2.4)(2.5) - (2.4)(4.9)}{1.6}$$

which yields $v_{1f} = 1.9$ m/s. The block continues going to the right after the collision.

(b) To see if the collision is elastic, we compare the total kinetic energy before the collision with the total kinetic energy after the collision. The total kinetic energy before is

$$K_i = \frac{1}{2}m_1v_{1i}^2 + \frac{1}{2}m_2v_{2i}^2 = \frac{1}{2}(1.6)(5.5)^2 + \frac{1}{2}(2.4)(2.5)^2 = 31.7\text{ J}\ .$$

The total kinetic energy after is

$$K_f = \frac{1}{2}m_1v_{1f}^2 + \frac{1}{2}m_2v_{2f}^2 = \frac{1}{2}(1.6)(1.9)^2 + \frac{1}{2}(2.4)(4.9)^2 = 31.7\text{ J}\ .$$

Since $K_i = K_f$ the collision is found to be elastic.

(c) Now $v_{2i} = -2.5$ m/s and

$$v_{1f} = \frac{m_1v_{1i} + m_2v_{2i} - m_2v_{2f}}{m_1} = \frac{(1.6)(5.5) + (2.4)(-2.5) - (2.4)(4.9)}{1.6}$$

which yields $v_{1f} = -5.6$ m/s. Thus, the velocity is opposite to the direction shown in Fig. 10-37.

36. We use m_1 for the mass of the electron and $m_2 = 1840m_1$ for the mass of the hydrogen atom. Using Eq. 10-31,

$$v_{2f} = \frac{2m_1}{m_1 + 1840m_1}v_{1i} = \frac{2}{1841}v_{1i}$$

we compute the final kinetic energy of the hydrogen atom:

$$K_{2f} = \frac{1}{2}(1840m_1)\left(\frac{2\,v_{1i}}{1841}\right)^2 = \frac{(1840)(4)}{1841^2}\left(\frac{1}{2}(1840m_1)v_{1i}^2\right)$$

so we find the fraction to be $(1840)(4)/1841^2 \approx 2.2 \times 10^{-3}$, or 0.22%.

37. (a) Let m_1 be the mass of the cart that is originally moving, v_{1i} be its velocity before the collision, and v_{1f} be its velocity after the collision. Let m_2 be the mass of the cart that is originally at rest and v_{2f} be its velocity after the collision. Then, according to Eq. 10-30,

$$v_{1f} = \frac{m_1 - m_2}{m_1 + m_2}v_{1i}\ .$$

Using SI units (so $m_1 = 0.34$ kg), we obtain

$$m_2 = \frac{v_{1i} - v_{1f}}{v_{1i} + v_{1f}}m_1 = \left(\frac{1.2 - 0.66}{1.2 + 0.66}\right)(0.34) = 0.099\text{ kg}\ .$$

(b) The velocity of the second cart is given by Eq. 10-31:

$$v_{2f} = \frac{2m_1}{m_1+m_2} v_{1i} = \left(\frac{2(0.34)}{0.34+0.099}\right)(1.2) = 1.9 \text{ m/s} .$$

(c) The speed of the center of mass is

$$v_{\text{com}} = \frac{m_1 v_{1i} + m_2 v_{2i}}{m_1+m_2} = \frac{(0.34)(1.2)+0}{0.34+0.099} = 0.93 \text{ m/s} .$$

Values for the initial velocities were used but the same result is obtained if values for the final velocities are used.

38. No mechanical energy is "lost" in this encounter, so we analyze it with the elastic collision equations, particularly Eq. 10-38. Thus,

$$v_{1f} = \frac{m_1 - m_2}{m_1+m_2} v_{1i} + \frac{2m_2}{m_1+m_2} v_{2i} \approx -v_{1i} + 2v_{2i}$$

where we have made the (certainly reasonable) approximation that $m_2 \gg m_1$ and simplified accordingly. Thus, $v_{1\,f} = -12 + 2(-13) = -38$, resulting in a final speed (relative to the Sun) of 38 km/s.

39. We use $m_1 = 4$ u for the mass of the alpha particle and $m_2 = 197$ u for the mass of the gold nucleus in Eq. 10-31:

$$v_{2f} = \frac{2(4)}{4+197} v_{1i} = \frac{8}{201} v_{1i}$$

we compute the final kinetic energy of the gold nucleus (which must be the same as the kinetic energy lost by the alpha particle – since this is an elastic collision)

$$K_{2f} = \frac{1}{2} m_2 v_{2f}^2 = \frac{1}{2}(197\,\text{u})\left(\frac{8\,v_{1i}}{201}\right)^2 .$$

We divide this by the initial alpha particle energy $K_i = \frac{1}{2}(4\,\text{u})v_{1i}^2$ to obtain

$$\frac{K_{2f}}{K_i} = \frac{(197)(8)^2}{(4)(201)^2} \approx 0.078$$

so we find the percentage is 7.8%.

40. First, we find the speed v of the ball of mass m_1 right before the collision (just as it reaches its lowest point of swing). Mechanical energy conservation (with $h = 0.700$ m) leads to

$$m_1 g h = \frac{1}{2} m_1 v^2 \implies v = \sqrt{2gh} = 3.7 \text{ m/s} .$$

(a) We now treat the elastic collision (with SI units) using Eq. 10-30:

$$v_{1f} = \frac{m_1 - m_2}{m_1+m_2} v = \frac{0.5-2.5}{0.5+2.5}(3.7) = -2.47$$

which means the final speed of the ball is 2.47 m/s.

(b) Finally, we use Eq. 10-31 to find the final speed of the block:

$$v_{2f} = \frac{2m_1}{m_1+m_2} v = \frac{2(0.5)}{0.5+2.5}(3.7) = 1.23 \text{ m/s} .$$

41. (a) Let m_1 be the mass of the body that is originally moving, v_{1i} be its velocity before the collision, and v_{1f} be its velocity after the collision. Let m_2 be the mass of the body that is originally at rest and v_{2f} be its velocity after the collision. Then, according to Eq. 10–30,

$$v_{1f} = \frac{m_1 - m_2}{m_1 + m_2} v_{1i} \,.$$

We solve for m_2 to obtain

$$m_2 = \frac{v_{1i} - v_{1f}}{v_{1f} + v_{1i}} m_1 \,.$$

We combine this with $v_{1f} = v_{1i}/4$ to obtain $m_2 = 3m_1/5 = 3(2.0)/5 = 1.2\,\text{kg}$.

(b) The speed of the center of mass is

$$v_{\text{com}} = \frac{m_1 v_{1i} + m_2 v_{2i}}{m_1 + m_2} = \frac{(2.0)(4.0)}{2.0 + 1.2} = 2.5 \text{ m/s} \,.$$

42. We refer to the discussion in the textbook (Sample Problem 10-4, which uses the same notation that we use here) for some important details in the reasoning. We choose rightward in Fig. 10-15 as our $+x$ direction. We use the notation $\vec{v}$ when we refer to velocities and v when we refer to speeds (which are necessarily positive). Since the algebra is fairly involved, we find it convenient to introduce the notation $\Delta m = m_2 - m_1$ (which, we note for later reference, is a positive-valued quantity).

(a) Since $\vec{v}_{1i} = +\sqrt{2gh_1}$ where $h_1 = 9.0$ cm, we have

$$\vec{v}_{1f} = \frac{m_1 - m_2}{m_1 + m_2} v_{1i} = -\frac{\Delta m}{m_1 + m_2} \sqrt{2gh_1}$$

which is to say that the *speed* of sphere 1 immediately after the collision is $v_{1f} = (\Delta m/(m_1 + m_2))\sqrt{2gh_1}$ and that $\vec{v}_{1f}$ points in the $-x$ direction. This leads (by energy conservation $m_1 g h_{1f} = \frac{1}{2} m_1 v_{1f}^2$) to

$$h_{1f} = \frac{v_{1f}^2}{2g} = \left(\frac{\Delta m}{m_1 + m_2}\right)^2 h_1 \,.$$

With $m_1 = 50$ g and $m_2 = 85$ g, this becomes $h_{1f} \approx 0.6$ cm.

(b) Eq. 10-31 gives

$$v_{2f} = \frac{2m_1}{m_1 + m_2} v_{1i} = \frac{2m_1}{m_1 + m_2} \sqrt{2gh_1}$$

which leads (by energy conservation $m_2 g h_{2f} = \frac{1}{2} m_2 v_{2f}^2$) to

$$h_{2f} = \frac{v_{2f}^2}{2g} = \left(\frac{2m_1}{m_1 + m_2}\right)^2 h_1 \,.$$

With $m_1 = 50$ g and $m_2 = 85$ g, this becomes $h_{2f} \approx 4.9$ cm.

(c) Fortunately, they hit again at the lowest point (as long as their amplitude of swing was "small" – this is further discussed in Chapter 16). At the risk of using cumbersome notation, we refer to the *next* set of heights as h_{1ff} and h_{2ff}. At the lowest point (before this second collision) sphere 1 has velocity $+\sqrt{2gh_{1f}}$ (rightward in Fig. 10-15) and sphere 2 has velocity $-\sqrt{2gh_{1f}}$ (that is, it points in the $-x$ direction). Thus, the velocity of sphere 1 immediately after the second collision is, using Eq. 10-38,

$$\begin{aligned}
\vec{v}_{1ff} &= \frac{m_1 - m_2}{m_1 + m_2} \sqrt{2gh_{1f}} + \frac{2m_2}{m_1 + m_2} \left(-\sqrt{2gh_{2f}}\right) \\
&= \frac{-\Delta m}{m_1 + m_2} \left(\frac{\Delta m}{m_1 + m_2} \sqrt{2gh_1}\right) - \frac{2m_2}{m_1 + m_2} \left(\frac{2m_1}{m_1 + m_2} \sqrt{2gh_1}\right) \\
&= -\frac{(\Delta m)^2 + 4m_1 m_2}{(m_1 + m_2)^2} \sqrt{2gh_1} \,.
\end{aligned}$$

This can be greatly simplified (by expanding $(\Delta m)^2$ and $(m_1 + m_2)^2$) to arrive at the conclusion that the speed of sphere 1 immediately after the second collision is simply $v_{1ff} = \sqrt{2gh_1}$ and that $\vec{v}_{1ff}$ points in the $-x$ direction. Energy conservation ($m_1 g h_{1ff} = \frac{1}{2} m_1 v_{1ff}^2$) leads to

$$h_{1ff} = \frac{v_{1ff}^2}{2g} = h_1 = 9.0 \text{ cm} .$$

(d) One can reason (energy-wise) that $h_{1\,ff} = 0$ simply based on what we found in part (c). Still, it might be useful to see how this shakes out of the algebra. Eq. 10-39 gives the velocity of sphere 2 immediately after the second collision:

$$\begin{aligned} v_{2\,ff} &= \frac{2m_1}{m_1 + m_2}\sqrt{2gh_{1\,f}} + \frac{m_2 - m_1}{m_1 + m_2}\left(-\sqrt{2gh_{2\,f}}\right) \\ &= \frac{2m_1}{m_1 + m_2}\left(\frac{\Delta m}{m_1 + m_2}\sqrt{2gh_1}\right) + \frac{\Delta m}{m_1 + m_2}\left(\frac{-2m_1}{m_1 + m_2}\sqrt{2gh_1}\right) \end{aligned}$$

which vanishes since $(2m_1)(\Delta m) - (\Delta m)(2m_1) = 0$. Thus, the second sphere (after the second collision) stays at the lowest point, which basically recreates the conditions at the start of the problem (so all subsequent swings-and-impacts, neglecting friction, can be easily predicted – as they are just replays of the first two collisions).

43. (a) Let m_1 be the mass of one sphere, v_{1i} be its velocity before the collision, and v_{1f} be its velocity after the collision. Let m_2 be the mass of the other sphere, v_{2i} be its velocity before the collision, and v_{2f} be its velocity after the collision. Then, according to Eq. 10–38,

$$v_{1f} = \frac{m_1 - m_2}{m_1 + m_2} v_{1i} + \frac{2m_2}{m_1 + m_2} v_{2i} .$$

Suppose sphere 1 is originally traveling in the positive direction and is at rest after the collision. Sphere 2 is originally traveling in the negative direction. Replace v_{1i} with v, v_{2i} with $-v$, and v_{1f} with zero to obtain $0 = m_1 - 3m_2$. Thus $m_2 = m_1/3 = (300\,\text{g})/3 = 100\,\text{g}$.

(b) We use the velocities before the collision to compute the velocity of the center of mass:

$$v_{\text{com}} = \frac{m_1 v_{1i} + m_2 v_{2i}}{m_1 + m_2} = \frac{(300\,\text{g})(2.0\,\text{m/s}) + (100\,\text{g})(-2.0\,\text{m/s})}{300\,\text{g} + 100\,\text{g}}$$

which yields $v_{\text{com}} = 1.0$ m/s.

44. The velocities of m_1 and m_2 just after the collision with each other are given by Eq. 10-38 and Eq. 10-39 (setting $v_{1i} = 0$).

$$\begin{aligned} v_{1f} &= \frac{2m_2}{m_1 + m_2} v_{2i} \\ v_{2f} &= \frac{m_2 - m_1}{m_1 + m_2} v_{2i} \end{aligned}$$

After bouncing off the wall, the velocity of m_2 becomes $-v_{2f}$ (see *a massive target* in §10-5). In these terms, the problem requires

$$\begin{aligned} v_{1f} &= -v_{2f} \\ \frac{2m_2}{m_1 + m_2} v_{2i} &= -\frac{m_2 - m_1}{m_1 + m_2} v_{2i} \end{aligned}$$

which simplifies to

$$2m_2 = -(m_2 - m_1) \implies m_2 = \frac{m_1}{3} .$$

45. (a) We use conservation of mechanical energy to find the speed of either ball after it has fallen a distance h. The initial kinetic energy is zero, the initial gravitational potential energy is Mgh, the final kinetic energy is $\frac{1}{2}Mv^2$, and the final potential energy is zero. Thus $Mgh = \frac{1}{2}Mv^2$ and $v = \sqrt{2gh}$. The collision of the ball of M with the floor is an elastic collision of a light object with a stationary massive object. The velocity of the light object reverses direction without change in magnitude. After the collision, the ball is traveling upward with a speed of $\sqrt{2gh}$. The ball of mass m is traveling downward with the same speed. We use Eq. 10-38 to find an expression for the velocity of the ball of mass M after the collision:

$$\begin{aligned} v_{Mf} &= \frac{M-m}{M+m}v_{Mi} + \frac{2m}{M+m}v_{mi} \\ &= \frac{M-m}{M+m}\sqrt{2gh} - \frac{2m}{M+m}\sqrt{2gh} \\ &= \frac{M-3m}{M+m}\sqrt{2gh}\ . \end{aligned}$$

For this to be zero, $M = 3m$.

(b) We use the same equation to find the velocity of the ball of mass m after the collision:

$$v_{mf} = -\frac{m-M}{M+m}\sqrt{2gh} + \frac{2M}{M+m}\sqrt{2gh} = \frac{3M-m}{M+m}\sqrt{2gh}$$

which becomes (upon substituting $M = 3m$) $v_{mf} = 2\sqrt{2gh}$. We next use conservation of mechanical energy to find the height h' to which the ball rises. The initial kinetic energy is $\frac{1}{2}mv_{mf}^2$, the initial potential energy is zero, the final kinetic energy is zero, and the final potential energy is mgh'. Thus

$$\frac{1}{2}mv_{mf}^2 = mgh' \implies h' = \frac{v_{mf}^2}{2g} = 4h$$

where $2\sqrt{2gh}$ is substituted for v_{mf}.

46. (a) Conservation of linear momentum implies $m_A\vec{v}_A + m_B\vec{v}_B = m_A\vec{v}\,'_A + m_B\vec{v}\,'_B$. Since $m_A = m_B = m = 2.0\,\text{kg}$, the masses divide out and we obtain (in m/s)

$$\begin{aligned} \vec{v}\,'_B &= \vec{v}_A + \vec{v}_B - \vec{v}\,'_A \\ &= (15\hat{\text{i}} + 30\hat{\text{j}}) + (-10\hat{\text{i}} + 5\hat{\text{j}}) - (-5\hat{\text{i}} + 20\hat{\text{j}}) \\ &= 10\hat{\text{i}} + 15\hat{\text{j}}\ . \end{aligned}$$

(b) The final and initial kinetic energies are

$$\begin{aligned} K_f &= \frac{1}{2}mv'^2_A + \frac{1}{2}mv'^2_B = \frac{1}{2}(2.0)\left((-5)^2 + 20^2 + 10^2 + 15^2\right) = 8.0\times10^2\ \text{J} \\ K_i &= \frac{1}{2}mv_A^2 + \frac{1}{2}mv_B^2 = \frac{1}{2}(2.0)\left(15^2 + 30^2 + (-10)^2 + 5^2\right) = 1.3\times10^3\ \text{J}\ . \end{aligned}$$

The change kinetic energy is then $\Delta K = -5.0\times10^2$ J (that is, 500 J of the initial kinetic energy is lost).

47. We orient our $+x$ axis along the initial direction of motion, and specify angles in the "standard" way – so $\theta = +64°$ for the alpha (α) particle (after collision) and $\phi = -51°$ for the oxygen nucleus (o) (which is going into the fourth quadrant, in our scenario). We apply the conservation of linear momentum to the x and y axes respectively.

$$\begin{aligned} m_\alpha v_\alpha &= m_\alpha v'_\alpha \cos\theta + m_o v'_o \cos\phi \\ 0 &= m_\alpha v'_\alpha \sin\theta + m_o v'_o \sin\phi \end{aligned}$$

We are given $v'_o = 1.2\times10^5$ m/s, which leaves us two unknowns and two equations, which is sufficient for solving.

(a) We solve for the final alpha particle speed using the y-momentum equation:

$$v'_\alpha = -\frac{m_\alpha v'_\alpha \sin\theta}{m_o \sin\phi} = -\frac{(16)\left(1.2\times 10^5\right)\sin(-51°)}{(4)\sin(64°)}$$

which yields $v'_\alpha = 4.15 \times 10^5$ m/s.

(b) Plugging our result from part (a) into the x-momentum equation produces the initial alpha particle speed:

$$\begin{aligned} m_\alpha v_\alpha &= \frac{m_\alpha v'_\alpha \cos\theta + m_o v'_o \cos\phi}{m_\alpha} \\ &= \frac{(4)\left(4.15\times 10^5\right)\cos(64°) + (16)\left(1.2\times 10^5\right)\cos(-51°)}{4} \\ &= 4.84\times 10^5 \text{ m/s} . \end{aligned}$$

48. We orient our $+x$ axis along the initial direction of motion, and specify angles in the "standard" way – so $\theta = +60°$ for the proton (1) which is assumed to scatter into the first quadrant and $\phi = -30°$ for the target proton (2) which scatters into the fourth quadrant (recall that the problem has told us that this is perpendicular to θ). We apply the conservation of linear momentum to the x and y axes respectively.

$$\begin{aligned} m_1 v_1 &= m_1 v'_1 \cos\theta + m_2 v'_2 \cos\phi \\ 0 &= m_1 v'_1 \sin\theta + m_2 v'_2 \sin\phi \end{aligned}$$

We are given $v_1 = 500$ m/s, which provides us with two unknowns and two equations, which is sufficient for solving. Since $m_1 = m_2$ we can cancel the mass out of the equations entirely.

(a) Combining the above equations and solving for v'_2 we obtain

$$v'_2 = \frac{v_1 \sin(\theta)}{\sin(\theta - \phi)} = \frac{500\sin(60°)}{\sin(90°)} = 433$$

in SI units (m/s). We used the identity $\sin(\theta)\cos(\phi) - \cos(\theta)\sin(\phi) = \sin(\theta - \phi)$ in simplifying our final expression.

(b) In a similar manner, we find

$$v'_1 = \frac{v_1 \sin(\phi)}{\sin(\phi - \theta)} = \frac{500\sin(-30°)}{\sin(-90°)} = 250 \text{ m/s} .$$

49. (a) We use Fig. 10-16 of the text (which treats both angles are positive-valued, even though one of them is in the fourth quadrant; this is why there is an explicit minus sign in Eq. 10-43 as opposed to it being implicitly in the angle). We take the cue ball to be body 1 and the other ball to be body 2. Conservation of the x component of the total momentum of the two-ball system leads to $mv_{1i} = mv_{1f}\cos\theta_1 + mv_{2f}\cos\theta_2$ and conservation of the y component leads to $0 = -mv_{1f}\sin\theta_1 + mv_{2f}\sin\theta_2$. The masses are the same and cancel from the equations. We solve the second equation for $\sin\theta_2$:

$$\sin\theta_2 = \frac{v_{1f}}{v_{2f}}\sin\theta_1 = \left(\frac{3.50\,\text{m/s}}{2.00\,\text{m/s}}\right)\sin 22.0° = 0.656 .$$

Consequently, the angle between the second ball and the initial direction of the first is $\theta_2 = 41.0°$.

(b) We solve the first momentum conservation equation for the initial speed of the cue ball.

$$\begin{aligned} v_{1i} &= v_{1f}\cos\theta_1 + v_{2f}\cos\theta_2 \\ &= (3.50\,\text{m/s})\cos 22.0° + (2.00\,\text{m/s})\cos 41.0° \\ &= 4.75 \text{ m/s} . \end{aligned}$$

(c) With SI units understood, the initial kinetic energy is

$$K_i = \frac{1}{2}mv_i^2 = \frac{1}{2}m(4.75)^2 = 11.3m$$

and the final kinetic energy is

$$K_f = \frac{1}{2}mv_{1f}^2 + \frac{1}{2}mv_{2f}^2 = \frac{1}{2}m\left((3.50)^2 + (2.00)^2\right) = 8.1m\ .$$

Kinetic energy is not conserved.

50. We orient our $+x$ axis along the initial direction of motion, and specify angles in the "standard" way – so $\theta = -90°$ for the particle B which is assumed to scatter "downward" and $\phi > 0$ for particle A which presumably goes into the first quadrant. We apply the conservation of linear momentum to the x and y axes respectively.

$$\begin{aligned} m_B v_B &= m_B v'_B \cos\theta + m_A v'_A \cos\phi \\ 0 &= m_B v'_B \sin\theta + m_A v'_A \sin\phi \end{aligned}$$

(a) Setting $v_B = v$ and $v'_B = v/2$, the y-momentum equation yields

$$m_A v'_A \sin\phi = m_B\,\frac{v}{2}$$

and the x-momentum equation yields

$$m_A v'_A \cos\phi = m_B\, v\ .$$

Dividing these two equations, we find $\tan\phi = \frac{1}{2}$ which yields $\phi = 27°$. If we choose to measure this from the final direction of motion for B, then this becomes $90° + 27° = 117°$.

(b) We can *formally* solve for v'_A (using the y-momentum equation and the fact that $\sin\phi = 1/\sqrt{5}$)

$$v'_A = \frac{\sqrt{5}}{2}\,\frac{m_B}{m_A}\,v$$

but lacking numerical values for v and the mass ratio, we cannot fully determine the final speed of A. Note: substituting $\cos\phi = 2/\sqrt{5}$, into the x-momentum equation leads to exactly this same relation (that is, no new information is obtained which might help us determine an answer).

51. Suppose the objects enter the collision along lines that make the angles $\theta > 0$ and $\phi > 0$ with the x axis, as shown in the diagram below. Both have the same mass m and the same initial speed v.

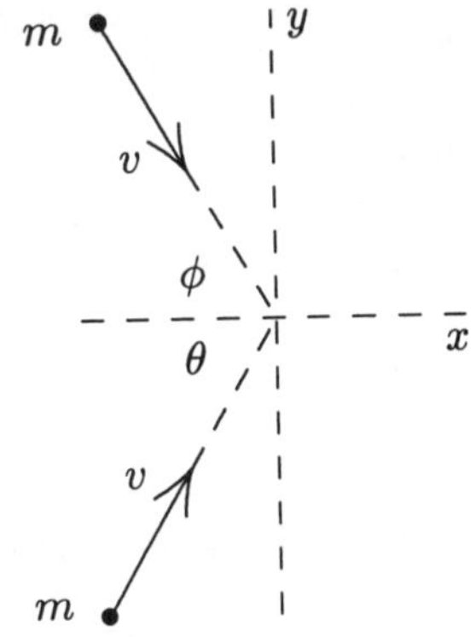

We suppose that after the collision the combined object moves in the positive x direction with speed V. Since the y component of the total momentum of the two-object system is conserved, $mv\sin\theta - mv\sin\phi = 0$. This means $\phi = \theta$. Since the x component is conserved, $2mv\cos\theta = 2mV$. We now use $V = v/2$ to find that $\cos\theta = 1/2$. This means $\theta = 60°$. The angle between the initial velocities is $120°$.

52. We orient our $+x$ axis along the initial direction of motion, and specify angles in the "standard" way – so $\theta = +60°$ for one ball (1) which is assumed to go into the first quadrant with speed $v_1' = 1.1$ m/s, and $\phi < 0$ for the other ball (2) which presumably goes into the fourth quadrant. The mass of each ball is m, and the initial speed of one of the balls is $v_0 = 2.2$ m/s. We apply the conservation of linear momentum to the x and y axes respectively.

$$\begin{aligned} mv_0 &= mv_1' \cos\theta + mv_2' \cos\phi \\ 0 &= mv_1' \sin\theta + mv_2' \sin\phi \end{aligned}$$

The mass m cancels out of these equations, and we are left with two unknowns and two equations, which is sufficient to solve.

(a) With SI units understood, the y-momentum equation can be rewritten as

$$v_2' \sin\phi = -v_1' \sin 60° = -0.95$$

and the x-momentum equation yields

$$v_2' \cos\phi = v_0 - v_1' \cos 60° = 1.65$$

Dividing these two equations, we find $\tan\phi = -0.577$ which yields $\phi = -30°$. If we choose to measure this as a positive-valued angle (as the textbook does in §10-6), then this becomes 30°. We plug $\phi = -30°$ into either equation and find $v_2' \approx 1.9$ m/s.

(b) One can check to see if this an elastic collision by computing

$$\frac{2K_i}{m} = v_0^2 \quad \text{and} \quad \frac{2K_f}{m} = v_1'^2 + v_2'^2$$

and seeing if they are equal (they are), but one must be careful not to use rounded-off values. Thus, it is useful to note that the answer in part (a) can be expressed "exactly" as $v_2' = \frac{1}{2}v_0\sqrt{3}$ (and of course $v_1' = \frac{1}{2}v_0$ "exactly" – which makes it clear that these two kinetic energy expressions are indeed equal).

53. The diagram below shows the situation as the incident ball (the left-most ball) makes contact with the other two. It exerts an impulse of the same magnitude on each ball, along the line that joins the centers of the incident ball and the target ball. The target balls leave the collision along those lines, while the incident ball leaves the collision along the x axis. The three dotted lines that join the centers of the balls in contact form an equilateral triangle, so both of the angles marked θ are 30°. Let v_0 be the velocity of the incident ball before the collision and V be its velocity afterward. The two target balls leave the collision with the same speed. Let v represent that speed. Each ball has mass m.

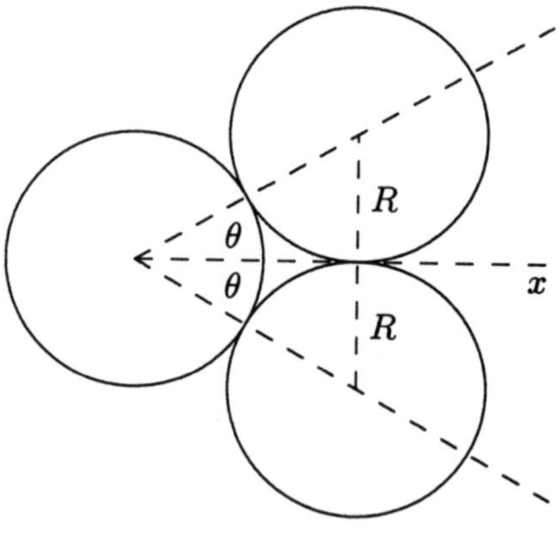

Since the x component of the total momentum of the three-ball system is conserved,

$$mv_0 = mV + 2mv\cos\theta$$

and since the total kinetic energy is conserved,

$$\frac{1}{2}mv_0^2 = \frac{1}{2}mV^2 + 2\left(\frac{1}{2}mv^2\right) .$$

We know the directions in which the target balls leave the collision so we first eliminate V and solve for v. The momentum equation gives $V = v_0 - 2v\cos\theta$, so $V^2 = v_0^2 - 4v_0 v\cos\theta + 4v^2\cos^2\theta$ and the energy equation becomes $v_0^2 = v_0^2 - 4v_0 v\cos\theta + 4v^2\cos^2\theta + 2v^2$. Therefore,

$$v = \frac{2v_0\cos\theta}{1+2\cos^2\theta} = \frac{2(10\,\text{m/s})\cos 30^\circ}{1+2\cos^2 30^\circ} = 6.93\ \text{m/s} .$$

(a) The discussion and computation above determines the final velocity of ball 2 (as labeled in Fig. 10-41) to be 6.9 m/s at 30° counterclockwise from the $+x$ axis.

(b) Similarly, the final velocity of ball 3 is 6.9 m/s at 30° clockwise from the $+x$ axis.

(c) Now we use the momentum equation to find the final velocity of ball 1:

$$V = v_0 - 2v\cos\theta = 10\,\text{m/s} - 2(6.93\,\text{m/s})\cos 30^\circ = -2.0\,\text{m/s} .$$

The minus sign indicates that it bounces back in the $-x$ direction.

54. The problem involves the completely inelastic collision of the two children followed by their completely inelastic collision with the (already moving) man. Speeds are given but no angles, so we are free to orient our $-x$ axis along the direction of motion of the man before his collision with the children (so his angle is 180°). The magnitude of the man's momentum before that collision is $(75\,\text{kg})(2.0\,\text{m/s}) = 150$ kg·m/s. Thus, with SI units understood, the second collision is described by momentum conservation:

$$\vec{p} + (150\ \angle\ 180^\circ) = 0$$

which yields the momentum of the stuck-together children $\vec{p} = (150\ \angle\ 0^\circ)$ in magnitude-angle notation. We now describe the first collision (of the two children) using momentum conservation:

$$\vec{p}_1 + \vec{p}_2 = (150\ \angle\ 0^\circ) \quad \text{or} \quad 150\,\hat{\text{i}}$$

where the unit-vector notation has also been used, in case the magnitude-angle notation is less familiar. Now, since $m_1 = m_2 = 30$ kg and $|\vec{p}_1| = |\vec{p}_2| = 120$ kg·m/s, we see that the y components of the children's initial velocities must be equal and opposite. Therefore, if child 1 has an initial velocity angle θ then child 2 has an initial velocity angle $-\theta$. The previous equation becomes

$$120\cos(\theta) + 120\cos(-\theta) = 150$$

which has the solution $\theta = 51^\circ$. The angle between the children (initially) is therefore $2\theta \approx 103^\circ$.

55. Let $m_n = 1.0$ u be the mass of the neutron and $m_d = 2.0$ u be the mass of the deuteron. In our manipulations we treat these masses as "exact", so, for instance, we write $m_n/m_d = \frac{1}{2}$ in our simplifying steps. We assume the neutron enters with a velocity $\vec{v}_\text{o}$ pointing in the $+x$ direction and leaves along the positive y axis with speed v_n. The deuteron goes into the fourth quadrant with velocity components $v_{dx} > 0$ and $v_{dy} < 0$. Conservation of the x component of momentum leads to

$$m_n v_\text{o} = m_d v_{dx} \implies v_{dx} = \frac{1}{2}v_\text{o}$$

and conservation of the y component leads to

$$0 = m_n v_n + m_d v_{fy} \implies v_{dy} = -\frac{1}{2}v_\text{n} .$$

Also, the collision is elastic, so kinetic energy "conservation" leads to

$$\frac{1}{2}m_n v_o^2 = \frac{1}{2}m_n v_n^2 + \frac{1}{2}m_d v_d^2$$

which we simplify by multiplying through with $2/m_n$ and using $v_d^2 = v_{dx}^2 + v_{dy}^2$

$$v_o^2 = v_n^2 + \frac{m_d}{m_n}\left(v_{dx}^2 + v_{dy}^2\right) .$$

Now we substitute in the relations found from the momentum conditions:

$$v_o^2 = v_n^2 + 2\left(\frac{v_o^2}{4} + \frac{v_n^2}{4}\right) \implies v_n = v_o\sqrt{\frac{1}{3}} .$$

Finally, we set up a ratio expressing the (relative) loss of kinetic energy (by the neutron).

$$\frac{K_o - K_n}{K_o} = 1 - \frac{v_n^2}{v_o^2} = 1 - \frac{1}{3} = \frac{2}{3} .$$

56. (a) Choosing upward as the positive direction, the momentum change of the foot is

$$\Delta\vec{p} = 0 - m_{\text{foot}}\,\vec{v}_i = -(0.003\,\text{kg})(-1.5\text{m/s})$$

which yields an impulse of 4.50×10^{-3} N·s.

(b) Using Eq. 10-8 and now treating *downward* as the positive direction, we have

$$\vec{J} = \vec{F}_{\text{avg}}\Delta t = m_{\text{lizard}}\,g\,\Delta t = (0.090)(9.8)(0.6)$$

which yields $\vec{J} = 0.529$ N·s.

(c) Considering the large difference between the answers for part (a) and part (b), we see that the slap cannot account for the support; we infer, then, that the push does the job.

57. From mechanical energy conservation (or simply using Eq. 2-16 with $\vec{a} = g$ downward) we obtain

$$v = \sqrt{2gh} = \sqrt{2(9.8)(1.5)} = 5.4 \text{ m/s}$$

for the speed just as the body makes contact with the ground.

(a) During the compression of the body, the center of mass must decelerate over a distance $d = 0.30$ m. Choosing $+y$ downward, the deceleration a is found using Eq. 2-16

$$0 = v^2 + 2ad \implies a = -\frac{v^2}{2d} = -\frac{5.4^2}{2(0.30)}$$

which yields $a = -49$ m/s^2. Thus, the magnitude of the net (vertical) force is $m|a| = 49m$ in SI units, which (since $49 = 5(9.8)$) can be expressed as $5mg$.

(b) During the deceleration process, the forces on the dinosaur are (in the vertical direction) $\vec{N}$ and $m\vec{g}$. If we choose $+y$ upward, and use the final result from part (a), we therefore have $N - mg = 5mg$, or $N = 6mg$. In the horizontal direction, there is also a deceleration (from $v_o = 19$ m/s to zero), in this case due to kinetic friction $f_k = \mu_k N = \mu_k(6mg)$. Thus, the net force exerted by the ground on the dinosaur is

$$F_{\text{ground}} = \sqrt{f_k^2 + N^2} \approx 7mg .$$

(c) We can applying Newton's second law in the horizontal direction (with the sliding distance denoted as Δx) and then use Eq. 2-16, or we can apply the general notions of energy conservation. The latter approach is shown:

$$\frac{1}{2}mv_o^2 = \mu_k(6mg)\Delta x \implies \Delta x = \frac{19^2}{2(6)(0.6)(9.8)} \approx 5 \text{ m} .$$

58. (a) As explained in the problem, the height of the nth domino is $h_n = 1.5^{n-1}$ in centimeters. Therefore, $h_{32} = 1.5^{31} = 2.9 \times 10^5$ cm $= 2.9$ km (!).

(b) When the center of the domino is directly over the corner, the height of the center-point is

$$h_c = \sqrt{\left(\frac{h}{2}\right)^2 + \left(\frac{d}{2}\right)^2} = \frac{d}{2}\sqrt{101}$$

where $h = 10d$ has been used in that last step. While the domino is in its usual resting position, the height of that point is only $h_o = h/2$ which can be written as $5d$. Since the answer is requested to be in terms of U_1 then

$$U_1 = mg(5d) \implies d = \frac{U_1}{5mg} .$$

Therefore, the energy needed to push over the domino is

$$\Delta U = mgh_c - U_1 = mg\left(\frac{d}{2}\sqrt{101}\right) - U_1 = \frac{U_1}{10}\sqrt{101} - U_1$$

which yields approximately $0.005U_1$; the problem refers to this as $\Delta E_{1,\text{in}}$.

(c) The "loss" of potential energy equal to

$$mgh_c - mg\left(\frac{h}{2}\sin\theta\right)$$

becomes the kinetic energy (denoted $\Delta E_{1,\text{out}}$ in the problem). Therefore, we obtain

$$\Delta E_{1,\text{out}} = mg\left(\frac{d}{2}\sqrt{101}\right) - mg\left(\frac{10d}{2}\sin\theta\right)$$

which (using $\theta = 45°$) simplifies to $1.49mgd$. Since $d = U_1/5mg$ this becomes roughly $\Delta E_{1,\text{out}} = 0.30U_1$.

(d) We see from part (b) that $\Delta E_{n,\text{in}}$ is directly proportional to $m_n d_n$ and consequently (since the density is assumed the same for all of them and the volume of a domino is hdw where w is the width) is proportional to $w_n h_n d_n^2$. The width also scales like the other quantities, so $\Delta E_{n,\text{in}}$ is proportional to $1.5^{4(n-1)}$. Therefore, $\Delta E_{2,\text{in}} = 1.5^4 \Delta E_{1,\text{in}}$ which implies $\Delta E_{2,\text{in}} = 0.025U_1$.

(e) Therefore,

$$\frac{\Delta E_{1,\text{out}}}{\Delta E_{2,\text{in}}} = \frac{0.30U_1}{0.025U_1} = 12 .$$

59. (a) We choose $+x$ to be away from the armor (pointing back towards the gun). The velocity is there negative-valued and the acceleration is positive-valued. Using Eq. 2-11,

$$0 = \vec{v}_0 + \vec{a}t \implies \vec{a} = -\frac{\vec{v}_0}{t} = -\frac{-300}{40 \times 10^{-6}} = 7.5 \times 10^6 \text{ m/s}^2 .$$

(b) Since the final momentum is zero, the momentum change is

$$\Delta\vec{p} = 0 - m\vec{v}_0 = -(0.0102\,\text{kg})(-300\,\text{m/s}) = 3.1\text{ kg}\cdot\text{m/s} .$$

(c) We compute $K_f - K_i = 0 - \frac{1}{2}mv_0^2$ and obtain $-\frac{1}{2}(0.0102)(300)^2 \approx -460$ J.

(d) If we assume uniform deceleration, Eq. 2-17 gives

$$\Delta x = \frac{1}{2}(\vec{v}_0 + 0)\,t = \frac{1}{2}(-300)\,(40 \times 10^{-6})$$

so that the distance is $|\Delta x| = 0.0060$ m.

(e) By the impulse-momentum theorem, the impulse of the armor on the bullet is $\vec{J} = \Delta\vec{p} = 3.1$ N·s. By Newton's third law, the impulse of the bullet on the armor must have that same magnitude.

(f) Using Eq. 10-8, we find the magnitude of the (average) force exerted by the bullet on the armor:

$$F_{\text{avg}} = \frac{J}{t} = \frac{3.1}{40 \times 10^{-6}} = 7.7 \times 10^4 \text{ N} .$$

(g) From Newton's second law, we find $a_p = F_{\text{avg}}/M$ (where $M = 65$ kg) to be 1.2×10^3 m/s^2.

(h) Momentum conservation leads to $V = mv_0/M = 0.047$ m/s. (This result can be gotten a number of ways, given the information available at this point in the problem.)

(i) Shortening the distance means decreasing the stopping time (Eq. 2-17 shows this clearly) which (recalling our calculation in part (a)) means the magnitude of the bullet's deceleration increases. It does not change the answer to part (b) (for the change in momentum), nor does it affect part (c) (the change in kinetic energy). Since $\vec{J}$ is determined by $\Delta\vec{p}$, part (e) is unchanged. But with t smaller, $J/t = F_{\text{avg}}$ is larger, as is a_p. Finally, v_p is the same as before since momentum conservation describes the input/output of the collision and not the inner dynamics of it.

60. From mechanical energy conservation (or simply using Eq. 2-16 with $\vec{a} = g$ downward) we obtain

$$v = \sqrt{2gh} = \sqrt{2(9.8)(6.0)} = 10.8 \text{ m/s}$$

for the speed just as the $m = 3000$-kg block makes contact with the pile. At the moment of "joining", they are a system of mass $M = 3500$ kg and speed V. With downward positive, momentum conservation leads to

$$mv = MV \implies V = \frac{(3000)(10.8)}{3500} = 9.3 \text{ m/s} .$$

Now this block-pile "object" must be rapidly decelerated over the small distance $d = 0.030$ m. Using Eq. 2-16 and choosing $+y$ downward, we have

$$0 = V^2 + 2ad \implies a = -\frac{9.3^2}{2(0.030)} = -1440$$

in SI units (m/s^2). Thus, the net force during the decelerating process has magnitude $M\,|a| = 5.0\times10^6$ N.

61. Using Eq. 10-31 with $m_1 = 3.0$ kg, $v_{1i} = 8.0$ m/s and $v_{2f} = 6.0$ m/s, then

$$v_{2f} = \frac{2m_1}{m_1 + m_2}v_{1i} \implies m_2 = m_1\left(\frac{2v_{1i}}{v_{2f}} - 1\right)$$

leads to $m_2 = M = 5.0$ kg.

62. In the momentum relationships, we could as easily work with weights as with masses, but because part (b) of this problem asks for kinetic energy – we will find the masses at the outset: $m_1 = 280 \times 10^3/9.8 = 2.86 \times 10^4$ kg and $m_2 = 210 \times 10^3/9.8 = 2.14 \times 10^4$ kg. Both cars are moving in the $+x$ direction: $v_{1i} = 1.52$ m/s and $v_{2i} = 0.914$ m/s.

(a) If the collision is completely elastic, momentum conservation leads to a final speed of

$$V = \frac{m_1v_{1i} + m_2v_{2i}}{m_1 + m_2} = 1.26 \text{ m/s} .$$

(b) We compute the total initial kinetic energy and subtract from it the final kinetic energy.

$$K_i - K_f = \frac{1}{2}m_1v_{1i}^2 + \frac{1}{2}m_2v_{2i}^2 - \frac{1}{2}(m_1 + m_2)V^2 = 2.25 \times 10^3 \text{ J} .$$

(c) and (d) Using Eq. 10-38 and Eq. 10-39, we find

$$\begin{aligned} v_{2\,f} &= \frac{2m_1}{m_1+m_2}v_{1\,i} + \frac{m_2-m_1}{m_1+m_2}v_{2\,i} = 1.61 \text{ m/s} \quad \text{and} \\ v_{1\,f} &= \frac{m_1-m_2}{m_1+m_2}v_{1\,i} + \frac{2m_2}{m_1+m_2}v_{2\,i} = 1.00 \text{ m/s} . \end{aligned}$$

63. We choose coordinates with $+x$ East and $+y$ North, with the standard conventions for measuring the angles. With SI units understood, we write the initial magnitude of the man's momentum as $(60)(6.0) = 360$ and the final momentum of the two of them together as $(98)(3.0) = 294$. Using magnitude-angle notation (quickly implemented using a vector capable calculator in polar mode), momentum conservation becomes

$$\begin{aligned} \vec{p}_{\text{man}} + \vec{p}_{\text{child}} &= \vec{p}_{\text{together}} \\ (360 \angle 90°) + \vec{p} &= (294 \angle 35°) \end{aligned}$$

Therefore, the momentum of the 38 kg child before the collision is $\vec{p} = (308 \angle -38°)$. Thus, the child's velocity has magnitude equal to $308/38 = 8.1$ m/s and direction of 38° south of east.

64. (a) We choose a coordinate system with $+x$ downriver and $+y$ in the initial direction of motion of the second barge. The velocities in component forms are $\vec{v}_{1i} = (6.2\,\text{m/s})\hat{\text{i}}$ and $\vec{v}_{2i} = (4.3\,\text{m/s})\hat{\text{j}}$ before collision. After the collision, barge 2 has velocity

$$\vec{v}_{2f} = (5.1\,\text{m/s})\left((\sin 18°)\hat{\text{i}} + (\cos 18°)\hat{\text{j}}\right) .$$

Writing $\vec{v}_{1f} = v_{1f}\left((\cos\theta)\hat{\text{i}} + (\sin\theta)\hat{\text{j}}\right)$, with θ we express the component form of the conservation of momentum:

$$\begin{aligned} m_1v_{1i} &= m_1v_{1f}\cos\theta + m_2v_{2f}\sin 18° \\ m_2v_{2i} &= m_1v_{1f}\sin\theta + m_2v_{2f}\cos 18° . \end{aligned}$$

Substituting $v_{1i} = 6.2\,\text{m/s}$, $v_{2i} = 4.3\,\text{m/s}$, and $v_{2f} = 5.1\,\text{m/s}$, we find: $v_{1f} = 3.4\,\text{m/s}$, $\theta = 17°$ (from the point of view of someone on that barge, this deflection is toward the left).

(b) The loss of kinetic energy is

$$K_i - K_f = \left(\frac{1}{2}m_1v_{1i}^2 + \frac{1}{2}m_2v_{2i}^2\right) - \left(\frac{1}{2}m_1v_{1f}^2 + \frac{1}{2}m_2v_{2f}^2\right)$$

which yields 9.5×10^5 J.

65. Let the mass of each ball be m. Conservation of (kinetic) energy in elastic collisions requires that $K_i = K_f$ which leads to

$$\frac{1}{2}mV^2 = \frac{1}{2}(16m)v^2$$

which yields $v = V/4$.

66. The speed of each particle of mass m upon impact with the scale is found from mechanical energy conservation (or simply using Eq. 2-16 with $\vec{a} = g$ downward): $v = \sqrt{2gh}$, where $h = 3.5\,\text{m}$. With $+y$ upward, the change in momentum for the particle is therefore

$$\Delta\vec{p} = m\Delta\vec{v} = 2mv = 2m\sqrt{2gh} .$$

During a time interval Δt, the number of collisions is $N = R\Delta t$ where $R = 42\,\mathrm{s}^{-1}$. Thus, using the impulse-momentum theorem and Eq. 10-8, the average force is

$$\begin{aligned}\vec{F}_{\text{avg}} &= \frac{N\Delta\vec{p}}{\Delta t}\\ &= 2mR\sqrt{2gh}\\ &= 2(0.110)(42)\sqrt{2(9.8)(3.5)}\\ &= 77\ \mathrm{N}\end{aligned}$$

which corresponds to a mass reading of $77/9.8 = 7.8\,\mathrm{kg}$.

67. The momentum before the collision (with $+x$ rightward) is

$$(6.0\ \mathrm{kg})(8.0\ \mathrm{m/s}) + (4.0\ \mathrm{kg})(2.0\ \mathrm{m/s}) = 56\ \mathrm{kg{\cdot}m/s}\ .$$

(a) The total momentum at this instant is $(6.0\,\mathrm{kg})(6.4\,\mathrm{m/s}) + (4.0\,\mathrm{kg})\vec{v}$. Since this must equal the initial total momentum (56, using SI units), then we find $\vec{v} = 4.4$ m/s.

(b) The initial kinetic energy was

$$\frac{1}{2}(6.0\ \mathrm{kg})(8.0\ \mathrm{m/s})^2 + \frac{1}{2}(4.0\ \mathrm{kg})(2.0\ \mathrm{m/s})^2 = 200\ \mathrm{J}\ .$$

The kinetic energy at the instant described in part (a) is

$$\frac{1}{2}(6.0\ \mathrm{kg})(6.4\ \mathrm{m/s})^2 + \frac{1}{2}(4.0\ \mathrm{kg})(4.4\ \mathrm{m/s})^2 = 162\ \mathrm{J}\ .$$

The "missing" 38 J is not dissipated since there is no friction; it is the energy stored in the spring at this instant when it is compressed. Thus, $U_e = 38$ J.

68. This is a completely inelastic collision, followed by projectile motion. In the collision, we use momentum conservation.

$$\begin{aligned}\vec{p}_{\text{shoes}} &= \vec{p}_{\text{together}}\\ (3.2\,\mathrm{kg})(3.0\,\mathrm{m/s}) &= (5.2\,\mathrm{kg})\vec{v}\end{aligned}$$

Therefore, $\vec{v} = 1.8$ m/s toward the right as the combined system is projected from the edge of the table. Next, we can use the projectile motion material from Ch. 4 or the energy techniques of Ch. 8; we choose the latter.

$$\begin{aligned}K_{\text{edge}} + U_{\text{edge}} &= K_{\text{floor}} + U_{\text{floor}}\\ \frac{1}{2}(5.2\,\mathrm{kg})(1.8\,\mathrm{m/s})^2 + (5.2\,\mathrm{kg})(9.8\ \mathrm{m/s^2})(0.40\,\mathrm{m}) &= K_{\text{floor}} + 0\end{aligned}$$

Therefore, the kinetic energy of the system right before hitting the floor is $K_{\text{floor}} = 29$ J.

69. We use the impulse-momentum theorem $\vec{J} = \Delta\vec{p}$ where $\vec{J} = \int \vec{F}\,dt$. Integrating the given expression for force from the moment it starts from rest up to a variable upper limit t, we have $\vec{J} = \left(16t - \frac{1}{3}t^3\right)\hat{\imath}$ with SI units understood.

(a) Since $\left(16t - \frac{1}{3}t^3\right)\hat{\imath} = m\vec{v}$ with $m = 1.6$, we obtain $\vec{v} = 24\hat{\imath}$ in meters-per-second, for $t = 3.0$ s.

(b) Setting $\left(16t - \frac{1}{3}t^3\right)\hat{\imath} = m\vec{v}$ equal to zero leads to $t = 6.9$ s as the positive root.

(c) We can work through the $\frac{d\vec{v}}{dt} = 0$ condition using our $\left(16t - \frac{1}{3}t^3\right)\hat{\imath} = m\vec{v}$ relation, or more simply observe, from the outset, that this is equivalent to finding when the acceleration, hence the force, is zero. We obtain $t = 4.0$ s as the positive root, which we plug into the $\left(16t - \frac{1}{3}t^3\right)\hat{\imath} = m\vec{v}$ relation and find $\vec{v}_{\text{max}} = 27\hat{\imath}$ m/s.

70. (a) We use coordinates with $+x$ rightward and $+y$ upward, with the usual conventions for measuring the angles (so that the final angle is written $90° - 40° = 50°$). With SI units understood, the magnitude of the diver's momentum before contact is $(60.0)(3.00) = 180$ and after contact is $(60.0)(5.00) = 300$. Using magnitude-angle notation (quickly implemented using a vector capable calculator in polar mode), the change in momentum is

$$(300 \angle 50°) - (180 \angle -90°) = (453 \angle 65°) .$$

This equals the *total* impulse delivered to the diver (by the board and by gravity). If F_{net} denotes the magnitude of the average *net* force exerted on the diver, then we have

$$F_{\text{net}}\Delta t = 453 \implies F_{\text{net}} = \frac{453}{1.2} = 377 \text{ N} .$$

(b) Since $\vec{F}_{\text{net}} = (377 \angle 65°)$ and the weight of the diver is $(588 \angle -90)$, we obtain

$$(377 \angle 65°) - (588 \angle -90°) = (943 \angle 80°) .$$

Therefore, the magnitude of the average force exerted by the board on the diver is 943 N.

71. The magnitude of the impulse exerted by the gunner on the gun per minute is $J = F_{\text{avg}}\Delta t$, where $F_{\text{avg}} = 180\,\text{N}$ and $\Delta t = 60\,\text{s}$. The impulse exerted on the gun by each bullet of mass m and speed v is $J' = mv$. The maximum number of bullets N that he could fire per minute satisfies $J = NJ'$. Thus

$$N = \frac{J}{J'} = \frac{F_{\text{avg}}\Delta t}{mv} = \frac{(180)(60)}{(50 \times 10^{-3})(1000)} = 216 .$$

72. (a) The magnitude of the force is

$$F = \frac{\Delta p}{\Delta t} = \frac{9.0 \times 10^3 \,\text{kg}\cdot\text{m/s}}{12\,\text{s}} = 750\,\text{N} .$$

(b) Assuming this is one-dimensional motion (so that any acceleration implies a change in the magnitude of the velocity), we find the speed increase to be

$$\Delta v = \frac{\Delta p}{m} = \frac{9.0 \times 10^3 \,\text{kg}\cdot\text{m/s}}{1500\,\text{kg}} = 6.0\,\text{m/s} .$$

73. (a) The momentum conservation equation (for this completely inelastic collision) $m_A\vec{v}_A + m_B\vec{v}_B = (m_A + m_B)\vec{V}$ can be written in terms of weights by multiplying through by g:

$$w_A\vec{v}_A + w_B\vec{v}_B = (w_A + w_B)\vec{V} .$$

Our $\hat{\imath}$ direction is West and $\hat{\jmath}$ is South, so we have (with weights in kN and speeds in km/h)

$$\begin{aligned} \vec{V} &= \frac{(12.0)\left(64.4\hat{\imath}\right) + (16.0)\left(96.6\hat{\jmath}\right)}{12.0 + 16.0} \\ &= 27.6\hat{\imath} + 55.2\hat{\jmath} \end{aligned}$$

which implies that the final speed is 61.7 km/h.

(b) And the angle for the final velocity is $\tan^{-1}(55.2/27.6) = 63.4°$ South of West.

74. We choose $\hat{\imath}$ East and $\hat{\jmath}$ North, and use SI units (kg for mass and m/s for speed). The initially moving tin cookie has mass $m_1 = 2.0$ and velocity $\vec{v}_o = 8.0\,\hat{\imath}$, and the initially stationary cookie tin has mass $m_2 = 4.0$.

(a) Momentum conservation leads to

$$\begin{aligned} m_1\vec{v}_o &= m_1\vec{v}_1 + m_2\vec{v}_2 \\ 16\,\hat{\imath} &= 8\cos(37°)\,\hat{\imath} + 8\sin(37°)\,\hat{\jmath} + (4.0)\vec{v}_2 \end{aligned}$$

which leads to

$$\vec{v}_2 = 2.4\hat{\imath} - 1.2\hat{\jmath} \implies \vec{v}_2 = (2.7 \angle\ 27°)$$

where magnitude-angle notation is used. Thus, the speed of the cookie tin is 2.7 m/s.

(b) And its angle is $\tan^{-1}(-1.2/2.4) = -27°$ which can be expressed as 27° south of east.

75. We choose $\hat{\imath}$ East and $\hat{\jmath}$ North, and use SI units. The ball initially moving eastward has mass $m_1 = 5.0$ kg and initial velocity $\vec{v}_{1i} = 4.0\,\hat{\imath}$ m/s, and the ball initially moving westward has mass $m_2 = 4.0$ kg and velocity $\vec{v}_{2i} = -3.0\,\hat{\imath}$ m/s. The final velocity of m_1 is $\vec{v}_{1f} = -1.2\,\hat{\jmath}$.

(a) Momentum conservation leads to

$$\begin{aligned} m_1\vec{v}_{1i} + m_2\vec{v}_{2i} &= m_1\vec{v}_1 + m_2\vec{v}_2 \\ 20\,\hat{\imath} - 12\,\hat{\imath} &= -6\,\hat{\jmath} + 4\vec{v}_2 \end{aligned}$$

which leads to

$$\vec{v}_2 = 2.0\,\hat{\imath} + 1.5\,\hat{\jmath} \implies \vec{v}_2 = (2.5 \angle\ 37°)$$

where magnitude-angle notation is used. Thus, the speed of the 4.0 kg ball just after the collision is 2.5 m/s.

(b) We compute the decrease in total kinetic energy:

$$K_i - K_f = \frac{1}{2}(5)(4)^2 + \frac{1}{2}(4)(3)^2 - \frac{1}{2}(5)(1.2)^2 - \frac{1}{2}(4)(2.5)^2$$

which gives the result 42 J.

76. Using mechanical energy conservation, we find the speed v of a pendulum at the bottom of its swing is related to the height h it was released from (or that it swings up to) by $v^2 = 2gh$. Thus, the conservation of momentum at the instant they collide can be expressed as

$$m_1\sqrt{2gd} = (m_1 + m_2)\sqrt{2gh_f}\ .$$

Therefore, the "final" height of the system (which it swings to shortly after the collision) is

$$h_f = \left(\frac{m_1}{m_1 + m_2}\right)^2 d\ .$$

77. If we neglect the time required for the spring to decelerate the leftward moving glider m_2 and re-accelerate it (rightward), then we are effectively assuming that glider bounces elastically off the wall (with the spring playing no dynamic role). Thus, we assume the time t required for m_2 to travel distance $d + x$ (to the wall and then rightward to position x, assuming the origin is at the wall) is simply $t = (d + x)/v$ where $v = |v_{2f}|$ is its speed resulting from the first elastic collision. This velocity is found from Eq. 10-31:

$$v_{2f} = \frac{2m_1}{m_1 + m_2}\,v_{1i} = \frac{2(590)}{940}(-75)$$

which yields −94 cm/s. Thus, with $d = 53$ cm, we have the relation $t = (53 + x)/94$ with x in cm and t in s. During that time, glider m_1 has a displacement $\Delta x = x - d$ due its velocity v_{1f} where

$$v_{1f} = \frac{m_1 - m_2}{m_1 + m_2}\,v_{1i} = \frac{240}{940}(-75)$$

which yields $v_{1f} = -19$ cm/s. This provides another relation between t and x: $t = (x - d)/v_{1f} = (53 - x)/19$. Equating these to relations, we obtain

$$\frac{53 + x}{94} = \frac{53 - x}{19} \implies x = 35 \text{ cm} .$$

78. Eq. 10-31, for situations where $m_1 \gg m_2$, reduces simply to $v_{2f} \approx 2v_{1i}$. Thus, the speed of the fly after the collision is $2(2.1) = 4.2$ m/s.

79. (a) We find the velocity $\vec{v}_{1f}$ of the 1200 kg car after the collision (taking the direction of motion as positive) using momentum conservation (with mass in kg and speed in km/h).

$$\begin{aligned} m_1\vec{v}_{1i} + m_2\vec{v}_{2i} &= m_1\vec{v}_{1f} + m_2\vec{v}_{2f} \\ (1200)(70) + (900)(60) &= (1200)\vec{v}_{1f} + (900)(70) \end{aligned}$$

This gives the result $\vec{v}_{1f} = 62.5$ km/h.

(b) We compute the reduction of total kinetic energy in the collision:

$$Q = K_i - K_f = \frac{1}{2}(1200)(70)^2 + \frac{1}{2}(900)(60)^2 - \frac{1}{2}(1200)(62.5)^2 - \frac{1}{2}(900)(70)^2$$

which gives the result 11250 in mixed units (kg·km^2/h^2). We set up the requested ratio (where $v_o = 5$ km/h):

$$\frac{Q}{\frac{1}{2}m_1v_o^2} = \frac{11250}{15000} = \frac{3}{4} .$$

80. We refer to the discussion in the textbook (see Sample Problem 10-2, which uses the same notation that we use here) for many of the important details in the reasoning. Here we only present the primary computational step (using SI units).

(a) The bullet's initial kinetic energy is

$$\frac{1}{2}mv^2 = \frac{1}{2}m\left(\frac{m+M}{m}\sqrt{2gh}\right)^2 = \frac{m+M}{m}U_f$$

where $U_f = (m + M)gh$ is the system's final potential energy (equal to its total mechanical energy since its speed is zero at height h). Thus,

$$\frac{U_f}{\frac{1}{2}mv^2} = \frac{m}{m+M} = \frac{0.008}{7.008} = 0.00114 .$$

(b) The fraction $m/(m + M)$ shown in part (a) has no v-dependence. The answer remains the same.

(c) As we found in part (a), the fraction is $m/(m+M)$. The numerical value of h given in the problem statement has not been used in this solution.

81. (a) Since $\vec{F}_{\text{net}} = \frac{d\vec{p}}{dt}$ (Eq. 9-23), we read from value of F_x (see graph) that the rate of change of momentum is 4.0 kg·m/s^2 at $t = 3.0$ s.

(b) The impulse, which causes the change in momentum, is equivalent to the area under the curve in this graph (see Eq. 10-3). We break the area into that of a triangle $\frac{1}{2}(2.0\,\text{s})(4.0\,\text{N})$ plus that of a rectangle $(1.0\,\text{s})(4.0\,\text{N})$, which yields a total of 8.0 N·s. Since the car started from rest, its momentum at $t = 3.0$ s must therefore be 8.0 kg·m/s.

82. We use $J = \int F dt = m\Delta v = mv_f$. The integral $\int F\, dt$ is estimated from the area under the curve in Fig. 10-61 as approximately 4 N·s. (If one doesn't want to "count squares" one can assume the curve to be a parabola, in which case $F = \xi(t-3.25)(t-0.35)$ (with t in milliseconds) will fit it once the parameter ξ is adjusted so that $F = 2200$ N when t is midway between 0.35 ms and 3.25 ms. Then the integral can be done explicitly.) Thus, the final speed of the ball is

$$v_f = \frac{J}{m} = \frac{4\,\text{N}\cdot\text{s}}{0.5\,\text{kg}} = 8\ \text{m/s}\ .$$

83. (a) The impulse on the ball is

$$\vec{J} = \Delta\vec{p} = m\vec{v} - 0 = (46\times10^{-3}\,\text{kg})(50\,\text{m/s})\hat{\imath} = (2.3\,\text{N}\cdot\text{s})\hat{\imath}$$

where we choose $\hat{\imath}$ to be in the direction of the velocity $\vec{v}$ of the ball as it leaves the club (at 30° above horizontal – so it is like the x axis of an inclined plane problem).

(b) The impulse on the club is, by Newton's third law, $\vec{J}' = -\vec{J} = -(2.3\,\text{N}\cdot\text{s})\hat{\imath}$. We note that it is directed opposite to the direction of motion.

(c) Using Eq. 10-8, the average force on the ball is

$$\vec{F}_{\text{avg}} = \frac{\vec{J}}{\Delta t} = \frac{(2.3)\hat{\imath}}{1.7\times10^{-3}} = 1400\,\hat{\imath}\ \text{N}\ .$$

(d) The work done on the ball is

$$W = \Delta K = \frac{1}{2}mv^2 = \frac{1}{2}(46\times10^{-3})(50)^2 = 58\ \text{J}\ .$$

84. We first note that when the the velocity of a projectile is simply reversed as a result of collision, its change in momentum (in magnitude) is $2mv$ (where v is its speed). If this collision takes time Δt, then the average force involved is (using Eq. 10-8) $F_{\text{avg}} = 2mv/\Delta t$. To relate this observation to the present situation, we replace m with Δm (representing just that amount of the water stream which is in contact with the blade during Δt, and since the impinging flow rate dm/dt is constant (and no water is lost or "splattered away" in the process) then we conclude $dm/dt = \Delta m/\Delta t$. Therefore,

$$F_{\text{avg}} = 2v\,\frac{dm}{dt}\ .$$

85. One could reason as in §9-7 (with the thrust concept) or proceed with Eq. 10-8. Choosing the latter approach, we note that (with the final momentum being zero) the average force is (in magnitude)

$$F_{\text{avg}} = v\,\frac{\Delta m}{\Delta t}$$

where Δm is the portion of the water that is decelerated (by the wall) from speed $v = 500$ cm/s to zero during time Δt. If the impinging mass flow rate dm/dt is constant, then we conclude $dm/dt = \Delta m/\Delta t$. Thus, $F_{\text{avg}} = v\,dm/dt$. We are given the volume flow rate $dV/dt = 300\ \text{cm}^3/\text{s}$, and we use the concept of density to relate mass and volume: $m = \rho V$ where $\rho = 1.0\ \text{g/cm}^3$ for water (most students have seen density in previous courses). Thus,

$$F_{\text{avg}} = v\,\frac{dm}{dt} = \rho\, v\,\frac{dV}{dt} = (1.0)(500)(300)$$

which yields $F_{\text{avg}} = 1.5\times10^5\ \text{g}\cdot\text{cm/s}^2$ which we convert to SI, giving the result $F_{\text{avg}} = 1.5$ N.

86. Although we do not present problems and solutions here, we share a few thoughts on the matter.

(a) This might be more like part (b) of problem 80, in which energy is "liberated" in the collision, but this depends on what particular sort of pinball collision one has in mind.

(b) This is a good example of an inelastic (but not completely so) collision and might be similar to part (a) of problem 80.

(c) This might be similar to problem 85, finding the average force on the car in the hailstorm. Instead of having the hail be halted completely by the collision (as is done with the water in problem 85) there should be some small rebound speed.

(d) An interesting comparison can be made here between the impact of fist with face with glove, and without the glove. The increase in contact time with the glove certainly decreases the force of impact.

(e) If baseball is chosen as one's example, it might be of interest to refer to the article by Howard Brody in the August 1990 issue of the *American Journal of Physics*, where he considers that the bat may be viewed as a relatively free body in the batting process.

87. (First problem in **Cluster 1**)
Instead of using V for final speed in completely inelastic collisions (as is used in Eq. 10-18), we use $v_{1f} = v_{2f}$, since that facilitates comparison of the results of parts (a) and (b). When we make comparisons, we assume $v_{1i} > 0$.

(a) When they stick together, we have

$$v_{1f} = v_{2f} = \frac{m_1}{m_1 + m_2} v_{1i} \; .$$

(b) Eq. 10-30 and Eq. 10-31 provide the elastic collision results:

$$\begin{aligned} v_{1f} &= \frac{m_1 - m_2}{m_1 + m_2} v_{1i} \\ v_{1f} &= \frac{2m_1}{m_1 + m_2} v_{1i} \end{aligned}$$

from which it is evident that $v_{1f\,\text{elastic}} < v_{1f\,\text{inelastic}}$ and $v_{2f\,\text{elastic}} > v_{2f\,\text{inelastic}}$.

88. (Second problem in **Cluster 1**)
We note that the problem has implicitly chosen the initial direction of motion (of m_1) as the positive direction. The questions to find "greatest" and "least" values are understood in terms of that axis choice (*greatest* = largest positive value, and *least* = the negative value of greatest magnitude or the smallest non-negative value). In addition to the assumptions mentioned in the problem, we also assume that m_1 cannot pass through m_2 (like a bullet might be able to). We are only able to use momentum conservation, since no assumptions about the total kinetic energy can be made.

$$m_1 v_{1i} = m_1 v_{1f} + m_2 v_{2f}$$

This (since $m_2 = 2.00 m_1$) simplifies to

$$v_{1i} = v_{1f} + 2.00 v_{2f} \; .$$

(a) Using $v_{1i} = 10.0$ m/s, we have

$$v_{2f} = (5.00\,\text{m/s}) - 0.500 v_{1f} \; .$$

(b) Ignoring physics considerations, our function is a line of infinite extent with negative slope.

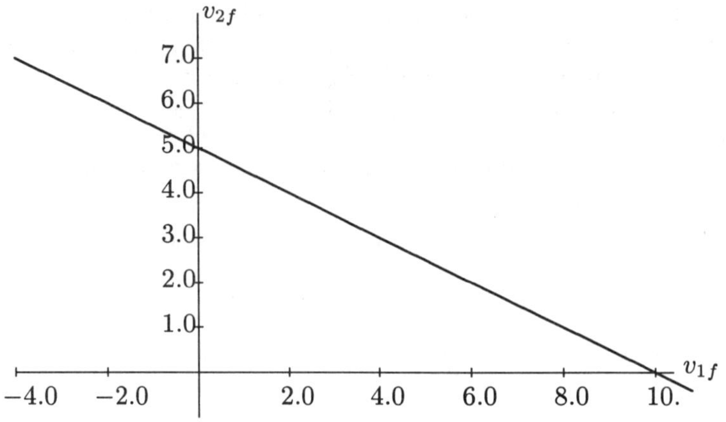

(c) The greatest possible value of v_{1f} occurs in the completely inelastic case (reasons mentioned in the next several parts) where (see solution to part (a) of previous problem) its value would be $(10.0)(1/3) \approx 3.33$ m/s.

(d) Clearly, this is also the value of v_{2f} in this case.

(e) They stick together (completely inelastic collision).

(f) As mentioned above, we assume m_1 does not pass through m_2 and the problem states that there's no energy production so that $K_{1f} \le K_{1i}$ which implies $v_{1f} \le v_{1i}$.

(g) The plot is shown below, in part (ℓ).

(h) With energy production not a possibility, then the "hardest rebound" m_1 can suffer is in an elastic collision, in which its final velocity (see part (b) of the previous problem) is $(10.0)(1-2)/3 \approx -3.33$ m/s.

(i) Eq. 10-31 gives the velocity of m_2 as $(10.0)(2/3) \approx 6.67$ m/s (see also part (b) of previous problem).

(j) As mentioned, this is an elastic collision (no "loss" of kinetic energy).

(k) The problem states that there's no energy production so that $K_{1i} - K_{1f} = K_{2f}$ and any greater value of $|v_{2f}|$ would violate this condition.

(l) The above graph is redrawn here, with the dark part representing the physically allowed region; the small circles bounding the dark segment correspond to the values calculated in the previous parts of this problem.

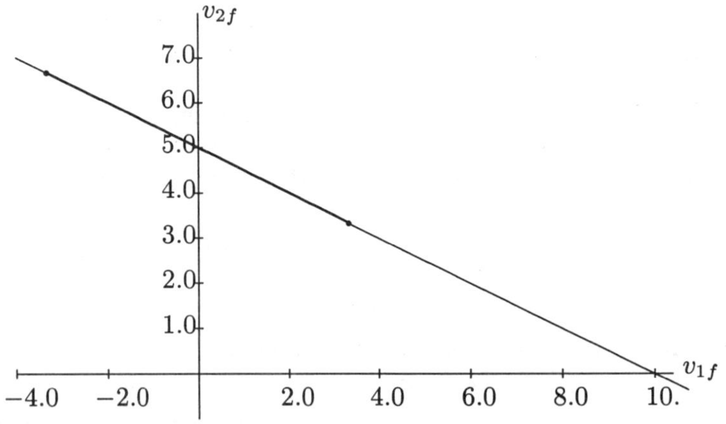

89. (Third problem in **Cluster 1**)
We note that the problem has implicitly chosen the initial direction of motion (of m_1) as the positive

direction. The questions to find "greatest" and "least" values are understood in terms of that axis choice (*greatest* = largest positive value, and *least* = the negative value of greatest magnitude or the smallest non-negative value). In addition to the assumptions mentioned in the problem, we also assume that m_1 cannot pass through m_2 (like a bullet might be able to). We are only able to use momentum conservation, since no assumptions about the total kinetic energy can be made.

$$m_1 v_{1i} = m_1 v_{1f} + m_2 v_{2f}$$

This (since $m_2 = 0.500 m_1$) simplifies to

$$v_{1i} = v_{1f} + 0.500 v_{2f} .$$

(a) Using $v_{1i} = 10.0$ m/s, we have

$$v_{2f} = (20.0\,\text{m/s}) - 2.00 v_{1f} .$$

(b) Ignoring physics considerations, our function is a line of infinite extent with negative slope.

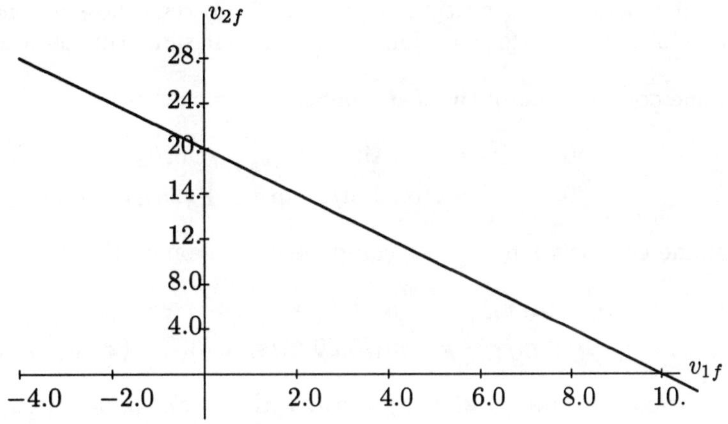

(c) The greatest possible value of v_{1f} occurs in the completely inelastic case (reasons mentioned in the next several parts) where (see solution to part (a) of previous problem) its value would be $(10.0)(2/3) \approx 6.67$ m/s.

(d) Clearly, this is also the value of v_{2f} in this case.

(e) They stick together (completely inelastic collision).

(f) As mentioned above, we assume m_1 does not pass through m_2 and the problem states that there's no energy production so that $K_{1f} \leq K_{1i}$ which implies $v_{1f} \leq v_{1i}$.

(g) The plot is shown below, in part (ℓ).

(h) With energy production not a possibility, then the "hardest rebound" m_1 can suffer is in an elastic collision, in which its final velocity (see part (b) of the previous problem) is $(10.0)(2-1)/3 \approx 3.33$ m/s.

(i) Eq. 10-31 gives the velocity of m_2 as $(10.0)(4/3) \approx 13.3$ m/s (see also part (b) of previous problem).

(j) As mentioned, this is an elastic collision (no "loss" of kinetic energy).

(k) The problem states that there's no energy production so that $K_{1i} - K_{1f} = K_{2f}$ and any greater value of $|v_{2f}|$ would violate this condition.

(l) The above graph is redrawn here, with the dark part representing the physically allowed region; the small circles bounding the dark segment correspond to the values calculated in the previous parts of this problem.

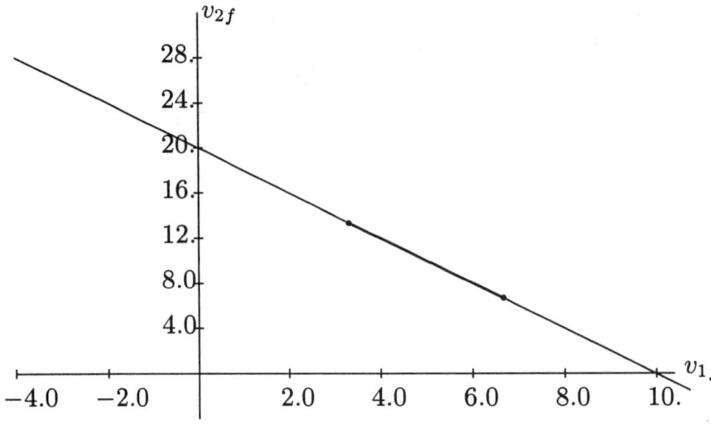

90. (First problem in **Cluster 2**)
The setup for this cluster refers to Fig. 10-16 in the chapter that assumes both angles are positive (at least, this is what is assumed in writing down Eq. 10-43) regardless of whether they are measured clockwise or counterclockwise. In this solution, we adopt that same convention.

(a) We first examine conservation of the y components of momentum:

$$\begin{aligned} 0 &= -m_1 v_{1f} \sin\theta_1 + m_2 v_{2f} \sin\theta_2 \\ 0 &= -m_1 (5.00 \text{ m/s}) \sin 30^\circ + (2m_1)\, v_{2f} \sin\theta_2 \end{aligned}$$

Next, we examine conservation of the x components of momentum.

$$\begin{aligned} m_1 v_{1i} &= m_1 v_{1f} \cos\theta_1 + m_2 v_{2f} \cos\theta_2 \\ m_1 (10.0 \text{ m/s}) &= m_1 (5.00 \text{ m/s}) \cos 30^\circ + (2m_1)\, v_{2f} \cos\theta_2 \end{aligned}$$

From the y equation, we obtain $1.25 = v_{2f} \sin\theta_2$ with SI units understood; similarly, the x equation yields $2.83 = v_{2f} \cos\theta_2$. Squaring these two relations and adding them leads to

$$1.25^2 + 2.83^2 = v_{2f}^2 \left(\sin^2\theta_2 + \cos^2\theta_2\right)$$

and consequently to $v_{2f} = \sqrt{1.25^2 + 2.83^2} = 3.10$ m/s. Plugging back in to either the x or y equation yields the angle $\theta_2 = 23.8^\circ$.

(b) We compute decrease in total kinetic energy:

$$K_i - K_f = 27.9\, m_1$$

so that the collision is seen to be inelastic. We find that

$$\frac{27.9\, m_1}{\frac{1}{2} m_1\, 10^2} = 0.558 \;,$$

or roughly 56%, of the initial energy has been "lost."

91. (Second problem in **Cluster 2**)
As explained in the previous solution, we take both angles θ_1 and θ_2 to be positive-valued.

(a) We first examine conservation of the y components of momentum.

$$\begin{aligned} 0 &= -m_1 v_{1f} \sin\theta_1 + m_2 v_{2f} \sin\theta_2 \\ 0 &= -m_1 v_{1f} \sin 30^\circ + 2m_1\, v_{2f} \sin\theta_2 \end{aligned}$$

Next, we examine conservation of the x components of momentum.

$$\begin{aligned} m_1 v_{1i} &= m_1 v_{1f} \cos\theta_1 + m_2 v_{2f} \cos\theta_2 \\ m_1(10.0 \text{ m/s}) &= m_1 v_{1f} \cos 30^\circ + 2m_1 v_{2f} \cos\theta_2 \end{aligned}$$

From the y equation, we obtain $v_{1f} = 4v_{2f} \sin\theta_2$; similarly, the x equation yields $20 - v_{1f}\sqrt{3} = 4v_{2f} \cos\theta_2$ with SI units understood (also, $\cos 30^\circ = \sqrt{3}/2$ has been used). Squaring these two relations and adding them leads to

$$v_{1f}^2 (1+3) - 40v_{1f}\sqrt{3} + 400 = 16\, v_{2f}^2 \left(\sin^2\theta_2 + \cos^2\theta_2\right)$$

and thus to $v_{2f}^2 = v_{1f}^2/4 - 5v_{1f}\sqrt{3}/2 + 25$. We plug this into the condition of total kinetic energy "conservation."

$$\begin{aligned} K_i &= K_f \\ \frac{1}{2} m_1 v_{1i}^2 &= \frac{1}{2} m_1 v_{1f}^2 + \frac{1}{2} m_2 v_{2f}^2 \\ \frac{1}{2} m_1 \left(10\ \frac{\text{m}}{\text{s}}\right)^2 &= \frac{1}{2} m_1 v_{1f}^2 + \frac{1}{2}(2m_1)\left(\frac{v_{1f}^2}{4} - \frac{5\sqrt{3}}{2} v_{1f} + 25\right) \end{aligned}$$

This leads to an equation of second degree (in the variable v_{1f}):

$$\frac{3}{4} v_{1f}^2 - \frac{5\sqrt{3}}{2} v_{1f} - 25 = 0$$

which has a positive root $v_{1f} = \frac{5}{3}\sqrt{3}\,(1+\sqrt{5}) \approx 9.34$ m/s.

(b) We plug our result for v_{1f} into the relation $v_{2f} = \sqrt{v_{1f}^2/4 - 5v_{1f}\sqrt{3}/2 + 25}$ derived above and obtain $v_{2f} = \frac{5}{6}\sqrt{6}\,(\sqrt{5} - 1) \approx 2.52$ m/s.

(c) Plugging these values of v_{1f} and v_{2f} into, say, the $v_{1f} = 4v_{2f} \sin\theta_2$ relation, we find $\theta_2 = 67.8^\circ$.

92. (Second problem in **Cluster 2**)
As explained in the first solution in this cluster, we take both angles θ_1 and θ_2 to be positive-valued.

(a) We first examine conservation of the y components of momentum.

$$\begin{aligned} 0 &= -m_1 v_{1f} \sin\theta_1 + m_2 v_{2f} \sin\theta_2 \\ 0 &= -m_1 v_{1f} \sin 30^\circ + 2m_1 v_{2f} \sin\theta_2 \end{aligned}$$

Next, we examine conservation of the x components of momentum.

$$\begin{aligned} m_1 v_{1i} &= m_1 v_{1f} \cos\theta_1 + m_2 v_{2f} \cos\theta_2 \\ m_1(10.0 \text{ m/s}) &= m_1 v_{1f} \cos 30^\circ + 2m_1 v_{2f} \cos\theta_2 \end{aligned}$$

From the y equation, we obtain $v_{1f} = 4v_{2f} \sin\theta_2$; similarly, the x equation yields $20 - v_{1f}\sqrt{3} = 4v_{2f} \cos\theta_2$ with SI units understood (and the fact that $\cos 30^\circ = \sqrt{3}/2$ has been used). Squaring these two relations and adding them leads to

$$v_{1f}^2 (1+3) - 40v_{1f}\sqrt{3} + 400 = 16\, v_{2f}^2 \left(\sin^2\theta_2 + \cos^2\theta_2\right)$$

and thus to

$$v_{2f}^2 = v_{1f}^2/4 - 5v_{1f}\sqrt{3}/2 + 25\ .$$

(b) The plot (v_{2f} versus v_{1f}) is shown below. The units for both axes are meters/second.

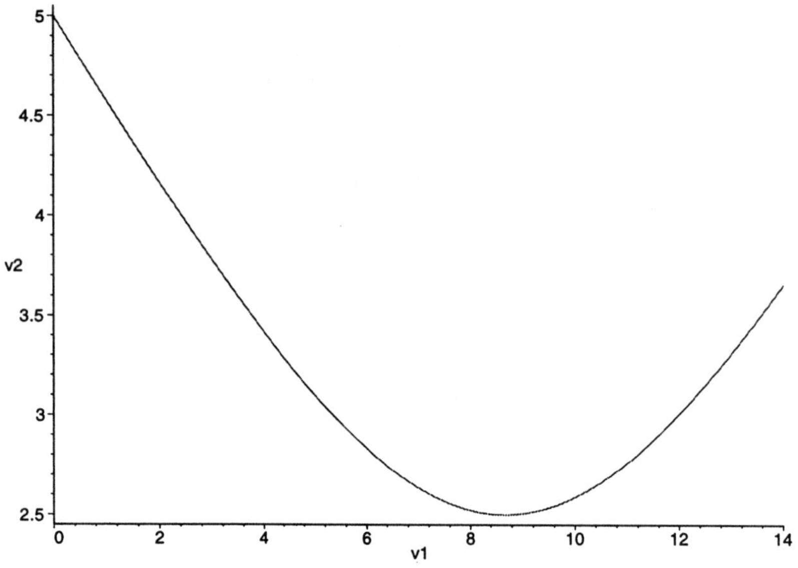

(c) Simply from the total kinetic energy requirement that $K_i \geq K_f$ we see immediately that $v_{1f} \leq v_{1i} = 10.0$ m/s (where the upper bound represents the trivial case where it passes m_2 by completely with $K_i = K_f$), and with the more stringent requirement that it does strike m_2 and scatters at $\theta_1 = 30°$ we again find that it is bounded by the $K_i = K_f$ case. The elastic collision scenario was worked in the previous problem with the result $v_{1f} = 9.34$ m/s.

(d) And we also found the result $v_{2f} = 2.52$ m/s.

(e) As mentioned, this is an elastic collision.

(f) A higher speed for v_{1f} would require energy conversion into kinetic form (say, from an explosion) since $K_i < K_f$ would be the result.

(g) To save space, a separate graph for this part is not shown.

(h) Returning to the x and y equations derived in part (a), we divide them to obtain

$$\frac{v_{1f}}{20 - v_{1f}\sqrt{3}} = \frac{4v_{2f}\sin\theta_2}{4v_{2f}\cos\theta_2} = \tan\theta_2$$

which leads to

$$\theta_2 = \tan^{-1}\left(\frac{v_{1f}}{20 - v_{1f}\sqrt{3}}\right).$$

(i) See part (k).

(j) The value for the elastic case was computed in the previous problem; we find $\theta_2 = 67.8°$ when $v_{1f} = 9.34$ m/s.

(k) This corresponds to the upper righthand point of the curve shown below.

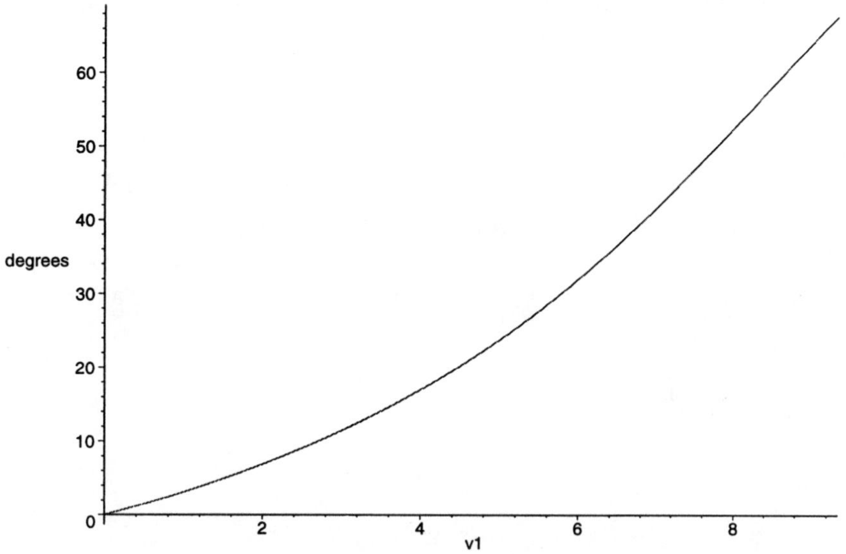

(l) , (m), (n), and (o)

Now, unlike the notation used in the one-dimensional collisions, this v_{1f} cannot be negative (it is the *magnitude* of the velocity). This suggests that its smallest value is zero, but the requirement that it scatter at $\theta_1 = 30°$ might seem to conflict with this. However, if one considers the (smooth) limit of $v_{1f} \to 0$, we find there is nothing inconsistent with $\theta_1 = 30°$ in setting $v_{1f} = 0$. It is certainly inelastic (but not completely so! A completely inelastic collision *would* be inconsistent with the $\theta_1 = 30°$ condition!); we find from $v_{2f} = 5.00$ m/s (see the graph for part (b)) that $K_i < K_f$ in this case. Clearly, $\theta_2 = 0°$ in this circumstance (see, e.g., the graph for part(i)).

Chapter 11

1. (a) Eq. 11-6 leads to

$$\omega = \frac{d}{dt}\left(at + bt^3 - ct^4\right) = a + 3bt^2 - 4ct^3 .$$

(b) And Eq. 11-8 gives

$$\alpha = \frac{d}{dt}\left(a + 3bt^2 - 4ct^3\right) = 6bt - 12ct^2 .$$

2. (a) The second hand of the smoothly running watch turns through 2π radians during 60 s. Thus,

$$\omega = \frac{2\pi}{60} = 0.105 \text{ rad/s} .$$

(b) The minute hand of the smoothly running watch turns through 2π radians during 3600 s. Thus,

$$\omega = \frac{2\pi}{3600} = 1.75 \times 10^{-3} \text{ rad/s} .$$

(c) The hour hand of the smoothly running 12-hour watch turns through 2π radians during 43200 s. Thus,

$$\omega = \frac{2\pi}{43200} = 1.45 \times 10^{-4} \text{ rad/s} .$$

3. (a) The time for one revolution is the circumference of the orbit divided by the speed v of the Sun: $T = 2\pi R/v$, where R is the radius of the orbit. We convert the radius:

$$R = \left(2.3 \times 10^4 \text{ ly}\right)\left(9.46 \times 10^{12} \text{ km/ly}\right) = 2.18 \times 10^{17} \text{ km}$$

where the ly $\leftrightarrow$ km conversion can be found in Appendix D or figured "from basics" (knowing the speed of light). Therefore, we obtain

$$T = \frac{2\pi\left(2.18 \times 10^{17} \text{ km}\right)}{250 \text{ km/s}} = 5.5 \times 10^{15} \text{ s} .$$

(b) The number of revolutions N is the total time t divided by the time T for one revolution; that is, $N = t/T$. We convert the total time from years to seconds and obtain

$$N = \frac{\left(4.5 \times 10^9 \text{ y}\right)\left(3.16 \times 10^7 \text{ s/y}\right)}{5.5 \times 10^{15} \text{ s}} = 26 .$$

4. If we make the units explicit, the function is

$$\theta = (4.0 \text{ rad/s})t - \left(3.0 \text{ rad/s}^2\right) t^2 + \left(1.0 \text{ rad/s}^3\right) t^3$$

but generally we will proceed as shown in the problem – letting these units be understood. Also, in our manipulations we will generally not display the coefficients with their proper number of significant figures.

(a) Eq. 11-6 leads to

$$\omega = \frac{d}{dt}\left(4t - 3t^2 + t^3\right) = 4 - 6t + 3t^2 \ .$$

Evaluating this at $t = 2$ s yields $\omega_2 = 4.0$ rad/s.

(b) Evaluating the expression in part (a) at $t = 4$ s gives $\omega_4 = 28$ rad/s.

(c) Consequently, Eq. 11-7 gives

$$\alpha_{\text{avg}} = \frac{\omega_4 - \omega_2}{4 - 2} = 12 \text{ rad/s}^2 \ .$$

(d) And Eq. 11-8 gives

$$\alpha = \frac{d\omega}{dt} = \frac{d}{dt}\left(4 - 6t + 3t^2\right) = -6 + 6t \ .$$

Evaluating this at $t = 2$ s produces $\alpha_2 = 6.0$ rad/s^2.

(e) Evaluating the expression in part (d) at $t = 4$ s yields $\alpha_4 = 18$ rad/s^2. We note that our answer for α_{avg} does turn out to be the arithmetic average of α_2 and α_4 but point out that this will not always be the case.

5. If we make the units explicit, the function is

$$\theta = 2\,\text{rad} + \left(4\,\text{rad/s}^2\right)t^2 + \left(2\,\text{rad/s}^3\right)t^3$$

but in some places we will proceed as indicated in the problem – by letting these units be understood.

(a) We evaluate the function θ at $t = 0$ to obtain $\theta_0 = 2$ rad.

(b) The angular velocity as a function of time is given by Eq. 11-6:

$$\omega = \frac{d\theta}{dt} = \left(8\,\text{rad/s}^2\right)t + \left(6\,\text{rad/s}^3\right)t^2$$

which we evaluate at $t = 0$ to obtain $\omega_0 = 0$.

(c) For $t = 4$ s, the function found in the previous part is $\omega_4 = (8)(4) + (6)(4)^2 = 128$ rad/s. If we round this to two figures, we obtain $\omega_4 \approx 130$ rad/s.

(d) The angular acceleration as a function of time is given by Eq. 11-8:

$$\alpha = \frac{d\omega}{dt} = 8\,\text{rad/s}^2 + \left(12\,\text{rad/s}^3\right)t$$

which yields $\alpha_2 = 8 + (12)(2) = 32\,\text{rad/s}^2$ at $t = 2$ s.

(e) The angular acceleration, given by the function obtained in the previous part, depends on time; it is not constant.

6. (a) To avoid touching the spokes, the arrow must go through the wheel in not more than

$$\Delta t = \frac{1/8 \text{ rev}}{2.5\,\text{rev/s}} = 0.050 \text{ s} \ .$$

The minimum speed of the arrow is then

$$v_{\text{min}} = \frac{20\,\text{cm}}{0.050\,\text{s}} = 400\,\text{cm/s} = 4.0 \text{ m/s} \ .$$

(b) No – there is no dependence on radial position in the above computation.

7. Applying Eq. 2-15 to the vertical axis (with $+y$ downward) we obtain the free-fall time:

$$\Delta y = v_{0y}t + \frac{1}{2}gt^2 \implies t = \sqrt{\frac{2(10)}{9.8}} = 1.4 \text{ s} .$$

Thus, by Eq. 11-5, the magnitude of the average angular velocity is

$$\omega_{\text{avg}} = \frac{(2.5)(2\pi)}{1.4} = 11 \text{ rad/s} .$$

8. (a) We assume the sense of rotation is positive. Applying Eq. 11-12, we obtain

$$\omega = \omega_0 + \alpha t \implies \alpha = \frac{3000 - 1200}{12/60} = 9000 \text{ rev/min}^2 .$$

(b) And Eq. 11-15 gives

$$\theta = \frac{1}{2}(\omega_0 + \omega)t = \frac{1}{2}(1200 + 3000)\left(\frac{12}{60}\right)$$

which yields $\theta = 420$ rev.

9. We assume the sense of initial rotation is positive. Then, with $\omega_0 > 0$ and $\omega = 0$ (since it stops at time t), our angular acceleration is negative-valued.

(a) The angular acceleration is constant, so we can apply Eq. 11-12 ($\omega = \omega_0 + \alpha t$). To obtain the requested units, we have $t = 30/60 = 0.50$ min. Thus,

$$\alpha = -\frac{33.33 \text{ rev/min}}{0.50 \text{ min}} = -66.7 \text{ rev/min}^2 .$$

(b) We use Eq. 11-13:

$$\theta = \omega_0 t + \frac{1}{2}\alpha t^2 = (33.33)(0.50) + \frac{1}{2}(-66.7)(0.50)^2 = 8.3 \text{ rev} .$$

10. We assume the sense of initial rotation is positive. Then, with $\omega_0 = +120$ rad/s and $\omega = 0$ (since it stops at time t), our angular acceleration ("deceleration") will be negative-valued: $\alpha = -4.0$ rad/s^2.

(a) We apply Eq. 11-12 to obtain t.

$$\omega = \omega_0 + \alpha t \implies t = \frac{0 - 120}{-4.0} = 30 \text{ s} .$$

(b) And Eq. 11-15 gives

$$\theta = \frac{1}{2}(\omega_0 + \omega)t = \frac{1}{2}(120 + 0)(30)$$

which yields $\theta = 1800$ rad. Alternatively, Eq. 11-14 could be used if it is desired to only use the given information (as opposed to using the result from part (a)) in obtaining θ. If using the result of part (a) is acceptable, then any angular equation in Table 11-1 (except Eq. 11-12) can be used to find θ.

11. We apply Eq. 11-12 twice, assuming the sense of rotation is positive. We have $\omega > 0$ and $\alpha < 0$. Since the angular velocity at $t = 1$ min is $\omega_1 = (0.90)(250) = 225$ rev/min, we have

$$\omega_1 = \omega_0 + \alpha t \implies \alpha = \frac{225 - 250}{1} = -25 \text{ rev/min}^2 .$$

Next, between $t = 1$ min and $t = 2$ min we have the interval $\Delta t = 1$ min. Consequently, the angular velocity at $t = 2$ min is

$$\omega_2 = \omega_1 + \alpha \Delta t = 225 + (-25)(1) = 200 \text{ rev/min} .$$

12. We assume the sense of rotation is positive, which (since it starts from rest) means all quantities (angular displacements, accelerations, etc.) are positive-valued.

   (a) The angular acceleration satisfies Eq. 11-13:

$$25\,\text{rad} = \frac{1}{2}\alpha(5.0\,\text{s})^2 \implies \alpha = 2.0\ \text{rad/s}^2 .$$

   (b) The average angular velocity is given by Eq. 11-5:

$$\omega_{\text{avg}} = \frac{\Delta\theta}{\Delta t} = \frac{25\,\text{rad}}{5.0\,\text{s}} = 5.0\ \text{rad/s} .$$

   (c) Using Eq. 11-12, the instantaneous angular velocity at $t = 5.0$ s is

$$\omega = \left(2.0\,\text{rad/s}^2\right)(5.0\,\text{s}) = 10\ \text{rad/s} .$$

   (d) According to Eq. 11-13, the angular displacement at $t = 10$ s is

$$\theta = \omega_0 + \frac{1}{2}\alpha t^2 = 0 + \frac{1}{2}(2.0)(10)^2 = 100\ \text{rad} .$$

   Thus, the displacement between $t = 5$ s and $t = 10$ s is $\Delta\theta = 100 - 25 = 75$ rad.

13. We take $t = 0$ at the start of the interval and take the sense of rotation as positive. Then at the end of the $t = 4.0$ s interval, the angular displacement is $\theta = \omega_0 t + \frac{1}{2}\alpha t^2$. We solve for the angular velocity at the start of the interval:

$$\omega_0 = \frac{\theta - \frac{1}{2}\alpha t^2}{t} = \frac{120\,\text{rad} - \frac{1}{2}\left(3.0\,\text{rad/s}^2\right)(4.0\,\text{s})^2}{4.0\,\text{s}} = 24\ \text{rad/s} .$$

We now use $\omega = \omega_0 + \alpha t$ (Eq. 11-12) to find the time when the wheel is at rest:

$$t = -\frac{\omega_0}{\alpha} = -\frac{24\,\text{rad/s}}{3.0\,\text{rad/s}^2} = -8.0\ \text{s} .$$

That is, the wheel started from rest 8.0 s before the start of the described 4.0 s interval.

14. The wheel starts turning from rest ($\omega_0 = 0$) at $t = 0$, and accelerates uniformly at $\alpha = 2.00\ \text{rad/s}^2$. Between t_1 and t_2 it turns through $\Delta\theta = 90.0$ rad, where $t_2 - t_1 = \Delta t = 3.00$ s.

   (a) We use Eq. 11-13 (with a slight change in notation) to describe the motion for $t_1 \le t \le t_2$:

$$\Delta\theta = \omega_1 \Delta t + \frac{1}{2}\alpha\,(\Delta t)^2 \implies \omega_1 = \frac{\Delta\theta}{\Delta t} - \frac{\alpha\Delta t}{2}$$

   which we plug into Eq. 11-12, set up to describe the motion during $0 \le t \le t_1$:

$$\begin{aligned} \omega_1 &= \omega_0 + \alpha t_1 \\ \frac{\Delta\theta}{\Delta t} - \frac{\alpha\Delta t}{2} &= \alpha t_1 \\ \frac{90.0}{3.00} - \frac{(2.00)(3.00)}{2} &= (2.00)t_1 \end{aligned}$$

   yielding $t_1 = 13.5$ s.

   (b) Plugging into our expression for ω_1 (in previous part) we obtain

$$\omega_1 = \frac{\Delta\theta}{\Delta t} - \frac{\alpha\Delta t}{2} = \frac{90.0}{3.00} - \frac{(2.00)(3.00)}{2} = 27.0\ \text{rad/s} .$$

15. The problem has (implicitly) specified the positive sense of rotation. The angular acceleration of magnitude $0.25\ \text{rad/s}^2$ in the negative direction is assumed to be constant over a large time interval, including negative values (for t).

(a) We specify θ_{max} with the condition $\omega = 0$ (this is when the wheel reverses from positive rotation to rotation in the negative direction). We obtain θ_{max} using Eq. 11-14:

$$\theta_{\text{max}} = -\frac{\omega_o^2}{2\alpha} = -\frac{4.7^2}{2(-0.25)} = 44\ \text{rad}\ .$$

(b) We find values for t_1 when the angular displacement (relative to its orientation at $t = 0$) is $\theta_1 = 22$ rad (or 22.09 rad if we wish to keep track of accurate values in all intermediate steps and only round off on the final answers). Using Eq. 11-13 and the quadratic formula, we have

$$\theta_1 = \omega_o t_1 + \frac{1}{2}\alpha t_1^2 \implies t_1 = \frac{-\omega_o \pm \sqrt{\omega_o^2 + 2\theta_1\alpha}}{\alpha}$$

which yields the two roots 5.5 s and 32 s.

(c) We find values for t_2 when the angular displacement (relative to its orientation at $t = 0$) is $\theta_2 = -10.5$ rad. Using Eq. 11-13 and the quadratic formula, we have

$$\theta_2 = \omega_o t_2 + \frac{1}{2}\alpha t_2^2 \implies t_2 = \frac{-\omega_o \pm \sqrt{\omega_o^2 + 2\theta_2\alpha}}{\alpha}$$

which yields the two roots -2.1 s and 40 s.

(d) With radians and seconds understood, the graph of θ versus t is shown below (with the points found in the previous parts indicated as small circles).

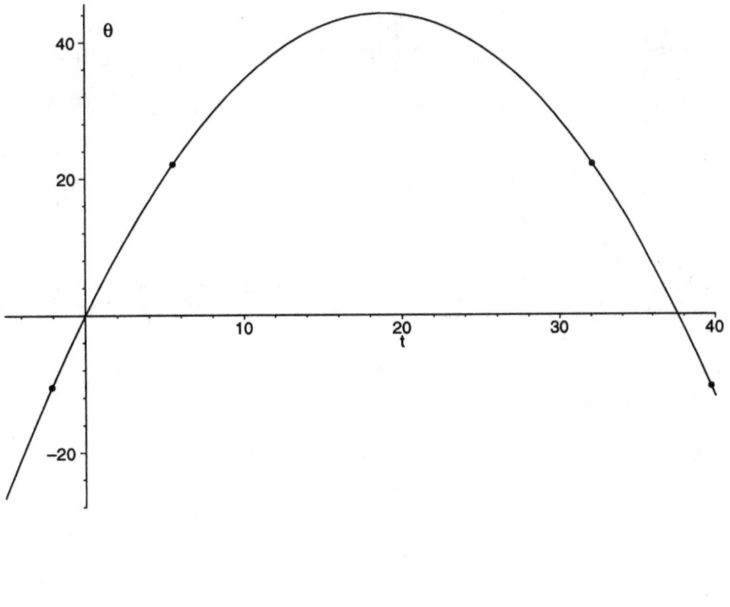

16. The wheel starts turning from rest ($\omega_o = 0$) at $t = 0$, and accelerates uniformly at $\alpha > 0$, which makes our choice for positive sense of rotation. At t_1 its angular velocity is $\omega_1 = +10$ rev/s, and at t_2 its angular velocity is $\omega_2 = +15$ rev/s. Between t_1 and t_2 it turns through $\Delta\theta = 60$ rev, where $t_2 - t_1 = \Delta t$.

(a) We find α using Eq. 11-14:

$$\omega_2^2 = \omega_1^2 + 2\alpha\Delta\theta \implies \alpha = \frac{15^2 - 10^2}{2(60)}$$

which yields $\alpha = 1.04$ rev/s^2 which we round off to 1.0 rev/s^2.

(b) We find Δt using Eq. 11-15:

$$\Delta\theta = \frac{1}{2}(\omega_1 + \omega_2)\,\Delta t \implies \Delta t = \frac{2(60)}{10+15} = 4.8 \text{ s} .$$

(c) We obtain t_1 using Eq. 11-12:

$$\omega_1 = \omega_0 + \alpha t_1 \implies t_1 = \frac{10}{1.04} = 9.6 \text{ s} .$$

(d) Any equation in Table 11-1 involving θ can be used to find θ_1 (the angular displacement during $0 \le t \le t_1$); we select Eq. 11-14.

$$\omega_1^2 = \omega_0^2 + 2\alpha\theta_1 \implies \theta_1 = \frac{10^2}{2(1.04)} = 48 \text{ rev} .$$

17. The wheel has angular velocity $\omega_0 = +1.5$ rad/s $= +0.239$ rev/s^2 at $t = 0$, and has constant value of angular acceleration $\alpha < 0$, which indicates our choice for positive sense of rotation. At t_1 its angular displacement (relative to its orientation at $t = 0$) is $\theta_1 = +20$ rev, and at t_2 its angular displacement is $\theta_2 = +40$ rev and its angular velocity is $\omega_2 = 0$.

(a) We obtain t_2 using Eq. 11-15:

$$\theta_2 = \frac{1}{2}(\omega_0 + \omega_2)\,t_2 \implies t_2 = \frac{2(40)}{0.239}$$

which yields $t_2 = 335$ s which we round off to $t_2 \approx 340$ s.

(b) Any equation in Table 11-1 involving α can be used to find the angular acceleration; we select Eq. 11-16.

$$\theta_2 = \omega_2 t_2 - \frac{1}{2}\alpha t_2^2 \implies \alpha = -\frac{2(40)}{335^2}$$

which yields $\alpha = -7.12 \times 10^{-4}$ rev/s^2 which we convert to $\alpha = -4.5 \times 10^{-3}$ rad/s^2.

(c) Using $\theta_1 = \omega_0 t_1 + \frac{1}{2}\alpha t_1^2$ (Eq. 11-13) and the quadratic formula, we have

$$t_1 = \frac{-\omega_0 \pm \sqrt{\omega_0^2 + 2\theta_1\alpha}}{\alpha} = \frac{-0.239 \pm \sqrt{0.239^2 + 2(20)\left(-7.12 \times 10^{-4}\right)}}{-7.12 \times 10^{-4}}$$

which yields two positive roots: 98 s and 572 s. Since the question makes sense only if $t_1 < t_2$ we conclude the correct result is $t_1 = 98$ s.

18. The wheel starts turning from rest ($\omega_0 = 0$) at $t = 0$, and accelerates uniformly at $\alpha = +4.0$ rad/s^2, which makes our choice for positive sense of rotation. At t_1 its angular displacement (relative to its orientation at $t = 0$) is θ_1, and at t_2 its angular velocity is θ_2, where $\theta_2 - \theta_1 = \Delta\theta = 80$ rad. Also, $t_2 - t_1 = \Delta t = 4.0$ s.

(a) We find the angular velocity at t_1 using Eq. 11-13 (set up to describe the interval $t_1 \le t \le t_2$).

$$\Delta\theta = \omega_1\Delta t + \frac{1}{2}\alpha\,(\Delta t)^2 \implies \omega_1 = \frac{80 - \frac{1}{2}(4.0)(4.0)^2}{4.0}$$

which yields $\omega_1 = 12$ rad/s.

(b) We obtain t_1 using Eq. 11-12:

$$\omega_1 = \omega_0 + \alpha t_1 \quad \Longrightarrow \quad t_1 = \frac{12}{4.0} = 3.0 \text{ s} .$$

19. The magnitude of the acceleration is given by $a = \omega^2 r$ (Eq. 11-23) where r is the distance from the center of rotation and ω is the angular velocity. We convert the given angular velocity to rad/s:

$$\omega = \frac{(33.33\,\text{rev/min})(2\pi\,\text{rad/rev})}{60\,\text{s/min}} = 3.49 \text{ rad/s} .$$

Therefore,

$$a = \left(3.49\,\text{rad/s}^2\right)^2 (0.15\,\text{m}) = 1.8 \text{ m/s}^2 .$$

The acceleration vector is toward the center of the record.

20. (a) We obtain

$$\omega = \frac{(33.33\,\text{rev/min})(2\pi\,\text{rad/rev})}{60\,\text{s/min}} = 3.49 \text{ rad/s} .$$

(b) Using Eq. 11-18, we have

$$v = r\omega = (15)(3.49) = 52 \text{ cm/s} .$$

(c) Similarly, when $r = 7.4$ cm we find $v = r\omega = 26$ cm/s. The goal of this exercise to observe what is and is not the same at different locations on a body in rotational motion (ω is the same, v is not), as well as to emphasize the importance of radians when working with equations such as Eq. 11-18.

21. With $v = 50(1000/3600) = 13.9$ m/s, Eq. 11-18 leads to

$$\omega = \frac{v}{r} = \frac{13.9}{110} = 0.13 \text{ rad/s} .$$

22. (a) We obtain

$$\omega = \frac{(200\,\text{rev/min})(2\pi\,\text{rad/rev})}{60\,\text{s/min}} = 20.9 \text{ rad/s} .$$

(b) With $r = 1.20/2 = 0.60$ m, Eq. 11-18 leads to

$$v = r\omega = (0.60)(20.9) = 12.6 \text{ m/s} .$$

(c) With $t = 1$ min, $\omega = 1000$ rev/min and $\omega_o = 200$ rev/min, Eq. 11-12 gives

$$\alpha = \frac{\omega - \omega_o}{t} = 800 \text{ rev/min}^2 .$$

(d) With the same values used in part (c), Eq. 11-15 becomes

$$\theta = \frac{1}{2}(\omega_o + \omega)\,t = \frac{1}{2}(200 + 1000)(1) = 600 \text{ rev} .$$

23. (a) Using Eq. 11-6, the angular velocity at $t = 5.0$ s is

$$\omega = \left.\frac{d\theta}{dt}\right|_{t=5.0} = \frac{d}{dt}\left(0.30t^2\right)\Big|_{t=5.0} = 2(0.30)(5.0) = 3.0 \text{ rad/s} .$$

(b) Eq. 11-18 gives the linear speed at $t = 5.0$ s:

$$v = \omega r = (3.0\,\text{rad/s})(10\,\text{m}) = 30 \text{ m/s} .$$

(c) The angular acceleration is, from Eq. 11-8,

$$\alpha = \frac{d\omega}{dt} = \frac{d}{dt}(0.60t) = 0.60 \text{ rad/s}^2 .$$

Then, the tangential acceleration at $t = 5.0\,\text{s}$ is, using Eq. 11-22,

$$a_t = r\alpha = (10\,\text{m})\left(0.60\,\text{rad/s}^2\right) = 6.0 \text{ m/s}^2 .$$

(d) The radial (centripetal) acceleration is given by Eq. 11-23:

$$a_r = \omega^2 r = (3.0\,\text{rad/s})^2(10\,\text{m}) = 90 \text{ m/s}^2 .$$

24. (a) Converting from hours to seconds, we find the angular velocity (assuming it is positive) from Eq. 11-18:

$$\omega = \frac{v}{r} = \frac{(2.90 \times 10^4\,\text{km/h})\left(\frac{1.00\,\text{h}}{3600\,\text{s}}\right)}{3.22 \times 10^3\,\text{km}} = 2.50 \times 10^{-3}\,\text{rad/s} .$$

(b) The radial (or centripetal) acceleration is computed according to Eq. 11-23:

$$a_r = \omega^2 r = \left(2.50 \times 10^{-3}\,\text{rad/s}\right)^2 \left(3.22 \times 10^6\,\text{m}\right) = 20.2 \text{ m/s}^2 .$$

(c) Assuming the angular velocity is constant, then the angular acceleration and the tangential acceleration vanish, since

$$\alpha = \frac{d\omega}{dt} = 0 \quad \text{and} \quad a_t = r\alpha = 0 .$$

25. (a) In the time light takes to go from the wheel to the mirror and back again, the wheel turns through an angle of $\theta = 2\pi/500 = 1.26 \times 10^{-2}\,\text{rad}$. That time is

$$t = \frac{2\ell}{c} = \frac{2(500\,\text{m})}{2.998 \times 10^8\,\text{m/s}} = 3.34 \times 10^{-6} \text{ s}$$

so the angular velocity of the wheel is

$$\omega = \frac{\theta}{t} = \frac{1.26 \times 10^{-2}\,\text{rad}}{3.34 \times 10^{-6}\,\text{s}} = 3.8 \times 10^3 \text{ rad/s} .$$

(b) If r is the radius of the wheel, the linear speed of a point on its rim is

$$v = \omega r = \left(3.8 \times 10^3\,\text{rad/s}\right)(0.05\,\text{m}) = 190 \text{ m/s} .$$

26. (a) The angular acceleration is

$$\alpha = \frac{\Delta\omega}{\Delta t} = \frac{0 - 150\,\text{rev/min}}{(2.2\,\text{h})(60\,\text{min/1\,h})} = -1.14 \text{ rev/min}^2 .$$

(b) Using Eq. 11-13 with $t = (2.2)(60) = 132$ min, the number of revolutions is

$$\begin{aligned}\theta &= \omega_0 t + \frac{1}{2}\alpha t^2 \\ &= (150\,\text{rev/min})(132\,\text{min}) + \frac{1}{2}\left(-1.14\,\text{rev/min}^2\right)(132\,\text{min})^2 \\ &= 9.9 \times 10^3 \text{ rev} .\end{aligned}$$

(c) With $r = 500$ mm, the tangential acceleration is

$$a_t = \alpha r = \left(-1.14\,\text{rev/min}^2\right)\left(\frac{2\pi\,\text{rad}}{1\,\text{rev}}\right)\left(\frac{1\,\text{min}}{60\,\text{s}}\right)^2 (500\,\text{mm})$$

which yields $a_t = -0.99\ \text{mm/s}^2$.

(d) With $r = 0.50$ m, the radial (or centripetal) acceleration is given by Eq. 11-23:

$$a_r = \omega^2 r = \left((75\,\text{rev/min})\left(\frac{2\pi\,\text{rad/rev}}{1\,\text{min}/60\,\text{s}}\right)\right)^2 (0.50\,\text{m})$$

which yields $a_r = 31$ in SI units – and is seen to be much bigger than a_t. Consequently, the magnitude of the acceleration is

$$|\vec{a}| = \sqrt{a_r^2 + a_t^2} \approx a_r = 31\ \text{m/s}^2 .$$

27. (a) Earth makes one rotation per day and $1\,d$ is $(24\,\text{h})(3600\,\text{s/h}) = 8.64 \times 10^4$ s, so the angular speed of Earth is

$$\omega = \frac{2\pi\,\text{rad}}{8.64 \times 10^4\,\text{s}} = 7.27 \times 10^{-5}\ \text{rad/s} .$$

(b) We use $v = \omega r$, where r is the radius of its orbit. A point on Earth at a latitude of 40° moves along a circular path of radius $r = R\cos 40°$, where R is the radius of Earth (6.37×10^6 m). Therefore, its speed is

$$v = \omega\,(R\cos 40°) = \left(7.27 \times 10^{-5}\,\text{rad/s}\right)\left(6.37 \times 10^6\,\text{m}\right)\cos 40° = 355\ \text{m/s} .$$

(c) At the equator (and all other points on Earth) the value of ω is the same (7.27×10^{-5} rad/s).

(d) The latitude is 0° and the speed is

$$v = \omega R = \left(7.27 \times 10^{-5}\,\text{rad/s}\right)\left(6.37 \times 10^6\,\text{m}\right) = 463\ \text{m/s} .$$

28. (a) The tangential acceleration, using Eq. 11-22, is

$$a_t = \alpha r = \left(14.2\,\text{rad/s}^2\right)(2.83\,\text{cm}) = 40.2\ \text{cm/s}^2 .$$

(b) In rad/s, the angular velocity is $\omega = (2760)(2\pi/60) = 289$, so

$$a_r = \omega^2 r = (289\,\text{rad/s})^2(0.0283\,\text{m}) = 2.36 \times 10^3\ \text{m/s}^2 .$$

(c) The angular displacement is, using Eq. 11-14,

$$\theta = \frac{\omega^2}{2\alpha} = \frac{289^2}{2(14.2)} = 2.94 \times 10^3\ \text{rad} .$$

Then, using Eq. 11-1, the distance traveled is

$$s = r\theta = (0.0283\ \text{m})\left(2.94 \times 10^3\,\text{rad}\right) = 83.2\ \text{m} .$$

29. Since the belt does not slip, a point on the rim of wheel C has the same tangential acceleration as a point on the rim of wheel A. This means that $\alpha_A r_A = \alpha_C r_C$, where α_A is the angular acceleration of wheel A and α_C is the angular acceleration of wheel C. Thus,

$$\alpha_C = \left(\frac{r_A}{r_C}\right)\alpha_A = \left(\frac{10\,\text{cm}}{25\,\text{cm}}\right)(1.6\,\text{rad/s}^2) = 0.64\ \text{rad/s}^2 .$$

Since the angular speed of wheel C is given by $\omega_C = \alpha_C t$, the time for it to reach an angular speed of $\omega = 100\,\text{rev/min} = 10.5\,\text{rad/s}$ starting from rest is

$$t = \frac{\omega_C}{\alpha_C} = \frac{10.5\,\text{rad/s}}{0.64\,\text{rad/s}^2} = 16\ \text{s} .$$

30. The function $\theta = \xi e^{\beta t}$ where $\xi = 0.40$ rad and $\beta = 2\,\mathrm{s}^{-1}$ is describing the angular coordinate of a line (which is marked in such a way that all points on it have the same value of angle at a given time) on the object. Taking derivatives with respect to time leads to $\frac{d\theta}{dt} = \xi\,\beta\,e^{\beta t}$ and $\frac{d^2\theta}{dt^2} = \xi\,\beta^2\,e^{\beta t}$.

(a) Using Eq. 11-22, we have

$$a_t = \alpha r = \frac{d^2\theta}{dt^2}\,r = \ 6.4\ \mathrm{cm/s^2}\ .$$

(b) Using Eq. 11-23, we have

$$a_r = \omega^2 r = \left(\frac{d\theta}{dt}\right)^2 r = \ 2.6\ \mathrm{cm/s^2}\ .$$

31. (a) A complete revolution is an angular displacement of $\Delta\theta = 2\pi$ rad, so the angular velocity in rad/s is given by $\omega = \Delta\theta/T = 2\pi/T$. The angular acceleration is given by

$$\alpha = \frac{d\omega}{dt} = -\frac{2\pi}{T^2}\frac{dT}{dt}\ .$$

For the pulsar described in the problem, we have

$$\frac{dT}{dt} = \frac{1.26 \times 10^{-5}\,\mathrm{s/y}}{3.16 \times 10^{7}\,\mathrm{s/y}} = 4.00 \times 10^{-13}\ .$$

Therefore,

$$\alpha = -\left(\frac{2\pi}{(0.033\,\mathrm{s})^2}\right)(4.00 \times 10^{-13}) = -2.3 \times 10^{-9}\,\mathrm{rad/s^2}\ .$$

The negative sign indicates that the angular acceleration is opposite the angular velocity and the pulsar is slowing down.

(b) We solve $\omega = \omega_0 + \alpha t$ for the time t when $\omega = 0$:

$$t = -\frac{\omega_0}{\alpha} = -\frac{2\pi}{\alpha T} = -\frac{2\pi}{(-2.3 \times 10^{-9}\,\mathrm{rad/s^2})(0.033\,\mathrm{s})} = 8.3 \times 10^{10}\,\mathrm{s}\ .$$

This is about 2600 years.

(c) The pulsar was born $1992 - 1054 = 938$ years ago. This is equivalent to $(938\,\mathrm{y})(3.16 \times 10^7\,\mathrm{s/y}) = 2.96 \times 10^{10}$ s. Its angular velocity at that time was

$$\omega = \omega_0 + \alpha t = \frac{2\pi}{T} + \alpha t = \frac{2\pi}{0.033\,\mathrm{s}} + (-2.3 \times 10^{-9}\,\mathrm{rad/s^2})(-2.96 \times 10^{10}\,\mathrm{s}) = 258\,\mathrm{rad/s}\ .$$

Its period was

$$T = \frac{2\pi}{\omega} = \frac{2\pi}{258\,\mathrm{rad/s}} = 2.4 \times 10^{-2}\,\mathrm{s}\ .$$

32. (a) The angular speed in rad/s is

$$\omega = \left(33\frac{1}{3}\,\mathrm{rev/min}\right)\left(\frac{2\pi\,\mathrm{rad/rev}}{60\,\mathrm{s/min}}\right) = 3.49\ \mathrm{rad/s}\ .$$

Consequently, the radial (centripetal) acceleration is (using Eq. 11-23)

$$a = \omega^2 r = (3.49\,\mathrm{rad/s})^2\,(6.0 \times 10^{-2}\,\mathrm{m}) = 0.73\,\mathrm{m/s^2}\ .$$

(b) Using Ch. 6 methods, we have $ma = f_s \le f_{s,\max} = \mu_s mg$, which is used to obtain the (minimum allowable) coefficient of friction:

$$\mu_{s,\min} = \frac{a}{g} = \frac{0.73}{9.8} = 0.075\ .$$

(c) The radial acceleration of the object is $a_r = \omega^2 r$, while the tangential acceleration is $a_t = \alpha r$. Thus

$$|\vec{a}| = \sqrt{a_r^2 + a_t^2} = \sqrt{(\omega^2 r)^2 + (\alpha r)^2} = r\sqrt{\omega^4 + \alpha^2} \ .$$

If the object is not to slip at any time, we require

$$f_{s,\max} = \mu_s mg = m a_{\max} = mr\sqrt{\omega_{\max}^4 + \alpha^2} \ .$$

Thus, since $\alpha = \omega/t$ (from Eq. 11-12), we find

$$\begin{aligned} \mu_{s,\min} &= \frac{r\sqrt{\omega_{\max}^4 + \alpha^2}}{g} \\ &= \frac{r\sqrt{\omega_{\max}^4 + (\omega_{\max}/t)^2}}{g} \\ &= \frac{(0.060)\sqrt{3.49^4 + (3.49/0.25)^2}}{9.8} \\ &= 0.11 \ . \end{aligned}$$

33. The kinetic energy (in J) is given by $K = \frac{1}{2}I\omega^2$, where I is the rotational inertia (in kg·m^2) and ω is the angular velocity (in rad/s). We have

$$\omega = \frac{(602\,\text{rev/min})(2\pi\,\text{rad/rev})}{60\,\text{s/min}} = 63.0\,\text{rad/s} \ .$$

Consequently, the rotational inertia is

$$I = \frac{2K}{\omega^2} = \frac{2(24400\,\text{J})}{(63.0\,\text{rad/s})^2} = 12.3\ \text{kg}\cdot\text{m}^2 \ .$$

34. The translational kinetic energy of the molecule is

$$K_t = \frac{1}{2}mv^2 = \frac{1}{2}\left(5.30\times10^{-26}\right)(500)^2 = 6.63\times10^{-21}\ \text{J} \ .$$

With $I = 1.94\times10^{-46}$ kg·m^2, we employ Eq. 11-27:

$$\begin{aligned} K_r &= \frac{2}{3}K_t \\ \frac{1}{2}I\omega^2 &= \frac{2}{3}\left(6.63\times10^{-21}\right) \end{aligned}$$

which leads to $\omega = 6.75\times10^{12}$ rad/s.

35. Since the rotational inertia of a cylinder is $I = \frac{1}{2}MR^2$ (Table 11-2(c)), its rotational kinetic energy is

$$K = \frac{1}{2}I\omega^2 = \frac{1}{4}MR^2\omega^2 \ .$$

For the first cylinder, we have $K = \frac{1}{4}(1.25)(0.25)^2(235)^2 = 1.1\times10^3$ J. For the second cylinder, we obtain $K = \frac{1}{4}(1.25)(0.75)^2(235)^2 = 9.7\times10^3$ J.

36. (a) Using Table 11-2(c), the rotational inertia is

$$I = \frac{1}{2}mR^2 = \frac{1}{2}(1210\,\text{kg})\left(\frac{1.21\,\text{m}}{2}\right)^2 = 221\ \text{kg}\cdot\text{m}^2 \ .$$

(b) The rotational kinetic energy is, by Eq. 11-27,

$$\begin{aligned} K &= \frac{1}{2} I\omega^2 \\ &= \frac{1}{2}\left(2.21\times 10^2\,\text{kg}\cdot\text{m}^2\right)\left((1.52\,\text{rev/s})(2\pi\,\text{rad/rev})\right)^2 \\ &= 1.10\times 10^4\ \text{J}\ . \end{aligned}$$

37. The particles are treated "point-like" in the sense that Eq. 11-26 yields their rotational inertia, and the rotational inertia for the rods is figured using Table 11-2(e) and the parallel-axis theorem (Eq. 11-29).

(a) With subscript 1 standing for the rod nearest the axis and 4 for the particle farthest from it, we have

$$\begin{aligned} I &= I_1 + I_2 + I_3 + I_4 \\ &= \left(\frac{1}{12}Md^2 + M\left(\frac{1}{2}d\right)^2\right) + md^2 + \left(\frac{1}{12}Md^2 + M\left(\frac{3}{2}d\right)^2\right) + m(2d)^2 \\ &= \frac{8}{3}Md^2 + 5md^2\ . \end{aligned}$$

(b) Using Eq. 11-27, we have

$$K = \frac{1}{2}I\omega^2 = \left(\frac{4}{3}Md^2 + \frac{5}{2}md^2\right)\omega^2\ .$$

38. (a) The rotational inertia of the three blades (each of mass m and length L) is

$$I = 3\left(\frac{1}{3}mL^2\right) = mL^2 = (240\,\text{kg})(5.2\,\text{m})^2 = 6.49\times 10^3\ \text{kg}\cdot\text{m}^2\ .$$

(b) The rotational kinetic energy is

$$\begin{aligned} K &= \frac{1}{2}I\omega^2 \\ &= \frac{1}{2}\left(6.49\times 10^3\,\text{kg}\cdot\text{m}^2\right)\left((350\,\text{rev/min})\left(\frac{2\pi\,\text{rad/rev}}{60\,\text{s/min}}\right)\right)^2 \\ &= 4.36\times 10^6\ \text{J} = 4.36\ \text{MJ}\ . \end{aligned}$$

39. We use the parallel axis theorem: $I = I_{\text{com}} + Mh^2$, where I_{com} is the rotational inertia about the center of mass (see Table 11-2(d)), M is the mass, and h is the distance between the center of mass and the chosen rotation axis. The center of mass is at the center of the meter stick, which implies $h = 0.50\,\text{m} - 0.20\,\text{m} = 0.30\,\text{m}$. We find

$$I_{\text{com}} = \frac{1}{12}ML^2 = \frac{1}{12}(0.56\,\text{kg})(1.0\,\text{m})^2 = 4.67\times 10^{-2}\ \text{kg}\cdot\text{m}^2\ .$$

Consequently, the parallel axis theorem yields

$$I = 4.67\times 10^{-2}\,\text{kg}\cdot\text{m}^2 + (0.56\,\text{kg})(0.30\,\text{m})^2 = 9.7\times 10^{-2}\ \text{kg}\cdot\text{m}^2\ .$$

40. (a) We show the figure with its axis of rotation (the thin horizontal line).

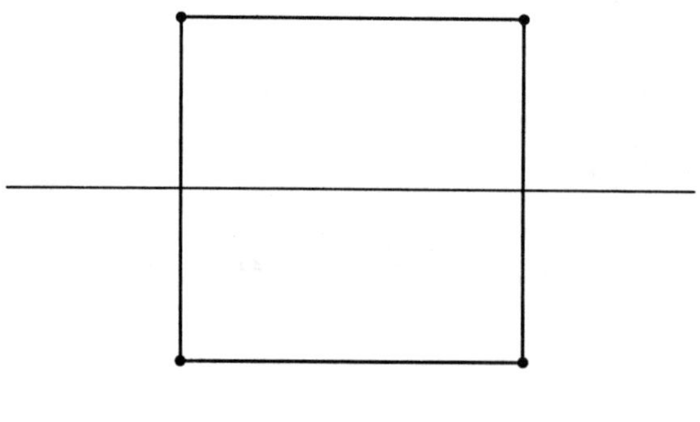

We note that each mass is $r = 1.0$ m from the axis. Therefore, using Eq. 11-26, we obtain

$$I = \sum m_i r_i^2 = 4(0.50 \text{ kg})(1.0 \text{ m})^2 = 2 \text{ kg}\cdot\text{m}^2 \ .$$

(b) In this case, the two masses nearest the axis are $r = 1.0$ m away from it, but the two furthest from the axis are $r = \sqrt{1.0^2 + 2.0^2}$ m from it. Here, then, Eq. 11-26 leads to

$$I = \sum m_i r_i^2 = 2(0.50 \text{ kg})(1.0 \text{ m}^2) + 2(0.50 \text{ kg})(5.0 \text{ m}^2) = 6.0 \text{ kg}\cdot\text{m}^2 \ .$$

(c) Now, two masses are on the axis (with $r = 0$) and the other two are a distance $r = \sqrt{1.0^2 + 1.0^2}$ m away. Now we obtain $I = 2.0 \text{ kg}\cdot\text{m}^2$.

41. We use the parallel-axis theorem. According to Table 11-2(i), the rotational inertia of a uniform slab about an axis through the center and perpendicular to the large faces is given by

$$I_{\text{com}} = \frac{M}{12}(a^2 + b^2) \ .$$

A parallel axis through the corner is a distance $h = \sqrt{(a/2)^2 + (b/2)^2}$ from the center. Therefore,

$$I = I_{\text{com}} + Mh^2 = \frac{M}{12}\left(a^2 + b^2\right) + \frac{M}{4}\left(a^2 + b^2\right) = \frac{M}{3}\left(a^2 + b^2\right) \ .$$

42. (a) We apply Eq. 11-26:

$$I_x = \sum_{i=1}^{4} m_i y_i^2 = 50(2.0)^2 + (25)(4.0)^2 + 25(-3.0)^2 + 30(4.0)^2 = 1.3 \times 10^3 \text{ g}\cdot\text{cm}^2 \ .$$

(b) For rotation about the y axis we obtain

$$I_y = \sum_{i=1}^{4} m_i x_i^2 = 50(2.0)^2 + (25)(0)^2 + 25(3.0)^2 + 30(2.0)^2 = 5.5 \times 10^2 \text{ g}\cdot\text{cm}^2 \ .$$

(c) And about the z axis, we find (using the fact that the distance from the z axis is $\sqrt{x^2 + y^2}$)

$$I_z = \sum_{i=1}^{4} m_i(x_i^2 + y_i^2) = I_x + I_y = 1.3 \times 10^3 + 5.5 \times 10^2 = 1.9 \times 10^2 \text{ g}\cdot\text{cm}^2 \ .$$

(d) Clearly, the answer to part (c) is $A + B$.

43. (a) According to Table 11-2, the rotational inertia formulas for the cylinder (radius R) and the hoop (radius r) are given by

$$I_C = \frac{1}{2}MR^2 \quad \text{and} \quad I_H = Mr^2 \ .$$

Since the two bodies have the same mass, then they will have the same rotational inertia if $R^2/2 = R_H^2$, or $R_H = R/\sqrt{2}$.

(b) We require the rotational inertia to be written as $I = Mk^2$, where M is the mass of the given body and k is the radius of the "equivalent hoop." It follows directly that $k = \sqrt{I/M}$.

44. (a) Using Table 11-2(c) and Eq. 11-27, the rotational kinetic energy is

$$\begin{aligned} K &= \frac{1}{2}I\omega^2 \\ &= \frac{1}{2}\left(\frac{1}{2}MR^2\right)\omega^2 \\ &= \frac{1}{4}(500\,\text{kg})(200\pi\,\text{rad/s})^2(1.0\,\text{m})^2 \\ &= 4.9 \times 10^7 \text{ J} \ . \end{aligned}$$

(b) We solve $P = K/t$ (where P is the average power) for the operating time t.

$$t = \frac{K}{P} = \frac{4.9 \times 10^7 \text{ J}}{8.0 \times 10^3 \text{ W}} = 6.2 \times 10^3 \text{ s}$$

which we rewrite as $t \approx 100$ min.

45. Two forces act on the ball, the force of the rod and the force of gravity. No torque about the pivot point is associated with the force of the rod since that force is along the line from the pivot point to the ball. As can be seen from the diagram, the component of the force of gravity that is perpendicular to the rod is $mg \sin\theta$. If ℓ is the length of the rod, then the torque associated with this force has magnitude $\tau = mg\ell \sin\theta = (0.75)(9.8)(1.25)\sin 30° = 4.6\,\text{N}\cdot\text{m}$. For the position shown, the torque is counterclockwise.

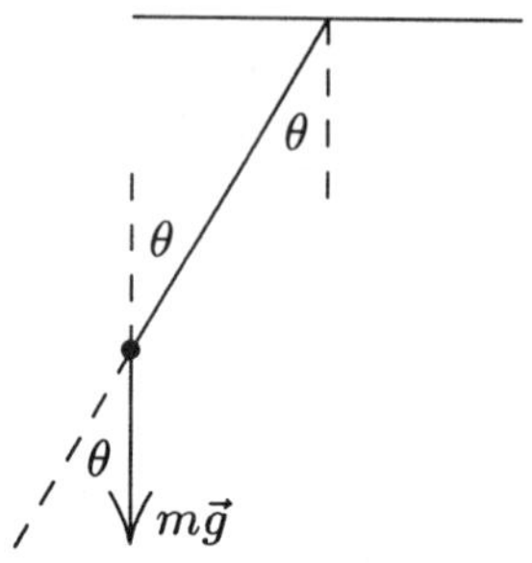

46. We compute the torques using $\tau = rF\sin\phi$.

$$\begin{aligned}\tau_a &= (0.152\,\text{m})(111\,\text{N})\sin 30° = 8.4\ \text{N}\cdot\text{m} \\ \tau_b &= (0.152\,\text{m})(111\,\text{N})\sin 90° = 17\ \text{N}\cdot\text{m} \\ \tau_c &= (0.152\,\text{m})(111\,\text{N})\sin 180° = 0\end{aligned}$$

47. (a) We take a torque that tends to cause a counterclockwise rotation from rest to be positive and a torque tending to cause a clockwise rotation to be negative. Thus, a positive torque of magnitude $r_1F_1\sin\theta_1$ is associated with $\vec{F}_1$ and a negative torque of magnitude $r_2F_2\sin\theta_2$ is associated with $\vec{F}_2$. The net torque is consequently

$$\tau = r_1F_1\sin\theta_1 - r_2F_2\sin\theta_2 \ .$$

(b) Substituting the given values, we obtain

$$\tau = (1.30\,\text{m})(4.20\,\text{N})\sin 75° - (2.15\,\text{m})(4.90\,\text{N})\sin 60° = -3.85\ \text{N}\cdot\text{m} \ .$$

48. The net torque is

$$\begin{aligned}\tau &= \tau_A + \tau_B + \tau_C \\ &= F_Ar_A\sin\phi_A - F_Br_B\sin\phi_B + F_Cr_C\sin\phi_C \\ &= (10)(8.0)\sin 135° - (16)(4.0)\sin 90° + (19)(3.0)\sin 160° \\ &= 12\ \text{N}\cdot\text{m} \ .\end{aligned}$$

49. (a) We use the kinematic equation $\omega = \omega_0 + \alpha t$, where ω_0 is the initial angular velocity, ω is the final angular velocity, α is the angular acceleration, and t is the time. This gives

$$\alpha = \frac{\omega - \omega_0}{t} = \frac{6.20\,\text{rad/s}}{220 \times 10^{-3}\,\text{s}} = 28.2\ \text{rad/s}^2 \ .$$

(b) If I is the rotational inertia of the diver, then the magnitude of the torque acting on her is

$$\tau = I\alpha = \left(12.0\,\text{kg}\cdot\text{m}^2\right)\left(28.2\,\text{rad/s}^2\right) = 3.38 \times 10^2\,\text{N}\cdot\text{m} \ .$$

50. The rotational inertia is found from Eq. 11-37.

$$I = \frac{\tau}{\alpha} = \frac{32.0}{25.0} = 1.28 \text{ kg}\cdot\text{m}^2$$

51. (a) We use $\tau = I\alpha$, where τ is the net torque acting on the shell, I is the rotational inertia of the shell, and α is its angular acceleration. Therefore,

$$I = \frac{\tau}{\alpha} = \frac{960\,\text{N}\cdot\text{m}}{6.20\,\text{rad/s}^2} = 155\,\text{kg}\cdot\text{m}^2 \ .$$

(b) The rotational inertia of the shell is given by $I = (2/3)MR^2$ (see Table 11-2 of the text). This implies

$$M = \frac{3I}{2R^2} = \frac{3(155\,\text{kg}\cdot\text{m}^2)}{2(1.90\,\text{m})^2} = 64.4\,\text{kg} \ .$$

52. According to the sign conventions used in the book, the magnitude of the net torque exerted on the cylinder of mass m and radius R_2 is

$$\begin{aligned}\tau_{\text{net}} &= F_1R_2 - F_2R_2 - F_3R_1 \\ &= (6.0\,\text{N})(0.12\,\text{m}) - (4.0\,\text{N})(0.12\,\text{m}) - (2.0\,\text{N})(0.05\,\text{m}) \\ &= 71 \text{ N}\cdot\text{m} \ .\end{aligned}$$

The resulting angular acceleration of the cylinder (with $I = \frac{1}{2}MR^2$ according to Table 11-2(c)) is

$$\begin{aligned}\alpha &= \frac{\tau_{\text{net}}}{I} \\ &= \frac{71 \text{ N}\cdot\text{m}}{\frac{1}{2}(2.0\,\text{kg})(0.12\,\text{m})^2} \\ &= 9.7 \text{ rad/s}^2\end{aligned}$$

and is counterclockwise (which is the positive sense of rotation).

53. We use $\tau = Fr = I\alpha$, where α satisfies $\theta = \frac{1}{2}\alpha t^2$ (Eq. 11-13). Here $\theta = 90° = \frac{\pi}{2}$ rad and $t = 30\,\text{s}$. The force needed is consequently

$$F = \frac{I\alpha}{r} = \frac{I\,(2\theta/t^2)}{r} = \frac{(8.7\times10^4)\,(2(\pi/2)/30^2)}{2.4} = 1.3\times10^2\,\text{N} \ .$$

54. With rightward positive for the block and clockwise negative for the wheel (as is conventional), then we note that the tangential acceleration of the wheel is of opposite sign from the block's acceleration (which we simply denote as a); that is, $a_t = -a$. Applying Newton's second law to the block leads to

$$P - T = ma \qquad \text{where } m = 2.0 \text{ kg} \ .$$

Applying Newton's second law (for rotation) to the wheel leads to

$$-TR = I\alpha \qquad \text{where } I = 0.050 \text{ kg}\cdot\text{m}^2 \ .$$

Noting that $R\alpha = a_t = -a$, we multiply this equation by R and obtain

$$-TR^2 = -Ia \implies T = a\frac{I}{R^2} \ .$$

Adding this to the above equation (for the block) leads to

$$P = \left(m + \frac{I}{R^2}\right)a \ .$$

Thus, $a = 0.92 \text{ m/s}^2$ and therefore $\alpha = -4.6 \text{ rad/s}^2$, where the negative sign should not be mistaken for a deceleration (it simply indicates the clockwise sense to the motion).

55. (a) We use constant acceleration kinematics. If down is taken to be positive and a is the acceleration of the heavier block, then its coordinate is given by $y = \frac{1}{2}at^2$, so

$$a = \frac{2y}{t^2} = \frac{2(0.750\,\text{m})}{(5.00\,\text{s})^2} = 6.00 \times 10^{-2}\ \text{m/s}^2 .$$

The lighter block has an acceleration of $6.00 \times 10^{-2}\,\text{m/s}^2$ upward.

(b) Newton's second law for the heavier block is $m_h g - T_h = m_h a$, where m_h is its mass and T_h is the tension force on the block. Thus,

$$T_h = m_h(g - a) = (0.500\,\text{kg})\left(9.8\,\text{m/s}^2 - 6.00 \times 10^{-2}\,\text{m/s}^2\right) = 4.87\ \text{N} .$$

(c) Newton's second law for the lighter block is $m_l g - T_l = -m_l a$, where T_l is the tension force on the block. Thus,

$$T_l = m_l(g + a) = (0.460\,\text{kg})\left(9.8\,\text{m/s}^2 + 6.00 \times 10^{-2}\,\text{m/s}^2\right) = 4.54\,\text{N} .$$

(d) Since the cord does not slip on the pulley, the tangential acceleration of a point on the rim of the pulley must be the same as the acceleration of the blocks, so

$$\alpha = \frac{a}{R} = \frac{6.00 \times 10^{-2}\,\text{m/s}^2}{5.00 \times 10^{-2}\,\text{m}} = 1.20\ \text{rad/s}^2 .$$

(e) The net torque acting on the pulley is $\tau = (T_h - T_l)R$. Equating this to $I\alpha$ we solve for the rotational inertia:

$$\begin{aligned} I &= \frac{(T_h - T_l)R}{\alpha} \\ &= \frac{(4.87\,\text{N} - 4.54\,\text{N})(5.00 \times 10^{-2}\,\text{m})}{1.20\,\text{rad/s}^2} \\ &= 1.38 \times 10^{-2}\ \text{kg}\cdot\text{m}^2 . \end{aligned}$$

56. Since the force acts tangentially at $r = 0.10$ m, the angular acceleration (presumed positive) is

$$\alpha = \frac{\tau}{I} = \frac{Fr}{I} = \frac{(0.5t + 0.3t^2)(0.10)}{1.0 \times 10^{-3}} = 50t + 30t^2$$

in SI units (rad/s^2).

(a) At $t = 3$ s, the above expression becomes $\alpha = 420\ \text{rad/s}^2$.

(b) We integrate the above expression, noting that $\omega_o = 0$, to obtain the angular speed at $t = 3$ s:

$$\omega = \int_0^3 \alpha\, dt = \left(25t^2 + 10t^3\right)\Big|_0^3 = 5.0 \times 10^2\,\text{rad/s} .$$

57. With counterclockwise positive, the angular acceleration α for both masses satisfies $\tau = mgL_1 - mgL_2 = I\alpha = (mL_1^2 + mL_2^2)\alpha$, by combining Eq. 11-37 with Eq. 11-32 and Eq. 11-26. Therefore, using SI units,

$$\alpha = \frac{g\,(L_1 - L_2)}{L_1^2 + L_2^2} = \frac{(9.8)(0.20 - 0.80)}{0.80^2 + 0.20^2} = -8.65\ \text{rad/s}^2$$

where the negative sign indicates the system starts turning in the clockwise sense. The magnitude of the acceleration vector involves no radial component (yet) since it is evaluated at $t = 0$ when the instantaneous velocity is zero. Thus, for the two masses, we apply Eq. 11-22 and obtain the respective answers for parts (a) and (b):

$$\begin{aligned} |\vec{a}_1| = |\alpha| L_1 &= \left(8.65\,\text{rad/s}^2\right)(0.80\,\text{m}) = 6.9\ \text{m/s}^2 \\ |\vec{a}_2| = |\alpha| L_2 & \\ &= \left(8.65\,\text{rad/s}^2\right)(0.20\,\text{m}) \\ &= 1.7\ \text{m/s}^2\ . \end{aligned}$$

58. (a) The speed of v of the mass m after it has descended $d = 50\,\text{cm}$ is given by $v^2 = 2ad$ (Eq. 2-16) where a is calculated as in Sample Problem 11-7 except that here we choose $+y$ downward (so $a > 0$). Thus, using $g = 980\ \text{cm/s}^2$, we have

$$v = \sqrt{2ad} = \sqrt{\frac{2(2mg)d}{M+2m}} = \sqrt{\frac{4(50)(980)(50)}{400+2(50)}} = 1.4 \times 10^2\ \text{cm/s}\ .$$

(b) The answer is still 1.4×10^2 cm/s = 1.4 m/s, since it is independent of R.

59. With $\omega = (1800)(2\pi/60) = 188.5$ rad/s, we apply Eq. 11-47:

$$P = \tau\omega \implies \tau = \frac{74600\ \text{W}}{188.5\,\text{rad/s}}$$

which yields $\tau = 396$ N·m.

60. The initial angular speed is $\omega = (280)(2\pi/60) = 29.3$ rad/s. We use Eq. 11-44 for the work and Eq. 7-42 for the average power.

(a) Since the rotational inertia is (Table 11-2(a)) $I = (32)(1.2)^2 = 46.1\ \text{kg}\cdot\text{m}^2$, the work done is

$$W = \Delta K = 0 - \frac{1}{2}I\omega^2 = -\frac{1}{2}(46.1)(29.3)^2$$

which yields $|W| = 19.8 \times 10^3$ J.

(b) The average power (in absolute value) is therefore

$$|P| = \frac{|W|}{\Delta t} = \frac{19.8 \times 10^3}{15} = 1.32 \times 10^3\ \text{W}\ .$$

61. (a) We apply Eq. 11-27:

$$K = \frac{1}{2}I\omega^2 = \frac{1}{2}\left(\frac{1}{3}mL^2\right)\omega^2 = \frac{1}{6}mL^2\omega^2\ .$$

(b) Simple conservation of mechanical energy leads to $K = mgh$. Consequently, the center of mass rises by

$$h = \frac{K}{mg} = \frac{mL^2\omega^2}{6mg} = \frac{L^2\omega^2}{6g}\ .$$

62. (a) The angular speed ω associated with Earth's spin is $\omega = 2\pi/T$, where $T = 86400\,\text{s}$ (one day). Thus

$$\omega = \frac{2\pi}{86400\,\text{s}} = 7.27 \times 10^{-5}\,\text{rad/s}$$

and the angular acceleration α required to accelerate the Earth from rest to ω in one day is $\alpha = \omega/T$. The torque needed is then

$$\tau = I\alpha = \frac{I\omega}{T} = \frac{\left(9.71 \times 10^{27}\right)\left(7.27 \times 10^{-5}\right)}{86400} = 8.17 \times 10^{28}\ \text{N}\cdot\text{m}$$

where we used

$$I = \frac{2}{5}MR^2 = \frac{2}{5}\left(5.98 \times 10^{24}\right)\left(6.37 \times 10^6\right)^2$$

for Earth's rotational inertia.

(b) Using the values from part (a), the kinetic energy of the Earth associated with its rotation about its own axis is $K = \frac{1}{2}I\omega^2 = 2.57 \times 10^{29}$ J. This is how much energy would need to be supplied to bring it (starting from rest) to the current angular speed.

(c) The associated power is

$$P = \frac{K}{T} = \frac{2.57 \times 10^{29}\,\text{J}}{86400\,\text{s}} = 2.97 \times 10^{24}\ \text{W}\ .$$

63. We use ℓ to denote the length of the stick. Since its center of mass is $\ell/2$ from either end, its initial potential energy is $\frac{1}{2}mg\ell$, where m is its mass. Its initial kinetic energy is zero. Its final potential energy is zero, and its final kinetic energy is $\frac{1}{2}I\omega^2$, where I is its rotational inertia about an axis passing through one end of the stick and ω is the angular velocity just before it hits the floor. Conservation of energy yields

$$\frac{1}{2}mg\ell = \frac{1}{2}I\omega^2 \implies \omega = \sqrt{\frac{mg\ell}{I}}\ .$$

The free end of the stick is a distance ℓ from the rotation axis, so its speed as it hits the floor is (from Eq. 11-18)

$$v = \omega\ell = \sqrt{\frac{mg\ell^3}{I}}\ .$$

Using Table 11-2 and the parallel-axis theorem, the rotational inertial is $I = \frac{1}{3}m\ell^2$, so

$$v = \sqrt{3g\ell} = \sqrt{3\left(9.8\,\text{m/s}^2\right)(1.00\,\text{m})} = 5.42\ \text{m/s}\ .$$

64. (a) We use the parallel-axis theorem to find the rotational inertia:

$$\begin{aligned} I &= I_{\text{com}} + Mh^2 = \frac{1}{2}MR^2 + Mh^2 \\ &= \frac{1}{2}(20\,\text{kg})(0.10\,\text{m})^2 + (20\,\text{kg})(0.50\,\text{m})^2 \\ &= 0.15\ \text{kg}\cdot\text{m}^2\ . \end{aligned}$$

(b) Conservation of energy requires that $Mgh = \frac{1}{2}I\omega^2$, where ω is the angular speed of the cylinder as it passes through the lowest position. Therefore,

$$\omega = \sqrt{\frac{2Mgh}{I}} = \sqrt{\frac{2(20)(9.8)(0.050)}{0.15}} = 11\ \text{rad/s}\ .$$

65. We use conservation of mechanical energy. The center of mass is at the midpoint of the cross bar of the **H** and it drops by $L/2$, where L is the length of any one of the rods. The gravitational potential energy decreases by $MgL/2$, where M is the mass of the body. The initial kinetic energy is zero and the final kinetic energy may be written $\frac{1}{2}I\omega^2$, where I is the rotational inertia of the body and ω is its angular velocity when it is vertical. Thus

$$0 = -MgL/2 + \frac{1}{2}I\omega^2 \implies \omega = \sqrt{MgL/I}\ .$$

Since the rods are thin the one along the axis of rotation does not contribute to the rotational inertia. All points on the other leg are the same distance from the axis of rotation, so that leg contributes

$(M/3)L^2$, where $M/3$ is its mass. The cross bar is a rod that rotates around one end, so its contribution is $(M/3)L^2/3 = ML^2/9$. The total rotational inertia is $I = (ML^2/3) + (ML^2/9) = 4ML^2/9$. Consequently, the angular velocity is

$$\omega = \sqrt{\frac{MgL}{I}} = \sqrt{\frac{MgL}{4ML^2/9}} = \sqrt{\frac{9g}{4L}} .$$

66. From Table 11-2, the rotational inertia of the spherical shell is $2MR^2/3$, so the kinetic energy (after the object has descended distance h) is

$$K = \frac{1}{2}\left(\frac{2}{3}MR^2\right)\omega_{\text{sphere}}^2 + \frac{1}{2}I\omega_{\text{pulley}}^2 + \frac{1}{2}mv^2 .$$

Since it started from rest, then this energy must be equal (in the absence of friction) to the potential energy mgh with which the system started. We substitute v/r for the pulley's angular speed and v/R for that of the sphere and solve for v.

$$v = \sqrt{\frac{mgh}{\frac{1}{2}m + \frac{1}{2}\frac{I}{r^2} + \frac{M}{3}}} = \sqrt{\frac{2gh}{1 + (I/mr^2) + (2M/3m)}}$$

67. (a) We use conservation of mechanical energy to find an expression for ω^2 as a function of the angle θ that the chimney makes with the vertical. The potential energy of the chimney is given by $U = Mgh$, where M is its mass and h is the altitude of its center of mass above the ground. When the chimney makes the angle θ with the vertical, $h = (H/2)\cos\theta$. Initially the potential energy is $U_i = Mg(H/2)$ and the kinetic energy is zero. The kinetic energy is $\frac{1}{2}I\omega^2$ when the chimney makes the angle θ with the vertical, where I is its rotational inertia about its bottom edge. Conservation of energy then leads to

$$MgH/2 = Mg(H/2)\cos\theta + \frac{1}{2}I\omega^2 \implies \omega^2 = (MgH/I)(1 - \cos\theta) .$$

The rotational inertia of the chimney about its base is $I = MH^2/3$ (found using Table 11-2(e) with the parallel axis theorem). Thus

$$\omega = \sqrt{\frac{3g}{H}(1 - \cos\theta)} .$$

(b) The radial component of the acceleration of the chimney top is given by $a_r = H\omega^2$, so $a_r = 3g(1 - \cos\theta)$.

(c) The tangential component of the acceleration of the chimney top is given by $a_t = H\alpha$, where α is the angular acceleration. We are unable to use Table 11-1 since the acceleration is not uniform. Hence, we differentiate $\omega^2 = (3g/H)(1 - \cos\theta)$ with respect to time, replacing $d\omega/dt$ with α, and $d\theta/dt$ with ω, and obtain

$$\frac{d\omega^2}{dt} = 2\omega\alpha = (3g/H)\omega\sin\theta \implies \alpha = (3g/2H)\sin\theta .$$

Consequently, $a_t = H\alpha = \frac{3g}{2}\sin\theta$.

(d) The angle θ at which $a_t = g$ is the solution to $\frac{3g}{2}\sin\theta = g$. Thus, $\sin\theta = 2/3$ and we obtain $\theta = 41.8°$.

68. (a) The longitudinal separation between Helsinki and the explosion site is $\Delta\theta = 102° - 25° = 77°$. The spin of the earth is constant at

$$\omega = \frac{1\,\text{rev}}{1\,\text{day}} = \frac{360°}{24\,\text{h}}$$

so that an angular displacement of $\Delta\theta$ corresponds to a time interval of

$$\Delta t = (77^\circ)\left(\frac{24\,\text{h}}{360^\circ}\right) = 5.1 \text{ h} .$$

(b) Now $\Delta\theta = 102^\circ - (-20^\circ) = 122^\circ$ so the required time shift would be

$$\Delta t = (122^\circ)\left(\frac{24\,\text{h}}{360^\circ}\right) = 8.1 \text{ h} .$$

69. Analyzing the forces tending to drag the $M = 5124$ kg stone down the oak beam, we find

$$F = Mg\,(\sin\theta + \mu_s\cos\theta)$$

where $\mu_s = 0.22$ (static friction is assumed to be at its maximum value) and the incline angle θ for the oak beam is $\sin^{-1}(3.9/10) = 23^\circ$ (but the incline angle for the spruce log is the complement of that). We note that the component of the weight of the workers (N of them) which is perpendicular to the spruce log is $Nmg\cos(90^\circ - \theta) = Nmg\sin\theta$, where $m = 85$ kg. The corresponding torque is therefore $Nmg\ell\sin\theta$ where $\ell = 4.5 - 0.7 = 3.8$ m (see figure). This must (at least) equal the magnitude of torque due to F, so with $r = 0.7$ m, we have

$$Mgr\,(\sin\theta + \mu_s\cos\theta) = Ngm\ell\sin\theta .$$

This expression yields $N \approx 17$ for the number of workers.

70. (a) We apply Eq. 11-18, using the subscript J for the Jeep.

$$\omega = \frac{v_J}{r_J} = \frac{114\,\text{km/h}}{0.100\,\text{km}}$$

which yields 1140 rad/h or (dividing by 3600) 0.32 rad/s for the value of the angular speed ω.

(b) Since the cheetah has the same angular speed, we again apply Eq. 11-18, using the subscript c for the cheetah.

$$v_c = r_c\omega = (92\,\text{m})(1140\,\text{rad/h})$$

which yields 1.05×10^5 m/h or 105 km/h for the cheetah's speed.

71. The *Hint* given in the problem would make the computation in part (a) very straightforward (without doing the integration as we show here), but we present this further level of detail in case that hint is not obvious or – simply – in case one wishes to see how the calculus supports our intuition.

(a) The (centripetal) force exerted on an infinitesimal portion of the blade with mass dm located a distance r from the rotational axis is (Newton's second law) $dF = (dm)\omega^2 r$, where dm can be written as $(M/L)dr$ and the angular speed is $\omega = (320)(2\pi/60) = 33.5$ rad/s. Thus for the entire blade of mass M and length L the total force is given by

$$\begin{aligned} F &= \int dF = \int \omega^2 r\, dm \\ &= \frac{M}{L}\int_0^L \omega^2 r\, dr \\ &= \left.\frac{M\omega^2 r^2}{2L}\right|_0^L = \frac{M\omega^2 L}{2} \\ &= \frac{(110\,\text{kg})(33.5\,\text{rad/s})^2(7.80\,\text{m})}{2} \\ &= 4.8 \times 10^5 \text{ N} . \end{aligned}$$

(b) About its center of mass, the blade has $I = ML^2/12$ according to Table 11-2(e), and using the parallel-axis theorem to "move" the axis of rotation to its end-point, we find the rotational inertia becomes $I = ML^2/3$. Using Eq. 11-37, the torque (assumed constant) is

$$\begin{aligned}\tau &= I\alpha \\ &= \left(\frac{1}{3}ML^2\right)\left(\frac{\Delta\omega}{\Delta t}\right) \\ &= \frac{1}{3}(110\,\mathrm{kg})(7.8\,\mathrm{m})^2\left(\frac{33.5\,\mathrm{rad/s}}{6.7\,\mathrm{s}}\right) \\ &= 1.1\times 10^4\ \mathrm{N{\cdot}m}\ .\end{aligned}$$

(c) Using Eq. 11-44, the work done is

$$\begin{aligned}W &= \Delta K = \frac{1}{2}I\omega^2 - 0 \\ &= \frac{1}{2}\left(\frac{1}{3}ML^2\right)\omega^2 \\ &= \frac{1}{6}(110\,\mathrm{kg})(7.80\,\mathrm{m})^2(33.5\,\mathrm{rad/s})^2 \\ &= 1.3\times 10^6\ \mathrm{J}\ .\end{aligned}$$

72. (a) Constant angular acceleration kinematics can be used to compute the angular acceleration α. If ω_0 is the initial angular velocity and t is the time to come to rest, then

$$0 = \omega_0 + \alpha t \implies \alpha = -\frac{\omega_0}{t}$$

which yields $-39/32 = -1.2$ rev/s or (multiplying by 2π) $-7.66\,\mathrm{rad/s^2}$ for the value of α.

(b) We use $\tau = I\alpha$, where τ is the torque and I is the rotational inertia. The contribution of the rod to I is $M\ell^2/12$ (Table 11-2(e)), where M is its mass and ℓ is its length. The contribution of each ball is $m(\ell/2)^2$, where m is the mass of a ball. The total rotational inertia is

$$I = \frac{M\ell^2}{12} + 2\frac{m\ell^2}{4} = \frac{(6.40\,\mathrm{kg})(1.20\,\mathrm{m})^2}{12} + \frac{(1.06\,\mathrm{kg})(1.20\,\mathrm{m})^2}{2}$$

which yields $I = 1.53\ \mathrm{kg{\cdot}m^2}$. The torque, therefore, is

$$\tau = (1.53\,\mathrm{kg\cdot m^2})\left(-7.66\,\mathrm{rad/s^2}\right) = -11.7\ \mathrm{N{\cdot}m}\ .$$

(c) Since the system comes to rest the mechanical energy that is converted to thermal energy is simply the initial kinetic energy

$$K_i = \frac{1}{2}I\omega_0^2 = \frac{1}{2}\left(1.53\,\mathrm{kg\cdot m^2}\right)\left((2\pi)(39)\,\mathrm{rad/s}\right)^2 = 4.59\times 10^4\ \mathrm{J}\ .$$

(d) We apply Eq. 11-13:

$$\theta = \omega_0 t + \frac{1}{2}\alpha t^2 = \left((2\pi)(39)\,\mathrm{rad/s}\right)(32.0\,\mathrm{s}) + \frac{1}{2}\left(-7.66\,\mathrm{rad/s^2}\right)(32.0\,\mathrm{s})^2$$

which yields 3920 rad or (dividing by 2π) 624 rev for the value of angular displacement θ.

(e) Only the mechanical energy that is converted to thermal energy can still be computed without additional information. It is 4.59×10^4 J no matter how τ varies with time, as long as the system comes to rest.

73. We assume the given rate of 1.2×10^{-3} m/y is the linear speed of the top; it is also possible to interpret it as just the horizontal component of the linear speed but the difference between these interpretations is arguably negligible. Thus, Eq. 11-18 leads to

$$\omega = \frac{1.2 \times 10^{-3}\,\text{m/y}}{55\,\text{m}} = 2.18 \times 10^{-5}\ \text{rad/y}$$

which we convert (since there are about 3.16×10^7 s in a year) to $\omega = 6.9 \times 10^{-13}$ rad/s.

74. The rotational inertia of the passengers is (to a good approximation) given by Eq. 11-26: $I = \sum mR^2 = NmR^2$ where N is the number of people and m is the (estimated) mass per person. We apply Eq. 11-44:

$$W = \frac{1}{2}I\omega^2 = \frac{1}{2}NmR^2\omega^2 \ .$$

where $R = 38$ m and $N = 36 \times 60 = 2160$ persons. The rotation rate is constant so that $\omega = \theta/t$ which leads to $\omega = 2\pi/120 = 0.052$ rad/s. The mass (in kg) of the average person is probably in the range $50 \le m \le 100$, so the work should be in the range

$$\begin{aligned} \frac{1}{2}(2160)(50)(38)^2(0.052)^2 \ \le\ W &\ \le\ \frac{1}{2}(2160)(100)(38)^2(0.052)^2 \\ 2 \times 10^5\,\text{J} \ \le\ W &\ \le\ 4 \times 10^5\,\text{J} \ . \end{aligned}$$

75. (a) The axis of rotation is at the bottom right edge of the rod along the ground, a horizontal distance of $d_3 + d_2 + d_1/2$ from the middle of the table assembly (mass $m = 90$ kg). The linebacker's center of mass at that critical moment was a horizontal distance of $d_4 + d_5$ from the axis of rotation. For the clockwise torque caused by the linebacker (mass M) to overcome the counterclockwise torque of the table assembly, we require (using Eq. 11-33)

$$Mg\,(d_4 + d_5) > mg\left(d_3 + d_2 + \frac{d_1}{2}\right) \ .$$

With the values given in the problem, we do indeed find the inequality is satisfied.

(b) Replacing our inequality with an equality and solving for M, we obtain

$$M = m\,\frac{d_3 + d_2 + \frac{1}{2}d_1}{d_4 + d_5} = 114\ \text{kg} \ .$$

76. We choose positive coordinate directions (different choices for each item) so that each is accelerating positively, which will allow us to set $a_1 = a_2 = R\alpha$ (for simplicity, we denote this as a). Thus, we choose upward positive for m_1, downward positive for m_2 and (somewhat unconventionally) clockwise for positive sense of disk rotation. Applying Newton's second law to m_1, m_2 and (in the form of Eq. 11-37) to M, respectively, we arrive at the following three equations.

$$\begin{aligned} T_1 - m_1 g &= m_1 a_1 \\ m_2 g - T_2 &= m_2 a_2 \\ T_2 R - T_1 R &= I\alpha \end{aligned}$$

(a) The rotational inertia of the disk is $I = \frac{1}{2}MR^2$ (Table 11-2(c)), so we divide the third equation (above) by R, add them all, and use the earlier equality among accelerations – to obtain:

$$m_2 g - m_1 g = \left(m_1 + m_2 + \frac{1}{2}M\right) a$$

which yields $a = \frac{4}{25}\,g = 1.6$ m/s^2.

(b) Plugging back in to the first equation, we find $T_1 = \frac{29}{24} m_1 g = 4.6$ N (where it is important in this step to have the mass in SI units: $m_1 = 0.40$ kg).

(c) Similarly, with $m_2 = 0.60$ kg, we find $T_2 = \frac{5}{6} m_2 g = 4.9$ N.

77. We employ energy methods in this solution; thus, considerations of positive versus negative sense (regarding the rotation of the wheel) are not relevant.

(a) The speed of the box is related to the angular speed of the wheel by $v = R\omega$, so that

$$K_{\text{box}} = \frac{1}{2} m_{\text{box}} v^2 \quad \Longrightarrow \quad v = \sqrt{\frac{2K_{\text{box}}}{m_{\text{box}}}} = 1.41 \text{ m/s}$$

implies that the angular speed is $\omega = 1.41/0.20 = 0.71$ rad/s. Thus, the kinetic energy of rotation is $\frac{1}{2} I\omega^2 = 10.0$ J.

(b) Since it was released from rest at what we will consider to be the reference position for gravitational potential, then (with SI units understood) energy conservation requires

$$\begin{aligned} K_0 + U_0 &= K + U \\ 0 + 0 &= (6.0 + 10.0) + m_{\text{box}} g(-h) \ . \end{aligned}$$

Therefore, $h = 16.0/58.8 = 0.27$ m.

78. The distances from P to the particles are as follows:

$$\begin{aligned} r_1 &= a \quad \text{for } m_1 = 2M \quad \text{(lower left)} \\ r_2 &= \sqrt{b^2 - a^2} \quad \text{for } m_2 = M \quad \text{(top)} \\ r_3 &= a \quad \text{for } m_1 = 2M \quad \text{(lower right)} \end{aligned}$$

The rotational inertia of the system about P is

$$I = \sum_{i=1}^{3} m_i r_i^2 = \left(3a^2 + b^2\right) M$$

which yields $I = 0.208$ kg·m^2 for $M = 0.40$ kg, $a = 0.30$ m and $b = 0.50$ m. Applying Eq. 11-44, we find

$$W = \frac{1}{2} I\omega^2 = \frac{1}{2}(0.208)(5.0)^2 = 2.6 \text{ J} \ .$$

79. We choose positive coordinate directions (different choices for each item) so that each is accelerating positively, which will allow us to set $a_2 = a_1 = R\alpha$ (for simplicity, we denote this as a). Thus, we choose rightward positive for $m_2 = M$ (the block on the table), downward positive for $m_1 = M$ (the block at the end of the string) and (somewhat unconventionally) clockwise for positive sense of disk rotation. This means that we interpret θ given in the problem as a positive-valued quantity. Applying Newton's second law to m_1, m_2 and (in the form of Eq. 11-37) to M, respectively, we arrive at the following three equations (where we allow for the possibility of friction f_2 acting on m_2).

$$\begin{aligned} m_1 g - T_1 &= m_1 a_1 \\ T_2 - f_2 &= m_2 a_2 \\ T_1 R - T_2 R &= I\alpha \end{aligned}$$

(a) From Eq. 11-13 (with $\omega_0 = 0$) we find

$$\theta = \omega_0 t + \frac{1}{2}\alpha t^2 \quad \Longrightarrow \quad \alpha = \frac{2\theta}{t^2} \ .$$

(b) From the fact that $a = R\alpha$ (noted above), we obtain $a = 2R\theta/t^2$.

(c) From the first of the above equations, we find

$$T_1 = m_1(g - a_1) = M\left(g - \frac{2R\theta}{t^2}\right) .$$

(d) From the last of the above equations, we obtain the second tension:

$$T_2 = T_1 - \frac{I\alpha}{R} = M\left(g - \frac{2R\theta}{t^2}\right) - \frac{2I\theta}{Rt^2}$$

80. (a) With $r = 0.780$ m, the rotational inertia is

$$I = Mr^2 = (1.30\,\text{kg})(0.780\,\text{m})^2 = 0.791\ \text{kg}\cdot\text{m}^2 .$$

(b) The torque that must be applied to counteract the effect of the drag is

$$\tau = rf = (0.780\,\text{m})(2.30 \times 10^{-2}\,\text{N}) = 1.79 \times 10^{-2}\ \text{N}\cdot\text{m} .$$

81. (a) The rotational inertia relative to the specified axis is

$$I = \sum m_i r_i^2 = (2M)L^2 + (2M)L^2 + M(2L)^2$$

which is found to be $I = 4.6$ kg·m². Then, with $\omega = 1.2$ rad/s, we obtain the kinetic energy from Eq. 11-27:

$$K = \frac{1}{2}I\omega^2 = 3.3\ \text{J} .$$

(b) In this case the axis of rotation would appear as a standard y axis with origin at P. Each of the $2M$ balls are a distance of $r = L\cos 30°$ from that axis. Thus, the rotational inertia in this case is

$$I = \sum m_i r_i^2 = (2M)r^2 + (2M)r^2 + M(2L)^2$$

which is found to be $I = 4.0$ kg·m². Again, from Eq. 11-27 we obtain the kinetic energy

$$K = \frac{1}{2}I\omega^2 = 2.9\ \text{J} .$$

82. We make use of Table 11-2(e) as well as the parallel-axis theorem, Eq. 11-27, where needed. We use ℓ (as a subscript) to refer to the long rod and s to refer to the short rod.

(a) The rotational inertia is

$$I = I_s + I_\ell = \frac{1}{12}m_s L_s^2 + \frac{1}{3}m_\ell L_\ell^2 = 0.019\ \text{kg}\cdot\text{m}^2 .$$

(b) We note that the center of the short rod is a distance of $h = 0.25$ m from the axis. The rotational inertia is

$$I = I_s + I_\ell = \frac{1}{12}m_s L_s^2 + m_s h^2 + \frac{1}{12}m_\ell L_\ell^2$$

which again yields $I = 0.019$ kg·m².

83. This may be derived from Eq. 11-28 or (suitably interpreted) from Eq. 11-26. Since every element of the hoop has the same radius $r = R$, the integration (or summation, if preferred) is trivial: $I = \int r^2\,dm = R^2\int dm = MR^2$.

84. (a) Using Eq. 11-15 with $\omega = 0$, we have

$$\theta = \frac{\omega_0 + \omega}{2} t = 2.8 \text{ rad} .$$

(b) One ingredient in this calculation is $\alpha = (0 - 3.5\,\text{rad/s})/(1.6\,\text{s}) = -2.2\,\text{rad/s}^2$, so that the tangential acceleration is $r\alpha = 0.33\ \text{m/s}^2$. Another ingredient is $\omega = \omega_0 + \alpha t = 1.3$ rad/s for $t = 1.0$ s, so that the radial (centripetal) acceleration is $\omega^2 r = 0.26\,\text{m/s}^2$. Thus, the magnitude of the acceleration is

$$|\vec{a}| = \sqrt{0.33^2 + 0.26^2} = 0.42\ \text{m/s}^2 .$$

85. (a) Using $T = 1$ yr $= 3.16 \times 10^7$ s for the time to make one full revolution (or 2π rad), we obtain

$$\omega = \frac{2\pi}{T} = \frac{2\pi}{3.16 \times 10^7} = 2.0 \times 10^{-7}\,\text{rad/s} .$$

(b) The radius r of Earth's orbit can be found in Appendix C or the inside front cover. Eq. 11-18 gives

$$v = \omega r = (2.0 \times 10^{-7}\,\text{rad/s})(1.49 \times 10^{11}\,\text{m}) = 3.0 \times 10^4\,\text{m/s} .$$

(c) The (radial, or centripetal) acceleration is

$$a = \omega^2 r = \left(2.0 \times 10^{-7}\,\text{rad/s}\right)^2 (1.49 \times 10^{11}\,\text{m}) = 5.9 \times 10^{-3}\,\text{m/s}^2 .$$

The direction of $\vec{a}$ is toward the sun.

86. Using Eq. 11-12, we have

$$\omega = \omega_0 + \alpha t \implies \alpha = \frac{2.6 - 8.0}{3.0}$$

which yields $\alpha = -1.8\ \text{rad/s}^2$. Using this value in Eq. 11-14 leads to

$$\omega^2 = \omega_0^2 + 2\alpha\theta \implies \theta = \frac{0^2 - 8.0^2}{2\,(-1.8)} = 18 \text{ rad} .$$

87. The motion consists of two stages. The first, the interval $0 \le t \le 20$ s, consists of constant angular acceleration given by

$$\alpha = \frac{5.0\,\text{rad/s}}{2.0\,\text{s}} = 2.5\ \text{rad/s}^2 .$$

The second stage, $20 < t \le 40$ s, consists of constant angular velocity $\omega = \Delta\theta/\Delta t$. Analyzing the first stage, we find

$$\begin{aligned} \theta_1 &= \left.\frac{1}{2}\alpha t^2\right|_{t=20} = 500 \text{ rad} \\ \omega &= \left.\alpha t\right|_{t=20} = 50 \text{ rad/s} . \end{aligned}$$

Analyzing the second stage, we obtain

$$\theta_2 = \theta_1 + \omega\Delta t = 500 + (50)(20) = 1500 \text{ rad} .$$

88. (a) Eq. 11-12 leads to $\alpha = -\omega_o/t = -25.0/20.0 = -1.25\,\text{rad/s}^2$.

(b) Eq. 11-15 leads to $\theta = \frac{1}{2}\omega_o t = \frac{1}{2}(25.0)(20.0) = 250\,\text{rad}$.

(c) Dividing the previous result by 2π we obtain $\theta = 39.8$ rev.

89. (a) We integrate the angular acceleration (as a function of τ) with respect to τ to find the angular velocity as a function of $t > 0$.

$$\omega = \omega_0 + \int_0^t \left(4a\tau^3 - 3b\tau^2\right) d\tau = \omega_0 + at^4 - bt^3 .$$

(b) We integrate the angular velocity (as a function of τ) with respect to τ to find the angular position as a function of $t > 0$.

$$\theta = \theta_0 + \int_0^t \left(4a\tau^3 - 3b\tau^2\right) d\tau = \theta_0 + \omega_0 t + \frac{a}{5}t^5 - \frac{b}{4}t^4 .$$

90. (a) The particle at A has $r = 0$ with respect to the axis of rotation. The particle at B is $r = L = 0.50$ m from the axis; similarly for the particle directly above A in the figure. The particle diagonally opposite A is a distance $r = \sqrt{2}\,L = 0.71$ m from the axis. Therefore,

$$I = \sum m_i r_i^2 = 2mL^2 + m\left(\sqrt{2}\,L\right)^2 = 0.20 \text{ kg}\cdot\text{m}^2 .$$

(b) One imagines rotating the figure (about point A) clockwise by 90° and noting that the center of mass has fallen a distance equal to L as a result. If we let our reference position for gravitational potential be the height of the center of mass at the instant AB swings through vertical orientation, then

$$\begin{aligned} K_0 + U_0 &= K + U \\ 0 + (4m)gh_0 &= K + 0 . \end{aligned}$$

Since $h_0 = L = 0.50$ m, we find $K = 3.9$ J. Then, using Eq. 11-27, we obtain

$$K = \frac{1}{2} I_A\, \omega^2 \implies \omega = 6.3 \ \frac{\text{rad}}{\text{s}} .$$

91. The center of mass is initially at height $h = \frac{L}{2}\sin 40°$ when the system is released (where $L = 2.0$ m). The corresponding potential energy Mgh (where $M = 1.5$ kg) becomes rotational kinetic energy $\frac{1}{2}I\omega^2$ as it passes the horizontal position (where I is the rotational inertia about the pin). Using Table 11-2(e) and the parallel axis theorem, we find $I = \frac{1}{12}ML^2 + M(L/2)^2 = \frac{1}{3}ML^2$. Therefore,

$$Mg\frac{L}{2}\sin 40° = \frac{1}{2}\left(\frac{1}{3}ML^2\right)\omega^2 \implies \omega = \sqrt{\frac{3g\sin 40°}{L}}$$

which yields $\omega = 3.1$ rad/s.

92. We choose $\pm$ directions such that the initial angular velocity is $\omega_0 = -317$ rad/s and the values for α, τ and F are positive.

(a) Combining Eq. 11-12 with Eq. 11-37 and Table 11-2(f) (and using the fact that $\omega = 0$) we arrive at the expression

$$\tau = \left(\frac{2}{5}MR^2\right)\left(-\frac{\omega_0}{t}\right) = -\frac{2}{5}\frac{MR^2\omega_0}{t} .$$

With $t = 15.5$ s, $R = 0.226$ m and $M = 1.65$ kg, we obtain $\tau = 0.689$ N·m.

(b) From Eq. 11-32, we find $F = \tau/R = 3.05$ N.

(c) Using again the expression found in part (a), but this time with $R = 0.854$ m, we get $\tau = 9.84$ N·m.

(d) Now, $F = \tau/R = 11.5$ N.

93. We choose positive coordinate directions so that each is accelerating positively, which will allow us to set $a_{\text{box}} = R\alpha$ (for simplicity, we denote this as a). Thus, we choose downhill positive for the $m = 2.0$ kg box and (as is conventional) counterclockwise for positive sense of wheel rotation. Applying Newton's second law to the box and (in the form of Eq. 11-37) to the wheel, respectively, we arrive at the following two equations (using θ as the incline angle 20°, not as the angular displacement of the wheel).

$$\begin{aligned} mg\sin\theta - T &= ma \\ TR &= I\alpha \end{aligned}$$

Since the problem gives $a = 2.0\ \text{m/s}^2$, the first equation gives the tension $T = m(g\sin\theta - a) = 2.7$ N. Plugging this and $R = 0.20$ m into the second equation (along with the fact that $\alpha = a/R$) we find the rotational inertia $I = TR^2/a = 0.054\,\text{kg}\cdot\text{m}^2$.

94. Eq. 11-32 leads to $\tau = mgr = (70)(9.8)(0.20)$ in SI units, which yields $\tau = 1.4 \times 10^2$ N.

95. The disk centered on A has $I = \frac{1}{2}MR^2$ (Table 11-2(c)) about that point, but the rotational inertia of the other disk is found using the parallel-axis theorem $I = \frac{1}{2}MR^2 + M(2R)^2 = \frac{9}{2}MR^2$ about point A. Adding these two results, we obtain

$$\frac{1}{2}MR^2 + \frac{9}{2}MR^2 = 5MR^2 = 5(4.0)(0.40)^2$$

which yields $3.2\,\text{kg}\cdot\text{m}^2$.

96. (a) One particle is on the axis, so $r = 0$ for it. For each of the others, the distance from the axis is $r = (0.60\ \text{m})\sin 60° = 0.52$ m. Therefore, the rotational inertia is $I = \sum m_i r_i^2 = 0.27\ \text{kg}\cdot\text{m}^2$.

(b) The two particles that are nearest the axis are each a distance of $r = 0.30$ m from it. The particle "opposite" from that side is a distance $r = (0.60\ \text{m})\sin 60° = 0.52$ m from the axis. Thus, the rotational inertia is $I = \sum m_i r_i^2 = 0.22\ \text{kg}\cdot\text{m}^2$.

(c) The distance from the axis for each of the particles is $r = \frac{1}{2}(0.60\,\text{m})\sin 60°$. Now, $I = 3(0.50\,\text{kg})(0.26\,\text{m})^2 = 0.10\,\text{kg}\cdot\text{m}^2$.

97. The parallel axis theorem gives $I = I_{\text{com}} + Mh^2$ for the rotational inertia about any axis (parallel to the axis used to compute I_{com}). Let us assume that an axis has already been chosen through the center of mass of the body such that I_{com} is as small as it possibly can be. Since $Mh^2 >$ for all nonzero values of h, then $I > I_{\text{com}}$ from the parallel axis theorem as long as $h \neq 0$. Thus, with $h = 0$ we get $I = I_{\text{com}}$ and therefore the smallest possible value of rotational inertia.

98. (a) The linear speed at $t = 15.0$ s is

$$v = a_t t = \left(0.500\,\text{m/s}^2\right)(15.0\,\text{s}) = 7.50\ \text{m/s}\ .$$

The radial (centripetal) acceleration at that moment is

$$a_r = \frac{v^2}{r} = \frac{(7.50\,\text{m/s})^2}{30.0\,\text{m}} = 1.875\ \text{m/s}^2\ .$$

Thus, the net acceleration has magnitude:

$$a = \sqrt{a_t^2 + a_r^2} = \sqrt{\left(0.500\,\text{m/s}^2\right)^2 + \left(1.875\,\text{m/s}^2\right)^2} = 1.94\ \text{m/s}^2\ .$$

(b) We note that $\vec{a}_t \parallel \vec{v}$. Therefore, the angle between $\vec{v}$ and $\vec{a}$ is

$$\tan^{-1}\left(\frac{a_r}{a_t}\right) = \tan^{-1}\left(\frac{1.875}{0.5}\right) = 75.1°$$

so that the vector is pointing more toward the center of the track than in the direction of motion.

99. First, we convert the angular velocity: $\omega = (2000)(2\pi/60) = 209$ rad/s. Also, we convert the plane's speed to SI units: $(480)(1000/3600) = 133$ m/s. We use Eq. 11-18 in part (a) and (implicitly) Eq. 4-39 in part (b).

(a) The speed of the tip as seen by the pilot is

$$v_t = \omega r = (209\,\text{rad/s})(1.5\,\text{m}) = 314\text{ m/s}$$

which (since the radius is given to only two significant figures) we write as $v = 3.1 \times 10^2$ m/s.

(b) The plane's velocity $\vec{v}_p$ and the velocity of the tip $\vec{v}_t$ (found in the plane's frame of reference), in any of the tip's positions, must be perpendicular to each other. Thus, the speed as seen by an observer on the ground is

$$v = \sqrt{v_p^2 + v_t^2} = \sqrt{(133\,\text{m/s})^2 + (314\,\text{m/s})^2} = 3.4 \times 10^2\text{ m/s}\ .$$

100. Using Eq. 11-7 and Eq. 11-18, the average angular acceleration is

$$\alpha_{\text{avg}} = \frac{\Delta\omega}{\Delta t} = \frac{\Delta v}{r\Delta t} = \frac{25-12}{(0.75/2)(6.2)} = 5.6\text{ rad/s}^2\ .$$

101. (a) Eq. 11-15 gives

$$90\,\text{rev} = \frac{1}{2}\left(\omega_0 + 10\,\text{rev/s}\right)(15\,\text{s})$$

which leads to $\omega_0 = 2.0\,\text{rev/s}$.

(b) From Eq. 11-12, the angular acceleration is

$$\alpha = \frac{10\,\text{rev/s} - 2.0\,\text{rev/s}}{15\,\text{s}} = 0.53\text{ rev/s}^2\ .$$

Using the equation again (with the same value for α) we seek a *negative* value of t (meaning an earlier time than that when $\omega_0 = 2.0$ rev/s) such that $\omega = 0$. Thus,

$$t = -\frac{\omega_0}{\alpha} = -\frac{2.0\,\text{rev/s}}{0.53\,\text{rev/s}^2} = -3.8\text{ s}$$

which means that the wheel was at rest 3.8 s before the 15 s interval began.

102. (a) Using Eq. 11-1, the angular displacement is

$$\theta = \frac{5.6\,\text{m}}{8.0 \times 10^{-2}\,\text{m}} = 1.4 \times 10^2\text{ rad}\ .$$

(b) We use $\theta = \frac{1}{2}\alpha t^2$ (Eq. 11-13) to obtain t:

$$t = \sqrt{\frac{2\theta}{\alpha}} = \sqrt{\frac{2(1.4\times 10^2\,\text{rad})}{1.5\,\text{rad/s}^2}} = 14\text{ s}\ .$$

103. The problem asks us to assume v_{com} and ω are constant. For consistency of units, we write

$$v_{\text{com}} = (85\,\text{mi/h})\left(\frac{5280\,\text{ft/mi}}{60\,\text{min/h}}\right) = 7480\text{ ft/min}\ .$$

Thus, with $\Delta x = 60$ ft, the time of flight is $t = \Delta x/v_{\text{com}} = 60/7480 = 0.00802$ min. During that time, the angular displacement of a point on the ball's surface is

$$\theta = \omega t = (1800\,\text{rev/min})(0.00802\,\text{min}) \approx 14\text{ rev}\ .$$

Chapter 12

1. The initial speed of the car is $v = (80.0)(1000/3600) = 22.2$ m/s. The tire radius is $R = 0.750/2 = 0.375$ m.

   (a) The initial speed of the car is the initial speed of the center of mass of the tire, so Eq. 12-2 leads to

   $$\omega_0 = \frac{v_{\text{com}\,0}}{R} = \frac{22.2}{0.375} = 59.3 \text{ rad/s} .$$

   (b) With $\theta = (30.0)(2\pi) = 188$ rad and $\omega = 0$, Eq. 11-14 leads to

   $$\omega^2 = \omega_0^2 + 2\alpha\theta \implies |\alpha| = \frac{59.3^2}{2(188)} = 9.31 \text{ rad/s}^2 .$$

   (c) Eq. 12-1 gives $R\theta = 70.7$ m for the distance traveled.

2. We define the direction of motion of the car as the $+x$ direction. The velocity of the car is a constant $\vec{v} = +(80)(1000/3600) = +22$ m/s, and the radius of the wheel is $r = 0.66/2 = 0.33$ m.

   (a) In the car's reference frame (where the lady perceives herself to be at rest) the road is moving towards the rear at $\vec{v}_{\text{road}} = -v = -22$ m/s, and the motion of the tire is purely rotational. In this frame, the center of the tire is "fixed" so $v_{\text{center}} = 0$.

   (b) This frame of reference is not accelerating, so "fixed" points within it have zero acceleration; thus, $a_{\text{center}} = 0$.

   (c) Since the tire's motion is only rotational (not translational) in this frame, Eq. 11-18 gives $\vec{v}_{\text{top}} = +v = +22$ m/s.

   (d) Not only is the motion purely rotational in this frame, but we also have $\omega =$ constant, which means the only acceleration for points on the rim is radial (centripetal). Therefore, the magnitude of the acceleration is

   $$a_{\text{top}} = \frac{v^2}{r} = \frac{22^2}{0.33} = 1.5 \times 10^3 \text{ m/s}^2 .$$

   (e) The bottom-most point of the tire is (momentarily) in firm contact with the road (not skidding) and has the same velocity as the road: $\vec{v}_{\text{bottom}} = -22$ m/s. This also follows from Eq. 11-18.

   (f) The magnitude of the acceleration is the same as in part (d): $a_{\text{bottom}} = 1.5 \times 10^3 \text{ m/s}^2$.

   (g) Now we examine the situation in the road's frame of reference (where the road is "fixed" and it is the car that appears to be moving). The center of the tire undergoes purely translational motion while points at the rim undergo a combination of translational and rotational motions. The velocity of the center of the tire is $\vec{v} = +v = +22$ m/s.

   (h) The translational motion of the center is constant; it does not accelerate.

(i) In part (c), we found $\vec{v}_{\text{top,car}} = +v$ and we use Eq. 4-39:

$$\begin{aligned}\vec{v}_{\text{top,ground}} &= \vec{v}_{\text{top,car}} + \vec{v}_{\text{car,ground}} \\ &= v + v\end{aligned}$$

which yields $2v = +44$ m/s. This is consistent with Fig. 12-3(c).

(j) Since we are transforming between constant-velocity frames of reference, the accelerations are unaffected. The answer is as it was in part (d): 1.5×10^3 m/s^2.

(k) We can proceed as in part (i) or simply recall that the bottom-most point is in firm contact with the (zero-velocity) road. Either way – the answer is zero.

(l) As explained in part (j), $a = 1.5 \times 10^3$ m/s^2.

3. By Eq. 11-44, the work required to stop the hoop is the negative of the initial kinetic energy of the hoop. The initial kinetic energy is $K = \frac{1}{2}I\omega^2 + \frac{1}{2}mv^2$ (Eq. 12-5), where $I = mR^2$ is its rotational inertia about the center of mass, $m = 140$ kg, and $v = 0.150$ m/s is the speed of its center of mass. Eq. 12-2 relates the angular speed to the speed of the center of mass: $\omega = v/R$. Thus,

$$K = \frac{1}{2}mR^2\left(\frac{v^2}{R^2}\right) + \frac{1}{2}mv^2 = mv^2 = (140)(0.150)^2$$

which implies that the work required is -3.15 J.

4. The rotational kinetic energy is $K = \frac{1}{2}I\omega^2$, where $I = mR^2$ is its rotational inertia about the center of mass (Table 11-2(a)), $m = 140$ kg, and $\omega = v_{\text{com}}/R$ (Eq. 12-2). The asked-for ratio is

$$\frac{K_{\text{transl}}}{K_{\text{rot}}} = \frac{\frac{1}{2}mv_{\text{com}}^2}{\frac{1}{2}\left(mR^2\right)\left(v_{\text{com}}/R\right)^2} = 1 .$$

5. Let M be the mass of the car (presumably including the mass of the wheels) and v be its speed. Let I be the rotational inertia of one wheel and ω be the angular speed of each wheel. The kinetic energy of rotation is

$$K_{\text{rot}} = 4\left(\frac{1}{2}I\omega^2\right)$$

where the factor 4 appears because there are four wheels. The total kinetic energy is given by $K = \frac{1}{2}Mv^2 + 4\left(\frac{1}{2}I\omega^2\right)$. The fraction of the total energy that is due to rotation is

$$\text{fraction} = \frac{K_{\text{rot}}}{K} = \frac{4I\omega^2}{Mv^2 + 4I\omega^2} .$$

For a uniform disk (relative to its center of mass) $I = \frac{1}{2}mR^2$ (Table 11-2(c)). Since the wheels roll without sliding $\omega = v/R$ (Eq. 12-2). Thus the numerator of our fraction is

$$4I\omega^2 = 4\left(\frac{1}{2}mR^2\right)\left(\frac{v}{R}\right)^2 = 2mv^2$$

and the fraction itself becomes

$$\text{fraction} = \frac{2mv^2}{Mv^2 + 2mv^2} = \frac{2m}{M + 2m} = \frac{2(10)}{1000} = \frac{1}{50} .$$

The wheel radius cancels from the equations and is not needed in the computation.

6. Interpreting h as the height increase for the center of mass of the body, then (using Eq. 12-5) mechanical energy conservation leads to

$$\begin{aligned} K_i &= U_f \\ \frac{1}{2}mv_{\text{com}}^2 + \frac{1}{2}I\omega^2 &= mgh \\ \frac{1}{2}mv^2 + \frac{1}{2}I\left(\frac{v}{R}\right)^2 &= mg\left(\frac{3v^2}{4g}\right) \end{aligned}$$

from which v cancels and we obtain $I = \frac{1}{2}mR^2$ (solid cylinder – Table 11-2(c)).

7. Rather than reproduce the analysis in §12-3, we simply use the results from that section.

(a) We substitute $I = \frac{2}{5}MR^2$ (Table 11-2(f)) and $a = -0.10g$ into Eq. 12-10:

$$-0.10g = -\frac{g\sin\theta}{1 + \left(\frac{2}{5}MR^2\right)/MR^2} = -\frac{g\sin\theta}{7/5}$$

which yields $\theta = \sin^{-1}(0.14) = 8.0°$.

(b) The acceleration would be more. We can look at this in terms of forces or in terms of energy. In terms of forces, the uphill static friction would then be absent so the downhill acceleration would be due only to the downhill gravitational pull. In terms of energy, the rotational term in Eq. 12-5 would be absent so that the potential energy it started with would simply become $\frac{1}{2}mv^2$ (without it being "shared" with another term) resulting in a greater speed (and, because of Eq. 2-16, greater acceleration).

8. We choose $+x$ rightward (so $\vec{F} = 10\hat{\imath}$ in Newtons) and apply Eq. 9-14 and Eq. 11-37.

(a) Newton's second law in the x direction leads to

$$F - f_s = ma \implies f_s = 10\,\text{N} - (10\,\text{kg})(0.60\,\text{m/s}^2)$$

which yields $f_s = 4.0$ N. As assumed in setting up the equation, $\vec{f}_s$ points leftward.

(b) With $R = 0.30$ m, we find the magnitude of the angular acceleration to be $|\alpha| = |a_{\text{com}}|/R = 2.0\,\text{rad/s}^2$, from Eq. 12-6. The only force not directed towards (or away from) the center of mass is $\vec{f}_s$, and the torque it produces is clockwise:

$$\begin{aligned} |\tau| &= I\,|\alpha| \\ (0.30\,\text{m})(4.0\,\text{N}) &= I\,(2.0\,\text{rad/s}^2) \end{aligned}$$

which yields the wheel's rotational inertia about its center of mass: $I = 0.60\ \text{kg}\cdot\text{m}^2$.

9. To find where the ball lands, we need to know its speed as it leaves the track (using conservation of energy). Its initial kinetic energy is $K_i = 0$ and its initial potential energy is $U_i = MgH$. Its final kinetic energy (as it leaves the track) is $K_f = \frac{1}{2}Mv^2 + \frac{1}{2}I\omega^2$ (Eq. 12-5) and its final potential energy is Mgh. Here we use v to denote the speed of its center of mass and ω is its angular speed – at the moment it leaves the track. Since (up to that moment) the ball rolls without sliding we can set $\omega = v/R$. Using $I = \frac{2}{5}MR^2$ (Table 11-2(f)), conservation of energy leads to

$$\begin{aligned} MgH &= \frac{1}{2}Mv^2 + \frac{1}{2}I\omega^2 + Mgh \\ &= \frac{1}{2}Mv^2 + \frac{2}{10}Mv^2 + Mgh \\ &= \frac{7}{10}Mv^2 + Mgh\,. \end{aligned}$$

The mass M cancels from the equation, and we obtain

$$v = \sqrt{\frac{10}{7}g(H-h)} = \sqrt{\frac{10}{7}\left(9.8\,\text{m/s}^2\right)(6.0\,\text{m} - 2.0\,\text{m})} = 7.48\,\text{m/s} .$$

Now this becomes a projectile motion of the type examined in Chapter 4. We put the origin at the position of the center of mass when the ball leaves the track (the "initial" position for this part of the problem) and take $+x$ rightward and $+y$ downward. Then (since the initial velocity is purely horizontal) the projectile motion equations become

$$x = vt \quad \text{and} \quad y = -\frac{1}{2}gt^2 .$$

Solving for x at the time when $y = h$, the second equation gives $t = \sqrt{2h/g}$. Then, substituting this into the first equation, we find

$$x = v\sqrt{\frac{2h}{g}} = (7.48)\sqrt{\frac{2(2.0)}{9.8}} = 4.8 \text{ m} .$$

10. (a) When the small sphere is released at the edge of the large "bowl" (the hemisphere of radius R), its center of mass is at the same height at that edge, but when it is at the bottom of the "bowl" its center of mass is a distance r above the the bottom surface of the hemisphere. Since the small sphere descends by $R - r$, its loss in gravitational potential energy is $mg(R - r)$, which, by conservation of mechanical energy, is equal to its kinetic energy at the bottom of the track.

(b) Using Eq. 12-5 for K, the asked-for fraction becomes

$$\frac{K_{\text{rot}}}{K} = \frac{\frac{1}{2}I\omega^2}{\frac{1}{2}I\omega^2 + \frac{1}{2}Mv_{\text{com}}^2} = \frac{1}{1 + \left(\frac{M}{I}\right)\left(\frac{v_{\text{com}}}{\omega}\right)^2} .$$

Substituting $v_{\text{com}} = R\omega$ (Eq. 12-2) and $I = \frac{2}{5}MR^2$ (Table 11-2(f)), we obtain

$$\frac{K_{\text{rot}}}{K} = \frac{1}{1 + \left(\frac{5}{2R^2}\right)R^2} = \frac{2}{7} .$$

(c) The small sphere is executing circular motion so that when it reaches the bottom, it experiences a radial acceleration upward (in the direction of the normal force which the "bowl" exerts on it). From Newton's second law along the vertical axis, the normal force N satisfies $N - mg = ma_{\text{com}}$ where $a_{\text{com}} = v_{\text{com}}^2/(R - r)$. Therefore,

$$N = mg + \frac{mv_{\text{com}}^2}{R - r} = \frac{mg(R - r) + mv_{\text{com}}^2}{R - r} .$$

But from part (a), $mg(R - r) = K$, and from Eq. 12-5, $\frac{1}{2}mv_{\text{com}}^2 = K - K_{\text{rot}}$. Thus,

$$N = \frac{K + 2\left(K - K_{\text{rot}}\right)}{R - r} = 3\frac{K}{R - r} - 2\frac{K_{\text{rot}}}{R - r} .$$

We now plug in $R - r = K/mg$ and use the result of part (b):

$$N = 3mg - 2mg\left(\frac{2}{7}\right) = \frac{17}{7}mg .$$

11. (a) We find its angular speed as it leaves the roof using conservation of energy. Its initial kinetic energy is $K_i = 0$ and its initial potential energy is $U_i = Mgh$ where $h = 6.0 \sin 30° = 3.0$ m (we are using the edge of the roof as our reference level for computing U). Its final kinetic energy (as it leaves

the roof) is $K_f = \frac{1}{2}Mv^2 + \frac{1}{2}I\omega^2$ (Eq. 12-5). Here we use v to denote the speed of its center of mass and ω is its angular speed – at the moment it leaves the roof. Since (up to that moment) the ball rolls without sliding we can set $v = R\omega = v$ where $R = 0.10$ m. Using $I = \frac{1}{2}MR^2$ (Table 11-2(c)), conservation of energy leads to

$$\begin{aligned} Mgh &= \frac{1}{2}Mv^2 + \frac{1}{2}I\omega^2 \\ &= \frac{1}{2}MR^2\omega^2 + \frac{1}{4}MR^2\omega^2 \\ &= \frac{3}{4}MR^2\omega^2 \ . \end{aligned}$$

The mass M cancels from the equation, and we obtain

$$\omega = \frac{1}{R}\sqrt{\frac{4}{3}gh} = \frac{1}{0.10\,\text{m}}\sqrt{\frac{4}{3}\left(9.8\,\text{m/s}^2\right)(3.0\,\text{m})} = 63\,\text{rad/s} \ .$$

(b) Now this becomes a projectile motion of the type examined in Chapter 4. We put the origin at the position of the center of mass when the ball leaves the track (the "initial" position for this part of the problem) and take $+x$ leftward and $+y$ downward. The result of part (a) implies $v_0 = R\omega = 6.3$ m/s, and we see from the figure that (with these positive direction choices) its components are

$$\begin{aligned} v_{0x} &= v_0 \cos 30° = 5.4 \text{ m/s} \quad \text{and} \\ v_{0y} &= v_0 \sin 30° = 3.1 \text{ m/s} \ . \end{aligned}$$

The projectile motion equations become

$$x = v_{0x}t \quad \text{and} \quad y = v_{0y}t + \frac{1}{2}gt^2 \ .$$

We first find the time when $y = 5.0$ m from the second equation (using the quadratic formula, choosing the positive root):

$$t = \frac{-v_{0y} + \sqrt{v_{0y}^2 + 2gy}}{g} = 0.74 \text{ s} \ .$$

Then we substitute this into the x equation and obtain

$$x = (5.4\,\text{m/s})(0.74\,\text{s}) = 4.0 \text{ m} \ .$$

12. Using the floor as the reference position for computing potential energy, mechanical energy conservation leads to

$$\begin{aligned} U_{\text{release}} &= K_{\text{top}} + U_{\text{top}} \\ mgh &= \frac{1}{2}mv_{\text{com}}^2 + \frac{1}{2}I\omega^2 + mg(2R) \ . \end{aligned}$$

Substituting $I = \frac{2}{5}mr^2$ (Table 11-2(f)) and $\omega = v_{\text{com}}/r$ (Eq. 12-2), we obtain

$$\begin{aligned} mgh &= \frac{1}{2}mv_{\text{com}}^2 + \frac{1}{2}\left(\frac{2}{5}mr^2\right)\left(\frac{v_{\text{com}}}{r}\right)^2 + 2mgR \\ gh &= \frac{7}{10}v_{\text{com}}^2 + 2gR \end{aligned}$$

where we have canceled out mass m in that last step.

(a) To be on the verge of losing contact with the loop (at the top) means the normal force is vanishingly small. In this case, Newton's second law along the vertical direction ($+y$ downward) leads to

$$mg = ma_r \implies g = \frac{v_{\text{com}}^2}{R-r}$$

where we have used Eq. 11-23 for the radial (centripetal) acceleration (of the center of mass, which at this moment is a distance $R-r$ from the center of the loop). Plugging the result $v_{\text{com}}^2 = g(R-r)$ into the previous expression stemming from energy considerations gives

$$gh = \frac{7}{10}(g)(R-r) + 2gR$$

which leads to

$$h = 2.7R - 0.7r \approx 2.7R \ .$$

(b) The energy considerations shown above (now with $h = 6R$) can be applied to point Q (which, however, is only at a height of R) yielding the condition

$$g(6R) = \frac{7}{10}v_{\text{com}}^2 + gR$$

which gives us $v_{\text{com}}^2 = 50gR/7$. Recalling previous remarks about the radial acceleration, Newton's second law applied to the horizontal axis at Q ($+x$ leftward) leads to

$$\begin{aligned} N &= m\frac{v_{\text{com}}^2}{R-r} \\ &= m\frac{50gR}{7(R-r)} \end{aligned}$$

which (for $R \gg r$) gives $N \approx 50mg/7$.

13. From $I = \frac{2}{3}MR^2$ (Table 11-2(g)) we find

$$M = \frac{3I}{2R^2} = \frac{3(0.040)}{2(0.15)^2} = 2.7 \text{ kg} \ .$$

It also follows from the rotational inertia expression that $\frac{1}{2}I\omega^2 = \frac{1}{3}MR^2\omega^2$. Furthermore, it rolls without slipping, $v_{\text{com}} = R\omega$, and we find

$$\frac{K_{\text{rot}}}{K_{\text{com}} + K_{\text{rot}}} = \frac{\frac{1}{3}MR^2\omega^2}{\frac{1}{2}mR^2\omega^2 + \frac{1}{3}MR^2\omega^2} \ .$$

(a) Simplifying the above ratio, we find $K_{\text{rot}}/K = 0.4$. Thus, 40% of the kinetic energy is rotational, or $K_{\text{rot}} = (0.4)(20) = 8.0$ J.

(b) From $K_{\text{rot}} = \frac{1}{3}MR^2\omega^2 = 8.0$ J (and using the above result for M) we find

$$\omega = \frac{1}{0.15\,\text{m}}\sqrt{\frac{3(8.0\,\text{J})}{2.7\,\text{kg}}} = 20 \text{ rad/s}$$

which leads to $v_{\text{com}} = (0.15)(20) = 3.0$ m/s.

(c) We note that the inclined distance of 1.0 m corresponds to a height $h = 1.0\sin 30° = 0.50$ m. Mechanical energy conservation leads to

$$\begin{aligned} K_i &= K_f + U_f \\ 20\,\text{J} &= K_f + Mgh \end{aligned}$$

which yields (using the values of M and h found above) $K_f = 6.9$ J.

(d) We found in part (a) that 40% of this must be rotational, so

$$\frac{1}{3}MR^2\omega_f^2 = (0.40)K_f \implies \omega_f = \frac{1}{0.15}\sqrt{\frac{3(0.40)(6.9)}{2.7}}$$

which yields $\omega_f = 12$ rad/s and leads to

$$v_{\text{com}f} = R\omega_f = (0.15)(12) = 1.8 \text{ m/s} .$$

14. (a) We choose clockwise as the negative rotational sense and rightwards as the positive translational direction. Thus, since this is the moment when it begins to roll smoothly, Eq. 12-2 becomes

$$v_{\text{com}} = -R\omega = (-0.11\,\text{m})\omega .$$

This velocity is positive-valued (rightward) since ω is negative-valued (clockwise) as shown in Fig. 12-34.

(b) The force of friction exerted on the ball of mass m is $-\mu_k mg$ (negative since it points left), and setting this equal to ma_{com} leads to

$$a_{\text{com}} = -\mu g = -(0.21)\left(9.8\,\text{m/s}^2\right) = -2.1 \text{ m/s}^2$$

where the minus sign indicates that the center of mass acceleration points left, opposite to its velocity, so that the ball is decelerating.

(c) Measured about the center of mass, the torque exerted on the ball due to the frictional force is given by $\tau = -\mu mgR$. Using Table 11-2(f) for the rotational inertia, the angular acceleration becomes (using Eq. 11-37)

$$\alpha = \frac{\tau}{I} = \frac{-\mu mgR}{2mR^2/5} = \frac{-5\mu g}{2R} = \frac{-5(0.21)(9.8)}{2(0.11)} = -47 \text{ rad/s}^2$$

where the minus sign indicates that the angular acceleration is clockwise, the same direction as ω (so its angular motion is "speeding up").

(d) The center-of-mass of the sliding ball decelerates from $v_{\text{com},0}$ to v_{com} during time t according to Eq. 2-11:

$$v_{\text{com}} = v_{\text{com},0} - \mu g t .$$

During this time, the angular speed of the ball increases (in magnitude) from zero to $|\omega|$ according to Eq. 11-12:

$$|\omega| = |\alpha|\, t = \frac{5\mu g t}{2R} = \frac{v_{\text{com}}}{R}$$

where we have made use of our part (a) result in the last equality. We have two equations involving v_{com}, so we eliminate that variable and find

$$t = \frac{2v_{\text{com},0}}{7\mu g} = \frac{2(8.5)}{7(0.21)(9.8)} = 1.2 \text{ s} .$$

(e) The skid length of the ball is (using Eq. 2-15)

$$\Delta x = v_{\text{com},0}t - \frac{1}{2}(\mu g)t^2 = (8.5)(1.2) - \frac{1}{2}(0.21)(9.8)(1.2)^2 = 8.6 \text{ m} .$$

(f) The center of mass velocity at the time found in part (d) is

$$v_{\text{com}} = v_{\text{com},0} - \mu g t = 8.5 - (0.21)(9.8)(1.2) = 6.1 \text{ m/s} .$$

15. (a) The derivation of the acceleration is found in §12-4; Eq. 12-13 gives

$$a_{\rm com} = -\frac{g}{1 + I_{\rm com}/MR_0^2}$$

where the positive direction is upward. We use $I_{\rm com} = 950\,{\rm g}\cdot{\rm cm}^2$, $M = 120\,{\rm g}$, $R_0 = 0.32\,{\rm cm}$ and $g = 980\,{\rm cm/s}^2$ and obtain

$$|a_{\rm com}| = \frac{980}{1 + (950)/(120)(0.32)^2} = 12.5\ {\rm cm/s}^2\ .$$

(b) Taking the coordinate origin at the initial position, Eq. 2-15 leads to $y_{\rm com} = \frac{1}{2}a_{\rm com}t^2$. Thus, we set $y_{\rm com} = -120\,{\rm cm}$, and find

$$t = \sqrt{\frac{2y_{\rm com}}{a_{\rm com}}} = \sqrt{\frac{2(-120\,{\rm cm})}{-12.5\,{\rm cm/s}^2}} = 4.38\ {\rm s}\ .$$

(c) As it reaches the end of the string, its center of mass velocity is given by Eq. 2-11: $v_{\rm com} = a_{\rm com}t = (-12.5\,{\rm cm/s}^2)(4.38\,{\rm s}) = -54.8\,{\rm cm/s}$, so its linear speed then is approximately 55 cm/s.

(d) The translational kinetic energy is $\frac{1}{2}mv_{\rm com}^2 = \frac{1}{2}(0.120\,{\rm kg})(0.548\,{\rm m/s})^2 = 1.8\times 10^{-2}\,{\rm J}$.

(e) The angular velocity is given by $\omega = -v_{\rm com}/R_0$ and the rotational kinetic energy is

$$\frac{1}{2}I_{\rm com}\omega^2 = \frac{1}{2}I_{\rm com}\frac{v_{\rm com}^2}{R_0^2} = \frac{1}{2}\frac{(9.50\times 10^{-5}\,{\rm kg}\cdot{\rm m}^2)(0.548\,{\rm m/s})^2}{(3.2\times 10^{-3}\,{\rm m})^2}$$

which yields $K_{\rm rot} = 1.4$ J.

(f) The angular speed is $\omega = |v_{\rm com}|/R_0 = (0.548\,{\rm m/s})/(3.2\times 10^{-3}\,{\rm m}) = 1.7\times 10^2\,{\rm rad/s} = 27\,{\rm rev/s}$.

16. (a) The acceleration is given by Eq. 12-13:

$$a_{\rm com} = -\frac{g}{1 + I_{\rm com}/MR_0^2}$$

where upward is the positive translational direction. Taking the coordinate origin at the initial position, Eq. 2-15 leads to

$$y_{\rm com} = v_{\rm com,0}\,t + \frac{1}{2}a_{\rm com}t^2 = v_{\rm com,0}\,t - \frac{\frac{1}{2}gt^2}{1 + I_{\rm com}/MR_0^2}$$

where $y_{\rm com} = -1.2\,{\rm m}$ and $v_{\rm com,0} = -1.3$ m/s. Substituting $I_{\rm com} = 0.000095\,{\rm kg}\cdot{\rm m}^2$, $M = 0.12\,{\rm kg}$, $R_0 = 0.0032\,{\rm m}$ and $g = 9.8\,{\rm m/s}^2$, we use the quadratic formula and find

$$\begin{aligned} t &= \frac{\left(1 + \frac{I_{\rm com}}{MR_0^2}\right)\left(v_{\rm com,0} \mp \sqrt{v_{\rm com,0}^2 - \frac{2gy_{\rm com}}{1+I_{\rm com}/MR_0^2}}\right)}{g} \\ &= \frac{\left(1 + \frac{0.000095}{(0.12)(0.0032)^2}\right)\left(-1.3 \mp \sqrt{1.3^2 - \frac{2(9.8)(-1.2)}{1+0.000095/(0.12)(0.0032)^2}}\right)}{9.8} \\ &= -21.7 \quad \text{or} \quad 0.885 \end{aligned}$$

where we choose $t = 0.89$ s as the answer.

(b) We note that the initial potential energy is $U_i = Mgh$ and $h = 1.2$ m (using the bottom as the reference level for computing U). The initial kinetic energy is as shown in Eq. 12-5, where the initial angular and linear speeds are related by Eq. 12-2. Energy conservation leads to

$$\begin{aligned} K_f &= K_i + U_i \\ &= \frac{1}{2} m v_{\text{com},0}^2 + \frac{1}{2} I \left(\frac{v_{\text{com},0}}{R_0}\right)^2 + Mgh \\ &= \frac{1}{2}(0.12)(1.3)^2 + \frac{1}{2}\left(9.5\times10^{-5}\right)\left(\frac{1.3}{0.0032}\right)^2 + (0.12)(9.8)(1.2) \\ &= 9.4 \text{ J} . \end{aligned}$$

(c) As it reaches the end of the string, its center of mass velocity is given by Eq. 2-11:

$$v_{\text{com}} = v_{\text{com},0} + a_{\text{com}} t = v_{\text{com},0} - \frac{gt}{1 + I_{\text{com}}/MR_0^2} .$$

Thus, we obtain

$$v_{\text{com}} = -1.3 - \frac{(9.8)(0.885)}{1 + \frac{0.000095}{(0.12)(0.0032)^2}} = -1.41 \text{ m/s}$$

so its linear speed at that moment is approximately 1.4 m/s.

(d) The translational kinetic energy is $\frac{1}{2} m v_{\text{com}}^2 = \frac{1}{2}(0.12)(1.41)^2 = 0.12\,\text{J}$.

(e) The angular velocity at that moment is given by

$$\omega = -\frac{v_{\text{com}}}{R_0} = -\frac{-1.41}{0.0032} = 441$$

or approximately 440 rad/s.

(f) And the rotational kinetic energy is

$$\frac{1}{2} I_{\text{com}} \omega^2 = \frac{1}{2}\left(9.50\times10^{-5}\,\text{kg}\cdot\text{m}^2\right)(441\,\text{rad/s})^2 = 9.2 \text{ J} .$$

17. One method is to show that $\vec{r}\cdot\left(\vec{r}\times\vec{F}\right) = \vec{F}\cdot\left(\vec{r}\times\vec{F}\right) = 0$, but we choose here a more pedestrian approach: without loss of generality we take $\vec{r}$ and $\vec{F}$ to be in the xy plane – and will show that $\vec{\tau}$ has no x and y components (that it is parallel to the $\hat{\text{k}}$ direction). We proceed as follows: in the general expression $\vec{r} = x\hat{\text{i}} + y\hat{\text{j}} + z\hat{\text{k}}$, we will set $z = 0$ to constrain $\vec{r}$ to the xy plane, and similarly for $\vec{F}$. Using Eq. 3-30, we find $\vec{r}\times\vec{F}$ is equal to

$$(yF_z - zF_y)\,\hat{\text{i}} + (zF_x - xF_z)\,\hat{\text{j}} + (xF_y - yF_x)\,\hat{\text{k}}$$

and once we set $z = 0$ and $F_z = 0$ we obtain

$$\vec{\tau} = \vec{r}\times\vec{F} = (xF_y - yF_x)\,\hat{\text{k}}$$

which demonstrates that $\vec{\tau}$ has no component in the xy plane.

18. If we write $\vec{r} = x\hat{\text{i}} + y\hat{\text{j}} + z\hat{\text{k}}$, then (using Eq. 3-30) we find $\vec{r}\times\vec{F}$ is equal to

$$(yF_z - zF_y)\,\hat{\text{i}} + (zF_x - xF_z)\,\hat{\text{j}} + (xF_y - yF_x)\,\hat{\text{k}} .$$

(a) In the above expression, we set (with SI units understood) $x = -2$, $y = 0$, $z = 4$, $F_x = 6$, $F_y = 0$ and $F_z = 0$. Then we obtain $\vec{\tau} = \vec{r}\times\vec{F} = 24\hat{\text{j}}$ N·m.

(b) The values are just as in part (a) with the exception that now $F_x = -6$. We find $\vec{\tau} = \vec{r}\times\vec{F} = -24\hat{\text{j}}$ N·m.

(c) In the above expression, we set $x = -2$, $y = 0$, $z = 4$, $F_x = 0$, $F_y = 0$ and $F_z = 6$. We get $\vec{\tau} = \vec{r} \times \vec{F} = 12\hat{\jmath}$ N·m.

(d) The values are just as in part (c) with the exception that now $F_z = -6$. We find $\vec{\tau} = \vec{r} \times \vec{F} = -12\hat{\jmath}$ N·m.

19. If we write $\vec{r} = x\hat{\imath} + y\hat{\jmath} + z\hat{k}$, then (using Eq. 3-30) we find $\vec{r} \times \vec{F}$ is equal to

$$(yF_z - zF_y)\,\hat{\imath} + (zF_x - xF_z)\,\hat{\jmath} + (xF_y - yF_x)\,\hat{k} .$$

(a) In the above expression, we set (with SI units understood) $x = 0$, $y = -4$, $z = 3$, $F_x = 2$, $F_y = 0$ and $F_z = 0$. Then we obtain $\vec{\tau} = \vec{r} \times \vec{F} = \left(6\,\hat{\jmath} + 8\,\hat{k}\right)$ N·m. This has magnitude $\sqrt{6^2 + 8^2} = 10$ N·m and is seen to be parallel to the yz plane. Its angle (measured counterclockwise from the $+y$ direction) is $\tan^{-1}(8/6) = 53°$.

(b) In the above expression, we set $x = 0$, $y = -4$, $z = 3$, $F_x = 0$, $F_y = 2$ and $F_z = 4$. Then we obtain $\vec{\tau} = \vec{r} \times \vec{F} = -22\,\hat{\imath}$ N·m. This has magnitude 22 N·m and points in the $-x$ direction.

20. We use the notation $\vec{r}\,'$ to indicate the vector pointing from the axis of rotation directly to the position of the particle. If we write $\vec{r}\,' = x'\,\hat{\imath} + y'\,\hat{\jmath} + z'\,\hat{k}$, then (using Eq. 3-30) we find $\vec{r}\,' \times \vec{F}$ is equal to

$$(y'F_z - z'F_y)\,\hat{\imath} + (z'F_x - x'F_z)\,\hat{\jmath} + (x'F_y - y'F_x)\,\hat{k} .$$

(a) Here, $\vec{r}\,' = \vec{r}$. Dropping the primes in the above expression, we set (with SI units understood) $x = 0$, $y = 0.5$, $z = -2.0$, $F_x = 2$, $F_y = 0$ and $F_z = -3$. Then we obtain $\vec{\tau} = \vec{r} \times \vec{F} = \left(-1.5\,\hat{\imath} - 4\,\hat{\jmath} - \hat{k}\right)$ N·m.

(b) Now $\vec{r}\,' = \vec{r} - \vec{r}_o$ where $\vec{r}_o = 2\,\hat{\imath} - 3\,\hat{k}$. Therefore, in the above expression, we set $x' = -2.0$, $y' = 0.5$, $z' = 1.0$, $F_x = 2$, $F_y = 0$ and $F_z = -3$. Thus, we obtain $\vec{\tau} = \vec{r}\,' \times \vec{F} = \left(-1.5\,\hat{\imath} - 4\,\hat{\jmath} - \hat{k}\right)$ N·m.

21. If we write $\vec{r} = x\hat{\imath} + y\hat{\jmath} + z\hat{k}$, then (using Eq. 3-30) we find $\vec{r} \times \vec{F}$ is equal to

$$(yF_z - zF_y)\,\hat{\imath} + (zF_x - xF_z)\,\hat{\jmath} + (xF_y - yF_x)\,\hat{k} .$$

(a) Plugging in, we find

$$\vec{\tau} = ((3.0\,\text{m})(6.0\,\text{N}) - (4.0\,\text{m})(-8.0\,\text{N}))\,\hat{k} = 50\,\hat{k}\ \text{N·m} .$$

(b) We use Eq. 3-27, $|\vec{r} \times \vec{F}| = rF\sin\phi$, where ϕ is the angle between $\vec{r}$ and $\vec{F}$. Now $r = \sqrt{x^2 + y^2} = 5.0\,\text{m}$ and $F = \sqrt{F_x^2 + F_y^2} = 10\,\text{N}$. Thus $rF = (5.0\,\text{m})(10\,\text{N}) = 50\,\text{N}\cdot\text{m}$, the same as the magnitude of the vector product calculated in part (a). This implies $\sin\phi = 1$ and $\phi = 90°$.

22. If we write $\vec{r}\,' = x'\,\hat{\imath} + y'\,\hat{\jmath} + z'\,\hat{k}$, then (using Eq. 3-30) we find $\vec{r}\,' \times \vec{F}$ is equal to

$$(y'F_z - z'F_y)\,\hat{\imath} + (z'F_x - x'F_z)\,\hat{\jmath} + (x'F_y - y'F_x)\,\hat{k} .$$

(a) Here, $\vec{r}\,' = \vec{r}$ where $\vec{r} = 3\,\hat{\imath} - 2\,\hat{\jmath} + 4\,\hat{k}$, and $\vec{F} = \vec{F}_1$. Thus, dropping the primes in the above expression, we set (with SI units understood) $x = 3$, $y = -2$, $z = 4$, $F_x = 3$, $F_y = -4$ and $F_z = 5$. Then we obtain $\vec{\tau} = \vec{r} \times \vec{F}_1 = \left(6.0\,\hat{\imath} - 3.0\,\hat{\jmath} - 6.0\,\hat{k}\right)$ N·m.

(b) This is like part (a) but with $\vec{F} = \vec{F}_2$. We plug in $F_x = -3$, $F_y = -4$ and $F_z = -5$ and obtain $\vec{\tau} = \vec{r} \times \vec{F}_2 = \left(26\,\hat{\imath} + 3.0\,\hat{\jmath} - 18\,\hat{k}\right)$ N·m.

(c) We can proceed in either of two ways. We can add (vectorially) the answers from parts (a) and (b), or we can first add the two force vectors and then compute $\vec{\tau} = \vec{r} \times \left(\vec{F}_1 + \vec{F}_2\right)$ (these total force components are computed in the next part). The result is $\left(32\,\hat{\mathrm{i}} - 24\,\hat{\mathrm{k}}\right)$ N·m.

(d) Now $\vec{r}\,' = \vec{r} - \vec{r}_o$ where $\vec{r}_o = 3\hat{\mathrm{i}} + 2\hat{\mathrm{j}} + 4\hat{\mathrm{k}}$. Therefore, in the above expression, we set $x' = 0$, $y' = -4$, $z' = 0$, $F_x = 3 - 3 = 0$, $F_y = -4 - 4 = -8$ and $F_z = 5 - 5 = 0$. We get $\vec{\tau} = \vec{r}\,' \times \left(\vec{F}_1 + \vec{F}_2\right) = 0$.

23. We could proceed formally by setting up an xyz coordinate system and using Eq. 3-30 for the vector cross product, or we can approach this less formally in the style of Sample Problem 12-4 (which is our choice). For the 3.1 kg particle, Eq. 12-21 yields

$$\ell_1 = r_{\perp 1} m v_1 = (2.8)(3.1)(3.6) = 31.2 \text{ kg}\cdot\text{m}^2/\text{s} .$$

Using the right-hand rule for vector products, we find this $(\vec{r}_1 \times \vec{p}_1)$ is out of the page, perpendicular to the plane of Fig. 12-35. And for the 6.5 kg particle, we find

$$\ell_2 = r_{\perp 2} m v_2 = (1.5)(6.5)(2.2) = 21.4 \text{ kg}\cdot\text{m}^2/\text{s} .$$

And we use the right-hand rule again, finding that this $(\vec{r}_2 \times \vec{p}_2)$ is into the page. Consequently, the two angular momentum vectors are in opposite directions, so their vector sum is the *difference* of their magnitudes:

$$L = \ell_1 - \ell_2 = 9.8 \text{ kg}\cdot\text{m}^2/\text{s} .$$

24. We note that the component of $\vec{v}$ perpendicular to $\vec{r}$ has magnitude $v \sin\phi$ where $\phi = 30°$. A similar observation applies to $\vec{F}$.

(a) Eq. 12-20 leads to

$$\ell = r m v_\perp = (3.0)(2.0)(4.0)\sin 30° = 12 \text{ kg}\cdot\text{m}^2/\text{s} .$$

Using the right-hand rule for vector products, we find $\vec{r} \times \vec{p}$ points out of the page, perpendicular to the plane of Fig. 12-36.

(b) Eq. 11-31 (which is the same as Eq. 12-15) leads to

$$\tau = rF\sin\phi = (3.0)(2.0)\sin 30° = 3.0 \text{ N}\cdot\text{m} .$$

Using the right-hand rule for vector products, we find $\vec{r} \times \vec{F}$ is also out of the page.

25. (a) We use $\vec{\ell} = m\vec{r} \times \vec{v}$, where $\vec{r}$ is the position vector of the object, $\vec{v}$ is its velocity vector, and m is its mass. Only the x and z components of the position and velocity vectors are nonzero, so Eq. 3-30 leads to $\vec{r} \times \vec{v} = (-xv_z + zv_x)\,\hat{\mathrm{j}}$. Therefore,

$$\begin{aligned} \vec{\ell} &= m\,(-xv_z + zv_x)\,\hat{\mathrm{j}} \\ &= (0.25\,\text{kg})\,(-(2.0\,\text{m})(5.0\,\text{m/s}) + (-2.0\,\text{m})(-5.0\,\text{m/s}))\,\hat{\mathrm{j}} \\ &= 0\,. \end{aligned}$$

(b) If we write $\vec{r} = x\hat{\mathrm{i}} + y\hat{\mathrm{j}} + z\hat{\mathrm{k}}$, then (using Eq. 3-30) we find $\vec{r} \times \vec{F}$ is equal to

$$(yF_z - zF_y)\,\hat{\mathrm{i}} + (zF_x - xF_z)\,\hat{\mathrm{j}} + (xF_y - yF_x)\,\hat{\mathrm{k}} .$$

With $x = 2.0$, $z = -2.0$, $F_y = 4.0$ and all other components zero (and SI units understood) the expression above yields $\vec{\tau} = \vec{r} \times \vec{F} = \left(8.0\,\hat{\mathrm{i}} + 8.0\,\hat{\mathrm{k}}\right)$ N·m.

26. If we write $\vec{r}\,' = x'\,\hat{\mathrm{i}} + y'\,\hat{\mathrm{j}} + z'\,\hat{\mathrm{k}}$, then (using Eq. 3-30) we find $\vec{r}\,' \times \vec{v}$ is equal to

$$(y'v_z - z'v_y)\,\hat{\mathrm{i}} + (z'v_x - x'v_z)\,\hat{\mathrm{j}} + (x'v_y - y'v_x)\,\hat{\mathrm{k}} .$$

(a) Here, $\vec{r}\,' = \vec{r}$ where $\vec{r} = 3\hat{\imath} - 4\hat{\jmath}$. Thus, dropping the primes in the above expression, we set (with SI units understood) $x = 3$, $y = -4$, $z = 0$, $v_x = 30$, $v_y = 60$ and $v_z = 0$. Then (with $m = 2.0$ kg) we obtain $\vec{\ell} = m\,(\vec{r}\times\vec{v}) = 600\,\hat{\mathrm{k}}$ kg·m²/s.

(b) Now $\vec{r}\,' = \vec{r} - \vec{r}_o$ where $\vec{r}_o = -2\hat{\imath} - 2\hat{\jmath}$. Therefore, in the above expression, we set $x' = 5$, $y' = -2$, $z' = 0$, $v_x = 30$, $v_y = 60$ and $v_z = 0$. We get $\vec{\ell} = m\,(\vec{r}\,'\times\vec{v}) = 720\,\hat{\mathrm{k}}$ kg·m²/s.

27. (a) The diagram below shows the particles and their lines of motion. The origin is marked O and may be anywhere. The angular momentum of particle 1 has magnitude $\ell_1 = mvr_1\sin\theta_1 = mv(d+h)$ and it is into the page. The angular momentum of particle 2 has magnitude $\ell_2 = mvr_2\sin\theta_2 = mvh$ and it is out of the page. The net angular momentum has magnitude $L = mv(d+h) - mvh = mvd$ and is into the page. This result is independent of the location of the origin.

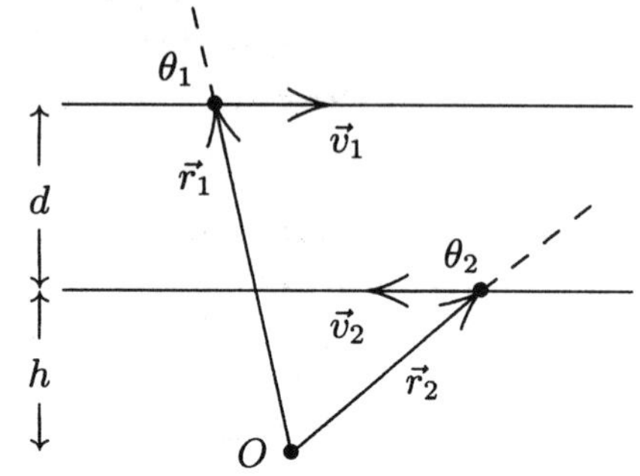

(b) As indicated above, the expression does not change.

(c) Suppose particle 2 is traveling to the right. Then $L = mv(d+h) + mvh = mv(d+2h)$. This result depends on h, the distance from the origin to one of the lines of motion. If the origin is midway between the lines of motion, then $h = -d/2$ and $L = 0$.

28. (a) With $\vec{p} = m\vec{v} = -16\hat{\jmath}$ kg · m/s, we take the vector cross product (using either Eq. 3-30 or, more simply, Eq. 12-20 and the right-hand rule):

$$\vec{\ell} = \vec{r}\times\vec{p} = -32\,\hat{\mathrm{k}}\ \mathrm{kg\cdot m^2/s}\ .$$

(b) Now the axis passes through the point $\vec{R} = 4.0\hat{\jmath}$ m, parallel with the z axis. With $\vec{r}\,' = \vec{r} - \vec{R} = 2.0\hat{\imath}$ m, we again take the cross product and arrive at the same result as before:

$$\vec{\ell}\,' = \vec{r}\,'\times\vec{p} = -32\,\hat{\mathrm{k}}\ \mathrm{kg\cdot m^2/s}\ .$$

(c) Torque is defined in Eq. 12-14: $\vec{\tau} = \vec{r}\times\vec{F} = 12\,\hat{\mathrm{k}}$ N·m.

(d) Using the notation from part (b),

$$\vec{\tau}\,' = \vec{r}\,'\times\vec{F} = 0\ .$$

29. If we write (for the general case) $\vec{r} = x\hat{\imath} + y\hat{\jmath} + z\hat{\mathrm{k}}$, then (using Eq. 3-30) we find $\vec{r}\times\vec{v}$ is equal to

$$(yv_z - zv_y)\,\hat{\imath} + (zv_x - xv_z)\,\hat{\jmath} + (xv_y - yv_x)\,\hat{\mathrm{k}}\ .$$

(a) The angular momentum is given by the vector product $\vec{\ell} = m\vec{r}\times\vec{v}$, where $\vec{r}$ is the position vector of the particle, $\vec{v}$ is its velocity, and $m = 3.0$ kg is its mass. Substituting (with SI units understood) $x = 3$, $y = 8$, $z = 0$, $v_x = 5$, $v_y = -6$ and $v_z = 0$ into the above expression, we obtain

$$\vec{\ell} = (3.0)\,((3)(-6) - (8.0)(5.0))\,\hat{\mathrm{k}} = -1.7\times10^2\,\hat{\mathrm{k}}\ \mathrm{kg\cdot m^2/s}\ .$$

(b) The torque is given by Eq. 12-14, $\vec{\tau} = \vec{r}\times\vec{F}$. We write $\vec{r} = x\hat{\imath} + y\hat{\jmath}$ and $\vec{F} = F_x\,\hat{\imath}$ and obtain

$$\vec{\tau} = \left(x\,\hat{\imath} + y\,\hat{\jmath}\right)\times\left(F_x\,\hat{\imath}\right) = -yF_x\,\hat{\mathrm{k}}$$

since $\hat{\imath}\times\hat{\imath} = 0$ and $\hat{\jmath}\times\hat{\imath} = -\hat{\mathrm{k}}$. Thus, we find $\vec{\tau} = -(8.0\,\mathrm{m})(-7.0\,\mathrm{N})\,\hat{\mathrm{k}} = 56\,\hat{\mathrm{k}}$ N·m.

(c) According to Newton's second law $\vec{\tau} = d\vec{\ell}/dt$, so the rate of change of the angular momentum is $56\,\text{kg}\cdot\text{m}^2/\text{s}^2$, in the positive z direction.

30. The rate of change of the angular momentum is

$$\frac{d\vec{\ell}}{dt} = \vec{\tau}_1 + \vec{\tau}_2 = 2.0\hat{\text{i}} - 4.0\hat{\text{j}}\ \text{N}\cdot\text{m}\ .$$

Consequently, the vector $d\vec{\ell}/dt$ has a magnitude $\sqrt{2.0^2 + (-4.0)^2} = 4.5\,\text{N}\cdot\text{m}$ and is at an angle θ (in the xy plane, or a plane parallel to it) measured from the positive x axis, where $\theta = \tan^{-1}\left(\frac{-4.0}{2.0}\right) = -63°$, the negative sign indicating that the angle is measured clockwise as viewed "from above" (by a person on the $+z$ axis).

31. We use a right-handed coordinate system with $+\hat{\text{k}}$ directed out of the xy plane so as to be consistent with counterclockwise rotation (and the right-hand rule). Thus, all the angular momenta being considered are along the $-\hat{\text{k}}$ direction; for example, in part (b) $\vec{\ell} = -4.0t^2\,\hat{\text{k}}$ in SI units. We use Eq. 12-23.

(a) The angular momentum is constant so its derivative is zero. There is no torque in this instance.

(b) Taking the derivative with respect to time, we obtain the torque:

$$\vec{\tau} = \frac{d\vec{\ell}}{dt} = (-4.0\,\hat{\text{k}})\,\frac{dt^2}{dt} = -8.0t\,\hat{\text{k}}$$

in SI units (N·m). This vector points in the $-\hat{\text{k}}$ direction (causing the clockwise motion to speed up) for all $t > 0$.

(c) With $\vec{\ell} = -4.0\sqrt{t}\,\hat{\text{k}}$ in SI units, the torque is

$$\vec{\tau} = \left(-4.0\hat{\text{k}}\right)\frac{d\sqrt{t}}{dt} = \left(-4.0\hat{\text{k}}\right)\left(\frac{1}{2\sqrt{t}}\right)$$

which yields $\vec{\tau} = -2.0/\sqrt{t}\,\hat{\text{k}}$ in SI units. This vector points in the $-\hat{\text{k}}$ direction (causing the clockwise motion to speed up) for all $t > 0$ (and it is undefined for $t < 0$).

(d) Finally, we have

$$\vec{\tau} = \left(-4.0\hat{\text{k}}\right)\frac{dt^{-2}}{dt} = \left(-4.0\hat{\text{k}}\right)\left(\frac{-2}{t^3}\right)$$

which yields $\vec{\tau} = 8.0/t^3\,\hat{\text{k}}$ in SI units. This vector points in the $+\hat{\text{k}}$ direction (causing the initially clockwise motion to slow down) for all $t > 0$.

32. Both $\vec{r}$ and $\vec{v}$ lie in the xy plane. The position vector $\vec{r}$ has an x component that is a function of time (being the integral of the x component of velocity, which is itself time-dependent) and a y component that is constant ($y = -2.0$ m). In the cross product $\vec{r} \times \vec{v}$, all that matters is the y component of $\vec{r}$ since $v_x \neq 0$ but $v_y = 0$:

$$\vec{r} \times \vec{v} = -yv_x\,\hat{\text{k}}\ .$$

(a) The angular momentum is $\vec{\ell} = m\,(\vec{r} \times \vec{v})$ where the mass is $m = 2.0$ kg in this case. With SI units understood and using the above cross-product expression, we have

$$\vec{\ell} = (2.0)\left(-(-2.0)\left(-6.0t^2\right)\right)\hat{\text{k}} = -24t^2\,\hat{\text{k}}$$

in $\text{kg}\cdot\text{m}^2/\text{s}$. This implies the particle is moving clockwise (as observed by someone on the $+z$ axis) for $t > 0$.

(b) The torque is caused by the (net) force $\vec{F} = m\vec{a}$ where

$$\vec{a} = \frac{d\vec{v}}{dt} = -12t\,\hat{\mathrm{i}}\ \mathrm{m/s^2}\ .$$

The remark above that only the y component of $\vec{r}$ still applies, since $a_y = 0$. We use $\vec{\tau} = \vec{r} \times \vec{F} = m\,(\vec{r} \times \vec{a})$ and obtain

$$\vec{\tau} = (2.0)\,(-(-2.0)(-12t))\,\hat{\mathrm{k}} = -48t\,\hat{\mathrm{k}}$$

in N·m. The torque on the particle (as observed by someone on the $+z$ axis) is clockwise, causing the particle motion (which was clockwise to begin with) to increase.

(c) We replace $\vec{r}$ with $\vec{r}\,'$ (measured relative to the new reference point) and note (again) that only its y component matters in these calculations. Thus, with $y' = -2.0 - (-3.0) = 1.0$ m, we find

$$\vec{\ell}\,' = (2.0)\left(-(1.0)\left(-6.0t^2\right)\right)\hat{\mathrm{k}} = 12t^2\,\hat{\mathrm{k}}$$

in kg·m²/s. The fact that this is positive implies that the particle is moving counterclockwise relative to the new reference point.

(d) Using $\vec{\tau}\,' = \vec{r}\,' \times \vec{F} = m\,(\vec{r}\,' \times \vec{a})$, we obtain

$$\vec{\tau} = (2.0)\,(-(1.0)\,(-12t))\,\hat{\mathrm{k}} = 24t\,\hat{\mathrm{k}}$$

in N·m. The torque on the particle (as observed by someone on the $+z$ axis) is counterclockwise, relative to the new reference point.

33. (a) Since $\tau = dL/dt$, the average torque acting during any interval Δt is given by $\tau_{\text{avg}} = (L_f - L_i)\,/\Delta t$, where L_i is the initial angular momentum and L_f is the final angular momentum. Thus

$$\tau_{\text{avg}} = \frac{0.800\,\mathrm{kg\cdot m^2/s} - 3.00\,\mathrm{kg\cdot m^2/s}}{1.50\,\mathrm{s}}$$

which yields $\tau_{\text{avg}} = -1.467 \approx= -1.47\,\mathrm{N\cdot m}$. In this case the negative sign indicates that the direction of the torque is opposite the direction of the initial angular momentum, implicitly taken to be positive.

(b) The angle turned is $\theta = \omega_0 t + \frac{1}{2}\alpha t^2$. If the angular acceleration α is uniform, then so is the torque and $\alpha = \tau/I$. Furthermore, $\omega_0 = L_i/I$, and we obtain

$$\begin{aligned}\theta &= \frac{L_i t + \frac{1}{2}\tau t^2}{I}\\ &= \frac{(3.00\,\mathrm{kg\cdot m^2/s})(1.50\,\mathrm{s}) + \frac{1}{2}(-1.467\,\mathrm{N\cdot m})(1.50\,\mathrm{s})^2}{0.140\,\mathrm{kg\cdot m^2}}\\ &= 20.4\ \mathrm{rad}\ .\end{aligned}$$

(c) The work done on the wheel is

$$W = \tau\theta = (-1.47\,\mathrm{N\cdot m})(20.4\,\mathrm{rad}) = -29.9\ \mathrm{J}$$

where more precise values are used in the calculation than what is shown here. An equally good method for finding W is Eq. 11-44, which, if desired, can be rewritten as $W = (L_f^2 - L_i^2)/2I$.

(d) The average power is the work done by the flywheel (the negative of the work done on the flywheel) divided by the time interval:

$$P_{\text{avg}} = -\frac{W}{\Delta t} = -\frac{-29.8\,\mathrm{J}}{1.50\,\mathrm{s}} = 19.9\ \mathrm{W}\ .$$

34. (a) Eq. 11-27 gives $\alpha = \tau/I$ and Eq. 11-12 leads to $\omega = \alpha t = \tau t/I$. Therefore, the angular momentum at $t = 0.033$ s is

$$I\omega = \tau t = (16\,\mathrm{N\cdot m})(0.033\,\mathrm{s}) = 0.53\ \mathrm{kg{\cdot}m^2/s}$$

where this is essentially a derivation of the angular version of the impulse-momentum theorem.

(b) We find

$$\omega = \frac{\tau t}{I} = \frac{(16)(0.033)}{1.2 \times 10^{-3}} = 440\ \mathrm{rad}$$

which we convert as follows: $\omega = (440)(60/2\pi) \approx 4200$ rev/min.

35. (a) A particle contributes mr^2 to the rotational inertia. Here r is the distance from the origin O to the particle. The total rotational inertia is

$$I = m(3d)^2 + m(2d)^2 + m(d)^2 = 14md^2\ .$$

(b) The angular momentum of the middle particle is given by $L_m = I_m\omega$, where $I_m = 4md^2$ is its rotational inertia. Thus $L_m = 4md^2\omega$.

(c) The total angular momentum is $I\omega = 14md^2\omega$.

36. We integrate Eq. 12-29 (for a single torque) over the time interval (where the angular speed at the beginning is ω_i and at the end is ω_f)

$$\int \tau\, dt = \int \frac{dL}{dt}\, dt = L_f - L_i = I\,(\omega_f - \omega_i)$$

and if we use the calculus-based notion of the average of a function f

$$f_{\text{avg}} = \frac{1}{\Delta t}\int f\, dt$$

then (using Eq. 12-16) we obtain

$$\int \tau\, dt = \tau_{\text{avg}}\Delta t = F_{\text{avg}}R\Delta t\ .$$

Inserting this into the top line proves the relationship shown in the problem.

37. Suppose cylinder 1 exerts a uniform force of magnitude F on cylinder 2, tangent to the cylinder's surface at the point of contact. The torque applied to cylinder 2 is $\tau_2 = R_2F$ and the angular acceleration of that cylinder is $\alpha_2 = \tau_2/I_2 = R_2F/I_2$. As a function of time its angular velocity is

$$\omega_2 = \alpha_2 t = \frac{R_2Ft}{I_2}\ .$$

The forces of the cylinders on each other obey Newton's third law, so the magnitude of the force of cylinder 2 on cylinder 1 is also F. The torque exerted by cylinder 2 on cylinder 1 is $\tau_1 = R_1F$ and the angular acceleration of cylinder 1 is $\alpha_1 = \tau_1/I_1 = R_1F/I_1$. This torque slows the cylinder. As a function of time, its angular velocity is $\omega_1 = \omega_0 - R_1Ft/I_1$. The force ceases and the cylinders continue rotating with constant angular speeds when the speeds of points on their rims are the same ($R_1\omega_1 = R_2\omega_2$). Thus,

$$R_1\omega_0 - \frac{R_1^2Ft}{I_1} = \frac{R_2^2Ft}{I_2}\ .$$

When this equation is solved for the product of force and time, the result is

$$Ft = \frac{R_1I_1I_2}{I_1R_2^2 + I_2R_1^2}\,\omega_0\ .$$

Substituting this expression for Ft in the ω_2 equation above, we obtain

$$\omega_2 = \frac{R_1R_2I_1}{I_1R_2^2 + I_2R_1^2}\,\omega_0\ .$$

38. (a) For the hoop, we use Table 11-2(h) and the parallel-axis theorem to obtain

$$I_1 = I_{\text{com}} + mh^2 = \frac{1}{2}mR^2 + mR^2 = \frac{3}{2}mR^2 \ .$$

Of the thin bars (in the form of a square), the member along the rotation axis has (approximately) no rotational inertia about that axis (since it is thin), and the member farthest from it is very much like it (by being parallel to it) except that it is displaced by a distance h; it has rotational inertia given by the parallel axis theorem:

$$I_2 = I_{\text{com}} + mh^2 = 0 + mR^2 = mR^2 \ .$$

Now the two members of the square perpendicular to the axis have the same rotational inertia (that is, $I_3 = I_4$). We find I_3 using Table 11-2(e) and the parallel-axis theorem:

$$I_3 = I_{\text{com}} + mh^2 = \frac{1}{12}mR^2 + m\left(\frac{R}{2}\right)^2 = \frac{1}{3}mR^2 \ .$$

Therefore, the total rotational inertia is

$$I_1 + I_2 + I_3 + I_4 = \frac{19}{6}mR^2 = 1.6 \text{ kg}\cdot\text{m}^2 \ .$$

(b) The angular speed is constant:

$$\omega = \frac{\Delta\theta}{\Delta t} = \frac{2\pi}{2.5} = 2.5 \text{ rad/s} \ .$$

Thus, $L = I_{\text{total}}\omega = 4.0 \text{ kg}\cdot\text{m}^2/\text{s}$.

39. (a) No external torques act on the system consisting of the man, bricks, and platform, so the total angular momentum of the system is conserved. Let I_i be the initial rotational inertia of the system and let I_f be the final rotational inertia. Then $I_i\omega_i = I_f\omega_f$ and

$$\begin{aligned} \omega_f &= \left(\frac{I_i}{I_f}\right)\omega_i \\ &= \left(\frac{6.0\,\text{kg}\cdot\text{m}^2}{2.0\,\text{kg}\cdot\text{m}^2}\right)(1.2\,\text{rev/s}) \\ &= 3.6 \text{ rev/s} \ . \end{aligned}$$

(b) The initial kinetic energy is $K_i = \frac{1}{2}I_i\omega_i^2$, the final kinetic energy is $K_f = \frac{1}{2}I_f\omega_f^2$, and their ratio is

$$\frac{K_f}{K_i} = \frac{I_f\omega_f^2}{I_i\omega_i^2} = \frac{(2.0\,\text{kg}\cdot\text{m}^2)(3.6\,\text{rev/s})^2}{(6.0\,\text{kg}\cdot\text{m}^2)(1.2\,\text{rev/s})^2} = 3.0 \ .$$

(c) The man did work in decreasing the rotational inertia by pulling the bricks closer to his body. This energy came from the man's store of internal energy.

40. We use conservation of angular momentum: $I_m\omega_m = I_p\omega_p$. The respective angles θ_m and θ_p by which the motor and probe rotate are therefore related by

$$\int I_m\omega_m dt = I_m\theta_m = \int I_p\omega_p dt = I_p\theta_p$$

which gives

$$\theta_m = \frac{I_p\theta_p}{I_m} = \frac{(12\,\text{kg}\cdot\text{m}^2)\,(30^\circ)}{2.0\times10^{-3}\,\text{kg}\cdot\text{m}^2} = 180000^\circ \ .$$

The number of revolutions for the rotor is then $1.8\times10^5/360 = 500\,\text{rev}$.

41. (a) No external torques act on the system consisting of the two wheels, so its total angular momentum is conserved. Let I_1 be the rotational inertia of the wheel that is originally spinning (at ω_i) and I_2 be the rotational inertia of the wheel that is initially at rest. Then $I_1\omega_i = (I_1 + I_2)\omega_f$ and

$$\omega_f = \frac{I_1}{I_1 + I_2}\,\omega_i$$

where ω_f is the common final angular velocity of the wheels. Substituting $I_2 = 2I_1$ and $\omega_i = 800\,\text{rev/min}$, we obtain $\omega_f = 267\,\text{rev/min}$.

(b) The initial kinetic energy is $K_i = \frac{1}{2}I_1\omega_i^2$ and the final kinetic energy is $K_f = \frac{1}{2}(I_1 + I_2)\omega_f^2$. We rewrite this as

$$K_f = \frac{1}{2}(I_1 + 2I_1)\left(\frac{I_1\omega_i}{I_1 + 2I_1}\right)^2 = \frac{1}{6}I\omega_i^2\ .$$

Therefore, the fraction lost, $(K_i - K_f)/K_i$, is

$$1 - \frac{K_f}{K_i} = 1 - \frac{\frac{1}{6}I\omega_i^2}{\frac{1}{2}I\omega_i^2} = \frac{2}{3}\ .$$

42. (a) We apply conservation of angular momentum: $I_1\omega_1 + I_2\omega_2 = (I_1 + I_2)\omega$. The angular speed after coupling is therefore

$$\omega = \frac{I_1\omega_1 + I_2\omega_2}{I_1 + I_2} = \frac{(3.3\,\text{kg}\cdot\text{m}^2)\,(450\,\text{rev/min}) + (6.6\,\text{kg}\cdot\text{m}^2)\,(900\,\text{rev/min})}{3.3\,\text{kg}\cdot\text{m}^2 + 6.6\,\text{kg}\cdot\text{m}^2} = 750\,\text{rev/min}\ .$$

(b) In this case, we obtain

$$\omega = \frac{I_1\omega_1 + I_2\omega_2}{I_1 + I_2} = \frac{(3.3)(450) + (6.6)(-900)}{3.3 + 6.6} = -450\,\text{rev/min}$$

where the minus sign indicates that $\vec{\omega}$ is in the direction of the second disk's initial angular velocity.

43. (a) In terms of the radius of gyration k, the rotational inertia of the merry-go-round is $I = Mk^2$. We obtain $I = (180\,\text{kg})(0.910\,\text{m})^2 = 149\,\text{kg}\cdot\text{m}^2$.

(b) An object moving along a straight line has angular momentum about any point that is not on the line. The magnitude of the angular momentum of the child about the center of the merry-go-round is given by Eq. 12-21, mvR, where R is the radius of the merry-go-round. Therefore,

$$\left|\vec{L}_{\text{child}}\right| = (44.0\,\text{kg})(3.00\,\text{m/s})(1.20\,\text{m}) = 158\,\text{kg}\cdot\text{m}^2/\text{s}\ .$$

(c) No external torques act on the system consisting of the child and the merry-go-round, so the total angular momentum of the system is conserved. The initial angular momentum is given by mvR; the final angular momentum is given by $(I + mR^2)\,\omega$, where ω is the final common angular velocity of the merry-go-round and child. Thus $mvR = (I + mR^2)\omega$ and

$$\omega = \frac{mvR}{I + mR^2} = \frac{158\,\text{kg}\cdot\text{m}^2/\text{s}}{149\,\text{kg}\cdot\text{m}^2 + (44.0\,\text{kg})(1.20\,\text{m})^2} = 0.744\,\text{rad/s}\ .$$

44. Angular momentum conservation $I_i\omega_i = I_f\omega_f$ leads to

$$\frac{\omega_f}{\omega_i} = \frac{I_i}{I_f}\omega_i = 3$$

which implies

$$\frac{K_f}{K_i} = \frac{\frac{1}{2}I_f\omega_f^2}{\frac{1}{2}I_i\omega_i^2} = \frac{I_f}{I_i}\left(\frac{\omega_f}{\omega_i}\right)^2 = 3\ .$$

45. No external torques act on the system consisting of the train and wheel, so the total angular momentum of the system (which is initially zero) remains zero. Let $I = MR^2$ be the rotational inertia of the wheel. Its final angular momentum is $= I\omega\hat{k} = -MR^2|\omega|\hat{k}$, where $\hat{k}$ is *up* in Fig. 12-40 and that last step (with the minus sign) is done in recognition that the wheel's clockwise rotation implies a negative value for ω. The linear speed of a point on the track is ωR and the speed of the train (going counterclockwise in Fig. 12-40 with speed v' relative to an outside observer) is therefore $v' = v - |\omega|R$ where v is its speed relative to the tracks. Consequently, the angular momentum of the train is $m(v-|\omega|R)R\,\hat{k}$. Conservation of angular momentum yields

$$0 = -MR^2|\omega|\,\hat{k} + m\,(v - |\omega|R)\,R\,\hat{k}\;.$$

When this equation is solved for the angular speed, the result is

$$|\omega| = \frac{mvR}{(M+m)R^2} = \frac{mv}{(M+m)R}\;.$$

46. We assume that from the moment of grabbing the stick onward, they maintain rigid postures so that the system can be analyzed as a symmetrical rigid body with center of mass midway between the skaters.

(a) The total linear momentum is zero (the skaters have the same mass and equal-and-opposite velocities). Thus, their center of mass (the middle of the 3.0 m long stick) remains fixed and they execute circular motion (of radius $r = 1.5$ m) about it. Using Eq. 11-18, their angular velocity (counterclockwise as seen in Fig. 12-41) is

$$\omega = \frac{v}{r} = \frac{1.4}{1.5} = 0.93 \text{ rad/s}\;.$$

(b) Their rotational inertia is that of two particles in circular motion at $r = 1.5$ m, so Eq. 11-26 yields

$$I = \sum mr^2 = 2(50)(1.5)^2 = 225 \text{ kg}\cdot\text{m}^2\;.$$

Therefore, Eq. 11-27 leads to

$$K = \frac{1}{2}I\omega^2 = \frac{1}{2}(225)(0.93)^2 = 98 \text{ J}\;.$$

(c) Angular momentum is conserved in this process. If we label the angular velocity found in part (a) ω_i and the rotational inertia of part (b) as I_i, we have

$$I_i\omega_i = (225)(0.93) = I_f\omega_f\;.$$

The final rotational inertia is $\sum mr_f^2$ where $r_f = 0.5$ m so $I_f = 25$ kg·m^2. Using this value, the above expression gives $\omega_f = 8.4$ rad/s.

(d) We find

$$K_f = \frac{1}{2}I_f\omega_f^2 = \frac{1}{2}(25)(8.4)^2 = 8.8\times10^2 \text{ J}\;.$$

(e) We account for the large increase in kinetic energy (part (d) minus part (b)) by noting that the skaters do a great deal of work (converting their internal energy into mechanical energy) as they pull themselves closer – "fighting" what appears to them to be large "centrifugal forces" trying to keep them apart.

47. So that we don't get confused about ± signs, we write the angular *speed* of the lazy Susan as $|\omega|$ and reserve the ω symbol for the angular velocity (which, using a common convention, is negative-valued when the rotation is clockwise). When the roach "stops" we recognize that it comes to rest relative to the lazy Susan (not relative to the ground).

(a) Angular momentum conservation leads to

$$mvR + I\omega_0 = \left(mR^2 + I\right)\omega_f$$

which we can write (recalling our discussion about angular speed versus angular velocity) as

$$mvR - I\,|\omega_0| = -\left(mR^2 + I\right)|\omega_f| \ .$$

We solve for the final angular speed of the system:

$$|\omega_f| = \frac{mvR - I\,|\omega_0|}{mR^2 + I} \ .$$

(b) No, $K_f \neq K_i$ and – if desired – we can solve for the difference:

$$K_i - K_f = \frac{m\,I}{2}\,\frac{v^2 + \omega_0^2 R^2 + 2Rv\,|\omega_0|}{mR^2 + I}$$

which is clearly positive. Thus, some of the initial kinetic energy is "lost" – that is, transferred to another form. And the culprit is the roach, who must find it difficult to stop (and "internalize" that energy).

48. The initial angular momentum of the system is zero. The final angular momentum of the girl-plus-merry-go-round is $\left(I + MR^2\right)\omega$ which we will take to be positive. The final angular momentum we associate with the thrown rock is negative: $-mRv$, where v is the speed (positive, by definition) of the rock relative to the ground.

(a) Angular momentum conservation leads to

$$0 = \left(I + MR^2\right)\omega - mRv \implies \omega = \frac{mRv}{I + MR^2} \ .$$

(b) The girl's linear speed is given by Eq. 11-18:

$$R\omega = \frac{mvR^2}{I + MR^2} \ .$$

49. For simplicity, we assume the record is turning freely, without any work being done by its motor (and without any friction at the bearings or at the stylus trying to slow it down). Before the collision, the angular momentum of the system (presumed positive) is $I_i\omega_i$ where $I_i = 5.0 \times 10^{-4}$ kg·m^2 and $\omega_i = 4.7$ rad/s. The rotational inertia afterwards is $I_f = I_i + mR^2$ where $m = 0.020$ kg and $R = 0.10$ m. The mass of the record (0.10 kg), although given in the problem, is not used in the solution. Angular momentum conservation leads to

$$I_i\omega_i = I_f\omega_f \implies \omega_f = \frac{I_i\omega_i}{I_i + mR^2} = 3.4 \text{ rad/s} \ .$$

50. The axis of rotation is in the middle of the rod, $r = 0.25$ m from either end. By Eq. 12-19, the initial angular momentum of the system (which is just that of the bullet, before impact) is $rmv\sin\phi$ where $m = 0.003$ kg and $\phi = 60°$. Relative to the axis, this is counterclockwise and thus (by the common convention) positive. After the collision, the moment of inertia of the system is $I = I_{\text{rod}} + mr^2$ where $I_{\text{rod}} = ML^2/12$ by Table 11-2(e), with $M = 4.0$ kg and $L = 0.5$ m. Angular momentum conservation leads to

$$rmv\sin\phi = \left(\frac{1}{12}ML^2 + mr^2\right)\omega \ .$$

Thus, with $\omega = 10$ rad/s, we obtain

$$v = \frac{\left(\frac{1}{12}(4.0)(0.5)^2 + (0.003)(0.25)^2\right)(10)}{(0.25)(0.003)\sin 60°} = 1.3 \times 10^3 \text{ m/s} \ .$$

51. (a) If we consider a short time interval from just before the wad hits to just after it hits and sticks, we may use the principle of conservation of angular momentum. The initial angular momentum is the angular momentum of the falling putty wad. The wad initially moves along a line that is $d/2$ distant from the axis of rotation, where $d = 0.500\,\text{m}$ is the length of the rod. The angular momentum of the wad is $mvd/2$ where $m = 0.0500\,\text{kg}$ and $v = 3.00$ m/s are the mass and initial speed of the wad. After the wad sticks, the rod has angular velocity ω and angular momentum $I\omega$, where I is the rotational inertia of the system consisting of the rod with the two balls and the wad at its end. Conservation of angular momentum yields $mvd/2 = I\omega$ where $I = (2M+m)(d/2)^2$ and $M = 2.00$ kg is the mass of each of the balls. We solve $mvd/2 = (2M+m)(d/2)^2\omega$ for the angular speed:

$$\omega = \frac{2mv}{(2M+m)d} = \frac{2(0.0500)(3.00)}{(2(2.00)+0.0500)(0.500)} = 0.148\ \text{rad/s}\ .$$

(b) The initial kinetic energy is $K_i = \frac{1}{2}mv^2$, the final kinetic energy is $K_f = \frac{1}{2}I\omega^2$, and their ratio is $K_f/K_i = I\omega^2/mv^2$. When $I = (2M+m)d^2/4$ and $\omega = 2mv/(2M+m)d$ are substituted, this becomes

$$\frac{K_f}{K_i} = \frac{m}{2M+m} = \frac{0.0500}{2(2.00)+0.0500} = 0.0123\ .$$

(c) As the rod rotates, the sum of its kinetic and potential energies is conserved. If one of the balls is lowered a distance h, the other is raised the same distance and the sum of the potential energies of the balls does not change. We need consider only the potential energy of the putty wad. It moves through a 90° arc to reach the lowest point on its path, gaining kinetic energy and losing gravitational potential energy as it goes. It then swings up through an angle θ, losing kinetic energy and gaining potential energy, until it momentarily comes to rest. Take the lowest point on the path to be the zero of potential energy. It starts a distance $d/2$ above this point, so its initial potential energy is $U_i = mgd/2$. If it swings up to the angular position θ, as measured from its lowest point, then its final height is $(d/2)(1-\cos\theta)$ above the lowest point and its final potential energy is $U_f = mg(d/2)(1-\cos\theta)$. The initial kinetic energy is the sum of that of the balls and wad: $K_i = \frac{1}{2}I\omega^2 = \frac{1}{2}(2M+m)(d/2)^2\omega^2$. At its final position, we have $K_f = 0$. Conservation of energy provides the relation:

$$mg\frac{d}{2} + \frac{1}{2}(2M+m)\left(\frac{d}{2}\right)^2\omega^2 = mg\frac{d}{2}(1-\cos\theta)\ .$$

When this equation is solved for $\cos\theta$, the result is

$$\begin{aligned}\cos\theta &= -\frac{1}{2}\left(\frac{2M+m}{mg}\right)\left(\frac{d}{2}\right)\omega^2 \\ &= -\frac{1}{2}\left(\frac{2(2.00\,\text{kg})+0.0500\,\text{kg}}{(0.0500\,\text{kg})(9.8\,\text{m/s}^2)}\right)\left(\frac{0.500\,\text{m}}{2}\right)(0.148\,\text{rad/s})^2 \\ &= -0.0226\ .\end{aligned}$$

Consequently, the result for θ is 91.3°. The total angle through which it has swung is $90° + 91.3° = 181°$.

52. We denote the cockroach with subscript 1 and the disk with subscript 2.

(a) Initially the angular momentum of the system consisting of the cockroach and the disk is

$$L_i = m_1v_{1i}r_{1i} + I_2\omega_{2i} = m_1\omega_0R^2 + \frac{1}{2}m_2\omega_0R^2\ .$$

After the cockroach has completed its walk, its position (relative to the axis) is $r_{1f} = R/2$ so the final angular momentum of the system is

$$L_f = m_1\omega_f\left(\frac{R}{2}\right)^2 + \frac{1}{2}m_2\omega_fR^2\ .$$

Then from $L_f = L_i$ we obtain

$$\omega_f \left(\frac{1}{4}m_1R^2 + \frac{1}{2}m_2R\right) = \omega_0 \left(m_1R^2 + \frac{1}{2}m_2R^2\right) .$$

Thus,

$$\begin{aligned}\omega_f - \omega_0 &= \omega_0 \left(\frac{m_1R^2 + m_2R^2/2}{m_1R^2/4 + m_2R^2/2}\right) - \omega_0 \\ &= \omega_0 \left(\frac{m + 10m/2}{m/4 + 10m/2} - 1\right) \\ &= \omega_0(1.14 - 1)\end{aligned}$$

which yields $\Delta\omega = 0.14\omega_0$. For later use, we note that $\omega_f/\omega_i = 1.14$.

(b) We substitute $I = L/\omega$ into $K = \frac{1}{2}I\omega^2$ and obtain $K = \frac{1}{2}L\omega$. Since we have $L_i = L_f$, the the kinetic energy ratio becomes

$$\frac{K}{K_0} = \frac{\frac{1}{2}L_f\omega_f}{\frac{1}{2}L_i\omega_i} = \frac{\omega_f}{\omega_i} = 1.14 .$$

(c) The cockroach does positive work while walking toward the center of the disk, increasing the total kinetic energy of the system.

53. If the polar cap melts, the resulting body of water will effectively increase the equatorial radius of the Earth from R_e to $R'_e = R_e + \Delta R$, thereby increasing the moment of inertia of the Earth and slowing its rotation (by conservation of angular momentum), causing the duration T of a day to increase by ΔT. We note that (in rad/s) $\omega = 2\pi/T$ so

$$\frac{\omega'}{\omega} = \frac{2\pi/T'}{2\pi/T} = \frac{T}{T'}$$

from which it follows that

$$\frac{\Delta\omega}{\omega} = \frac{\omega'}{\omega} - 1 = \frac{T}{T'} - 1 = -\frac{\Delta T}{T'} .$$

We can approximate that last denominator as T so that we end up with the simple relationship $|\Delta\omega|/\omega = \Delta T/T$. Now, conservation of angular momentum gives us

$$\Delta L = 0 = \Delta(I\omega) \approx I(\Delta\omega) + \omega(\Delta I)$$

so that $|\Delta\omega|/\omega = \Delta I/I$. Thus, using our expectation that rotational inertia is proportional to the equatorial radius squared (supported by Table 11-2(f) for a perfect uniform sphere, but then this isn't a perfect uniform sphere) we have

$$\begin{aligned}\frac{\Delta T}{T} &= \frac{\Delta I}{I} \\ &= \frac{\Delta(R_e^2)}{R_e^2} \approx \frac{2\Delta R_e}{R_e} \\ &= \frac{2(30\,\text{m})}{6.37 \times 10^6\,\text{m}}\end{aligned}$$

so with $T = 86400\,\text{s}$ we find (approximately) that $\Delta T = 0.8$ s. The radius of the earth can be found in Appendix C or on the inside front cover of the textbook.

54. The initial rotational inertia of the system is $I_i = I_{\text{disk}} + I_{\text{student}}$ where $I_{\text{disk}} = 300$ kg·m^2 (which, incidentally, does agree with Table 11-2(c)) and $I_{\text{student}} = mR^2$ where $m = 60$ kg and $R = 2.0$ m. The rotational inertia when the student reaches $r = 0.5$ m is $I_f = I_{\text{disk}} + mr^2$. Angular momentum conservation leads to

$$I_i\omega_i = I_f\omega_f \implies \omega_f = \omega_i \frac{I_{\text{disk}} + mR^2}{I_{\text{disk}} + mr^2}$$

which yields, for $\omega_i = 1.5$ rad/s, a final angular velocity of $\omega_f = 2.6$ rad/s.

55. Their angular velocities, when they are stuck to each other, are equal, regardless of whether they share the same central axis. The initial rotational inertia of the system is

$$I_0 = I_{\text{big disk}} + I_{\text{small disk}} \quad \text{where} \quad I_{\text{big disk}} = \frac{1}{2}MR^2$$

using Table 11-2(c). Similarly, since the small disk is initially concentric with the big one, $I_{\text{small disk}} = \frac{1}{2}mr^2$. After it slides, the rotational inertia of the small disk is found from the parallel axis theorem (using $h = R - r$). Thus, the new rotational inertia of the system is

$$I = \frac{1}{2}MR^2 + \frac{1}{2}mr^2 + m(R-r)^2 .$$

(a) Angular momentum conservation, $I_0\omega_0 = I\omega$, leads to the new angular velocity:

$$\omega = \omega_0 \frac{\frac{1}{2}MR^2 + \frac{1}{2}mr^2}{\frac{1}{2}MR^2 + \frac{1}{2}mr^2 + m(R-r)^2} .$$

Substituting $M = 10m$ and $R = 3r$, this becomes $\omega = \omega_0(91/99)$. Thus, with $\omega_0 = 20$ rad/s, we find $\omega = 18$ rad/s.

(b) From the previous part, we know that

$$\frac{I_0}{I} = \frac{91}{99} \quad \text{and} \quad \frac{\omega}{\omega_0} = \frac{91}{99} .$$

Plugging these into the ratio of kinetic energies, we have

$$\frac{\frac{1}{2}I\omega^2}{\frac{1}{2}I_0\omega_0^2} = \frac{I}{I_0}\left(\frac{\omega}{\omega_0}\right)^2 = \frac{99}{91}\left(\frac{91}{99}\right)^2$$

which yields $K/K_0 = 0.92$.

56. This is a completely inelastic collision which we analyze using angular momentum conservation. Let m and v_0 be the mass and initial speed of the ball and R the radius of the merry-go-round. The initial angular momentum is

$$\vec{\ell}_0 = \vec{r}_0 \times \vec{p}_0 \implies \ell_0 = R(mv_0)\sin 53°$$

where 53° is the angle between the radius vector pointing to the child and the direction of $\vec{v}_0$. Thus, $\ell_0 = 19\,\text{kg}\cdot\text{m}^2/\text{s}$. Now, with SI units understood,

$$\begin{aligned} \ell_0 &= L_f \\ 19 &= I\omega \\ &= \left(150 + (30)R^2 + (1.0)R^2\right)\omega \end{aligned}$$

so that $\omega = 0.070$ rad/s.

57. (a) With $r = 0.60$ m, we obtain $I = 0.060 + (0.501)r^2 = 0.24\ \text{kg}\cdot\text{m}^2$.

(b) Invoking angular momentum conservation, with SI units understood,

$$\begin{aligned} \ell_0 &= L_f \\ mv_0r &= = I\omega \\ (0.001)v_0(0.60) &= (0.24)(4.5) \end{aligned}$$

which leads to $v_0 = 1.8 \times 10^3$ m/s.

58. We make the unconventional choice of *clockwise* sense as positive, so that the angular velocities in this problem are positive. With $r = 0.60$ m and $I_0 = 0.12\ \text{kg}\cdot\text{m}^2$, the rotational inertia of the putty-rod system (after the collision) is $I = I_0 + (0.20)r^2 = 0.19\ \text{kg}\cdot\text{m}^2$. Invoking angular momentum conservation, with SI units understood, we have

$$\begin{aligned} L_0 &= L_f \\ I_0\omega_0 &= I\omega \\ (0.12)(2.4) &= (0.19)\omega \end{aligned}$$

which yields $\omega = 1.5$ rad/s.

59. We make the unconventional choice of *clockwise* sense as positive, so that the angular velocities (and angles) in this problem are positive. Mechanical energy conservation applied to the particle (before impact) leads to

$$mgh = \frac{1}{2}mv^2 \implies v = \sqrt{2gh}$$

for its speed right before undergoing the completely inelastic collision with the rod. The collision is described by angular momentum conservation:

$$mvd = \left(I_{\text{rod}} + md^2\right)\omega$$

where I_{rod} is found using Table 11-2(e) and the parallel axis theorem:

$$I_{\text{rod}} = \frac{1}{12}Md^2 + M\left(\frac{d}{2}\right)^2 = \frac{1}{3}Md^2 .$$

Thus, we obtain the angular velocity of the system immediately after the collision:

$$\omega = \frac{md\sqrt{2gh}}{\frac{1}{3}Md^2 + md^2}$$

which means the system has kinetic energy $\frac{1}{2}(I_{\text{rod}} + md^2)\omega^2$ which will turn into potential energy in the final position, where the block has reached a height H (relative to the lowest point) and the center of mass of the stick has increased its height by $H/2$. From trigonometric considerations, we note that $H = d(1 - \cos\theta)$, so we have

$$\begin{aligned} \frac{1}{2}\left(I_{\text{rod}} + md^2\right)\omega^2 &= mgH + Mg\frac{H}{2} \\ \frac{1}{2}\frac{m^2d^2(2gh)}{\frac{1}{3}Md^2 + md^2} &= \left(m + \frac{M}{2}\right)gd\,(1 - \cos\theta) \end{aligned}$$

from which we obtain

$$\theta = \cos^{-1}\left(1 - \frac{m^2h}{\left(m + \frac{1}{2}M\right)\left(m + \frac{1}{3}M\right)}\right) .$$

60. (a) Since the motorcycle is going leftward across our field of view, then when its wheels are rolling they must be going counterclockwise (which we take as the positive sense of rotation, which is the usual convention).

(b) Just before the rear wheel spins up to ω_{wf} it has the angular velocity necessary for rolling $\omega_{wR} = v/R$ where $v = 32$ m/s and $R = 0.30$ m. Since $\omega_{wf} > \omega_{wR}$ the system would seem to have suddenly acquired an increase in (positive) angular momentum – without the action of external torques! Since this is not possible, then the other constituents of the system (the man and the motorcycle body, which the problem just refers to as "the motorcycle") must have acquired some (negative) angular momentum. Thus, the motorcycle rotated clockwise.

(c) Assuming the system's (translational) projectile motion is symmetrical (as in Fig. 4-34 in the textbook) then (with $+y$ upward) it starts with $v_{0y} = v \sin 15°$ and returns with $v_y = -v \sin 15°$. Substituting these into Eq. 2-11 (with $a = -g$) leads to

$$-v \sin 15° = v \sin 15° - gt \implies t = \frac{2v \sin 15°}{g} = 1.7 \text{ s} .$$

(d) As noted in our solution of part (b), $\omega_{wR} = v/R$ which yields the value $\omega_{wR} = 32/0.30 = 106.7$ rad/s. In keeping with the significant figures rules, we round this to 1.1×10^2 rad/s.

(e) We have $L_w = I_w \omega_{wR} = (0.40)(106.7) = 43 \text{ kg}\cdot\text{m}^2/\text{s}$.

(f) Recalling our discussion in part (b), we apply angular momentum conservation:

$$I_w \omega_{wR} = I_w \omega_{wf} + I_c \omega_c \implies \omega_c = -\frac{I_w (\omega_{wf} - \omega_{wR})}{I_c}$$

which yields $\omega_c = -1.067$ rad/s or $|\omega_c| \approx 1.1$ rad/s.

(g) The problem states that the spin up occurs immediately – the moment this becomes a projectile motion problem (for the center of mass). We assume the motorcycle turns at the (constant) rate $|\omega_c|$ for the duration of the motion. Using the more precise values from our previous results, we are led to

$$\theta = \omega_c t = -1.80 \text{ rad}$$

which we convert (multiplying by $180/\pi$) to $-103°$. Rounding off, we find $|\theta| \approx 100°$.

61. (a) The derivation of the acceleration is found in §12-4; Eq. 12-13 gives

$$a_{\text{com}} = -\frac{g}{1 + I_{\text{com}}/MR_0^2}$$

where the positive direction is upward. We use $I_{\text{com}} = \frac{1}{2}MR^2$ where the radius is $R = 0.32$ m and $M = 116$ kg is the *total* mass (thus including the fact that there are two disks) and obtain

$$a = -\frac{g}{1 + \frac{1}{2}MR^2/MR_0^2} = \frac{g}{1 + \frac{1}{2}\left(\frac{R}{R_0}\right)^2}$$

which yields $a = -g/51$ upon plugging in $R_0 = R/10 = 0.032$ m. Thus, the magnitude of the center of mass acceleration is 0.19 m/s^2 and the direction of that vector is down.

(b) As observed in §12-4, our result in part (a) applies to both the descending and the rising yoyo motions.

(c) The external forces on the center of mass consist of the cord tension (upward) and the pull of gravity (downward). Newton's second law leads to

$$T - Mg = ma \implies T = M\left(g - \frac{g}{51}\right)$$

which yields $T = 1.1 \times 10^3$ N.

(d) Our result in part (c) indicates that the tension is well below the ultimate limit for the cord.

(e) As we saw in our acceleration computation, all that mattered was the ratio R/R_0 (and, of course, g). So if it's a scaled-up version, then such ratios are unchanged and we obtain the same result.

(f) Since the tension also depends on mass, then the larger yoyo will involve a larger cord tension.

62. We denote the wheel with subscript 1 and the whole system with subscript 2. We take clockwise as the negative sense for rotation (as is the usual convention). Conservation of angular momentum gives $L = I_1\omega_1 = I_2\omega_2$, where $I_1 = m_1R_1^2$. Thus

$$\omega_2 = \omega_1\frac{I_1}{I_2} = (-57.7\,\text{rad/s})\frac{\left(37\,\text{N}/9.8\,\text{m/s}^2\right)(0.35\,\text{m})^2}{2.1\,\text{kg}\cdot\text{m}^2}$$

which yields $\omega_2 = -12.7\,\text{rad/s}$. The system therefore rotates clockwise (as seen from above) at the rate of 12.7 rad/s.

63. We use $L = I\omega$ and $K = \frac{1}{2}I\omega^2$ and observe that the speed of points on the rim (corresponding to the speed of points on the belt) of wheels A and B must be the same (so $\omega_A R_A = \omega_B r_B$).

(a) If $L_A = L_B$ (call it L) then the ratio of rotational inertias is

$$\frac{I_A}{I_B} = \frac{L/\omega_A}{L/\omega_B} = \frac{\omega_A}{\omega_B} = \frac{R_A}{R_B} = \frac{1}{3}\,.$$

(b) If we have $K_A = K_B$ (call it K) then the ratio of rotational inertias becomes

$$\frac{I_A}{I_B} = \frac{2K/\omega_A^2}{2K/\omega_B^2} = \left(\frac{\omega_B}{\omega_A}\right)^2 = \left(\frac{R_A}{R_B}\right)^2 = \frac{1}{9}\,.$$

64. Since we will be taking the vector cross product in the course of our calculations, below, we note first that when the two vectors in a cross product $\vec{A}\times\vec{B}$ are in the xy plane, we have $\vec{A} = A_x\hat{\imath} + A_y\hat{\jmath}$ and $\vec{B} = B_x\hat{\imath} + B_y\hat{\jmath}$, and Eq. 3-30 leads to

$$\vec{A}\times\vec{B} = (A_xB_y - A_yB_x)\,\hat{k}\,.$$

(a) We set up a coordinate system with its origin at the firing point, the positive x axis in the horizontal direction of motion of the projectile and the positive y axis vertically upward. The projectile moves in the xy plane, and if $+x$ is to our right then the "rotation" sense will be clockwise. Thus, we expect our answer to be negative. The position vector for the projectile (as a function of time) is given by

$$\vec{r} = (v_{0x}t)\hat{\imath} + \left(v_{0y}t - \frac{1}{2}gt^2\right)\hat{\jmath} = (v_0\cos\theta_0 t)\hat{\imath} + (v_0\sin\theta_0 - gt)\hat{\jmath}$$

and the velocity vector is

$$\vec{v} = v_x\hat{\imath} + v_y\hat{\jmath} = (v_0\cos\theta_0)\,\hat{\imath} + (v_0\sin\theta_0 - gt)\hat{\jmath}\,.$$

Thus (using the above observation about the cross product of vectors in the xy plane) the angular momentum of the projectile as a function of time is

$$\vec{\ell} = m\vec{r}\times\vec{v} = -\frac{1}{2}mv_0\cos\theta_0 gt^2\hat{k}\,.$$

(b) We take the derivative of our result in part (a): $\frac{d\vec{\ell}}{dt} = -v_0 mgt\cos\theta_0\hat{k}$.

(c) Again using the above observation about the cross product of vectors in the xy plane, we find

$$\vec{r}\times\vec{F} = \left((v_0\cos\theta_0 t)\hat{\imath} + r_y\hat{\jmath}\right)\times(-mg\hat{\jmath}) = -v_0 mgt\cos\theta_0\,\hat{k}$$

which is the same as the result in part (b).

(d) They are the same because $d\vec{\ell}/dt = \tau = \vec{r}\times\vec{F}$.

65. The problem asks that we put the origin of coordinates at point O but compute all the angular momenta and torques relative to point A. This requires some care in defining $\vec{r}$ (which occurs in the angular momentum and torque formulas). If $\vec{r}_O$ locates the point (where the block is) in the prescribed coordinates, and $\vec{r}_{OA} = -1.2\hat{\jmath}$ points from O to A, then $\vec{r} = \vec{r}_O - \vec{r}_{OA}$ gives the position of the block relative to point A. SI units are used throughout this problem.

(a) Here, the momentum is $\vec{p}_0 = m\vec{v}_0 = 1.5\hat{\imath}$ and $\vec{r}_0 = 1.2\hat{\jmath}$, so that

$$\vec{\ell}_0 = \vec{r}_0 \times \vec{p}_0 = -1.8\,\hat{k}\ \text{kg}\cdot\text{m}^2/\text{s}\ .$$

(b) The horizontal component of momentum doesn't change in projectile motion (without friction), and its vertical component depends on how far its fallen. From either the free-fall equations of Ch. 2 or the energy techniques of Ch. 8, we find the vertical momentum component after falling a distance h to be $-m\sqrt{2gh}$. Thus, with $m = 0.50$ and $h = 1.2$, the momentum just before the block hits the floor is $\vec{p} = 1.5\,\hat{\imath} - 2.4\hat{\jmath}$. Now, $\vec{r} = R\hat{\imath}$ where R is figured from the projectile motion equations of Ch. 4 to be $R = v_0\sqrt{\frac{2h}{g}} = 1.5$ m. Consequently,

$$\vec{\ell} = \vec{r} \times \vec{p} = -3.6\,\hat{k}\ \text{kg}\cdot\text{m}^2/\text{s}\ .$$

(c) and (d) The only force on the object is its weight $m\vec{g} = -4.9\,\hat{\jmath}$. Thus,

$$\begin{aligned} \vec{\tau}_0 &= \vec{r}_0 \times \vec{F} = 0 \\ \vec{\tau} &= \vec{r} \times \vec{F} = -7.3\,\hat{k}\ \text{N}\cdot\text{m}\ . \end{aligned}$$

66. Since we will be taking the vector cross product in the course of our calculations, below, we note first that when the two vectors in a cross product $\vec{A} \times \vec{B}$ are in the xy plane, we have $\vec{A} = A_x\hat{\imath} + A_y\hat{\jmath}$ and $\vec{B} = B_x\hat{\imath} + B_y\hat{\jmath}$, and Eq. 3-30 leads to

$$\vec{A} \times \vec{B} = (A_xB_y - A_yB_x)\,\hat{k}\ .$$

Now, we choose coordinates centered on point O, with $+x$ rightwards and $+y$ upwards. In unit-vector notation, the initial position of the particle, then, is $\vec{r}_0 = s\,\hat{\imath}$ and its later position (halfway to the ground) is $\vec{r} = s\,\hat{\imath} - \frac{1}{2}h\,\hat{\jmath}$. Using either the free-fall equations of Ch. 2 or the energy techniques of Ch. 8, we find the speed at its later position to be $v = \sqrt{2g|\Delta y|} = \sqrt{gh}$. Its momentum there is $\vec{p} = -M\sqrt{gh}\,\hat{\jmath}$. We find the angular momentum using Eq. 12-18 and our observation, above, about the cross product of two vectors in the xy plane.

$$\vec{\ell} = \vec{r} \times \vec{p} = -sM\sqrt{gh}\ \hat{k}$$

Therefore, its magnitude is $|\vec{\ell}| = sM\sqrt{gh}$.

67. We may approximate the planets and their motions as particles in circular orbits, and use Eq. 12-26

$$L = \sum_{i=1}^{9} \ell_i = \sum_{i=1}^{9} m_i r_i^2 \omega_i$$

to compute the total angular momentum. Since we assume the angular speed of each one is constant, we have (in rad/s) $\omega_i = 2\pi/T_i$ where T_i is the time for that planet to go around the Sun (this and related information is found in Appendix C but there, the T_i are expressed in years and we'll need to convert with 3.156×10^7 s/y, and the M_i are expressed as multiples of M_{earth} which we'll convert by multiplying by 5.98×10^{24} kg).

(a) Using SI units, we find (with $i = 1$ designating Mercury)

$$\begin{aligned}
L &= \sum_{i=1}^{9} m_i r_i^2 \left(\frac{2\pi}{T_i}\right) \\
&= 2\pi \frac{3.34\times10^{23}}{7.61\times10^{6}} \left(57.9\times10^{9}\right)^2 + 2\pi \frac{4.87\times10^{24}}{19.4\times10^{7}} \left(108\times10^{9}\right)^2 + \\
&\quad 2\pi \frac{5.98\times10^{24}}{3.156\times10^{7}} \left(150\times10^{9}\right)^2 + 2\pi \frac{6.40\times10^{23}}{5.93\times10^{7}} \left(228\times10^{9}\right)^2 + \\
&\quad 2\pi \frac{1.9\times10^{27}}{3.76\times10^{8}} \left(778\times10^{9}\right)^2 + 2\pi \frac{5.69\times10^{26}}{9.31\times10^{8}} \left(1430\times10^{9}\right)^2 + \\
&\quad 2\pi \frac{8.67\times10^{25}}{2.65\times10^{9}} \left(2870\times10^{9}\right)^2 + 2\pi \frac{1.03\times10^{26}}{5.21\times10^{9}} \left(4500\times10^{9}\right)^2 + \\
&\quad 2\pi \frac{1.2\times10^{22}}{7.83\times10^{9}} \left(5900\times10^{9}\right)^2 \\
&= 3.14\times10^{43}\ \text{kg}\cdot\text{m}^2/\text{s}\ .
\end{aligned}$$

(b) The fractional contribution of Jupiter is

$$\frac{\ell_5}{L} = \frac{2\pi \frac{1.9\times10^{27}}{3.76\times10^{8}} \left(778\times10^{9}\right)^2}{3.14\times10^{43}} = 0.61\ .$$

68. If we write $\vec{r} = x\hat{\imath} + y\hat{\jmath} + z\hat{k}$, then (using Eq. 3-30) we find $\vec{r}\times\vec{F}$ is equal to

$$(yF_z - zF_y)\,\hat{\imath} + (zF_x - xF_z)\,\hat{\jmath} + (xF_y - yF_x)\,\hat{k}\ .$$

With (using SI units) $x = 0$, $y = -4.0$, $z = 5.0$, $F_x = 0$, $F_y = -2.0$ and $F_z = 3.0$ (these latter terms being the individual forces that contribute to the net force), the expression above yields

$$\vec{\tau} = \vec{r}\times\vec{F} = -2.0\,\hat{\imath}\ \text{N}\cdot\text{m}\ .$$

69. We make the unconventional choice of *clockwise* sense as positive, so that the angular acceleration are positive (as is the linear acceleration of the center of mass, since we take rightwards as positive).

(a) We approach this in the manner of Eq. 12-3 (*pure rotation* about point P) but use torques instead of energy:

$$\tau = I_P\,\alpha \quad \text{where } I_P = \frac{1}{2}MR^2 + MR^2$$

where the parallel-axis theorem and Table 11-2(c) has been used. The torque (relative to point P) is due to the $F = 12$ N force and is $\tau = F(2R)$. In this way, we find

$$\alpha = \frac{(12)(0.20)}{0.05 + 0.10} = 16\ \text{rad/s}^2\ .$$

Hence, $a_{\text{com}} = R\alpha = 1.6\ \text{m/s}^2$.

(b) As shown above, $\alpha = 16\ \text{rad/s}^2$.

(c) Applying Newton's second law in its linear form yields

$$(12\ \text{N}) - f = Ma_{\text{com}}\ .$$

Therefore, $f = -4.0$ N. Contradicting what we assumed in setting up our force equation, the friction force is found to point *rightward* (with magnitude 4.0 N).

70. The speed of the center of mass of the car is $v = (40)(1000/3600) = 11$ m/s. The angular speed of the wheels is given by Eq. 12-2: $\omega = v/R$ where the wheel radius R is not given (but will be seen to cancel in these calculations).

   (a) For one wheel of mass $M = 32$ kg, Eq. 11-27 gives (using Table 11-2(c))

$$K_{\text{rot}} = \frac{1}{2}I\omega^2 = \frac{1}{2}\left(\frac{1}{2}MR^2\right)\left(\frac{v}{R}\right)^2 = \frac{1}{4}Mv^2$$

   which yields $K_{\text{rot}} = 9.9 \times 10^2$ J. The time given in the problem (10 s) is not used in the solution.

   (b) Adding the above to the wheel's translational kinetic energy, $\frac{1}{2}Mv^2$, leads to

$$K_{\text{wheel}} = \frac{1}{2}Mv^2 + \frac{1}{4}Mv^2 = \frac{3}{4}(32)(11)^2 = 3.0 \times 10^3 \text{ J} .$$

   (c) With $M_{\text{car}} = 1700$ kg and the fact that there are four wheels, we have

$$\frac{1}{2}M_{\text{car}}v^2 + 4\left(\frac{3}{4}Mv^2\right) = 1.2 \times 10^5 \text{ J} .$$

71. Information relevant to this calculation can be found in Appendix C. We apply angular momentum conservation using Table 11-2(f):

$$I_i\omega_i = I_f\omega_f \implies \frac{\omega_i}{\omega_f} = \frac{I_f}{I_i} = \frac{\frac{2}{5}MR_f^2}{\frac{2}{5}MR_i^2}$$

and we note that $\omega = 2\pi/T$ in rad/min if T is the period in minutes. Plugging this into to our expression above (and simplifying) yields

$$\frac{T_f}{T_i} = \left(\frac{R_f}{R_i}\right)^2 .$$

Substituting $T_i = 25(24)(60) = 36000$ min, $R_f = 6.37 \times 10^6$ m and $R_i = 6.96 \times 10^8$ m into this relation, we obtain $T_f = 3.0$ min.

72. (a) We use Table 11-2(e) and the parallel-axis theorem to obtain the rod's rotational inertia about an axis through one end:

$$I = I_{\text{com}} + Mh^2 = \frac{1}{12}ML^2 + M\left(\frac{L}{2}\right)^2 = \frac{1}{3}ML^2$$

   where $L = 6.00$ m and $M = 10.0/9.8 = 1.02$ kg. Thus, $I = 12.2$ kg·m^2.

   (b) Using $\omega = (240)(2\pi/60) = 25.1$ rad/s, Eq. 12-31 gives the magnitude of the angular momentum as $I\omega = (12.2)(25.1) = 308$ kg·m^2/s. Since it is rotating clockwise as viewed from above, then the right-hand rule indicates that its direction is down.

73. This problem involves the vector cross product of vectors lying in the xy plane. For such vectors, if we write $\vec{r}\,' = x'\,\hat{\text{i}} + y'\,\hat{\text{j}}$, then (using Eq. 3-30) we find

$$\vec{r}\,' \times \vec{v} = (x'v_y - y'v_x)\,\hat{\text{k}} .$$

   (a) Here, $\vec{r}\,'$ points in either the $+\hat{\text{i}}$ or the $-\hat{\text{i}}$ direction (since the particle moves along the x axis). It has no y' or z' components, and neither does $\vec{v}$, so it is clear from the above expression (or, more simply, from the fact that $\hat{\text{i}} \times \hat{\text{i}} = 0$) that $\vec{\ell} = m(\vec{r}\,' \times \vec{v}) = 0$ in this case.

   (b) The net force is in the $-\hat{\text{i}}$ direction (as one finds from differentiating the velocity expression, yielding the acceleration), so, similar to what we found in part (a), we obtain $\tau = \vec{r}\,' \times \vec{F} = 0$.

(c) Now, $\vec{r}\,' = \vec{r} - \vec{r}_o$ where $\vec{r}_o = 2.0\,\hat{\text{i}} + 5.0\,\hat{\text{j}}$ (with SI units understood) and points from $(2.0, 5.0, 0)$ to the instantaneous position of the car (indicated by $\vec{r}$ which points in either the $+x$ or $-x$ directions, or nowhere (if the car is passing through the origin)). Since $\vec{r} \times \vec{v} = 0$ we have (plugging into our general expression above)

$$\vec{\ell} = m\,(\vec{r}\,' \times \vec{v}) = -m\,(\vec{r}_o \times \vec{v}) = -(3.0)\,\left((2.0)(0) - (5.0)\left(-2.0t^3\right)\right)\hat{\text{k}}$$

which yields $\vec{\ell} = -30t^3\,\hat{\text{k}}$ in SI units ($\text{kg}\cdot\text{m}^2/\text{s}$).

(d) The acceleration vector is given by $\vec{a} = \frac{d\vec{v}}{dt} = -6.0t^2\,\hat{\text{i}}$ in SI units, and the net force on the car is $m\vec{a}$. In a similar argument to that given in the previous part, we have

$$\vec{\tau} = m\,(\vec{r}\,' \times \vec{a}) = -m\,(\vec{r}_o \times \vec{a}) = -(3.0)\,\left((2.0)(0) - (5.0)\left(-6.0t^2\right)\right)\hat{\text{k}}$$

which yields $\vec{\tau} = -90t^2\,\hat{\text{k}}$ in SI units (N·m).

(e) In this situation, $\vec{r}\,' = \vec{r} - \vec{r}_o$ where $\vec{r}_o = 2.0\,\hat{\text{i}} - 5.0\,\hat{\text{j}}$ (with SI units understood) and points from $(2.0, -5.0, 0)$ to the instantaneous position of the car (indicated by $\vec{r}$ which points in either the $+x$ or $-x$ directions, or nowhere (if the car is passing through the origin)). Since $\vec{r} \times \vec{v} = 0$ we have (plugging into our general expression above)

$$\vec{\ell} = m\,(\vec{r}\,' \times \vec{v}) = -m\,(\vec{r}_o \times \vec{v}) = -(3.0)\,\left((2.0)(0) - (-5.0)\left(-2.0t^3\right)\right)\hat{\text{k}}$$

which yields $\vec{\ell} = 30t^3\,\hat{\text{k}}$ in SI units ($\text{kg}\cdot\text{m}^2/\text{s}$).

(f) Again, the acceleration vector is given by $\vec{a} = -6.0t^2\,\hat{\text{i}}$ in SI units, and the net force on the car is $m\vec{a}$. In a similar argument to that given in the previous part, we have

$$\vec{\tau} = m\,(\vec{r}\,' \times \vec{a}) = -m\,(\vec{r}_o \times \vec{a}) = -(3.0)\,\left((2.0)(0) - (-5.0)\left(-6.0t^2\right)\right)\hat{\text{k}}$$

which yields $\vec{\tau} = 90t^2\,\hat{\text{k}}$ in SI units (N·m).

74. This problem involves the vector cross product of vectors lying in the xy plane. For such vectors, if we write $\vec{r} = x\,\hat{\text{i}} + y\,\hat{\text{j}}$, then (using Eq. 3-30) we find

$$\vec{r} \times \vec{p} = (\Delta x p_y - \Delta y p_x)\,\hat{\text{k}}\ .$$

The momentum components are $p_x = p\cos\theta$ and $p_y = p\sin\theta$ where $p = 2.4$ (SI units understood) and $\theta = 115°$. The mass (0.80 kg) given in the problem is not used in the solution. Thus, with $x = 2.0$, $y = 3.0$ and the momentum components described above, we obtain

$$\vec{\ell} = \vec{r} \times \vec{p} = 7.4\,\hat{\text{k}}\ \text{kg}\cdot\text{m}^2/\text{s}\ .$$

75. Information relevant to this calculation can be found in Appendix C or on the inside front cover of the textbook. The angular speed is constant so

$$\omega = \frac{2\pi}{T} = \frac{2\pi}{86400} = 7.3 \times 10^{-5}\ \text{rad/s}\ .$$

Thus, with $m = 84$ kg and $R = 6.37 \times 10^6$ m, we find $\ell = mR^2\omega = 2.5 \times 10^{11}\ \text{kg}\cdot\text{m}^2/\text{s}$.

76. With $r_\perp = 1300$ m, Eq. 12-21 gives

$$\ell = r_\perp m v = (1300)(1200)(80) = 1.2 \times 10^8\ \text{kg}\cdot\text{m}^2/\text{s}\ .$$

77. The result follows immediately from Eq. 3-30. If what is desired to show here is basically a derivation of Eq. 3-30, then (with the slight change to position and force notation) that is shown in some detail in our solution to problem 32 of Chapter 3.

78. (a) Using Eq. 2-16 for the translational (center-of-mass) motion, we find

$$v^2 = v_0^2 + 2a\Delta x \implies a = -\frac{v_0^2}{2\Delta x}$$

which yields $a = -4.11$ for $v_0 = 43$ and $\Delta x = 225$ (SI units understood). The magnitude of the linear acceleration of the center of mass is therefore 4.11 m/s^2.

(b) With $R = 0.250$ m, Eq. 12-6 gives $|\alpha| = |a|/R = 16.4$ rad/s^2. If the wheel is going rightward, it is rotating in a clockwise sense. Since it is slowing down, this angular acceleration is counterclockwise (opposite to ω) so (with the usual convention that counterclockwise is positive) there is no need for the absolute value signs for α.

(c) Eq. 12-8 applies with Rf_s representing the magnitude of the frictional torque. Thus, $Rf_s = I\alpha = (0.155)(16.4) = 2.55\,\text{N}\cdot\text{m}$.

79. We note that its mass is $M = 36/9.8 = 3.67$ kg and its rotational inertia is $I_{\text{com}} = \frac{2}{5}MR^2$ (Table 11-2(f)).

(a) Using Eq. 12-2, Eq. 12-5 becomes

$$\begin{aligned} K &= \frac{1}{2}I_{\text{com}}\omega^2 + \frac{1}{2}Mv_{\text{com}}^2 \\ &= \frac{1}{2}\left(\frac{2}{5}MR^2\right)\left(\frac{v_{\text{com}}}{R}\right)^2 + \frac{1}{2}Mv_{\text{com}}^2 \\ &= \frac{7}{10}Mv_{\text{com}}^2 \end{aligned}$$

which yields $K = 61.7$ J for $v_{\text{com}} = 4.9$ m/s.

(b) This kinetic energy turns into potential energy Mgh at some height $h = d\sin\theta$ where the sphere comes to rest. Therefore, we find the distance traveled up the $\theta = 30°$ incline from energy conservation:

$$\frac{7}{10}Mv_{\text{com}}^2 = Mgd\sin\theta \implies d = \frac{7v_{\text{com}}^2}{10g\sin\theta} = 3.43 \text{ m} .$$

(c) As shown in the previous part, M cancels in the calculation for d. Since the answer is independent of mass, then, it is also independent of the sphere's weight.

80. Although we will not be "working" this problem, we do – briefly – share a few thoughts about it.

(a) A figure in the textbook that may be referred to is Fig. 8-16. The idea, crudely stated, is to show that although all bodies will return to the same height they're released from (in the absence of dissipative effects), the one with the least rotational inertia (say, a sphere) will get there the fastest because its speed is greatest at every point inbetween.

(b) Several people might be pulling on ropes attached to a merry-go-round to set it into motion. The ropes should be at different angles (measured relative to tangent lines at the appropriate points). The idea is to calculate the net torque using Eq. 12-15 and then to find the angular acceleration (using Eq. 11-37) of the merry-go-round.

(c) This might require particular care in the wording, especially regarding a clown "falling off." If he falls off in what might described as the "natural way" (simply letting go and pursuing a straight-line trajectory tangent to the merry-go-round) then there is no change in the angular momentum. It's easier to see that there'd be a change in angular momentum in the case of a clown (initially at rest) stepping onto the moving merry-go-round.

(d) This is an important astrophysical application of the angular momentum concept (angular momentum is conserved in gravitational-dominated situations such as binary star systems). When the masses of the stars are similar and the mass transfer is relatively steady, they are often known as Algol binaries, and realistic numerical values can be found in many astronomy textbooks (and, probably, on the Web).

81. (First problem in **Cluster 1**)

(a) Applying Newton's second law in its linear form yields

$$(200\ \text{N}) - f = M_{\text{cart}}\, a \quad .$$

Therefore, $f = 200 - (50.0)(3.00) = 50$ N.

(b) The torque associated with the friction is $\tau_f = fR = (50)(0.200) = 10$ N·m. (We make the unconventional choice of the clockwise sense as positive, so that the frictional torque and this angular acceleration are positive.)

(c) Applying the rotational form of Newton's second law (relative to the axle) yields

$$\tau_f = I\alpha \qquad \text{where } \alpha = \frac{a}{R} = 15.0\ \text{rad/s}^2 \ .$$

Therefore, $I = 0.667$ kg·m^2.

82. (Second problem in **Cluster 1**)

(a) If we interpret this "one-wheel cart" which has a wheel that is a "long cylinder" as simply the cylinder itself, then an appropriate picture for this problem is Fig. 12-30 in the textbook. We make the unconventional choice of *clockwise* sense as positive, so that the angular velocity in this problem is positive; we choose *downhill* positive for the x axis (which is parallel to the incline surface) so that $a_{\text{com}} = R\alpha$ holds. We can combine the rotational (about the center of mass) and linear forms of Newton's second law, or we can more simply adopt the view of pure rotation (see, for example, Eq. 12-3) and examine torques about the bottom-most point P:

$$MgR\sin\theta = I_P\,\alpha = I_P\,\frac{a_{\text{com}}}{R}$$

We have assumed that the center of mass of the cart-wheel system is at the center of the wheel (the axle), although this is not stated in the problem. Now, $\theta = 30.0°$, $R = 0.200$ m, $M = 50.0$ kg, and $I_P = 0.667\ \text{kg}\cdot\text{m}^2 + MR^2 = 2.67\ \text{kg}\cdot\text{m}^2$ (using the parallel-axis theorem and the result of the previous problem). Thus, we find $a_{\text{com}} = 3.68\,\text{m/s}^2$.

(b) If we apply the linear form of Newton's law, we have

$$\begin{aligned}\sum F_x &= Mg\sin\theta - f_{s,\,\text{max}} = Ma_{\text{com}}\\ \sum F_y &= N - Mg\sin\theta = 0\end{aligned}$$

Solving for $f_{s,\,\text{max}}$ and N and dividing, we obtain

$$\mu_s = \frac{f_{s,\,\text{max}}}{N} = 0.14 \quad .$$

83. (Third problem in **Cluster 1**)
An appropriate picture for this problem is Fig. 12-7 in the textbook. We make the unconventional choice of *clockwise* sense as positive, so that the angular velocity in this problem is positive; we choose *downhill* positive for the x axis (which is parallel to the incline surface) so that $a_{\text{com}} = R\alpha$ holds. For simplicity, we refer to a_{com} as a. We examine the rotational (about the center of mass) and linear forms of Newton's second law:

$$\begin{aligned}\sum \tau_z &= f_sR = I\alpha = I\frac{a}{R}\\ \sum F_x &= Mg\sin\theta - f_s = Ma\\ \sum F_y &= N - Mg\cos\theta = 0\end{aligned}$$

Combining the first two of these equations, we obtain

$$f_s = \frac{Mg\sin\theta}{1+\frac{MR^2}{I}} .$$

We now let $f_s = f_{s,\max} = \mu_s N$ and combine this with the third equation above:

$$\mu_s Mg\cos\theta = \frac{Mg\sin\theta}{1+\frac{MR^2}{I}} \implies \theta = \tan^{-1}\left(\mu_s + \frac{MR^2\mu_s}{I}\right) .$$

84. (Fourth problem in **Cluster 1**)

(a) We take the tangential acceleration of the bottom-most point on the (positively) accelerating disk to equal $R\alpha + a_{\text{com}}$. This in turn must equal the (forward) acceleration of the truck $a_{\text{truck}} = a > 0$. Since the disk is rolling toward the back of the truck, $a_{\text{com}} < a$ which implies that α is positive. If the forward direction is *rightward*, then this makes it consistent to choose counterclockwise as the positive rotational sense, which is the usual convention. Thus, $\sum \tau = I\alpha$ becomes

$$f_s R = I\alpha \qquad \text{where } I = \frac{1}{2}MR^2$$

and $\sum F_x = Ma_{\text{com}}$ becomes

$$f_s = M(a - R\alpha) .$$

Combining these two equations, we find $R\alpha = \frac{2}{3}a$. From the previous discussion, we see acceleration of the disk relative to the truck bed is $a_{\text{com}} - a = -R\alpha$, so this has a magnitude of $\frac{2}{3}$ and is directed leftward.

(b) Returning to $R\alpha + a_{\text{com}} = a$ with our result that $R\alpha = \frac{2}{3}a$, we find $a_{\text{com}} = \frac{1}{3}a$. This is positive, hence rightward.

85. (First problem in **Cluster 2**)
The last line of the problem indicates our choice of positive directions: up for m_2, down for m_1 and counterclockwise for the two-pulley device. This allows us to write $R_2\alpha = a_2$ and $R_1\alpha = a_1$ with all terms positive. We apply Newton's second law to the elements of this system:

$$\begin{aligned} T_2 - m_2 g &= m_2 a_2 = m_2 R_2 \alpha \\ m_1 g - T_1 &= m_1 a_1 = m_1 R_1 \alpha \\ T_1 R_1 - T_2 R_2 &= I\alpha \end{aligned}$$

Multiplying the first equation by R_2, the second by R_1 and adding the equations leads to

$$\alpha = \frac{m_1 g R_1 - m_2 g R_2}{I + m_1 R_1^2 + m_2 R_2^2} .$$

(a) Therefore, again using $R_1\alpha = a_1$, we obtain

$$a_1 = \frac{m_1 g R_1^2 - m_2 g R_1 R_2}{I + m_1 R_1^2 + m_2 R_2^2} .$$

(b) Once more, we use $R_2\alpha = a_2$ and find

$$a_2 = \frac{m_1 g R_1 R_2 - m_2 g R_2^2}{I + m_1 R_1^2 + m_2 R_2^2} .$$

86. (Second problem in **Cluster 2**)
This system is extensively discussed in §12-4. Rather than repeat those steps here, we refer to their conclusion, Eq. 12-13.

(a) The magnitude of the result in Eq. 12-13 is

$$|a| = g\,\frac{1}{1+\frac{I}{MR^2}}$$

(b) The relation $a = a_{\text{com}} = -R\alpha$ used in §12-3 must now be modified to read $a_f - a_{\text{com}} = R\alpha$ where a_f is the acceleration of the finger. With this in mind, the linear and angular versions of Newton's second law become

$$\begin{aligned} T - Mg &= Ma_{\text{com}} \\ TR &= I\alpha \qquad \text{where } \alpha = \frac{a_f - a_{\text{com}}}{R} \end{aligned}$$

If we require $a_{\text{com}} = 0$ then these equations yield

$$a_f = g\,\frac{MR^2}{I} \; .$$

87. (Third problem in **Cluster 2**)
Our analysis of spool 2 is exactly as in the solution of part (b) of the previous problem, but with a_f replaced with $-a_s$. The negative sign is due to the wording of the problem (which refers to a "downward acceleration a_s"):

$$\begin{aligned} T - Mg &= Ma_1 \\ TR_1 &= I_1\alpha_1 = I_1\left(\frac{-a_s - a_1}{R_1}\right) \end{aligned}$$

In our analysis of spool 1, we pay close attention to signs: positive (downward) a_s corresponds to clockwise (conventionally taken to be negative) rotation of spool 1; hence, $R_2\alpha_2 = -a_s$. For spool 1, we therefore have

$$\sum \tau_z = -TR_2 = I_2\alpha_1 = I_2\left(\frac{-a_s}{R_2}\right) \; .$$

(a) Simultaneous solution (certainly non-trivial) of these three equations yields

$$a_1 = -\frac{g}{1+\dfrac{1}{\dfrac{MR_1^2}{I_1}+\dfrac{MR_2^2}{I_2}}} \; .$$

The problem asks for the magnitude of this (which eliminates the negative sign).

(b) This amounts to eliminating the $\frac{MR_2^2}{I_2}$ term in the expression for a_1.

NOTES

NOTES

NOTES

NOTES

NOTES

NOTES

NOTES

NOTES

NOTES

NOTES

NOTES

NOTES